LINEAR DIFFERENTIAL EQUATIONS WITH PERIODIC COEFFICIENTS

V. A. Yakubovich and V. M. Starzhinskii

LINEAR DIFFERENTIAL EQUATIONS WITH PERIODIC COEFFICIENTS

2

Translated from Russian by D. Louvish

A HALSTED PRESS BOOK

JOHN WILEY & SONS
New York · Toronto

ISRAEL PROGRAM FOR SCIENTIFIC TRANSLATIONS
Jerusalem · London

Sole distributors for the Western Hemisphere
HALSTED PRESS, a division of
JOHN WILEY & SONS, INC., NEW YORK

Library of Congress Cataloging in Publication Data

I͡Akubovich, Vladimir Andreevich.
Linear differential equations with periodic coefficients.

Translation of *Lineĭnye different͡sial'nye uravnenii͡a s periodicheskimi koėffit͡sientami i ikh prilozhenii͡a.*
"A Halsted Press book."
Bibliography: p.
Includes index.
1. Differential equations, Linear. I. Starzhinskiĭ, V͡iacheslav Mikhaĭlovich, joint author. II. Title.
QA372.I213 515'.35 75-4982
ISBN 0-470-96953-9

Distributors for the U.K., Europe, Africa and the Middle East
JOHN WILEY & SONS, LTD., CHICHESTER

Distributors for Japan, Southeast Asia, and India
TOPPAN COMPANY, LTD., TOKYO AND SINGAPORE

Distributed in the rest of the world by
KETER PUBLISHING HOUSE JERUSALEM LTD.
ISBN 0 7065 1480 7
IPST cat. no. 22123 2

This book is a translation from Russian of
LINEINYE DIFFERENTSIAL'NYE URAVNENIYA
S PERIODICHESKIMI KOEFFITSIENTAMI I IKH
PRILOZHENIYA
Izdatel'stvo "Nauka,"
Moscow, 1972

Printed and bound by Keterpress Enterprises, Jerusalem
Printed in Israel

CONTENTS

CONTENTS OF VOLUME ONE

Chapter V

THEORY OF PARAMETRIC RESONANCE

Parametric resonance differs from resonance in the familiar sense of the term in two essential respects.

1. In ordinary resonance, the spectrum of frequencies at which vibrations with indefinitely increasing amplitudes may build up is discrete — it is the set of natural frequencies of the system; in parametric resonance the spectrum is the union of several small intervals. The lengths of the intervals depend on the amplitude of the perturbations and go to zero as the amplitude approaches zero. The frequencies to which these intervals shrink are known as the *critical frequencies*.

2. In ordinary resonance, the amplitudes of a dynamic system increase in accordance with a power law (generally linear), whereas in parametric resonance the increase is exponential.

Both these features render parametric resonance more "dangerous" (or "useful," depending on the situation) than ordinary resonance.

It turns out that the principal characteristics (lengths of instability intervals, vibrational modes, relative build-up intensities) are given, in the first approximation, by extremely simple formulas, so that it is often possible to draw important conclusions without any computations, by simply studying the differential equations.

We now proceed to a rigorous exposition of the problem and a derivation of these formulas.

§ 1. FIRST-APPROXIMATION FORMULAS FOR BOUNDARIES OF DYNAMIC INSTABILITY DOMAINS

1.1. Dynamic instability domains for canonical systems. Critical frequencies

Let S be some dynamic system, capable of performing only vibrational motion, described by the vector equation

$$\frac{d\mathbf{x}}{dt} = \mathbf{C}\mathbf{x}. \tag{1.1}$$

We assume that the constant real $2k \times 2k$ matrix $\mathbf{C}$ is reducible to diagonal form with pure imaginary elements $\pm i\omega_\varkappa$ ($\varkappa = 1, \ldots, k$; $|\omega_\varkappa|$ are the natural frequencies of the system).

We shall refer to (1.1) as the unperturbed equation. Suppose now that certain parameters of the system S begin to vary periodically with frequency ϑ and small amplitudes depending on a small parameter ε. The perturbed motion of the system is described by a vector equation

$$\frac{d\mathbf{x}}{dt}=[\mathbf{C}+\varepsilon\mathbf{B}_1(\vartheta t)+\varepsilon^2\mathbf{B}_2(\vartheta t)+\ldots]\,\mathbf{x}, \tag{1.2}$$

where

$$\mathbf{B}_j(s+2\pi)=\mathbf{B}_j(s) \qquad (j=1,\ 2,\ \ldots).$$

We assume that the matrices $\mathbf{B}_j(\vartheta t)$ have the properties listed in the introduction to Chapter IV, and that the series converges in the same sense. We refer to (1.2) as the perturbed equation. Unlike (1.1), the perturbed equation may have unbounded solutions for arbitrarily small ε and suitable ϑ. In that case we shall say that the system S is parametrically excited or that parametric resonance sets in.

Consider the half-plane $\vartheta \geqslant 0$ on the $\varepsilon\vartheta$-plane. Let us say that a point $(\varepsilon_0,\ \vartheta_0)$ is unstable if it has the property that for all nonnegative ε and ϑ sufficiently close to ε_0 and ϑ_0 equation (1.2) has a solution which is not bounded as $t\rightarrow+\infty$. We shall see later that the set of all unstable points in the neighborhood of the ϑ-axis is the union of countably many wedge-shaped connected domains, as shown by the hatched regions in Figure 22, known as dynamic instability domains.

The values of ϑ_0 corresponding to the points $(0,\ \vartheta_0)$ on the ϑ-axis where dynamic instability domains are "attached" are called the critical frequencies. To be precise: ϑ_0 is a critical frequency if for any $\delta>0$ there exist $\varepsilon,\ \vartheta,\ |\varepsilon|<\delta,\ \ |\vartheta-\vartheta_0|<\delta$, for which equation (1.2) has solutions which are unbounded as $t\rightarrow+\infty$.

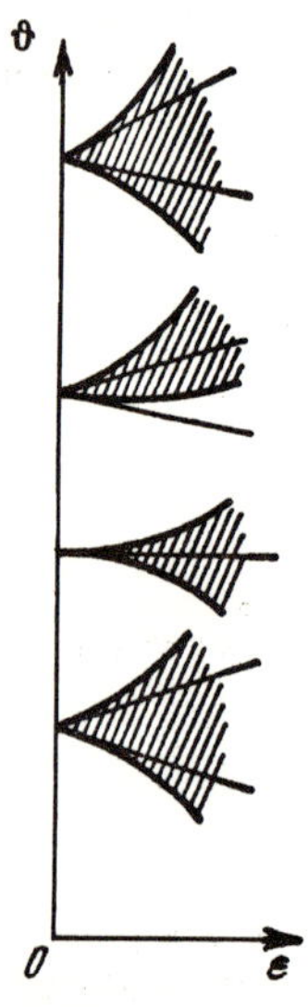

FIGURE 22.

It is easily shown that, if the matrices $\mathbf{B}_j(\vartheta t)$ in equation (1.2) are "too general" in form, any frequency is critical. In applications, however, the matrices $\mathbf{C},\ \mathbf{B}_j(\vartheta t)$ in (1.1), (1.2) have a special form. The most important case is that in which equations (1.1) and (1.2) are canonical. We shall first consider a more general case, assuming only that for any $\varepsilon>0$ the multipliers of equation (1.2) are symmetric about the unit circle.

Theorem I. Assume that for any $\varepsilon>0$ the spectrum (set of multipliers) of equation (1.2) is symmetric about the unit circle. Then the critical frequencies must be numbers of the form

$$\vartheta=\frac{1}{m}\,|\,\omega_j\pm\omega_h\,| \tag{1.3}$$

$$(j,\ h=1,\ \ldots,\ k;\ m=1,\ 2,\ \ldots).$$

Proof. Let ϑ_0 be a critical frequency. We shall show that for $\varepsilon=0,\ \vartheta=\vartheta_0$ equation (1.2) has a multiplier

ρ_0 of multiplicity at least 2 on the unit circle. It follows from the definition of critical frequencies that there is a sequence

$$\{\varepsilon_l, \vartheta_l\} \to (0, \vartheta_0)$$

such that equation (1.2) with parameters $\varepsilon=\varepsilon_l$, $\vartheta=\vartheta_l$ has a multiplier ρ_l either outside the unit circle or on the unit circle but with a multiple elementary divisor. There are two possibilities: 1) there is an infinite set of multipliers ρ_l outside the unit circle; 2) there are finitely many such multipliers but there is an infinite set of multipliers ρ_l on the unit circle with multiple elementary divisors. Consider the first case. The multipliers of equation (1.2) are continuous functions of ε and ϑ. Hence, as $l\to\infty$ the multipliers ρ_l tend to some multiplier $\rho^{(0)}$ of equation (1.1) with period $T_0=2\pi/\vartheta_0$. Since the spectrum is symmetric, equation (1.2) with $\varepsilon=\varepsilon_l$, $\vartheta=\vartheta_l$ has a multiplier $\rho_l'=\overline{\rho_l}^{-1}$ symmetric to ρ_l with respect to the unit circle. Since $|\rho^{(0)}|=1$ it is also true that $\rho_l'\to\rho^{(0)}$ $(l\to\infty)$. This means that $\rho^{(0)}$ is at least a twofold multiplier of equation (1.1). (If $\rho^{(0)}$ were simple, it would follow by continuity that $\rho^{(0)}$ has a neighborhood in which equation (1.1), $\varepsilon=\varepsilon_l$, $\vartheta=\vartheta_l$, has only one multiplier for sufficiently large l.) In the second case, one proves similarly that $\rho_l\to\rho_0$, where ρ_0 is a multiplier of multiplicity at least 2.

Equation (1.1) with period $T_0=2\pi/\vartheta_0$ has the monodromy matrix

$$\mathbf{X}(T_0, 0)=\exp\left(\frac{2\pi}{\vartheta_0}\mathbf{C}\right),$$

and so its multipliers are

$$\rho^{(\pm\varkappa)}=\exp\left(\pm\frac{2\pi i}{\vartheta_0}\omega_\varkappa\right)\ (\varkappa=1,\dots, k).$$

Two multipliers $\rho^{(j)}$ and $\rho^{(h)}$ $(j, h=\pm1,\dots, \pm k)$ coincide if and only if, for some natural number m,

$$\frac{2\pi}{\vartheta_0}\omega_j=\pm2\pi m\pm\frac{2\pi}{\vartheta_0}\omega_h,$$

whence, since $\vartheta_0>0$, we deduce (1.3), completing the proof.

We now assume that equations (1.1), (1.2) (for real ε) are canonical. In other words,

$$\left.\begin{aligned}&\mathbf{C}=\mathbf{J}^{-1}\mathbf{H}_0,\quad \mathbf{J}=\begin{pmatrix}\mathbf{0} & -\mathbf{I}_k\\ \mathbf{I}_k & \mathbf{0}\end{pmatrix},\quad \mathbf{H}_0^\tau=\mathbf{H}_0,\\ &\mathbf{B}_l(s)=\mathbf{J}^{-1}\mathbf{H}_l(s),\quad \mathbf{H}_l(s)^\tau=\mathbf{H}_l(s)\ (l=1, 2,\dots),\end{aligned}\right\}\tag{1.4}$$

where $\mathbf{H}_0$, $\mathbf{H}_1(s)$, $\mathbf{H}_2(s)$, ... are real matrices.

We recall the main definitions and results of Chapter III. Let $i\omega_\nu$ be a simple eigenvalue and $\mathbf{c}_\nu$ a corresponding eigenvector of $\mathbf{C}$:

$$\mathbf{C}\mathbf{c}_\nu=i\omega_\nu\mathbf{c}_\nu.$$

We call $i\omega_\nu$ an **eigenvalue of the first kind** (§III. 1.2) if

$$\langle \mathbf{c}_\nu, \mathbf{c}_\nu \rangle = i(\mathbf{J}\mathbf{c}_\nu, \mathbf{c}_\nu) > 0,$$

and of the **second kind** if $\langle \mathbf{c}_\nu, \mathbf{c}_\nu \rangle < 0$. Now let $i\omega_\nu$ be an eigenvalue of multiplicity r. Its elementary divisors must be simple, since by assumption $\mathbf{C}$ is reducible to diagonal form. Consequently, there are r linearly independent eigenvectors

$$\mathbf{C}\mathbf{c}_\rho = i\omega_\nu \mathbf{c}_\rho \quad (\rho = 1, \ldots, r).$$

These vectors may be chosen so that

$$\langle \mathbf{c}_\sigma, \mathbf{c}_\rho \rangle = 0 \quad (\sigma, \rho = 1, \ldots, r;\ \sigma \neq \rho),$$

and they may be normalized so that

$$\langle \mathbf{c}_\sigma, \mathbf{c}_\sigma \rangle = \pm 1 \ (\delta = 1, \ldots, r).$$

Let p be the number of positive ones and q the number of negative ones in the diagonal matrix $\| \langle \mathbf{c}_\sigma, \mathbf{c}_\rho \rangle \|_1^r$:

$$\| \langle \mathbf{c}_\sigma, \mathbf{c}_\rho \rangle \|_1^r = \begin{pmatrix} \mathbf{I}_p & 0 \\ 0 & -\mathbf{I}_q \end{pmatrix} \quad (p+q=r).$$

We shall then say that p eigenvalues of the first kind and q of the second **coincide at the point** $i\omega_\nu$.

If $q=0\,(p=0)$ the p-fold (q-fold) eigenvalue $i\omega_\nu$ of the matrix $\mathbf{C}$ is called a **p-fold (q-fold) eigenvalue of the first (second) kind**. In each of these two cases the eigenvalue $i\omega_\nu$ is said to be **definite**.

If $p \neq 0$, $q \neq 0$, the $(p+q)$-fold eigenvalue $i\omega_\nu$ is said to be **of mixed kind** or **indefinite**.

By Lemma I in §III. 2. 7, the matrix $\mathbf{C} = \mathbf{J}^{-1}\mathbf{H}_0$ has k eigenvalues of the first kind

$$i\omega_1, i\omega_2, \ldots, i\omega_k \tag{1.5}$$

(some of which may coincide). Assume that they are indexed so that

$$\omega_1 \leqslant \omega_2 \leqslant \ldots \leqslant \omega_k.$$

Then by Lemma III of §III.2. 7 the numbers

$$-i\omega_1, -i\omega_2, \ldots, -i\omega_k \tag{1.6}$$

are the eigenvalues of the second kind. Note that each number $\pm i\omega_\varkappa$ appears in (1.5) or (1.6) a number of times corresponding to its multiplicity. If the matrix $\mathbf{C}$ has an eigenvalue $i\omega^{(0)}$ of mixed kind (and p eigenvalues of the first kind and q of the second coincide at $i\omega^{(0)}$), the number $i\omega^{(0)}$ appears p times in the sequence (1.5) and q times in (1.6).

Theorem II. Let (1.1) and (1.2) be canonical equations and let the numbers (1.5) be the eigenvalues of the first kind for the matrix $\mathbf{C}=\mathbf{J}^{-1}\mathbf{H}_0$. *Suppose that* $\mathbf{C}$ *has no eigenvalues of mixed kind, i.e.,*

$$\omega_j+\omega_h\neq 0 \qquad (j, h=1, \ldots, k).$$

Then the critical frequencies must be numbers of the form

$$\vartheta=\frac{|\omega_j+\omega_h|}{m} \qquad (j, h=1, \ldots, k;\ m=1, 2, \ldots). \tag{1.7}$$

Proof. The assertion follows at once from Krein's strong stability theorem (§III. 1.4). Indeed, the numbers

$$\rho_j=\exp\frac{2\pi i\omega_j}{\vartheta} \qquad (j=1, \ldots, k)$$

are the multipliers of the first kind for the unperturbed equation (1.1) with period $T=2\pi/\vartheta$, and the numbers $\bar{\rho}_j$ (see Lemma III, §III. 2.7) are its multipliers of the second kind. If $\rho_j\neq\bar{\rho}_h$, it follows from the strong stability theorem that for sufficiently small ε all solutions of equation (1.2) are bounded. Thus the only possible critical frequencies are the values of $\vartheta>0$ such that for some $j,\ h=1, \ldots, k$,

$$\rho_j=\bar{\rho}_h,$$

whence follows (1.7). This proves Theorem II.

Remark. We shall show below that any of the frequencies (1.7) is indeed a critical frequency for some perturbed equation (1.2). We shall also show that if $\omega_j+\omega_h=0$ for some j and h then any frequency $\vartheta>0$ is critical for suitable perturbations.

We emphasize that for any fixed m formula (1.7) gives only half as many values as formula (1.3) (some of these values may be repeated).

The frequencies (1.7) for $j=h$, i.e., $\vartheta=2|\omega_j|/m$ $(j=1, \ldots, k;\ m=1, 2, \ldots)$, will be called principal-resonance frequencies, and for $j\neq h$ combination-resonance frequencies.

1.2. Eigenvalues of first and second kinds of stationary canonical systems

We again consider a stable canonical system (1.1) $(\mathbf{C}=\mathbf{J}^{-1}\mathbf{H}_0,\ \mathbf{H}_0^{\tau}=\mathbf{H}_0)$. As before, let the numbers (1.5) be the eigenvalues of the first kind of the matrix $\mathbf{C}=\mathbf{J}^{-1}\mathbf{H}_0$ and (1.6) its eigenvalues of the second kind. From the $2k$ numbers (1.5) and (1.6) we choose k numbers with nonnegative imaginary parts and denote them by $i\tilde{\omega}_j$ $(j=1, \ldots, k)$. Let $0\leqslant\tilde{\omega}_1\leqslant\tilde{\omega}_2\leqslant\ldots\leqslant\tilde{\omega}_k$. In this new notation, the eigenvalues of $\mathbf{C}$ will be the numbers $\pm i\tilde{\omega}_j$ $(j=1, \ldots, k)$. The numbers $\tilde{\omega}_j>0$ are the natural frequencies of the system S described by the vector equation (1.1). If $\det \mathbf{H}_0 = 0$, then $\tilde{\omega}_1 = 0$. In that case, in the interests of a unified presentation, we shall also call the numbers $\tilde{\omega}_j = 0$ natural frequencies. A frequency $\tilde{\omega}_j \geqslant 0$ will be called a natural frequency of the first (second) kind if $i\tilde{\omega}_j$ is an eigenvalue of the first (second) kind of the matrix $\mathbf{C} = \mathbf{J}^{-1}\mathbf{H}_0$.

The following rephrased version of Theorem II in §1.1 (and the remark following it) is often more convenient.

Theorem. Suppose that the system S has no coinciding natural frequencies of different kinds. Then the critical frequencies must have the form

$$\vartheta = \frac{\tilde{\omega}_j + \tilde{\omega}_h}{m} \quad (m = 1, 2, \ldots), \tag{1.8}$$

where $\tilde{\omega}_j$ and $\tilde{\omega}_h$ are natural frequencies of one kind, or

$$\vartheta = \frac{\tilde{\omega}_j - \tilde{\omega}_h}{m} \quad (m = 1, 2, \ldots; j > h), \tag{1.9}$$

where $\tilde{\omega}_j$ and $\tilde{\omega}_h$ are natural frequencies of different kinds. Moreover, any of the frequencies (1.8), (1.9) is actually critical for a suitable parametric excitation.

If the system S has equal natural frequencies of different kinds (i.e., $\omega_j + \omega_h = 0$ for some j and h), any frequency $\vartheta \neq 0$ is critical for a suitable parametric excitation.

The nature of the parametric excitation with respect to which a frequency (1.8) or (1.9) (or any frequency, if $\omega_j + \omega_h = 0$) is critical will be described later in this section. For the moment, we assert only that if $i\tilde{\omega}_j$ and $i\tilde{\omega}_h$ are simple eigenvalues, then the frequencies (1.8) or (1.9) are certainly critical if $(\mathbf{H}_1^{(m)}\mathbf{c}_j, \mathbf{c}_h) \neq 0$, where $\mathbf{c}_j, \mathbf{c}_h$ are corresponding eigenvectors of $\mathbf{C} = \mathbf{J}^{-1}\mathbf{H}_0$ and $\mathbf{H}_1^{(m)}$ are the Fourier coefficients of the matrix $\mathbf{H}_1(t)$:

$$\mathbf{H}_1(t) \sim \sum_m \exp\left(\frac{2\pi i t}{T}\right) \mathbf{H}_1^{(m)}.$$

We claim that if $\mathbf{H}_0 > 0$ all the natural frequencies are of the same — first — kind. Indeed, in that case, for the eigenvalues (1.5) of the first kind,

$$\langle \mathbf{c}_j, \mathbf{c}_j \rangle > 0, \quad \langle \mathbf{C}\mathbf{c}_j, \mathbf{c}_j \rangle = i\omega_j \langle \mathbf{c}_j, \mathbf{c}_j \rangle,$$
$$\langle \mathbf{C}\mathbf{c}_j, \mathbf{c}_j \rangle = i(\mathbf{J}\mathbf{J}^{-1}\mathbf{H}_0\mathbf{c}_j, \mathbf{c}_j) = i(\mathbf{H}_0\mathbf{c}_j, \mathbf{c}_j),$$

whence $\omega_j = (\mathbf{H}_0\mathbf{c}_j, \mathbf{c}_j)/\langle \mathbf{c}_j, \mathbf{c}_j \rangle > 0$, so that $\tilde{\omega}_1 = \omega_1, \ldots, \tilde{\omega}_k = \omega_k$ are the natural frequencies of the first kind. Similarly one proves that if $\mathbf{H}_0 < 0$ all the natural frequencies are of the second kind.

Let $\mathbf{H}_0$ be a sign definite matrix with l negative eigenvalues. We claim that then l is even, $l = 2q$, and of the natural frequencies $\tilde{\omega}_1, \ldots, \tilde{\omega}_k$ exactly $k-q$ are of the first and exactly q of the second kind. In fact, by Lemma II of §III.6.5, there exists a symplectic matrix $\mathbf{U}$ such that $\mathbf{U}^\tau\mathbf{H}_0\mathbf{U} = \mathbf{H}_0'$, where $\mathbf{H}_0' = \operatorname{diag}(\omega_1, \ldots, \omega_k, \omega_1, \ldots, \omega_k)$. The matrices $\mathbf{H}_0$ and $\mathbf{H}_0'$ have the same number of negative eigenvalues (by the law of inertia for quadratic forms; see §I.1.14). It follows from the form of the matrix $\mathbf{H}_0'$ that this number is even,

$l=2q$. We have $\tilde{\omega}_j=-\omega_j$ if $\omega_j<0$, and $\tilde{\omega}_j=\omega_j$ if $\omega_j>0$; thus the numbers of natural frequencies of the first and second kinds are respectively $k-q$ and q, as claimed.

Note that the above argument makes essential use of the assumption that system (1.1) is stable.

We now consider the special case that the system S is described by the vector equation

$$\mathbf{M}\frac{d^2\mathbf{y}}{dt^2}+\mathbf{Q}\frac{d\mathbf{y}}{dt}+\mathbf{P}\mathbf{y}=0, \tag{1.10}$$

where $\mathbf{M}^\tau=\mathbf{M}>0$, $\mathbf{Q}^\tau=-\mathbf{Q}$, $\mathbf{P}^\tau=\mathbf{P}$ are real $k\times k$ matrices. Recall that this system is stable (any solution $y=y(t)$ is bounded on $(-\infty, +\infty)$). As shown in § II. 3.2, the substitution $\mathbf{z}=\mathbf{y}+\mathbf{M}\frac{d\mathbf{y}}{dt}$ reduces equation (1.10) to a canonical equation $\tilde{\mathbf{J}}\frac{d\mathbf{z}}{dt}=\tilde{\mathbf{H}}\mathbf{z}$, where

$$\tilde{\mathbf{J}}=\begin{pmatrix}-\mathbf{Q} & -\mathbf{I}_k\\ \mathbf{I}_k & 0\end{pmatrix},\quad \tilde{\mathbf{H}}=\begin{pmatrix}\mathbf{P} & 0\\ 0 & \mathbf{M}^{-1}\end{pmatrix}.$$

This equation may be reduced by a suitable substitution $\mathbf{z}=\mathbf{S}\mathbf{x}$, where $\mathbf{S}$ is a real matrix such that $\det\mathbf{S}=1$, to the canonical equation $\mathbf{J}\frac{d\mathbf{x}}{dt}=\mathbf{H}_0\mathbf{x}$, where $\mathbf{H}_0=\mathbf{S}^*\tilde{\mathbf{H}}\mathbf{S}$ (see § II. 3.2 and § III. 2. 3). It is clear from the above formulas that the matrices $\mathbf{H}_0$, $\tilde{\mathbf{H}}$ and $\mathbf{P}$ have the same number l of negative eigenvalues. The statement just proved, that l is an even number, implies the well-known theorem of Kelvin (for equation (1.10)). We have proved the following proposition.

The number of natural frequencies of the first (second) kind for the system S described by equation (1.10) is equal to the number of positive (negative) eigenvalues of the matrix $\mathbf{P}$.

1.3. Notation. Auxiliary results

Consider the canonical system (1.2), assuming (1.4). The numbers (1.5) are eigenvalues of the first kind for $\mathbf{C}=\mathbf{J}^{-1}\mathbf{H}_0$. We introduce negative subscripts, letting

$$i\omega_{-\varkappa}=-i\omega_\varkappa \qquad (\varkappa=1,\ldots,k) \tag{1.11}$$

denote the eigenvalues of the second kind. Let us assume that the matrix $\mathbf{C}$ has no eigenvalues of mixed kind, i. e.,

$$\omega_j+\omega_h\neq 0 \qquad (j,h=1,\ldots,k). \tag{1.12}$$

Let $\mathbf{c}_\nu$ $(\nu=\pm1,\ldots\pm k)$ be eigenvectors of the matrix $\mathbf{C}$:

$$\mathbf{C}\mathbf{c}_\nu=i\omega_\nu\mathbf{c}_\nu \qquad (\nu=\pm1,\ldots,\pm k). \tag{1.13}$$

By the lemma of § III. 2. 4, we may assume that the eigenvectors are J-orthogonal:

$$\langle \mathbf{c}_j, \mathbf{c}_h \rangle = \delta_{jh} \operatorname{sign} j \qquad (j, h = \pm 1, \ldots, \pm k). \tag{1.14}$$

If all the multipliers of system (1.1) are distinct, it follows from the Lyapunov-Poincaré theorem (§ II. 3. 5) that for sufficiently small $|\varepsilon|$ the trivial solution of system (1.2) is stable. For critical frequencies ϑ, system (1.1) has multiple multipliers. Let ϑ_0 be some such frequency. The multipliers of the unperturbed system (1.1) are

$$\rho_\nu^{(0)} = \exp \frac{2\pi i \omega_\nu}{\vartheta_0} \qquad (\nu = \pm 1, \ldots, \pm k).$$

Even though the unperturbed system is independent of t, in view of the perturbations we shall assign it the period $T = 2\pi/\vartheta_0$; this explains the preceding formula.

Divide the eigenvalues $i\omega_\nu$ $(\nu = \pm 1, \ldots, \pm k)$ of $\mathbf{C}$ into congruence classes modulo $i\vartheta_0 = 2\pi i/T$. Each class determines a unique $\rho_\nu^{(0)}$. Let $i\omega_{j_1}, \ldots, i\omega_{j_r}$ be one of these congruence classes, determining an r-fold multiplier

$$\rho^{(0)} = \exp \frac{2\pi i \alpha_0}{\vartheta_0}.$$

Define integers m_j by

$$\omega_j = \alpha_0 + m_j \vartheta_0 \qquad (j = j_1, \ldots, j_r). \tag{1.15}$$

Formula (1.7) may be written

$$\vartheta_0 = \frac{\omega_j + \omega_h}{m} \quad (j, h = \pm 1, \ldots, \pm k; \ m = 1, 2, \ldots) \tag{1.16}$$

for suitable j and h. Moreover, we then have $\rho_j^{(0)} = \rho_{-h}^{(0)}$, so that $r \geqslant 2$. Note that the most general case is that of $r = 2$. The case $r > 2$, when

$$\vartheta_0 = \frac{\omega_j + \omega_h}{m_1} = \frac{\omega_{j_1} + \omega_{h_1}}{m} = \ldots, \tag{1.17}$$

is in a certain sense exceptional,* since then the numbers ω_j must be interrelated in some way.

Thus, at the frequencies $\vartheta = \vartheta_0$ of interest, there must be multiple multipliers on the upper closed unit semicircle of the complex ρ-plane. The case in which the multiplicity of each of these multipliers is 2 will be referred to as *fundamental*; otherwise, we shall speak of the *singular* case.

* The equalities (1.17) imply that either some r is greater than 2 or there are several multipliers on the upper closed unit semicircle, of multiplicity $r \geqslant 2$, for which the equalities (1.16) are satisfied.

1.4. Characteristic exponents on a ray $\vartheta=\vartheta_0+\mu\varepsilon$

Introducing a new independent variable $s=\vartheta t$ in equation (1.2), we obtain

$$\frac{d\mathbf{x}}{ds}=\frac{1}{\vartheta}\left[\mathbf{C}+\varepsilon\mathbf{B}_1(s)+\varepsilon^2\mathbf{B}_2(s)+\ldots\right]\mathbf{x}. \tag{1.18}$$

We shall first derive formulas for the characteristic exponents for fixed $\vartheta=\vartheta_0\ (\mu=0)$. As before, let $i\omega_{j_1}, \ldots, i\omega_{j_r}$ be one of the classes (1.15) of eigenvalues of $\mathbf{C}$. Then

$$i\frac{\omega_j}{\vartheta_0}=i\frac{\alpha_0}{\vartheta_0}+im_j \qquad (j=j_1, \ldots, j_r)$$

is the corresponding class of eigenvalues of the matrix $\frac{1}{\vartheta_0}\mathbf{C}$, corresponding to an r-fold characteristic exponent $i\alpha_0/\vartheta_0$ of the unperturbed equation (1.18) at $\vartheta=\vartheta_0$, i. e., the equation

$$\frac{d\mathbf{x}}{ds}=\frac{1}{\vartheta_0}\mathbf{C}\mathbf{x}.$$

By §IV. 3. 2, for each r-fold characteristic exponent of this equation we have r characteristic exponents of equation (1.18) at $\vartheta=\vartheta_0$:

$$\alpha_j(\varepsilon)=i\frac{\alpha_0}{\vartheta_0}+i\nu_j\varepsilon+O(\varepsilon^{1+\delta}) \qquad (j=j_1, \ldots, j_r), \tag{1.19}$$

where ν_j are the roots of the equation (IV. 3.10)

$$\det\|\sigma_{jh}+\nu\gamma_{jh}\|_{j,\,h=j_1,\,\ldots,\,j_r}=0, \tag{1.20}$$

in which (see formulas (IV. 3. 7), (IV. 3. 9))

$$\left.\begin{aligned}
&\gamma_{jh}=\langle\mathbf{c}_j, \mathbf{c}_h\rangle\equiv i(\mathbf{J}\mathbf{c}_j, \mathbf{c}_h)=\begin{cases}0, & j\neq h,\\ 1, & j=h>0,\\ -1, & j=h<0,\end{cases}\\
&\sigma_{jh}=-\frac{1}{\vartheta_0}(\mathbf{H}_1^{(m_h-m_j)}\mathbf{c}_j, \mathbf{c}_h) \quad (j, h=j_1, \ldots, j_r),\\
&\mathbf{H}_1(s)=\mathbf{J}\mathbf{B}_1(s)\sim\sum_m e^{ims}\mathbf{H}_1^{(m)}.
\end{aligned}\right\} \tag{1.21}$$

Note that the subscripts $j_1, \ldots, j_r$ are numbers in the set $\pm1, \ldots, \pm k$, corresponding to the class under consideration.

We write the last formula as

$$\sigma_{jh}=\sigma_{jh}(0)=-\frac{1}{\vartheta_0}\chi_{jh}(0), \quad \chi_{jh}=\chi_{jh}(0)=(\mathbf{H}_1^{(m_h-m_j)}\mathbf{c}_j, \mathbf{c}_h) \qquad (j, h=j_1, \ldots, j_r). \tag{1.22}$$

The zero argument emphasizes that this formula is true for $\vartheta=\vartheta_0$ $(\mu=0)$; recall that ϑ_0 is a critical frequency.

Now consider a ray through the point $(0, \vartheta_0)$ in the first quadrant of the $\varepsilon\vartheta$-plane ($\varepsilon \geqslant 0$, $\vartheta > 0$):

$$\vartheta = \vartheta_0 + \mu\varepsilon$$

(like ϑ_0, μ is held fixed). We shall determine the characteristic exponents of system (1.2) on this ray for sufficiently small ε. We use the expansion

$$\frac{1}{\vartheta} = \frac{1}{\vartheta_0 + \mu\varepsilon} = \frac{1}{\vartheta_0}\left[1 - \frac{\mu}{\vartheta_0}\varepsilon + \frac{\mu^2}{\vartheta_0^2}\varepsilon^2 - \ldots\right]$$

and write equation (1.18) as

$$\frac{d\mathbf{x}}{ds} = \frac{1}{\vartheta_0}\left[\mathbf{C} + \varepsilon\left(\mathbf{B}_1(s) - \frac{\mu}{\vartheta_0}\mathbf{C}\right) + \varepsilon^2 \ldots\right]\mathbf{x}. \tag{1.23}$$

To adapt formula (IV. 3.9) to this equation, we must replace $\mathbf{H}_1$ by

$$\mathbf{J}\left[\frac{1}{\vartheta_0}\mathbf{B}_1(s) - \frac{\mu}{\vartheta_0^2}\mathbf{C}\right] = \frac{1}{\vartheta_0}\mathbf{H}_1(s) - \frac{\mu}{\vartheta_0^2}\mathbf{H}_0,$$

where

$$\mathbf{H}_1(s) = \mathbf{J}\mathbf{B}_1(s), \quad \mathbf{H}_0 = \mathbf{J}\mathbf{C}.$$

Then, by (IV. 3. 9),

$$\sigma_{jh}(\mu) = -\left(\left[\frac{1}{\vartheta_0}\mathbf{H}_1 - \frac{\mu}{\vartheta_0^2}\mathbf{J}\mathbf{C}\right]^{(m_h - m_j)}\mathbf{c}_j,\ \mathbf{c}_h\right) =$$
$$= -\frac{1}{\vartheta_0}\left[(\mathbf{H}_1^{(m_h - m_j)}\mathbf{c}_j,\ \mathbf{c}_h) - \frac{\mu}{\vartheta_0}(\mathbf{J}\mathbf{C}\mathbf{c}_j,\ \mathbf{c}_h)^{(m_h - m_j)}\right] \qquad (j,\ h = j_1,\ \ldots,\ j_r).$$

Since $\mathbf{C}\mathbf{c}_j = i\omega_j\mathbf{c}_j$, we can write the last term as

$$(\mathbf{J}\mathbf{C}\mathbf{c}_j,\ \mathbf{c}_h) = i\omega_j(\mathbf{J}\mathbf{c}_j,\ \mathbf{c}_h) = \omega_j\langle\mathbf{c}_j,\ \mathbf{c}_h\rangle = \omega_j\gamma_{jh}.$$

Using (1.22), we get

$$\sigma_{jh}(\mu) = \sigma_{jh}(0) + \frac{\mu}{\vartheta_0^2}\omega_j\gamma_{jh},$$

or, in other words

$$\sigma_{jh}(\mu) = -\frac{1}{\vartheta_0}\chi_{jh}(\mu), \text{ where } \chi_{jh}(\mu) = \chi_{jh}(0) - \frac{\mu}{\vartheta_0}\omega_j\gamma_{jh}. \tag{1.24}$$

Thus the values of ν_j for the characteristic exponents (1.19) of equation (1.18) with $\vartheta = \vartheta_0 + \mu\varepsilon$ are defined by equation (1.20), where $\sigma_{jh} = \sigma_{jh}(\mu)$ is given by (1.24).

1.5. Case of twofold multiplier of mixed kind $(\omega_j+\omega_h\neq 0)$

Consider the case that for $\varepsilon=0$ two multipliers of different kinds coincide at a point $\rho^{(0)}$ on the unit circle:

$$\rho^{(0)}=\exp\frac{2\pi i\omega_{-j}}{\vartheta_0}=\exp\frac{2\pi i\omega_h}{\vartheta_0}\qquad(\omega_{-j}=-\omega_j).$$

Assuming that $\omega_j+\omega_h\neq 0$ and (to fix ideas) that $\omega_h>\omega_{-j}$, we obtain

$$\vartheta_0=\frac{1}{m}(\omega_j+\omega_h)\qquad(m=1,\ 2,\ \ldots).\tag{1.25}$$

We thus have a class of two numbers $\{\omega_{-j},\ \omega_h\}$, and by (1.15)

$$\left.\begin{aligned}\omega_{-j}&=\alpha_0+m_{-j}\vartheta_0,\\ \omega_h&=\alpha_0+m_h\vartheta_0.\end{aligned}\right\}\tag{1.26}$$

Hence

$$\omega_j+\omega_h\equiv\omega_h-\omega_{-j}=(m_h-m_{-j})\,\vartheta_0,$$

and so

$$m_h-m_{-j}=m.$$

In view of the fact that $i\omega_h$ is an eigenvalue of $\mathbf{C}$ of the first kind and $i\omega_{-j}$ an eigenvalue of the second kind, i. e., $\gamma_{hh}=+1,\ \gamma_{-j,\,-j}=-1$, we see that the equation (1.20) for ν is

$$\begin{vmatrix}\sigma_{-j,\,-j}(\mu)-\nu & \sigma_{-jh}(\mu)\\ \sigma_{h,\,-j}(\mu) & \sigma_{hh}(\mu)+\nu\end{vmatrix}=0,$$

or, in the notation of (1.24),

$$\nu^2-\frac{1}{\vartheta_0}[\chi_{hh}(\mu)-\chi_{-j,\,-j}(\mu)]\,\nu-\frac{1}{\vartheta_0^2}[\chi_{hh}(\mu)\,\chi_{-j,\,-j}(\mu)-\chi_{h,\,-j}(\mu)\,\chi_{-jh}(\mu)]=0.$$

Hence the corrections to the characteristic exponents are

$$\nu_{1,2}=\frac{1}{\vartheta_0}\left\{\frac{1}{2}[\chi_{hh}(\mu)-\chi_{-j,\,-j}(\mu)]\mp\sqrt{\Delta(\mu)}\right\},\tag{1.27}$$

where

$$\Delta(\mu)=\frac{1}{4}[\chi_{-j,-j}(\mu)+\chi_{hh}(\mu)]^2-\chi_{h,-j}(\mu)\,\chi_{-jh}(\mu).$$

We transform this expression as follows:

$$\chi_{-j,-j}(\mu)+\chi_{hh}(\mu)=\chi_{-j,-j}(0)+\chi_{hh}(0)-\frac{\mu}{\vartheta_0}(\omega_h-\omega_{-j})=$$
$$=\chi_{-j,-j}(0)+\chi_{hh}(0)-m\mu,$$
$$\chi_{h-j}(\mu)=\chi_{h,-j}(0)=(\mathbf{H}_1^{(m-j-m_h)}\mathbf{c}_h,\ \mathbf{c}_{-j})=(\mathbf{c}_h,\ \mathbf{H}_1^{(m_h-m-j)}\mathbf{c}_{-j})=\overline{\chi}_{-jh}(0),$$
$$\chi_{h,-j}(\mu)\,\chi_{-jh}(\mu)=\overline{\chi}_{-jh}(0)\,\chi_{-jh}(0)=|\chi_{-jh}(0)|^2.$$

Thus,

$$\Delta(\mu)=\frac{1}{4}(\chi_{-j,-j}(0)+\chi_{hh}(0)-m\mu)^2-|\chi_{-jh}(0)|^2. \tag{1.28}$$

Instability holds if $\Delta(\mu)<0$, i. e.,

$$\frac{1}{m}(\chi_{-j,-j}+\chi_{hh}-2|\chi_{-jh}|)<\mu<\frac{1}{m}(\chi_{-j,-j}+\chi_{hh}+2|\chi_{-jh}|).$$

The final result is as follows.

Suppose that when $\varepsilon=0,\ \vartheta=\vartheta_0=(\omega_j+\omega_h)/m$ $(m=1,\ 2,\ \ldots)$ *exactly two multipliers of different kinds coincide at the point* $\rho^{(0)}$. *Define*

$$\left.\begin{aligned}&\chi_{-j,-j}=(\mathbf{H}_1^{(0)}\overline{\mathbf{c}}_j,\ \overline{\mathbf{c}}_j),\quad \chi_{hh}=(\mathbf{H}_1^{(0)}\mathbf{c}_h,\ \mathbf{c}_h),\\&\chi_{-j,h}=(\mathbf{H}_1^{(m)}\overline{\mathbf{c}}_j,\ \mathbf{c}_h),\quad \mu_{1,2}=\frac{1}{m}(\chi_{-j,-j}+\chi_{hh}\mp2|\chi_{-jh}|),\end{aligned}\right\} \tag{1.29}$$

where $\mathbf{c}_j,\ \mathbf{c}_h$ *are eigenvectors of* $\mathbf{C}$ *belonging to the eigenvalues* $i\omega_j$ *and* $i\omega_h$ *and normalized by conditions (1.21). Then, for sufficiently small* $\varepsilon>0$ *and parameters* ε *and* ϑ *in the region*

$$\vartheta_0+\varepsilon\mu_1+\ldots<\vartheta<\vartheta_0+\varepsilon\mu_2+\ldots \tag{1.30}$$

the multipliers of system (1.2) near $\rho^{(0)}$ *will lie on different sides of the unit circle, one inside and the other outside; outside this region, however, the multipliers will be distinct and lie on the unit circle.*

The characteristic exponents of system (1.2) corresponding to these two multipliers (as functions of t, not of s) are

$$\alpha_{1,2}(\varepsilon)=i\alpha_0+i\vartheta_0\nu_{1,2}\varepsilon+O(\varepsilon^2).$$

By Theorem II of § IV. 2. 7, $\alpha_{1,2}$ are analytic functions of ε. In the region (1.30),

$$\operatorname{Re}\alpha_{1,2}=\mp\sqrt{-\Delta(\mu)}\varepsilon+O(\varepsilon^2).$$

The instability domain (1.30) will be called w i d e if $\mu_1<\mu_2$, n a r r o w if $\mu_1=\mu_2$. It follows from (1.29) that an instability domain attached at the point $(0,\vartheta_0)$ $(\vartheta_0=(\omega_j+\omega_h)/m)$ will be wide if

$$\chi_{-jh}=(\mathbf{H}_1^{(m)}\overline{\mathbf{c}}_j,\ \mathbf{c}_h)\neq0,$$

and narrow if $\chi_{-jh}=0$. The following should be kept in mind, however. If $\varepsilon=0$ and $\vartheta=\vartheta_0$ and there are $q\geqslant1$ pairs of multiple multipliers on the upper closed unit semicircle, then to each such pair corresponds a specific sector (1.30) and the instability domain is the (set-theoretic) union of these sectors.

Given a wide instability domain, let us consider the locus in this domain of the points (ε, ϑ) such that $\mathrm{Re}\,\alpha=\zeta$, where ζ is some small positive number. Inverting the series

$$\sqrt{-\Delta(\mu)}\,\varepsilon+O(\varepsilon^2)=\vartheta,$$

we obtain

$$\varepsilon=\frac{\zeta+O(\zeta^2)}{\sqrt{-\Delta(\mu)}}\,.$$

The corresponding curve is illustrated in Figure 23. The hatched region is the open domain in which the solution increases more rapidly than $e^{\zeta t}$ $(\mathrm{Re}\,\alpha>\zeta)$.

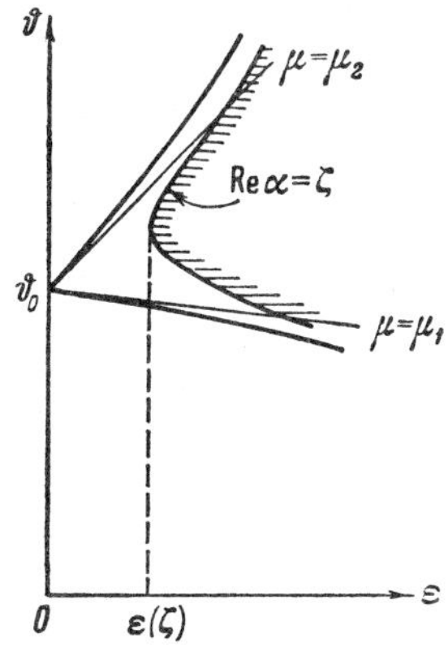

FIGURE 23.

Set

$$\inf_{\mathrm{Re}\,\alpha>\zeta}\varepsilon=\varepsilon(\zeta).$$

Then it follows from (1.28) that

$$\varepsilon(\zeta)=\frac{\zeta}{|\chi_{-jh}|}+O(\zeta^2).$$

1.6. On vibrational modes in instability domains

The eigenvector of the matrix $\tilde{K}(\varepsilon)$ (see §IV. 4) belonging to the characteristic exponent $\alpha(\varepsilon)$ is

$$\mathbf{a}(\varepsilon)=\mathbf{a}_0+\varepsilon\mathbf{b}+\ldots,$$

where

$$\mathbf{a}_0=\zeta_1\bar{\mathbf{c}}_j+\zeta_2\mathbf{c}_h \tag{1.31}$$

and the numbers ζ_1, ζ_2 are determined by the vector equation

$$\begin{Vmatrix}\sigma_{-j,-j}(\mu)-\nu & \sigma_{-jh}(\mu)\\ \sigma_{h,-j}(\mu) & \sigma_{hh}(\mu)+\nu\end{Vmatrix}\begin{pmatrix}\zeta_1\\ \zeta_2\end{pmatrix}=0. \tag{1.32}$$

Using (1.27), we write equation (1.32) as

$$\begin{Vmatrix}\frac{1}{2}[\chi_{-j,-j}(\mu)+\chi_{hh}(\mu)]+\sqrt{\Delta(\mu)} & \chi_{-jh}(0)\\ \chi_{h,-j}(0) & \frac{1}{2}[\chi_{-j,-j}(\mu)+\chi_{hh}(\mu)]-\sqrt{\Delta(\mu)}\end{Vmatrix}\begin{pmatrix}\zeta_1\\ \zeta_2\end{pmatrix}=0. \tag{1.33}$$

Assuming that $\chi_{-jh}=\chi_{-jh}(0)\neq 0$, we can set

$$\zeta_1^{(1)}=\chi_{-jh},\quad \zeta_2^{(1)}=-\frac{1}{2}[\chi_{-j,-j}(\mu)+\chi_{hh}(\mu)-m\mu]-\sqrt{\Delta(\mu)}, \tag{1.34}$$

since the vector $\mathbf{a}_0$ is determined uniquely up to a multiplicative constant.

Since $\alpha(\varepsilon)$ and $\mathbf{a}(\varepsilon)$ are respectively an eigenvalue and eigenvector of $\tilde{\mathbf{K}}(\varepsilon)$, we also have

$$e^{s\tilde{\mathbf{K}}(\varepsilon)}\mathbf{a}(\varepsilon)=e^{s\alpha(\varepsilon)}\mathbf{a}(\varepsilon).$$

The vector solution satisfying the initial conditions $\mathbf{x}(0,\varepsilon)=\mathbf{a}(\varepsilon)$ may be written

$$\mathbf{x}(s,\varepsilon)=\mathbf{X}(s,\varepsilon)\mathbf{a}(\varepsilon)=e^{s\mathbf{C}_0}[\mathbf{I}_{2k}+\varepsilon\tilde{\mathbf{F}}_1(s)+\ldots]e^{\alpha(\varepsilon)s}(\mathbf{a}_0+\varepsilon\mathbf{b}+\ldots).$$

Thus, if $\vartheta=\vartheta_0+\varepsilon\mu$, $\varepsilon>0$ is sufficiently small, and $\mu_1<\mu<\mu_2$, we have an unbounded increasing solution corresponding to the first root, $i\nu_1$:

$$\mathbf{x}^{(1)}(s,\mu,\varepsilon)=e^{s\mathbf{C}_0}e^{\left[i\frac{\alpha_0}{\vartheta_0}+\varepsilon i\nu_1(\mu)+\ldots\right]s}\{\mathbf{a}_0+\varepsilon[\mathbf{b}_1+\tilde{\mathbf{F}}_1(s)\mathbf{a}_0]+\ldots\},$$

which becomes at $\varepsilon=0$ a solution of the unperturbed system

$$\mathbf{x}^{(1)}(s,\mu,0)=e^{\left(\mathbf{C}_0+i\frac{\alpha_0}{\vartheta_0}\mathbf{I}_{2k}\right)s}\mathbf{a}_0.$$

In view of (1.15), we have

$$\left(\mathbf{C}_0+i\frac{\alpha_0}{\vartheta_0}\mathbf{I}_{2k}\right)\mathbf{c}_{-j}=-i\frac{\omega_j}{\vartheta_0}\bar{\mathbf{c}}_j,\quad \left(\mathbf{C}_0+i\frac{\alpha_0}{\vartheta_0}\mathbf{I}_{2k}\right)\mathbf{c}_h=i\frac{\omega_h}{\vartheta_0}\mathbf{c}_h$$

and by (1.31)

$$\mathbf{x}^{(1)}(s,\mu,0)=\zeta_1^{(1)}e^{-i\frac{\omega_j}{\vartheta_0}s}\bar{\mathbf{c}}_j+\zeta_2^{(1)}e^{i\frac{\omega_h}{\vartheta_0}s}\mathbf{c}_h, \tag{1.35}$$

and the corresponding unbounded solution is

$$\mathbf{x}^{(1)}(s, \mu, \varepsilon) = e^{\varepsilon[\beta_1(\mu)+O(\varepsilon)s]}[\mathbf{x}^{(1)}(s, \mu, 0)+O(\varepsilon)]. \tag{1.36}$$

Recall that $\zeta_1^{(1)}$ and $\zeta_2^{(1)}$ are the roots of equation (1.33), where $\beta=\beta_1(\mu)=i\nu_1(\mu)$ is the root (1.27) with the minus sign before the radical. For a wide instability domain, when $\chi_{-jh}\neq 0$, we have

$$\left.\begin{aligned}
&\zeta_1^{(1} = \chi_{-jh} = (\mathbf{H}_1^{(m)}\bar{\mathbf{c}}_j, \mathbf{c}_h),\\
&\zeta_2^{(1)} = -\frac{1}{2}(\chi_{-j,-j}+\chi_{hh}-m\mu)-\sqrt{\Delta(\mu)} =\\
&= -\frac{1}{2}[(\mathbf{H}_1^{(0)}\bar{\mathbf{c}}_j, \bar{\mathbf{c}}_j)+(\mathbf{H}_1^{(0)}\mathbf{c}_h, \mathbf{c}_h)-m\mu]-\\
&-\sqrt{\frac{1}{4}[(\mathbf{H}_1^{(0)}\bar{\mathbf{c}}_j, \bar{\mathbf{c}}_j)+(\mathbf{H}_1^{(0)}\mathbf{c}_h, \mathbf{c}_h)-m\mu]^2-|(\mathbf{H}_1^{(m)}\bar{\mathbf{c}}_j, \mathbf{c}_h)|^2}.
\end{aligned}\right\} \tag{1.37}$$

Formulas analogous to (1.35) and (1.36) may be derived for the vector solution corresponding to the second root $\nu_2(\mu)$, but now $\zeta_1^{(2)}$ and $\zeta_2^{(2)}$ are defined by the formulas obtained from (1.37) when the sign of the radical is reversed. This solution tends to zero as $t\to+\infty$.

In the case of combination resonance, we also have the complex conjugate solution $\bar{\mathbf{x}}$, increasing without bound as $t\to+\infty$. The solution (1.35) will be called the g e n e r a t i n g s o l u t i o n of the unperturbed system (1.2).

To summarize:

If $\vartheta=\vartheta_0+\mu\varepsilon$, $\vartheta_0=(\omega_j+\omega_h)/m$, $\mu_1<\mu<\mu_2$ and ε is sufficiently small, the effect of the perturbation on the general solution of the unperturbed system

$$\mathbf{x} = \sum_{l=-k}^{k} \pi_l \exp(i\omega_l t)\,\mathbf{c}_l$$

is to initiate a build-up of the two harmonics (principal resonance) of indices j and $-j$, or the four harmonics (combination resonance) of indices j, $-j$, h, $-h$.

If there are other coinciding multipliers of different kinds at $\varepsilon=0$, other harmonics begin to build up.

These harmonics are natural vibrations of the system. For example, if ω_j is the frequency of flexural vibrations of a shaft, corresponding to j nodes, and ω_h is the frequency of torsional vibrations corresponding to h nodes, then at $\vartheta=(\omega_j+\omega_h)/m$ intensive flexural-torsional vibrations will build up at precisely the above harmonics, with amplitudes that increase exponentially with growth factor ζ_1/ζ_2.

Let us examine the phenomenon in greater detail.

a) P r i n c i p a l r e s o n a n c e. $\rho^{(0)}=+1$ (or -1) is the only point on the unit circle at which two multipliers of different kinds coincide. The general solution for $\varepsilon=0$ is

$$\mathbf{x}\Big|_{\varepsilon=0} = \sum_{l=-m}^{m} \pi_l e^{i\omega_l t}\mathbf{c}_l. \tag{1.38}$$

If $\varepsilon\neq 0$, $\vartheta=2\omega_h/m+\mu\varepsilon$ $(\mu_1<\mu<\mu_2)$, the terms with $l\neq h, -h$ in (1.38) vary "weakly"—they remain bounded functions of time. The terms with $l=h, -h$

split into two sums, one of which begins to increase as $\exp(\beta(\mu)\varepsilon t)$ $(\beta>0)$, while the other begins to decrease as $\exp(-\beta(\mu)\varepsilon t)$:

$$\begin{aligned} \mathbf{x}= & \sum_{\substack{l=-k\\(l\neq h,\,-h)}}^{k} \pi_l \exp\{i[\omega_l+\varepsilon\beta_l(\mu)+O(\varepsilon^2)]t\}[\mathbf{c}_l+\varepsilon\mathbf{u}_l(\vartheta t)+\dots]+\\ & +\pi' e^{[\beta(\mu)\varepsilon+O(\varepsilon^2)]t}\{\zeta_1 e^{i\omega_h t}\mathbf{c}_h+\zeta_2 e^{-i\omega_h t}\bar{\mathbf{c}}_h+\dots\}+\\ & +\pi'' e^{[-\beta(\mu)\varepsilon+O(\varepsilon^2)]t}\{\zeta_1' e^{i\omega_h t}\mathbf{c}_h+\zeta_2' e^{-i\omega_h t}\bar{\mathbf{c}}_h+\dots\}. \end{aligned}$$

b) Combination resonance. Let $\rho^{(0)}$, $\bar{\rho}^{(0)}=\pm 1$ be the only points on the unit circle at which two multipliers of different kinds coincide. If

$$\varepsilon\neq 0,\quad \vartheta=(\omega_j+\omega_h)/m+\mu\varepsilon \qquad (\mu_1<\mu<\mu_2)$$

then four terms in (1.38), namely those indexed j, $-j$, h, $-h$, will split into terms, two of which begin to increase as $\exp[\beta(\mu)\varepsilon t]$, while the other two decrease as $\exp[-\beta(\mu)\varepsilon t]$.

The remaining terms vary "weakly," remaining bounded functions of time. We have

$$\begin{aligned} \mathbf{x}= & \sum_{\substack{l=-k\\(l\neq\pm j,\,\pm h)}}^{k} \pi_l \exp\{i[\omega_l+\varepsilon\beta_l(\mu)+O(\varepsilon^2)]t\}[\mathbf{c}_l+\varepsilon\mathbf{u}_l(\vartheta t)+\dots]+\\ & +\pi' e^{[\beta(\mu)\varepsilon+O(\varepsilon^2)]t}\{\zeta_j e^{i\omega_j t}[\mathbf{c}_j+\varepsilon\mathbf{u}_j(\vartheta t)+\dots]+\\ & \qquad +\zeta_h e^{i\omega_h t}[\mathbf{c}_h+\varepsilon\mathbf{u}_h(\vartheta t)+\dots]\}+\\ & +\pi'' e^{-[\beta(\mu)\varepsilon+O(\varepsilon^2)]t}\{\zeta_{-j} e^{-i\omega_j t}[\bar{\mathbf{c}}_j+\varepsilon\mathbf{u}_j(-\vartheta t)+\dots]+\\ & \qquad +\zeta_{-h} e^{-i\omega_h t}[\bar{\mathbf{c}}_h+\varepsilon\mathbf{u}_h(-\vartheta t)+\dots]\}. \end{aligned}$$

c) If there are several points on the upper closed unit semicircle such that two multipliers of different kinds coincide at each, the overall pattern is obtained by superimposing the individual patterns of the above type independently (in the first approximation).

Consider a system with several degrees of freedom, capable of performing two types of vibration, say flexural and torsional (or flexural in different planes, etc.). Let $\omega_j^{(1)}$ and $\omega_h^{(2)}$ be the frequencies of the first and second types. The subscripts j and h indicate the number of nodes (mode number). According to the above discussion, if $\vartheta\approx(\omega_j^{(1)}+\omega_h^{(2)})/m$, flexural-torsional vibrations corresponding to frequencies $\omega_j^{(1)}$ and $\omega_h^{(2)}$ begin to stand out from the overall superposition of small vibrations, their amplitudes increase and they begin to predominate.

1.7. Case of r-fold multiplier of mixed kind $(\omega_j+\omega_h\neq 0)$

Now let r multipliers $(r\geqslant 2)$ coincide at a point $\rho^{(0)}$ on the unit circle. At $\varepsilon=0$ we have an r-fold pure imaginary characteristic exponent α_0. The corresponding r characteristic exponents for sufficiently small ε and $\vartheta=\vartheta_0+\mu\varepsilon$ are given by (1.19), (1.20). Substituting (1.24) and setting

$$\nu=\frac{1}{\vartheta_0}\xi \qquad \left(\beta=\frac{i}{\vartheta_0}\xi\right),$$

we write equation (1.20) as

$$\det \left\| \chi_{jh} - \mu \frac{\omega_j}{\vartheta_0} \gamma_{jh} - \xi \gamma_{jh} \right\|_{j,\, h=j_1,\, \ldots,\, j_r} = 0, \tag{1.39}$$

where $\chi_{jh} = \chi_{jh}(0)$,

$$\chi_{jh} = (\mathbf{H}_1^{(m_h - m_j)} \mathbf{c}_j,\ \mathbf{c}_h),\ \gamma_{jh} = \langle \mathbf{c}_j,\ \mathbf{c}_h \rangle = \begin{cases} 0 & \text{for } j \neq h, \\ +1 & \text{for } j = h > 0, \\ -1 & \text{for } j = h < 0. \end{cases} \tag{1.40}$$

The subscripts j, h take on the values $j_1, \ldots, j_r$ as in (1.15).

Developing the determinant in (1.39), we write the equation as

$$\xi^r + \delta_1(\mu)\xi^{r-1} + \ldots + \delta_r(\mu) = 0. \tag{1.41}$$

We claim that although the numbers χ_{jh} are generally complex, the coefficients $\delta_1(\mu), \ldots, \delta_r(\mu)$ are real for real μ. The matrix $\mathbf{H}_1(t)$ is hermitian and therefore

$$[\mathbf{H}^{(m)}]^* = \mathbf{H}^{(-m)}$$

so that the matrix $\|\chi_{jh}\|$ is hermitian. The matrix

$$\mathbf{G} = \|\gamma_{jh}\|_{j,\, h=j_1,\, \ldots,\, j_r}$$

is also hermitian.

Set

$$\mathbf{Q}(\mu) \equiv -\frac{1}{\vartheta_0} \|\sigma_{jh}(\mu)\| = \left\| \chi_{jh} - \mu \frac{\omega_j}{\vartheta_0} \gamma_{jh} \right\|_{j,\, h=j_1,\, \ldots,\, j_r}.$$

Equation (1.39) may be written

$$\det [\mathbf{G}^{-1}\mathbf{Q}(\mu) - \xi \mathbf{I}_r] = 0.$$

The matrix $\mathbf{G}^{-1}\mathbf{Q}(\mu)$ is G-hermitian, and so its spectrum is symmetric with respect to the real axis. In fact, this is also apparent from the equality

$$(\mathbf{G}^{-1}\mathbf{Q})^* = \mathbf{G}(\mathbf{G}^{-1}\mathbf{Q})\mathbf{G}^{-1}.$$

Thus, for every root $\xi(\mu)$, equation (1.39) also has the root $\bar{\xi}(\mu)$. It follows that the coefficients of equation (1.41) are real.

For sufficiently large $|\mu|$, all the roots of equation (1.41) are real, since in this case the trivial solution of system (1.2) is stable. Let $(\mu_1^{(-)}, \mu_1^{(+)})$, $(\mu_2^{(-)}, \mu_2^{(+)})$, ... be the intervals in which equation (1.39) has at least one nonreal root whereas outside these intervals all the roots are real and distinct. Then, for sufficiently small μ, the sectors attached at the point $(0, \vartheta_0)$, defined by the inequalities

$$\left.\begin{array}{l}\vartheta_0+\mu_1^{(-)}\varepsilon+\ldots<\vartheta<\vartheta_0+\mu_1^{(+)}\varepsilon+\ldots,\\ \vartheta_0+\mu_2^{(-)}\varepsilon+\ldots<\vartheta<\vartheta_0+\mu_2^{(+)}\varepsilon+\ldots,\\ \ldots\ldots\ldots\ldots\ldots\ldots\ldots\ldots\end{array}\right\} \qquad (1.42)$$

will contain multipliers outside the unit circle in the neighborhood of $\rho^{(0)}$. By the corollary to the Lyapunov-Poincaré theorem (§ II. 3.6), it follows that outside the union of these sectors, for sufficiently small ε, all the multipliers lie on the unit circle. Thus for sufficiently small ε the dynamic stability domain is the union of all sectors (1.42) for all points $\rho^{(0)}$ of the above type.

The condition for all roots of equation (1.41) to be real is (see, e. g., /56/)

$$A_2>0,\quad A_3>0,\ \ldots,\ A_r>0, \qquad (1.43)$$

where $A_0=1,\ A_1=r,\ A_2,\ \ldots,\ A_r$ are the leading coefficients in a chain of Sturm functions for equation (1.41).

Here $A_2,\ \ldots,\ A_r$ are polynomials in μ and the slopes $\mu^{(\pm)}$ of the tangents to the boundaries of the dynamic instability domains are found from the equation

$$\Delta(\mu)=0, \qquad (1.44)$$

where $\Delta(\mu)$ is the discriminant of equation (1.41). However, this equation has "superfluous" roots, other than $\mu_1^{(\pm)},\ \mu_2^{(\pm)},\ \ldots.$ To determine the intervals (1.42), one can proceed as follows. Let $\mu_1,\ \ldots,\ \mu_N$ be the roots of equation (1.44), arranged in decreasing order. On each of the intervals

$$(-\infty,\ \mu_1),\ (\mu_1,\ \mu_2),\ \ldots,\ (\mu_N,\ +\infty) \qquad (1.45)$$

the coefficients $A_2,\ \ldots,\ A_r$ have a constant sign. Thus, choosing an arbitrary point in each of these intervals and determining the signs of $A_2,\ \ldots,\ A_r$ there, we can choose those of the intervals in which at least one of inequalities (1.43) fails to hold. These are precisely the instability intervals — in our notation, the intervals $(\mu_\nu^{(-)},\ \mu_\nu^{(+)})$. In fact, as the roots of equation (1.41) are real, they may become complex only provided they "first" coincide. Thus equation (1.41) has multiple roots on the boundary of the instability sectors, i. e., at $\mu=\mu^{(\pm)}$.

1.8. Case of r-fold definite multiplier

Suppose that multipliers of only one kind coincide at the point $\rho^{(0)}$. Then

$$\gamma_{jh}=\delta_{|j|\,|h|}\ \mathrm{sign}\,j$$

in (1.40), where $\mathrm{sign}\,j=1$ for all j if the coinciding multipliers are of the first kind and $\mathrm{sign}\,j=-1$ if they are of the second. Equation (1.39) is

$$\det[\mathbf{Q}(\mu)-\xi\mathbf{I}_r]=0,$$

where $\mathbf{Q}(\mu)$ is hermitian for any real μ. Hence all roots $\xi(\mu)$ of the equation are real. The intervals $(\mu_1^{(-)}, \mu_1^{(+)}), (\mu_2^{(-)}, \mu_2^{(+)}), \ldots$ contract to points, and so in the first approximation the domains (1.42) degenerate into curves. As yet, however, we cannot state that there are no instability domains attached at the point $(0, \vartheta_0)$ in the $\varepsilon\vartheta$-plane — there may (a priori) be "narrow" instability domains. We shall now show that in fact there are not even narrow domains attached at $(0, \vartheta_0)$.

Indeed, by Krein's theorem (§ III. 1. 4) all multipliers that tend to $\rho^{(0)}$ at $\varepsilon=0$ lie on the unit circle for sufficiently small positive ε, and the corresponding elementary divisors are simple. In the first quadrant of the $\varepsilon\vartheta$-plane, the dynamic instability domain degenerates into a curve, on some initial segment of which the trivial solution of system (1.2) is certainly stable.

In particular, if $r=2$ this means that the frequencies

$$\vartheta_0=\frac{\omega_j-\omega_h}{m} \qquad (j,\ h=1,\ \ldots,\ k;\ m=1,\ 2,\ \ldots)$$

are not critical (where $i\omega_1, \ldots, i\omega_k$ are the eigenvalues of the first kind of the matrix $\mathbf{C}$). We state this proposition as a corollary to Krein's theorem.

Corollary. Suppose that for fixed j, h, m $(j,\ h=1,\ \ldots,\ k;$ $m=1,\ 2,\ \ldots)$

$$\frac{\omega_j+\omega_h}{m}\neq\frac{\omega_{j'}+\omega_{h'}}{m'},$$

where j', h', m' run through the same values $j', h'=1, \ldots, k;\ m'=1, 2, \ldots$. Then $\vartheta_0=(\omega_j-\omega_h)/m$ is not a critical frequency.

1.9. The case $\omega_j+\omega_h=0$

It was assumed in §1.4 that $\omega_j+\omega_h\neq 0$ $(j,\ h=\pm 1,\ \ldots,\ \pm k)$. We now consider the singular case: for some j_0, h_0,

$$\omega_{j_0}+\omega_{h_0}=0.$$

Recall that $i\omega_1, \ldots, i\omega_k$ are the eigenvalues of the first kind of the matrix $\mathbf{C}$. We have

$$\rho^{(0)}=\exp\frac{2\pi i\omega_{-j_0}}{\vartheta_0}=\exp\frac{2\pi i\omega_{h_0}}{\vartheta_0},$$

i. e., for any $\vartheta_0\neq 0$ there are coincident multipliers of different kinds. By (1.26), $m_{-j_0}=m_{h_0}$. Repeating the reasoning of §1.4 $(m=0)$, we find

$$\Delta(\mu)\equiv\Delta=\frac{1}{4}\left[\chi_{-j_0,-j_0}(0)+\chi_{h_0h_0}(0)\right]^2-|\chi_{-j_0h_0}|^2, \tag{1.46}$$

where

$$\chi_{\alpha\beta} = (\mathbf{H}_1^{(0)}\mathbf{c}_\alpha,\ \mathbf{c}_\beta) \qquad (\alpha, \beta = -j_0, h_0).$$

If $\Delta < 0$, the multipliers of system (1.2) on any ray $\vartheta = \vartheta_0 + \mu\varepsilon$ $(\vartheta_0 \neq 0)$ leave the unit circle. In this situation, a sufficiently small strip $(0 \leqslant \varepsilon < \varepsilon^*)$, containing the half-axis $\vartheta > 0$, is an instability domain, i. e., every frequency is critical. But if $\Delta > 0$, no frequency is critical. * The value of Δ then depends only on $\mathbf{H}_1^{(0)} = [\mathbf{H}_1(s)]_{av}$. If $\mathbf{H}_1^{(0)} > 0$, then

$$\Delta = \frac{1}{4}[\chi^2_{-j_0,\,-j_0} + \chi^2_{h_0 h_0}] + \left(\frac{1}{2}\chi_{-j_0,\,-j_0}\chi_{h_0 h_0} - |\chi_{-j_0 h_0}|^2\right) > 0,$$

since by Schwarz's inequality the term in parentheses is nonnegative.

1.10. Formulas for second-order canonical vector equations

Let us write a system of k second-order canonical differential equations as a vector differential equation:

$$\mathbf{M}\frac{d^2\mathbf{y}}{dt^2} + [\mathbf{P}_0 + \varepsilon\mathbf{P}_1(\vartheta t) + \varepsilon^2\mathbf{P}_2(\vartheta t) + \ldots]\,\mathbf{y} = 0, \qquad (1.47)$$

where $\mathbf{M} = \mathbf{M}^\tau > 0$, $\mathbf{P}_0 = \mathbf{P}_0^\tau > 0$ are constant real symmetric positive definite matrices, the real matrix-functions are periodic and symmetric:

$$\mathbf{P}_l(s) = \mathbf{P}_l(s + 2\pi) = \mathbf{P}_l(s)^\tau \qquad (0 \leqslant s \leqslant 2\pi,\ l = 1, 2, \ldots),$$

and the bracketed series in (1.47) is convergent in the usual sense (see introduction to Chapter IV). All the matrices are of order $k \times k$.

We substitute

$$\mathbf{x} = \begin{pmatrix} \mathbf{y} \\ \mathbf{M}\dot{\mathbf{y}} \end{pmatrix}.$$

Denoting

$$\mathbf{C} = -\mathbf{J}\mathbf{H}_0 = \begin{pmatrix} 0 & \mathbf{M}^{-1} \\ -\mathbf{P}_0 & 0 \end{pmatrix}, \quad \mathbf{B}_1(\vartheta t) = -\mathbf{J}\mathbf{H}_1(\vartheta t) = \begin{pmatrix} 0 & 0 \\ -\mathbf{P}_1(\vartheta t) & 0 \end{pmatrix},$$

we obtain a system equivalent to (1.2):

$$\frac{d\mathbf{x}}{dt} = [\mathbf{C} + \varepsilon\mathbf{B}_1(\vartheta t) + \ldots]\,\mathbf{x}.$$

* Provided that $\rho^{(0)}$ is the only point at which multipliers coincide on the upper closed unit semicircle. Otherwise one must also investigate other points $\rho^{(0)}$.

Following §IV. 3.3, let $\omega_\varkappa^2 (\omega_\varkappa > 0)$ and $\mathbf{a}_\varkappa$ denote corresponding eigenvalues and eigenvectors of the matrix $\mathbf{M}^{-1}\mathbf{P}_0$:

$$\mathbf{P}_0\mathbf{a}_\varkappa = \omega_\varkappa^2 \mathbf{M}\mathbf{a}_\varkappa \qquad (\varkappa = 1, \ldots, k),$$

the eigenvectors being normalized so that

$$(\omega_\varkappa + \omega_\lambda)(\mathbf{M}\mathbf{a}_\varkappa, \mathbf{a}_\lambda) = \delta_{\varkappa\lambda} \qquad (\varkappa, \lambda = 1, \ldots, k),$$

where $\delta_{\varkappa\lambda}$ is the Kronecker symbol. Then the vectors

$$\mathbf{c}_\nu = \begin{pmatrix} \mathbf{a}_{|\nu|} \\ i\omega_\nu \mathbf{M}\mathbf{a}_{|\nu|} \end{pmatrix} \qquad (\nu = \pm 1, \ldots, \pm k),$$

where $\omega_{-\nu} = -\omega_\nu$, are readily seen to be eigenvectors of $\mathbf{C}$:

$$\mathbf{C}\mathbf{c}_\nu = i\omega_\nu \mathbf{c}_\nu \qquad (\nu = \pm 1, \ldots, \pm k).$$

Here

$$\gamma_{\nu\iota} = \langle \mathbf{c}_\nu, \mathbf{c}_\iota \rangle \equiv i(\mathbf{J}\mathbf{c}_\nu, \mathbf{c}_\iota) = \begin{cases} 0, & \text{if } \nu \neq \iota, \\ +1, & \text{if } \nu = \iota > 0, \\ -1, & \text{if } \nu = \iota < 0 \end{cases} \qquad (\nu, \iota = \pm 1, \ldots, \pm k),$$

so that $i\omega_1, \ldots, i\omega_k$ are the eigenvalues of the first kind for $\mathbf{C}$, while $-i\omega_1, \ldots, -i\omega_k$ are the eigenvalues of the second kind. A few simple manipulations yield

$$(\mathbf{H}_1(\vartheta t)\mathbf{c}_\nu, \mathbf{c}_\iota) = (\mathbf{P}_1(\vartheta t)\mathbf{a}_{|\nu|}, \mathbf{a}_{|\iota|}).$$

Formulas (1.22) for χ_{jh} $(r \geqslant 2)$ become

$$\chi_{jh} = (\mathbf{P}_1^{(m_h - m_j)}\mathbf{a}_{|j|}, \mathbf{a}_{|h|}) \qquad (j, h = \pm 1, \ldots, \pm k),$$

and formulas (1.29) $(r = 2)$ are

$$\chi_{jj} = (\mathbf{P}_1^{(0)})\mathbf{a}_{|j|}, \mathbf{a}_{|j|}) \qquad (j = \pm 1, \ldots, \pm k),$$
$$\chi_{-jh} = (\mathbf{P}_1^{(m)}\mathbf{a}_j, \mathbf{a}_h) \qquad (i, h = 1, \ldots, k).$$

1.11. Summary of final results

We consider a system (1.2) of equations of canonical type, with real coefficients which are 2π-periodic functions of $s = \vartheta t$. Consequently,

$$\mathbf{B}_j(s) = -\mathbf{J}\mathbf{H}_j(s), \quad \mathbf{H}_j(s)^\tau = \mathbf{H}_j(s) \qquad (0 \leqslant s \leqslant 2\pi, \ j = 1, 2, \ldots).$$

The matrix $\mathbf{C}$ is reducible to diagonal form with pure imaginary diagonal elements $\pm i\omega_1, \ldots, \pm i\omega_k$. The multipliers of the unperturbed system (1.1) (which is system (1.2) with $\varepsilon=0$, $\vartheta=\vartheta_0$) are

$$\rho_\nu^{(0)} = \exp\frac{2\pi i\omega_\nu}{\vartheta_0} \qquad (\nu=\pm 1, \ldots, \pm k).$$

Number the eigenvalues of $\mathbf{C}$ so that $i\omega_1, \ldots, i\omega_k$ $(\omega_1 \leqslant \omega_2 \leqslant \ldots \leqslant \omega_k)$ are the eigenvalues of the first kind (§1.1). Set $\omega_{-\varkappa} = -\omega_\varkappa (\varkappa=1, \ldots, k)$; then

$$i\omega_{-1}, \ldots, i\omega_{-k}$$

are the eigenvalues of the second kind. Corresponding eigenvectors $\mathbf{c}_\nu$ of $\mathbf{C}$ such that

$$\mathbf{C}\mathbf{c}_\nu = i\omega_\nu \mathbf{c}_\nu \qquad (\nu=\pm 1, \ldots, \pm k),$$

are chosen in such a way that (§1.1)

$$\langle \mathbf{c}_\nu, \mathbf{c}_\iota \rangle = \gamma_{\nu\iota} \qquad (\nu, \iota=\pm 1, \ldots, \pm k).$$

Divide the eigenvalues $i\omega_\nu (\nu=\pm 1, \ldots, \pm k)$ into congruence classes modulo $i\vartheta_0$ so that each class

$$i\omega_{j_1}, \ldots i\omega_{j_r}$$

contains one r-fold multiplier of system (1.2) for $\varepsilon=0$, $\vartheta=\vartheta_0$:

$$\rho^{(0)} = \exp\frac{2\pi i\omega^{(0)}}{\vartheta_0}.$$

Let Γ denote the upper closed unit semicircle.

a) At $\vartheta=\vartheta_0$, there is one twofold multiplier $\rho^{(0)}$ of mixed kind on Γ. This means that for some $j, h=1, \ldots, k$ and $m=1, 2, \ldots$,

$$\vartheta_0 = \frac{\omega_j+\omega_h}{m},$$

and the multiplier $\rho^{(0)}$ determines a class of eigenvalues $\{i\omega_{-j}, i\omega_h\}$. Expand the matrix $\mathbf{H}_1(\vartheta t) = \mathbf{J}\mathbf{B}_1(\vartheta t)$ in Fourier series:

$$\mathbf{H}_1(\vartheta t) \sim \sum_{l=-\infty}^{\infty} e^{il\vartheta t}\mathbf{H}_1^{(l)}.$$

Define the numbers $\chi_{-jh}, \chi_{-j,-j}, \chi_{hh}, \mu_1, \mu_2$ by (1.29). If $\chi_{-jh} \neq 0$, there is a wide dynamic instability domain attached at the point $(0, \vartheta_0)$ in the first quadrant of the $\varepsilon\vartheta$-plane; if $\chi_{-jh}=0$, this instability domain is narrow. The width of the instability domain $(\chi_{-jh} \neq 0)$ is given by

$$\Psi_{jhm} = \frac{1}{\pi}\operatorname{arcctg}\frac{1+\mu_1\mu_2}{\mu_2-\mu_1},$$

and the domain itself is defined by

$$\vartheta_0+\mu_1\varepsilon+\ldots<\vartheta<\vartheta_0+\mu_2\varepsilon+\ldots,$$

where the dots stand for terms of second order in ε. If

$$[\mathbf{H}_1(\vartheta t)]_{\text{av}}\equiv\mathbf{H}_1^{(0)}=0,$$

then

$$\chi_{-j,\,-j}=0,\quad \chi_{hh}=0,\quad \chi_{-jh}=(\mathbf{H}_1^{(m)}\bar{\mathbf{c}}_j,\ \mathbf{c}_h),$$
$$\mu_{1,2}=\mp\frac{2}{m}|\chi_{-jh}|,\quad \Psi_{jhm}=\psi\left(\frac{2}{m}|\chi_{-j,h}|\right),$$

where $\psi(x)$ is a monotone function of x $(0<x<\infty)$:

$$\psi(x)=\frac{1}{\pi}\operatorname{arcctg}\frac{1-x^2}{2x}.$$

The dynamic instability domain attached at the point $(0,\vartheta_0)$ contains a region in which the solutions increase more rapidly than $e^{\zeta t}$ $(\zeta>0)$. This region has the shape illustrated in Figure 23. The value of $\varepsilon(\zeta)$ is given by

$$\varepsilon(\zeta)=\frac{\zeta}{|\chi_{-jh}|}+O(\zeta^2).$$

If (ε,ϑ) is a point in the dynamic instability domain, there exist one $(j=h)$ or two $(j\neq h)$ solutions which increase as $\exp[\zeta(\varepsilon)t]$, and one $(j=h)$ or two $(j\neq h)$ solutions decreasing as $\exp[-\zeta(\varepsilon)t]$. If $\vartheta=\vartheta_0+\varepsilon\mu$ $(\mu_1<\mu<\mu_2)$, the value of $\zeta(\varepsilon)$ is given by

$$\zeta(\varepsilon)=\sqrt{-\Delta(\mu)\,\varepsilon}+O(\varepsilon^2),$$

where

$$\Delta(\mu)=\frac{1}{4}(\chi_{-j,\,-j}+\chi_{hh}-m\vartheta)^2-|\chi_{-jh}|^2.$$

The remaining linearly independent solutions are almost-periodic functions. For fixed μ, the unboundedly increasing solution tends as $\varepsilon\to 0$ to a "generating" solution

$$\zeta_1^{(1)}e^{-i\omega_j\vartheta_0 t}\bar{\mathbf{c}}_j+\zeta_2^{(1)}e^{i\omega_h\vartheta_0 t}\mathbf{c}_h,$$

where $\zeta_1^{(1)}$, $\zeta_2^{(1)}$ are defined by (1.37):

$$\zeta_1^{(1)}=\chi_{-jh},\quad \zeta_2^{(1)}=-\frac{1}{2}(\chi_{-j,-j}+\chi_{hh}-m\mu)-\sqrt{\Delta(\mu)}.$$

If $j\neq h$, a second "generating" solution is obtained when h is replaced by $-h$, $-j$ by j, $\zeta_1^{(1)}$ by $\zeta_1^{(2)}$, and $\zeta_2^{(1)}$ by $\zeta_2^{(2)}$ (§1.6).

b) Let the system have the form

$$\mathbf{M}\frac{d^2\mathbf{y}}{dt^2}+[\mathbf{P}_0+\varepsilon\mathbf{P}_1(\vartheta t)+\varepsilon^2\mathbf{P}_2(\vartheta t)+\ldots]\,\mathbf{y}=0, \tag{1.48}$$

where $\mathbf{M}=\mathbf{M}^{\tau}>0$, $\mathbf{P}_0=\mathbf{P}_0^{\tau}>0$ are symmetric positive definite matrices and $\mathbf{P}_j(s)^{\tau}=\mathbf{P}_j(s)=\mathbf{P}_j(s+2\pi)$ $(0\leqslant s\leqslant 2\pi;\ \ j=1,\ 2,\ \ldots)$. Define eigenvectors and eigenvalues of the matrix $\mathbf{M}^{-1}\mathbf{P}_0$ by

$$\mathbf{P}_0\mathbf{a}_\varkappa=\omega_\varkappa^2\mathbf{M}\mathbf{a}_\varkappa,\quad (\omega_\varkappa+\omega_\lambda)(\mathbf{M}\mathbf{a}_\varkappa,\ \mathbf{a}_\lambda)=\delta_{\varkappa\lambda}\qquad (\varkappa,\ \lambda=1,\ \ldots,\ k).$$

Expand $\mathbf{P}_1(\vartheta t)$ in Fourier series:

$$\mathbf{P}_1(\vartheta t)\sim\sum_{l=-\infty}^{\infty}e^{il\vartheta t}\mathbf{P}_1^{(l)}.$$

Suppose that when $\vartheta=\vartheta_0$ there is only one point $\rho^{(0)}$ on Γ of the type described in (a); then there is exactly one triple $j,\ h,\ m$ such that

$$\vartheta_0=\frac{\omega_j+\omega_h}{m}.$$

Set

$$\chi_j=(\mathbf{P}_1^{(0)}\mathbf{a}_j,\ \mathbf{a}_j),\quad \chi_{jh}=(\mathbf{P}_1^{(m)}\mathbf{a}_j,\ \mathbf{a}_h).$$

Everything said above in (a) remains true here, provided we replace $\chi_{-j,\,-j}$ by χ_j, χ_{hh} by χ_h, and χ_{-jh} by χ_{jh}. In particular, relative to a basis diagonalizing $\mathbf{P}_0$ and $\mathbf{M}$:

$$\mathbf{P}_0=\begin{pmatrix}p_1 & & 0\\ & \ddots & \\ 0 & & p_k\end{pmatrix},\quad \mathbf{M}=\begin{pmatrix}q_1 & & 0\\ & \ddots & \\ 0 & & q_k\end{pmatrix},\quad \mathbf{P}_1(\vartheta t)\sim\sum_{l=-\infty}^{\infty}e^{il\vartheta t}\left\|\pi_{\varkappa\lambda}^{(l)}\right\|_1^k, \tag{1.49}$$

we have

$$\omega_\varkappa=\sqrt{\frac{p_\varkappa}{q_\varkappa}},\quad \mathbf{a}_\varkappa=\frac{1}{\sqrt{2\omega_\varkappa q_\varkappa}}\mathbf{e}_\varkappa\qquad (\varkappa=1,\ \ldots,\ k),$$

and so

$$\chi_j=\frac{1}{2\omega_j q_j}\pi_{jj}^{(0)},\quad \chi_{jh}=\frac{1}{2\sqrt{\omega_j\omega_h q_j q_h}}\pi_{jh}^{(m)}\qquad (j,\ h=1,\ \ldots,\ k). \tag{1.50}$$

These coefficients are thus expressed directly in terms of the coefficients of the original system of differential equations. In particular, the domain attached at the point $(0,\ \vartheta_0)$, where

$$\vartheta_0=\frac{\omega_j+\omega_h}{m},$$

is wide if $\pi_{jh}^{(m)}\neq 0$ and narrow if $\pi_{jh}^{(m)}=0$. If $\mathbf{P}_1^{(0)}\equiv[\mathbf{P}_1(\vartheta t)]_{\mathrm{av}}=0$, the width of the domain is a monotone function of the quantity

$$\frac{2}{m}|\chi_{jh}|.$$

c) If $\vartheta=\vartheta_0=(\omega_j+\omega_h)/m$ there are several multipliers $\rho^{(0)}$ of the desired type on Γ (twofold multipliers of mixed kind). For each such multiplier we have a sector on the $\varepsilon\vartheta$-plane:

$$\vartheta_0+\mu_1\varepsilon+\ldots<\vartheta<\vartheta_0+\mu_2\varepsilon+\ldots$$

The whole dynamic instability domain is the union of these sectors. The formulas of (a) and (b) above are valid in each sector.

d) If $\vartheta=\vartheta_0=(\omega_j+\omega_h)/m$ there is a multiplier $\rho^{(0)}$ of mixed kind on Γ which has multiplicity $r>2$. Let $i\omega_{j_1}, \ldots, i\omega_{j_r}$ be the corresponding class of eigenvalues of $\mathbf{C}$. Define integers m_j by

$$\omega_j=\alpha_0+m_j\vartheta_0 \qquad (j=j_1, \ldots, j_r),$$

where α_0 is a fixed number, defined up to a multiple of ϑ_0. Define the coefficients for $j,\ h=j_1, \ldots, j_r$ by

$$\gamma_{jh}=\langle \mathbf{c}_j, \mathbf{c}_h\rangle=\begin{cases} 0, & \text{if } j\neq h, \\ +1, & \text{if } j=h>0, \\ -1, & \text{if } j=h<0, \end{cases} \qquad \chi_{jh}=(\mathbf{H}_1^{(m_h-m_j)}\mathbf{c}_j, \mathbf{c}_h).$$

Consider the equation

$$\det\left\| \chi_{jh}-\mu\frac{\omega_j}{\vartheta_0}\gamma_{jh}-\xi\gamma_{jh}\right\|_{j,\,h=j_1, \ldots, j_r}=0$$

written in developed form as

$$\xi^r+\delta_1(\mu)\,\xi^{r-1}+\ldots+\delta_r(\mu)=0.$$

The coefficients $\delta_1(\mu), \ldots, \delta_r(\mu)$ are real for real μ (though χ_{jh} are complex). Let $(\mu_1^{(-)}, \mu_1^{(+)}), (\mu_2^{(-)}, \mu_2^{(+)}), \ldots$ be the intervals on which this equation has at least one nonreal root ξ, whereas outside the intervals all the roots are real and distinct. Then the dynamic instability domain is the union of the sectors

$$\begin{gathered}\vartheta_0+\mu_1^{(-)}\varepsilon+\ldots<\vartheta<\vartheta_0+\mu_1^{(+)}\varepsilon+\ldots, \\ \vartheta_0+\mu_2^{(-)}\varepsilon+\ldots<\vartheta<\vartheta_0+\mu_2^{(+)}\varepsilon+\ldots, \\ \ldots\ldots\ldots\ldots\ldots\ldots\end{gathered}$$

The characteristic exponents of the corresponding r linearly independent solutions may be expressed (for fixed $\mu,\ \varepsilon$) in terms of the roots $\xi\in(\mu_j^{(-)},\mu_j^{(+)})$ by

$$\alpha(\varepsilon)=i\,(\alpha_0+\xi\varepsilon+\ldots)\,\vartheta_0,$$

where the dots stand for terms of the order of $\varepsilon^{1+\delta}\,(\delta>0)$. For a second-order system (1.48), the coefficients χ_{jh} are defined by

$$\chi_{jh}=(\mathbf{P}_1^{(m_h-m_j)}\mathbf{a}_{|j|}, \mathbf{a}_{|h|}).$$

If $\mathbf{P}_0$, $\mathbf{M}$ are diagonal, then

$$\chi_{jh} = \frac{1}{2\sqrt{\omega_{|j|}\omega_{|h|}q_{|j|}q_{|h|}}}\pi^{(m_h - m_j)}_{|j||h|}$$

(some of the indices j, h may be negative). Here again

$$\mathbf{P}_1(\vartheta t) \sim \sum_{l=-\infty}^{\infty} e^{il\vartheta t} \| \pi^{(l)}_{\varkappa\lambda} \|^k_1.$$

Remark. We have assumed that $\omega_j + \omega_h \neq 0$. This is surely true for system (1.1) if $\omega_1 > 0, \ldots, \omega_k > 0$. The case $\omega_j + \omega_h = 0$ was studied in §1.8.

1.12. Example

Consider the system of three differential equations

$$\left.\begin{aligned} &q_1 \frac{d^2\eta_1}{dt^2} + p_1\eta_1 + [a\eta_1 \cos\vartheta t + (b\cos 2\vartheta t + c\sin 2\vartheta t)\,\eta_3] = 0, \\ &q_2 \frac{d^2\eta_2}{dt^2} + p_2\eta_2 + [g\eta_3 \sin 5\vartheta t] = 0, \\ &q_3 \frac{d^2\eta_3}{dt^2} + p_3\eta_3 + [(b\cos 2\vartheta t + c\sin 2\vartheta t)\,\eta_1 + g\eta_2 \sin 5\vartheta t] = 0, \end{aligned}\right\} \tag{1.51}$$

where a, b, c, g are small parameters of the same order of magnitude. We assume in addition that the natural frequencies

$$\omega_\varkappa = \sqrt{\frac{p_\varkappa}{q_\varkappa}} \qquad (\varkappa = 1, 2, 3)$$

of the unperturbed system are such that if

$$\frac{\omega_j + \omega_h}{m} = \frac{\omega_{j'} + \omega_{h'}}{m'} \qquad (m, m' = 1, 2, \ldots)$$

then $j = j'$, $h = h'$, $m = m'$. Then for any ϑ_0 there is at most a twofold multiplier of mixed kind on Γ.

By §1.11 (b), we have:

1) The nonvanishing coefficients $\pi^{(m)}_{jh}$ $(m \neq 0)$ are

$$\pi^{(1)}_{11}, \quad \pi^{(2)}_{13}, \quad \pi^{(5)}_{23}.$$

Consequently, there are wide instability domains only at the following frequencies ϑ_0:

$$2\omega_1, \quad \frac{\omega_1 + \omega_3}{2}, \quad \frac{\omega_2 + \omega_3}{5}.$$

2) Using the formulas of §1.11 (b), we find the coefficients χ_{jh} for the above subscripts:

$\frac{\omega_j+\omega_h}{m}$	$2\omega_1$	$\frac{1}{2}(\omega_1+\omega_3)$	$\frac{1}{5}(\omega_2+\omega_3)$
$\frac{2}{m}\vert\chi_{jh}\vert$	$\frac{1}{2\omega_1 q_1}\vert a\vert$	$\frac{1}{4}\sqrt{\frac{b^2+c^2}{\omega_1\omega_3 q_1 q_3}}$	$\frac{1}{10}\sqrt{\frac{1}{\omega_2\omega_3 q_2 q_3}}\vert g\vert$

The larger the coefficient $2|\chi_{jh}|/m$, the more "dangerous" is the corresponding frequency $(\omega_j+\omega_h)/m$; more precisely, the wider is the corresponding dynamic instability domain. The three wide instability domains are defined in the first approximation by the formulas of §1.12(a) for $\varepsilon=1$:

$$2\omega_1-\frac{1}{2\omega_1 q_1}|a|+\ldots<\vartheta<2\omega_1+\frac{1}{2\omega_1 q_1}|a|+\ldots,$$

$$\frac{1}{2}(\omega_1+\omega_3)-\frac{1}{4}\sqrt{\frac{b^2+c^2}{\omega_1\omega_3 q_1 q_3}}+\ldots<\vartheta<\frac{1}{2}(\omega_1+\omega_3)+\frac{1}{4}\sqrt{\frac{b^2+c^2}{\omega_1\omega_3 q_1 q_3}}+\ldots,$$

$$\frac{1}{5}(\omega_2+\omega_3)-\frac{|g|}{10\sqrt{\omega_2\omega_3 q_2 q_3}}+\ldots<\vartheta<\frac{1}{5}(\omega_2+\omega_3)+\frac{|g|}{10\sqrt{\omega_2\omega_3 q_2 q_3}}+\ldots$$

(dots stand for second-order quantities in a, b, c, g).

All the other domains attached to points $(0, \vartheta_{jhm})$, where

$$\vartheta_{jhm}=\frac{\omega_j+\omega_h}{m},$$

are narrow. Their boundaries are

$$\vartheta=\vartheta_{jhm}+\ldots,$$

where the dots again stand for second-order terms with respect to a, b, c, g.

Suppose we are interested in a value of $\varepsilon=\varepsilon(\zeta)$ such that for $|\varepsilon|<\varepsilon(\zeta)$ the solutions of system (1.51) increase at a lower rate than $e^{\zeta t}$. By the formula in §1.11 (a), we have

$$\varepsilon(\zeta)=\frac{4\omega_1 q_1}{|a|}\zeta+\ldots$$

in the domain corresponding to frequency $2\omega_1$,

$$\varepsilon(\zeta)=4\sqrt{\frac{\omega_1\omega_3 q_1 q_3}{b^2+c^2}}\zeta+\ldots$$

in the domain corresponding to frequency $\frac{1}{2}(\omega_1+\omega_3)$, and

$$\varepsilon(\zeta)=\frac{4\sqrt{\omega_2\omega_3 q_2 q_3}}{|g|}\zeta+\ldots$$

in the domain corresponding to frequency $\frac{1}{5}(\omega_2+\omega_3)$. The dots stand for terms of the order of ζ^2.

§ 2. FIRST-APPROXIMATION FORMULAS FOR BOUNDARIES OF DYNAMIC INSTABILITY DOMAINS WHEN THE COEFFICIENTS OF THE EQUATIONS ARE NONLINEAR FUNCTIONS OF THE PARAMETER $1/\vartheta$

§1 was devoted to a detailed account of the most familiar case, in which the motion of the system is described by a canonical system of linear differential equations (1.2), reduced by the substitution $t=\gamma\tau$ ($\gamma=1/\vartheta$) to the form

$$\frac{d\mathbf{x}}{d\tau}=\gamma[\mathbf{C}+\varepsilon\mathbf{B}(\tau,\varepsilon)]\mathbf{x}.$$

In this section we shall study the case in which the right-hand side of the equation is a nonlinear function of the parameter γ:

$$\frac{d\mathbf{x}}{d\tau}=[\mathbf{C}(\gamma)+\varepsilon\mathbf{B}(\tau,\varepsilon,\gamma)]\mathbf{x}.$$

This complication involves no essential difficulties; the exposition largely duplicates that of §1, and so we shall be far more concise. The first-approximation formulas for the boundaries of the dynamic instability domains are indeed more cumbersome. We shall therefore confine ourselves to cases which, though easily tractable from the mathematical standpoint, are the most important in applications.

2.1. Statement of the problem

Consider a canonical system (§II. 3.1) of $2k$ linear differential equations with periodic coefficients, in vector notation:

$$\tilde{\mathbf{J}}\frac{d\mathbf{x}}{d\tau}=\mathbf{H}(\tau,\varepsilon,\gamma)\mathbf{x}, \tag{2.1}$$

where $\tilde{\mathbf{J}}$ is a real nonsingular skew-symmetric matrix, $\mathbf{H}(\tau,\varepsilon,\gamma)$ a real symmetric matrix-function of period 2π in τ; ε and γ are real parameters. The precise assumptions made concerning the function $\mathbf{H}(\tau,\varepsilon,\gamma)$ will be specified below. It is assumed that the matrix $\mathbf{H}(\tau,0,\gamma)=\mathbf{H}_0(\gamma)$ is independent of τ and all solutions of the unperturbed system

$$\tilde{\mathbf{J}}\frac{d\mathbf{x}}{d\tau}=\mathbf{H}_0(\gamma)\mathbf{x} \tag{2.2}$$

are bounded as $\tau\to\infty$ for all $\gamma\in[\gamma_1,\gamma_2]$.

In the accepted terminology, γ_0 is called a critical value of γ if, for any $\delta_1 > 0$, $\delta_2 > 0$, the domain

$$|\gamma-\gamma_0| < \delta_1, \quad |\varepsilon| < \delta_2$$

contains a point (ε, γ) for which the perturbed system (2.1) has solutions which are unbounded as $\tau \to \infty$. Consider the set of all points on the $\varepsilon\gamma$-plane for which system (2.1) has solutions which are unbounded as $\tau \to \infty$. The interior of this set is the union of "dynamic instability" domains attached at the points $(0, \gamma_0)$, where γ_0 are the critical values of γ.

Let us write the unperturbed system (2.2) as

$$\frac{d\mathbf{x}}{d\tau} = \mathbf{C}(\gamma)\mathbf{x} \qquad (\mathbf{C}(\gamma) = \tilde{\mathbf{J}}^{-1}\mathbf{H}_0(\gamma)).$$

By assumption, for any $\gamma \in [\gamma_1, \gamma_2]$ the real $2k \times 2k$ matrix $\mathbf{C}(\gamma)$ is reducible to diagonal form with pure imaginary diagonal elements $\pm i\omega_j$, which are the roots of the equation

$$\det[\tilde{\mathbf{J}}^{-1}\mathbf{H}_0(\gamma) - i\omega\mathbf{I}_{2k}] = 0.$$

The next definition is the same as in §1.1. We normalize and index the eigenvectors $\mathbf{c}_\nu(\gamma)$ of the matrix $\mathbf{C}(\gamma)$ for $\gamma \in [\gamma_1, \gamma_2]$, i. e.,

$$\tilde{\mathbf{J}}^{-1}\mathbf{H}_0(\gamma)\mathbf{c}_\nu(\gamma) = i\omega_\nu(\gamma)\mathbf{c}_\nu(\gamma) \qquad (\nu = \pm 1, \ldots, \pm k), \tag{2.3}$$

in such a way that $\omega_{-j}(\gamma) = -\omega_j(\gamma)$ and

$$\|\gamma_{\nu\mu}\|_{-k}^{k} = \|\langle\mathbf{c}_\nu(\gamma), \mathbf{c}_\mu(\gamma)\rangle\|_{-k}^{k} = \|i(\tilde{\mathbf{J}}\mathbf{c}_\nu(\gamma), \mathbf{c}_\mu(\gamma))\|_{-k}^{k} = \begin{pmatrix} -\mathbf{I}_k & 0 \\ 0 & \mathbf{I}_k \end{pmatrix}. \tag{2.4}$$

Then

$$i\omega_1(\gamma), \ldots, i\omega_k(\gamma) \qquad (\omega_1(\gamma) \leqslant \ldots \leqslant \omega_k(\gamma))$$

are the eigenvalues of the first kind, and

$$-i\omega_1(\gamma), \ldots, -i\omega_k(\gamma)$$

those of the second kind for the matrix $\mathbf{C}(\gamma)$.

By the corollary to Krein's theorem (§§ III. 1.4 and 1.2), the critical values of γ for system (2.1) must be roots of the equations

$$\omega_j(\gamma) + \omega_h(\gamma) = m \qquad (j, h = 1, \ldots, k;\ m = 0, \pm 1, \pm 2, \ldots). \tag{2.5}$$

The dynamic instability domain attached at $(0, \gamma_0)$, where γ_0 is a root of equation (2.5), will be called a principal-resonance domain if $j = h$ and a combination-resonance domain if $j \neq h$.

The case

$$\mathbf{H}(\tau, \varepsilon, \gamma) = \gamma \mathbf{H}(\tau, \varepsilon)$$

was first studied, under certain additional assumptions (incongruence of the natural frequencies, inter alia) by Malkin /62b/. Under general conditions, this case was considered in §1 of this chapter. In the present section we are assuming that $\mathbf{H}(\tau, \varepsilon, \gamma)$ is a more complicated function of the parameter γ, satisfying the following assumptions.

We shall assume that there exist $\varepsilon_0 > 0$, $\delta_0 > 0$ such that for $|\varepsilon| \leqslant \varepsilon_0$, $|\gamma - \gamma_0| \leqslant \delta_0$, where γ_0 is any fixed point in the interval $[\gamma_1, \gamma_2]$,

$$\mathbf{H}(\tau, \varepsilon, \gamma) = \sum_{r, s=0}^{\infty} \mathbf{H}_{rs}(\tau, \gamma_0)\, \varepsilon^r (\gamma - \gamma_0)^s \qquad (\mathbf{H}_{0s} = \mathbf{H}_{0s}(\gamma_0)). \tag{2.6}$$

The functions $\mathbf{H}_{rs}(\tau, \gamma_0)$ are assumed here to be integrable, piecewise-continuous, 2π-periodic matrix-functions, whose norms satisfy estimates

$$\int_0^{2\pi} |\mathbf{H}_{rs}(\tau, \gamma_0)|\, d\tau \leqslant h_{rs}(\gamma_0),$$

where the series

$$\sum_{r, s=0}^{\infty} h_{rs}(\gamma_0)\, \varepsilon^r (\gamma - \gamma_0)^s$$

is convergent for $|\varepsilon| \leqslant \varepsilon_0$, $|\gamma - \gamma_0| \leqslant \delta_0$.

2.2. Fundamental theorem

Theorem. Let γ_0 be a root of equation (2.5) for some triple j, h, m. Assume that the "fundamental" case holds, i.e., if

$$\omega_j(\gamma_0) + \omega_{h'}(\gamma_0) = m' \quad (h' = 1, \ldots, k;\ m' = 0, \pm 1, \pm 2, \ldots)$$

then

$$h' = h, \quad m' = m,$$

and also

$$\frac{d}{d\gamma}(\omega_j + \omega_h)\big|_{\gamma=\gamma_0} \neq 0. \tag{2.7}$$

Define numbers $\chi_{-j, -j}$, χ_{hh}, χ_{-jh}, $\chi^{(1)}$, $\chi^{(2)}$ by

$$\left.\begin{aligned} \chi_{-j, -j} &= (\mathbf{H}_{10}^{(0)}(\gamma_0)\, \bar{\mathbf{c}}_j(\gamma_0), \bar{\mathbf{c}}_j(\gamma_0)), \quad \chi_{hh} = (\mathbf{H}_{10}^{(0)}(\gamma_0)\, \mathbf{c}_h(\gamma_0), \mathbf{c}_h(\gamma_0)), \\ \chi_{-jh} &= (\mathbf{H}_{10}^{(m)}(\gamma_0)\, \bar{\mathbf{c}}_j(\gamma_0), \mathbf{c}_h(\gamma_0)), \quad \chi^{(1,2)} = \frac{-\chi_{-j, -j} - \chi_{hh} \mp 2|\chi_{-jh}|}{\frac{d}{d\gamma}(\omega_j + \omega_h)\big|_{\gamma=\gamma_0}} \end{aligned}\right\} \tag{2.8}$$

where $\mathbf{H}_{10}^{(m)}(\gamma_0)$ *are the Fourier coefficients of the matrix-function* $\mathbf{H}_{10}(\tau, \gamma_0)$ *of (2.6). Then the dynamic instability domains attached at the point* $(0, \gamma_0)$ *on the* (ε, γ)*-plane are*

$$\gamma_0+\chi^{(1)}\varepsilon+o(\varepsilon)<\gamma<\gamma_0+\chi^{(2)}\varepsilon+o(\varepsilon) \quad \text{for } \varepsilon>0$$

and

$$\gamma_0+\chi^{(2)}\varepsilon+o(\varepsilon)<\gamma<\gamma_0+\chi^{(1)}\varepsilon+o(\varepsilon) \quad \text{for } \varepsilon<0.$$

Remarks. 1. Later (§3.2) we shall see that the remainder term in these formulas is in fact $o(\varepsilon)=O(\varepsilon^2)$.

2. In the "fundamental" case when $\varepsilon=0$ there are only twofold multipliers of mixed kind on the unit circle. The proof will in fact imply a procedure for constructing instability domains when these multipliers are of arbitrary multiplicity.

Proof. Let γ_0 be a root of some equation (2.5) as stipulated. Set

$$\gamma=\gamma_0+\chi\varepsilon;$$

we are going to investigate the characteristic exponents of system (2.1) on the ray $\gamma=\gamma_0+\chi\varepsilon$. Suppose that for some χ and arbitrarily small $|\varepsilon|$ system (2.1) has solutions unbounded as $\tau\to\infty$; then we have a dynamic instability domain on the (ε, γ)-plane, attached at the point $(0, \gamma_0)$, which contains a sufficiently small segment of the ray $\gamma=\gamma_0+\chi\varepsilon\ (\varepsilon>0)$, and γ_0 will be a critical value of γ. Divide the eigenvalues of the matrix $\mathbf{C}(\gamma)=\tilde{\mathbf{J}}^{-1}\mathbf{H}_0(\gamma)$ into congruence classes modulo i and pick out the classes $\{\omega_{j_1}, \ldots, \omega_{j_r}\}$ which contain eigenvalues of different kinds. In order to determine whether γ_0 is a critical value of γ, we must investigate (see §1.4) the equation

$$\det\|\sigma_{jh}(\gamma_0)+\nu\gamma_{jh}\|_{j,\,h=j_1,\,\ldots,\,j_r}=0 \tag{2.9}$$

for each of these classes. To evaluate the quantities σ_{jh}, we write system (2.1), using (2.6), as

$$\tilde{\mathbf{J}}\frac{d\mathbf{x}}{d\tau}=[\mathbf{H}_{00}(\gamma_0)+(\gamma-\gamma_0)\mathbf{H}_{01}(\gamma_0)+\ldots+\varepsilon\mathbf{H}_{10}(\tau, \gamma_0)+\varepsilon(\gamma-\gamma_0)\mathbf{H}_{11}(\tau, \gamma_0)+\ldots]\mathbf{x}$$

and then, substituting $\gamma=\gamma_0+\chi\varepsilon$, as

$$\tilde{\mathbf{J}}\frac{d\mathbf{x}}{dt}=\{\mathbf{H}_{00}(\gamma_0)+\varepsilon[\mathbf{H}_{10}(\tau, \gamma_0)+\chi\mathbf{H}_{01}(\gamma_0)]+O(\varepsilon^2)\}\mathbf{x}.$$

To treat this equation, we must replace $\mathbf{H}_1$ in formula (IV.3.9) by

$$\mathbf{H}_{10}(\tau, \gamma_0)+\chi\mathbf{H}_{01}(\gamma_0).$$

The result is

$$\begin{aligned}\sigma_{jh}(\chi)&=-([\mathbf{H}_{10}(\tau, \gamma_0)+\chi\mathbf{H}_{01}(\gamma_0)]^{(m_h-m_j)}\mathbf{c}_j(\gamma_0), \mathbf{c}_h(\gamma_0))=\\&=-\chi_{jh}-\chi\psi_{jh}\delta_{m_j,m_h} \qquad (j, h=j_1, \ldots, j_r),\end{aligned} \tag{2.10}$$

where

$$\chi_{jh}=(\mathbf{H}_{10}^{(m_h-m_j)}\mathbf{c}_j(\gamma_0),\ \mathbf{c}_h(\gamma_0)),\quad \psi_{jh}=(\mathbf{H}_{01}(\gamma_0)\,\mathbf{c}_j(\gamma_0),\ \mathbf{c}_h(\gamma_0)), \tag{2.11}$$

and $\mathbf{H}_{10}^{(m_h-m_j)}$ are the Fourier coefficients of $\mathbf{H}_{10}(\tau,\ \gamma_0)$:

$$\mathbf{H}_{10}(\tau,\ \gamma_0)\sim\sum_{l=-\infty}^{\infty}e^{il\tau}\mathbf{H}_{10}^{(l)}.$$

Let $\nu_\rho\ (\rho=1,\ \ldots,\ r)$ be the roots of the r-th order equation (2.9), whose coefficients are real polynomials in χ. Using familiar techniques (see §1.7), we must find intervals $(\chi_1^{(-)},\ \chi_1^{(+)}),\ (\chi_2^{(-)},\ \chi_2^{(+)}),\ \ldots$ on the χ-axis, in each of which equation (2.9) has at least one nonreal root. In the first approximation, the dynamic instability domain is the union of the sectors

$$\begin{gathered}\gamma_0+\chi_1^{(-)}\varepsilon+\ldots<\gamma<\gamma_0+\chi_1^{(+)}\varepsilon+\ldots,\\ \gamma_0+\chi_2^{(-)}\varepsilon+\ldots<\gamma<\gamma_0+\chi_2^{(+)}\varepsilon+\ldots,\\ \ldots\ldots\ldots\ldots\ldots\ldots\ldots\ldots,\end{gathered}$$

taken over all classes $\{\omega_{j_1},\ \ldots,\ \omega_{j_r}\}$ that contain eigenvalues of different kinds.

We now transform the second of formulas (2.11), multiplying both sides of (2.3) on the left by the matrix $\tilde{\mathbf{J}}$ and expanding the quantities figuring in the formula as series in powers of $\gamma-\gamma_0$:

$$\begin{aligned}&[\mathbf{H}_{00}(\gamma_0)+(\gamma-\gamma_0)\,\mathbf{H}_{01}(\gamma_0)+\ldots]\left[\mathbf{c}_j(\gamma_0)+(\gamma-\gamma_0)\left(\frac{d\mathbf{c}_j}{d\gamma}\right)_0+\ldots\right]=\\ &\qquad=i\left[\omega_j(\gamma_0)+\left(\frac{d\omega_j}{d\gamma}\right)_0(\gamma-\gamma_0)+\ldots\right]\left[\tilde{\mathbf{J}}\mathbf{c}_j(\gamma_0)+(\gamma-\gamma_0)\,\tilde{\mathbf{J}}\left(\frac{d\mathbf{c}_j}{d\gamma}\right)_0+\ldots\right].\end{aligned}$$

Comparing coefficients of like powers of $\gamma-\gamma_0$, we get

$$\begin{gathered}[\mathbf{H}_{00}(\gamma_0)-i\omega_j(\gamma_0)\,\tilde{\mathbf{J}}]\,\mathbf{c}_j(\gamma_0)=0,\\ \mathbf{H}_{01}(\gamma_0)\,\mathbf{c}_j(\gamma_0)+[\mathbf{H}_{00}(\gamma_0)-i\omega_j(\gamma_0)\,\tilde{\mathbf{J}}]\left(\frac{d\mathbf{c}_j}{d\gamma}\right)_0=i\left(\frac{d\omega_j}{d\gamma}\right)_0\tilde{\mathbf{J}}\mathbf{c}_j(\gamma_0).\end{gathered}$$

Multiply both sides of the second equality by $\mathbf{c}_j(\gamma_0)$ and use formula (I.1.14):

$$\begin{aligned}&(\mathbf{H}_{01}(\gamma_0)\,\mathbf{c}_j(\gamma_0),\ \mathbf{c}_j(\gamma_0))+\left(\left(\frac{d\mathbf{c}_j}{d\gamma}\right)_0,\ [\mathbf{H}_{00}(\gamma_0)-i\omega_j(\gamma_0)\,\tilde{\mathbf{J}}]\,\mathbf{c}_j(\gamma_0)\right)=\\ &\qquad\qquad=i\left(\frac{d\omega_j}{d\gamma}\right)_0(\tilde{\mathbf{J}}\mathbf{c}_j(\gamma_0),\ \mathbf{c}_j(\gamma_0)).\end{aligned}$$

By the first of the above equalities, the second term on the left vanishes, and by (2.4)

$$\psi_{jj}=(\mathbf{H}_{01}(\gamma_0)\,\mathbf{c}_j(\gamma_0),\ \mathbf{c}_j(\gamma_0))=\left(\frac{d\omega_j}{d\gamma}\right)_0\gamma_{jj}.$$

Since it follows from $m_h=m_j$ that $\omega_h(\gamma_0)=\omega_j(\gamma_0)$ and $\mathbf{c}_h(\gamma_0)=\mathbf{c}_j(\gamma_0)$, we may write formula (2.10) as

$$\sigma_{jh}(\chi) = -\chi_{jh} - \chi\left(\frac{d\omega_j}{d\gamma}\right)_0 \gamma_{jh} \qquad (j,\ h = j_1,\ \ldots,\ j_r), \tag{2.12}$$

where, as we recall, χ_{jh} and γ_{jh} are defined by (2.11), (2.4). Note that $\omega_j(\gamma)$ are continuously differentiable functions on $[\gamma_1,\ \gamma_2]$, since for $\gamma \in [\gamma_1,\ \gamma_2]$ the matrix $\mathbf{C}(\gamma) = \tilde{\mathbf{J}}^{-1}\mathbf{H}_0(\gamma)$ has simple elementary divisors (see § IV. 1).

In the second part of the proof, we examine in greater detail the most important "general case," in which the matrix $\mathbf{C}(\gamma_0)$ has only pairs of eigenvalues of different kinds congruent modulo i, i. e., $j_1 = h > 0$, $j_2 = -j < 0$, and

$$\omega_h(\gamma_0) = \alpha_0 + m_h,\quad \omega_{-j}(\gamma_0) = \alpha_0 + m_{-j},\quad \omega_h(\gamma_0) - \omega_{-j}(\gamma_0) = m_h - m_{-j}.$$

Since the last equality may be written as

$$\omega_j(\gamma_0) + \omega_h(\gamma_0) = m_h - m_{-j},$$

we see, comparing this with (2.5), that

$$m_h - m_{-j} = m.$$

Equation (2.9) becomes

$$\begin{vmatrix} \sigma_{-j,\,-j}(\chi) - \nu & \sigma_{-jh}(\chi) \\ \sigma_{h,\,-j}(\chi) & \sigma_{hh}(\chi) + \nu \end{vmatrix} = 0.$$

The discriminant of this equation is

$$\Delta(\chi) = \frac{1}{4}(\sigma_{hh} + \sigma_{-j,\,-j})^2 - \sigma_{h,\,-j}\sigma_{-jh}.$$

By (2.11) and (2.12), we have

$$\sigma_{h,\,-j} = -\chi_{h,\,-j},\qquad \sigma_{-jh} = -\chi_{-jh},\qquad \chi_{h,\,-j} = \bar{\chi}_{-jh}$$

and finally

$$\Delta(\chi) = \frac{1}{4}\left[\chi_{-j,\,-j} + \chi_{hh} + \chi\frac{d}{d\gamma}(\omega_j + \omega_h)\Big|_{\gamma=\gamma_0}\right]^2 - |\chi_{-jh}|^2.$$

If $\Delta(\chi) > 0$, the equation for ν has distinct real roots; if $\Delta(\chi) < 0$ the roots are complex conjugates. The formulas for the slopes of the tangents to the boundaries of the dynamic instability domain are obtained from the equation $\Delta(\chi) = 0$. If

$$\frac{d}{d\gamma}(\omega_j + \omega_h)\Big|_{\gamma=\gamma_0} \neq 0,$$

we have

$$\chi^{(1,2)} = \frac{1}{\frac{d}{d\gamma}(\omega_j + \omega_h)\Big|_{\gamma=\gamma_0}}(\mp 2|\chi_{-jh}| - \chi_{-j,\,-j} - \chi_{hh}). \tag{2.13}$$

The rays $\gamma=\gamma_0+\chi_{1,2}\varepsilon$ are tangent to the boundaries of the instability domain. Let $\chi^{(1)}<\chi^{(2)}$; then instability holds for

$$\varepsilon>0,\ \gamma_0+\chi^{(1)}\varepsilon+\ldots<\gamma<\gamma_0+\chi^{(2)}\varepsilon+\ldots$$

and for

$$\varepsilon<0,\qquad \gamma_0+\chi^{(1)}\varepsilon+\ldots<\gamma<\gamma_0+\chi^{(2)}\varepsilon+\ldots.$$

The dots stand for higher-order terms. This completes the proof.

Remark. If

$$\frac{d}{d\gamma}(\omega_j+\omega_h)\,|_{\gamma=\gamma_0}=0,$$

then $\Delta(\chi)$ is independent of χ, $\Delta(\chi)=\Delta(0)$. Hence, if $\Delta(0)>0$, there is no dynamic instability domain attached at $(0,\ \gamma_0)$; if $\Delta(0)<0$, any ray $\gamma=\gamma_0+\chi\varepsilon$ $(-\infty<\chi<+\infty)$ lies inside the instability domain.

Definition. As in §1.5, we shall say that an instability domain attached at the point $(0,\ \gamma_0)$ is *wide* if it contains a sector with vertex at $(0,\ \gamma_0)$ $(\chi_1\neq\chi_2)$, and *narrow* otherwise $(\chi_1=\chi_2)$.

2.3. Formulas for second-order canonical vector equation

The equation

$$\frac{d}{d\tau}\left[\mathbf{M}(\tau,\ \varepsilon,\ \gamma)\frac{d\mathbf{y}}{d\tau}\right]+\mathbf{P}(\tau,\ \varepsilon,\ \gamma)\,\mathbf{y}=\mathbf{0}, \tag{2.14}$$

which occurs quite frequently in applications, may be reduced to equation (2.1). Here $\mathbf{y}$ is a k-vector, $\mathbf{M}(\tau,\ \varepsilon,\ \gamma)$ and $\mathbf{P}(\tau,\ \varepsilon,\ \gamma)$ are real symmetric $k\times k$ matrices, 2π-periodic in τ; $\mathbf{M}(\tau,\ \varepsilon,\ \gamma)$ is assumed to be nonsingular for the relevant values of $\tau,\ \varepsilon,\ \gamma$. It is also assumed that the matrices $\mathbf{M}(\tau,\ 0,\ \gamma)=\mathbf{M}_0(\gamma)$, $\mathbf{P}(\tau,\ 0,\ \gamma)=\mathbf{P}_0(\gamma)$ are independent of τ and $\mathbf{M}_0(\gamma)>\mathbf{0}$, $\mathbf{P}_0(\gamma)>\mathbf{0}$, i. e., these matrices are positive in the sense of the quadratic forms $(\mathbf{M}_0(\gamma)\,\mathbf{c},\ \mathbf{c})$ and $(\mathbf{P}_0(\gamma)\,\mathbf{c},\ \mathbf{c})$ for all $\gamma\in[\gamma_1,\ \gamma_2]$. Concerning the matrix-functions $\mathbf{M}$ and $\mathbf{P}$, we assume that there exist $\varepsilon_0>0$, $\delta_0>0$ such that for $|\varepsilon|\leqslant\varepsilon_0$, $|\gamma-\gamma_0|\leqslant\delta_0$, where γ_0 is any fixed point in $[\gamma_1,\ \gamma_2]$,

$$\left.\begin{aligned}\mathbf{M}(\tau,\ \varepsilon,\ \gamma)&=\sum_{r,\ s=0}^{\infty}\mathbf{M}_{rs}(\tau,\ \gamma_0)\,\varepsilon^r(\gamma-\gamma_0)^s,\\ \mathbf{P}(\tau,\ \varepsilon,\ \gamma)&=\sum_{r,\ s=0}^{\infty}\mathbf{P}_{rs}(\tau,\ \gamma_0)\,\varepsilon^r(\gamma-\gamma_0)^s,\end{aligned}\right\} \tag{2.15}$$

where $\mathbf{M}_{rs}(\tau,\ \gamma_0)$, $\mathbf{P}_{rs}(\tau,\ \gamma_0)$ are integrable piecewise-continuous 2π-periodic matrix-functions with norms satisfying the estimates

$$\int_0^{2\pi}|\mathbf{M}_{rs}(\tau,\ \gamma_0)|\,d\tau\leqslant m_{rs}(\gamma_0),\quad \int_0^{2\pi}|\mathbf{P}_{rs}(\tau,\ \gamma_0)|\,d\tau\leqslant p_{rs}(\gamma_0)$$

such that the series

$$\sum_{r,\, s=0}^{\infty} m_{rs}(\gamma_0)\,\varepsilon^r(\gamma-\gamma_0)^s, \qquad \sum_{r,\, s=0}^{\infty} p_{rs}(\gamma_0)\,\varepsilon^r(\gamma-\gamma_0)^s$$

are convergent for $|\varepsilon|\leqslant\varepsilon_0$, $|\gamma-\gamma_0|\leqslant\delta_0$.

We extend the results of §2.1 to equation (2.14). The substitution

$$\mathbf{x}=\begin{pmatrix}\mathbf{y}\\ \mathbf{M}(\tau,\ \varepsilon,\ \gamma)\dfrac{d\mathbf{y}}{d\tau}\end{pmatrix}$$

reduces equation (2.14) to equation (2.1):

$$\mathbf{J}\frac{d\mathbf{x}}{d\tau}=\mathbf{H}(\tau,\ \varepsilon,\ \gamma)\,\mathbf{x}, \quad \mathbf{J}=\begin{pmatrix}\mathbf{0} & -\mathbf{I}_k\\ \mathbf{I}_k & \mathbf{0}\end{pmatrix},$$

where the matrix-function

$$\mathbf{H}(\tau,\ \gamma,\ \varepsilon)=\begin{pmatrix}\mathbf{P}(\tau,\ \varepsilon,\ \gamma) & \mathbf{0}\\ \mathbf{0} & \mathbf{M}^{-1}(\tau,\ \varepsilon,\ \gamma)\end{pmatrix}$$

satisfies all the assumptions of §2.1. Let $\omega_\varkappa^2(\gamma)$ $(\omega_\varkappa>0)$ and $\mathbf{a}_\varkappa(\gamma)$ denote the eigenvalues and eigenvectors of the matrix $\mathbf{M}_0(\gamma)^{-1}\cdot\mathbf{P}_0(\gamma)$, i. e.,

$$\mathbf{P}_0(\gamma)\,\mathbf{a}_\varkappa=\omega_\varkappa^2(\gamma)\,\mathbf{M}_0(\gamma)\,\mathbf{a}_\varkappa(\gamma) \qquad (\varkappa=1,\ \ldots,\ k), \tag{2.16}$$

with the eigenvectors normalized so that

$$[\omega_\varkappa(\gamma)+\omega_\lambda(\gamma)]\,(\mathbf{M}_0(\gamma)\,\mathbf{a}_\varkappa(\gamma),\ \mathbf{a}_\lambda(\gamma))=\delta_{\varkappa\lambda} \qquad (\varkappa,\ \lambda=1,\ \ldots,\ k). \tag{2.17}$$

Then the vectors

$$\mathbf{c}_\varkappa(\gamma)=\begin{pmatrix}\mathbf{a}_{|\varkappa|}(\gamma)\\ i\omega_\varkappa(\gamma)\,\mathbf{M}_0(\gamma)\,\mathbf{a}_{|\varkappa|}(\gamma)\end{pmatrix},$$

where $\omega_{-\varkappa}=-\omega_\varkappa$ are eigenvectors of the matrix

$$\mathbf{C}(\gamma)=\mathbf{J}^{-1}\mathbf{H}(\tau,\ \gamma,\ 0)=\mathbf{J}^{-1}\mathbf{H}_0(\gamma)=\begin{pmatrix}\mathbf{0} & \mathbf{M}_0(\gamma)^{-1}\\ -\mathbf{P}_0(\gamma) & \mathbf{0}\end{pmatrix},$$

i. e.,

$$\mathbf{J}^{-1}\mathbf{H}_0(\gamma)\,\mathbf{c}_\varkappa(\gamma)=i\omega_\varkappa(\gamma)\,\mathbf{c}_\varkappa(\gamma) \qquad (\varkappa=\pm 1,\ \ldots,\ \pm k).$$

Here

$$\gamma_{\varkappa\lambda}=\langle\mathbf{c}_\varkappa(\gamma),\ \mathbf{c}_\lambda(\gamma)\rangle\equiv i\,(\mathbf{J}\mathbf{c}_\varkappa(\gamma),\ \mathbf{c}_\lambda(\gamma))=\begin{cases}0, & \varkappa\neq\lambda,\\ +1, & \varkappa=\lambda>0,\\ -1, & \varkappa=\lambda<0\end{cases}$$

$(\varkappa, \lambda = \pm 1, \ldots, \pm k)$, so that $i\omega_1(\gamma), \ldots, i\omega_k(\gamma)$ are the eigenvalues of $\mathbf{C}$ of the first kind and $-i\omega_1(\gamma), \ldots, -i\omega_k(\gamma)$ its eigenvalues of the second kind.

Let γ_0 be a critical value of γ, that is, a root of equation (2.5). It follows from (2.15) that

$$\mathbf{M}(\tau, \gamma_0, \varepsilon) = \mathbf{M}_0(\gamma_0) + \varepsilon\mathbf{M}_{10}(\tau, \gamma_0) + \varepsilon^2 \ldots \qquad (\mathbf{M}_0(\gamma_0) \equiv \mathbf{M}_{00}(\gamma_0))$$

and the inverse matrix is

$$\mathbf{M}(\tau, \gamma_0, \varepsilon)^{-1} = \mathbf{M}_0(\gamma_0)^{-1} - \varepsilon\mathbf{M}_0(\gamma_0)^{-1}\mathbf{M}_{10}(\tau, \gamma_0)\mathbf{M}_0(\tau, \gamma_0)^{-1} + \ldots$$

Thus the Fourier coefficients are

$$\mathbf{H}_{10}^{(l)}(\gamma_0) = \begin{pmatrix} \mathbf{P}_{10}^{(l)}(\gamma_0) & 0 \\ 0 & -\mathbf{M}_0(\gamma_0)^{-1}\mathbf{M}_{10}^{(l)}(\gamma_0)\mathbf{M}_0(\gamma_0)^{-1} \end{pmatrix}.$$

A few additional manipulations finally yield the following formulas for (2.8):

$$\left.\begin{array}{c} \chi_{jj} = ([\mathbf{P}_{10}^{(0)}(\gamma_0) - \omega_j(\gamma_0)^2\mathbf{M}_{10}^{(0)}(\gamma_0)]\,\mathbf{a}_{|j|}(\gamma_0), \mathbf{a}_{|j|}(\gamma_0)), \\ \chi_{-jh} = ([\mathbf{P}_{10}^{(m)}(\gamma_0) + \omega_j(\gamma_0)\,\omega_h(\gamma_0)\mathbf{M}_{10}^{(m)}(\gamma_0)]\,\mathbf{a}_{|j|}(\gamma_0), \mathbf{a}_{|h|}(\gamma_0)) \\ (j, h = \pm 1, \ldots, \pm k). \end{array}\right\} \tag{2.18}$$

Here $\mathbf{P}_{10}^{(m)}(\gamma_0)$ and $\mathbf{M}_{10}^{(m)}(\gamma_0)$ are the Fourier coefficients of the matrix-functions $\mathbf{P}_{10}(\tau, \gamma_0)$ and $\mathbf{M}_{10}(\tau, \gamma_0)$:

$$\mathbf{P}_{10}(\tau, \gamma_0) \sim \sum_{l=-\infty}^{\infty} e^{il\tau}\mathbf{P}_{10}^{(l)}(\gamma_0), \quad \mathbf{M}_{10}(\tau, \gamma_0) \sim \sum_{l=-\infty}^{\infty} e^{il\tau}\mathbf{M}_{10}^{(l)}(\gamma_0).$$

For the reader's convenience, we summarize the final result as a theorem, analogous to the theorem of §2.2.

Theorem. Let γ_0 be a root of equation (2.5), where $\omega_1(\gamma)^2, \ldots, \omega_k(\gamma)^2$ $(\omega_\varkappa > 0)$ are the eigenvalues of the matrix $\mathbf{M}_0(\gamma)^{-1}\mathbf{P}_0(\gamma)$. Suppose that the general case holds (see theorem in §2.1) and also that

$$\frac{d}{d\gamma}(\omega_j + \omega_h)\,|_{\gamma=\gamma_0} \neq 0. \tag{2.19}$$

Define eigenvectors of $\mathbf{M}_0(\gamma)^{-1}\mathbf{P}_0(\gamma)$ by formulas (2.16) and (2.17) and numbers $\chi_j, \chi_{jh}, \mu_1, \mu_2$ by

$$\left.\begin{array}{c} \chi_j = ([\mathbf{P}_{10}^{(0)}(\gamma_0) - \omega_j(\gamma_0)^2\mathbf{M}_{10}^{(0)}(\gamma_0)]\,\mathbf{a}_j(\gamma_0), \mathbf{a}_j(\gamma_0)), \\ \chi_{jh} = ([\mathbf{P}_{10}^{(m)}(\gamma_0) + \omega_j(\gamma_0)\,\omega_h(\gamma_0)\mathbf{M}_{10}^{(m)}(\gamma_0)]\,\mathbf{a}_j(\gamma_0), \mathbf{a}_h(\gamma_0)) \\ (j, h = 1, \ldots, k), \end{array}\right\} \tag{2.20}$$

$$\mu_{1,2} = \frac{1}{\frac{d}{d\gamma}(\omega_j + \omega_h)\,|_{\gamma=\gamma_0}}(-\chi_j - \chi_h \mp 2\,|\chi_{jh}|), \tag{2.21}$$

where $\mathbf{M}_{10}^{(0)}$, $\mathbf{M}_{10}^{(m)}$, $\mathbf{P}_{10}^{(0)}$, $\mathbf{P}_{10}^{(m)}$ *are the Fourier coefficients of the matrix-functions* $\mathbf{M}_{10}(\tau, \gamma_0)$ *and* $\mathbf{P}_{10}(\tau, \gamma_0)$. *Then there is a dynamic instability domain*

$$\gamma_0+\mu_1\varepsilon+\ldots<\gamma<\gamma_0+\mu_2\varepsilon+\ldots,$$

attached at the point $(0, \gamma_0)$ *in the first quadrant of the* $\varepsilon\gamma$*-plane* $(\varepsilon>0, \gamma>0)$. *The omitted terms here are of the order of* ε^2 *and higher.*

Remark I. Condition (2.19) will certainly hold if the matrices

$$\left[\frac{d\mathbf{P}_0}{d\gamma}-\omega_j^2\frac{d\mathbf{M}_0}{d\gamma}\right]_{\gamma=\gamma_0}, \quad \left[\frac{d\mathbf{P}_0}{d\gamma}-\omega_h^2\frac{d\mathbf{M}_0}{d\gamma}\right]_{\gamma=\gamma_0}$$

are both positive definite or both negative definite.

Proof. Consider the expression

$$\begin{aligned}\left(\left[\frac{d\mathbf{P}_0}{d\gamma}-\omega_j^2\frac{d\mathbf{M}_0}{d\gamma}\right]\mathbf{a}_j, \mathbf{a}_j\right)&=\frac{d}{d\gamma}(\mathbf{P}_0\mathbf{a}_j, \mathbf{a}_j)-\left(\mathbf{P}_0\frac{d\mathbf{a}_j}{d\gamma}, \mathbf{a}_j\right)-\\ &-\left(\mathbf{P}_0\mathbf{a}_j, \frac{d\mathbf{a}_j}{d\gamma}\right)-\omega_j^2\left[\frac{d}{d\gamma}(\mathbf{M}_0\mathbf{a}_j, \mathbf{a}_j)-\left(\mathbf{M}_0\frac{d\mathbf{a}_j}{d\gamma}, \mathbf{a}_j\right)-\left(\mathbf{M}_0\mathbf{a}_j, \frac{d\mathbf{a}_j}{d\gamma}\right)\right]=\\ &=\frac{d}{d\gamma}(\mathbf{P}_0\mathbf{a}_j, \mathbf{a}_j)-\omega_j^2\frac{d}{d\gamma}(\mathbf{M}_0\mathbf{a}_j, \mathbf{a}_j)-2\operatorname{Re}\left([\mathbf{P}_0-\omega_j^2\mathbf{M}_0]\mathbf{a}_j, \frac{d\mathbf{a}_j}{d\gamma}\right).\end{aligned}$$

The last term vanishes, as follows from (2.16). As for the first term, by (2.16) and (2.17),

$$\frac{d}{d\gamma}(\omega_j^2\mathbf{M}_0\mathbf{a}_j, \mathbf{a}_j)=\frac{1}{2}\frac{d\omega_j}{d\gamma}.$$

By (2.17), the second term gives

$$-\omega_j^2\frac{d}{d\gamma}\frac{1}{2\omega_j}=\frac{1}{2}\frac{d\omega_j}{d\gamma}.$$

Finally, we have

$$\left(\left[\frac{d\mathbf{P}_0}{d\gamma}-\omega_j^2\frac{d\mathbf{M}_0}{d\gamma}\right]\mathbf{a}_j, \mathbf{a}_j\right)=\frac{d\omega_j}{d\gamma}.$$

If, say,

$$\frac{d\mathbf{P}_0}{d\gamma}-\omega^2\frac{d\mathbf{M}_0}{d\gamma}>0 \qquad (\omega^2=\omega_j^2, \ \omega_h^2),$$

it follows that

$$\frac{d\omega_j}{d\gamma}>0, \quad \frac{d\omega_h}{d\gamma}>0,$$

Q. E. D.

Remark II. If $\mathbf{M}(\tau, \gamma, \varepsilon) \equiv \mathbf{I}_k$, formula (2.20) is simplified:

$$\left.\begin{aligned} \chi_j &= (\mathbf{P}_{10}^{(0)}(\gamma_0)\,\mathbf{a}_j(\gamma_0),\ \mathbf{a}_j(\gamma_0)),\\ \chi_{jh} &= (\mathbf{P}_{10}^{(m)}(\gamma_0)\,\mathbf{a}_j(\gamma_0),\ \mathbf{a}_h(\gamma_0)) \qquad (j, h = 1, \ldots, k), \end{aligned}\right\} \tag{2.22}$$

where

$$\left.\begin{aligned} &\mathbf{P}_0(\gamma_0)\,\mathbf{a}_\varkappa(\gamma_0) = \omega_\varkappa^2(\gamma_0)\,\mathbf{a}_\varkappa(\gamma_0),\\ &[(\omega_\varkappa(\gamma_0) + \omega_\lambda(\gamma_0)]\,(\mathbf{a}_\varkappa(\gamma_0),\ \mathbf{a}_\lambda(\gamma_0)) = \delta_{\varkappa\lambda} \qquad (\varkappa, \lambda = 1, \ldots, k). \end{aligned}\right\} \tag{2.23}$$

Remark III. If in addition to $\mathbf{M} = \mathbf{I}_k$ it is also true that

$$\mathbf{P}_{10}^{(0)} = \left[\frac{\partial}{\partial\varepsilon}\mathbf{P}(\tau, 0, \gamma_0)\right]_{\mathrm{av}} = 0,$$

then the formula (2.21) for $\mu_{1,2}$ is also simplified:

$$\mu_{1,2} = \mp \left.\frac{2\,(\mathbf{P}_{10}^{(m)}\mathbf{a}_j,\ \mathbf{a}_h)}{\dfrac{d\omega_j}{d\gamma} + \dfrac{d\omega_h}{d\gamma}}\right|_{\gamma=\gamma_0}. \tag{2.24}$$

Remark IV. If $\mathbf{M}(\tau, \gamma, \varepsilon) \equiv \mathbf{I}_k$ and $\mathbf{P}_0(\gamma)$ is diagonal,

$$\mathbf{P}_0(\gamma) = \begin{pmatrix} p_1(\gamma) & & 0 \\ & \ddots & \\ 0 & & p_k(\gamma) \end{pmatrix} \qquad (p_\varkappa(\gamma) > 0),$$

then

$$\omega_\varkappa(\gamma) = \sqrt{p_\varkappa(\gamma)}, \quad \mathbf{a}_\varkappa(\gamma) = \frac{1}{\sqrt{2\omega_\varkappa(\gamma)}}\,\mathbf{e}_\varkappa \qquad (\varkappa = 1, \ldots, k),$$

where $\mathbf{e}_1, \ldots, \mathbf{e}_k$ are the standard basis vectors (columns of the identity matrix). Let us express $\mathbf{P}_{10}(\tau, \gamma)$ as

$$\mathbf{P}_{10}(\tau, \gamma) = \sum_{l=-\infty}^{\infty} e^{il\tau}\,\|\pi_{\varkappa\lambda}^{(l)}(\gamma)\|_1^k.$$

The formulas for χ_j and χ_{jh} become

$$\chi_j = \frac{1}{2\sqrt{p_j(\gamma_0)}}\,\pi_{jj}^{(0)}(\gamma_0), \quad \chi_{jh} = \frac{1}{2\sqrt[4]{p_j(\gamma_0)\,p_h(\gamma_0)}}\,\pi_{jh}^{(m)}(\gamma_0).$$

In this case, the dynamic instability domain will be wide if and only if the element of the matrix

$$\mathbf{P}_{10}^{(m)}(\gamma_0) = \|\pi_{\varkappa\lambda}^{(m)}(\gamma_0)\|_1^k$$

at the intersection of the j-th row and h-th column does not vanish.

§ 3. PROPERTIES OF BOUNDARIES OF DYNAMIC INSTABILITY DOMAINS. SECOND AND HIGHER APPROXIMATIONS FOR THE BOUNDARIES

3.1. General characterization of equations. Statement of the problem

We shall again consider a canonical equation (2.1), in which the Hamiltonian $\mathbf{H}(\tau, \varepsilon, \gamma) = \mathbf{H}_0(\gamma) + \varepsilon\mathbf{H}_1(\tau, \varepsilon, \gamma)$ has the properties listed in §2.1: it is an analytic function of γ and ε in the sense described there and a 2π-periodic function of τ.

The following frequently occurring special cases of equation (2.1) are worthy of mention:

$$\mathbf{J}\frac{d\mathbf{x}}{dt} = [\mathbf{H}_0 + \varepsilon\mathbf{H}_1(\vartheta t) + \varepsilon^2\mathbf{H}_2(\vartheta t) + \ldots]\,\mathbf{x} \tag{3.1}$$

and

$$\mathbf{M}\frac{d^2\mathbf{y}}{dt^2} + [\mathbf{P}_0 + \varepsilon\mathbf{P}_1(\vartheta t) + \varepsilon^2\mathbf{P}_2(\vartheta t) + \ldots]\,\mathbf{y} = 0. \tag{3.2}$$

Here, as usual, the coefficients of the bracketed series are real and symmetric for real values of all the parameters and 2π-periodic in $\tau = \vartheta t$; $\mathbf{M} = \mathbf{M}^{\tau} > 0$ is a real skew-symmetric matrix. Equations (3.1) and (3.2) are reduced to the form of (2.1) by the substitution

$$\tau = \vartheta t, \quad \gamma = \frac{1}{\vartheta}$$

(concerning the reduction of equation (3.2) to (3.1), see § II. 3.2).

We shall assume that the natural frequencies $\omega_1(\gamma), \ldots, \omega_k(\gamma)$ of the unperturbed system (see §2.1) are analytic functions of γ for $\gamma_1 \leqslant \gamma \leqslant \gamma_2$. This condition is clearly valid for equations (3.1) and (3.2), provided the numbers γ_1 and γ_2 are of the same sign (since in these cases $\gamma = 1/\vartheta$, where ϑ is the perturbation frequency, it is natural to take $\gamma_2 > \gamma_1 > 0$). The eigenvectors $\mathbf{c}_\nu(\gamma)$, normalized by (2.4), will also be analytic functions of γ for $\gamma_1 \leqslant \gamma \leqslant \gamma_2$.

In many cases, an additional condition is satisfied: the symmetric matrix

$$\mathbf{H}_0'(\gamma_0) = \left(\frac{d\mathbf{H}_0(\gamma)}{d\gamma}\right)_{\gamma=\gamma_0}$$

is positive (or negative) definite. (For example, this is the case for system (3.1) with a Hamiltonian $\mathbf{H}_0 > 0$.) We claim that in this case the numbers

$$\omega_j'(\gamma_0) = \left(\frac{d\omega_j(\gamma)}{d\gamma}\right)_{\gamma=\gamma_0} \qquad (j = 1, \ldots, k)$$

are of the same sign: positive if $\mathbf{H}_0'(\gamma_0) > 0$ and negative if $\mathbf{H}_0'(\gamma_0) < 0$. Indeed, differentiating the identity (2.3) for $\nu = j$, taking the scalar product of the resulting equation by $\mathbf{c}_j(\gamma)$, and using §2.2, we get

$$\frac{d\omega_j}{d\gamma}=\frac{(\mathbf{H}_0'(\gamma)\,\mathbf{c}_j(\gamma),\ \mathbf{c}_j(\gamma))}{\langle \mathbf{c}_j(\gamma),\ \mathbf{c}_j(\gamma)\rangle} \qquad (j=1,\ \ldots,\ k), \tag{3.3}$$

and this, together with (2.4), yields the assertion.

As mentioned in §2.1, a dynamic instability domain may be attached at points $(0, \gamma_0)$ only for values of γ_0 which are roots of equations (2.5). Let γ_0 be a root of some equation (2.5) such that when $\varepsilon=0$, $\gamma=\gamma_0$ equation (2.1) has only a twofold multiplier ρ_0 of mixed kind on the upper closed unit semicircle ("general case"). On the boundaries of the dynamic instability domain, equation (2.1) has a twofold multiplier of mixed kind. If $\rho_0=\pm 1$ (principal resonance), it is readily seen (for details, see below) that $\rho(\varepsilon)=\pm 1$. This means that when $\rho_0=\pm 1$ there is either a 2π-periodic ($\rho_0=1$) or a 2π-antiperiodic ($\rho_0=-1$) solution on the boundaries of the instability domain. If $\rho_0\neq\pm 1$ (combination resonance) then, generally speaking, $\rho(\varepsilon)\not\equiv\rho_0$. If there is a wide dynamic instability domain attached at $(0, \gamma_0)$ (i. e., the tangents to the boundaries of the domain at the point $(0, \gamma_0)$ do not coincide), the existence of this domain may be established by the methods of §2. In that case, $\chi_{-jh}\neq 0$ (see (2.11) and (2.13)). The boundaries themselves may be determined to any approximation (in principle, at least) by utilizing the above property of the existence of a twofold multiplier $\rho(\varepsilon)$ tending to ρ_0 at $\varepsilon=0$ (combination resonance), or the existence of a 2π-periodic or 2π-antiperiodic solution (principal resonance). If $\chi_{-jh}=0$, the tangents coincide. In that case only a narrow instability domain may be attached at the point $(0, \gamma_0)$, and there may be none at all. In the neighborhood of $(0, \gamma_0)$ one can then find curves on which there exists a twofold multiplier $\rho(\varepsilon)$ (for principal resonance $\rho(\varepsilon)\equiv\rho_0$).

Now, however, it is not clear whether these curves are boundaries of a dynamic instability domain or not. We shall show in this section that under quite broad assumptions, to be formulated below, (a) the curves in question are defined by analytic functions $\gamma=\gamma^{(1)}(\varepsilon)$, $\gamma=\gamma^{(2)}(\varepsilon)$ of ε; (b) if $\gamma^{(1)}(\varepsilon)<\gamma^{(2)}(\varepsilon)$ for $\varepsilon>0$, then for small $\varepsilon>0$ the domain $\gamma^{(1)}(\varepsilon)<\gamma<\gamma^{(2)}(\varepsilon)$ is invariably a dynamic instability domain.

Property (a) substantially simplifies the practical derivation of approximate formulas for $\gamma^{(1,2)}(\varepsilon)$. Property (b) releaves us of the need to analyze the generally highly complicated position of the multipliers in the domain $\gamma^{(1)}(\varepsilon)<\gamma<\gamma^{(2)}(\varepsilon)$.

Properties (a) and (b) are also valid if when $\varepsilon=0$, $\gamma=0$ equation (2.1) has an arbitrary number of twofold multipliers of mixed kind on the unit circle. The more complicated cases will not be considered in this section (the necessary arguments and results are considerably more tedious).

We need some more details concerning the analyticity of the functions $\gamma^{(1,2)}(\varepsilon)$.

Let $\mathbf{X}(\tau, \varepsilon, \gamma)$ be the matrizant of equation (2.1), $\mathbf{U}(\varepsilon, \gamma)=\mathbf{X}(2\pi, \varepsilon, \gamma)$ its monodromy matrix and $\Delta(\varepsilon, \gamma)$ the discriminant of the characteristic equation:

$$\det[\mathbf{U}(\varepsilon, \gamma)-\rho\mathbf{I}]=0. \tag{3.4}$$

(Here and below, $\mathbf{I}$ is the $2k \times 2k$ identity matrix.) Since multipliers of different kinds must coincide on the boundary of a dynamic instability domain, it follows that the functions $\gamma = \gamma(\varepsilon)$ defining its boundary satisfy the equation

$$\Delta(\varepsilon, \gamma) = 0$$

and the initial condition $\gamma(0) = \gamma_0$. The matrix-functions $\mathbf{X}(\tau, \gamma, \varepsilon)$, $\mathbf{U}(\gamma, \varepsilon)$, and hence also $\Delta(\gamma, \varepsilon)$, are analytic functions of ε and γ at $(0, \gamma_0)$. We may thus expand the functions $\gamma = \gamma^{(j)}(\varepsilon)$ as

$$\gamma_j = \gamma_0 + \gamma_1^{(j)} \varepsilon^{p_j/q_j} + \gamma_2^{(j)} \varepsilon^{(p_j+1)/q_j} + \dots \qquad (j = 1, 2) \tag{3.5}$$

where p_j, q_j are integers. The most logical procedure for determining the coefficients of (3.5) is the following. We must successively determine the matrices $\mathbf{X}_h(\tau, \gamma)$, $\mathbf{U}_h(\gamma)$ in the expansions

$$\mathbf{X}(\tau, \varepsilon, \gamma) = \mathbf{X}_0(\tau, \gamma) + \varepsilon \mathbf{X}_1(\tau, \gamma) + \varepsilon^2 \mathbf{X}_2(\tau, \gamma) + \dots,$$
$$\mathbf{U}(\varepsilon, \gamma) = \mathbf{U}_0(\gamma) + \varepsilon \mathbf{U}_1(\gamma) + \varepsilon^2 \mathbf{U}_2(\gamma) + \dots,$$

then the coefficients $\Delta_h(\gamma)$ in

$$\Delta(\varepsilon, \gamma) = \Delta_0(\gamma) + \Delta_1(\gamma)\varepsilon + \Delta_2(\gamma)\varepsilon^2 + \dots$$

and finally the coefficients of (3.5). This procedure, however, is quite unacceptable, owing to the enormously messy algebraic manipulations needed, most of which have nothing to do with the intrinsic essence of the problem. Indeed, in determining the coefficients $\Delta_h(\gamma)$ we are preparing the ground for determination of all the roots of the equation $\Delta(\varepsilon, \gamma) = 0$, which is quite superfluous.

Below we shall present far more convenient methods to determine the coefficients of (3.5). We shall first show, in accordance with the foregoing, that under quite broad assumptions only integral powers of ε will appear in the expansions (3.5).

Note that this is a characteristic property of Hamiltonian systems. For t-invariant systems (§ II. 3. 9), for which the Lyapunov-Poincaré theorem is also true and so the whole context of dynamic instability domains applies to them as well, the functions $\gamma^{(1,2)}(\varepsilon)$ are no longer necessarily analytic. There are examples of such systems in which the functions $\gamma^{(1,2)}(\varepsilon)$ may be expanded only in fractional powers of ε. Even for such systems, though, one can develop criteria for the functions $\gamma^{(1,2)}(\varepsilon)$ to be analytic.

3.2. Analyticity of $\gamma^{(1,2)}(\varepsilon)$ as functions of ε

Theorem. Suppose that for $\varepsilon = 0$, $\gamma = \gamma_0$, *two multipliers of different kinds of equation (2.1) coincide at a point* $\rho^{(0)}$ *on the unit circle, in other words, for some* j *and* h *of the same sign, equation (2.5) holds and*

$$\rho^{(0)} = \exp[2\pi i \omega_j(\gamma_0)] = \exp[-2\pi i \omega_h(\gamma_0)].$$

Assume that

$$\frac{d}{d\gamma}\Big[\omega_j(\gamma)+\omega_h(\gamma)\Big]\Big|_{\gamma=\gamma_0}\neq 0. \tag{3.6}$$

Then the locus of the points on the (ε, γ)*-plane in the neighborhood of* $\varepsilon=0,\ \gamma=\gamma_0$ *for which equation (2.1) has a twofold multiplier of mixed kind*

$$\rho(\varepsilon)=e^{2\pi i\alpha(\varepsilon)},\quad \rho(0)=\rho^{(0)}$$

in the neighborhood of $\rho^{(0)}$ *consists of two (not necessarily distinct) curves* $\gamma=\gamma^{(l)}(\varepsilon)$ $(l=1, 2)$ *through the point* $(0, \gamma_0)$. *The functions* $\gamma^{(l)}(\varepsilon)$, $\alpha(\varepsilon)$ *are analytic in* ε *at* $\varepsilon=0$*:*

$$\left.\begin{aligned}\gamma^{(l)}(\varepsilon)&=\gamma_0+\gamma_1^{(l)}\varepsilon+\gamma_2^{(l)}\varepsilon^2+\ldots\qquad (l=1, 2),\\ \alpha(\varepsilon)&=\alpha_0+\alpha_1\varepsilon+\alpha_2\varepsilon^2+\ldots\end{aligned}\right\} \tag{3.7}$$

If $\rho^{(0)}=\pm 1$, *then* $\rho(\varepsilon)\equiv\pm 1$.

Remark 1. It follows from the previous discussion that system (2.1) satisfies condition (3.6) if $\mathbf{H}_0'(\gamma_0)$ is sign definite, and system (3.1) satisfies the condition if $\omega_j+\omega_h\neq 0$.

Proof. Let $\delta,\ \varepsilon_1,\ \delta_1$ be positive numbers, so small that for

$$|\varepsilon|<\varepsilon_1,\quad |\gamma-\gamma_0|<\delta_1 \tag{3.8}$$

(ε and γ are complex) the disk $|\rho-\rho^{(0)}|<\delta$ contains exactly two roots of equation (3.4), say $\rho_1(\varepsilon, \gamma)$ and $\rho_2(\varepsilon, \gamma)$. We claim that

$$\Delta_0(\varepsilon, \gamma)=[\rho_1(\varepsilon, \gamma)-\rho_2(\varepsilon, \gamma)]^2$$

is an analytic function of ε and γ in the domain (3.8). Let γ be held fixed, $|\gamma-\gamma_0|<\delta_1$, and let ε describe a circle $|\varepsilon|=\text{const}<\varepsilon_1$. In the process, the roots ρ_1 and ρ_2 can only be interchanged. Thus $\Delta_0(\varepsilon, \gamma)$ is a single-valued function for fixed γ, $|\gamma-\gamma_0|<\delta_1$, and so it is an analytic function of ε for $|\varepsilon|<\varepsilon_1$. Similarly, for fixed ε, $|\varepsilon|<\varepsilon_1$, $\Delta_0(\varepsilon, \gamma)$ is an analytic function of γ for $|\gamma-\gamma_0|<\delta_1$. Hence, by Hartogs' theorem /95/, $\Delta_0(\varepsilon, \gamma)$ is an analytic function in the domain (3.8). Since

$$\rho_1(0, \gamma)=\exp[2\pi i\omega_j(\gamma)],\quad \rho_2(0, \gamma)=\exp[-2\pi i\omega_h(\gamma)], \tag{3.9}$$

it follows that as $\gamma\to\gamma_0$,

$$\Delta_0(\varepsilon, \gamma)=\{2\pi i\rho^{(0)}[\omega_j'(\gamma_0)+\omega_h'(\gamma_0)]\}^2(\gamma-\gamma_0)^2+O[(\gamma-\gamma_0)^3]. \tag{3.10}$$

The function $\gamma(\varepsilon)$ is a root of the equation $\Delta_0(\varepsilon, \gamma)=0$. Since by (3.10) and (3.6)

$$\frac{\partial^2\Delta_0(0, \gamma_0)}{\partial\gamma^2}\neq 0, \tag{3.11}$$

it follows that in (3.5) we have $q_1+q_2=2$, so that $\gamma(\varepsilon)$ is either an analytic function of $\sqrt{\varepsilon}$ or has two branches $\gamma^{(1)}(\varepsilon)$, $\gamma^{(2)}(\varepsilon)$ which are analytic functions of ε and coincide in the limiting case.

We claim that the first alternative is impossible. Consider the behavior of the multipliers $\rho_1(\varepsilon, \gamma), \rho_2(\varepsilon, \gamma)$ on the real $\varepsilon\gamma$-plane in the neighborhood of $(0, \gamma_0)$. By assumption, when $\varepsilon=0$, $\gamma=\gamma_0$, two multipliers of different kinds coincide at $\rho^{(0)}$. Hence, for all real ε, γ in the neighborhood of $(0, \gamma_0)$, the roots $\rho_1(\varepsilon, \gamma)$, $\rho_2(\varepsilon, \gamma)$ are multipliers of different kinds. (Recall that, by the theorem of § III. 2.10, the multipliers together with their kinds are continuous functions of real ε, γ.) It follows from condition (3.6) and formulas (3.9) that as one describes the γ-axis from $\gamma<\gamma_0$ to $\gamma>\gamma_0$ the function $\arg\rho_1-\arg\rho_2$ changes sign, i. e., the order in which the multipliers of first and second kinds appear on the unit circle is reversed (Figure 24).

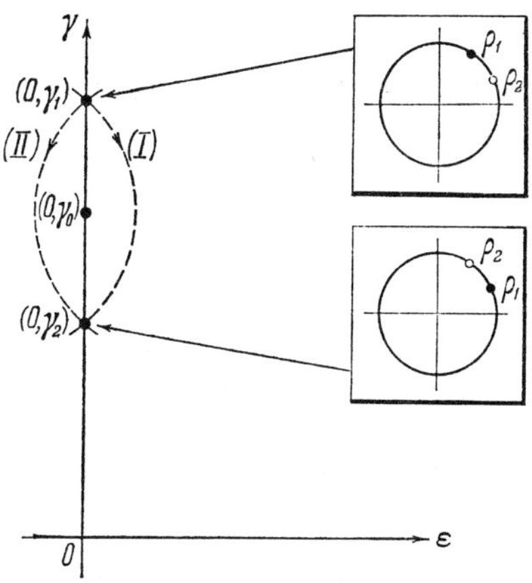

FIGURE 24.

Suppose that the point (ε, γ) moves from $(0, \gamma_0)$ to $(0, \gamma_1)$ in a sufficiently small neighborhood $(0, \gamma_2)$, along the dashed curve (I) in Figure 24. If curve (I) intersects neither of the curves $\gamma=\gamma^{(1,2)}(\varepsilon)$, the multipliers ρ_1, ρ_2 cannot meet, i. e., they move only along the unit circle and the position of the multipliers of both points $(0, \gamma_1)$, $(0, \gamma_2)$ is of the same type. Since these points correspond to multiplier positions which cannot be interchanged continuously if the multipliers ρ_1, ρ_2 move along the unit circle without meeting and remain in the neighborhood of $\rho^{(0)}$, it follows that curve (I) must intersect at least one of the curves $\gamma=\gamma^{(1,2)}(\varepsilon)$. For the same reason, the dashed curve (II) in Figure 24 must also intersect at least one of these curves. If $\gamma(\varepsilon)$ is an analytic function of $\sqrt{\varepsilon}$, at least one of the half-planes $\varepsilon>0$, $\varepsilon<0$ must contain no real values of $\gamma(\varepsilon)$. Therefore this case is impossible, so that the expansions (3.7) hold for $\gamma^{(1,2)}(\varepsilon)$ with real coefficients $\gamma_j^{(1,2)}$.

Let ε be real and $\gamma=\gamma^{(l)}(\varepsilon)$ $(l=1, 2)$. By the definition of $\gamma^{(l)}(\varepsilon)$, equation (2.1) has a twofold multiplier $\rho(\varepsilon)$ on the unit circle. For sufficiently small ε, equation (2.1) cannot have any other multipliers in the neighborhood of $\rho^{(0)}$. The function $\rho(\varepsilon)$ is a root of equation (3.4) for $\gamma=\gamma^{(l)}(\varepsilon)$, and the left-hand side of this equation is an analytic function at $\varepsilon=0, \rho=\rho^{(0)}$. Moreover, all values of this function for sufficiently small real ε lie on the unit circle. By the Lemma of § III. 1.5, $\rho(\varepsilon)$ and (hence) $\alpha(\varepsilon)$ are analytic functions of ε at $\varepsilon=0$.

Now suppose that $\rho^{(0)}=\pm 1$. For real ε and γ, the monodromy matrix $\mathbf{U}(\varepsilon, \gamma)$ is real, its spectrum is symmetric with respect to the real axis. If $\rho(\varepsilon)\not\equiv\pm 1$, it would follow that for sufficiently small ε a neighborhood of the point $\rho^{(0)}=\pm 1$ would contain at least two twofold roots, which is impossible. This contradiction completes the proof.

Remark 2. Suppose that condition (3.6) fails to hold and $\gamma=\gamma_0$ is a p-fold zero of the function $\omega_j(\gamma)+\omega_h(\gamma)-\omega_j(\gamma_0)-\omega_h(\gamma_0)$. Then, repeating the proof, one sees that there are $2p$ curves in the neighborhood of $(0, \gamma_0)$ on which equation (2.1) has coinciding multipliers of different kinds.

3.3. Position of multipliers on different sides of curves $\gamma=\gamma^{(1,2)}(\varepsilon)$

Consider a neighborhood of the point $(0, \gamma_0)$ on the (ε, γ)-plane, where γ_0 is a critical value. If equation (2.1) ($\varepsilon=0$, $\gamma=\gamma_0$) has only one twofold multiplier $\rho^{(0)}$ of mixed kind on the upper closed unit semicircle, it follows from the theorem of §3.2 that there are two curves, $\gamma=\gamma^{(1)}(\varepsilon)$ and $\gamma=\gamma^{(2)}(\varepsilon)$, emanating from $(0, \gamma_0)$, on which the equation has a twofold multiplier of mixed kind, but outside these curves all multipliers in the neighborhood of $\rho^{(0)}$ are distinct (Figure 25). Since by assumption there are no multiple multipliers of mixed kind on $\varepsilon=0$ in the neighborhood of $(0, \gamma_0)$, it follows by continuity that this is the situation in the open domains (I) and (II) shown in Figure 25. In these domains, then, all the solutions of equation (2.1) are bounded. As one crosses the curve $\gamma=\gamma^{(1)}(\varepsilon)$ or $\gamma=\gamma^{(2)}(\varepsilon)$ into domain (III), the multipliers may either leave the unit circle (in which case (III) is a dynamic instability domain) or move apart, remaining on the unit circle Γ (in which case (III) is a stability domain). We shall now show that the latter case is impossible. Moreover, if for $\varepsilon=0$, $\gamma=\gamma_0$ there are several twofold multipliers of mixed kind on Γ, each of them generates a different sector of type (III), and the dynamic instability domain is the union of all such sectors. Thus, each pair of twofold multipliers of mixed kind, $\rho^{(0)}=e^{2\pi i\alpha}\neq\pm 1$ and $\bar{\rho}^{(0)}$, and each of the twofold multipliers $\rho^{(0)}=1$ and $\rho^{(0)}=-1$, generates an instability sector on the (ε, γ)-plane. For example, if equation (2.1) with $\varepsilon=0, \gamma=\gamma_0$ has the spectrum shown in Figure 26 (filled circles — multipliers of first kind, empty circles — multipliers of second kind), there will be three instability sectors near the point $(0, \gamma_0)$ on the (ε, γ)-plane (Figure 27). Their relative positions are determined by the perturbation matrix $\mathbf{H}(\tau, \varepsilon, \gamma)$. The dynamic instability domain is the union of these sectors.

Theorem. Let

$$\omega_j(\gamma_0)+\omega_h(\gamma_0)=m \qquad (j, h=1, \ldots, k;\ m=\pm 1, \pm 2, \ldots) \qquad (3.12)$$

and suppose that for $\varepsilon=0$, $\gamma=\gamma_0$ *two multipliers of equation (2.1) of different kinds coincide at a point* $\rho^{(0)}$ *on the upper closed unit semicircle.*

Suppose that

$$\omega_j'(\gamma_0)\,\omega_h'(\gamma_0)>0 \qquad (3.13)$$

and that $\gamma^{(1)}(\varepsilon) < \gamma^{(2)}(\varepsilon)$ *for sufficiently small* $\varepsilon > 0$, *where* $\gamma^{(1)}(\varepsilon)$ *and* $\gamma^{(2)}(\varepsilon)$ *are the functions defined in §3.2.*

Then for sufficiently small $\varepsilon > 0$ *and* $|\gamma-\gamma_0|$ *equation (2.1) has two distinct multipliers on the unit circle in the neighborhood of* $\rho^{(0)}$ *in the domains* $\gamma < \gamma^{(1)}(\varepsilon)$ *and* $\gamma > \gamma^{(2)}(\varepsilon)$ *and in the domain* $\gamma^{(1)}(\varepsilon) < \gamma < \gamma^{(2)}(\varepsilon)$ *it has one multiplier outside the unit circle and one multiplier inside it.*

A similar statement holds for $\varepsilon < 0$.

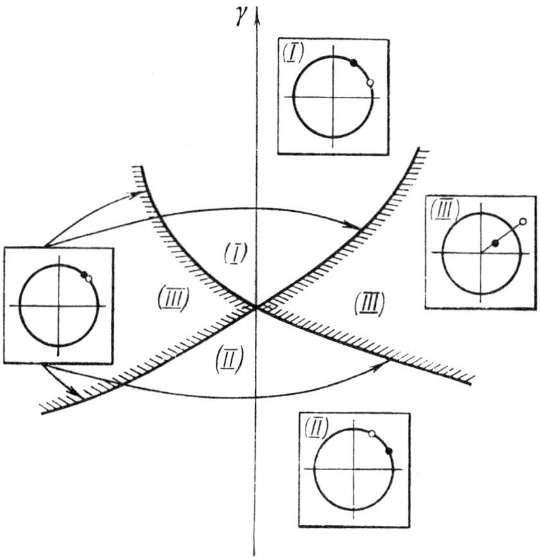

FIGURE 25.

P r o o f. When $\varepsilon = 0$, the multipliers of first and second kinds which coincide when $\gamma = \gamma_0$ are the numbers (3.9); we may assume without loss of generality that $j > 0$, $h > 0$. By (3.13), as γ increases the multipliers ρ_1 and ρ_2 move along the unit circle in different directions. The positions of the multipliers in domains (I) and (II) are illustrated in Figure 25.

Suppose that in the domain $\varepsilon > 0$, $\gamma^{(1)}(\varepsilon) < \gamma < \gamma^{(2)}(\varepsilon)$ the multipliers lie on the unit circle; in other words, in this domain the distribution of multipliers is of type (I) or (II) (Figure 25). We claim that this leads to a contradiction.

Let $\rho = \exp[i\varphi(\varepsilon, \gamma)]$ be a multiplier of the first kind in the domain $\varepsilon > 0$, $\gamma^{(1)}(\varepsilon) < \gamma < \gamma^{(2)}(\varepsilon)$. Since ρ is an isolated eigenvalue, it follows that ρ and hence also φ are analytic functions of γ in domain (III). We have (see §III. 1.7)

$$\frac{\partial\varphi}{\partial\gamma} = \frac{1}{\langle \mathbf{a}, \mathbf{a}\rangle} \int_0^{2\pi} \left(\left[\frac{d\mathbf{H}_0}{d\gamma} + \varepsilon \frac{\partial \mathbf{H}_1}{\partial\gamma} \right] \mathbf{x}, \mathbf{x} \right) dt, \tag{3.14}$$

where $\mathbf{a} = \mathbf{a}(\varepsilon, \gamma)$ is an eigenvector belonging to the multiplier $\rho = \rho(\varepsilon, \gamma)$, $\mathbf{x} = \mathbf{X}(\tau, \varepsilon, \gamma)\,\mathbf{a}$.

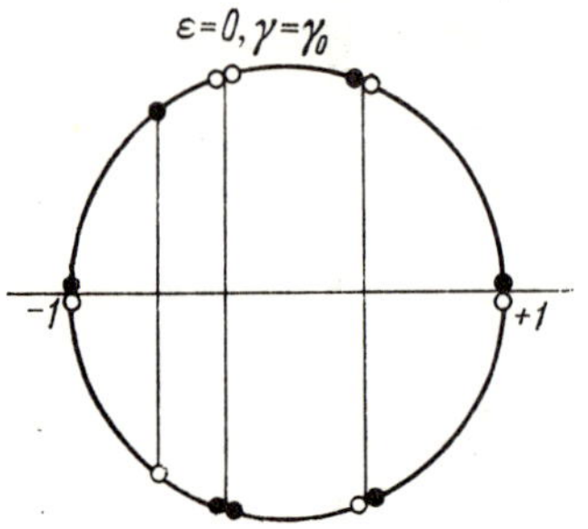

FIGURE 26.

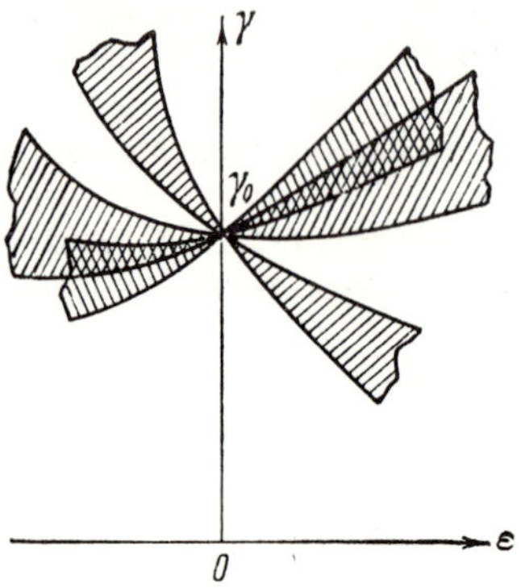

FIGURE 27.

We may assume that $\mathbf{a}(\varepsilon, \gamma)$ is a continuous function of ε, γ.

Let us evaluate the integral in (3.14) for $\varepsilon=0, \gamma=\gamma_0$. At this point the integrand is $(\mathbf{H}_0'(\gamma_0)\mathbf{x}^{(0)}, \mathbf{x}^{(0)})$, where $\mathbf{x}^{(0)}=\mathbf{X}(\tau, 0, \gamma_0)\,\mathbf{a}(0, \gamma_0)$, i. e.,

$$\mathbf{x}^{(0)}=\xi_j \exp[2\pi i\omega_j(\gamma_0)t]\,\mathbf{c}_j+\xi_{-h}\exp[2\pi i\omega_{-h}(\gamma_0)t]\,\mathbf{c}_{-h},$$

where ξ_j, ξ_{-h} are complex numbers and $\mathbf{c}_\nu=\mathbf{c}_\nu(\gamma_0)$ $(\nu=\pm1, \ldots, \pm k)$ (see (2.3)). Consequently,

$$(\mathbf{H}_0'(\gamma_0)\mathbf{x}^{(0)}, \mathbf{x}^{(0)})=|\xi_j|^2(\mathbf{H}_0'(\gamma_0)\mathbf{c}_j, \mathbf{c}_j)+|\xi_{-h}|^2(\mathbf{H}_0'(\gamma_0)\mathbf{c}_{-h}, \mathbf{c}_{-h})+ \\ +2\,\mathrm{Re}\{\xi_j\bar{\xi}_{-h}(\mathbf{H}_0'(\gamma_0)\mathbf{c}_j, \mathbf{c}_{-h})\exp[2\pi i(\omega_j(\gamma_0)+\omega_h(\gamma_0))t]\}. \quad (3.15)$$

It follows from (3.3) that $(\mathbf{H}_0'(\gamma_0)\mathbf{c}_j, \mathbf{c}_j)=\langle\mathbf{c}_j, \mathbf{c}_j\rangle\,\omega_j'(\gamma_0)$. Using (3.12), we see that the integral of the last term in (3.15) vanishes. Thus the integral of (3.14) at $\varepsilon=0, \gamma=\gamma_0$ is

$$\varkappa=|\xi_j|^2\langle\mathbf{c}_j, \mathbf{c}_j\rangle\,\omega_j'(\gamma_0)-|\xi_{-h}|^2\langle\mathbf{c}_{-h}, \mathbf{c}_{-h}\rangle\,\omega_h'(\gamma_0). \quad (3.16)$$

By (2.4) and (3.13), $\varkappa\neq0$. Thus the integral in (3.14) does not vanish and its sign is the same as that of $\varkappa$ in a sufficiently small neighborhood of $(0, \gamma_0)$. Since a multiplier of the first kind is characterized by the property $\langle\mathbf{a}, \mathbf{a}\rangle>0$, it follows from (3.14) that $\partial\varphi/\partial\gamma$ does not vanish and its sign is that of $\varkappa$ in a sufficiently small neighborhood of $(0, \gamma_0)$. Formula (3.14) is also valid for multipliers of the second kind in domain (III) (Figure 25), except that now $\langle\mathbf{a}, \mathbf{a}\rangle<0$. Consequently, for a multiplier of the second kind $\partial\varphi/\partial\gamma$ does not vanish and its sign is opposite to that of $\varkappa$. Thus, for fixed ε and variable γ, the multipliers on the unit circle move in different directions, i. e., they move apart with either increasing or decreasing γ. Let us vary γ (for sufficiently small fixed ε) in such a way that the multipliers move apart. Since the multipliers coincide on the curves $\gamma=\gamma^{(1)}(\varepsilon)$, $\gamma=\gamma^{(2)}(\varepsilon)$, it follows that in a sufficiently small neighborhood of the curve containing the point (ε, γ) the multipliers are arbitrarily close to each other. This contradiction shows that in domain (III) multipliers in the neighborhood of the point $\rho^{(0)}$ cannot lie on the unit circle, Q. E. D.

Remark. It follows from (3.3) that conditions (3.6) are satisfied for systems (2.1) and (3.1) if $\mathbf{H}_0'(\gamma_0)$ and $\mathbf{H}_0$ are sign definite matrices.

3.4. Computational procedure

Many other equations besides (3.1) and (3.2) may be reduced to the form of (2.1) (see §§II. 3.1, II. 3.2). The procedure whereby one determines the boundaries of the dynamic instability domains depends on the specific case at hand, but the general scheme of computations is nevertheless the same for all cases.

To fix ideas, consider the equation

$$\frac{d^2\mathbf{y}}{d\tau^2}+\gamma\mathbf{Q}\frac{d\mathbf{y}}{d\tau}+\gamma^2\left[\mathbf{P}_0+\varepsilon\mathbf{P}_1(\tau)+\varepsilon^2\mathbf{P}_2(\tau)+\ldots\right]\mathbf{y}=\mathbf{0}, \tag{3.17}$$

where $\mathbf{P}_j$ are symmetric matrices, $\mathbf{Q}^\tau=-\mathbf{Q}$, and $\gamma=\frac{1}{\vartheta}>0$ (ϑ is the perturbation frequency). The equation for the natural frequencies $\omega=\omega_\nu(\gamma)$ of the unperturbed system is $\det[\lambda^2\mathbf{I}+\lambda\gamma\mathbf{Q}+\gamma^2\mathbf{P}_0]=0$, where $\lambda=i\omega(\gamma)$. Let us assume that this equation has only pure imaginary roots

$$\lambda_\nu=i\gamma\omega_\nu^0 \qquad (\nu=\pm 1,\ldots,\pm k) \tag{3.18}$$

(so that $\omega_\nu(\gamma)=\gamma\omega_\nu^0$) and that equation (3.17) is stable at $\varepsilon=0$. As we know (§II. 3.3), this will certainly be true if $\mathbf{P}_0>\mathbf{0}$. If $\mathbf{Q}\neq\mathbf{0}$, however, there is a case (gyroscopic stabilization) in which $\mathbf{P}_0$ is a sign definite matrix but all the roots are pure imaginary and we have stability at $\varepsilon=0$ (see §II. 3.3). Introducing the vector

$$\mathbf{x}=\begin{pmatrix}\mathbf{y}\\ \gamma^{-1}\frac{d\mathbf{y}}{d\tau}\end{pmatrix},$$

we transform equation (3.17) to the form of (2.1) with

$$\tilde{\mathbf{J}}=\begin{pmatrix}-\mathbf{Q} & -\mathbf{I}_k\\ \mathbf{I}_k & \mathbf{0}\end{pmatrix},\quad \mathbf{H}_0(\gamma)\equiv\mathbf{H}(\tau,0,\gamma)=\gamma\begin{pmatrix}\mathbf{P}_0 & \mathbf{0}\\ \mathbf{0} & \mathbf{I}_k\end{pmatrix}.$$

It was shown in §III. 2.7 that if $\mathbf{P}_0>\mathbf{0}$ the eigenvalues of the first kind are the numbers (3.18) with $\omega_\nu^0>0$,* i. e., according to our notation, $\omega_\nu^0>0$ for $\nu=1,\ldots,k$. In the general case, one must determine the kinds of the numbers (3.18) as shown in §1.1 and provide the eigenvalues of the first kind with the subscripts $1,\ldots,k$.

We shall assume that

$$\omega_j^0+\omega_h^0\neq 0 \qquad (j,h=1,\ldots,k).$$

* Recall that to fix ideas we are assuming that $\gamma>0$.

(As in §1.9, it can be shown that if $\omega_j^0+\omega_h^0=0$ for some j, h then the entire strip $|\varepsilon|\leqslant\varepsilon_0$ is a dynamic instability domain under suitable perturbations.) By (2.5), the critical values of the parameter γ (critical perturbation frequencies ϑ) are the numbers*

$$\gamma_0^{-1}=\vartheta_0=\frac{|\omega_j^0+\omega_0^h|}{m} \qquad (j, h=1, \ldots, k;\ m=1, 2, \ldots). \tag{3.19}$$

As before, we shall assume that only two multipliers coincide at the point

$$\rho_0=\exp(2\pi i\gamma_0\omega_j^0)=\exp(-2\pi i\gamma_0\omega_h^0)=\exp(2\pi i\alpha_0). \tag{3.20}$$

The sequence of computations depend on whether one is concerned with principal or combination resonance.

a) Principal resonance: $\rho_0=(-1)^m$. The procedure is exactly the same as in /62b/. By the theorem of §3.2, the boundaries of the instability domain for the multiplier $\rho^{(0)}$ are two curves $\gamma=\gamma^{(j)}(\varepsilon)$, $\gamma^{(j)}(0)=$ $=\gamma_0$ $(j=1, 2)$.

When $\gamma=\gamma^{(j)}(\varepsilon)$, it follows from the above-mentioned theorem that equation (3.17) has a solution $\mathbf{y}=\mathbf{y}(\tau, \varepsilon)$ satisfying the condition

$$\mathbf{y}(\tau+2\pi, \varepsilon)\equiv(-1)^m\mathbf{y}(\tau, \varepsilon). \tag{3.21}$$

Since the corresponding eigenvector of the monodromy matrix $\mathbf{U}(\varepsilon, \gamma^{(j)}(\varepsilon))$ may be assumed an analytic function of ε, we may assume that $\mathbf{y}(\tau, \varepsilon)$ is an analytic function of ε.

Substituting the series

$$\mathbf{y}=\mathbf{y}_0(\tau)+\varepsilon\mathbf{y}_1(\tau)+\varepsilon^2\mathbf{y}_2(\tau)+\ldots, \quad \gamma=\gamma_0+\gamma_1\varepsilon+\gamma_2\varepsilon^2+\ldots$$

into equation (3.17) and comparing coefficients of like powers of ε, we obtain the equations

$$\left.\begin{aligned} &\mathbf{L}_0\mathbf{y}_0=\mathbf{0},\\ &\mathbf{L}_0\mathbf{y}_l=\mathbf{g}_l(\tau)-\gamma_l\left(\mathbf{Q}\frac{d\mathbf{y}_0}{d\tau}+2\gamma_0\mathbf{P}_0\mathbf{y}_0\right)-\gamma_0^2\mathbf{P}_l\mathbf{y}_0, \end{aligned}\right\} \tag{3.22}$$

where

$$\mathbf{L}_0=\frac{d^2}{d\tau^2}+\gamma_0\mathbf{Q}\frac{d}{d\tau}+\gamma_0^2\mathbf{P}_0,$$

$l=1, 2, \ldots$ and $\mathbf{g}_l(\tau)$ are expressed in terms of $\mathbf{y}_1, \ldots, \mathbf{y}_{l-1}, \gamma_1, \ldots, \gamma_{l-1}$.

Let us consider the derivation of equation (3.22) in greater detail.

Let $\mathbf{L}=\mathbf{L}(\varepsilon, \gamma)$ be any linear operator that depends analytically on ε and γ: $\gamma=\gamma_0+\gamma_1\varepsilon+\ldots$, $\mathbf{y}=\mathbf{y}_0+\varepsilon\mathbf{y}_1+\ldots$, $\mathbf{L}[\varepsilon, \gamma(\varepsilon)]\mathbf{y}(\varepsilon)=\mathbf{z}_0+\varepsilon\mathbf{z}_1+\varepsilon^2\mathbf{z}_2+\ldots$ are power series in ε; what is the relationship between the coefficients $\mathbf{z}_l$ and the coefficients $\gamma_0, \gamma_1, \ldots, \mathbf{y}_0, \mathbf{y}_1, \ldots$? Denoting $D=\frac{d}{d\varepsilon}$, $(\ )_0=(\ )_{\varepsilon=0}$, we have

$$\mathbf{z}_l=D^l\{\mathbf{L}[\varepsilon, \gamma(\varepsilon)]\mathbf{y}(\varepsilon)\}_0=(D^l\mathbf{L})_0\mathbf{y}_0+\mathbf{L}_0\mathbf{y}_l+ \\ +\mathbf{f}_1[(D\mathbf{L})_0, \ldots, (D^{l-1}\mathbf{L})_0, \mathbf{y}_1, \ldots, \mathbf{y}_{l-1}],$$

* Remember that $\vartheta_0>0$.

where $\mathbf{f}_1$ is a polynomial with zero free term. Since $D^l = (\bar{D}+\partial)^l$, where $\bar{D}$ is the derivative on the assumption that only γ depends on ε and ∂ denotes partial differentiation with respect to ε, it follows that

$$(D^l\mathbf{L})_0 = (\bar{D}^l\mathbf{L})_0 + (\partial^l\mathbf{L})_0 + f_2(\gamma_1, \ldots, \gamma_{l-1}),$$

where f_2 is a polynomial with zero free term. Since

$$\bar{D}^l\mathbf{L} = \left(\frac{\partial \mathbf{L}}{\partial \lambda}\right)_0 \gamma_l + f_3(\gamma_1, \ldots, \gamma_{l-1}),$$

where f_3 is again a polynomial with zero free term, we finally obtain

$$\mathbf{z}_l = \mathbf{L}(\gamma_0, 0)\,\mathbf{y}_l + \left[\left(\frac{\partial^l \mathbf{L}}{\partial \varepsilon^l}\right)_0 + \left(\frac{\partial \mathbf{L}}{\partial \gamma}\right)_0 \gamma_l\right]\mathbf{y}_0 + \mathbf{f}(\gamma_1, \ldots, \gamma_{l-1}, \mathbf{y}_1, \ldots, \mathbf{y}_{l-1}), \quad (3.23)$$

where $\mathbf{f}$ is a polynomial with zero free term.

If $\mathbf{L}[\varepsilon, \gamma(\varepsilon)]\,\mathbf{y}(\varepsilon) \equiv \mathbf{0}$, the equation $\mathbf{z}_l = 0$ yields an equation for $\mathbf{y}_l$. Taking

$$\mathbf{L} = \frac{d^2}{d\tau^2} + \gamma\mathbf{Q}\frac{d}{d\tau} + \gamma^2[\mathbf{P}_0 + \varepsilon\mathbf{P}_1(\tau) + \ldots],$$

we now obtain the second equation of (3.22) from (3.23).

The solutions $\mathbf{y}_l(\tau)$ of equations (3.22) must satisfy the conditions

$$\mathbf{y}_l(\tau+2\pi) \equiv (-1)^m\,\mathbf{y}_l(\tau) \qquad (l = 0, 1, 2, \ldots). \quad (3.24)$$

Since by assumption ρ_0 is a twofold multiplier with simple elementary divisors, the general solution $\mathbf{y}_0$ of the first equation (3.22) satisfying condition (3.24) depends on two parameters. Inserting $\mathbf{y}_0$ into equation (3.22) for $l=1$, we see that a solution $\mathbf{y}_1$ satisfying conditions (3.24) will exist provided γ_1 is a root of a certain quadratic equation. If the roots $\gamma_1^{(1)}, \gamma_1^{(2)}$ of this equation are distinct, each of the subsequent equations (3.22) uniquely (uniqueness follows from the theorem of §3.2) determines a sequence of coefficients, $\gamma_2^{(1)}, \gamma_3^{(1)}, \gamma_4^{(1)}, \ldots$ and $\gamma_2^{(2)}, \gamma_3^{(2)}, \gamma_4^{(2)}, \ldots$. If $\gamma_1^{(1)} = \gamma_1^{(2)}$, we have an analogous quadratic equation for γ_2. If the roots $\gamma_2^{(1)}, \gamma_2^{(2)}$ of this equation are distinct, we again have two uniquely determined series of coefficients, $\gamma_3^{(1)}, \gamma_4^{(1)}, \ldots, \gamma_3^{(2)}, \gamma_4^{(2)}, \ldots$, and so on. In applications, the matrices $\mathbf{P}_l(\tau)$ are usually trigonometric polynomials involving a relatively small number of harmonics. In such cases the work is much easier if one looks for solutions $\mathbf{y}_l$ of equations (3.22) expanded in Fourier series:

$$\mathbf{y}_l = \sum_n e^{in\tau}\mathbf{c}_n^{(l)}$$

if m is even, or

$$\mathbf{y}_l = \sum_{n\neq 0} e^{\frac{1}{2}in\tau}\,\mathbf{c}_n^{(l)}$$

if m is odd.

b) Combination resonance: $\vartheta_0=\gamma_0^{-1}=\frac{1}{m}|\omega_j^0+\omega_h^0|,\ j\neq h$. Two multipliers of different kinds of the unperturbed equation (3.17) ($\varepsilon=0,\ \gamma=\gamma_0$) coincide at the point (3.20).

The boundaries of the dynamic instability domain are two curves $\gamma=\gamma^{(1)}(\varepsilon),\ \gamma=\gamma^{(2)}(\varepsilon)$. The functions $\gamma^{(1,2)}(\varepsilon)$ are analytic at $\varepsilon=0$. By the theorem of §3.2, if $\gamma=\gamma^{(1,2)}(\varepsilon)$ equation (3.17) has a solution $\mathbf{y}(\tau,\varepsilon)$ satisfying the condition

$$\mathbf{y}(\tau+2\pi,\varepsilon)=\exp[2\pi i\alpha(\varepsilon)]\,\mathbf{y}(\tau,\varepsilon), \tag{3.25}$$

where $\alpha(\varepsilon)$ is an unknown analytic function of ε, and $\alpha(0)=\alpha_0$ is the number in (3.20). The multiplier $\exp[2\pi i\alpha(\varepsilon)]$ is twofold (there are automatically multipliers of different kinds that coincide at $\exp[2\pi i\alpha(\varepsilon)]$). We may assume that $\mathbf{y}(\tau,\varepsilon)$ is an analytic function.

It follows from (3.25) that

$$\mathbf{y}(\tau,\varepsilon)=e^{i\alpha(\varepsilon)\tau}\mathbf{u}(\tau,\varepsilon), \tag{3.26}$$

where

$$\mathbf{u}(\tau+2\pi,\varepsilon)\equiv\mathbf{u}(\tau,\varepsilon)$$

is an analytic function of ε at $\varepsilon=0$, 2π-periodic in τ. We substitute (3.26) into equation (3.17), letting

$$\left.\begin{aligned}\alpha(\varepsilon)&=\alpha_0+\alpha_1\varepsilon+\alpha_2\varepsilon^2+\dots,\\ \mathbf{u}(\tau,\varepsilon)&=\mathbf{u}_0(\tau)+\varepsilon\mathbf{u}_1(\tau)+\varepsilon^2\mathbf{u}_2(\tau)+\dots,\\ \gamma=\gamma(\varepsilon)&=\gamma_0+\gamma_1\varepsilon+\gamma_2\varepsilon^2+\dots\end{aligned}\right\} \tag{3.27}$$

Comparing coefficients of like powers of ε, we obtain the equations*

$$\left.\begin{aligned}\mathbf{L}_0\mathbf{u}_0&=\frac{d^2\mathbf{u}_0}{d\tau^2}+(2i\alpha_0\mathbf{I}+\gamma_0\mathbf{Q})\frac{d\mathbf{u}_0}{d\tau}+(\gamma_0^2\mathbf{P}_0-\alpha_0^2\mathbf{I}+i\alpha_0\gamma_0\mathbf{Q})\,\mathbf{u}_0=0,\\ \mathbf{L}_0\mathbf{u}_l&=\mathbf{g}_l-\gamma_l\left[\mathbf{Q}\frac{d\mathbf{u}_0}{d\tau}+(2\gamma_0\mathbf{P}_0+i\alpha_0\gamma_0\mathbf{Q})\,\mathbf{u}_0\right]+\\ &+2i\alpha_l\frac{d\mathbf{u}_0}{d\tau}+(\gamma_0^2\mathbf{P}_l+i\alpha_l\gamma_0\mathbf{Q}-\{\alpha\}_l\,\mathbf{I})\,\mathbf{u}_0\qquad(l=1,2,\dots),\end{aligned}\right\} \tag{3.28}$$

where $\mathbf{g}_1=0$, $\mathbf{g}_l$ depends on $\mathbf{y}_1,\dots,\mathbf{y}_{l-1},\alpha_1,\dots,\alpha_{l-1},\gamma_1,\dots,\gamma_{l-1}$ and

$$\{\alpha\}_l=\left(\frac{\partial^l\alpha^2}{\partial\varepsilon^l}\right)_{\varepsilon=0}=\alpha_0\alpha_l+l\alpha_1\alpha_{l-1}+\dots+l\alpha_{l-1}\alpha_1+\alpha_l\alpha_0.$$

The solutions $\mathbf{u}_l(\tau)$ must be 2π-periodic:

$$\mathbf{u}_l(\tau+2\pi)=\mathbf{u}_l(\tau)\qquad(l=1,2,\dots). \tag{3.29}$$

* The second equation of (3.28) follows easily from the equation $\mathbf{z}_l=0$ via (3.23).

We see from the first equation of (3.28) that the general solution $\mathbf{u}_0$ satisfying (3.29) is of the form $\mathbf{u}_0=\xi\mathbf{v}+\eta\mathbf{w}$, where ξ, η are arbitrary constants. (Recall that ρ_0 is a twofold multiplier.) Inserting this into equation (3.28) for $l=1$, we deduce that a necessary condition for the existence of a nontrivial solution ($\mathbf{u}_0\not\equiv 0$) is a certain equation whose left-hand side is a quadratic polynomial in α_1, γ_1. We require that α_1 be a twofold root of this equation (otherwise the multipliers of equation (3.17) at $\varepsilon\neq 0$, $\gamma=\gamma^{(1,2)}(\varepsilon)$ in the neighborhood of ρ_0 would be distinct). Equating the discriminant to zero, we obtain a quadratic equation in γ_1. Let $\gamma_1^{(1)}$, $\gamma_1^{(2)}$ be its roots. The equation for α_1 yields the required value of α_1.

Using the equation for $\mathbf{u}_2$, one similarly finds $\gamma_2^{(1)}$, $\gamma_2^{(2)}$, α_2, and so on. Thus one successively determines the coefficients of (3.27). We shall not provide more details about the computations, for since we have proved the existence of a solution of type (3.26) and justified the expansions (3.27) the coefficients may always be determined by simple substitution.

Note that in the exceptional case we have $\gamma^{(1)}(\varepsilon)\equiv\gamma^{(2)}(\varepsilon)$ for both principal and combination resonance. Only in this case is the value of γ_0 not critical in the one-parameter class of equations (3.17).

3.5. Second-approximation formulas for boundaries of instability domains of a canonical system

We begin with a system of type (3.1). As in §1.3, we let $\mathbf{c}_{-k},\ \ldots,\ \mathbf{c}_{-1}$, $\mathbf{c}_1,\ \ldots,\ \mathbf{c}_k$ be an orthonormal system of eigenvectors of the matrix $-\mathbf{J}\mathbf{H}_0$:

$$\begin{aligned} -\mathbf{J}\mathbf{H}_0\mathbf{c}_\nu &= i\omega_\nu\mathbf{c}_\nu, \quad \langle\mathbf{c}_\nu,\ \mathbf{c}_\mu\rangle=0 \qquad (\nu\neq\mu),\\ \langle\mathbf{c}_\nu,\ \mathbf{c}_\nu\rangle &= \operatorname{sign}\nu \qquad (\nu,\ \mu=\pm1,\ \ldots,\ \pm k), \end{aligned} \tag{3.30}$$

where, as we recall,

$$\langle\mathbf{x},\ \mathbf{y}\rangle\equiv i(\mathbf{J}\mathbf{x},\ \mathbf{y}),\quad \mathbf{J}=\begin{pmatrix}0 & -\mathbf{I}_k\\ \mathbf{I}_k & 0\end{pmatrix},\quad \omega_{-\varkappa}=-\omega_\varkappa \qquad (\varkappa=1,\ \ldots,\ k).$$

For convenience, we recall formula (1.7) for the critical frequencies:

$$\vartheta_0=\frac{\omega_j+\omega_h}{m} \qquad (j,\ h=1,\ \ldots,\ k;\ m=1,\ 2,\ \ldots). \tag{3.31}$$

This means that at $\varepsilon=0$, $\vartheta=\vartheta_0$ system (3.1) has two coinciding multipliers of different kinds, i.e., the congruence class of natural frequencies $\{\omega_{j_1},\ \ldots,\ \omega_{j_r}\}$ modulo ϑ_0 contains only ω_{-j} and ω_h:

$$\begin{aligned} \omega_{-j}&=\alpha_0+m_{-j}\vartheta_0,\quad \omega_h=\alpha_0+m_h\vartheta_0,\\ \omega_\nu&\neq\alpha_0+m_\nu\vartheta_0 \qquad (\nu=\pm1,\ \ldots,\ \pm k;\ \nu\neq -j,\ \nu\neq h), \end{aligned} \tag{3.32}$$

where m_{-j} and m_h are certain fixed integers (or zero), and the m_ν are arbitrary integers (or zero). Comparison of formulas (3.31) and (3.32) shows that

$$m_h - m_{-j} = m. \qquad (3.33)$$

By Theorem I of §3.2, the boundaries of the dynamic instability domain are analytic functions of ε at $\varepsilon=0$:

$$\vartheta = \vartheta_0 + \vartheta_1\varepsilon + \vartheta_2\varepsilon^2 + \dots.$$

Consequently,

$$\frac{1}{\vartheta} = \frac{1}{\vartheta_0}\left[1 - \frac{\vartheta_1}{\vartheta_0}\varepsilon + \left(\frac{\vartheta_1^2}{\vartheta_0^2} - \frac{\vartheta_2}{\vartheta_0}\right)\varepsilon^2 + \dots\right]$$

and system (3.1) takes the form $(s=\vartheta t)$

$$\frac{d\mathbf{x}}{ds} = [\mathbf{C}_0 + \varepsilon\mathbf{C}_1(s) + \varepsilon^2\mathbf{C}_2(s) + \dots]\,\mathbf{x} \quad (\mathbf{C}_l(s+2\pi) = \mathbf{C}_l(s),\ l=1,\ 2,\ \dots), \qquad (3.34)$$

where

$$\mathbf{C}_0 = -\frac{1}{\vartheta_0}\mathbf{JH}_0,\ \mathbf{C}_1 = -\frac{1}{\vartheta_0}\mathbf{JH}_1 + \frac{\vartheta_1}{\vartheta_0^2}\mathbf{JH}_0,$$
$$\mathbf{C}_2 = -\frac{1}{\vartheta_0}\mathbf{JH}_2 + \frac{\vartheta_1}{\vartheta_0^2}\mathbf{JH}_1 - \left(\frac{\vartheta_1^2}{\vartheta_0^3} - \frac{\vartheta_2}{\vartheta_0^2}\right)\mathbf{JH}_0.$$

By §§II. 2.5 and 3.2, there is a solution $\mathbf{x}(s,\ \varepsilon) = e^{i\alpha(\varepsilon)s}\mathbf{u}(s,\ \varepsilon)$ $(\mathbf{u}(s+2\pi,\ \varepsilon) = \mathbf{u}(s,\ \varepsilon))$ of system (3.34) on the boundary of the instability domain; the multiplier $\rho(\varepsilon) = \exp[2\pi i\alpha(\varepsilon)]$ is twofold and the series

$$\alpha(\varepsilon) = \alpha_0 + \alpha_1(\varepsilon) + \alpha_2\varepsilon^2 + \dots, \qquad \mathbf{u}(s,\ \varepsilon) = \mathbf{u}_0(s) + \varepsilon\mathbf{u}_1(s) + \varepsilon^2\mathbf{u}_2(s) + \dots$$

are convergent for small $|\varepsilon|$. Inserting these series in (3.34) and comparing coefficients of like powers of ε, we get

$$\left.\begin{aligned}
\frac{d\mathbf{u}_0}{ds} &= (\mathbf{C}_0 - i\alpha_0\mathbf{I}_{2k})\,\mathbf{u}_0,\\
\frac{d\mathbf{u}_1}{ds} &= (\mathbf{C}_0 - i\alpha_0\mathbf{I}_{2k})\,\mathbf{u}_1 + (\mathbf{C}_1 - i\alpha_1\mathbf{I}_{2k})\,\mathbf{u}_0,\\
\frac{d\mathbf{u}_2}{ds} &= (\mathbf{C}_0 - i\alpha_0\mathbf{I}_{2k})\,\mathbf{u}_2 + (\mathbf{C}_1 - i\alpha_1\mathbf{I}_{2k})\,\mathbf{u}_1 + (\mathbf{C}_2 - i\alpha_2\mathbf{I}_{2k})\,\mathbf{u}_0.\\
&\dots\dots\dots\dots\dots\dots\dots\dots
\end{aligned}\right\} \qquad (3.35)$$

The problem thus reduces to finding 2π-periodic solutions of these equations under the above multiplicity conditions. The general form of a 2π-periodic solution of the first equation of (3.35) is, by (3.30) and (3.32),

$$\mathbf{u}_0 = \beta_1 e^{im_h s}\mathbf{c}_h + \beta_2 e^{im_{-j}s}\mathbf{c}_{-j}.$$

The equation for $\mathbf{u}_l$ is

$$\frac{d\mathbf{u}_l}{ds} = (\mathbf{C}_0 - i\alpha_0\mathbf{I}_{2k})\,\mathbf{u}_l + \mathbf{g}_l(s) \qquad (\mathbf{g}_l(s+2\pi) = \mathbf{g}_l(s);\ l=1,\ 2,\ \dots).$$

By Lemma I of § II. 2.10, $\mathbf{u}_l(s)$ is periodic if and only if $((\mathbf{g}_l, \mathbf{z}_\varkappa))=0$, where $\mathbf{z}_\varkappa = e^{-im_\varkappa s}\,\mathbf{d}_\varkappa$ are linearly independent solutions of the homogeneous adjoint system. It was shown in § IV. 3.2 that we may take $\mathbf{d}_\varkappa = \mathbf{J}\mathbf{c}_\varkappa$. Therefore,

$$((\mathbf{g}_l, \mathbf{z}_\varkappa)) = (\mathbf{g}_l^{(m_\varkappa)}, \mathbf{d}_\varkappa) = i\langle \mathbf{g}_l^{(m_\varkappa)}, \mathbf{c}_\varkappa \rangle,$$

and in our case the conditions for periodicity of $\mathbf{u}_l(s)$ are

$$\langle \mathbf{g}_l^{(m_h)}, \mathbf{c}_h\rangle = \langle \mathbf{g}_l^{(m_{-j})}, \mathbf{c}_{-j}\rangle = 0 \qquad \left(\mathbf{g}_l(s) \sim \sum_{n=-\infty}^{\infty} e^{ins}\mathbf{g}_l^{(n)}\right), \tag{3.36}$$

where m_h and m_{-j} are defined in (3.32).

The first condition of (3.36) for $l=1$ is

$$\begin{aligned}(\{\mathbf{J}[\mathbf{C}_1 - i\alpha_1 \mathbf{I}_{2k}]\,\mathbf{u}_0\}^{(m_h)}, \mathbf{c}_h) &= \\ = \left(\left\{\left[\frac{1}{\vartheta_0}\mathbf{H}_1 - \frac{\vartheta_1}{\vartheta_0^2}\mathbf{H}_0 - i\alpha_1\mathbf{J}\right]\right.\right.&\left.\left.[\beta_1 e^{im_h s}\mathbf{c}_h + \beta_2 e^{im_{-j}s}\mathbf{c}_{-j}]\right\}^{(m_h)}, \mathbf{c}_h\right) = 0.\end{aligned}$$

Proceeding in this way for the second condition of (3.36), we get

$$\left[(\mathbf{H}_1^{(0)}\mathbf{c}_h, \mathbf{c}_h) - \frac{\vartheta_1}{\vartheta_0}(\mathbf{H}_0\mathbf{c}_h, \mathbf{c}_h) - i\alpha_1\vartheta_0(\mathbf{J}\mathbf{c}_h, \mathbf{c}_h)\right]\beta_1 + (\mathbf{H}_1^{(m_h - m_{-j})}\mathbf{c}_{-j}, \mathbf{c}_h)\,\beta_2 = 0,$$

$$(\mathbf{H}_1^{(m_{-j} - m_h)}\mathbf{c}_h, \mathbf{c}_{-j})\,\beta_1 + \left[(\mathbf{H}_1^{(0)}\mathbf{c}_{-j}, \mathbf{c}_{-j}) - \frac{\vartheta_1}{\vartheta_0}(\mathbf{H}_0\mathbf{c}_{-j}, \mathbf{c}_{-j}) - i\alpha_1\vartheta_0(\mathbf{J}\mathbf{c}_{-j}, \mathbf{c}_{-j})\right]\beta_2 = 0.$$

By (3.30),

$$(\mathbf{H}_0\mathbf{c}_\nu, \mathbf{c}_\nu) = (i\omega_\nu \mathbf{J}\mathbf{c}_\nu, \mathbf{c}_\nu) = \omega_\nu\,\mathrm{sign}\,\nu.$$

Introducing the notation of §1.4,

$$\chi_{pq}^{(r)} = (\mathbf{H}_1^{(r)}\mathbf{c}_p, \mathbf{c}_q) \qquad (\chi_{qp}^{(-r)} = \overline{\chi_{pq}^{(r)}}) \tag{3.37}$$

and using (3.33), we write the existence condition for a 2π-periodic solution of the second equation in (3.35) as

$$\left.\begin{aligned}\left(\chi_{hh}^{(0)} - \frac{\omega_h}{\vartheta_0}\vartheta_1 - \alpha_1\vartheta_0\right)\beta_1 + \chi_{-jh}^{(m)}\beta_2 = 0,\\ \chi_{h,-j}^{(-m)}\beta_1 + \left(\chi_{-j,-j}^{(0)} + \frac{\omega_{-j}}{\vartheta_0}\vartheta_1 + \alpha_1\vartheta_0\right)\beta_2 = 0.\end{aligned}\right\} \tag{3.38}$$

To determine the nonzero solutions β_1 and β_2, we demand that the determinant of this algebraic system vanish:

$$\begin{aligned}(\alpha_1\vartheta_0)^2 - \left(\chi_{hh}^{(0)} - \chi_{-j,-j}^{(0)} - \frac{\omega_h + \omega_{j-}}{\vartheta_0}\vartheta_1\right)\alpha_1\vartheta_0 - \\ - \left(\chi_{hh}^{(0)} - \frac{\omega_h}{\vartheta_0}\vartheta_1\right)\left(\chi_{-j,-j}^{(0)} + \frac{\omega_{-j}}{\vartheta_0}\vartheta_1\right) + \chi_{h,-j}^{(-m)}\chi_{-jh}^{(m)} = 0.\end{aligned} \tag{3.39}$$

On the boundary of the instability domain, α_1 is a multiple root, and therefore the discriminant Δ of equation (3.39) must vanish:

$$\Delta = \frac{1}{4}(\chi_{hh}^{(0)} + \chi_{-j,-j}^{(0)} - m\vartheta_1)^2 - |\chi_{-jh}^{(m)}|^2 = 0.$$

We thus obtain the formula (1.29) for the slope ϑ_1 of the tangent to the boundary of the instability domain (recall that in our new notation $\mu_{1,2}=\vartheta_1^{(1,2)}$)

$$\vartheta_1^{(1,2)}=\frac{1}{m}(\chi_{-j,-j}^{(0)}+\chi_{hh}^{(0)}\mp 2|\chi_{-jh}^{(m)}|). \tag{3.40}$$

The minus sign here gives the lower boundary, the plus sign the upper. The first correction α_1 to the characteristic exponent on the boundary is determined from (3.39):

$$\alpha_1=\frac{1}{2\vartheta_0}\left[\chi_{hh}^{(0)}-\chi_{-j,-j}^{(0)}-\frac{\omega_h+\omega_{-j}}{m\vartheta_0}(\chi_{-j,-j}^{(0)}+\chi_{hh}^{(0)}\mp 2|\chi_{-jh}^{(m)}|)\right], \tag{3.41}$$

and by (3.38)

$$\beta_1=\chi_{-jh}^{(m)},\quad \beta_2=\mp|\chi_{-jh}^{m}|. \tag{3.42}$$

To determine a periodic solution of the second equation of (3.35) we transform its inhomogeneous part:

$$\mathbf{g}_1(s)=-\frac{1}{\vartheta_0}\left(\mathbf{JH}_1-\frac{\vartheta_1}{\vartheta_0}\mathbf{JH}_0+i\alpha_1\vartheta_0\mathbf{I}_{2k}\right)(\beta_1e^{im_hs}\mathbf{c}_h+\beta_2e^{im_{-j}s}\mathbf{c}_{-j}).$$

Since $\mathbf{c}_{-k},\ldots,\mathbf{c}_{-1},\mathbf{c}_1,\ldots,\mathbf{c}_k$ is an orthonormal basis, it follows that

$$\mathbf{JH}_1^{(l)}\mathbf{c}_\mu=-i\sum_{\nu=-k}^{k}\operatorname{sign}\nu\,\chi_{\mu\nu}^{(l)}\mathbf{c}_\nu \qquad (\mu=\pm 1,\ldots,\pm k),$$

as can be verified by multiplying both sides of the equality by $i\mathbf{J}\mathbf{c}_\lambda$. Therefore,

$$\begin{aligned}\mathbf{g}_1(s)=\frac{i}{\vartheta_0}\Bigg\{&\sum_{\nu=-k}^{k}\sum_{l=-\infty}^{\infty}\operatorname{sign}\nu\,\beta_1\chi_{h\nu}^{(l)}e^{i(m_h+l)s}\mathbf{c}_\nu+\\&+\sum_{\nu=-k}^{k}\sum_{l=-\infty}^{\infty}\operatorname{sign}\nu\,\beta_2\chi_{-j\nu}^{(l)}e^{i(m_{-j}+l)s}\mathbf{c}_\nu-\\&-\left(\frac{\vartheta_1}{\vartheta_0}\omega_h+\alpha_1\vartheta_0\right)\beta_1e^{im_hs}\mathbf{c}_h-\left(\frac{\vartheta_1}{\vartheta_0}\omega_{-j}+\alpha_1\vartheta_0\right)\beta_2e^{-im_{-j}s}\mathbf{c}_{-j}\Bigg\}.\end{aligned}$$

We isolate two of the terms in the first sum: $\nu=h,\ l=0;\ \nu=-j,\ l=-m$, and two in the second sum: $\nu=h,\ l=m;\ \nu=-j,\ l=0$. We then note that the coefficients of $\exp(im_hs)\mathbf{c}_h$ and $\exp(im_{-j}s)\mathbf{c}_{-j}$ vanish by (3.38). Changing the running index in the first sum to $n=l+m_h$ and that in the second to $n=l+m_{-j}$, we write the second equation of (3.35) as

$$\frac{d\mathbf{u}_1}{ds}=\frac{1}{\vartheta_0}(-\mathbf{JH}_0-i\alpha_0\mathbf{I}_{2k})\mathbf{u}_1+\frac{i}{\vartheta_0}\sum_{\nu,n}{}'\operatorname{sign}\nu\,[\chi_{h\nu}^{(n-m_h)}\beta_1+\chi_{-j}^{(n-m-j)}\beta_2]\,e^{ins}\mathbf{c}_\nu, \tag{3.43}$$

where the prime indicates that the summation omits the terms with indices $\nu=h,\ n=m_h$ and $\nu=-j,\ n=m_{-j}$. By Lemma I of §II. 2.10, any 2π-periodic solution of this equation may be written

$$\mathbf{u}_1=\mathbf{u}_{10}(s)+\gamma_1e^{im_hs}\mathbf{c}_h+\gamma_2e^{im_{-j}s}\mathbf{c}_{-j}.$$

A particular 2π-periodic solution $\mathbf{u}_{10}(s)$ will be sought as a sum

$$\mathbf{u}_{10}(s)=\sum_{\nu,\,n}{}' p_\nu^{(n)} e^{ins}\mathbf{c}_\nu.$$

Inserting this sum into equation (3.43), we obtain

$$p_\nu^{(n)}=\frac{\operatorname{sign}\nu}{\alpha_0+n\vartheta_0-\omega_\nu}\left(\chi_{h\nu}^{(n-m_h)}\beta_1+\chi_{-j\nu}^{(n-m_{-j})}\beta_2\right). \tag{3.44}$$

We now determine $\mathbf{J}\mathbf{g}_2(s)$, where $\mathbf{g}_2(s)$ is the inhomogeneous part of the third equation of (3.35):

$$\begin{aligned}\mathbf{J}\mathbf{g}_2(s)=\frac{1}{\vartheta_0}\left[\mathbf{H}_1-\frac{\vartheta_1}{\vartheta_0}\mathbf{H}_0-i\alpha_1\vartheta_0\mathbf{J}\right]\left[\gamma_1 e^{im_h s}\mathbf{c}_h+\gamma_2 e^{im_{-j}s}\mathbf{c}_{-j}+\sum_{\nu,\,n}{}' p_\nu^{(n)}e^{ins}\mathbf{c}_\nu\right]+\\+\frac{1}{\vartheta_0}\left[\mathbf{H}_2-\frac{\vartheta_1}{\vartheta_0}\mathbf{H}_1+\left(\frac{\vartheta_1^2}{\vartheta_0^2}-\frac{\vartheta_2}{\vartheta_0}\right)\mathbf{H}_0-i\alpha_2\vartheta_0\mathbf{J}\right]\left(\beta_1 e^{im_h s}\mathbf{c}_h+\beta_2 e^{im_{-j}s}\mathbf{c}_{-j}\right).\end{aligned}$$

The existence conditions for a 2π-periodic solution (3.36) for the third equation are

$$\left(\mathbf{J}\mathbf{g}_2^{(m_h)},\ \mathbf{c}_h\right)=\left(\mathbf{J}\mathbf{g}_2^{(m_{-j})},\ \mathbf{c}_{-j}\right)=0.$$

In expanded form, these equalities are

$$\begin{aligned}&\left(\chi_{hh}^{(0)}-\frac{\vartheta_1}{\vartheta_0}\omega_h-\alpha_1\vartheta_0\right)\gamma_1+\chi_{-jh}^{(m)}\gamma_2+\sum{}' p_\nu^{(n)}\chi_{\nu h}^{(m_h-n)}+\\&\quad+\left[\sigma_{hh}^{(0)}-\frac{\vartheta_1}{\vartheta_0}\chi_{hh}^{(0)}+\left(\frac{\vartheta_1^2}{\vartheta_0^2}-\frac{\vartheta_2}{\vartheta_0}\right)\omega_h-\alpha_2\vartheta_0\right]\beta_1+\left(\sigma_{-jh}^{(m)}-\frac{\vartheta_1}{\vartheta_0}\chi_{-jh}^{(m)}\right)\beta_2=0,\\&\chi_{h,-j}^{(-m)}\gamma_1+\left(\chi_{-j,-j}^{(0)}+\frac{\vartheta_1}{\vartheta_0}\omega_{-j}+\alpha_1\vartheta_0\right)\gamma_2+\sum{}' p_\nu^{(n)}\chi_{\nu,-j}^{(m_{-j}-n)}+\left(\sigma_{h,-j}^{(-m)}-\frac{\vartheta_1}{\vartheta_0}\chi_{h,-j}^{(-m)}\right)\beta_1+\\&\quad+\left[\sigma_{-j,-j}^{(0)}-\frac{\vartheta_1}{\vartheta_0}\chi_{-j,-j}^{(0)}-\left(\frac{\vartheta_1^2}{\vartheta_0^2}-\frac{\vartheta_2}{\vartheta_0}\right)\omega_{-j}+\alpha_2\vartheta_0\right]\beta_2=0,\end{aligned} \tag{3.45}$$

where

$$\sigma_{pq}^{(r)}=\left(\mathbf{H}_2^{(r)}\mathbf{c}_p,\ \mathbf{c}_q\right)\quad\left(\sigma_{qp}^{(-r)}=\overline{\sigma_{pq}^{(r)}}\right). \tag{3.46}$$

The determinant of the inhomogeneous algebraic system (3.45) is the same as that of system (3.38) and vanishes because of (3.39). System (3.45) will have a solution if and only if the determinant of the augmented matrix also vanishes:

$$\begin{vmatrix}\begin{aligned}&\sum{}' p_\nu^{(n)}\chi_{\nu h}^{(m_h-n)}+\left[\sigma_{hh}^{(0)}-\frac{\vartheta_1}{\vartheta_0}\chi_{hh}^{(0)}+\left(\frac{\vartheta_1^2}{\vartheta_0^2}-\frac{\vartheta_2}{\vartheta_0}\right)\omega_h-\alpha_2\vartheta_0\right]\beta_1+\\&\qquad+\left(\sigma_{-jh}^{(m)}-\frac{\vartheta_1}{\vartheta_0}\chi_{-jh}^{(m)}\right)\beta_2\end{aligned} & \beta_1\\ \begin{aligned}&\sum{}' p_\nu^{(n)}\chi_{\nu,-j}^{(m_{-j}-n)}+\left(\sigma_{h,-j}^{(-m)}-\frac{\vartheta_1}{\vartheta_0}\chi_{h,-j}^{(-m)}\right)\beta_1+\\&\qquad+\left[\sigma_{-j,-j}^{(0)}-\frac{\vartheta_1}{\vartheta_0}\chi_{-j,-j}^{(0)}-\left(\frac{\vartheta_1^2}{\vartheta_0^2}-\frac{\vartheta_2}{\vartheta_0}\right)\omega_{-j}+\alpha_2\vartheta_0\right]\beta_1\end{aligned} & -\beta_2\end{vmatrix}=0.$$

To evaluate the elements of the second column in this determinant one uses formulas (3.40) – (3.42). Isolate the term of interest (that involving ϑ_2) in the last equality:

$$\frac{\omega_h-\omega_{-j}}{\vartheta_0}-\beta_1\beta_2\vartheta_2=\left[\sigma_{hh}^{(0)}+\sigma_{-j,\,-j}^{(0)}-\frac{\vartheta_1}{\vartheta_0}(\chi_{hh}^{(0)}+\chi_{-j,\,-j}^{(0)})+\frac{\vartheta_1^2}{\vartheta_0^2}(\omega_h-\omega_{-j})\right]\beta_1\beta_2+$$
$$+\sigma_{-jh}^{(m)}\beta_2^2+\sigma_{h,\,-j}^{(-m)}\beta_1^2-\frac{\vartheta_1}{\vartheta_0}(\chi_{-jh}^{(m)}\beta_2^2+\chi_{h,\,-j}^{(-m)}\beta_1^2)+\beta_2{\sum}'p_\nu^{(n)}\chi_{\nu h}^{(m_h-n)}+\beta_1{\sum}'p_\nu^{(n)}\chi_{\nu,\,-j}^{(m_{-j}-n)}.$$

Using formulas (3.31) and (3.42), we have

$$\vartheta_2=\frac{1}{m}\left\{\sigma_{hh}^{(0)}+\sigma_{-j,\,-j}^{(0)}-\frac{\vartheta_1}{\vartheta_0}(\chi_{-j,\,-j}^{(0)}+\chi_{hh}^{(0)}-m\vartheta_1\mp 2|\chi_{-jh}^{(m)}|)+\right.$$
$$\left.+\frac{\beta_2}{\beta_1}\sigma_{-jh}^{(m)}+\frac{\beta_1}{\beta_2}\sigma_{h,\,-j}^{(-m)}+{\sum}'p_\nu^{(n)}\left[\frac{1}{\beta_1}\chi_{\nu h}^{(m_h-n)}+\frac{1}{\beta_2}\chi_{\nu,\,-j}^{(m_{-j}-n)}\right]\right\}.$$

The expression in parentheses vanishes by (3.40), the last two terms are complex conjugates; we transform the last sum using (3.44):

$$\sum_{\nu,n}{}'\frac{\operatorname{sign}\nu}{\alpha_0+n\vartheta_0-\omega_\nu}\left[\chi_{h\nu}^{(n-m_h)}\chi_{\nu h}^{(m_h-n)}+\frac{\beta_2}{\beta_1}\chi_{-j\nu}^{(n-m_{-j})}\chi_{\nu h}^{(m_h-n)}+\right.$$
$$\left.+\frac{\beta_1}{\beta_2}\chi_{h\nu}^{(n-m_h)}\chi_{\nu,\,-j}^{(m_{-j}-n)}\right]+\sum_{\nu,n}{}'\frac{\operatorname{sign}\nu}{\alpha_0+n\vartheta_0-\omega_\nu}\chi_{-j\nu}^{(n-m_{-j})}\chi_{\nu,\,-j}^{(m_{-j}-n)}.$$

Now change the running index in the first sum from n to m_h+l, and that in the second sum from n to $m_{-j}+l$, and substitute the entire sum into the formula for ϑ_2. Using (3.32), (3.37), (3.42) and (3.46), we get

$$\vartheta_2=\frac{1}{m}\left[\sigma_{-j,\,-j}^{(0)}+\sigma_{hh}^{(0)}\mp 2\operatorname{Re}\left(\frac{\chi_{-jh}^{(m)}}{|\chi_{-jh}^{(m)}|}\sigma_{h,\,-j}^{(-m)}\right)+\sum_{\nu,\,l}{}'\frac{\operatorname{sign}\nu\,|\chi_{h\nu}^{(l)}|^2}{\omega_h-\omega_\nu+l\vartheta_0}\mp\right.$$
$$\left.\mp 2\operatorname{Re}\left(\frac{\chi_{-jh}^{(m)}}{|\chi_{-jh}^{(m)}|}\sum_{\nu,\,l}{}'\frac{\operatorname{sign}\nu\,\chi_{h\nu}^{(l)}\overline{\chi_{-j\nu}^{(m+l)}}}{\omega_h-\omega_\nu+l\vartheta_0}\right)+\sum_{\nu,\,l}{}'\frac{\operatorname{sign}\nu\,|\chi_{-j\nu}^{(l)}|^2}{-\omega_j-\omega_\nu+l\vartheta_0}\right]. \qquad (3.47)$$

Here $-k\leqslant\nu\leqslant k$ and $-\infty<l<\infty$, and the prime indicates that the summation omits all terms whose denominators vanish. Formulas (3.40) and (3.47) define the second approximation to the boundary of a wide ($\chi_{-jh}^{(m)}\neq 0$) instability domain of system (3.31),

$$\vartheta=\vartheta_0+\vartheta_1\varepsilon+\vartheta_2\varepsilon^2+\ldots.$$

The upper sign in formulas (3.40) and (3.47) defines the lower boundary, and vice versa.

3.6. Second-approximation formulas for boundaries of instability domains of a second-order vector equation

Formulas (3.40) and (3.47) are also applicable to equation (3.2), if we use the results of §1.10. Namely, the natural frequencies of the first kind $\omega_\varkappa>0$ ($\varkappa=1,\ldots,k$) are defined by the equation

$$\det[\mathbf{P}_0-\omega^2\mathbf{M}]=0,$$

and those of the second kind by setting $\omega_{-\varkappa}=-\omega_{\varkappa}\,(\varkappa=1,\ldots,k)$. Then

$$\chi_{pq}^{(r)}=(\mathbf{P}_1^{(r)}\mathbf{a}_{|p|},\ \mathbf{a}_{|q|}),\quad \sigma_{pq}^{(r)}=(\mathbf{P}_2^{(r)}\mathbf{a}_{|p|},\ \mathbf{a}_{|q|}\ )\qquad (p,\ q=\pm 1,\ \ldots,\ \pm k),\quad (3.48)$$

where $\mathbf{a}_1,\ \ldots,\ \mathbf{a}_k$ is an orthonormal system of eigenvectors of the matrix $\mathbf{M}^{-1}\mathbf{P}_0$:

$$\mathbf{P}_0\mathbf{a}_{\varkappa}=\omega_{\varkappa}^2\mathbf{M}\mathbf{a}_{\varkappa},\ (\omega_{\varkappa}+\omega_{\lambda})(\mathbf{M}\mathbf{a}_{\varkappa},\ \mathbf{a}_{\lambda})=\delta_{\varkappa\lambda}\qquad (\varkappa,\ \lambda=1,\ \ldots,\ k).\qquad (3.49)$$

Note that now

$$\chi_{\pm\varkappa,\ \pm\lambda}^{(r)}=\chi_{\varkappa\lambda}^{(r)},\quad \sigma_{\pm\varkappa,\ \pm\lambda}^{(r)}=\sigma_{\varkappa\lambda}^{(r)}\qquad (\varkappa,\ \lambda=1,\ \ldots,\ k).$$

3.7. Example. Dynamic stability of plane flexural vibrations

Bolotin (/13a/, § 58) examines a system of two second-order equations to which many dynamic stability problems may be reduced, in particular, the problem of dynamic stability of plane flexural vibrations of a plate. In vector notation, the system is

$$\mathbf{C}_2\frac{d^2\mathbf{f}}{dt^2}+[\mathbf{I}_2-\varphi_0(t)\,\mathbf{A}]\,\mathbf{f}=0,$$

where

$$\mathbf{C}_2=\begin{Vmatrix}\omega_1^{-2} & 0\\ 0 & \omega_2^{-2}\end{Vmatrix},\qquad \mathbf{A}=\begin{Vmatrix}0 & a_{12}\\ a_{21} & 0\end{Vmatrix},\qquad \varphi_0(t)=\alpha_0+\beta_0\cos\vartheta t,$$

α_0,β_0 are small parameters, $\omega_2>\omega_1>0$ and $a_{12}/\omega_1^2=a_{21}/\omega_2^2=\delta>0$. Multiplying by $\mathbf{C}_2^{-1}$, we obtain a vector equation

$$\frac{d^2\mathbf{f}}{dt^2}+[\mathbf{P}_0-\varphi(t)\,\mathbf{N}]\,\mathbf{f}=0,$$

where

$$\mathbf{P}_0=\begin{Vmatrix}\omega_1^2 & 0\\ 0 & \omega_2^2\end{Vmatrix},\quad \mathbf{N}=\begin{Vmatrix}0 & 1\\ 1 & 0\end{Vmatrix},\ \varphi(t)=\alpha+\beta\cos\vartheta t,\quad \alpha=\alpha_0\delta,\quad \beta=\beta_0\delta$$

(α, β are small parameters). To reduce this to the form of (3.2), we consider instead the equation

$$\frac{d^2\mathbf{v}}{dt^2}+[\mathbf{P}_0+\varepsilon\,(e^{-i\vartheta t}\mathbf{P}_1^{(-1)}+\mathbf{P}_1^{(0)}+e^{i\vartheta t}\mathbf{P}_1^{(1)})]\,\mathbf{y}=0,$$

where

$$\mathbf{P}_1^{(-1)}=\mathbf{P}_1^{(1)}=-\frac{1}{2}\begin{Vmatrix}0 & \beta\\ \beta & 0\end{Vmatrix},\quad P_1^{(0)}=-\begin{Vmatrix}0 & \alpha\\ \alpha & 0\end{Vmatrix}.$$

In the final formulas, one must set $\varepsilon=1$. The eigenvectors of $\mathbf{P}_0$, normalized to satisfy conditions (3.49), are

$$\mathbf{a}_\varkappa=\frac{1}{\sqrt{2\omega_\varkappa}}\mathbf{e}_\varkappa \quad (\varkappa=1,\,2),\ \text{where}\ \mathbf{e}_1=\begin{pmatrix}1\\0\end{pmatrix},\quad \mathbf{e}_2=\begin{pmatrix}0\\1\end{pmatrix}.$$

Of the numbers (3.48) the only nonzero ones are

$$\chi_{12}^{(0)}=\chi_{21}^{(0)}=-\frac{\alpha}{2\sqrt{\omega_1\omega_2}},\quad \chi_{12}^{(\pm1)}=\chi_{21}^{(\pm1)}=-\frac{\beta}{4\sqrt{\omega_1\omega_2}},$$

and so the only wide instability domain for this example corresponds to combination resonance at frequency $\vartheta_0=\omega_1+\omega_2$ ($h=1$, $j=2$, $m=1$). By formulas (3.40) and (3.47), we have

$$\vartheta_1^{(1,\,2)}=\mp\frac{\beta}{2\sqrt{\omega_1\omega_2}},$$
$$\vartheta_2=-\frac{1}{2(\omega_1+\omega_2)\,\omega_1\omega_2}\left(\alpha^2+\frac{1}{8}\beta^2\right).$$

Thus the only wide dynamic instability domain is

$$\omega_1+\omega_2-\frac{\beta}{2\sqrt{\omega_1\omega_2}}+\vartheta_2+O(|\alpha|^4+|\beta|^4)<\vartheta<$$
$$<\omega_1+\omega_2+\frac{\beta}{2\sqrt{\omega_1\omega_2}}+\vartheta_2+O(|\alpha|^4+|\beta|^4). \quad (3.50)$$

For the first principal resonance $\vartheta_0=2\omega_1$, $\vartheta_1=0$, and by (3.47) (with $j=h=1$, $m=1$) we see that the corresponding narrow dynamic instability domain is

$$2\omega_1+\vartheta_2^{(1)}+O(|\alpha|^4+|\beta|^4)<\vartheta<2\omega_1+\vartheta_2^{(2)}+O(|\alpha|^4+|\beta|^4), \quad (3.51)$$

where

$$\vartheta_2^{(1,2)}=\frac{1}{\omega_1^3}\left[-\frac{\left(\alpha\mp\frac{1}{2}\beta\right)^2}{\gamma^{-1}-1}+\frac{\beta^2}{4(9-\gamma^{-1})}\right]$$

and

$$\gamma=\frac{\omega_1^2}{\omega_2^2}<1. \quad (3.52)$$

Similarly, for the second principal resonance we obtain the narrow instability domain

$$2\omega_2+\tilde{\vartheta}_2^{(1)}+O(|\alpha|^4+|\beta|^4)<\vartheta<2\omega_2+\tilde{\vartheta}_2^{(2)}+O(|\alpha|^4+|\beta|^4), \quad (3.53)$$

where

$$\tilde{\vartheta}^{(1,2)}_{2} = \frac{1}{\omega_2^3}\left[\frac{\left(\alpha \mp \frac{1}{2}\beta\right)^2}{1-\gamma} + \frac{\beta^2}{4(9-\gamma)}\right]. \tag{3.54}$$

Note that formulas (3.51) and (3.53) are not quite the same as the relevant approximate formulas established in /13a/, pp. 311 — 317, using the method of harmonic balance. (The domain (3.50) cannot be constructed by the method of /13a/, since the latter is based on the existence of a T-periodic or T-antiperiodic solution on the boundary of the instability domain, and such solutions exist only on the boundaries of principal-resonance domains.) Moreover, the results of /13a/ by no means imply that the domains are actually dynamic instability domains — they are only "likely candidates" for this role. By §3.2, they are indeed dynamic instability domains. Thus the results of this subsection complement and improve the results of §58 in /13a/.

3.8. Example. Stability of a gyrocompass on a vessel describing a circular course

Discussing the theory of gyrocompasses, V. N. Koshlyakov (Prikl. Mat. Mekh. 23, No. 5, 1969) showed that the perturbed motion of a gyrocompass on a vessel traveling a circular course is described by the following system of differential equations:

$$\left.\begin{aligned} &\ddot{x}_1 + (\nu^2 - \Omega^2)x_1 = 2\Omega\dot{x}_2 \operatorname{tg}\varphi + \dot{\Omega}x_2 \operatorname{tg}\varphi, \\ &\ddot{x}_2 + (p^2 - \Omega^2)x_2 = -2\Omega\dot{x}_1 \operatorname{ctg}\varphi - \dot{\Omega}x_1 \operatorname{ctg}\varphi, \end{aligned}\right\} \tag{3.55}$$

where

$$\Omega = -\varepsilon\Theta \sin\theta t + \varepsilon^2\Theta \cos 2\Theta t + O(\varepsilon^3),$$
$$p^2 = p_0^2[1 - 2\varepsilon \operatorname{ctg}^2\varphi \sin\Theta t + \varepsilon^2(4\operatorname{ctg}^4\varphi \sin^2\Theta t + \operatorname{ctg}^2\varphi) + O(\varepsilon^3)],$$
$$\nu = \sqrt{\frac{g}{R}}, \quad p_0^2 = \frac{\delta}{\sin^2\varphi}, \quad \varepsilon = \frac{v}{Ru\cos\varphi},$$

v is the speed of the vessel, φ the latitude, θ the angular frequency of the vessel, g the acceleration of gravity, R and u the radius and angular velocity of rotation of the Earth, δ a quantity depending on the gyrocompass parameters.

We introduce a vector $\mathbf{x}$ of canonical variables q_1, q_2, p_1, p_2:

$$q_1 = \sqrt{\operatorname{ctg}\varphi}\,x_1, \quad p_1 = \sqrt{\operatorname{ctg}\varphi}\,\dot{x}_1 - \Omega\sqrt{\operatorname{tg}\varphi}\,x_2,$$
$$q_2 = \sqrt{\operatorname{tg}\varphi}\,x_2, \quad p_2 = \sqrt{\operatorname{tg}\varphi}\,\dot{x}_2 - \Omega\sqrt{\operatorname{ctg}\varphi}\,x_1.$$

In these variables, system (3.46) becomes*

* Note that system (3.56) is not exactly the same as (3.1). However, it is readily seen that formulas (3.40) and (3.47) of § 3.5 are applicable, provided we put

$$\mathbf{H}_1(\tau) = \tilde{\mathbf{H}}_1(\tau) + \Theta_0\hat{\mathbf{H}}_1(\tau), \quad \mathbf{H}(\tau) = \tilde{\mathbf{H}}(\tau) + \Theta_0\hat{\mathbf{H}}_2(\tau) + \Theta_1\hat{\mathbf{H}}_1(\tau).$$

$$\mathbf{J}\frac{d\mathbf{x}}{d\tau}=(\mathbf{H}_0+\varepsilon\mathbf{H}_1+\varepsilon^2\mathbf{H}_2+\dots)\,\mathbf{x}\qquad(\tau=\Theta t),\tag{3.56}$$

where

$$\mathbf{H}_j=\tilde{\mathbf{H}}_j(\tau)+\Theta\hat{\mathbf{H}}_j(\tau),\quad \tilde{\mathbf{H}}_j(\tau)=\alpha_j(\tau)\,\mathbf{A},\quad \hat{\mathbf{H}}_j(\tau)=\beta_j(\tau)\,\mathbf{B}\qquad(j=1,\ 2),$$
$$\alpha_1(\tau)=-2p_0^2\operatorname{ctg}^2\varphi\sin\tau,\quad \alpha_2(\tau)=p_0^2(4\operatorname{ctg}^4\varphi\sin^2\tau+\operatorname{ctg}^2\varphi),$$
$$\beta_1(\tau)=-\sin\tau,\quad \beta_2(\tau)=\cos 2\tau,$$
$$\mathbf{J}=\begin{pmatrix}\mathbf{0}&-\mathbf{I}_2\\ \mathbf{I}_2&\mathbf{0}\end{pmatrix},\quad \mathbf{H}_0=\begin{Vmatrix}v^2&0&0&0\\0&p_0^2&0&0\\0&0&1&0\\0&0&0&1\end{Vmatrix},\quad \mathbf{A}=\begin{Vmatrix}0&0&0&0\\0&1&0&0\\0&0&0&0\\0&0&0&0\end{Vmatrix},\quad \mathbf{B}=\begin{Vmatrix}0&0&0&-1\\0&0&1&0\\0&1&0&0\\-1&0&0&0\end{Vmatrix}.$$

The eigenvalues $i\omega_{\pm j}$ and eigenvectors of the matrix $\mathbf{J}^{-1}\mathbf{H}_0$ are defined by

$$\omega_{\pm 1}=\pm\nu,\quad \omega_{\pm 2}=\pm p_0,\quad \mathbf{f}_j=\frac{1}{\sqrt{2\omega_{|j|}}}\begin{pmatrix}\mathbf{e}_j\\ i\omega_j\mathbf{e}_j\end{pmatrix}\qquad(j=\pm 1,\ \pm 2).$$

Using the formulas of §3.5, one readily shows that in the $\varepsilon\Theta$-plane there can be wide instability domains attached only at points $(0,\ \Theta_0)$ with $\Theta_0=2\omega_2=2p_0$ for principal resonance and $\Theta_0=\omega_1+\omega_2=p_0+\nu$ for combination resonance. A few calculations yield the following equations for the boundaries $\Theta^{(1)}(\varepsilon)$, $\Theta^{(2)}(\varepsilon)$ of the instability domains $\Theta^{(1)}(\varepsilon)<\Theta<\Theta^{(2)}(\varepsilon)$, accurate to within ε^2 inclusive:

a) principal resonance, where

$$\Theta^{(1,2)}=2p_0\pm\varepsilon p_0\operatorname{ctg}^2\varphi+\varepsilon^2\left[\frac{2p_0(\nu^2-p_0^2)}{gp_0^2-\nu^2}+p_0\left(\frac{15}{8}\operatorname{ctg}^4\varphi+\operatorname{ctg}^2\varphi\right)\right]+O(\varepsilon^3);\tag{3.57}$$

b) combination resonance, $\Theta^{(1)}(\varepsilon)<\Theta<\Theta^{(2)}(\varepsilon)$, where

$$\Theta^{(1,2)}=p_0+\nu\pm\varepsilon\frac{\Theta_0(p_0-\nu)}{2\sqrt{p_0\nu}}+$$
$$+\varepsilon^2\left[\frac{(p_0-\nu)^2(2p_0+\nu)}{4p_0\Theta_0}+\frac{\operatorname{ctg}^4\varphi\, p_0(3p_0^2-\Theta_0^2)}{4p_0^2-\Theta_0^2}+\frac{p_0\operatorname{ctg}^2\varphi}{2}\right]+O(\varepsilon^3).\tag{3.58}$$

For example, let the speed of the vessel be $v=15$ m/sec at latitude 80°, and let

$$p_0^2=2{,}03\cdot 10^{-5}\ \text{sec}^{-2},\ \nu^2=0{,}154\cdot 10^{-5}\ \text{sec}^{-2}.$$

We then see from (3.57), (3.58) that the "resonant" frequencies lie in the following intervals:

$8.96\cdot 10^{-3}\ \text{sec}^{-1}<\Theta<9.02\cdot 10^{-3}\ \text{sec}^{-1}$ (principal resonance),
$5.85\cdot 10^{-3}\ \text{sec}^{-1}<\Theta<7.71\cdot 10^{-3}\ \text{sec}^{-1}$ (combination resonance).

In terms of the period T, we finally obtain:

11.6 min $<T<$ 11.7 min (principal resonance),
13.6 min $<T<$ 17.9 min (combination resonance).

Note that, as predicted by §1.11, the most dangerous phenomenon is combination parametric resonance, since it may appear in a wider range of periods and moreover the value of $\varepsilon(\zeta)$ for small ζ in this case is approximately one thirtieth of the value for principal resonance. (Concerning $\varepsilon(\zeta)$, see the end of §1. 5.)

§4. PARAMETRIC RESONANCE IN ALMOST-CANONICAL SYSTEMS

In this section we shall examine almost-canonical equations, that is to say, equations which are canonical only in the unperturbed state. The formulas developed in §1 for the critical frequencies and the boundaries of the instability domains in the parameter space are no longer applicable.

We shall show that the required formulas follow from the results of § IV. 3. In § VI. 4 we shall illustrate the application of these formulas to a fairly interesting applied problem, concerning the stability of flexural vibrations of two coaxial cantilever shafts.

4.1. Statement of the problem

Consider the following system of k linear second-order differential equations with periodic coefficients:

$$\mathbf{M}\ddot{\mathbf{y}}+\varepsilon\mathbf{Q}(\vartheta t,\ \varepsilon)\dot{\mathbf{y}}+[\mathbf{P}_0+\varepsilon\mathbf{P}(\vartheta t,\ \varepsilon)]\,\mathbf{y}=0, \tag{4.1}$$

where $\mathbf{y}$ is a k-vector, $\mathbf{P}(\vartheta t,\ \varepsilon)$ and $\mathbf{Q}(\vartheta t,\ \varepsilon)$ are $k\times k$ matrix-functions, analytic in ε in the neighborhood of $\varepsilon=0$, whose elements are periodic integrable complex-valued piecewise-continuous functions of t:

$$\mathbf{P}(\vartheta t+2\pi,\ \varepsilon)=\mathbf{P}(\vartheta t,\ \varepsilon),\quad \mathbf{Q}(\vartheta t+2\pi,\ \varepsilon)=\mathbf{Q}(\vartheta t,\ \varepsilon)$$

($\vartheta>0$ is the perturbation frequency); $\mathbf{M}$ and $\mathbf{P}_0$ are real constant symmetric $k\times k$ matrices, positive in the sense of the quadratic forms $(\mathbf{Mc},\ \mathbf{c})$ and $(\mathbf{P}_0\mathbf{c},\ \mathbf{c})$ (where $\mathbf{c}$ is a real k-vector). We shall also assume that the positive eigenvalues $\omega_1^2,\ \ldots,\ \omega_k^2$ of the matrix $\mathbf{M}^{-1}\mathbf{P}_0$ are arranged in nondecreasing order. The eigenvectors $\mathbf{a}_1,\ \ldots,\ \mathbf{a}_k$ of $\mathbf{M}^{-1}\mathbf{P}_0$,

$$\mathbf{P}_0\mathbf{a}_\varkappa=\omega_\varkappa^2\mathbf{M}\mathbf{a}_\varkappa,$$

are normalized so that

$$\frac{2}{\vartheta_0}\omega_\varkappa(\mathbf{M}\mathbf{a}_\varkappa,\ \mathbf{a}_\lambda)=\delta_{\varkappa\lambda}=\begin{cases}0, & \text{if}\quad \varkappa\neq\lambda,\\ 1, & \text{if}\quad \varkappa=\lambda\end{cases}\qquad (\varkappa,\ \lambda=1,\ \ldots,\ k). \tag{4.2}$$

Assume that for $\varepsilon=0$, $\vartheta=\vartheta_0$ equation (4.1) has an r-fold characteristic exponent $i\omega^{(0)}$. Consider a ray in the quadrant $\varepsilon\geqslant 0$, $\vartheta>0$ emanating from the point $(0, \vartheta_0)$:

$$\vartheta=\vartheta_0+\mu\varepsilon, \tag{4.3}$$

where μ, like ϑ_0, is held fixed. Let us determine the characteristic exponents of system (4.1) on this ray for sufficiently small ε. We go over to dimensionless time $\tau=\vartheta t$ and use the expansion

$$\frac{1}{\vartheta}=\frac{1}{\vartheta_0+\mu\varepsilon}=\frac{1}{\vartheta_0}\left(1-\frac{\mu}{\vartheta_0}\varepsilon+\dots\right),$$

which is convergent provided

$$|\mu|\varepsilon<\vartheta_0, \text{ i.e., } \vartheta<2\vartheta_0. \tag{4.4}$$

System (4.1) becomes

$$\mathbf{M}\frac{d^2\mathbf{y}}{d\tau^2}+\frac{1}{\vartheta_0}[\varepsilon\mathbf{Q}_1(\tau)+\dots]\frac{d\mathbf{y}}{d\tau}+\left\{\frac{1}{\vartheta_0^2}\mathbf{P}_0+\varepsilon\left[\frac{1}{\vartheta_0^2}\mathbf{P}_1(\tau)-2\frac{\mu}{\vartheta_0^3}\mathbf{P}_0+\dots\right]\right\}\mathbf{y}=0, \tag{4.5}$$

where $\mathbf{Q}_1(\tau)=\mathbf{Q}(\tau, 0)$, $\mathbf{P}_1(\tau)=\mathbf{P}(\tau, 0)$, and the dots stand for higher-order terms. Divide the numbers $\omega_\varkappa$ and $\omega_{-\varkappa}=-\omega_\varkappa$ $(\varkappa=1, \dots, k)$ into congruence classes modulo ϑ_0. Let $\{\omega_{j_1}, \dots, \omega_{j_r}\}$ be one of these classes, belonging to a characteristic exponent $i\omega^{(0)}/\vartheta_0$ of system (4.5) at $\varepsilon=0$. Define integers m_j by

$$\omega_j=\omega^{(0)}+m_j\vartheta_0 \qquad (j=j_1, \dots, j_r). \tag{4.6}$$

For small ε, the corresponding r characteristic exponents of system (4.5) are defined by formulas (IV. 3. 9):

$$\alpha_j(\varepsilon)=i\frac{\omega^{(0)}}{\vartheta_0}+i\nu_j\varepsilon+O(\varepsilon^{1+\delta}) \qquad (\delta>0;\ j=j_1, \dots, j_r); \tag{4.7}$$

the ν_j's are the roots of equation (IV. 3. 30):

$$\det\|\sigma_{jh}+\nu\gamma_{jh}\|_{j,\,h=j_1,\,\dots,\,j_r}=0, \tag{4.8}$$

where, by (IV. 3. 27) and (IV. 3. 31),

$$\gamma_{jh}=\begin{cases}0, & \text{if } j\neq h,\\ 1, & \text{if } j=h>0,\\ -1, & \text{if } j=h<0,\end{cases}$$

$$\sigma_{jh}=-\frac{1}{\vartheta_0^2}\left(\mathbf{P}_1^{(m_h-m_j)}\mathbf{a}_{|j|},\ \mathbf{a}_{|h|}\right)+\frac{\mu\omega_j}{\vartheta_0^2}\gamma_{jh}-\frac{i\omega_j}{\vartheta_0^3}(\mathbf{Q}_1^{(m_h-m_j)}\mathbf{a}_{|j|},\ \mathbf{a}_{|h|}), \tag{4.9}$$

and $\mathbf{P}_1^{(m)}$ and $\mathbf{Q}_1^{(m)}$ are the Fourier coefficients of the matrix-functions

$$\mathbf{P}(\tau, 0)\sim\sum_{m=-\infty}^{\infty}e^{im\tau}\mathbf{P}_1^{(m)}, \quad \mathbf{Q}(\tau, 0)\sim\sum_{m=-\infty}^{\infty}e^{im\tau}\mathbf{Q}_1^{(m)}.$$

4.2. Fundamental and supplementary cases

In contradistinction to parametric resonance in canonical systems, the "fundamental" case for almost-canonical systems is the following.

a) The class $\{\omega_{j_1}, \ldots, \omega_{j_r}\}$ consists of a single number. In other words, for some $j = \pm 1, \ldots, \pm k$ the number ω_j is incongruent modulo ϑ_0 to all the other numbers $\omega_{-k}, \ldots, \omega_{-1}, \omega_1, \ldots, \omega_k$. Equation (4.8) is now

$$\nu_j = -\sigma_{jj} \operatorname{sign} j, \tag{4.9a}$$

and the real part of the characteristic exponent (4.6) will be positive for sufficiently small $\varepsilon > 0$ if

$$\operatorname{Im}(\sigma_{jj} \operatorname{sign} j) > 0. \tag{4.10}$$

Conditions (4.10) are the first conditions for instability of the trivial solution of system (4.1) for sufficiently small $\varepsilon > 0$.

We must then examine the *supplementary case*, which under certain added assumptions was the "fundamental" case for canonical systems.

b) The class $\{\omega_{j_1}, \ldots, \omega_{j_r}\}$ consists of two numbers ω_{j_1} and ω_{j_2}, where $j_1, j_2 = \pm 1, \ldots, \pm k$ (to fix ideas, we assume that $j_1 < j_2$). Then by (4.6)

$$\omega_{j_1} = \omega^{(0)} + m_{j_1}\vartheta_0, \quad \omega_{j_2} = \omega^{(0)} + m_{j_2}\vartheta_0 \qquad (m_{j_2} > m_{j_1})$$

and so

$$\vartheta_0 = \frac{1}{m}(\omega_{j_2} - \omega_{j_1}) \qquad (m = m_{j_2} - m_{j_1} > 0). \tag{4.11}$$

Equation (4.8) becomes

$$\begin{vmatrix} \sigma_{j_1 j_1} + \nu \operatorname{sign} j_1 & \sigma_{j_1 j_2} \\ \sigma_{j_2 j_1} & \sigma_{j_2 j_2} + \nu \operatorname{sign} j_2 \end{vmatrix} = 0,$$

or, developing the determinant,

$$\nu^2 \vartheta_0^4 - 2(\alpha + i\beta)\nu\vartheta_0^2 - (\gamma + i\delta) = 0, \tag{4.12}$$

where

$$\left.\begin{aligned} \alpha + i\beta &= -\frac{1}{2}(\sigma_{j_1 j_1} \operatorname{sign} j_1 + \sigma_{j_2 j_2} \operatorname{sign} j_2)\,\vartheta_0^2, \\ \gamma + i\delta &= (\sigma_{j_1 j_2}\sigma_{j_2 j_1} - \sigma_{j_1 j_1}\sigma_{j_2 j_2})\,\vartheta_0^4 \operatorname{sign}(j_1 j_2). \end{aligned}\right\} \tag{4.13}$$

By (4.7), the real part of at least one of the characteristic exponents will be positive for sufficiently small $\varepsilon > 0$ if and (except for "borderline" cases) only if the imaginary part of at least one of the roots is negative. We emphasize that since the coefficients of system (4.1) are complex-valued the coefficients of equation (4.13) will also generally be complex. It is readily shown that the imaginary part of at least one root of equation (4.12) is negative if and only if one of the following conditions holds:

$$\left.\begin{array}{lll} \text{(a)} & \beta<0, & \\ \text{(b)} & \beta=0 \quad \text{and} \quad \delta\neq 0, & \\ \text{(c)} & \beta=\delta=0 \quad \text{and} \quad \gamma<-\alpha^2, & \\ \text{(d)} & \beta>0 \quad \text{and} \quad \delta^2+4\alpha\beta\delta-4\beta^2\gamma>0. & \end{array}\right\} \tag{4.14}$$

These inequalities determine the values of μ in (4.3) at which the trivial solution of system (4.1) is unstable for sufficiently small $\varepsilon>0$, i. e., they determine to a first approximation the dynamic stability domains attached at the point $(0, \vartheta_0)$. These domains may intersect the domains (4.10); in particular, they may even be subdomains of (4.10).

Note that in canonical systems case (a) never occurs, while by the theorem of §1.2 case (b) is possible only for "half" of the frequencies (4.11) — the critical frequencies.

4.3. A distinctive feature of combination resonance: transition from stability to instability with increasing friction

Consider a mechanical system which is described, not counting friction, by equation (1.48), where $\mathbf{M}$ and $\mathbf{P}_0$ are diagonal matrices (1.49). Then at $\varepsilon=0$ equation (1.48) describes a collection of independent subsystems

$$q_\varkappa\ddot{\eta}_\varkappa+p_\varkappa\eta_\varkappa=0 \qquad (\varkappa=1, \ldots, k). \tag{4.15}$$

It was shown in §1 that at a critical parametric perturbation frequency $\vartheta_0=(\omega_j+\omega_h)/m$ a wide combination resonance domain appears (for $\varepsilon\neq 0$) if and only if $\pi_{jh}^{(m)}\neq 0$ (see (1.50)). The physical meaning of this condition is that when $\varepsilon\neq 0$ the subsystems j and h are "coupled" in some way. Let (ε, ϑ) be a point in the stability domain near the ϑ-axis. Consider a system similar to (4.15) but involving a small friction force (of the order of ε). One might expect that the equation with parameters ε, ϑ will now be asymptotically stable, that friction has a stabilizing effect. This is indeed the case in many systems. Under certain conditions, however, the inclusion of friction causes instability. Even more: with increasing friction an asymptotically stable system may become unstable. It turns out that this may happen at combination resonance

$$\vartheta_0\approx(\omega_j+\omega_h)/m \qquad (j\neq h),$$

provided certain special relations hold among the dissipation forces.

The physical explanation for this phenomenon is as follows. On the one hand, friction stabilizes a system, but on the other it may increase the forces coupling, say, subsystems j and h. If $\pi_{jh}^{(m)}\neq 0$ the latter situation creates a favorable background for combination resonance, i. e., for instability. Whether the system is stable or unstable will depend on which of these tendencies predominates.

We now proceed to a rigorous account. To simplify the exposition, we consider a system

$$\mathbf{M}\frac{d^2\mathbf{y}}{dt^2}+\varepsilon\mathbf{Q}\frac{d\mathbf{y}}{dt}+[\mathbf{I}_2-\varepsilon\varphi(\vartheta t)\mathbf{P}]\,\mathbf{y}=0, \tag{4.16}$$

where $\mathbf{y}$ is a 2-vector,

$$\varphi(\vartheta t)=a+b\cos\vartheta t, \tag{4.17}$$

and

$$\mathbf{M}=\begin{Vmatrix}\frac{1}{\omega_1^2} & 0\\ 0 & \frac{1}{\omega_2^2}\end{Vmatrix},\quad \mathbf{Q}=\begin{Vmatrix}\varkappa_1 & 0\\ 0 & \varkappa_2\end{Vmatrix},\quad \mathbf{P}=\begin{Vmatrix}0 & a_{12}\\ a_{21} & 0\end{Vmatrix},\quad \mathbf{I}_2=\begin{Vmatrix}1 & 0\\ 0 & 1\end{Vmatrix}.$$

The number $\varepsilon>0$ is a small parameter. (We may assume that $\varepsilon=1$ and that the numbers $\varkappa_1$, $\varkappa_2$, a, b are small positive parameters of the same order of magnitude. In that case one should set $\varepsilon=1$ in the formulas derived below.) Equations (4.16) characterize many dynamic stability problems (see, e.g., /13a, b/). Let $\omega_2>\omega_1>0$. The system is clearly stable at $\varepsilon=0$. The characteristic exponents of system (4.16) at $\varepsilon=0$ (with t as independent variable) are

$$\alpha_{1,2}=\pm i\omega_1,\quad \alpha_{3,4}=\pm i\omega_2.$$

We consider the following cases.

a) Nonresonant case:

$$\vartheta_0\neq\frac{2\omega_1}{m},\quad \vartheta_0\neq\frac{\omega_2\pm\omega_1}{m},\quad \vartheta_0\neq\frac{2\omega_2}{m}\qquad (m=1,\ 2,\ \ldots).$$

In other words, $\omega_j\not\equiv\omega_h\ (\mathrm{mod}\,\vartheta)$ $(j,\ h=\pm1,\ \pm2;\ \omega_{-j}=-\omega_j)$, so that all the multipliers are distinct at $\varepsilon=0$. The numbers ω_j fall into four congruence classes mod ϑ_0: $\{\omega_1\}$, $\{\omega_2\}$, $\{\omega_{-1}\}$, $\{\omega_{-2}\}$ (as usual, $\omega_{-j}=-\omega_j$). Each class contains a single number, so that we have case (a) of §4.1. Applying formula (4.9), we obtain

$$\sigma_{jj}=\frac{\mu\omega_{|j|}}{\vartheta_0^2}-i\frac{\omega_j^2}{2\vartheta_0}\varkappa_j\,\mathrm{sign}\,j\qquad (j=\pm1,\ \pm2). \tag{4.18}$$

Since by (4.9a)

$$\nu_j=-\sigma_{jj}\,\mathrm{sign}\,j,$$

the real part of the first correction to the characteristic exponents (4.7) is

$$\mathrm{Re}\,(i\nu_j)=\mathrm{Im}\,(\sigma_{jj}\,\mathrm{sign}\,j)=-\frac{\omega_j^2}{2\vartheta_0}\varkappa_j\qquad (j=\pm1,\ \pm2).$$

Consequently, if $\vartheta=\vartheta_0+\mu\varepsilon$, for any μ $(-\infty<\mu<+\infty)$ and any sufficiently small $\varepsilon>0$, the system is asymptotically stable. These values are not critical — there are no dynamic instability domains attached at any of the above points $(0,\ \vartheta_0)$.

b) Principal resonance. Let $\vartheta_0 = 2\omega_1/m$, where m is some integer. Equation (4.16) at $\varepsilon=0$, $\vartheta=\vartheta_0$ has at least a twofold multiplier $\exp(i\omega_1)=\exp(-i\omega_1)$. Suppose that the multiplicity of this multiplier is at most 2 and that there are no other multiple multipliers, so that

$$\frac{2\omega_1}{m} \neq \frac{2\omega_2}{m_1}, \qquad \frac{2\omega_1}{m} \neq \frac{\omega_2 \pm \omega_1}{m_2}$$

for integers m_1, m_2. Then the numbers ω_j fall into classes $\{\omega_1, \omega_{-1}\}$, $\{\omega_2\}$, $\{\omega_{-2}\}$. Equation (4.8) ($j_1=1$, $j_2=-1$) becomes

$$\begin{vmatrix} \sigma_{11}+\nu & \sigma_{1,-1} \\ \sigma_{-11} & \sigma_{-1,-1}-\nu \end{vmatrix} = 0.$$

By (4.9),

$$\sigma_{1,-1} = \sigma_{-11} = 0.$$

Consequently, by (4.18),

$$i\nu_1 = -\frac{\varkappa_1\omega_1^2}{2\vartheta_0} - i\mu\frac{\omega_1}{\vartheta_0^2}, \quad i\nu_{-1} = -\frac{\varkappa_1\omega_1^2}{2\vartheta_0} + i\mu\frac{\omega_1}{\vartheta_0^2}.$$

As in case (a), we obtain

$$i\nu_2 = -\frac{\varkappa_2\omega_2^2}{2\vartheta_0} - i\mu\frac{\omega_2}{\vartheta_0^2}, \quad i\nu_{-2} = -\frac{\varkappa_2\omega_2^2}{2\vartheta_0} + i\mu\frac{\omega_2}{\vartheta_0^2}.$$

Since

$$\operatorname{Re}(i\nu_j) < 0 \qquad (j = \pm 1, \pm 2),$$

it follows that in this case too, for any $-\infty < \mu < +\infty$ and sufficiently small $\varepsilon > 0$, equation (4.16) with $\vartheta = \vartheta_0 + \mu\varepsilon$ is asymptotically stable. Thus the values of $\vartheta_0 = 2\omega_1/m$ are not critical.

Similarly one considers the case in which the ω_j's fall into classes $\{\omega_1, \omega_{-1}\}$, $\{\omega_2, \omega_{-2}\}$ (double principal resonance). In this case again we have asymptotic stability for small $\varepsilon > 0$.

The case in which the classes are $\{\omega_1\}$, $\{\omega_{-1}\}$, $\{\omega_2, \omega_{-2}\}$ is obtained from those just treated by notational changes.

c) Combination resonance at difference frequency: $\vartheta_0 = \frac{\omega_2-\omega_1}{m}$. When $\varepsilon=0$, $\vartheta=\vartheta_0$ equation (4.16) has two multiple multipliers $\exp(i\omega_1)=\exp(i\omega_2)$, $\exp(-i\omega_1)=\exp(-i\omega_2)$. Suppose that they are distinct:

$$\frac{\omega_2-\omega_1}{m} \neq \frac{2\omega_1}{m_1}, \ \frac{2\omega_2}{m_2}, \ \frac{\omega_1+\omega_2}{m_3},$$

where m_1, m_2, m_3 are integers. Then the congruence classes of ω_j modulo ϑ_0 are $\{\omega_1, \omega_2\}$, $\{\omega_{-1}, \omega_{-2}\}$.

For the first class, by (4.9),

$$\begin{Vmatrix} \sigma_{11} & \sigma_{12} \\ \sigma_{21} & \sigma_{22} \end{Vmatrix} = \begin{Vmatrix} \frac{\mu\omega_1}{\vartheta_0^2} - i\frac{\omega_1^2}{2\vartheta_0}\varkappa_1 & \frac{1}{2\vartheta_0}\sqrt{\omega_1\omega_2}\,a_{21}\varphi^{(m)} \\ \frac{1}{2\vartheta_0}\sqrt{\omega_1\omega_2}\,a_{12}\varphi^{(m)} & \frac{\mu\omega_2}{\vartheta_0^2} - i\frac{\omega_2^2}{2\vartheta_0}\varkappa_2 \end{Vmatrix},$$

where $\varphi^{(m)}$ are the Fourier coefficients of $\varphi(\tau)$:

$$\varphi(\tau) = \sum_l e^{il\tau}\varphi^{(l)}.$$

Equation (4.12) ($j_1 = 1,\ j_2 = 2$) is

$$\alpha + i\beta = -\frac{1}{2}(\omega_1+\omega_2)\,\mu + \frac{1}{4}i\vartheta_0(\omega_1^2\varkappa_1 + \omega_2^2\varkappa_2),$$

$$\gamma + i\delta = \frac{1}{4}\omega_1\omega_2\vartheta_0^2\varphi^{(m)^2}a_{12}a_{21} + \frac{1}{4}\omega_1^2\omega_2^2\vartheta_0^2\varkappa_1\varkappa_2 - \mu^2\omega_1\omega_2 + \frac{1}{2}i\mu\vartheta_0\omega_1\omega_2(\omega_1\varkappa_1 + \omega_2\varkappa_2).$$

Since $\beta > 0$ and $\delta^2 + 4\alpha\beta\delta - 4\beta^2\gamma < 0$ (as may be checked directly), it follows from (4.14) that for any μ $(-\infty < \mu < +\infty)$ and sufficiently small $\varepsilon > 0$ equation (4.16) with $\vartheta = \vartheta_0 + \mu\varepsilon$ is asymptotically stable.

d) Combination resonance at sum frequency: $\vartheta_0 = \frac{\omega_1+\omega_2}{m}$ (where m is an integer).

Suppose that

$$\frac{\omega_1+\omega_2}{m} \neq \frac{2\omega_1}{m_1},\quad \frac{2\omega_2}{m_2},\quad \frac{\omega_2-\omega_1}{m_3}$$

with integers $m_1,\ m_2,\ m_3$. Then the ω_j's fall into two congruence classes $\operatorname{mod}\vartheta_0$: $\{\omega_1,\ \omega_{-2}\}$, $\{\omega_{-1},\ \omega_2\}$. When $\varepsilon = 0$, there are two twofold complex-conjugate multipliers. By (4.9), for the pair $\{\omega_1,\ \omega_{-2}\}$ we have

$$\begin{Vmatrix} \sigma_{11} & \sigma_{1,-2} \\ \sigma_{-21} & \sigma_{-2,-2} \end{Vmatrix} = \begin{Vmatrix} \frac{\mu\omega_1}{\vartheta_0^2} - i\frac{\omega_1^2}{2\vartheta_0}\varkappa_1 & \frac{1}{2\vartheta_0}\sqrt{\omega_1\omega_2}\,a_{21}\varphi^{(m)} \\ \frac{1}{2\vartheta_0}\sqrt{\omega_1\omega_2}\,a_{12}\varphi^{(m)} & \frac{\mu\omega_2}{\vartheta_0^2} + i\frac{\omega_2^2}{2\vartheta_0}\varkappa_2 \end{Vmatrix}.$$

Referring to (4.17), we have $\varphi^{(m)} = 0$ for $m \neq 0,\ \pm 1$. Therefore, if $\vartheta_0 = (\omega_1+\omega_2)/m$ $(m = 2,\ 3,\ \ldots)$ the first corrections to the characteristic exponents $i\nu_1,\ i\nu_{-2}$ have negative real parts as in case (b). Let $\vartheta_0 = \omega_1 + \omega_2$ $(m = 1)$. The coefficients (4.13) are

$$\alpha + i\beta = \frac{1}{2}\mu(\omega_2 - \omega_1) + \frac{1}{4}i(\omega_1^2\varkappa_1 + \omega_2^2\varkappa_2),$$

$$\gamma + i\delta = \mu^2\omega_1\omega_2 + \frac{1}{4}\omega_1^2\omega_2^2\vartheta_0^2\varkappa_1\varkappa_2 - \frac{1}{4}\omega_1\omega_2\vartheta_0^2a_{12}a_{21}\varphi^{(1)^2} + \frac{1}{2}i\mu\omega_1\omega_2(\omega_2\varkappa_2 - \omega_1\varkappa_1)\,\vartheta_0.$$

Since $\beta > 0$, the real part of at least one of the corrections $i\nu\varepsilon$ to the characteristic exponents (4.7) is positive if and only if condition (4.14) holds, i. e.,

$$\omega_1\omega_2(\omega_1 + \omega_1)^2\varkappa_1\varkappa_2\mu^2 < \frac{1}{4}\vartheta_0^2(\omega_1^2\varkappa_1 + \omega_2^2\varkappa_2)^2(a_{12}a_{21}\varphi^{(1)^2} - \omega_1\omega_2\varkappa_1\varkappa_2).$$

Substituting $\varphi^{(1)}={}^1/_2 b$ (see (4.17)), we see that this inequality holds for $\mu_1<\mu<\mu_2$, where

$$\mu_{1,2}=\mp\frac{1}{4}\,\frac{\omega_1^2\varkappa_1+\omega_2^2\varkappa_2}{\sqrt{\omega_1\omega_2\varkappa_1\varkappa_2}}\sqrt{b^2a_{12}a_{21}-4\omega_1\omega_2\varkappa_1\varkappa_2}. \qquad (4.19)$$

It is assumed here that the radicand is positive, i. e.,

$$\beta^2a_{12}a_{21}>4\omega_1\omega_2\varkappa_1\varkappa_2, \qquad (4.20)$$

and also that $\varkappa_1\neq 0$, $\varkappa_2\neq 0$. If $\varkappa_1=\varkappa_2=0$, we obtain instead of (4.19)

$$\mu_{1,2}=\mp\frac{1}{2}\beta\sqrt{\omega_1\omega_2a_{12}a_{21}}. \qquad (4.21)$$

If

$$\beta^2a_{12}a_{21}<4\omega_1\omega_2\varkappa_1\varkappa_2, \qquad (4.22)$$

the numbers μ_1, μ_2 are complex and $\mathrm{Re}(i\nu_1)<0$, $\mathrm{Re}(i\nu_{-2})<0$ for any $-\infty<\mu<+\infty$. Since the multipliers of equation (4.16) must be complex conjugates, it follows that $\nu_{-1}=\bar{\nu}_1$, $\nu_2=\bar{\nu}_{-2}$ (as is readily seen by similar calculations). Thus, for the pair $\{\omega_{-1},\ \omega_2\}$ we obtain the same relations for $\mu_{1,2}$.

To sum up: if (4.22) is satisfied, then $\vartheta_0=\omega_1+\omega_2$ is not a critical frequency. Since none of the other frequences are critical, it follows that then equation (4.16) is asymptotically stable for any ϑ and sufficiently small $\varepsilon>0$. *

If inequality (4.20) holds, there is a unique critical frequency $\vartheta_0=\omega_1+\omega_2$. The dynamic instability domain is defined by

$$\vartheta_0+\varepsilon\mu_1+O(\varepsilon^{1+\delta})<\vartheta<\vartheta_0+\varepsilon\mu_2+O(\varepsilon^{1+\delta}) \qquad \left(\delta\geqslant\frac{1}{2}\right), \qquad (4.23)$$

where $\mu_{1,2}$ are defined by (4.19) $(\varkappa_1\neq 0,\ \varkappa_2\neq 0)$ or (4.21) $(\varkappa_1=\varkappa_2=0)$.

With the above results established, we can now see how the instability domain "expands" when friction is introduced. Indeed, formula (4.19) involves the factor

$$\frac{\omega_1^2\varkappa_1+\omega_2^2\varkappa_2}{\sqrt{\varkappa_1\varkappa_2}},$$

which may be made arbitrarily large for sufficiently small $\varkappa_1$, $\varkappa_2$ (for example, let $\varkappa_1=\varkappa_2^2$ be sufficiently small). Thus, for suitably chosen $\varkappa_1\neq 0$ and $\varkappa_2\neq 0$ the width of the domain (4.23) may be made arbitrarily large. In particular, for certain $\varkappa_1>0$, $\varkappa_2>0$ the domain (4.23) will certainly

* Recall that in addition we are assuming the frequencies ω_1, ω_2 to satisfy the above incongruence conditions. If the frequencies are congruent, an additional analysis, also based on the formulas of §4.2, is necessary; we shall not do this here.

contain (in the strip $0<\varepsilon\leqslant\varepsilon_0$, where $\varepsilon_0>0$ is small enough) the domain (4.23) with $\varkappa_1=\varkappa_2=0$. This will hold provided $\varkappa_1$ and $\varkappa_2$ satisfy the inequality

$$\mu_2(\varkappa_1, \varkappa_2)>\mu_2(0, 0),$$

where $\mu_2(\varkappa_1, \varkappa_2)$ is defined by (4.19) and $\mu_2(0, 0)$ by (4.21). Consequently, there are points on the (ε, ϑ)-plane arbitrarily close to $(0, \vartheta_0)$ for which equation (4.16) is stable (though not asymptotically) for $\varkappa_1=\varkappa_2=0$ (no friction) and exponentially unstable for certain $\varkappa_1>0$, $\varkappa_2>0$ (friction present).

Similarly, one can find points (ε, ϑ) which are in an asymptotic stability domain for certain $\varkappa_1^0$, $\varkappa_2^0$ but in an exponential instability domain for some $\varkappa_1>\varkappa_1^0$, $\varkappa_2>\varkappa_2^0$. The numbers $\varkappa_j$, $\varkappa_j^0$ must satisfy the inequality

$$\mu_1(\varkappa_1, \varkappa_2)>\mu_1(\varkappa_1^0, \varkappa_2^0).$$

Thus, with regard to the corresponding equation (4.16), increase of friction converts an asymptotically stable system into an unstable one.

We emphasize that this situation can occur (for this equation) *only at combination resonance* $\vartheta_0=\omega_1+\omega_2$.

Chapter VI

PARAMETRIC RESONANCE IN MECHANICAL AND PHYSICAL SYSTEMS

In this chapter the results established in Chapter V will be applied to the investigation of real mechanical (§§2—4) and physical (§§5, 6) systems. The main problem treated in the examples will be techniques for constructing first-approximation instability domains in the parameter space. (In §§5, 6, studying waveguides, these correspond to the stop-bands.) The results of Chapter V also allow a more detailed analysis in all cases, among other things to find approximate solution formulas.

Most of the problems considered in this chapter lead to partial differential equations, a rigorous mathematical treatment of which is beyond the scope of the book. Partial differential equations will be studied directly only in a very simple case, in §1.2, and as the reader will see the exposition will be rather cumbersome, making more demands than usual on the reader. In all other cases the original partial differential equations will be reduced by traditional methods (usually the Bubnov-Galerkin method /20, 47/) to a system of n ordinary differential equations, which will then be investigated with complete rigor. As is usual in the applied literature, we shall make no attempts in §§ 1—6 to justify the limiting procedure as $n \to \infty$ (i. e., the passage back to the original problem). In some cases the rigorous mathematical justification is quite simple (though, of course, the sense of the convergence must be made clear; one is usually dealing with convergence of domains of α-exponential growth for small $\alpha > 0$ in the parameter space of the system). Frequently, however, the rigorous arguments present substantial difficulties.

Many of the mathematical problems that arise here have been solved only recently; in certain cases of practical importance they are still unsolved. The main results known to date concerning justification of the Bubnov-Galerkin method to parametric resonance problems (justifying, in particular, the results of §§1, 3) are presented in §7. They are due to Derguzov and Fomin.

§1. VIBRATIONS OF SUSPENSION BRIDGES

The collapse of the Tacoma Narrows suspension bridge (Washington, DC, 1940) was one of the worst catastrophes in bridge-building history. In a wind blowing at about 60 km/hr, the entire central span of the bridge between the towers, a length of 853 m, collapsed. There is a voluminous literature on the reasons for the disaster (see /81/ for a bibliography, and also /28a, b, 37/, /13a, 18/). Some authors, such as Rocard (1954) and Steinman (1947) blamed nonlinear flutter; this viewpoint was developed most consistently and fully by Rocard in his interesting book /81/. His method, however, is not rigorous, and the logical inferences at some points are doubtful (reduction to a linear system, utilization of data on the true pattern of the collapse, etc.). Boloton /13a/ pointed out that recent wind-tunnel tests of models of stiffening girders give good grounds for the conjecture that the intense vibrations of the bridge recall the phenomenon of parametric resonance. Bolotin also derived approximate formulas for the dynamic instability domains of principal resonance. The first treatment of the problem of parametric vibrations of the stiffening girders of a suspension bridge was due to Gol'denblat /28a, b/.

Below we shall present a stability analysis of suspension bridges, adopting two alternative assumptions concerning the wind load. As a first approach, we assume that the wind load is steady, starting out from the physical premises and equations of Vlasov /18/. Rigorous mathematical analysis shows, however, that according to the steady wind-load assumption the Tacoma Narrows bridge should have been stable in a wind-speed range of $v < 130$ km/hr. Thus this hypothesis gives no satisfactory explanation of the collapse of the bridge. The qualitative pattern of the bridge vibrations is also inconsistent with this hypothesis, since it is known that, though it had withstood considerable wind speeds, the bridge experienced strong vibrations at certain wind speeds beginning with 5 — 6.5 km/hr. This pattern is typical of parametric resonance.

The second approach, which utilizes the parametric-resonance theory of Chapter V, is also based on Vlasov's differential equations /18/ (see also /13a/, p. 479), but now the wind load is assumed to be periodic. This periodicity is due to the phenomenon, familiar in aeromechanics, of periodic shedding of alternating eddies (with frequency ϑ) from the sharp edges of the streamlined cross-section of the bridge (Kármán vortex streets*). The Bubnov-Galerkin method /20, 47/ reduces the problem to investigating the stability of a canonical linear system of ordinary differential equations with periodic coefficients. The results are in good qualitative and quantitative agreement with the actual vibrations of the bridge just before the disaster (wind speed 60.5 km/hr, 8 nodes along the principal span; see /81/). The calculation shows that the most dangerous phenomenon for the stability of suspension bridges is combination parametric resonance.

1.1. Differential equations of flexural-torsional vibrations of suspension bridges

Let η be the vertical deflection of the bridge axis, Θ the angle of torsion of the cross-section (Figure 28). The differential equations of flexural-torsional vibrations of a suspension bridge, with boundary conditions, are /18/

* It is known from aeromechanics that there are periodic components acting both perpendicular to the flow direction /28/ and along the flow. The origin of the periodic forces is immaterial for the theory of parametric resonance; a similar effect is caused by pulsations of a real wind flow.

$$\left.\begin{aligned}
&\rho F\left(\frac{\partial^2\eta}{\partial t^2}-r_1^2\frac{\partial^4\eta}{\partial z^2\partial t^2}\right)+EFr_1^2\frac{\partial^4\eta}{\partial z^4}-2H\frac{\partial^2\eta}{\partial z^2}+\frac{\partial^2(M_y\Theta)}{\partial z^2}=\overline{q}_y,\\
&\rho\left(Fr^2\frac{\partial^2\Theta}{\partial t^2}-J_\omega\frac{\partial^4\Theta}{\partial z^2\partial t^2}\right)+M_y\frac{\partial^2\eta}{\partial z^2}+EJ_\omega\frac{\partial^4\Theta}{\partial z^4}-\left(GJ_d+\frac{1}{2}Hb^2\right)\frac{\partial^2\Theta}{\partial z^2}=\overline{m},\\
&\Theta(0,\ t)=\Theta(l,\ t)=\eta(0,\ t)=\eta(l,\ t)=0,\\
&\frac{\partial^2\Theta(0,\ t)}{\partial z^2}=\frac{\partial^2\Theta(l,\ t)}{\partial z^2}=\frac{\partial^2\eta(0,\ t)}{\partial z^2}=\frac{\partial^2\eta(l,\ t)}{\partial z^2}=0.
\end{aligned}\right\}\quad(1.1)$$

The meanings of the parameters figuring in the equations are as follows. E is Young's modulus, G the shear modulus, ρ the density of the material of the stiffening girder, F the cross-sectional area, $J_x=Fr_1^2$ the moment of inertia of the cross-section about the x-axis, $J_p=Fr^2$ the polar moment of inertia of the cross-section, J_d the moment of inertia for pure torsion, J_ω a certain geometric parameter of the cross-section (see /28b/), b the width of the bridge between the cable supports, l the length of the bridge, H the tension in each cable, q the uniformly distributed horizontal wind load, $M_y=\frac{1}{2}qz(l-z)$ the bending moment, $\overline{m}=\frac{1}{2}bq\Theta$ and $\overline{q}_y$ the torque and supporting force* produced by the wind per unit length of the bridge. The load q is assumed to be

$$q=0.9h\rho_a v^2,\qquad(1.2)$$

where v is the speed of the oncoming flow, ρ_a the density of air, h the height of the stiffening girder.

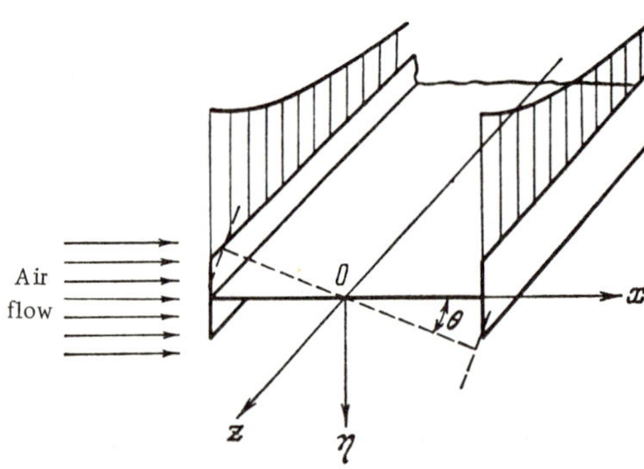

FIGURE 28.

Introducing dimensionless variables

$$u_1=\frac{\eta}{l},\quad u_2=\frac{r}{l}\Theta,\quad \zeta=\frac{z}{l},$$

* Henceforth, as in /28b/, the supporting force will be neglected.

we write system (1.1) and the boundary conditions as

$$\left.\begin{aligned}\frac{\partial^2 u_1}{\partial t^2}-\alpha_1\frac{\partial^4 u_1}{\partial\zeta^2\partial t^2}+\frac{g}{l}\left(\beta_1\frac{\partial^4 u_1}{\partial\zeta^4}-\gamma_1\frac{\partial^2 u_1}{\partial\zeta^2}+\Lambda\frac{\partial^2(f_0u_2)}{\partial\zeta^2}\right)=0,\\ \frac{\partial^2 u_2}{\partial t^2}-\alpha_2\frac{\partial^4 u_2}{\partial\zeta^2\partial t^2}+\frac{g}{l}\left(\Lambda f_0\frac{\partial^2 u_1}{\partial\zeta^2}+\beta_2\frac{\partial^4 u_2}{\partial\zeta^4}-\gamma_2\frac{\partial^2 u_2}{\partial\zeta^2}-\Lambda\delta u_2\right)=0,\end{aligned}\right\}\tag{1.3}$$

$$\left.\begin{aligned}&u_j(0,t)=u_j(1,t)=0,\\ &\frac{\partial^2 u_j(0,t)}{\partial\zeta^2}=\frac{\partial^2 u_j(1,t)}{\partial\zeta^2}=0\qquad(j=1,2).\end{aligned}\right\}\tag{1.4}$$

Here

$$\left.\begin{aligned}&\Lambda=\frac{ql}{pr},\quad \alpha_1=\frac{r_1^2}{l^2},\quad \beta_1=\frac{EFr_1^2}{pl^3},\quad \gamma_1=\frac{2H}{pl},\quad \alpha_2=\frac{I_\omega}{Fr^2l^2},\\ &\beta_2=\frac{EJ_\omega}{pr^2l^3},\quad \gamma_2=\frac{GJ_d+\frac{1}{2}Hb^2}{pr^2l},\quad \delta=\frac{b}{2r},\quad f_0=\frac{1}{2}\zeta(1-\zeta),\end{aligned}\right\}\tag{1.5}$$

p is the weight per unit length of the bridge, g the acceleration of gravity.

1.2. Stability analysis assuming steady wind load

Let $\mathbf{y}$ denote the column vector $\begin{pmatrix}u_1\\u_2\end{pmatrix}$ and $C^4[0,1]$ the set of vector-functions $\mathbf{y}(\zeta)$ defined and four times continuously differentiable on the interval $[0,1]$. Let $\mathbf{M}_0$, $\mathbf{C}_0$, $\mathbf{C}_1$ be the operators defined on vector-functions $\mathbf{y}(\zeta)\in C^4[0,1]$ satisfying conditions (1.4) by

$$\mathbf{M}_0\mathbf{y}=\begin{pmatrix}u_1-\alpha_1\frac{d^2u_1}{d\zeta^2}\\ u_2-\alpha_2\frac{d^2u_2}{d\zeta^2}\end{pmatrix},\quad \mathbf{C}_0\mathbf{y}=\frac{g}{l}\begin{pmatrix}\beta_1\frac{d^4u_1}{d\zeta^4}-\gamma_1\frac{d^2u_1}{d\zeta^2}\\ \beta_2\frac{d^4u_2}{d\zeta^4}-\gamma_2\frac{d^2u_2}{d\zeta^2}\end{pmatrix},\quad \mathbf{C}_1\mathbf{y}=\frac{g}{l}\begin{pmatrix}\frac{d^2(f_0u_2)}{d\zeta^2}\\ f_0\frac{d^2u_1}{d\zeta^2}-\delta u_2\end{pmatrix}.\tag{1.6}$$

Then system (1.3) becomes

$$\frac{d^2}{dt^2}(\mathbf{M}_0\mathbf{y})+\mathbf{C}\mathbf{y}=0\qquad(\mathbf{C}=\mathbf{C}_0+\Lambda\mathbf{C}_1);\tag{1.7}$$

it follows from (1.4), (1.6), that $\mathbf{M}_0$ and $\mathbf{C}$ are symmetric operators and $\mathbf{M}_0$ is positive definite. Using the general theory of selfadjoint differential operators /49/, one readily shows that system (1.7), with $\mathbf{M}_0$ and $\mathbf{C}$ defined as in (1.6), is stable if and only if all the eigenvalues of $\mathbf{C}$ are positive. This is certainly true if $\Lambda=0$ $(q=0)$, for then the eigenvalues $\rho_{1,2}^{(k)}$ and eigenvectors $u_{1,2}^{(k)}$ are defined by

$$\left.\begin{aligned}&\rho_j^{(k)}=\frac{g}{l}(\beta_jk^4\pi^4+\gamma_jk^2\pi^2),\\ &u_j^{(k)}=\sqrt{2}\sin k\pi\zeta\qquad(j=1,2;\ k=1,2,3,\dots).\end{aligned}\right\}\tag{1.8}$$

Let Λ_{cr} denote the smallest positive Λ for which $\mathbf{C}$ has a zero eigenvalue. The existence of Λ_{cr} follows from the general theory. It can be shown that system (1.7) is stable if $0 \leqslant \Lambda < \Lambda_{cr}$ and unstable if $\Lambda \geqslant \Lambda_{cr}$. This implies that system (1.7) is dynamically stable provided the equilibrium position $\mathbf{y}=\mathbf{0}$ is statically stable in Euler's sense (see also /13a/, p. 402).

It is difficult, however, to express Λ_{cr} as an elementary function of the system parameters. We shall therefore try to find a lower bound for Λ_{cr}. To this end, we consider the differential operator $\tilde{\mathbf{C}}$ defined on $\mathbf{y}(\zeta) \in C^2[0, 1]$ by

$$\tilde{\mathbf{C}}\mathbf{y} = \frac{g}{l}\begin{pmatrix} -\gamma_1 \dfrac{d^2u_1}{d\zeta^2} + \Lambda \dfrac{d^2(f_0u_2)}{d\zeta^2} \\ -\gamma_2 \dfrac{d^2u_2}{d\zeta^2} + \Lambda f_0 \dfrac{d^2u_1}{d\zeta^2} - \Lambda\delta u_2 \end{pmatrix}, \tag{1.9}$$

where

$$u_j(0) = u_j(1) = 0 \qquad (j = 1, 2). \tag{1.10}$$

It follows from (1.9), (1.10), that $\tilde{\mathbf{C}}$ is symmetric. The fundamental theorems of the theory of selfadjoint boundary-value problems (/49/, Chapter VII) will hold for the operator $\tilde{\mathbf{C}}$, provided the matrix of coefficients of the second derivatives is nonsingular for $\zeta \in [0, 1]$, i.e.,

$$\begin{vmatrix} -\gamma_1 & \Lambda f_0(\zeta) \\ \Lambda f_0(\zeta) & -\gamma_2 \end{vmatrix} = \gamma_1\gamma_2 - \Lambda^2 f_0^2(\zeta) \neq 0 \text{ for } \zeta \in [0, 1].$$

Hence, since $f_0(0) = 0$, we have $\gamma_1\gamma_2 > \Lambda^2 f_0^2(\zeta)$. We may therefore restrict attention to the interval

$$0 \leqslant \Lambda < \sqrt{\frac{\gamma_1\gamma_2}{\max(f_0^2)}} = 8\sqrt{\gamma_1\gamma_2}. \tag{1.11}$$

Using the explicit form of $\mathbf{C}$ and $\tilde{\mathbf{C}}$, one readily sees that on their common domain of definition $\mathbf{C} > \tilde{\mathbf{C}}$ (in the sense of inequality of the corresponding quadratic forms). Therefore, since $\tilde{\mathbf{C}}$ is positive definite, so is $\mathbf{C}$:

$$\Lambda_{cr} > \tilde{\Lambda}_{cr}, \tag{1.12}$$

where $\tilde{\Lambda}_{cr}$ is the smallest positive Λ for which $\tilde{\mathbf{C}}$ has a zero eigenvalue, i.e., for which the system

$$-\gamma_1 \frac{d^2u_1}{d\zeta^2} + \Lambda \frac{d^2(f_0u_2)}{d\zeta^2} = 0, \quad \Lambda f_0 \frac{d^2u_1}{d\zeta^2} - \gamma_2 \frac{d^2u_2}{d\zeta^2} - \Lambda\delta u_2 = 0 \tag{1.13}$$

has a nontrivial solution satisfying (1.10). Eliminating the function $u_1(\zeta)$, we get

$$\frac{d^2u_2}{d\zeta^2} = \frac{\Lambda^2}{\gamma_1\gamma_2} f_0 \frac{d^2(f_0u_2)}{d\zeta^2} - \frac{\Lambda\delta}{\gamma_2} u_2.$$

Hence

$$u_2(\zeta)=\int_0^1 G(\zeta,\ s)\left\{\frac{\Lambda^2}{\gamma_1\gamma_2}f_0(s)\frac{d^2}{ds^2}[f_0(s)u_2(s)]-\frac{\Lambda\delta}{\gamma_2}u_2(s)\right\}ds, \tag{1.14}$$

where $G(\zeta,\ s)$ is the Green's function of the operator $d^2/d\zeta^2$ for boundary conditions (1.10):

$$G(\zeta,\ s)=\begin{cases}-\zeta(1-s) & (\zeta<s),\\ -s(1-\zeta) & (\zeta>s).\end{cases}$$

Integrating by parts on the right and using the properties of $G(\zeta,\ s)$, we obtain an integral equation

$$u_2(\zeta)=\int_0^1 K(\zeta,\ s,\ \Lambda)\,u_2(s)\,ds, \tag{1.15}$$

$$K(\zeta,\ s,\ \Lambda)=\frac{1}{\gamma_1\gamma_2-\Lambda^2 f_0^2(\zeta)}\left\{\Lambda^2 f_0(s)\frac{\partial^2}{\partial s^2}[G(\zeta,\ s)f_0(s)]-\Lambda\delta\gamma_1 G(\zeta,\ s)\right\}. \tag{1.16}$$

By (1.11) and the properties of the Green's function, the kernel $K(\zeta,\ s,\ \Lambda)$ is piecewise-continuous in ζ and s. Denote

$$K(\Lambda)=\sqrt{\int_0^1\int_0^1 K^2(\zeta,\ s,\ \Lambda)\,d\zeta\,ds}. \tag{1.17}$$

It is known /46/ that if $K(\Lambda)<1$ the integral equation (1.15) has no nontrivial solution. Since direct calculation of $K(\Lambda)$ involves considerable labor, we proceed as follows. It is evident from (1.16) that

$$\left.\begin{aligned}&K(\zeta,\ s,\ \Lambda)=\frac{1}{\gamma_1\gamma_2-\Lambda^2 f_0^2(\zeta)}[K_1(\zeta,\ s,\ \Lambda)+K_2(\zeta,\ s,\ \Lambda)],\\ &K_1(\zeta,\ s,\ \Lambda)=\Lambda^2 f_0(s)\frac{\partial^2}{\partial s^2}[G(\zeta,\ s)f_0(s)],\ K_2(\zeta,\ s,\ \Lambda)=-\Lambda\delta\gamma_1 G(\zeta,\ s).\end{aligned}\right\} \tag{1.18}$$

Using (1.11) and Minkowski's integral inequality /46/, we get

$$K(\Lambda)<\hat{K}(\Lambda)=\frac{1}{\gamma_1\gamma_2-\frac{1}{64}\Lambda^2}[K_1(\Lambda)+K_2(\Lambda)], \tag{1.19}$$

where $K_1(\Lambda)$, $K_2(\Lambda)$ are defined for $K_1(\zeta,\ s,\ \Lambda)$, $K_2(\zeta,\ s,\ \Lambda)$ in the same way as $K(\Lambda)$. The condition $K(\Lambda)<1$ will hold if $\hat{K}(\Lambda)<1$. Fairly lengthy but simple calculations of $K_1(\Lambda)$, $K_2(\Lambda)$ yield condition (1.19) in the following equivalent form:

$$F(\Lambda)=\Lambda^2+3.54\,\delta\gamma_1\Lambda-33.6\,\gamma_1\gamma_2<0. \tag{1.20}$$

Letting Λ_0 denote the positive root of the equation $F(\Lambda)=0$, we thus obtain

$$\tilde{\Lambda}_{cr}>\Lambda_0=\frac{5.8\sqrt{\gamma_1\gamma_2}}{\sqrt{1+\psi^2}+\psi},\quad \psi=0.305\,\delta\sqrt{\frac{\gamma_1}{\gamma_2}}. \tag{1.21}$$

Note that Λ_0 satisfies condition (1.11). Using (1.12), we finally obtain (going back to the original variables)

$$q_{cr} > \frac{Hb}{l^2}\left(\sqrt{36.8+67.2\frac{GJ_d}{Hb^2}}-1.77\right)$$

or, by (1.2),

$$v_{cr}^2 > \frac{Hb}{0.9\rho_a hl^2}\left(\sqrt{36.8+67.2\frac{GJ_d}{Hb^2}}-1.77\right). \tag{1.22}$$

We now apply these estimates to the actual data of the Tacoma Narrows bridge. By /28b/, the parameters figuring in (1.22) are

$$H=5.65\cdot 10^6 \text{ kg}, \quad l=853 \text{ m}, \quad h=2.44 \text{ m}, \; b=11.9 \text{ m}, \; J_d=0.$$

Then

$$q_{cr} > 390 \text{ kg/m}, \; v_{cr} > 130 \text{ km/hr}. \tag{1.23}$$

In other words, the steady wind-load assumption implies that the Tacoma Narrows bridge should have been stable over the wind-speed range $v < 130$ km/hr, whereas the actual wind speed just before the collapse was about 60 km/hr. Moreover, the conclusion that a suspension bridge is stable at $v < v_{cr}$ and unstable at $v \geqslant v_{cr}$ is not in agreement with many observations. For example, before its collapse the Tacoma Narrows bridge was capable of withstanding wind speeds exceeding 60 km/hr, but on the other hand violent vibrations were observed at far lower speeds /37/. We must therefore reject the steady wind-load assumption.

1.3. Dynamic stability analysis according to parametric resonance theory

When a flow of air blows past various objects which are not fully streamlined, Kármán vortex streets are shed in their wake. This phenomenon has been studied both theoretically and experimentally in aeromechanics. The vortex shedding frequency at each of the sharp edges is given approximately by /81/

$$\vartheta = kv, \quad k=\frac{2\pi S}{h}, \tag{1.24}$$

where h is the characteristic transverse dimension of the streamline profile and S is the Strouhal number. It may be assumed in our case /81/ that

$$S \approx 0.16. \tag{1.25}$$

It is known /81/ that eddies are shed in alternating fashion from both edges, doubling the frequency of the forces acting on the bridge. It is therefore natural to assume that the wind load q due to the eddies is given by (see also /13a/)

$$q=q_0+q_1\cos 2\vartheta t. \tag{1.26}$$

The bridge vibrations are described by equations (1.3) with

$$\Lambda=\Lambda_0+\Lambda_1\cos 2\vartheta t,$$

where

$$\Lambda_0=\frac{q_0 l}{pr},\quad \Lambda_1=\frac{q_1 l}{pr}. \tag{1.27}$$

The other notation is the same as in (1.5). Using the Bubnov-Galerkin method, /20, 47/, we reduce system (1.3) to a system of ordinary differential equations. To do this, we set

$$\mathbf{y}(\zeta,\ t)=\sum_{k=1}^{n}u_k(\zeta)\begin{pmatrix}y_1^{(k)}(t)\\ y_2^{(k)}(t)\end{pmatrix},\quad u_k(\zeta)=\sqrt{2}\sin k\pi\zeta.$$

Then the coefficients $y_1^{(k)}$, $y_2^{(k)}$ are determined by a system of $2n$ differential equations. Introducing the $2n$-vector

$$\mathbf{y}=\begin{Vmatrix}y_1^{(1)}\\ \cdots\\ y_1^{(n)}\\ y_2^{(1)}\\ \cdots\\ y_2^{(n)}\end{Vmatrix},$$

we can write the system as

$$\mathbf{M}\frac{d^2\mathbf{y}}{dt^2}+(\mathbf{P}_0+\Lambda\mathbf{P}_1)\mathbf{y}=0, \tag{1.28}$$

where

$$\mathbf{M}=\begin{Vmatrix}m_1 & & 0\\ & \ddots & \\ 0 & & m_{2n}\end{Vmatrix},\quad \mathbf{P}_0=\begin{Vmatrix}p_{01} & & 0\\ & \ddots & \\ 0 & & p_{02n}\end{Vmatrix},\quad \mathbf{P}_1=\begin{pmatrix}\mathbf{0} & \mathbf{A}\\ \mathbf{A}^\tau & \mathbf{B}\end{pmatrix}=\|p_{1jh}\|_{j,h=1}^{2n} \tag{1.28a}$$

are symmetric $2n\times 2n$ matrices. The elements of the matrices $\mathbf{M}, \mathbf{P}_0, \mathbf{A}=\|a_{\mu\nu}\|_{\mu,\nu=1}^{n}$, $\mathbf{B}=\|b_{\mu\nu}\|_{\mu,\nu=1}^{n}$ are

$$\left.\begin{aligned}
&m_\nu = 1+\alpha_\nu \nu^2\pi^2, \quad p_{0\nu} = \frac{g}{l}(\beta_\nu \nu^4\pi^4 + \gamma_\nu \nu^2\pi^2) \qquad (\nu = 1, \ldots, 2n),\\
&\alpha_\nu = \alpha_1, \quad \beta_\nu = \beta_1, \quad \gamma_\nu = \gamma_1 \quad \text{for} \quad \nu \leqslant n,\\
&\alpha_\nu = \alpha_2, \quad \beta_\nu = \beta_2, \quad \gamma_\nu = \gamma_2 \quad \text{for} \quad \nu > n,\\
&a_{\mu\nu} = \begin{cases} -\frac{g}{4l}\left(1+\frac{1}{3}\nu^2\pi^2\right) & (\mu = \nu),\\ \frac{2g\nu\mu^3\,|(-1)^\mu + (-1)^\nu|}{l(\nu^2-\mu^2)^2} & (\mu \neq \nu),\end{cases}\\
&b_{\mu\nu} = -\frac{g\delta}{l}\delta_{\mu\nu},
\end{aligned}\right\} \tag{1.29}$$

$\delta_{\mu\nu}$ is the Kronecker symbol (μ, $\nu = 1, \ldots, 2n$). The quantities $\max f_0\Lambda/\gamma_1$, $\max f_0\Lambda/\gamma_2$ in equations (1.3) are of the order of $\varepsilon = 0.9\rho_a h v^2 l^2/16rH \approx 0.5$ at $v = 100$ km/hr, so that it is legitimate to use the method of small parameters We set

$$\varepsilon = \frac{0.9\rho_a h v^2 l^2}{16rH}, \quad \Lambda_0 = \varepsilon d_0, \quad \Lambda_1 = \varepsilon d_1 \qquad (v = 100 \text{ km/hr}). \tag{1.30}$$

Then system (1.28) becomes

$$\left.\begin{aligned}
&\mathbf{M}\frac{d^2\mathbf{y}}{dt^2} + [\mathbf{P}_0 + \varepsilon\mathbf{P}_1(\vartheta t)]\,\mathbf{y} = 0,\\
&\mathbf{P}_1(\vartheta t) = \varphi(\vartheta t)\,\mathbf{P}_1, \quad \varphi(\vartheta t) = \alpha_0 + \alpha_1\cos 2\vartheta t.
\end{aligned}\right\} \tag{1.31}$$

Clearly, at $\varepsilon = 0$ ($q = 0$) the trivial solution of system (1.31) is stable (all solutions bounded as $t \to \infty$). However, there are critical frequencies ϑ_0 such that for any $\delta_1 > 0$, $\delta_2 > 0$ the domain $|\varepsilon| < \delta_1$, $|\vartheta - \vartheta_0| < \delta_2$ contains values ε, ϑ for which system (1.31) has solutions which are unbounded as $t \to \infty$. The critical frequencies ϑ_0 are defined by Krein's formula (V. 1.7)

$$\vartheta_0 = \frac{\omega_j + \omega_h}{N} \qquad (j,\ h = 1, \ldots, 2n;\ N = 1, 2, \ldots), \tag{1.32}$$

where ω_j are the natural frequencies of the "unperturbed" system

$$\mathbf{M}\frac{d^2\mathbf{y}}{dt^2} + \mathbf{P}_0\mathbf{y} = 0,$$

i. e., by (1.29),

$$\omega_j = \sqrt{p_{0j}/m_j} \qquad (j = 1, \ldots, 2n).$$

Thus ω_j are flexural vibration frequencies for $j \leqslant n$ and torsional vibration frequencies for $j > n$. We shall call the frequencies (1.32) ($j \neq h$) the *combination-resonance frequencies* and those with $j = h$ the *principal-resonance frequencies*. The dynamic instability domains in the (ε, ϑ)-plane attached at the points (0, ϑ_0) are, in the first approximation,

$$\vartheta_0 + \mu_1\varepsilon + \ldots < \vartheta < \vartheta_0 + \mu_2\varepsilon + \ldots. \tag{1.33}$$

The coefficients $\mu_{1,2}$ are given by (V. 1.29)

$$\mu_{1,2}=\frac{1}{N}(\chi_j+\chi_h\mp 2|\chi_{jh}^{(N)}|), \tag{1.34}$$

where the expressions for χ_j, χ_h, $\chi_{jh}^{(N)}$ are in this case quite simple, as the matrices **M** and $\mathbf{P}_0$ are diagonal (see § V. 1.11):

$$\chi_j=\frac{\pi_{jj}^{(0)}}{2m_j\omega_j},\quad \chi_h=\frac{\pi_{hh}^{(0)}}{2m_h\omega_h},\quad \chi_{jh}^{(N)}=\frac{\pi_{jh}^{(N)}}{2\sqrt{m_jm_h\omega_j\omega_h}}. \tag{1.35}$$

Here $\|\pi_{jh}^{(l)}\|$ are the Fourier coefficients of the matrix

$$\mathbf{P}_1(\vartheta t)\sim\sum_{l=-\infty}^{\infty}\|\pi_{jh}^{(l)}\|e^{il\vartheta t}, \pi_{jh}^{(l)}=p_{1jh}\varphi^{(l)}.$$

Using (1.31), we deduce from (1.34)

$$\mu_{1,2}=\frac{1}{2N}\left(\frac{\alpha_0p_{1jj}}{m_j\omega_j}+\frac{\alpha_0p_{1hh}}{m_h\omega_h}\mp\frac{2|p_{1jh}||\varphi^{(N)}|}{\sqrt{m_jm_h\omega_j\omega_h}}\right), \tag{1.36}$$

where $\varphi^{(N)}$ is the N-th Fourier coefficient of the function $\varphi(\vartheta t)$. Following the terminology of § V. 1.5, we shall say that a dynamic instability domain is *wide* if the tangents to its boundaries at $(0, \vartheta_0)$ are different, i. e., $\mu_1\neq\mu_2$. It is immediate from (1.31), (1.36) that wide instability domains may be attached only at the points $\left(0, \frac{\omega_j+\omega_h}{2}\right)$, since $\varphi^{(N)}=0$ for $N\neq 0, \pm 2$, and in fact there are even fewer wide domains, as it follows from (1.28a) that $p_{1jh}\neq 0$ $(\mu_1\neq\mu_2)$ if either $j=h>n$ or $\min(j, h)\leqslant n$, $\max(j, h)>n$ and these two numbers are of the same parity. Let $\omega_1^{(f)}, \ldots, \omega_n^{(f)}$ denote the frequencies of flexural vibrations and $\omega_1^{(t)}, \ldots, \omega_n^{(t)}$ those of torsional vibrations. Then the above statements may be rephrased as follows: wide instability domains in the (ε, ϑ)-plane are attached only at the following points: $(0, \omega_j^{(t)})$ $(j=1, \ldots, n)$ — principal-resonance domains, $\left(0, \frac{\omega_j^{(f)}+\omega_h^{(t)}}{2}\right)$— combination-resonance domains. In the latter case, $j, h\leqslant n$ are of the same parity. The frequencies $\frac{1}{N}(\omega_j^{(f)}+\omega_h^{(t)})$, $\frac{1}{N}(\omega_j^{(f)}+\omega_h^{(f)})$, $\frac{1}{N}(\omega_j^{(t)}+\omega_h^{(t)})$, where $N=1, 2, \ldots$, and $j\leqslant n$, $h\leqslant n$ are of like parity, are in general not critical This is apparent from the fact that system (1.31) separates into two independent systems, one describing flexural-torsional vibrations of even harmonics, the other for odd harmonics. In the wide principal-resonance domains $(\vartheta\approx\omega_j^{(t)})$ one obtains strong torsional vibrations at the j-th harmonic, and in the combination-resonance domains $\left(\vartheta\approx\frac{\omega_j^{(f)}+\omega_h^{(t)}}{2}\right.$, j and h of the same parity) flexural-torsional vibrations at the j-th and h-th harmonics, respectively. Following /28b/, we shall not consider the first flexural nodeless mode, since the only reason for its appearance is that equations (1.3) neglect the almost zero elongation of the cables. Higher modes lead to an insignificant elongation of the cables, mainly modifying the shape of their sag.

The quantity

$$|\chi_{jh}^{(2)}| = \frac{|p_{1jh}||\alpha_1|}{4\sqrt{m_j m_h \omega_j \omega_h}} \tag{1.37}$$

is a good measure of the "danger" of the wide instability domain. Namely, for fixed ε, if (ε, ϑ) is a point within the instability domain attached at the point $\left(0, \frac{\omega_j+\omega_h}{2}\right)$, the corresponding system (1.31) will have unbounded solutions growing exponentially (as $e^{\alpha t}$). The exponent α is a maximum at some $\vartheta(\varepsilon)$, and (§ V. 1.5)

$$\alpha_{\max} = |\chi_{jh}^{(2)}|\varepsilon + O(\varepsilon^2).$$

On the other hand, it follows from (1.34) that the larger the coefficient $|\chi_{jh}^{(2)}|$ the larger (for fixed $\chi_j+\chi_h$) will be the angle between the tangents to the boundaries of the instability domain at $\left(0, \frac{\omega_j+\omega_h}{2}\right)$. In our case,

$$\varepsilon = \varepsilon_0 = \frac{0.9\rho_a h v_0^2 l^2}{16rH}, \quad \vartheta = kv. \tag{1.38}$$

The instability domains in the (ε, ϑ)-plane cut out instability intervals on the parabola (1.38) $(0 < v < 100$ km/hr). Correspondingly, we have instability intervals on the v-axis, located near the points

$$v_{jh} = \frac{\omega_j+\omega_h}{2k}. \tag{1.39}$$

The intervals cut out by the instability domains may be classified into more and less dangerous, depending on the magnitude of $|\chi_{jh}^{(2)}|$, which by (1.27), (1.30), (1.37) is proportional to χ_{jh} as defined by

$$\chi_{jh} = \frac{|p_{1jh}||q_1(v_{jh})|}{\sqrt{m_j m_h \omega_j \omega_h}}. \tag{1.40}$$

The dependence of the amplitude $q_1(v)$ of resistance vibrations on the speed v may be determined experimentally, say in wind-tunnel tests. Assuming a quadratic function,

$$q_1(v) = \text{const}\cdot v^2,$$

we see that the danger measure χ_{jh} of the critical speed $v_{jh} = (\omega_j+\omega_h)/2k$ is proportional to

$$\Psi_{jh} = \frac{|p_{1jh}|(\omega_j+\omega_h)^2}{\sqrt{m_j m_h \omega_j \omega_h}}. \tag{1.41}$$

We shall call Ψ_{jh} the *relative intensity of amplitude increase* of the mode in question. The larger Ψ_{jh}, the larger will be the corresponding instability interval on the v-axis near v_{jh}, and the faster the increase in the amplitude of the vibrations.

Using the actual data of the Tacoma Narrows bridge /28b/, we construct two diagrams which will clarify the subsequent reasoning. Figure 29 is a table of the coefficients Ψ_{jh} for combination resonance. The figure at the intersection of the j-th row and the h-th column is the relative intensity of amplitude increase for vibrations with j flexural nodes and h torsional nodes, at

$$v \approx v_{jh} = \frac{\omega_j^{(f)} + \omega_h^{(t)}}{2k}.$$

Formulas for low harmonics for the Tacoma data are

$$\omega_j^{(f)} = 0.54j, \quad \omega_h^{(t)} = 0.78h.$$

Figure 29 shows the family of straight lines

$$0.54j + 0.78h = 2kv, \tag{1.42}$$

for $v = 10, 20, \ldots, 100$ km/hr.

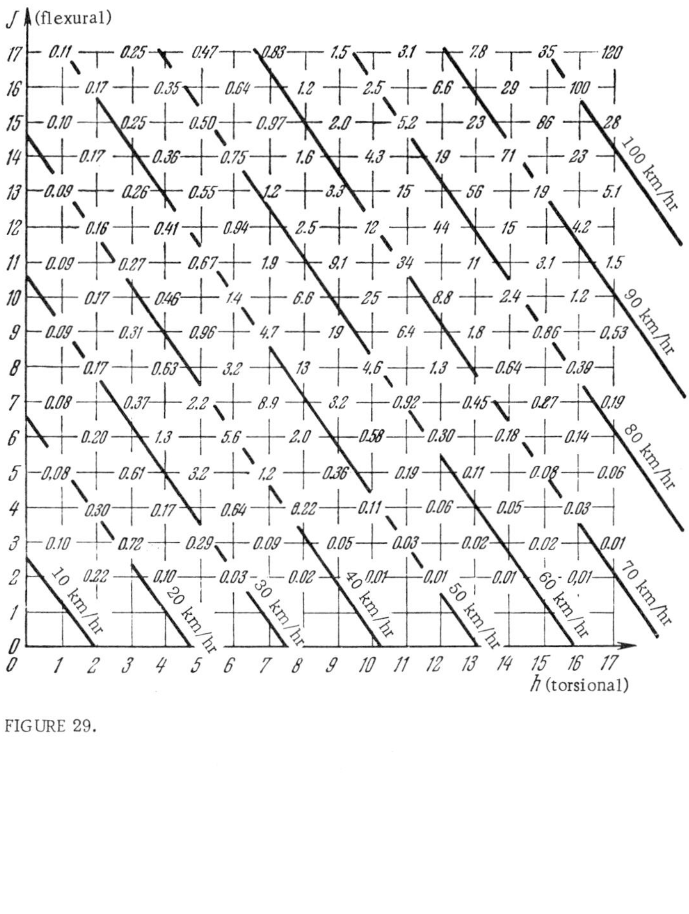

FIGURE 29.

If one of the points (j, h) of this integral lattice is near the straight line (1.42) for some v_0, this means that strong flexural-torsional vibrations at the corresponding harmonics may set in at a wind speed of v_0 km/hr. It is quite clear from the relative positions of the lines and the intensity figures at the points of the lattice that for fixed v the most dangerous effect is that of vibrations with the same number of flexural and torsional nodes. In this case the danger measure increases with increasing wind speed.

Along the horizontal axis in Figure 30 we have plotted the wind speed (and the proportional eddy shedding frequency). The vertical segments are equal in length to the relative intensity of the corresponding torsional vibrations at the principal resonance, numbered in accordance with the mode number 1, 2, . . . , 14. It is clear from Figure 30 that the danger of the principal resonance also increases with the wind speed.

Examining Figures 29 and 30 together, one sees that for fixed wind speeds the most dangerous phenomenon is combination resonance; parametric resonance sets in at quite low speeds: from 7 to 15 km/hr. As repeatedly mentioned, the wind speed just before the disaster was 60 km/hr. It is evident from Figure 29 that at this speed flexural-torsional vibrations with nine flexural and nine torsional nodes are most dangerous. This is in good agreement with the actual situation on the bridge just before it collapsed /81/.

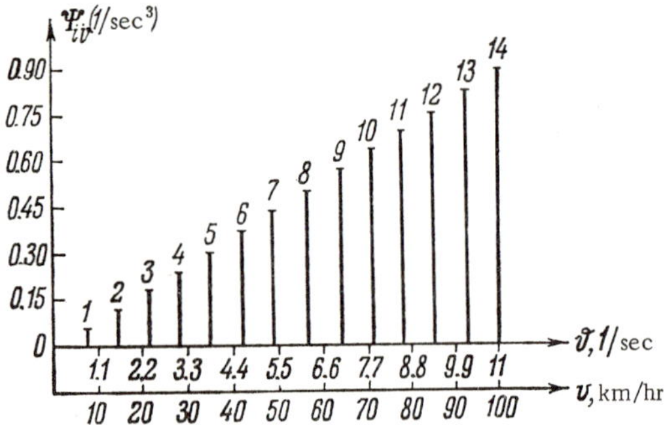

FIGURE 30.

§2. PROBLEMS IN WHICH THE COEFFICIENTS OF THE EQUATIONS ARE NONLINEAR IN $\gamma = 1/\vartheta$

2.1. Parametric resonance of rotating disks

According to /222/, small transverse vibrations of circular saws due to periodically varying stresses produced by the cutting forces are described by the following differential equation:

$$D\Delta\,\Delta w - h\left[\sigma_r\frac{\partial^2 w}{\partial r^2}+\sigma_\varphi\left(\frac{1}{r}\frac{\partial w}{\partial r}+\frac{1}{r^2}\frac{\partial^2 w}{\partial\varphi^2}\right)+\right.$$
$$\left.+2\tau_{r\varphi}\left(\frac{1}{r}\frac{\partial^2 w}{\partial r\,\partial\varphi}-\frac{1}{r^2}\frac{\partial w}{\partial\varphi}\right)+\frac{\partial V}{\partial r}\frac{\partial w}{\partial r}+\frac{1}{r^2}\frac{\partial V}{\partial\varphi}\frac{\partial w}{\partial\varphi}+\rho\frac{\partial^2 w}{\partial t^2}\right]=0, \quad (2.1)$$

where w is the deflection of the disk perpendicular to its plane, σ_r, σ_φ, $\tau_{r\varphi}$ the stresses in the middle plane of the disk (owing to the rotation of the disk, these stresses are periodic functions of time), $D=\frac{1}{12}Eh^3/(1-\mu^2)$ the cylindrical stiffness, h the thickness of the disk, E the modulus of elasticity of the first kind, μ Poisson's ratio, ρ the density of the material, V the potential of body forces; the other symbols are explained in Figure 31. The disk is rigidly attached at its inner circumference and completely free at its outer circumference.

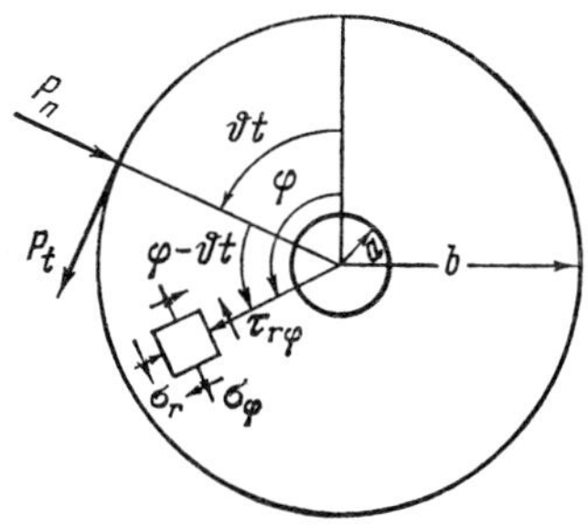

FIGURE 31.

The usual Bubnov-Galerkin reduction applied to (2.1) yields an infinite system of linear differential equations with periodic coefficients:

$$\frac{d^2\mathbf{y}}{d\tau^2}+\mathbf{C}(\tau,\vartheta,\varepsilon)\,\mathbf{y}=0, \quad (2.2)$$

where $\mathbf{y}$ is an infinite vector (whose terms are the coefficients of the Galerkin series), $\mathbf{C}(\tau,\vartheta,\varepsilon)$ an infinite symmetric matrix, $\varepsilon=bP_n/D$ a dimensionless scalar parameter, ϑ the angular velocity of the disk. Truncating the Galerkin series at the fourth term, we obtain system (2.2) with $\mathbf{C}(\tau,\vartheta,\varepsilon)=$ $=\mathbf{C}_0(\vartheta)+\varepsilon\mathbf{C}_1(\tau,\vartheta)$,

$$\mathbf{y}=\begin{pmatrix}\eta_1\\ \eta_2\\ \eta_3\\ \eta_4\end{pmatrix}, \quad \mathbf{C}_0(\vartheta)=\begin{Vmatrix} c_1(\vartheta)^2 & 0 & 0 & 0\\ 0 & c_2(\vartheta)^2 & 0 & 0\\ 0 & 0 & c_3(\vartheta)^2 & 0\\ 0 & 0 & 0 & c_4(\vartheta)^2\end{Vmatrix},$$

where

$$c_\varkappa(\vartheta)^2=\Omega_\varkappa(\vartheta)^2/\vartheta^2=p_\varkappa+\omega_\varkappa^2/\vartheta^2, \quad (2.3)$$

$\omega_\varkappa$ are the natural frequencies of transverse modes of the nonrotating disk, $\Omega^2_\varkappa(\vartheta)=\omega^2_\varkappa+p_\varkappa\vartheta^2$ the squared natural frequencies of transverse modes of the rotating disk ($p_\varkappa>0$; $\varkappa=1, 2, 3, 4$). We expand the perturbation matrix $\mathbf{C}_1(\tau, \vartheta)$ in Fourier series, confining attention to the approximation

$$\mathbf{C}_1(\tau, \vartheta)=\sum_{=-4}^{4}\mathbf{C}_1^{(l)}(\vartheta)e^{il\tau}.$$

The calculations give

$$\left.\begin{aligned}&\mathbf{C}_1^{(0)}=[\mathbf{C}_1(\tau, \vartheta)]_{\text{av}}=\frac{1}{\vartheta^2}\begin{Vmatrix}c_{11}^{(0)}&0&0&0\\0&c_{22}^{(0)}&c_{23}^{(0)}&0\\0&c_{32}^{(0)}&c_{33}^{(0)}&0\\0&0&0&c_{44}^{(0)}\end{Vmatrix},\quad \mathbf{C}_1^{(1)}=\frac{1}{\vartheta^2}\begin{Vmatrix}0&c_{12}^{(1)}&0&0\\c_{21}^{(1)}&0&0&0\\0&0&0&0\\0&0&0&0\end{Vmatrix},\\&\mathbf{C}_1^{(2)}=\frac{1}{\vartheta^2}\begin{Vmatrix}0&0&c_{13}^{(2)}&0\\0&c_{22}^{(2)}&0&0\\c_{31}^{(2)}&0&0&0\\0&0&0&0\end{Vmatrix},\quad \mathbf{C}_1^{(3)}=\frac{1}{\vartheta^2}\begin{Vmatrix}0&0&0&c_{14}^{(3)}\\0&0&0&0\\0&0&0&0\\c_{41}^{(3)}&0&0&0\end{Vmatrix},\quad \mathbf{C}_1^{(4)}=\frac{1}{\vartheta^2}\begin{Vmatrix}0&0&0&0\\0&0&0&0\\0&0&c_{33}^{(4)}&0\\0&0&0&0\end{Vmatrix}\end{aligned}\right\}. \tag{2.4}$$

Since $\mathbf{C}_0(\vartheta)$ is a diagonal matrix and $\mathbf{M}(\tau, \vartheta, \varepsilon)=\mathbf{I}_4$, it follows at once from Remark IV in § V. 2.3 that there may be wide dynamic instability domains on the (ε, ϑ)-plane only at points determined by the roots ϑ_0 of the following equations, involving the nonvanishing elements of $\mathbf{C}_1^{(m)}$ in (2.4):

$$\left.\begin{aligned}c_1(\vartheta)+c_2(\vartheta)&=1\text{—combination resonance,}\\c_1(\vartheta)+c_3(\vartheta)&=2\text{—combination resonance,}\\2c_2(\vartheta)&=2\text{— principal resonance,}\\c_1(\vartheta)+c_4(\vartheta)&=3\text{—combination resonance,}\\2c_3(\vartheta)&=4\text{—principal resonance.}\end{aligned}\right\} \tag{2.5}$$

(We are assuming that $c_j(\vartheta)>0$.) In view of (2.3), we obtain from (2.5) a formula for the critical frequency of rotation in the case of principal resonance:

$$\vartheta_0=\frac{2\omega_j}{\sqrt{m_j^2-4p_j}}\qquad (j=2, 3;\quad m_2=2,\quad m_3=4) \tag{2.6}$$

where

$$m_j>2\sqrt{p_j}\qquad (j=2, 3;\quad m_2=2,\quad m_3=4). \tag{2.7}$$

Similarly, we can derive from (2.5) formulas for the critical frequencies at combination resonance. It is immediately apparent from (2.3) that inclusion of the centrifugal forces of inertia raises the critical frequencies in comparison with the case of a nonrotating disk with the same m. This represents the physically evident fact that the stiffness of the disk increases upon rotation. For combination resonance, the analog of condition (2.7) is

$$m_j>\sqrt{p_1}+\sqrt{p_j}\qquad (j=2, 3, 4;\quad m_2=1,\quad m_3=2,\quad m_4=3). \tag{2.8}$$

For a circular saw with a typical ratio of inner to outer radii, $\delta = 0.2$, the estimates (2.8) are false, while the only valid case of (2.7) is $j = 3$. It follows that in the above approximation the circular saw has only one wide instability domain on the (ε, ϑ)-plane, corresponding to principal resonance, and attached at the point

$$\vartheta_0 = \frac{2\omega_3}{\sqrt{m_3^2 - 4p_3}} = \frac{\omega_3}{2\sqrt{1 - p_3/4}} = \frac{\omega_3}{1.2}$$

on the axis $\varepsilon = 0$ (for this disk, $p_3 = 2.56$).

The formulas of § V.2.3 yield the following instability domain in the first approximation:

$$\vartheta_0 + \mu_1\varepsilon + \ldots < \vartheta < \vartheta_0 + \mu_2\varepsilon + \ldots,$$

where

$$\mu_{1,2} = \frac{\vartheta_0}{2\omega_3^2}\left[c_{33}^0 \mp |c_{33}^{(4)}|\right] \qquad (\mu_1 < \mu_2). \tag{2.9}$$

The angle α between the tangents may be found (with sufficient accuracy for our case) from

$$\operatorname{tg} \alpha = \frac{|c_{33}^{(4)}|}{\omega_3\sqrt{4 - p_3}}. \tag{2.10}$$

This implies an interesting conclusion: inclusion of inertial forces will, on the one hand, increase the critical frequency but, on the other, increase the width of the dynamic instability domain.

Let us calculate the range of critical angular velocities for a circular saw with the following parameters: ratio of inner to outer radius $\delta = a/b = 0.2$, ratio of tangential component of cutting force to normal component $P_t/P_n = 2$, outer radius $b = 250$ mm, thickness of disk $h = 2$ mm, normal component of cutting force $P_n = 15$ kg. The calculations give

$$\mu_1 = -0.155\sqrt{\frac{D}{\rho h b^4}}, \quad \mu_2 = 0.0218\sqrt{\frac{D}{\rho h b^4}}, \quad \vartheta_0 = 5.82\sqrt{\frac{D}{\rho h b^4}}.$$

The critical angular velocities lie in the interval

$$(5.82 - 0.155\varepsilon + \ldots)\sqrt{\frac{D}{\rho h b^4}} < \vartheta < (5.82 + 0.0218\varepsilon + \ldots)\sqrt{\frac{D}{\rho h b^4}}.$$

For $\varepsilon = bP_n/D = 0.256$, we have 2674 rpm $< \vartheta <$ 2697 rpm. At these revolutions parametric resonance sets in, and so the saw may break.

2.2. Loss of stability of vertical vibrations in a plane simple pendulum

Consider a mathematical pendulum consisting of a mass m hanging on a weightless spring capable of vibrating both vertically (when the spring is deformed) and "transversely" in a fixed plane about a hinge O (see Figure 32). Let c denote the stiffness of the spring and l its normal length. Experiments show that for certain parameter values vertical vibrations are unstable and become transverse. In other words, vertical vibrations are damped and transverse ones build up. This is the phenomenon of energy transfer from vertical to transverse vibrations.* We shall study the phenomenon and find the parameter values at which it takes place.

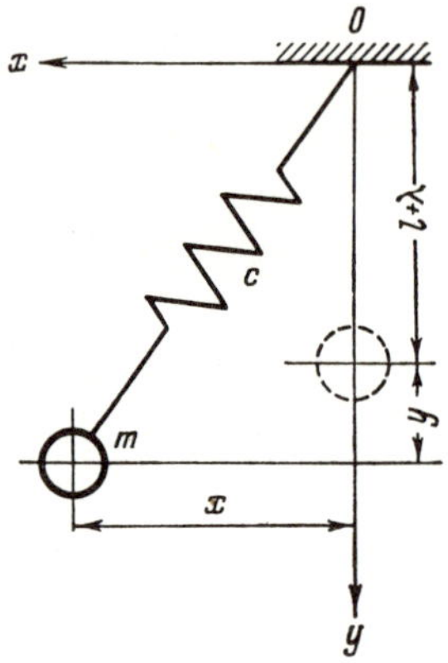

FIGURE 32.

As Lagrangian coordinates we take the abscissa x of the mass m and the ordinate y, measured along the vertical from the lower equilibrium position. The kinetic and potential energies are

$$T=\frac{1}{2}\,m\,(\dot{x}^2+\dot{y}^2),\quad \Pi=-\,mgy+\frac{1}{2}c\,[x^2+(l+\lambda+y)^2]-cl\sqrt{x^2+(l+\lambda+y)^2}.$$

The Lagrangian equations of motion are ($\lambda=mg/c$ is the static elongation of the spring)

$$m\ddot{x}=-\,cx+\frac{1}{\sqrt{x^2+(l+\lambda+y)^2}}\,clx,$$
$$m\ddot{y}=-\,c\,(l+y)+\frac{1}{\sqrt{x^2+(l+\lambda+y)^2}}\,cl\,(l+\lambda+y).$$

The solution (vertical vibrations) is

$$x\equiv 0,\quad y\equiv Y\cos\Omega t\qquad \left(\Omega=\sqrt{\frac{c}{m}}\right), \tag{2.11}$$

and this we treat as the unperturbed motion. We wish to analyze the stability of the vertical vibrations (2.11) with respect to the variables $x,\ y,\ \dot{x},\ \dot{y}$.

* The reverse phenomenon is also possible, but it will not be studied here.

Let the perturbed motion be

$$x=0+\xi, \quad y=Y\cos\Omega t+\eta.$$

The differential equation for the first approximation to the perturbed motion (variational equation) is

$$\left.\begin{aligned}\frac{d^2\xi}{d\tau^2}+\frac{\gamma+\varepsilon\cos\tau}{1+\gamma+\varepsilon\cos\tau}\xi=0,\\ \frac{d^2\eta}{d\tau^2}+\eta=0,\end{aligned}\right\} \tag{2.12}$$

where τ is dimensionless time and ε, γ dimensionless parameters:

$$\tau=\omega t, \quad \varepsilon=\frac{Y}{l}, \quad \gamma=\frac{\lambda}{l}.$$

Stability or instability of the trivial solution to the variational equations depends on the first of equations (2.12). In the present conservative case, however, the stability of the trivial solution of system (2.12) does not necessarily imply that the unperturbed motion (2.11) is stable with respect to x, y, $\dot{x}$, $\dot{y}$. However, if the trivial solution of system (2.12) is exponentially unstable, this does imply instability of the unperturbed motion with respect to x, y, $\dot{x}$, $\dot{y}$ (/102/, subsection 70). We therefore proceed to analyze the trivial solution of the first equation in (2.12) for instability:

$$\frac{d^2\xi}{d\tau^2}+[p_0(\gamma)+\varepsilon p_1(\tau, \gamma)+\varepsilon^2 p_2(\tau, \gamma)+\ldots]\xi=0, \tag{2.13}$$

where

$$p_0(\gamma)=\frac{\gamma}{1+\gamma}, \quad p_1(\tau, \gamma)=2p_1^{(1)}\cos\tau, \quad p_1^{(1)}=\frac{1}{2(1+\gamma)^2}.$$

The natural frequency ω_1 of the unperturbed system is

$$\omega_1=\sqrt{p_0(\gamma)}=\sqrt{\frac{\gamma}{1+\gamma}}.$$

The critical values of γ are determined from equations (V. 2.5), which are in this case $2\omega_1=m$, $m=1, 2 \ldots$ Hence

$$\gamma_m=\frac{m^2}{4-m^2} \qquad (m=1, 2, \ldots).$$

Since $\gamma>0$, we have a single critical value γ_1:

$$\gamma_1=\frac{1}{4-1}=\frac{1}{3}.$$

Since $p_1^{(0)}=(p_1)_{\mathrm{av}}=0$, we can use formula (V. 2.24) to evaluate $\mu_{1,2}$, with $j=h=1$, $m=1$, $\mathbf{a}_1=\mathbf{e}_1/\sqrt{2\omega_1}$; thus

$$\mu_{1,2} = \mp \left. \frac{p_1^{(1)}}{2\omega_1 \dfrac{d\omega_1}{d\gamma}} \right|_{\gamma=\gamma_0} = \mp \left. \frac{\dfrac{1}{2(1+\gamma)^2}}{\dfrac{1}{(1+\gamma)^2}} \right|_{\gamma=\gamma_0} = \mp \frac{1}{2}.$$

Thus, in the first approximation, the instability domain of the trivial solution of equation (2.13) is defined by the inequalities

$$\frac{1}{3} - \frac{1}{2}\varepsilon + \ldots < \gamma < \frac{1}{3} + \frac{1}{2}\varepsilon + \ldots \qquad (2.14)$$

Since $d\omega_1/d\gamma \neq 0$ for $\gamma = 1/3$, the boundaries of the dynamic instability domain are defined by analytic functions of ε (see § V. 3.2). Thus the omitted terms in (2.14) are of second order, $O(\varepsilon^2)$.

Finally, then: if the parameters of the pendulum satisfy the condition $\gamma \approx 1/3$, i. e. $mg/cl \approx 1/3$, small vertical vibrations are unstable. More precisely: if (2.14) holds, i. e.,

$$\frac{1}{3} - \frac{1}{2}\left(\frac{Y}{l}\right) + \ldots < \frac{mg}{cl} < \frac{1}{3} + \frac{1}{2}\left(\frac{Y}{l}\right) + \ldots,$$

the vertical vibrations (2.11) with amplitude Y bounded by a certain quantity will be unstable.

This loss of stability of certain vibrational modes may be found in arbitrary perfectly elastic vibrating chains /86d, 204m, n/, the simplest example of which is the simple pendulum just considered.

2.3. Loss of stability of vertical vibrations in a plane double pendulum

Consider a double pendulum with one arm absolutely rigid and the second elastic, with stiffness coefficient c. The notation is clear from Figure 33 a. We again study the stability of vertical vibrations, determining the relationships between the system parameters at which they are unstable, i. e., energy transfer occurs from one mode to another.

Let $\lambda_2 = m_2 g/c$ denote the static elongation of the spring. The kinetic and potential energies are

$$T = \frac{1}{2} m_2 (\dot{x}_2^2 + \dot{y}_2^2) + \frac{1}{2} m_1 l_1^2 \dot{\varphi}_1^2,$$

$$V = -m_1 g l_1 \cos\varphi_1 - m_2 g y_2 + \frac{1}{2} c [(x_2 - l_1 \sin\varphi_1)^2 + (l_2 + \lambda_2 + y_2 + l_1 - l_1 \cos\varphi_1)^2] -$$
$$- c l_2 \sqrt{(x_2 - l_1 \sin\varphi_1)^2 + (l_2 + \lambda_2 + y_2 + l_1 - l_1 \cos\varphi_1)^2}.$$

The Lagrange equations of the second kind (not given here; see /86d, 204m/), under the natural assumption that $l_2 + \lambda_2 + y_2 > 0$, have the solution

$$\varphi_1 = 0, \; x_2 = 0, \quad y_2 = Y_2 \cos \Omega_2 t \qquad \left(\Omega_2 = \sqrt{\frac{c}{m_2}}\right),$$

which describes vertical vibrations of the mass m_2 and is the unperturbed motion for our example.

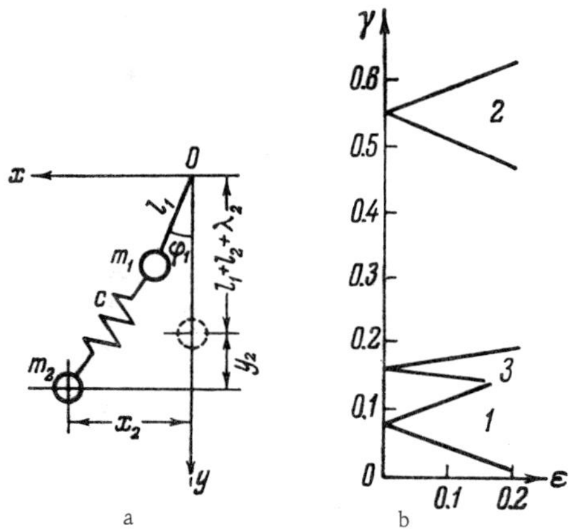

FIGURE 33.

The first group of variational equations ($\varphi_1 = 0+\Phi$, $x_2 = 0+\varkappa_2$) may be written in dimensionless variables as

$$\left.\begin{aligned}\frac{d^2\Phi}{d\tau^2}+\left[k\gamma(1+\mu)+\mu k\varepsilon\cos\tau+\mu\frac{\gamma+\varepsilon\cos\tau}{1+\gamma+\varepsilon\cos\tau}\right]\Phi-\\ -\sqrt{\mu}\,\frac{\gamma+\varepsilon\cos\tau}{1+\gamma+\varepsilon\cos\tau}\,\xi=0,\\ \frac{d^2\xi}{d\tau^2}-\sqrt{\mu}\,\frac{\gamma+\varepsilon\cos\tau}{1+\gamma+\varepsilon\cos\tau}\,\Phi+\frac{\gamma+\varepsilon\cos\tau}{1+\gamma+\varepsilon\cos\tau}\,\xi=0,\end{aligned}\right\} \tag{2.15}$$

where

$$\tau=\Omega_2 t,\ \xi=\sqrt{\mu}\,\frac{\varkappa_2}{l_1},\ \mu=\frac{m_2}{m_1},\ k=\frac{l_2}{l_1},\ \gamma=\frac{\lambda_2}{l_2},\ \varepsilon=\frac{Y_2}{l_2}.$$

Restricting ourselves to the case $m_1=m_2$, $l_1=l_2$ ($\mu=k=1$), we write system (2.15) in matrix notation:

$$\frac{d^2\mathbf{y}}{d\tau^2}+[\mathbf{P}_0(\gamma)+\varepsilon\mathbf{P}_1(\tau,\ \gamma)+\ldots]\,\mathbf{y}=\mathbf{0},$$

where

$$\mathbf{y}=\begin{pmatrix}\Phi\\ \xi\end{pmatrix},\quad \mathbf{P}_0(\gamma)=\begin{Vmatrix}2\gamma+\frac{\gamma}{1+\gamma} & -\frac{\gamma}{1+\gamma}\\ -\frac{\gamma}{1+\gamma} & \frac{\gamma}{1+\gamma}\end{Vmatrix},\quad \mathbf{P}_1(\tau,\ \gamma)=2\mathbf{P}_1^{(1)}\cos\tau,$$

$$\mathbf{P}_1^{(1)}(\gamma)=\begin{Vmatrix}\frac{1}{2}+\frac{1}{2(1+\gamma)^2} & -\frac{1}{2(1+\gamma)^2}\\ -\frac{1}{2(1+\gamma)^2} & \frac{1}{2(1+\gamma)^2}\end{Vmatrix}.$$

The matrix

$$\frac{d\mathbf{P}_0}{d\gamma} = \left\| \begin{matrix} 2+\frac{1}{(1+\gamma)^2} & -\frac{1}{(1+\gamma)^2} \\ -\frac{1}{(1+\gamma)^2} & \frac{1}{(1+\gamma)^2} \end{matrix} \right\|$$

is positive definite. Let us evaluate $\omega_1(\gamma)$ and $\omega_2(\gamma)$ — the natural frequencies of the unperturbed system $d^2\mathbf{y}/d\tau^2+\mathbf{P}_0(\gamma)\mathbf{y}=\mathbf{0}$:

$$\omega_1 = \sqrt{\gamma\left[1+\frac{1}{1+\gamma}-\sqrt{1+\frac{1}{(1+\gamma)^2}}\,\right]},$$
$$\omega_2 = \sqrt{\gamma\left[1+\frac{1}{1+\gamma}+\sqrt{1+\frac{1}{(1+\gamma)^2}}\,\right]}.$$

By Remark IV in § V. 2. 3, there may be wide dynamic instability domains only at points $(0,\gamma_0)$ on the axis $\varepsilon=0$, $\gamma>0$, for which

$$\omega_j(\gamma_0)+\omega_h(\gamma_0)=1 \qquad (j,\ h=1,\ 2).$$

Since $2\omega_1(\gamma)$, $2\omega_2(\gamma)$ and $\omega_1(\gamma)+\omega_2(\gamma)$ are in this example increasing functions of γ, there will be wide instability domains for a unique value of γ in each of the following three cases: 1) $2\omega_1(\gamma)=1$, 2) $2\omega_2(\gamma)=1$ — principal resonance, 3) $\omega_1(\gamma)+\omega_2(\gamma)=1$ — combination resonance. Since $(\mathbf{P}_1)_{av}=\mathbf{0}$, the slopes $\mu_{1,2}$ of the tangents are determined from formula (V. 2.24)

$$\mu_{1,2} = \mp \left. \frac{2\,(\mathbf{P}_1^{(1)}\mathbf{a}_j,\ \mathbf{a}_h)}{\frac{d\omega_j}{d\gamma}+\frac{d\omega_h}{d\gamma}} \right|_{\gamma=\gamma_0},$$

where γ_0 is a root of the equation

$$\omega_j(\gamma)+\omega_h(\gamma)=1 \qquad (j,\ h=1,\ 2),$$

and $\mathbf{a}_j$, $\mathbf{a}_h$ are appropriate eigenvectors of the matrix $\mathbf{P}_0$:

$$\mathbf{P}_0(\gamma)\,\mathbf{a}_j(\gamma)=\omega_j^2(\gamma)\,\mathbf{a}_j(\gamma), \qquad \mathbf{P}_0(\gamma)\,\mathbf{a}_h(\gamma)=\omega_h^2(\gamma)\,\mathbf{a}_h(\gamma),$$

with $2\omega_j(\mathbf{a}_j,\ \mathbf{a}_j)=1$, $2\omega_h(\mathbf{a}_h,\ \mathbf{a}_h)=1$.

The results are as follows. The principal-resonance domains are determined up to $O(\varepsilon^2)$ by the inequalities

$$0.076-0.364\varepsilon+\ldots<\gamma<0.076+0.364\varepsilon+\ldots,$$
$$0.549-0.385\varepsilon+\ldots<\gamma<0.549+0.385\varepsilon+\ldots,$$

and the combination-resonance domain by

$$0.157-0.088\varepsilon+\ldots<\gamma<0.157+0.088\varepsilon+\ldots$$

These domains are shown in Figure 33b.

It follows from § V. 3.2 that since

$$\frac{d\omega}{d\gamma} \neq 0,$$

the equations of the boundaries are analytic in ε, and therefore the neglected terms are of order ε^2 and higher. This was also the case in the examples of §§ 2.1, 2.2.

2.4. Orbital stability in a special case of the three-body problem

Let M, m and m' be point masses; let r, r' and R denote the distances Mm, Mm' and mm', respectively, and ψ, φ and φ' the angles of the triangle formed by the masses M, m, m', at the vertices M, m and m', respectively. Let Ω be the angular velocity of the fixed system defined by the point M, the direction Mm and the plane of the triangle Mmm'. We shall confine the discussion to the case of Newtonian attraction, when the mutual attraction of any two masses μ and μ' at a distance ρ is given by $\Gamma\mu\mu'/\rho^2$, where Γ is the gravitational constant. The three-body problem (in general, for any law of attraction) admits a particular periodic solution — Laplace's solution, in which $\psi = \frac{\pi}{3}$, $r = r'$ and r, Ω are determined by equations which in the case of Newtonian attraction are

$$\left.\begin{aligned} &\frac{d^2r}{dt^2} - r\Omega^2 + (M+m+m')\frac{\Gamma}{r^2} = 0,\\ &r^2\Omega = C, \end{aligned}\right\} \tag{2.16}$$

where C is an arbitrary constant. If $r = \text{const}$ the Laplacian motion is said to be *constant*. We now introduce a new independent variable τ, called the *true anomaly*, by setting $d\tau = \Omega\, dt$. By the first equation of (2.16),

$$d\tau = \pm \frac{d\left(\frac{1}{r}\right)}{\sqrt{h - \frac{1}{r^2} + \frac{2g\Gamma}{r}}},$$

where h is an arbitrary constant and $g = (M+m+m')/C^2$.

If the criterion for stability is that in perturbed motion the sides of the triangle should differ by an infinitesimal amount from their lengths in unperturbed motion at the same instant of time, then nonconstant Laplacian motions are generally unstable, as shown by Lyapunov /180a/. For nonconstant Laplacian motions, therefore, we can consider only orbital stability.

In any periodic Laplacian motion, the relative trajectory of each point with respect to one of the other two is an ellipse, with the reference point at one of the foci. We take the eccentricity ε of this ellipse as a small parameter. Following Lyapunov /180a/, we shall say that a periodic Laplacian motion is *L-stable* if after any infinitesimal perturbation the triangle remains throughout the motion infinitesimally close to equilateral, and the ranges over which its sides vary differs infinitesimally from the corresponding ranges in unperturbed motion. After appropriate manipulations, the perturbed system of equations may be written

$$\frac{d^2\mathbf{y}}{d\tau^2}+\mathbf{Q}_0\frac{d\mathbf{y}}{d\tau}+\mathbf{P}(\tau,\ \gamma,\ \varepsilon)\,\mathbf{y}=0 \qquad (2.17)$$

(see /180a/, subsections 2, 3, 12), where $\mathbf{y}$ is a 2-vector,

$$\mathbf{Q}_0=2\mathbf{J}_2=\begin{Vmatrix}0 & -2\\ 2 & 0\end{Vmatrix},\quad \mathbf{P}(\tau,\ \gamma,\ \varepsilon)=\begin{Vmatrix}\frac{-3\,[1-\lambda(\gamma)]}{1-\varepsilon\cos\tau} & 0\\ 0 & \frac{-3\lambda(\gamma)}{1-\varepsilon\cos\tau}\end{Vmatrix},$$

$\lambda(\gamma)$ a root of the equation

$$\lambda(1-\lambda)=\frac{\gamma}{4}, \qquad (2.18)$$

and γ a positive parameter depending only on the masses:

$$\gamma=3\,\frac{Mm+Mm'+mm'}{(M+m+m')^2}.$$

We confine attention to the interval $0<\gamma<1/9$.

The natural frequencies $\omega_1(\gamma)$, $\omega_2(\gamma)$ of system (2.17) at $\varepsilon=0$, i. e., the system

$$\mathbf{y}''+\mathbf{Q}_0\mathbf{y}'+\mathbf{P}_0(\gamma)\,\mathbf{y}=0 \qquad (\mathbf{P}_0(\gamma)=\mathbf{P}(\tau,\ \gamma,\ 0))$$

are solutions of the equation

$$\det\,[-\omega^2\mathbf{I}_2+i\mathbf{Q}_0\omega+\mathbf{P}_0(\gamma)]=\begin{vmatrix}-3(1-\lambda)-\omega^2 & -2i\omega\\ 2i\omega & -3\lambda-\omega^2\end{vmatrix}=0.$$

Hence

$$\omega_{1,2}(\gamma)=\sqrt{\frac{1}{2}\left[1\mp\sqrt{1-36\lambda(1-\lambda)}\right]}=\sqrt{\frac{1}{2}\left[1\mp\sqrt{1-9\gamma}\right]}.$$

Since $\mathbf{P}_0(\gamma)<0$, it follows from § V. 1.2 that the number l of natural frequencies of the first kind is zero, so that $\omega_1(\gamma)$, $\omega_2(\gamma)$ are the natural frequencies of the second kind of the matrix $\mathbf{C}(\gamma)$ and (V. 2.5) is the equation for the critical values of γ. Of all the equations (V. 2.5), only that with $j=h=1$ and $m=1$ is solvable, i. e., only the equation $2\omega_1=1$, so that we have a single critical value

$$\gamma_0=\frac{1}{12}.$$

The eigenvectors $\mathbf{c}_1(\gamma)$ and $\mathbf{c}_2(\gamma)$ must be normalized now not by conditions (V. 2.17), but by (V. 2.4), which in our case are the same as (IV. 3.27):

$$([(\omega_j+\omega_h)\,\mathbf{I}_2-i\mathbf{Q}_0]\,\mathbf{c}_j,\ \mathbf{c}_h)=\delta_{jh} \qquad (j,\ h=1,\ 2).$$

The calculations give

$$c_j = \sqrt{\frac{2\omega_j}{[\omega_j^2+3(1-\lambda)]^2-12(1-\lambda)}}\begin{pmatrix} 1 \\ \frac{\omega_j^2+3(1-\lambda)}{2\omega_j}\, i \end{pmatrix} \qquad (t=1,\ 2).$$

Since

$$\mathbf{P}_{10}^{(0)} = \left[\frac{\partial}{\partial \varepsilon}\mathbf{P}(\tau,\ \gamma_0,\ \varepsilon)\big|_{\varepsilon=0}\right]_{av} = 0,$$

it follows that the formula (V. 2.8) for $\mu_{1,2}$ is now

$$\mu_{1,2} = \mp\frac{(\mathbf{P}_{10}^{(1)}(\gamma_0)\bar{c}_1,\ c_1)}{\frac{d\omega_1(\gamma_0)}{d\gamma}},$$

where $\mathbf{P}_{10}^{(1)}$ is defined as in §IV. 3.4. In the present case,

$$\mathbf{P}_{10} = \frac{\partial}{\partial \varepsilon}\mathbf{P}(\tau,\ \gamma_0,\ \varepsilon)\big|_{\varepsilon=0} = -3\begin{Vmatrix} (1-\lambda_0)\cos\tau & 0 \\ 0 & \lambda_0\cos\tau \end{Vmatrix},\quad \mathbf{P}_{10}^{(1)} = -\frac{3}{2}\begin{Vmatrix} 1-\lambda_0 & 0 \\ 0 & \lambda_0 \end{Vmatrix},$$

where $\lambda_0 = \lambda(\gamma_0)$. This gives

$$\mu_{1,2} = \mp\frac{1}{3}\,\frac{-\lambda_0\left[\frac{1}{4}+3(1-\lambda_0)\right]^2+1-\lambda_0}{\left[\frac{1}{4}+3(1-\lambda_0)\right]^2-12(1-\lambda_0)}. \tag{2.19}$$

The values of λ_0 are determined from (2.18) with $\gamma_0 = 1/12$: $\lambda_0^{(1)} = 0.0213\ldots$, $\lambda_0^{(2)} = 0.9787\ldots$ Formula (2.19) now gives

$$\mu_{1,2} = \mp\frac{4}{25}.$$

Finally, the only L-instability domain is

$$\frac{1}{12}-\frac{4}{25}\varepsilon+\ldots<\gamma<\frac{1}{12}+\frac{4}{25}\varepsilon+\ldots$$

§3. DYNAMIC STABILITY OF THIN-WALLED BARS WITH LONGITUDINAL PERIODIC LOAD

3.1. Differential equations of flexural-torsional vibrations of a thin-walled bar

According to Vlasov (/18/, Chapter IX, §1), the static problem of equilibrium of thin-walled bars may be reduced to the system of equations

$$\left.\begin{aligned} EF\frac{\partial^2\zeta}{\partial t^2}+\int_L p_z\,ds=0,\quad EJ_y\frac{\partial^4\xi}{\partial z^4}-\int_L\frac{\partial p_z}{\partial z}x\,ds-q_x=0,\\ EJ_x\frac{\partial^4\eta}{\partial z^4}-\int_L\frac{\partial p_z}{\partial z}y\,ds-q_y=0,\quad EJ_\omega\frac{\partial^4\theta}{\partial z^4}-GJ_d\frac{\partial^2\theta}{\partial t^2}-\int_L\frac{\partial p_z}{\partial z}\omega\,ds=0.\end{aligned}\right\} \tag{3.1}$$

where z, s are coordinates measured along the generatrix and the directrix L, $\zeta(z, t)$, $\xi(z, t)$, $\eta(z, t)$ the components of the center of deflection along the generatrix of the cylindrical middle surface and the principal axes x, y of the cross-section, $\Theta = \Theta(z)$ the angle of torsion, F the cross-sectional area of the bar, J_x, J_y the principal moments of inertia, J_d the torsional resistance, J_ω the sectorial moment of inertia, E Young's modulus, G the shear modulus, $p_z(z, s)$ the projection of the surface load on the generatrix, $q_x(z)$, $q_y(z)$ the projections of the lateral load.

Introducing inertial forces in system (3.1) and assuming that the bar is subjected to a time-periodic longitudinal force

$$p(\vartheta t) = \alpha_0 + \beta_0 \cos \vartheta t = \alpha_0 + \frac{1}{2} \beta_0 (e^{-i\tau} + e^{i\tau}), \tag{3.2}$$

we obtain (/18/, Chapter IX, §3) a system of differential equations for the flexural-torsional vibrations of a thin-walled bar:

$$\left.\begin{aligned}
&EJ_y \frac{\partial^4 \xi}{\partial z^4} - \frac{\gamma J_y}{g} \frac{\partial^4 \xi}{\partial z^2 \partial t^2} + \frac{\gamma F}{g} \frac{\partial^2 \xi}{\partial t^2} + p(\vartheta t) \frac{\partial^2 \xi}{\partial z^2} + \frac{\gamma F a_y}{g} \frac{\partial^2 \Theta}{\partial t^2} - \\
&\qquad - p(\vartheta t)(e_y - a_y) \frac{\partial^2 \Theta}{\partial z^2} = 0, \\
&EJ_x \frac{\partial^4 \eta}{\partial z^4} - \frac{\gamma J_x}{g} \frac{\partial^4 \eta}{\partial z^2 \partial t^2} + \frac{\gamma F}{g} \frac{\partial^2 \eta}{\partial t^2} + p(\vartheta t) \frac{\partial^2 \eta}{\partial z^2} - \\
&\qquad \frac{-\gamma F a_x}{g} \frac{\partial^2 \Theta}{\partial t^2} + p(\vartheta t)(e_x - a_x) \frac{\partial^2 \Theta}{\partial z^2} = 0, \\
&\frac{\gamma F a_y}{g} \frac{\partial^2 \xi}{\partial t^2} - p(\vartheta t)(e_y - a_y) \frac{\partial^2 \xi}{\partial z^2} - \frac{\gamma F a_x}{g} \frac{\partial^2 \eta}{\partial t^2} + \\
&\qquad + p(\vartheta t)(e_x - a_x) \frac{\partial^2 \eta}{\partial z^2} + EJ_\omega \frac{\partial^4 \Theta}{\partial z^4} - \frac{\gamma J_\omega}{g} \frac{\partial^4 \Theta}{\partial z^2 \partial t^2} + \\
&\qquad + \frac{\gamma F r^2}{g} \frac{\partial^2 \Theta}{\partial t^2} - GJ_d \frac{\partial^2 \Theta}{\partial z^2} + p(\vartheta t)(r^2 + 2\beta_x e_x + 2\beta_y e_y) \frac{\partial^2 \Theta}{\partial z^2} = 0.
\end{aligned}\right\} \tag{3.3}$$

Here a_x, a_y are the coordinates of the center of deflection, e_x, e_y the coordinates of the point of application of the force $p(\vartheta t)$, γ the specific weight of the bar material, β_x, β_y geometric parameters of the cross-section,

$$r^2 = \frac{1}{F}(J_x + J_y) + a_x^2 + a_y^2.$$

We shall study the vibrations of a bar with freely supported ends, so that the boundary conditions are

$$\xi = \eta = \Theta = 0, \quad \frac{\partial^2 \xi}{\partial z^2} = \frac{\partial^2 \eta}{\partial z^2} = \frac{\partial^2 \Theta}{\partial z^2} = 0 \quad \text{for} \quad z = 0 \quad \text{and} \quad z = l.$$

Using the Bubnov-Galerkin method /20, 47/, one expresses the solution of system (3.3) as

$$\xi = \sum_{n=1}^{\infty} A_n(t) \sin \frac{n\pi z}{l}, \quad \eta = \sum_{n=1}^{\infty} B(t) \sin \frac{n\pi z}{l}, \quad \Theta = \sum_{n=1}^{\infty} C_n(t) \sin \frac{n\pi z}{l}.$$

Substituting these series into (3.3), we obtain an infinite sequence of systems of ordinary differential equations with periodic coefficients, in the unknowns A_n, B_n, C_n $(n = 1, 2, \ldots)$:

$$\frac{\gamma}{g}(F+J_y\lambda_n^2)\ddot{A}_n+\frac{\gamma F a_y}{g}\ddot{C}_n+[EJ_y\lambda_n^2-p(\vartheta t)]\lambda_n^2 A_n+(e_y-a_y)\lambda_n^2 p(\vartheta t)C_n=0,$$
$$\frac{\gamma}{g}(F+J_x\lambda_n^2)\ddot{B}_n+\frac{\gamma F a_x}{g}\ddot{C}_n+[EJ_x\lambda_n^2-p(\vartheta t)]\lambda_n^2 B_n-(e_x-a_x)\lambda_n^2 p(\vartheta t)C_n=0,$$
$$\frac{\gamma F a_y}{g}\ddot{A}_n-\frac{\gamma F a_x}{g}\ddot{B}_n+\frac{\gamma}{g}(Fr^2+J_\omega\lambda_n^2)\ddot{C}_n+(e_y-a_y)\lambda_n^2 p(\vartheta t)A_n-$$
$$-(e_x-a_x)\lambda_n^2 p(\vartheta t)B_n+[EJ_\omega\lambda_n^2+GJ_d-(r^2+2\beta_x e_x+2\beta_y e_y)p(\vartheta t)]\lambda_n^2 C_n=0,$$

where $\lambda_n=n\pi/l$ $(n=1,\ 2,\ \ldots)$. The coefficients of each system depend on the geometric and physical properties of the bar, on the magnitude of the longitudinal force and the coordinates of its point of application, and on the integer n, which is the number of half-waves extending over the length l of the bar. We write each of the systems in vector notation:

$$\mathbf{M}\frac{d^2\mathbf{f}_n}{d\tau^2}+\frac{1}{\vartheta^2}[\mathbf{P}_0-p(\vartheta t)\mathbf{N}]\mathbf{f}_n=0 \qquad (n=1,\ 2,\ \ldots), \tag{3.4}$$

introducing the notation $\tau=\vartheta t$ and

$$\left.\begin{aligned}
&\mathbf{M}=\frac{\gamma}{g}\begin{Vmatrix} F+J_y\lambda_n^2 & 0 & Fa_y \\ 0 & F+J_x\lambda_n^2 & -Fa_x \\ Fa_y & -Fa_x & Fr^2+J_\omega\lambda_n^2 \end{Vmatrix}=\begin{Vmatrix} a & 0 & 0 \\ 0 & b & 0 \\ 0 & 0 & c\end{Vmatrix}+\begin{Vmatrix} 0 & 0 & \delta \\ 0 & 0 & -\sigma \\ \delta & -\sigma & 0\end{Vmatrix},\\
&\mathbf{f}_n=\begin{pmatrix} A_n \\ B_n \\ C_n\end{pmatrix},\quad \mathbf{P}_0=\lambda_n^2\begin{Vmatrix} EJ_y\lambda_n^2 & 0 & 0 \\ 0 & EJ_x\lambda_n^2 & 0 \\ 0 & 0 & EJ_\omega\lambda_n^2+GJ_d\end{Vmatrix}=\begin{Vmatrix}\alpha & 0 & 0 \\ 0 & \beta & 0 \\ 0 & 0 & \gamma\end{Vmatrix},\\
&\mathbf{N}=\lambda_n^2\begin{Vmatrix} 1 & 0 & a_y-e_y \\ 0 & 1 & e_x-a_x \\ a_y-e_y & e_x-a_x & r^2+2\beta_x e_x+2\beta_y e_y\end{Vmatrix}=\begin{Vmatrix} 1 & 0 & e \\ 0 & 1 & d \\ e & d & h\end{Vmatrix},\\
&a=\frac{\gamma}{g}(F+J_y\lambda_n^2),\quad b=\frac{\gamma}{g}(F+J_x\lambda_n^2),\quad c=\frac{\gamma}{g}(Fr^2+J_\omega\lambda_n^2),\\
&\delta=\frac{\gamma F a_y}{g},\quad \sigma=\frac{\gamma F a_x}{g},\quad \alpha=EJ_y\lambda_n^4,\quad \beta=EJ_x\lambda_n^4,\quad \gamma=(EJ_\omega\lambda_n^2+GJ_d)\lambda_n^2,\\
&e=\lambda_n^2(a_y-e_y),\quad d=\lambda_n^2(e_x-a_x),\quad h=\lambda_n^2(r^2+2\beta_x e_x+2\beta_y e_y).
\end{aligned}\right\} \tag{3.5}$$

3.2. Bar with two cross-sectional axes of symmetry

In this frequently occurring case,

$$a_x=a_y=0.$$

Then

$$\mathbf{M}=\mathbf{M}_0=\begin{Vmatrix} a & 0 & 0 \\ 0 & b & 0 \\ 0 & 0 & c\end{Vmatrix}.$$

We provide system (3.4) with a small parameter ε representing the small magnitude of the longitudinal load:

$$\mathbf{M}_0\frac{d^2\mathbf{y}}{dt^2}+[\mathbf{P}_0+\varepsilon\mathbf{P}_1(\vartheta t)]\,\mathbf{y}=0 \qquad (\mathbf{P}_1=-p(\vartheta t)\,\mathbf{N}), \tag{3.6}$$

with a view to setting $\varepsilon=1$ in the final formulas. The squared natural frequencies of the unperturbed system ($\varepsilon=0$)

$$\mathbf{M}_0\frac{d^2\mathbf{y}}{dt^2}+\mathbf{P}_0\mathbf{y}=0$$

are

$$\omega_1^2=\frac{\alpha}{a},\quad \omega_2^2=\frac{\beta}{b},\quad \omega_3^2=\frac{\gamma}{c}. \tag{3.7}$$

The orthonormal system (V. 3.49) of eigenvectors of the matrix $\mathbf{M}_0^{-1}\mathbf{P}_0$ is

$$\mathbf{a}_1=\frac{1}{\sqrt{2\omega_1 a}}\mathbf{e}_1,\ \mathbf{a}_2=\frac{1}{\sqrt{2\omega_2 b}}\mathbf{e}_2,\ \mathbf{a}_3=\frac{1}{\sqrt{2\omega_3 b}}\mathbf{e}_3,\ \text{where } \mathbf{e}_1=\begin{pmatrix}1\\0\\0\end{pmatrix},\ \mathbf{e}_2=\begin{pmatrix}0\\1\\0\end{pmatrix},\ \mathbf{e}_3=\begin{pmatrix}0\\0\\1\end{pmatrix}.$$

We have

$$\mathbf{P}_1^{(0)}=-\alpha_0\mathbf{N},\quad \mathbf{P}_1^{(-1)}=\mathbf{P}_1^{(1)}=-\frac{1}{2}\beta_0\mathbf{N}, \tag{3.8}$$

and all the other Fourier coefficients $\mathbf{P}_q^{(l)}$ vanish. The quantities (V.3.48) may be arranged in a matrix

$$\|\chi_{\varkappa\lambda}^{(0)}\|_1^3=\left\|\begin{matrix} -\dfrac{\alpha_0}{2\omega_1 a}\lambda_n^2 & 0 & -\dfrac{\alpha_0 e}{2\sqrt{\omega_1\omega_3 ac}} \\ 0 & -\dfrac{\alpha_0}{2\omega_2 b}\lambda_n^2 & -\dfrac{\alpha_0 d}{2\sqrt{\omega_2\omega_3 bc}} \\ -\dfrac{\alpha_0 e}{2\sqrt{\omega_1\omega_2 ac}} & -\dfrac{\alpha_0 d}{2\sqrt{\omega_2\omega_3 bc}} & -\dfrac{\alpha_0 h}{2\omega_3 c}\end{matrix}\right\|,\quad \|\chi_{\varkappa\lambda}^{(\mp 1)}\|_1^3=\frac{\beta_0}{2\alpha_0}\|\chi_{\varkappa\lambda}^{(0)}\|_1^3,$$

since it is clear from (3.8) that

$$\chi_{\varkappa\lambda}^{(-1)}=\chi_{\varkappa\lambda}^{(+1)}=\frac{\beta_0}{2\alpha_0}\chi_{\varkappa\lambda}^{(0)} \qquad (\varkappa,\ \lambda=1,\ 2,\ 3).$$

It also follows from (V. 3.48) that

$$\begin{aligned}\chi_{\varkappa\lambda}^{(l)}&=0 \qquad (\varkappa,\ \lambda=1,\ 2,\ 3;\ l=\pm 2,\ \pm 3,\ \ldots),\\ \sigma_{\varkappa\lambda}^{(r)}&=0 \qquad (\varkappa,\ \lambda=1,\ 2,\ 3;\ r=0,\ \pm 1,\ \pm 2,\ \ldots).\end{aligned}$$

3.3. Wide instability domains in the second approximation

The only nonvanishing quantities $\chi_{\varkappa\lambda}^{(-m)}$ $(\varkappa,\ \lambda=1,\ 2,\ 3;\ m=1,\ 2,\ \ldots)$ are

$$\chi_{11}^{(-1)},\ \chi_{22}^{(-1)},\ \chi_{33}^{(-1)},\ \chi_{13}^{(-1)},\ \chi_{23}^{(-1)},$$

and so we have wide instability domains (see (V. 3.40)) only for three principal resonances, at the following critical frequencies of the longitudinal force:

$$\vartheta_0 = 2\omega_1, \quad \vartheta_0 = 2\omega_2, \quad \vartheta_0 = 2\omega_3$$

and two combination resonances, at critical frequencies

$$\vartheta_0 = \omega_1 + \omega_3, \quad \vartheta_0 = \omega_2 + \omega_3.$$

a) First principal resonance, $\vartheta_0 = 2\omega_1$. By formula (V. 3.40),

$$\vartheta_1^{(1,\,2)} = -\frac{1}{\omega_1 a}\left(\alpha_0 \pm \frac{1}{2}\beta_0\right)\lambda_n^2.$$

By (V. 3.47),

$$\vartheta_2^{(1,\,2)} = -\frac{1}{\omega_1 a}\left[\frac{e^2\left(\alpha_0 \pm \frac{\beta_0}{2}\right)^2}{c(\omega_3^2-\omega_1^2)} + \frac{\beta_0^2 e^2}{4c(\omega_3^2-9\omega_1^2)} + \frac{\left(\alpha_0 \pm \frac{1}{2}\beta_0\right)^2\lambda_n^4}{4\omega_1^2 a} - \frac{\beta_0^2\lambda_n^4}{32\omega_1^2 a}\right].$$

b) Second principal resonance, $\vartheta_0 = 2\omega_2$. Similarly:

$$\vartheta_1^{(1,\,2)} = -\frac{1}{\omega_2 b}\left(\alpha_0 \pm \frac{1}{2}\beta_0\right)\lambda_n^2,$$

$$\vartheta_2^{(1,\,2)} = -\frac{1}{\omega_2 b}\left[\frac{d^2\left(\alpha_0 \pm \frac{1}{2}\beta_0\right)^2}{c(\omega_3^2-\omega_2^2)} + \frac{\beta_0^2 d^2}{4c(\omega_3^2-9\omega_2^2)} + \frac{\left(\alpha_0 \pm \frac{1}{2}\beta_0\right)^2\lambda_n^4}{4\omega_2^2 b} - \frac{\beta_0^2\lambda_n^4}{32\omega_2^2 b}\right].$$

c) Third principal resonance, $\vartheta_0 = 2\omega_3$. The calculations give:

$$\vartheta_1^{(1,\,2)} = -\frac{h}{\omega_3 c}\left(\alpha_0 \pm \frac{1}{2}\beta_0\right),$$

$$\vartheta_2^{(1,\,2)} = \frac{1}{\omega_3 c}\left[\frac{e^2\left(\alpha_0 \pm \frac{1}{2}\beta_0\right)^2}{a(\omega_3^2-\omega_1^2)} + \frac{d^2\left(\alpha_0 \pm \frac{1}{2}\beta_0\right)^2}{b(\omega_3^2-\omega_2^2)} + \frac{\beta_0^2 e^2}{4a(9\omega_3^2-\omega_1^2)} + \right.$$
$$\left. + \frac{\beta_0^2 d^2}{4b(9\omega_3^2-\omega_1^2)} - \frac{h^2(\beta_0^2 + 8\alpha_0^2 \mp 8\alpha_0\beta_0)}{32c\omega_3^2}\right].$$

d) Combination resonance $\vartheta_0 = \omega_1 + \omega_3$:

$$\vartheta_1^{(1,\,2)} = -\frac{1}{2\omega_1\omega_3 ac}\left[\alpha_0(h\omega_1 a + \omega_3 c)\lambda_n^2 \pm \beta_0 e\sqrt{\omega_1\omega_3 ac}\right],$$

$$\vartheta_2^{(1,\,2)} = \frac{d^2}{2\omega_3 bc}\left\{\frac{\alpha_0^2}{\omega_3^2-\omega_2^2} + \frac{1}{4}\beta_0^2\left[\frac{1}{(2\omega_3+\omega_1)^2-\omega_2^2} - \frac{1}{\omega_2^2-\omega_1^2}\right]\right\} -$$
$$-\frac{h^2}{16\omega_3^2 c^2}\left[2\alpha_0^2 + \frac{\beta_0^2}{3\omega_3+\omega_1} + \frac{\beta_0^2}{\omega_3-\omega_1}\right] - \frac{e^2}{16\omega_1\omega_3 ac}\,\frac{\beta_0^2 + 8\alpha_0^2}{\omega_3+\omega_1} +$$
$$+\frac{\lambda_n^4}{16\omega_1^2 a^2}\left(\frac{\beta_0^2}{\omega_3-\omega_1} - \frac{2\alpha_0^2}{\omega_1} - \frac{\beta_0^2}{\omega_3+3\omega_1}\right) \pm \frac{\alpha_0\beta_0 e}{2\sqrt{\omega_1\omega_3 ac}}\left[\frac{\lambda_n^2}{a(\omega_3^2-\omega_1^2)} - \frac{h}{4\omega_3^2 c} - \frac{\lambda_n^2}{4\omega_1^2 a} - \frac{h}{c(\omega_3^2-\omega_1^2)}\right].$$

e) Combination resonance $\vartheta_0 = \omega_2 + \omega_3$:

$$\vartheta_1^{(1,2)} = -\frac{1}{2\omega_2\omega_3 bc}\left[\alpha_0 (h\omega_2 b + \omega_3 c)\lambda_n^2 \pm \beta_0 d\sqrt{\omega_2\omega_3 bc}\right],$$

$$\vartheta_2^{(1,2)} = \frac{e^2}{2\omega_3 ac}\left\{\frac{\alpha_0^2}{\omega_3^2-\omega_1^2} + \frac{1}{4}\beta_0^2\left[\frac{1}{(2\omega_3+\omega_2)^2-\omega_1^2} + \frac{1}{\omega_2^2-\omega_1^2}\right]\right\} -$$
$$-\frac{h^2}{16\omega_3^2 c^2}\left(2\alpha_0^2 + \frac{\beta_0^2}{3\omega_3+\omega_2} + \frac{\beta_0^2}{\omega_3-\omega_2}\right) - \frac{d^2}{16\omega_2\omega_3 bc}\,\frac{8\alpha_0^2+\beta_0^2}{\omega_3+\omega_2} +$$
$$+\frac{\lambda_n^4}{16\omega_2^2 b^2}\left(-\frac{2\alpha_0^2}{\omega_2} + \frac{\beta_0^2}{\omega_3-\omega_2} - \frac{\beta_0^2}{\omega_3+3\omega_2}\right) \mp \frac{\alpha_0\beta_0 d}{2\sqrt{\omega_2\omega_3 bc}}\left[\frac{\lambda_n^2}{b(\omega_3^2-\omega_2^2)} - \frac{h}{4\omega_3^2 c} - \frac{\lambda_n^2}{4\omega_2^2 b} - \frac{h}{c(\omega_3^2-\omega_2^2)}\right].$$

Each of the five wide instability domains is determined (up to second-order quantities) by the inequalities

$$\vartheta_0 + \vartheta_1^{(1)} + \vartheta_2^{(1)} < \vartheta < \vartheta_0 + \vartheta_1^{(2)} + \vartheta_2^{(2)}. \tag{3.9}$$

The parameters figuring in $\vartheta_1^{(1,2)}$ and $\vartheta_2^{(1,2)}$, are determined by formulas (3.2), (3.5) and (3.7). Each of the five domains (3.9) refers to an equation (3.4) with fixed n. The wide instability domains for system (3.3) are all the domains (3.9) with $n = 1, 2, 3, \ldots$, five domains for each n.

§4. STABILITY OF FLEXURAL VIBRATIONS OF ROTATING COAXIAL SHAFTS

In this section we use the formulas of § V.4 to analyze the stability of flexural vibrations of two rotating coaxial cantilever shafts with disks attached at their ends (Figure 34a).

4.1. Differential equations of flexural vibrations of coaxial shafts

Consider the steady rotation of shafts at constant independent angular velocities $\dot{\varphi}_1$ and $\dot{\varphi}_2$, when the external torsional moments exerted on the shafts by the drive are balanced by moments of frictional forces. The disks are assumed to be weightless; the masses and torsional vibrations of the shafts, as well as the gyroscopic effects of the disks, are neglected.* Under these assumptions and the additional assumption of small elastic displacements of the shafts and constant axial coordinates ζ_1 and of the centers of the disks, the equations of flexural vibrations (see /210/) may be written in the form of (V. 4.1):

$$\frac{d^2\mathbf{y}}{dt^2} + \varepsilon\mathbf{Q}_1(\vartheta t)\frac{d\mathbf{y}}{dt} + [\mathbf{P}_0 + \varepsilon\mathbf{P}_1(\vartheta t)]\,\mathbf{y} = 0, \tag{4.1}$$

where $\vartheta = \dot{\varphi}_1 - \dot{\varphi}_2 \geqslant 0$**, $\dot{\varphi}_1 > 0$,

* All these cases may be treated in similar but more cumbersome fashion.

** If $\vartheta = 0$ all the coefficients of system (4.1) are constants. This is regarded here as a limiting case.

$$\mathbf{P}_0=\begin{Vmatrix}\gamma_{11} & \gamma_{12}\\ \gamma_{21} & \gamma_{22}\end{Vmatrix},\quad \begin{aligned}\mathbf{P}_1(\vartheta t)&=\mathbf{P}_1^{(-1)}e^{-i\vartheta t}+\mathbf{P}_1^{(0)}+\mathbf{P}_1^{(1)}e^{i\vartheta t},\\ \mathbf{Q}_1(\vartheta t)&=\mathbf{Q}_1^{(-1)}e^{-i\vartheta t}+\mathbf{Q}_1^{(0)}+\mathbf{Q}_1^{(1)}e^{i\vartheta t},\end{aligned}$$

$$\mathbf{P}_1^{(-1)}=\begin{Vmatrix}0 & 0\\ -ik\dot{\varphi}_1\gamma_{21} & 0\end{Vmatrix},\quad \mathbf{P}_1^{(0)}=\begin{Vmatrix}-ik\dot{\varphi}_1\gamma_{11} & 0\\ 0 & -ik\dot{\varphi}_2\gamma_{22}\end{Vmatrix},\quad \mathbf{P}_1^{(1)}=\begin{Vmatrix}0 & -ik\dot{\varphi}_2\gamma_{12}\\ 0 & 0\end{Vmatrix},$$

$$\mathbf{Q}_1^{(-1)}=\begin{Vmatrix}0 & 0\\ k\gamma_{21} & 0\end{Vmatrix},\quad \mathbf{Q}_1^{(0)}=\begin{Vmatrix}(\varkappa+k)\gamma_{11} & \varkappa\gamma_{12}\\ \varkappa\gamma_{21} & (\varkappa+k)\gamma_{22}\end{Vmatrix},\quad \mathbf{Q}_1^{(1)}=\begin{Vmatrix}0 & k\gamma_{12}\\ 0 & 0\end{Vmatrix},$$

$\gamma_{ij}=c_{ij}/m_j$ $(i,\ j=1,\ 2;\ \gamma_{11}\gamma_{22}-\gamma_{12}\gamma_{21}>0)$, m_1, m_2 are the masses of the disks, c_{ij} the coefficients of the Maxwell effect for elastic forces applied at the disks' centers, εk is the coefficient of internal friction and $\varepsilon\varkappa$ the coefficient of external friction,

$$\mathbf{y}=\begin{pmatrix}z_1\\ z_2\end{pmatrix},\quad z_1=\xi_1+i\eta_1,\quad z_2=\xi_2+i\eta_2$$

and $(\xi_1,\ \eta_1)$, $(\xi_2,\ \eta_2)$ are the coordinates of the disks' mass centers. Finally, ε is a small positive parameter, representing the small magnitude of the internal and external friction. In the final result one must put $\varepsilon=1$; the result will be valid for small k and $\varkappa$. The unperturbed motion is described by the trivial solution $\mathbf{y}=\mathbf{0}$ of system (4.1).

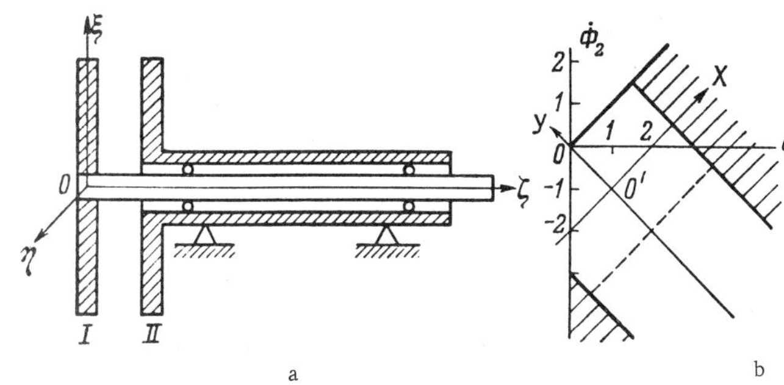

FIGURE 34.

The eigenvalues ω_1^2 and ω_2^2 of the matrix $\mathbf{P}_0$ are

$$\omega_{1,\,2}^2=\frac{1}{2}\left[\gamma_{11}+\gamma_{22}\mp\sqrt{(\gamma_{11}-\gamma_{22})^2+4\gamma_{12}\gamma_{21}}\right].$$

The eigenvectors $\mathbf{a}_1$ and $\mathbf{a}_2$, normalized according to (V. 4.2), are

$$\mathbf{a}_j=\frac{\sqrt{\vartheta_0}}{\sqrt{2\omega_j\left[1+\left(\frac{\omega_j^2-\gamma_{11}}{\gamma_{12}}\right)^2\right]}}\begin{pmatrix}1\\ \frac{\omega_j^2-\gamma_{11}}{\gamma_{12}}\end{pmatrix}\qquad (j=1,\ 2).$$

The vectors figuring in formulas (V. 4.9) are

$$\mathbf{P}_1^{(0)}\mathbf{a}_j = -i\frac{k}{\sqrt{R_j}}\begin{pmatrix}\gamma_{11} & \dot{\varphi}_1\\ \gamma_{22} & r_j\dot{\varphi}_2\end{pmatrix}, \quad \mathbf{P}_1^{(1)}\mathbf{a}_j = -i\frac{k}{\sqrt{R_j}}(\omega_j^2-\gamma_{11})\dot{\varphi}_2\begin{pmatrix}1\\0\end{pmatrix},$$

$$\mathbf{P}_1^{(-1)}\mathbf{a}_j = -i\frac{k}{\sqrt{R_j}}\gamma_{21}\dot{\varphi}_1\begin{pmatrix}0\\1\end{pmatrix}, \quad \mathbf{Q}_1^{(0)}\mathbf{a}_j = \frac{1}{\sqrt{R_j}}\begin{pmatrix}k\gamma_{11}+\varkappa\omega_j^2\\ \varkappa\gamma_{21}+(\varkappa+k)\gamma_{22}r_j\end{pmatrix},$$

$$\mathbf{Q}_1^{(1)}\mathbf{a}_j = \frac{k}{\sqrt{R_j}}(\omega_j^2-\gamma_{11})\begin{pmatrix}1\\0\end{pmatrix}, \quad \mathbf{Q}_1^{(-1)}\mathbf{a}_j = \frac{k}{\sqrt{R_j}}\gamma_{21}\begin{pmatrix}0\\1\end{pmatrix},$$

$$r_j = \frac{\omega_j^2-\gamma_{11}}{\gamma_{12}}, \quad R_j = \frac{2}{\vartheta_0}\omega_j(1+r_j^2) \qquad (j=1,\ 2).$$

4.2. Dynamic instability domains

We begin with the "general" case (a) of § V. 4.2. Using formulas (V. 4.9), we calculate $\sigma_{-2,-2}$, $\sigma_{-1,-1}$, σ_{11}, σ_{22}. Inequalities (V. 4.10) are

$$\omega_j[k\gamma_{11}+\varkappa\omega_j^2+\varkappa\gamma_{21}r_j+(\varkappa+k)\gamma_{22}r_j^2] < \pm k(\gamma_{11}\dot{\varphi}_1+\gamma_{22}r_j^2\dot{\varphi}_2) \qquad (j=1,\ 2). \quad (4.2)$$

For the mechanical system under consideration, with the values of the parameters γ_{11}, γ_{12}, γ_{21}, γ_{22}, k and $\varkappa$ held fixed, conditions (4.2) define two half-planes in the $(\dot{\varphi}_1, \dot{\varphi}_2)$-plane. The intersection of these half-planes with the half-planes $\dot{\varphi}_1 \geqslant \dot{\varphi}_2$, $\dot{\varphi}_1 \geqslant 0$ is the instability domain corresponding to the fundamental case.

Consider the supplementary case (b) of § V. 4.2. Since only the first harmonics of the matrix-functions $\mathbf{P}_1(\tau)$ and $\mathbf{Q}_1(\tau)$ are different from zero, the natural number m in (V. 4.11) must be unity, and we have the following critical frequencies:

$$\vartheta_0^{(1)} = 2\omega_1, \quad \vartheta_0^{(2)} = 2\omega_2,$$
$$\vartheta_0^{(3)} = \omega_1+\omega_2, \quad \vartheta_0^{(4)} = \omega_2-\omega_1.$$

First principal resonance, $\vartheta_0 = 2\omega_1$ $(j_1=-1,\ j_2=1)$:

By (V. 4.9) with $m_1 - m_{-1} = 1$ (see (V. 4.11)), we calculate $\sigma_{-1,1}$, $\sigma_{1,-1}$, and then

$$\left.\begin{aligned}
\alpha+i\beta &= \frac{1}{2}(\sigma_{-1,-1}-\sigma_{11})\vartheta_0^2 = \\
&= i\frac{\omega_1}{R_1}[k\gamma_{11}+\varkappa\omega_1^2+\varkappa\gamma_{21}r_1+(\varkappa+k)\gamma_{22}r_1^2],\\
\gamma+i\delta &= (\sigma_{-1,-1}\sigma_{11}-\sigma_{-1,1}\sigma_{1,-1})\vartheta_0^4 = \frac{1}{4}\mu^2-\frac{k^2}{R_1^2}(\dot{\varphi}_1\gamma_{11}+\gamma_{22}\dot{\varphi}_2r_1^2)^2+\\
&+\frac{\omega_1^2}{R_1^2}[k\gamma_{11}+\varkappa\omega_1^2+\varkappa\gamma_{21}r_1+(\varkappa+k)\gamma_{22}r_1^2]^2-\\
&-\frac{k^2\gamma_{12}r_1^2\gamma_{21}}{R_1^2}(\dot{\varphi}_2+\omega_1)(\omega_1-\dot{\varphi}_1)+i\frac{k}{R_1}(\dot{\varphi}_1\gamma_{11}+\dot{\varphi}_2\gamma_{22}r_1^2)\mu.
\end{aligned}\right\} \quad (4.3)$$

If conditions (V. 4.14) can be satisfied, they lead to inequalities

$$\mu_1 < \mu < \mu_2.$$

Then, by (V. 4.3), the dynamic instability domains at the first principal resonance are defined by

$$\vartheta_0+\mu_1\varepsilon+\ldots<\vartheta<\vartheta_0+\mu_2\varepsilon+\ldots$$

Substituting $\varepsilon=1$, $\vartheta_0=2\omega_1$, $\vartheta=\dot{\varphi}_1-\dot{\varphi}_2$, we see that the first approximation is

$$2\omega_1+\mu_1<\dot{\varphi}_1-\dot{\varphi}_2<2\omega_1+\mu_2.$$

Apart from the constraints stemming from the mechanical meaning of the problem ($\dot{\varphi}_1-\dot{\varphi}_2\geqslant 0$, $\dot{\varphi}_1>0$), one must add condition (V. 4.4), which is, for the first principal resonance,

$$\dot{\varphi}_1-\dot{\varphi}_2<4\omega_1. \tag{4.4}$$

Second principal resonance, $\vartheta_0=2\omega_2$ ($j_1=-2$, $j_2=2$). Same formulas as for first principal resonance, with ω_1 replaced by ω_2.

Combination resonance, $\vartheta_0=\omega_1+\omega_2$. Unlike the case of a canonical system, here we must consider two cases: $j_1=-1$, $j_2=2$ and $j_1=-2$, $j_2=1$. We write the coefficients of equation (V. 4.13) in a condensed notation: the upper sign and subscript before the comma refer to the first case, the lower sign and index after the comma to the second:

$$\begin{aligned}\alpha+i\beta=&\mp\frac{1}{2}(\omega_2-\omega_1)\mu+\frac{1}{2}i\bigg\{\frac{\omega_1}{R_1}\bigg[k\gamma_{11}\left(1\pm\frac{\dot{\varphi}_1}{\omega_1}\right)+\varkappa(\omega_1^2+\gamma_{21}r_1)+\\&+\gamma_{22}r_1^2\left(\varkappa+k\pm k\frac{\dot{\varphi}_2}{\omega_1}\right)\bigg]+\frac{\omega_2}{R_2}\bigg[k\gamma_{11}\left(1\mp\frac{\dot{\varphi}_1}{\omega_2}\right)+\varkappa(\omega_2^2+\gamma_{21}r_2)+\\&+\gamma_{22}r_2^2\left(\varkappa+k\mp k\frac{\dot{\varphi}_2}{\omega_2}\right)\bigg]\bigg\},\\ \gamma+i\beta=&\bigg\{\omega_1\mu\pm\frac{i}{R_1}[k\gamma_{11}(\omega_1\pm\dot{\varphi}_1)+\varkappa\omega_1(\omega_1^2+\gamma_{21}r_1)+\\&+\gamma_{22}r_1^2(k\omega_1+\varkappa\omega_1\pm k\dot{\varphi}_2)]\bigg\}\bigg\{\omega_2\mu\pm\frac{i}{R_2}[-k\gamma_{11}(\omega_2\mp\dot{\varphi}_1)-\varkappa\omega_2(\omega_2^2+\gamma_{21}r_2)-\\&-\gamma_{22}r_2^2(k\omega_2+\varkappa\omega_2\mp k\dot{\varphi}_2)]\bigg\}-\frac{1}{R_1R_2\gamma_{12}}\gamma_{21}k^2(\omega_{1,\,2}^2-\gamma_{11})^2(\omega_{1,\,2}+\dot{\varphi}_2)(\omega_{1,\,2}-\dot{\varphi}_1).\end{aligned}$$

Combination resonance at difference frequency, $\vartheta_0=\omega_2-\omega_1$. We consider two cases: $j_1=1$, $j_2=2$ and $j_1=-2$, $j_2=-1$. We distinguish between these cases as above; the resulting condensed formulas for the coefficients of equation (V. 4.13) are

$$\begin{aligned}\alpha+i\beta=&\mp\frac{1}{2}(\omega_1+\omega_2)\mu+\frac{1}{2}i\bigg\{\mp\frac{k}{R_1}[\dot{\varphi}_1\gamma_{11}+\dot{\varphi}_2\gamma_{22}r_1^2]\mp\\&\mp\frac{k}{R_2}(\dot{\varphi}_1\gamma_{11}+\dot{\varphi}_2\gamma_{22}r_2^2)+\frac{\omega_1}{R_1}[k\gamma_{11}+\varkappa\omega_1^2+\varkappa\gamma_{21}r_1+(\varkappa+k)\gamma_{22}r_1^2]+\\&+\frac{\omega_2}{R_2}[k\gamma_{11}+\varkappa\omega_2^2+\varkappa\gamma_{21}r_2+(\varkappa+k)\gamma_{22}r_2^2]\bigg\},\\ \gamma+i\delta=&-\bigg\{\omega_1\mu+i\frac{k}{R_1}(\dot{\varphi}_1\gamma_{11}+\dot{\varphi}_2\gamma_{22}r_1^2)\mp\\&\mp\frac{i\omega_1}{R_1}[k\gamma_{11}+\varkappa\omega_1^2+\varkappa\gamma_{21}r_1+(\varkappa+k)\gamma_{22}r_1^2]\bigg\}\bigg\{\omega_2\mu+i\frac{k}{R_2}(\dot{\varphi}_1\gamma_{11}+\dot{\varphi}_2\gamma_{22}r_2^2)\mp\\&\mp\frac{i\omega_2}{R_2}[k\gamma_{11}+\varkappa\omega_2^2+\varkappa\gamma_{21}r_2+(\varkappa+k)\gamma_{22}r_2^2]\bigg\}-\\&-\frac{1}{R_1R_2\gamma_{12}}k^2\gamma_{21}(\omega_{1,\,2}^2-\gamma_{11})^2(\omega_1\mp\dot{\varphi}_{2,\,1})(\omega_2\mp\dot{\varphi}_{1,\,2}).\end{aligned}$$

The instability domains for combination resonance at the sum $(\vartheta_0=\omega_1+\omega_2)$. and difference $(\vartheta_0=\omega_2-\omega_1)$ frequencies are defined by inequalities (V. 4.14), as in the case of first principal resonance.

4.3. Coaxial shafts with equivalent dynamic stiffness

We illustrate the construction of dynamic instability domains in the parameter space for coaxial shafts with equivalent dynamic stiffness, which is the special case with $\gamma_{22}=\gamma_{11}$ and $\gamma_{21}=\gamma_{12}$. Here

$$\omega_{1,2}^2=\gamma_{11}\mp\gamma_{12}.$$

We begin with the fundamental case (a) of §4.2. Inequalities (4.2) reduce to a single inequality

$$2\omega_1(k\gamma_{11}+\varkappa\omega_1^2)<k\gamma_{11}|\dot{\varphi}_1+\dot{\varphi}_2|.$$

Introducing positive dimensionless parameters

$$\Gamma_{11}=\frac{k\gamma_{11}}{2\omega_1},\ K=\frac{k\gamma_{11}+\varkappa\omega_1^2}{\omega_1}\qquad\left(2\Gamma_{11}<K<2\left(1+\frac{\varkappa}{k}\right)\Gamma_{11}\right),\tag{4.5}$$

we write this inequality as

$$|\dot{\Phi}_1+\dot{\Phi}_2|>\frac{K}{\Gamma_{11}}\qquad\left(\dot{\Phi}_1=\frac{\dot{\varphi}_1}{\omega_1},\ \dot{\Phi}_2=\frac{\dot{\varphi}_2}{\omega_1}\right).\tag{4.6}$$

In the first approximation, the instability domain of flexural vibrations corresponding to case (a) of §4.2 is the exterior of the strip hatched in Figure 34b for the relevant part of the $\dot{\Phi}_1\dot{\Phi}_2$-plane $(\dot{\Phi}_1-\dot{\Phi}_2\geqslant0,\ \dot{\Phi}_1>0)$.

Now for the supplementary case (b) of §4.2. For the first principal resonance, we have from (4.3)

$$\alpha=0,\quad \beta=\frac{1}{2}\omega_1K>0,$$
$$\gamma=\frac{1}{4}\omega_1^2\left[M^2-\Gamma_{11}^2(\dot{\Phi}_1+\dot{\Phi}_2)^2+K^2+\Gamma_{12}^2(\dot{\Phi}_1-1)(\dot{\Phi}_2+1)\right],$$
$$\delta=\frac{1}{2}\omega_1^2\Gamma_{11}(\dot{\Phi}_1+\dot{\Phi}_2)M,$$

where

$$M=\frac{\mu}{\omega_1},\quad \Gamma_{12}=\frac{k\gamma_{12}}{2\omega_1}\qquad(\Gamma_{12}<\Gamma_{11}).$$

Since $\beta>0$, there remains the condition (V. 4.14d)

$$\left[K^2-\Gamma_{11}^2(\dot{\Phi}_1+\dot{\Phi}_2)^2\right]M^2<K^2\left[\Gamma_{11}^2(\dot{\Phi}_1+\dot{\Phi}_2)^2-\Gamma_{12}^2(\dot{\Phi}_1-1)(\dot{\Phi}_2+1)-K^2\right].$$

Transforming from the $\dot{\Phi}_1\dot{\Phi}_2$-plane to the XY-plane by

$$\dot{\Phi}_1 = 1 + \frac{\sqrt{2}}{2}(X - Y), \quad \dot{\Phi}_2 = -1 + \frac{\sqrt{2}}{2}(X + Y)$$

(see Figure 34b), we write the inequality as

$$(K^2 - 2\Gamma_{11}^2 X^2)\, M^2 < \frac{1}{2} K^2 \left[(4\Gamma_{11}^2 - \Gamma_{12}^2)\, X^2 + \Gamma_{12}^2 Y^2 - 2K^2\right]. \tag{4.7}$$

We must investigate this inequality first inside the strip (4.6) (for outside the strip instability in the first approximation is guaranteed by the "general" case (a)), and then inside the strip corresponding to the convergence condition (4.4), i. e., inside the rectangle whose base is shown by a dashed line in Figure 34b:

$$|X| < \frac{1}{\sqrt{2}} \frac{K}{\Gamma_{11}}, \quad |Y| < \sqrt{2}. \tag{4.8}$$

Inside the strip (4.6), $|X| < K/\sqrt{2}\Gamma_{11}$, the left-hand side of inequality (4.7) is positive. The right-hand side of (4.7) is negative inside the ellipse

$$(4\Gamma_{11}^2 - \Gamma_{12}^2)\, X^2 + \Gamma_{12}^2 Y^2 - 2K^2 = 0.$$

The rectangle (4.8) lies entirely inside the ellipse, since

$$(4\Gamma_{11}^2 - \Gamma_{12}^2)\frac{K^2}{2\Gamma_{11}^2} + 2\Gamma_{12}^2 - 2K^2 = 2\left(1 - \frac{1}{4}\frac{K^2}{\Gamma_{11}^2}\right)\Gamma_{12}^2 < 0$$

by (4.5). This means that inequality (4.7) cannot hold inside the rectangle for any values of M, and thus the first principal resonance cannot expand the instability domain for any parameters of the mechanical system.

It is more difficult to draw similar conclusions with regard to the second principal resonance, combination resonance at sum or difference frequency, or, conversely, to expand the instability domain corresponding to the fundamental case for arbitrary parameter values (i. e., in a general setting). This kind of investigation in the $(\dot{\varphi}_1, \dot{\varphi}_2)$-plane, using numerical values of the parameters, would be based on investigation of inequalities (V. 4.14).

§5. PARAMETRIC RESONANCE IN WAVE-PROPAGATION PROBLEMS. ACOUSTIC WAVEGUIDES

5.1. Statement of the problem

The propagation of a scalar (e. g., acoustic) wave in a waveguide is described by the wave equation

$$\Delta U - \frac{1}{v_0^2(\mathbf{r})} \frac{\partial^2 U}{\partial t^2} = 0, \tag{5.1}$$

where Δ is the three-dimensional Laplacian, $v_0(\mathbf{r})$ the propagation velocity of the wave in infinite space, $U(\mathbf{r}, t)$ the potential defining the instantaneous velocity of particles in the vibrating medium, $\mathbf{v}=\operatorname{grad} U$ ($\mathbf{r}=\{x, y, z\}$ is the radius-vector of the observation point). We shall consider a wave which is a harmonic function of time:

$$U(\mathbf{r}, t)=u(\mathbf{r})\exp(-i\omega t).$$

The function $u(\mathbf{r})$ must satisfy the scalar Helmholtz equation

$$\Delta u+k^2u=0, \tag{5.2}$$

where $k=\omega/v_0$ is the wave number, which generally depends on $\mathbf{r}$.

Let us consider the propagation of waves in a three-dimensional region W containing the z-axis and bounded by a surface Π and planes $z=z_0$, $z=z_1$ (Figure 35). We assume that all cross-sections $S(z_i)$ $(z_0\leqslant z_i\leqslant z_1)$ of the interior of the waveguide W by planes $z=z_i$ are simply-connected and bounded. Propagation of a monochromatic wave at frequency ω in the interior of the waveguide is described by equation (5.2), and the boundary condition for a perfectly rigid boundary is

$$\left.\frac{\partial u}{\partial n}\right|_{\Pi}=0, \tag{5.3}$$

where $\mathbf{n}$ is the outer normal to the surface Π. Below we shall investigate either the more general, mixed boundary-value problem

$$\left[\frac{\partial u}{\partial n}+\alpha(\mathbf{r})\,u\right]_{\Pi}=0, \tag{5.4}$$

where $\alpha(\mathbf{r}_\Pi)$ is some real function defined on Π $(\mathbf{r}_\Pi\in\Pi)$ or the boundary-value problem

$$u|_{\Pi}=0, \tag{5.5}$$

corresponding to a perfectly soft boundary.

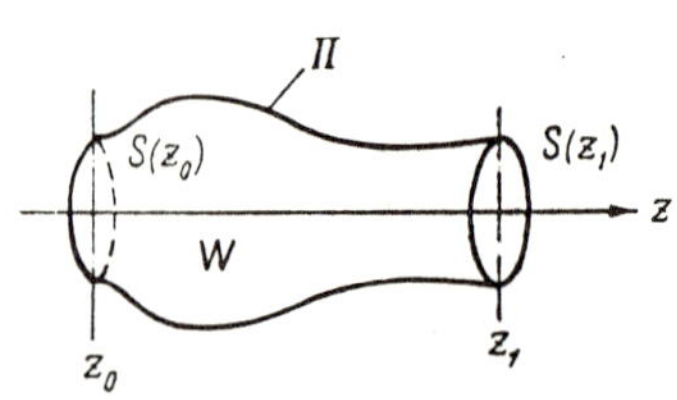

FIGURE 35.

Practical interest attaches to waveguides which are periodic along the z-axis, in which either α or k^2 is a periodic function of z, of period D, or the boundary is repeated at regular intervals D. The solutions of the boundary-value problem (5.2) – (5.4) (or (5.2)–(5.5)) satisfying the condition

$$u(\mathbf{r}+D\mathbf{z}^0)=\rho u(\mathbf{r}), \tag{5.6}$$

where ρ is a complex number and $\mathbf{z}^0$ a unit vector along the z-axis, are known as Floquet solutions. The multiplier ρ and Floquet solution u will depend on the parameters defining the boundary, and possibly also on those figuring in the functions k^2 and α. Under certain assumptions, u and ρ will be continuous functions of these parameters. If ρ lies on the unit circle throughout some range of variation of one of the parameters, but this is not true for any larger range, the interval will be called the pass-band for u with respect to the parameter in question. Determination of pass-bands is an important practical problem. In §§5.4 and 5.5 we shall develop formulas for pass-bands in the first (and sometimes also second) approximation. Another important problem, to be studied in §5.6, is to determine the phase of the multiplier (Floquet parameter), since it is linked directly to the phase and group velocities of the waves.

5.2. Reduction to ordinary differential equations

The boundary-value problems (5.2) – (5.4) and (5.2) – (5.5) need not be uniquely solvable; this may be due, for example, to the fact that the surface Π $(z_0\leqslant z\leqslant z_1)$ on which the boundary conditions (5.4) and (5.5) are specified is not closed.

Let us consider the problems in the variational setting. For problem (5.2)–(5.4), we define a functional

$$J(u)=\iiint\limits_W [(\operatorname{grad} u)^2-k^2u^2]\,dV+\iint\limits_\Pi \alpha u^2\,d\sigma. \tag{5.7}$$

The variation of this functional is

$$\delta J=\left[\frac{\partial}{\partial\varepsilon}J(u+\varepsilon\xi(x,y,z))\right]_{\varepsilon=0}=$$
$$=2\iiint\limits_W [\operatorname{div}(\xi\operatorname{grad} u)-(\Delta u+k^2u)\xi]\,dV+2\iint\limits_\Pi \alpha u\xi\,d\sigma$$

or, by the Gauss-Ostrogradskii formula,

$$\delta J=-2\iiint\limits_W(\Delta u+k^2u)\xi\,dV+2\iint\limits_\Pi\left(\frac{\partial u}{\partial n}+\alpha u\right)\xi\,d\sigma+$$
$$+2\iint\limits_{S(z_0)}\frac{\partial u}{\partial n}\xi\,dS+2\iint\limits_{S(z_1)}\frac{\partial u}{\partial n}\xi\,dS. \tag{5.8}$$

We shall assume that the "fixed-end" conditions are satisfied:

$$\xi(x,y,z_0)=\xi(x,y,z_1)\equiv 0. \tag{5.9}$$

Then $\delta J=0$ for any solution of problem (5.2) — (5.4). The boundary condition (5.4) turns out to be natural for this functional.

For the boundary-value problem (5.2)—(5.5), we consider the functional (5.7) with $\alpha=0$, defined on functions u satisfying the boundary condition (5.5), which is not natural for this functional. For any solution of problem (5.2)—(5.5), we again have $\delta J=0$ provided the variation is subject to conditions (5.9).

According to the method of Galerkin-Ritz-Kantorovich /20, 47/, we look for the solution to the original problem as a finite sum*

$$u(x, y, z)=\sum_{m=0}^{M-1}\sum_{n=0}^{N-1}\eta_n^m(x, y, z)f_n^m(z), \tag{5.10}$$

where $\eta_n^m(x, y, z)$ $(m=0, 1, \ldots, M-1;\ n=0, 1, \ldots, N-1)$ is some system of functions defined in W, satisfying the following additional condition in the case of problem (5.2)—(5.4):

$$\eta_n^m(x, y, z)|_{\Pi}=0 \qquad (m=0, 1, \ldots, M-1;\ n=0, 1, \ldots, N-1). \tag{5.11}$$

The functions $f_n^m(z)$ are unknown and are to be determined in such a way that the functional (5.7) attains an extremum.

Substituting (5.10) into (5.7), we obtain

$$J(u)=\sum_{m,\,n}\sum_{m',\,n'}\left(\int_{z_0}^{z_1}dz\iint_{S(z)}\left\{\left[(\operatorname{grad}\eta_n^m,\ \operatorname{grad}\eta_{n'}^{m'})-k^2\eta_n^m\eta_{n'}^{m'}\right]f_n^m f_{n'}^{m'}+\right.\right.$$
$$\left.\left.+\eta_n^m\eta_{n'}^{m'}\frac{df_n^m}{dz}\frac{df_{n'}^{m'}}{dz}+\frac{\partial\eta_n^m}{\partial z}\eta_{n'}^{m'}f_n^m\frac{df_{n'}^{m'}}{dz}+\frac{\partial\eta_{n'}^{m'}}{\partial z}\eta_n^m\frac{df_n^m}{dz}f_{n'}^{m'}\right\}dS+\iint_{\Pi}\alpha\eta_n^m\eta_{n'}^{m'}f_n^m f_{n'}^{m'}d\sigma\right).$$

Varying this expression for $J(u)$, subject to the condition that the values of $f_n^m(z)$ remain unchanged at z_0 and z_1, we obtain a finite system of ordinary differential equations for $f_n^m(z)$, which we shall now investigate.

We shall assume throughout that the waveguide boundary Π is defined by an equation in cylindrical coordinates

$$\rho=R(z, \varphi) \qquad (z_0\leqslant z\leqslant z_1,\ 0\leqslant\varphi\leqslant 2\pi,\ \rho=\sqrt{x^2+y^2}), \tag{5.12}$$

and is moreover smooth. Then the condition $\delta J(u)=0$ for variation with fixed ends (5.9) leads to the equation system

$$\frac{d}{dz}\sum_{m'=0}^{M-1}\sum_{n'=0}^{N-1}\left[G_{nn'}^{mm'}\frac{df_{n'}^{m'}}{dz}+Q_{nn'}^{mm'}f_{n'}^{m'}\right]-\sum_{m',\,n'}Q_{n'n}^{m'm}\frac{df_{n'}^{m'}}{dz}+\sum_{m',\,n'}P_{nn'}^{mm'}f_{n'}^{m'}=0 \tag{5.13}$$
$$(m=0, 1, \ldots, M-1;\ n=0, 1, \ldots, N-1),$$

where

$$\left.\begin{aligned}&G_{nn'}^{mm'}(z)=\iint_{S(z)}\eta_n^m\eta_{n'}^{m'}dS, \quad Q_{nn'}^{mm'}(z)=\iint_{S(z)}\eta_n^m\frac{\partial\eta_{n'}^{m'}}{\partial z}dS,\\ &P_{nn'}^{mm'}(z)=k^2G_{nn'}^{mm'}-\iint_{S(z)}(\operatorname{grad}\eta_n^m,\ \operatorname{grad}\eta_{n'}^{m'})dS-\int_0^{2\pi}(\eta_n^m\eta_{n'}^{m'})_{\Pi}\frac{\alpha R(z,\varphi)}{c_z c_\varphi}d\varphi,\end{aligned}\right\} \tag{5.14}$$

* We have chosen a two-index notation only for reasons of convenience. The problem with which we started is obtained from this one by letting $M\to\infty$, $N\to\infty$, but we shall not go into this here.

with $c_z = |\cos(\tau_z, \mathbf{e}_z)|$, $c_\varphi = |\cos(\tau_\varphi, \mathbf{e}_\varphi)|$; τ_z and τ_φ are unit vectors tangent to Π, τ_z in a plane through the z-axis and τ_φ in a plane perpendicular to the z-axis. System (5.13) may be written in vector notation as

$$\frac{d}{dz}\left[\mathbf{G}\frac{d\mathbf{f}}{dz}+\mathbf{Q}\mathbf{f}\right]-\mathbf{Q}^{\tau}\frac{d\mathbf{f}}{dz}+\mathbf{P}\mathbf{f}=0, \tag{5.15}$$

where the superscript τ denotes transposition. The components of the vector $\mathbf{f}$ will be arranged henceforth as follows:

$$\mathbf{f}=(f_0^0, f_1^0, \ldots, f_{N-1}^0, f_0^1, f_1^1, \ldots, f_{N-1}^1, \ldots, f_0^{M-1}, f_1^{M-1}, \ldots, f_{N-1}^{M-1})^{\tau}.$$

The arrangement of the elements in each of the matrices will be

$$\mathbf{A}=\|\mathbf{A}_{mm'}\|=\begin{pmatrix}\mathbf{A}_{00} & \cdots & \mathbf{A}_{0M-1}\\ \cdot & \cdot\cdot\cdot & \cdot \\ \mathbf{A}_{M-1,0} & \cdots & \mathbf{A}_{M-1\,M-1}\end{pmatrix},$$

where

$$\mathbf{A}_{mm'}=\|A_{nn'}^{mm'}\|=\begin{Vmatrix}A_{00}^{mm'} & \cdots & A_{0N-1}^{mm'}\\ \cdot & \cdot\cdot\cdot & \cdot \\ A_{N-1,0}^{mm'} & \cdots & A_{N-1\,N-1}^{mm'}\end{Vmatrix}.$$

Here $\mathbf{A}$ is any of the matrices $\mathbf{G}$, $\mathbf{Q}$, $\mathbf{P}$, i.e., $A_{nn'}^{mm'}$ is either $G_{nn'}^{mm'}$, $Q_{nn'}^{mm'}$ or $P_{nn'}^{mm'}$. It is readily seen that

$$G_{nn'}^{mm'}=G_{n'n}^{m'm}, \quad P_{nn'}^{mm'}=P_{n'n}^{m'm} \qquad (m, m'=0, 1, \ldots, M-1;\ n, n'=0, 1, \ldots, N-1) \tag{5.16}$$

so that $\mathbf{G}$ and $\mathbf{P}$ are symmetric matrices.

In accordance with the discussion in §5.1, we shall be interested only in Floquet solutions of equation (5.15), to be precise, in its multipliers, since, as mentioned in §5.1, stop-bands correspond to multipliers lying outside the unit circle.

Setting

$$\mathbf{g}=\mathbf{G}\frac{d\mathbf{f}}{dz}+\mathbf{Q}\mathbf{f} \quad \text{and} \quad \mathbf{h}=\begin{pmatrix}\mathbf{f}\\ \mathbf{g}\end{pmatrix},$$

we obtain a canonical equation for the vector $\mathbf{h}$:

$$\mathbf{J}\frac{d\mathbf{h}}{dz}=\mathbf{H}(z)\,\mathbf{h}, \tag{5.17}$$

where the skew-symmetric constant matrix $\mathbf{J}$ and the symmetric matrix-function $\mathbf{H}(z)$ are real square matrices of order $2MN$:

$$\mathbf{J}=\begin{pmatrix}0 & -\mathbf{I}_{MN}\\ \mathbf{I}_{MN} & 0\end{pmatrix}, \quad \mathbf{H}(z)=\begin{pmatrix}\mathbf{P}+\mathbf{Q}^{\tau}\mathbf{G}^{-1}\mathbf{Q} & -\mathbf{Q}^{\tau}\mathbf{G}^{-1}\\ -\mathbf{G}^{-1}\mathbf{Q} & \mathbf{G}^{-1}\end{pmatrix}, \tag{5.18}$$

and $\mathbf{I}_{MN}$ is the identity matrix of order MN.

Throughout the sequel (except in the first half of §5.3) we shall restrict ourselves to an azimuthally symmetric waveguide, where $k^2 = k^2(z)$, $\alpha = \alpha(z)$, and Π is a surface of revolution:

$$\rho = R(z) \qquad (z_0 \leqslant z \leqslant z_1). \tag{5.19}$$

To simplify the account, we shall choose the functions η_n^m in (5.10) so that the following orthogonality conditions hold identically in z:

$$G_{nn'}^{mm'}(z) = \delta_{mm'}\delta_{nn'}F(z) \qquad (m, m' = 0, 1, \ldots, M-1;\ n, n' = 0, 1, \ldots, N-1),$$

where δ_{pq} is the Kronecker symbol and $F(z)$ $(z_0 \leqslant z \leqslant z_1)$ some scalar function. Then $\mathbf{G} = F(z)\,\mathbf{I}_{MN}$. For example, these conditions hold for the functions*

$$\eta_n^m(\mathbf{r}) = [(1+\delta_{0m})\pi]^{-1/2}\,\psi_n^m(\rho, z)\cos m\varphi$$
$$(m = 0, 1, \ldots, M-1;\ n = 0, 1, \ldots, N-1), \tag{5.20}$$

where

$$\psi_n^m(\rho, z) = \begin{cases} \sqrt{2}\,\mu_n^m J(\mu_n^m)^{-1}[(\mu_n^m)^2 - m^2]^{-1/2} J_m\left(\mu_n^m \dfrac{\rho}{R(z)}\right) \text{for problem (5.2)—(5.4)}, \\ \sqrt{2}\,J_{m+1}(\nu_n^m)^{-1} J_m\left(\nu_n^m \dfrac{\rho}{R(z)}\right) \text{for problem (5.2)—(5.5)}. \end{cases} \tag{5.21}$$

Here $\rho = \sqrt{x^2+y^2}$, $J_m(x)$ are cylindrical harmonics (Bessel functions) of the first kind and m-th order, μ_n^m and ν_n^m $(m=0, 1, \ldots, M-1;\ n=0, 1, \ldots, N-1)$ are the $(n+1)$-th positive roots of the equations $J_m'(x) = 0$ and $J_m(x) = 0$, respectively. By the first of formulas (5.14), (5.21),

$$G_{nn'}^{mm'}(z) = \frac{1}{(1+\delta_{0m})\pi}\int_0^{R(z)} \psi_n^m(\rho, z)\,\psi_{n'}^{m'}(\rho, z)\,\rho\,d\rho \int_0^{2\pi} \cos m\varphi \cos m'\varphi\,d\varphi =$$
$$= \delta_{mm'} R^2(z)\frac{2\mu_n^m \mu_{n'}^m}{J_m(\mu_n^m) J_m(\mu_{n'}^m)}[(\mu_n^m)^2 - m^2]^{-1/2}[(\mu_{n'}^m)^2 - m^2]^{-1/2}\int_0^1 J_m(\mu_n^m x)\,J_m(\mu_{n'}^m x)\,x\,dx.$$

Using formulas /84/ for the integrals of cylindrical harmonics, we find, for problem (5.2)—(5.4),

$$G_{nn'}^{mm'}(z) = \delta_{mm'}\delta_{nn'}R(z)^2 \qquad (m, m' = 0, 1, \ldots, M-1;\ n, n' = 0, 1, \ldots, N-1). \tag{5.22}$$

The same formulas (5.22) may be established similarly for problem (5.2)—(5.5).

5.3. Simple special cases

1) Suppose that the shape of the boundary Π and the function $\alpha(\mathbf{r}_\Pi)$ are independent of z, so that the waveguide boundary is the surface of a cylinder

* To the functions (5.20) one should add the functions $\pi^{-1/2}\psi_n^m(\rho, z)\sin m\varphi$. However, these may be derived from (5.20) by the simple substitution $\varphi = \varphi' - \dfrac{\pi}{2m}$.

with an arbitrary, smooth, non-self-intersecting directrix. Then $R=R(\varphi)$, $\alpha=\alpha(\varphi)$ and the functions η_n^m in (5.20) are independent of z. Therefore, by (5.14), $\mathbf{Q}=\mathbf{0}$, and the matrices $\mathbf{G}$ and $\mathbf{P}$ are constant. Equation (5.15) becomes

$$\frac{d^2\mathbf{f}}{dz^2}+\mathbf{P}_1\mathbf{f}=0 \qquad (\mathbf{P}_1=\mathbf{G}^{-1}\mathbf{P}), \tag{5.23}$$

and the general solution is

$$\mathbf{f}=\cos(\mathbf{P}_1^{1/2}z)\,\mathbf{f}_0+\mathbf{P}_1^{-1/2}\sin(\mathbf{P}_1^{1/2}z)\left(\frac{d\mathbf{f}}{dz}\right)_0, \tag{5.24}$$

where $\cos(\mathbf{P}_1^{1/2}z)$ and $\mathbf{P}_1^{-1/2}\sin(\mathbf{P}_1^{1/2}z)$ are entire functions of $\mathbf{P}_1$ (depending on z as a parameter). Since $\mathbf{P}_1$ is symmetric, the general solution (5.24) is a linear combination of functions $\cos(h_jz)$, $\sin(h_jz)$, where h_j are real or pure imaginary numbers — the square roots of the eigenvalues of $\mathbf{P}_1$. The real h_j's will be called the *propagation constants* of the waves. The pure imaginary h_j's correspond to attenuating or increasing vibrations; the latter alternative may be rejected on the basis of physical considerations. The explicit form of the solution $\mathbf{f}(z)$ is conveniently determined in the usual manner (§ II. 1.6). Let $\mathbf{P}_1$ denote the Jordan form of $\mathbf{P}_1$: $\tilde{\mathbf{P}}_1=\mathbf{S}\tilde{\mathbf{P}}_1\mathbf{S}^{-1}$, where $\tilde{\mathbf{P}}_1$ is real diagonal and $\mathbf{S}$ a real nonsingular matrix. The linear transformation $\mathbf{f}=\mathbf{S}\tilde{\mathbf{f}}$ reduces equation (5.23) to

$$\frac{d^2\tilde{\mathbf{f}}}{dz^2}+\tilde{\mathbf{P}}_1\tilde{\mathbf{f}}=0.$$

The general solution of this equation is (§ II. 1.6)

$$\tilde{\mathbf{f}}=\cos(\tilde{\mathbf{P}}_1^{1/2}z)\,\tilde{\mathbf{f}}_0+\tilde{\mathbf{P}}_1^{-1/2}\sin(\tilde{\mathbf{P}}_1^{1/2}z)\left(\frac{d\tilde{\mathbf{f}}}{dz}\right)_0,$$

where $\cos(\tilde{\mathbf{P}}_1^{1/2}z)$ and $\tilde{\mathbf{P}}_1^{-1/2}\sin(\tilde{\mathbf{P}}_1^{1/2}z)$ are diagonal matrices. Substituting $\tilde{\mathbf{f}}=\mathbf{S}^{-1}\mathbf{f}$, we get an expression for $\mathbf{S}^{-1}\mathbf{f}$ as a combination of the natural modes of the waveguide, varying as harmonic functions of z:

$$\mathbf{S}^{-1}\mathbf{f}=\cos(\tilde{\mathbf{P}}_1^{1/2}z)\,\mathbf{S}^{-1}\mathbf{f}_0+\tilde{\mathbf{P}}_1^{-1/2}\sin(\tilde{\mathbf{P}}_1^{1/2}z)\,\mathbf{S}^{-1}\left(\frac{d\mathbf{f}}{dz}\right)_0.$$

2) Let the boundary Π be a surface of revolution and the function $\alpha(\mathbf{r}_\Pi)$ in (5.4) independent of the azimuth φ (azimuthal symmetry). Then the function R in (5.12) is independent of φ: $R=R(z)$ and the coefficients

$$G_{nn'}^{mm'},\ Q_{nn'}^{mm'},\ P_{nn'}^{mm'}$$

are proportional to $\delta_{mm'}$. System (5.13) thus separates into M independent subsystems

$$\frac{d}{dz}\sum_{n'=0}^{N-1}\left[G_{nn'}^{mm}\frac{df_{n'}^m}{dz}+Q_{nn'}^{mm}f_{n'}^m\right]-\sum_{n'=0}^{N-1}Q_{n'n}^{mm}\frac{df_{n'}^m}{dz}+\sum_{n'=0}^{N-1}P_{nn'}^{mm}f_{n'}^m=0 \qquad (5.25)$$
$$(n=0,\ 1,\ \ldots,\ N-1),$$

corresponding to fixed values of m ($m=0,1,\ \ldots,\ M-1$), pointing to the independence of waves with different azimuthal numbers m.

As an example combining cases 1 and 2, consider a uniform cylindrical waveguide of circular cross-section: $R=a=\text{const},\ \alpha=\text{const}$. In conformity with the previous discussion, we consider the subsystem (5.25) for which $m=0$, corresponding to propagation of an azimuthally symmetric wave. By (5.14), (5.20), we have $G_{nn'}^{00}=a^2\delta_{nn'}$, $Q_{nn'}^{00}=0$,

$$P_{nn'}^{00}=\begin{cases}[k^2a^2-(\mu_n^0)^2]\,\delta_{nn'}-2\alpha a & \text{for problem (5.2)}-\text{(5.4)},\\ [k^2a^2-(\nu_n^0)^2]\,\delta_{nn'} & \text{for problem (5.2)}-\text{(5.5)}.\end{cases}$$

We confine attention to the first problem, for small α. Determine the eigenvalues of the matrix

$$\mathbf{G}^{-1}\mathbf{P}(\alpha)=\frac{1}{a^2}\|P_{nn'}^{00}\|=k^2\mathbf{I}_N-\text{diag}\left(\frac{\mu_0^n}{a}\right)^2-2\frac{\alpha}{a}\mathbf{E}_N,$$

where $\mathbf{I}_N$ and diag are $N\times N$ identity and diagonal matrices, and $\mathbf{E}_N$ is the $N\times N$ matrix all of whose elements are equal to 1. In the notation introduced at the end of §IV. 1.1, we have

$$\beta_0=k^2-\left(\frac{\mu_n^0}{a}\right)^2,\quad \mathbf{L}=\mathbf{E}_N,\quad \varepsilon=-2\frac{\alpha}{n},\quad \mathbf{a}_1=\mathbf{e}_{n+1}=\mathbf{b}_1,$$
$$(\mathbf{La}_1,\ \mathbf{b}_1)=(\mathbf{E}_N\mathbf{e}_{n+1},\ \mathbf{e}_{n+1})=1,$$

where $\mathbf{e}_{n+1}$ is the $(n+1)$-th vector of the standard basis. By formula (IV. 1.27), the eigenvalues of the matrix $\mathbf{G}^{-1}\mathbf{P}(\alpha)$ are

$$\beta_n\equiv(h_n^0)^2=k^2-\left(\frac{\mu_n^0}{a}\right)^2-2\frac{\alpha}{a}+O(\alpha^2) \qquad (n=0,\ 1,\ \ldots,\ N-1).$$

Hence, up to $O(\alpha^2)$, we obtain the propagation constants in problem (5.2)—(5.4) as

$$h_n^0=\sqrt{k^2-\left(\frac{\mu_n^0}{a}\right)^2-2\frac{\alpha}{a}} \qquad (n=0,\ 1,\ \ldots,\ N-1), \qquad (5.26)$$

which coincide at $\alpha=0$ with the familiar expressions for a circularly cylindrical waveguide with rigid walls. For each n, we have a critical wave number k defined by

$$k_{\text{cr}}^{(n)}=\sqrt{\left(\frac{\mu_n^0}{a}\right)^2+2\frac{\alpha}{a}},$$

which in turn determines a critical frequency $\omega_{cr}^{(n)}=k_{cr}^{(n)}v_0$, such that at $\omega>\omega_{cr}^{(n)}$ the wave can propagate in the waveguide (h_n^0 is real), but not at $\omega<\omega_{cr}^{(n)}$.

5.4. Waveguide with periodic boundary and constant wave number. Pass-bands in first approximation

Let us assume that the coefficient α in the boundary condition (5.4) is constant, or that the boundary condition (5.5) obtains. We can distinguish two principal cases in which the matrix $\mathbf{H}$ of (5.17) depends on z.

I) Noncylindrical waveguide. In this case the function R in (5.12) depends on z. Thus the functions η_n^m defined in (5.20), (5.21) also depend on z, as does the domain of integration $S(z)$ in (5.14).

II) Inhomogeneously filled waveguide. The propagation velocity $v_0(\mathbf{r})$ of the wave in infinite space, and hence also k^2 in (5.2), depend on z.

Throughout the sequel we shall assume that the waveguide has a periodic structure, with period D. Then $\mathbf{H}(z)$ is D-periodic. In this subsection we consider case (I); case (II) will be studied in §5.5.

We are assuming, then, that the boundary Π has a period D along the z-axis, and that it is a surface of revolution: equation (5.12) is

$$\rho=R(z) \qquad (z_0\leqslant z\leqslant z_1,\ R(z+D)\equiv R(z)). \tag{5.27}$$

Let us evaluate the coefficients (5.14) in view of (5.20), (5.21) and (5.22). Using the formula for the gradient in cylindrical coordinates, formula (5.22), and also the fact that in our case $c_\varphi=1$, $c_z^{-1}=\sqrt{1+R'(z)^2}$, we obtain

a) for problem (5.2)—(5.4):

$$\begin{aligned} P_{nn'}^{mm'}={}&k^2R^2(z)\,\delta_{mm'}\delta_{nn'}-2\delta_{mm'}\mu_n^m\mu_{n'}^{m'}\,[(\mu_n^m)^2-m^2]^{-1/2}\,[(\mu_{n'}^{m'})^2-m'^2]^{-1/2}\times\\ &\times J_m(\mu_n^m)^{-1}J_{m'}(\mu_{n'}^{m'})^{-1}\Big\{\mu_n^m\mu_{n'}^{m'}\int_0^1 J_m'(\mu_n^m x)\,J_{m'}'(\mu_{n'}^{m'}x)\,x\,dx+\\ &+mm'\int_0^1 J_m(\mu_n^m x)\,J_{m'}(\mu_{n'}^{m'}x)\,\frac{dx}{x}+R'(z)^2\mu_n^m\mu_{n'}^{m'}\int_0^1 J_m'(\mu_n^m x)\,J_{m'}'(\mu_{n'}^{m'}x)\,x^3\,dx\Big\}-\\ &-2\delta_{mm'}\mu_n^m\mu_{n'}^{m'}\,[(\mu_n^m)^2-m^2]^{-1/2}\,[(\mu_{n'}^{m'})^2-m'^2]^{-1/2}\,\alpha R(z)\sqrt{1+R'(z)^2},\\ Q_{nn'}^{mm'}={}&-2\delta_{mm'}\mu_n^m(\mu_{n'}^{m'})^2\,[(\mu_n^m)^2-m^2]^{-1/2}\,[(\mu_{n'}^{m'})^2-m'^2]^{-1/2}\,J_m(\mu_n^m)^{-1}\,J_{m'}(\mu_{n'}^{m'})^{-1}\times\\ &\times R(z)\,R'(z)\int_0^1 J_m(\mu_n^m x)\,J_{m'}'(\mu_{n'}^{m'}x)\,x^2\,dx. \end{aligned}$$

Consequently, in this case (azimuthal symmetry) system (5.13) separates into subsystems with $m=0,1,\dots,M-1$. Using the formulas for integrals of cylindrical harmonics (see appendix at the end of this section), we determine the coefficients of each of the subsystems (fixed m):

$$\begin{aligned} &G_{nn'}^{mm}=R(z)^2\delta_{nn'},\quad P_{nn'}^{mm}=[k^2R(z)^2-(\mu_n^m)^2]\delta_{nn'}-R'(z)^2s_{nn'}^{m\mu}-\\ &\quad-\alpha R(z)\sqrt{1+R'(z)^2}\,t_{nn'}^{m\mu},\quad Q_{nn'}^{mm}=R(z)\,R'(z)\,v_{nn'}^{m\mu} \qquad (m=0,1,\dots,M-1),\\ &G_{nn'}^{mm'}=P_{nn'}^{mm'}=Q_{nn'}^{mm'}=0 \qquad (m,\ m'=0,1,\dots,M-1;\ m'\neq m) \end{aligned}$$

for $n, n'=0, 1, \ldots, N-1$. We have used the notation

$$\left.\begin{aligned}
s_{nn'}^{m\mu} &= 4\left(\frac{\mu_n^m}{\mu_{n'}^m}-\frac{\mu_{n'}^m}{\mu_n^m}\right)^{-2}\left[1-\frac{m^2}{(\mu_n^m)^2}\right]^{-1/2}\left[1-\frac{m^2}{(\mu_{n'}^m)^2}\right]^{-1/2}\times\\
&\quad\times\{2-m^2\,[(\mu_n^m)^{-2}+(\mu_{n'}^m)^{-2}]\} \qquad (n'\neq n),\\
s_{nn}^{m\mu} &= 2\,(\mu_n^m)^4\,[(\mu_n^m)^2-m^2]^{-1}J_m(\mu_n^m)^{-2}\int_0^1 J_m'(\mu_n^m x)^2\,x^3\,dx,\\
t_{nn'}^{m\mu} &= 2\mu_n^m\mu_{n'}^m\,[(\mu_n^m)^2-m^2]^{-1/2}\,[(\mu_{n'}^m)^2-m^2]^{-1/2},\\
v_{nn'}^{m\mu} &= 2\mu_n^m\mu_{n'}^m\,[(\mu_n^m)^2-(\mu_{n'}^m)^2]^{-1}\,[(\mu_{n'}^m)^2-m^2)^{1/2}\,[(\mu_n^m)^2-m^2]^{-1/2}\\
&\qquad (n'\neq n),\\
v_{nn}^{m\mu} &= -m^2\,[(\mu_n^m)^2-m^2]^{-1}.
\end{aligned}\right\}\qquad(5.28)$$

For each of the subsystems we have a vector equation (5.15) with

$$\left.\begin{aligned}
&\mathbf{G}(z)=R(z)^2\,\mathbf{I}_N,\\
&\mathbf{P}(z)=k^2R(z)^2\mathbf{I}_N-\operatorname{diag}(\mu_n^m)^2-R'(z)^2\mathbf{S}-\alpha R(z)\sqrt{1+R'(z)^2}\,\mathbf{T},\\
&\mathbf{Q}(z)=R(z)\,R'(z)\,\mathbf{V},\\
&\quad\mathbf{S}=\|s_{nn'}^{m\mu}\|,\quad \mathbf{T}=\|t_{nn'}^{m\mu}\|,\quad \mathbf{V}=\|v_{nn'}^{m\mu}\|,
\end{aligned}\right\}\qquad(5.29)$$

where all the matrices are of order $N\times N$ $(n, n'=0, 1, \ldots, N-1)$ and m is some fixed number from the set $0, 1, \ldots, M-1$; the superscript μ emphasizes that (5.28) and (5.29) are valid for problem (5.2)—(5.4).

b) For problem (5.2) — (5.5), we again have a vector equation (5.15) for each $m=0, 1, \ldots, M-1$, with

$$\left.\begin{aligned}
&\mathbf{G}(z)=R(z)^2\,\mathbf{I}_N,\quad \mathbf{P}(z)=k^2R(z)\,\mathbf{I}_N-\operatorname{diag}(\nu_n^m)^2-R'(z)^2\,\|s_{nn'}^{m\nu}\|,\\
&\mathbf{Q}(z)=R(z)\,R'(z)\,\|v_{nn'}^{m\nu}\|.
\end{aligned}\right\}\qquad(5.30)$$

The notation here is $(n, n'=0, 1, \ldots, N-1)$

$$\left.\begin{aligned}
&s_{nn'}^{m\nu}=\nu_n^m\nu_{n'}^m\,[(\nu_n^m)^2+(\nu_{n'}^m)^2]\,[(\nu_n^m)^2-(\nu_{n'}^m)^2]^{-1}\qquad(n'\neq n),\\
&s_{nn}^{m\nu}=2\,(\nu_n^m)^2\,J_m'(\nu_n^m)^{-2}\int_0^1 J_m'(\nu_n^m x)^2\,x^3\,dx,\\
&v_{nn'}^{m\nu}=2\nu_n^m\nu_{n'}^m\,[(\nu_n^m)^2-(\nu_{n'}^m)^2]^{-1}\qquad(n'\neq n),\quad v_{nn}^{m\nu}=1,
\end{aligned}\right\}\qquad(5.31)$$

and the superscript ν indicates that these formulas refer to problem (5.2)—(5.5).

Let us assume that the surface of revolution Π differs by a small amount from a right circular cylinder. Then equation (5.27) in cylindrical coordinates will be

$$\rho=R(z)\equiv a\,[1+\varepsilon g(\vartheta z)]\quad\left(g(s+2\pi)\equiv g(s),\quad g_{\mathrm{av}}\equiv\frac{1}{2\pi}\int_0^{2\pi}g(s)\,ds=0\right),\qquad(5.32)$$

where $\varepsilon>0$ is a small parameter, $\vartheta=2\pi/D$ is the perturbation frequency of the waveguide surface. We shall need various functions related to $R(z)$, which we collect here for future reference:

$$\left.\begin{aligned}
R'(z)^2 &= a^2\varepsilon^2\vartheta^2\dot{g}(\vartheta z)^2,\\
R(z)^{-2} &= a^{-2}\left[1-2\varepsilon g(\vartheta z)+3\varepsilon^2 g(\vartheta z)^2+O(\varepsilon^3)\right],\\
R(z)^2 &= a^2\left[1+2\varepsilon g(\vartheta z)+\varepsilon^2 g(\vartheta z)^2\right],\\
R(z)\sqrt{1+R'(z)^2} &= a\left[1+\varepsilon g(\vartheta z)+\frac{1}{2}a^2\varepsilon^2\vartheta^2\dot{g}(\vartheta z)^2+O(\varepsilon)^3\right],\\
\frac{R'(z)}{R(z)} &= \varepsilon\vartheta\dot{g}(\vartheta z)-\varepsilon^2\vartheta g(\vartheta z)\dot{g}(\vartheta z)+O(\varepsilon^3)
\end{aligned}\right\} \tag{5.33}$$

(the dot denotes differentiation with respect to ϑz).

From here until further notice we confine the discussion to problem (5.2)—(5.4). Expand the matrix coefficients of equations (5.15), (5.17) in powers of ε, using (5.29) and (5.33):

$$\left.\begin{aligned}
&\mathbf{P}(\vartheta z,\ \varepsilon)=\mathbf{P}_0+\varepsilon\mathbf{P}_1(\vartheta z)+\varepsilon^2\mathbf{P}_2(\vartheta z)+\ldots,\\
&\mathbf{H}(\vartheta z,\ \varepsilon)=\mathbf{H}_0+\varepsilon\mathbf{H}_1(\vartheta z)+\varepsilon^2\mathbf{H}_2(\vartheta z)+\ldots=\\
&=\begin{pmatrix}\mathbf{P}_0 & 0\\ 0 & a^{-2}\mathbf{I}_N\end{pmatrix}+\varepsilon\begin{pmatrix}\mathbf{P}_1 & -\vartheta\dot{g}(\vartheta z)\mathbf{V}^{\tau}\\ -\vartheta\dot{g}(\vartheta z)\mathbf{V} & -2a^{-2}g(\vartheta z)\mathbf{I}_N\end{pmatrix}+\\
&+\varepsilon^2\begin{pmatrix}\mathbf{P}_2+a^2\vartheta^2\dot{g}(\vartheta z)^2\mathbf{V}^{\tau}\mathbf{V} & \vartheta g(\vartheta z)\dot{g}(\vartheta z)\mathbf{V}^{\tau}\\ \vartheta g(\vartheta z)\dot{g}(\vartheta z)\mathbf{V} & 3a^{-2}g(\vartheta z)^2\mathbf{I}_N\end{pmatrix}+O(\varepsilon^3),
\end{aligned}\right\} \tag{5.34}$$

where

$$\left.\begin{aligned}
&\mathbf{P}_0=k^2a^2\mathbf{I}_N-\operatorname{diag}(\mu_n^m)^2-\alpha a\mathbf{T},\\
&\mathbf{P}_1=2k^2a^2g(\vartheta z)\,\mathbf{I}_N-\alpha a g(\vartheta z)\,\mathbf{T},\\
&\mathbf{P}_2=k^2a^2g(\vartheta z)^2\,\mathbf{I}_N-\frac{1}{2}\alpha a^3\vartheta^2\dot{g}(\vartheta z)^2\,\mathbf{T}-a^2\vartheta^2\dot{g}(\vartheta z)^2\mathbf{S}.
\end{aligned}\right\} \tag{5.35}$$

Consider the canonical equation (5.17)

$$\mathbf{J}\frac{d\mathbf{h}}{dz}=[\mathbf{H}_0+\varepsilon\mathbf{H}_1(\vartheta z)+\varepsilon^2\mathbf{H}_2(\vartheta z)+\ldots]\,\mathbf{h}, \tag{5.36}$$

in which the coefficients are defined by (5.34) and (5.35). For the reader's convenience, we briefly review the arguments of § V. 1.5 for the case $(\mathbf{H}_1)_{\mathrm{av}}=\mathbf{0}$. The instability domain for equation (5.36) in the strip $0<\varepsilon<\varepsilon_0$, attached at the point

$$\vartheta_{jhl}^{m\mu 0}=\frac{\omega_j+\omega_h}{l}\qquad (j,\ h=0,\ 1,\ \ldots,\ s;\ l=1,\ 2,\ \ldots), \tag{5.37}$$

is defined to first order in ε by the inequalities

$$\vartheta_{jhl}^{m\mu 0}-\vartheta_{jhl}^{m\mu 1}\varepsilon+\ldots<\vartheta<\vartheta_{jhl}^{m\mu 0}+\vartheta_{jhl}^{m\mu 1}\varepsilon+\ldots, \tag{5.38}$$

where

$$\vartheta_{jhl}^{m\mu 1}=\frac{2}{l}\left|(\mathbf{H}_1^{(l)}\mathbf{c}_{-j},\ \mathbf{c}_h)\right|. \tag{5.39}$$

Here $\mathbf{H}_1^{(l)}$ are the Fourier coefficients in

$$\mathbf{H}_1(\vartheta z)\sim\sum_{\lambda=-\infty}^{\infty}e^{i\lambda\vartheta z}\mathbf{H}_1^{(\lambda)},$$

$i\omega_p$ and $i\omega_{-p}=-i\omega_p$ are the eigenvalues of the matrix $\mathbf{J}^{-1}\mathbf{H}_0$, $\mathbf{c}_p$ and $\mathbf{c}_{-p}$ are corresponding eigenvectors

$$\mathbf{J}^{-1}\mathbf{H}_0\mathbf{c}_p=i\omega_p\mathbf{c}_p \qquad (p=\pm 0,\ \pm 1,\ \ldots,\ \pm s),$$

normalized so that

$$i(\mathbf{J}\mathbf{c}_p,\ \mathbf{c}_q)=\delta_{pq}\operatorname{sign} p \qquad (p,\ q=\pm 0,\ \pm 1,\ \ldots,\ \pm s).$$

For problem (5.2)—(5.3) (perfectly rigid boundary)

$$\omega_p^2=k^2-\left(\frac{\mu_{|p|}^m}{a}\right)^2,\quad \mathbf{c}_p=\frac{1}{a\sqrt{2\omega_{|p|}}}\begin{pmatrix}\mathbf{e}_{|p|}\\ i\omega_p a^2\mathbf{e}_{|p|}\end{pmatrix}\ (p=\pm 0,\ \pm 1,\ \ldots,\ \pm s) \tag{5.40}$$

($\mathbf{e}_0,\ \mathbf{e}_1,\ \ldots,\ \mathbf{e}_s$ are the standard basis vectors), and the condition $\omega_p^2>0$ implies the additional restriction $s\leqslant N-1$. To determine the numbers $\pm\vartheta_{jhl}^{m\mu 1}$ (slopes of the tangents to the instability domains), we calculate

$$(\mathbf{H}_1^{(l)}\mathbf{c}_{-j},\ \mathbf{c}_h)=$$
$$=\frac{1}{2a^2\sqrt{\omega_j\omega_h}}\{(\mathbf{P}_1^{(l)}\mathbf{e}_j,\ \mathbf{e}_h)+2a^2g^{(l)}\omega_j\omega_h\delta_{jh}+ia^2\vartheta_{jhl}^{m\mu 0}\dot{g}^{(l)}([\omega_j\mathbf{V}^{\mathsf{T}}+\omega_h\mathbf{V}]\,\mathbf{e}_j,\ \mathbf{e}_h)\},$$

where $g^{(l)}$ and $\dot{g}^{(l)}$ are the complex Fourier coefficients of the functions $g(\vartheta z)$ and $\dot{g}(\vartheta z)$. If we now use the usual Fourier expansion of $g(\vartheta z)$,

$$g(\vartheta z)\sim\sum_{\lambda=1}^{\infty}(\alpha_\lambda\cos\lambda\vartheta z+\beta_\lambda\sin\lambda\vartheta z) \qquad (\alpha_0=0), \tag{5.41}$$

we obtain, for the principal-resonance domains,

$$\vartheta_{jjl}^{m\mu 1}=\frac{1}{l}\sqrt{\alpha_l^2+\beta_l^2}\left|\frac{1}{\omega_j}\left[2k^2-\left(\frac{\mu_j^m}{a}\right)^2\right]-2\omega_j v_{jj}^{m\mu}\right|, \tag{5.42}$$

and for the combination-resonance domains ($h\neq j$),

$$\vartheta_{jhl}^{m\mu 1}=\frac{1}{2l}\sqrt{\alpha_l^2+\beta_l^2}\,\frac{\omega_j+\omega_h}{\sqrt{\omega_j\omega_h}}\,|\,\omega_j v_{jh}^{m\mu}+\omega_h v_{hj}^{m\mu}\,|. \tag{5.43}$$

We can now determine the pass-band of a periodic waveguide (see end of §5.1). Each solution of problem (5.2)—(5.4), or, briefly, each wave, may be expanded in terms of waveguide modes, each of which corresponds to one multiplier of equation (5.36). Fix one of the modes and let ρ_j be the corresponding multiplier. Looking at inequalities (5.38) for the fixed j-value (the first subscript), we obtain the full range of stop-bands for the wave in question (each of these bands corresponds to a multiplier ρ_j outside the unit circle).

The final result is as follows.

The pass-bands of a periodic waveguide, for a wave with multiplier ρ_j *and* m *azimuthal nodes, where the boundary surface is perfectly rigid (problem (5.2) — (5.3)) and defined by equation (5.32), are determined to first order in* ε *as follows. Given the relevant wave numbers* k *(assumed to be independent of* $\mathbf{r}$*), use formula (5.40) to determine a finite number of frequencies* $\omega_0, \omega_1, \ldots, \omega_s$ $(s \leqslant N-1$; *recall that* $\mu_\varkappa^m$ *is the* $(\varkappa+1)$*-th positive zero of the function* $J'_m(x))$. *Let the interval of relevant waveguide periods be* $D_1 < D < D_2$; *consider the interval*

$$\frac{2\pi}{D_2} < \vartheta < \frac{2\pi}{D_1} \qquad (5.44)$$

and find all values of $\vartheta_{jhl}^{m\mu 0}$ *(fixed* j*) in this interval (see (5.37)). For each of these quantities, use formulas (5.42)* $(h=j)$ *and (5.43)* $(h \neq j)$ *to determine* $\vartheta_{jhl}^{m\mu 1}$ *(the necessary quantities* $v_{jh}^{m\mu}$ *are evaluated from formulas (5.28)). Then construct the set of all intervals (5.38) for some fixed* ε*, defined from (5.32). Delete all these intervals (hatched in Figure 36) from the interval (5.44). The remaining part of the interval (5.44) is the set of pass-bands of the waveguide for the wave in question (to first order in* ε*).*

Remark. a) For azimuthally symmetric waves $(m=0)$ $v_{jj}^{0\mu}=0$, $v_{jh}^{0\mu}+v_{hj}^{0\mu}=-2$, and formulas (5.42), (5.43) become

$$\left.\begin{aligned} \vartheta_{jjl}^{0\mu 1} &= \frac{1}{l\omega_j}\sqrt{\alpha_l^2+\beta_l^2}\left|2k^2-\left(\frac{\mu_j^0}{a}\right)^2\right|, \\ \vartheta_{jhl}^{0\mu 1} &= \frac{1}{2l\sqrt{\omega_j\omega_h}}(\omega_j+\omega_h)\sqrt{\alpha_l^2+\beta_l^2}\left|-2\omega_j+(\omega_h-\omega_j)v_{hj}^{0\mu}\right|. \end{aligned}\right\} \qquad (5.45)$$

b) In the case of problem (5.2)—(5.4) and azimuthally symmetric waves $(\alpha \neq 0,\ m=0)$, the pass-bands of a waveguide with periodic boundary (5.32) were calculated in /162a/.

c) For a waveguide with perfectly soft boundary (boundary-value problem (5.2)— (5.5)) everything remains the same, subject to the following changes in formulas (5.40), (5.42), (5.43): replace μ_p^m by ν_p^m, μ_j^m by ν_j^m, and also $v_{jj}^{m\mu}$, $v_{jh}^{m\mu}$, $v_{hj}^{m\mu}$ by $v_{jj}^{m\nu}$, $v_{jh}^{m\nu}$, $v_{hj}^{m\nu}$, respectively (see (5.31)).

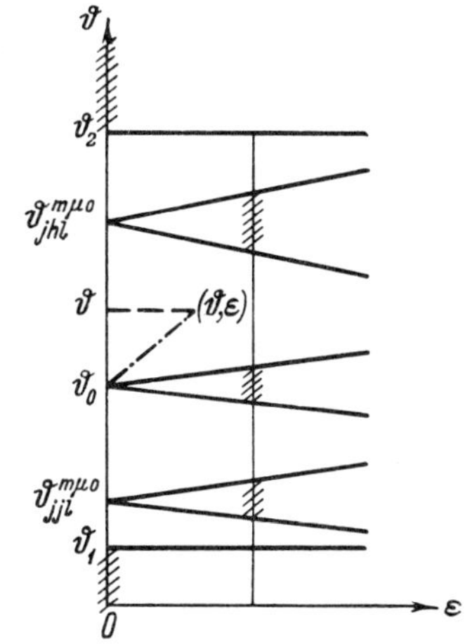

FIGURE 36.

5.5. Cylindrical periodically filled waveguide. Pass-bands in first and second approximations

Consider a waveguide whose boundary Π is the surface of a right circular cylinder with axis of symmetry along the z-axis and radius a, whose filling varies periodically in such a way that the wave number k obeys the law (see (II) at the beginning of §5.4)

$$k^2 = k^2(\Theta z) = k_0^2 [1 + \varepsilon \varkappa(\Theta z)] \qquad (\varkappa(s + 2\pi) = \varkappa(s),\ \varkappa_{av} = 0), \tag{5.46}$$

where D is the period and $\Theta = 2\pi/D$ is the frequency of the filling.

The coefficients (5.14) for the boundary-value problem (5.2)—(5.4) are

$$P_{nn'}^{mm} = [k(\Theta z)^2 a^2 - (\mu_n^m)^2]\,\delta_{nn'} - \alpha a t_{nn'}^{m\mu},$$
$$G_{nn'}^{mm} = a^2 \delta_{nn'},\ Q_{nn'}^{mm} = 0,\ G_{nn'}^{mm'} = P_{nn'}^{mm'} = Q_{nn'}^{mm'} = 0$$
$$(m,\ m' = 0,\ 1,\ \ldots,\ M-1;\ m' \neq m;\ n,\ n' = 0, 1,\ \ldots,\ N-1).$$

For each subsystem (fixed m) we have the vector equation (5.15)

$$\frac{d^2\mathbf{f}}{dz^2} + [\mathbf{P}_0 + \varepsilon \mathbf{P}_1(\Theta z)]\,\mathbf{f} = \mathbf{0}, \tag{5.47}$$

where

$$\mathbf{P}_0 = k_0^2 \mathbf{I}_N - \operatorname{diag}\left(\frac{\mu_n^m}{a}\right)^2 - \frac{\alpha}{a}\mathbf{T}, \qquad \mathbf{P}_1 = k_0^2 \varkappa(\Theta z)\,\mathbf{I}_N, \tag{5.48}$$

and the matrix $\mathbf{T} = \| t_{nn'}^{m\mu} \|$ is defined in (5.28). We confine the discussion to a perfectly rigid boundary (problem (5.2)—(5.3)), so that we must set $\alpha = 0$ in (5.48). The eigenvalues ω_n^2 of $\mathbf{P}_0$ are

$$\omega_n^2 = k_0^2 - \left(\frac{\mu_n^m}{a}\right)^2 \qquad (n = 0,\ 1,\ \ldots,\ s \leqslant N-1). \tag{5.49}$$

From (5.39) we derive the formula for the slope of the tangent to the instability domain in the first quadrant of the $(\varepsilon,\ \Theta)$-plane; by (5.48),

$$\Theta_{jhl}^{m\mu 1} = \frac{k_0^2 |\varkappa^{(l)}|}{l \omega_j}\,\delta_{jh} \qquad (j,\ h = 0,\ 1,\ \ldots,\ s;\ l = 1,\ 2,\ \ldots), \tag{5.50}$$

where $\varkappa^{(l)}$ denotes the complex Fourier coefficient of $\varkappa(\Theta z)$. It follows from (5.50) that the quantities $\Theta_{jhl}^{m\mu 1}$ are different from zero only if $h = j$, so that there are "wide" instability domains only of principal resonance, attached at points corresponding to critical frequencies

$$\Theta_{jjl}^{m\mu 0} = \frac{2\omega_j}{l}\sqrt{k_0^2 - \left(\frac{\mu_j^m}{a}\right)^2} \tag{5.51}$$

for $j = 0$, natural numbers $j\ (j = 0,\ 1,\ 2,\ \ldots,\ p(k_0))$ for which the radicand is positive, and values of l for which the Fourier coefficients of $\varkappa(\Theta z)$ do not vanish. Writing the Fourier expansion as

$$\varkappa(\Theta z) \sim \sum_{\lambda=1}^{\infty} (\gamma_\lambda \cos\lambda\Theta z + \delta_\lambda \sin\lambda\Theta z) \qquad (\gamma_0 = 0), \tag{5.52}$$

we can evaluate

$$\Theta_{jjl}^{m\mu 1} = \frac{k_0^2}{l}\left[k_0^2 - \left(\frac{\mu_j^m}{a}\right)^2\right]^{-1/2} \sqrt{\gamma_l^2 + \delta_l^2}$$

and obtain the principal-resonance instability domain (in the first approximation) by

$$\Theta_{jjl}^{m\mu 0} - \Theta_{jjl}^{m\mu 1}\varepsilon + \ldots < \Theta < \Theta_{jjl}^{m\mu 0} + \Theta_{jjl}^{m\mu 1}\varepsilon + \ldots \tag{5.53}$$

We now proceed to the second approximation for the boundaries of the principal-resonance instability domains. Let us express the equation of the boundary as

$$\Theta = \Theta_{jjl}^{m\mu 0} \mp \Theta_{jjl}^{m\mu 1}\varepsilon + \Theta_{jjl}^{m\mu 2}\varepsilon^2 + \ldots,$$

where the upper (lower) sign defines the lower (upper) boundary.

Formula (V. 3.47) for equation (5.47) is (see § V. 3.6)

$$\Theta_{jjl}^{m\mu 2} = \frac{1}{l}\left\{-\frac{|\chi_{jj}^{(l)}|^2}{2\omega_j} + \sum_{\substack{\lambda=-\infty \\ (\lambda \neq \pm l)}}^{\infty} \frac{4\omega_j |\chi_{jj}^{(\lambda)}|^2}{\lambda^2 (\Theta_{jjl}^{m\mu 0})^2 - 4\omega_j^2} \mp \right.$$
$$\left. \mp 2\,\mathrm{Re}\left[\frac{\chi_{jj}^{(l)}}{|\chi_{jj}^{(l)}|}\left(\sum_{\substack{\lambda=-\infty \\ (\lambda\neq 0)}}^{\infty} \frac{\chi_{jj}^{(\lambda)}\overline{\chi_{jj}^{(l+\lambda)}}}{\lambda\Theta_{jjl}^{m\mu 0}} - \sum_{\substack{\lambda=-\infty \\ (\lambda\neq -l)}}^{\infty} \frac{\chi_{jj}^{(\lambda)}\overline{\chi_{jj}^{(l+\lambda)}}}{2\omega_j + \lambda\Theta_{jjl}^{m\mu 0}}\right)\right]\right\}.$$

We have used the fact that $\sigma_{\varkappa\lambda}^{(r)} \equiv 0$, since in equation (5.47) $\mathbf{P}_2(\Theta z) \equiv \mathbf{0}$ and $\chi_{\varkappa\lambda}^{(r)} = \chi_{\varkappa\varkappa}^{(r)}\delta_{\varkappa\lambda}$ by (5.50). Using thc expansion (5.52), we obtain

$$\Theta_{jjl}^{m\mu 2} = \frac{k_0^4}{16 l \omega_j^3}\left\{-\frac{1}{2}(\gamma_l^2 + \delta_l^2) + 2\sum_{\substack{\lambda=1 \\ (\lambda\neq l)}}^{\infty} \frac{\gamma_\lambda^2 + \delta_\lambda^2}{\frac{\lambda^2}{l^2} - 1} \mp \right.$$
$$\mp \frac{1}{\sqrt{\gamma_l^2 + \delta_l^2}}\left[\gamma_l \sum_{\lambda=1}^{\infty} \frac{\gamma_\lambda\gamma_{l+\lambda} + \delta_\lambda\delta_{l+\lambda}}{\frac{\lambda}{l}\left(1 + \frac{\lambda}{l}\right)} + \delta_l \sum_{\lambda=1}^{\infty} \frac{\gamma_\lambda\delta_{l+\lambda} - \delta_\lambda\gamma_{l+\lambda}}{\frac{\lambda}{l}\left(1 + \frac{\lambda}{l}\right)} - \gamma_l \sum_{\lambda=1}^{l-1} \frac{\gamma_\lambda\gamma_{l-\lambda} - \delta_\lambda\delta_{l-\lambda}}{\frac{\lambda}{l}\left(1 - \frac{\lambda}{l}\right)} - \right.$$
$$\left.\left. - \delta_l \sum_{\lambda=1}^{l-1} \frac{\gamma_\lambda\delta_{l-\lambda} + \delta_\lambda\gamma_{l-\lambda}}{\frac{\lambda}{l}\left(1 - \frac{\lambda}{l}\right)} + \gamma_l \sum_{\lambda=l+1}^{\infty} \frac{\gamma_\lambda\gamma_{\lambda-l} + \delta_\lambda\delta_{l-\lambda}}{\frac{\lambda}{l}\left(\frac{\lambda}{l} - 1\right)} + \delta_l \sum_{\lambda=l+1}^{\infty} \frac{\delta_\lambda\gamma_{\lambda-l} - \gamma_\lambda\delta_{\lambda-l}}{\frac{\lambda}{l}\left(\frac{\lambda}{l} - 1\right)}\right]\right\}. \tag{5.54}$$

R e m a r k . All of the foregoing remains valid for a waveguide with perfectly soft boundary, provided one replaces μ_j^m by ν_j^m in formulas (5.49), (5.50), (5.51) and (5.54) (see end of §5.2).

5.6. Dispersion characteristics. Phase and group velocities of waves

The Floquet solution defined by (5.6),

$$u(\mathbf{r}+D\mathbf{z}^0)=\rho u(\mathbf{r}) \tag{5.55}$$

may be defined equivalently by

$$u(\mathbf{r})=e^{ihz}w(\mathbf{r}), \tag{5.56}$$

in which $w(\mathbf{r})\equiv w(\mathbf{r}+D\mathbf{z}^0)$ is a D-periodic function of z and h is defined by*

$$h=-\frac{i}{D}\ln\rho. \tag{5.57}$$

Expanding $w(\mathbf{r})$ in Fourier series as a function of z,

$$w(\mathbf{r})\sim\sum_{n=-\infty}^{\infty}u_n(\mathbf{r}_\perp)\exp\left(i\frac{2\pi n}{D}z\right)$$

and inserting this into (5.56), we obtain the expansion

$$u(\mathbf{r})\sim\sum_{n=-\infty}^{\infty}u_n(\mathbf{r}_\perp)\exp(ih_nz), \tag{5.58}$$

where

$$h_n=h+\frac{2\pi n}{D}=h+\vartheta n. \tag{5.59}$$

Formula (5.58) is the expansion in *space harmonics* (as the terms of (5.58) are called).

Recalling that $U(\mathbf{r},t)=u(\mathbf{r})\exp(-i\omega t)$, we deduce from (5.58) that

$$U(\mathbf{r},t)=\sum_{n=-\infty}^{\infty}u_n(\mathbf{r}_\perp)\exp[i(h_nz-\omega t)]. \tag{5.60}$$

In the pass-band, the multiplier ρ lies on the unit circle, and (5.57), (5.59) yield real values of h_n $(n=0,\pm1,\pm2,\ldots)$. According to (5.60), in this case the space harmonics are pure plane traveling waves with *phase velocities*

$$v_{\text{ph}\,n}=\frac{\omega}{h_n}=\frac{\omega}{h+\vartheta n}\qquad\left(\vartheta=\frac{2\pi}{D};\ n=0,\pm1,\pm2,\ldots\right). \tag{5.61}$$

It is at once apparent from (5.61) that there exist space harmonics with arbitrarily small phase velocities (for sufficiently large n).

An important property of the Floquet solution (natural mode of a periodic structure) in the pass-band is its *group velocity*, defined as

* This equation defines h up to an additive constant $2\pi n/D$. If necessary, therefore, one can always assume that $|\operatorname{Re}h|\leqslant\pi/D$.

$$v_{\mathrm{gr}}=\frac{1}{\frac{dh}{d\omega}}. \tag{5.62}$$

Thus the entire range of phase velocities (5.61) of space harmonics and the group velocity of the wave (5.62) are defined by the dispersivity $h(\omega)$ of the wave in the pass-band in question. As for the dispersitivities $h(\omega)$, these may be derived from (5.57) provided the values of $\rho(\omega)$ are known. Below we present a few sample calculations of $\rho=\exp(ihD)$.

We now return to a waveguide with periodic boundary (5.32) and constant wave number (§5.4), again considering only a perfectly rigid boundary $(\alpha=0)$, i.e., the boundary-value problem (5.2)—(5.3). We shall calculate a simple characteristic exponent of equation (5.36), following §§ IV.2.6 and IV.2.7. Namely, we assume that for some fixed $\vartheta\neq\vartheta_{jhl}^{m\mu 0}$ (see (5.37)) (dashed line in Figure 36) and $\varepsilon=0$, equation (5.36) has a characteristic exponent $i\omega_j$ $(j\geqslant 0)$ which is simple, i.e., ϑ is such that (see (5.40))

$$\omega_p\neq\omega_j+m_p\vartheta \tag{5.63}$$

for any integer $p\neq j$. Setting $\vartheta z=\zeta$, we write equation (5.36) as

$$\frac{d\mathbf{h}}{d\zeta}=[\mathbf{C}+\varepsilon\mathbf{B}_1(\zeta)+\varepsilon^2\mathbf{B}_2(\zeta)+\ldots]\,\mathbf{h}, \tag{5.64}$$

where

$$\left.\begin{aligned}
\mathbf{C}&=\begin{pmatrix}0 & a^{-2}\mathbf{I}_N\\ -k^2\vartheta^{-2}a^2\mathbf{I}_N+\vartheta^{-2}\operatorname{diag}(\mu_n^m)^2 & 0\end{pmatrix},\\
\mathbf{B}_1&=\begin{pmatrix}-\dot{g}\mathbf{V} & -2ga^{-2}\mathbf{I}_N\\ -2k^2\vartheta^{-2}a^2g\mathbf{I}_N & \dot{g}\mathbf{V}^{\tau}\end{pmatrix},\\
\mathbf{B}_2&=\begin{pmatrix}g\dot{g}\mathbf{V} & 3g^2a^{-2}\mathbf{I}_N\\ -k^2\vartheta^{-2}a^2g^2\mathbf{I}_N+a^2\dot{g}^2\mathbf{S}-a^2\dot{g}^2\mathbf{V}^{\tau}\mathbf{V} & -g\dot{g}\mathbf{V}^{\tau}\end{pmatrix}.
\end{aligned}\right\} \tag{5.65}$$

The required characteristic exponent of equation (5.64) at $\varepsilon=0$, i.e., the eigenvalue of the matrix $\mathbf{C}$, is $i\Omega_j(0)=i\omega_j/\vartheta$, and the corresponding eigenvectors $\mathbf{c}_j$ of $\mathbf{C}$ and $\mathbf{d}_j$ of $\mathbf{C}^{\tau}$, normalized so that $(\mathbf{c}_j,\ \mathbf{d}_j)=(\mathbf{d}_j,\ \mathbf{d}_j)=1$, are

$$\mathbf{c}_j=-i\frac{\sqrt{1+a^4\Omega_j^2}}{2a^2\Omega_j}\begin{pmatrix}\mathbf{e}_j\\ ia^2\Omega_j\mathbf{e}_j\end{pmatrix},\quad \mathbf{d}_j=\frac{1}{\sqrt{1+a^4\Omega_j^2}}\begin{pmatrix}-ia^2\Omega_j\mathbf{e}_j\\ \mathbf{e}_j\end{pmatrix},$$

where $\mathbf{e}_j$ are the standard basis vectors. It follows from formula (IV.2.67) that the characteristic exponent of equation (5.64) corresponding to $i\omega_j$ is defined by

$$\begin{aligned}
i\omega_j(\varepsilon)=i\omega_j+\varepsilon\vartheta(\mathbf{B}_1^{(0)}\mathbf{c}_j,\ \mathbf{d}_j)+\\
+\varepsilon^2\vartheta\left[(\mathbf{B}_2^{(0)}\mathbf{c}_j,\ \mathbf{d}_j)+\frac{1}{2\pi}\int_0^{2\pi}(\mathbf{B}_1(\zeta)\tilde{\mathbf{S}}\mathbf{B}_1(\zeta)\,\mathbf{c}_j,\ \mathbf{d}_j)\,d\zeta\right]+O(\varepsilon^3).
\end{aligned}$$

Here $\mathbf{B}_1^{(0)}$ and $\mathbf{B}_2^{(0)}$ are the mean values of the matrices $\mathbf{B}_1(\zeta)$ and $\mathbf{B}_2(\zeta)$, and $\tilde{\mathbf{S}}\mathbf{B}_1(\zeta)\mathbf{c}_j$ denotes a 2π-periodic solution of the equation

$$\frac{d\mathbf{y}}{d\zeta}-(\mathbf{C}-i\Omega_j\mathbf{I}_N)\,\mathbf{y}=\mathbf{B}_1(\zeta)\,\mathbf{c}_j.$$

For equation (5.64), we have $\mathbf{B}_1^{(0)}=0$, and so the correction of first order in ε vanishes and

$$i\omega_j(\varepsilon)=i\omega_j+i\Delta_j\varepsilon^2\vartheta+O(\varepsilon^3). \tag{5.66}$$

The calculations give

$$\begin{gathered}\Delta_j=\frac{1}{\Omega_j}\sum_{l=1}^{\infty}\left(3\Omega_j^2+\frac{k^2}{\vartheta^2}+l^2(v_{0j}^2+\ldots+v_{sj}^2-s_{jj})\right]|g^{(l)}|^2+\\ +\sum_{\substack{l=-\infty\\(l\neq 0)}}^{\infty}\left\{-\frac{1}{2}l^2\Omega_j(v_{j0}^2b_0+\ldots+v_{js}^2b_s)+l^2(\Omega_j+l)(v_{j0}b_0v_{0j}+\ldots+v_{js}b_sv_{sj})-\right.\\ -\frac{1}{2}\frac{l^2}{\Omega_j}(v_{0j}^2c_0+\ldots+v_{sj}^2c_s)-2\Omega_jc_j-2lv_{jj}c_j+\\ \left.+2\frac{k^2}{\vartheta^2}\left[2(\Omega_j+l)-\frac{k^2}{\vartheta^2}\frac{1}{\Omega_j}\right]b_j+2l\left[\Omega_j(\Omega_j+l)+\frac{k^2}{\vartheta^2}\frac{l}{\Omega_j}\right]v_{jj}b_j\right\}|g^{(l)}|^2.\end{gathered} \tag{5.67}$$

The notation used here is

$$\Omega_n=\frac{\omega_n}{\vartheta},\quad b_n=[\Omega_n^2-(\Omega_j+l)^2]^{-1},\quad c_n=\Omega_n^2b_n\qquad(n=0,\ 1,\ \ldots,\ s),$$

where $v_{nn'}=v_{nn'}^{m\mu}$, $s_{nn'}=s_{nn'}^{m\mu}$ $(n,\ n'=0,\ 1,\ \ldots,\ s)$ are defined by (5.28).

Remark. All the foregoing remains valid for a waveguide with perfectly soft boundary (problem (5.2)–(5.5)) provided we replace Ω_n (and accordingly Ω_j) by

$$\frac{1}{\vartheta}\sqrt{k^2-\left(\frac{\nu_n^m}{a}\right)^2}\qquad(n=0,\ 1,\ \ldots,\ s\leqslant N-1),$$

and $v_{nn'}^{m\mu}$, $s_{nn'}^{m\mu}$, by $v_{nn'}^{m\nu}$, $s_{nn'}^{m\nu}$, respectively (see (5.31)).

Using (5.66) and (5.67), we obtain the quantity h, defined by (5.57) as

$$h=\Omega_j(\varepsilon)\,\vartheta=\omega_j(\varepsilon)=\omega_j(0)+\vartheta\varepsilon^2\Delta_j+O(\varepsilon^3)\qquad(\omega_j=\omega_j(0)). \tag{5.68}$$

Inserting (5.68) into (5.61) and (5.62), we determine the phase and group velocities.

The various characteristics that we need may be evaluated differently. As before, consider a waveguide with periodic boundary (5.32) and constant wave number, confining attention to a perfectly rigid boundary $(\alpha=0)$ (problem (5.2)–(5.3)). We shall calculate the characteristic exponents of equation (5.36) near the critical point $\vartheta_0=\vartheta_{jhl}^{m\mu 0}$ (see (5.37)), to be precise, on the ray $\vartheta=\vartheta_0+\vartheta_1\varepsilon$ (dash-dot line in Figure 36), where $|\vartheta_1|>\vartheta_{jhl}^{m\mu 1}$ (see (5.43)). We denote $\zeta=\vartheta z$ and, using the expansion

$$\frac{1}{\vartheta}=\frac{1}{\vartheta_0+\vartheta_1\varepsilon}=\frac{1}{\vartheta_0}\left(1-\frac{\vartheta_1}{\vartheta_0}\varepsilon+\ldots\right),$$

which converges for $|\vartheta_1|\varepsilon<\vartheta_0$, i. e., for $\vartheta<2\vartheta_0$ (which we assume to be true), we obtain

$$\mathbf{J}\frac{d\mathbf{h}}{d\zeta}=[\tilde{\mathbf{H}}_0+\varepsilon\tilde{\mathbf{H}}_1(\zeta)+\ldots]\,\mathbf{h},$$

where

$$\tilde{\mathbf{H}}_0=\begin{pmatrix}\vartheta_0^{-2}\mathbf{P}_0 & 0\\ 0 & a^{-2}\mathbf{I}_N\end{pmatrix},\quad \tilde{\mathbf{H}}_1=\begin{pmatrix}\vartheta_0^{-2}\mathbf{P}_1(\zeta)-2\vartheta_1\vartheta_0^{-3}\mathbf{P}_0 & -\dot{g}(\zeta)\,\mathbf{V}^{\tau}\\ -\dot{g}(\zeta)\,\mathbf{V} & -2a^{-2}g(\zeta)\,\mathbf{I}_N\end{pmatrix}.$$

Here the matrices $\mathbf{V}$, $\mathbf{P}_0$, $\mathbf{P}_1(\zeta)$ are defined by formulas (5.28), (5.29) and (5.35) with $\alpha=0$, and $g(\zeta)$ is the function figuring in (5.32) and expanded as in (5.41). By the formulas of § IV. 3.2, the characteristic exponents of equation (5.36) that tend to $i\omega_j=i\omega_j(0)$ $(j=0,\ \pm1,\ \ldots,\ \pm s)$ at $\varepsilon=0$ (see (5.40)) will be

$$i\frac{\omega_j(\varepsilon)}{\vartheta_0}\vartheta=i\frac{\omega_j}{\vartheta_0}\vartheta+i\nu\vartheta\varepsilon+\ldots \qquad (5.69)$$

The results of the computations are as follows.

a) If $\vartheta_0=\vartheta_{jjl}^{m\mu 0}$, so that this is a critical frequency of principal resonance, then

$$\nu=\mp\frac{\omega_j}{\vartheta_0^2}\sqrt{\vartheta_1^2-\frac{\alpha_l^2+\beta_l^2}{l^2}\left|\frac{1}{\omega_j}\left[2k^2-\left(\frac{\mu_l^m}{a}\right)^2\right]-2\omega_j v_{jj}^{m\mu}\right|^2}. \qquad (5.70)$$

This formula may be written

$$\nu=\mp\frac{\omega_j}{\vartheta_0^2}\sqrt{\vartheta_1^2-(\vartheta_{jjl}^{m\mu 1})^2},$$

which provides a check on its validity, since when $|\vartheta_1|<\vartheta_{jjl}^{m\mu 1}$ the real part of one of the characteristic exponents of equation (5.36) becomes positive (the point $(\vartheta,\ \varepsilon)$ lies inside the instability domain).

b) If $\vartheta_0=\vartheta_{jhl}^{m\mu 0}$ $(h\neq j)$, so that we have combination resonance, then

$$\nu=\vartheta_1\frac{\omega_j-\omega_h}{2\vartheta_0^2}\mp\frac{l^2}{2(\omega_j+\omega_h)}\sqrt{\vartheta_1^2-\frac{1}{4}\frac{\alpha_l^2+\beta_l^2}{l^2}\frac{(\omega_j+\omega_h)^2}{\omega_j\omega_h}(\omega_j v_{jh}^{m\mu}+\omega_h v_{hj}^{m\mu})^2}=$$

$$=\vartheta_1\frac{\omega_j-\omega_h}{2\vartheta_0^2}\mp\frac{l^2}{2(\omega_j+\omega_h)}\sqrt{\vartheta_1^2-(\vartheta_{jhl}^{m\mu 1})^2}. \qquad (5.71)$$

5.7. Appendix. Certain integrals of cylindrical harmonics $J_m(x)$ (Bessel functions of the first kind)

Let μ_n^m and ν_n^m $(m=0,\ 1,\ \ldots,\ M-1;\ n=0,\ 1,\ \ldots,\ N-1)$ denote the $(n+1)$-th positive roots of the equations $J_m'(x)=0$ and $J_m(x)=0$, respectively.

a) By formulas 4.413, *

* Throughout this subsection the numerals without parentheses refer to the formulas in /84/.

$$\int_0^1 J_m(\nu_n^m x)\, J_m(\nu_{n'}^m x)\, x\, dx = \frac{1}{2} J_{m+1}(\nu_n^m)^2 \delta_{nn'}. \qquad \text{(I)}$$

Applying formulas 2.854 and 6.482, we get

$$\int_0^1 J_m(\mu_n^m x)\, J_m(\mu_{n'}^m x)\, x\, dx =$$

$$= [(\mu_n^m)^2 - (\mu_{n'}^m)^2]^{-1} [\mu_{n'}^m J_{m-1}(\mu_{n'}^m)\, J_m(\mu_n^m) - \mu_n^m J_{m-1}(\mu_n^m)\, J_m(\mu_{n'}^m)] = 0 \qquad (n' \neq n),$$

$$\int_0^1 J_m(\mu_n^m x)^2\, x\, dx = \frac{1}{2} [J_m(\mu_n^m)^2 - J_{m-1}(\mu_n^m)\, J_{m+1}(\mu_n^m)] = \frac{1}{2} [1 - m^2 (\mu_n^m)^{-2}]\, J_m(\mu_n^m)^2.$$

Thus

$$\int_0^1 J_m(\mu_n^m x)\, J_m(\mu_{n'}^m x)\, x\, dx = \frac{1}{2} [1 - m^2 (\mu_n^m)^{-2}]\, J_m(\mu_n^m)^2\, \delta_{nn'}. \qquad \text{(II)}$$

b) To abbreviate the notation, we write $\varkappa$ and $\varkappa'$ instead of μ_n^m and $\mu_{n'}^m$ or ν_n^m and $\nu_{n'}^m$. Using formulas 6.482-2 and 2.854, we find that

$$\varkappa\varkappa' \int_0^1 J_m'(\varkappa x)\, J_m'(\varkappa' x)\, x\, dx + m^2 \int_0^1 J_m(\varkappa x)\, J_m(\varkappa' x) \frac{dx}{x} =$$

$$= 2m^2 \int_0^1 J_m(\varkappa x)\, J_m(\varkappa' x) \frac{dx}{x} - m\varkappa \int_0^1 J_m(\varkappa' x)\, J_{m+1}(\varkappa x)\, dx -$$

$$- m\varkappa' \int_0^1 J_m(\varkappa x)\, J_{m+1}(\varkappa' x)\, dx + \varkappa\varkappa' \int_0^1 J_{m+1}(\varkappa x)\, J_{m+1}(\varkappa' x)\, x\, dx =$$

$$= m \int_0^1 d\, [J_m(\varkappa x)\, J_m(\varkappa' x)] + \varkappa\varkappa' \int_0^1 J_{m+1}(\varkappa x)\, J_{m+1}(\varkappa' x)\, x\, dx =$$

$$= m J_m(\varkappa)\, J_m(\varkappa') + \varkappa\varkappa' (\varkappa^2 - \varkappa'^2)^{-1} [\varkappa' J_m(\varkappa')\, J_{m+1}(\varkappa) - \varkappa J_m(\varkappa)\, J_{m+1}(\varkappa')] \quad (\varkappa' \neq \varkappa),$$

and for $n' = n$,

$$\varkappa^2 \int_0^1 J_m'(\varkappa x)^2\, x\, dx + m^2 \int_0^1 J_m(\varkappa x)^2 \frac{dx}{x} = m J_m(\varkappa)^2 + \frac{1}{2} \varkappa^2 [J_{m+1}(\varkappa)^2 - J_m(\varkappa)\, J_{m+2}(\varkappa)].$$

Hence

$$\nu_n^m \nu_{n'}^m \int_0^1 J_m'(\nu_n^m x)\, J_m'(\nu_{n'}^m x)\, x\, dx +$$

$$+ m^2 \int_0^1 J_m(\nu_n^m x)\, J_m(\nu_{n'}^m x) \frac{dx}{x} = \frac{1}{2} (\nu_n^m)^2\, J_m'(\nu_n^m)^2\, \delta_{nn'}. \qquad \text{(III)}$$

Again applying formulas 6.481-1 and 6.482-2, we have also

$$\mu_n^m \mu_{n'}^m \int_0^1 J'_m(\mu_n^m x) J'_m(\mu_{n'}^m x)\, x\, dx +$$
$$+ m^2 \int_0^1 J_m(\mu_n^m x) J_m(\mu_{n'}^m x) \frac{dx}{x} = \frac{1}{2}\left[(\mu_n^m)^2 - m^2\right] J_m(\mu_n^m)^2 \delta_{nn'}. \quad \text{(IV)}$$

c) Basing ourselves on formula 2.854-1, we see that for $\beta \neq \alpha$

$$\int_0^1 J_m(\alpha x) J'_m(\beta x)\, x^2\, dx = \frac{\partial}{\partial \beta} \int_0^1 J_m(\alpha x) J_m(\beta x)\, x\, dx =$$
$$= 2\beta(\alpha^2 - \beta^2)^{-2}\left[\beta J_m(\alpha) J'_m(\beta) - \alpha J'_m(\alpha) J_m(\beta)\right] +$$
$$+ (\alpha^2 - \beta^2)^{-1}\left[J_m(\alpha) J'_m(\beta) + \beta J_m(\alpha) J''_m(\beta) - \alpha J'_m(\alpha) J'_m(\beta)\right].$$

Hence it follows at once that

$$\int_0^1 J_m(\nu_n^m x) J'_m(\nu_{n'}^m x)\, x^2\, dx =$$
$$= -\nu_n^m\left[(\nu_n^m)^2 - (\nu_{n'}^m)^2\right]^{-1} J_{m-1}(\nu_n^m) J_{m-1}(\nu_{n'}^m) \qquad (n' \neq n), \quad \text{(V)}$$

and, via formula 6.482,

$$\int_0^1 J_m(\mu_n^m x) J'_m(\mu_{n'}^m x)\, x^2\, dx =$$
$$= (\mu_{n'}^m)^{-1}\left[m^2 - (\mu_{n'}^m)^2\right]\left[(\mu_n^m)^2 - (\mu_{n'}^m)^2\right]^{-1} J_m(\mu_n^m) J_m(\mu_{n'}^m) \qquad (n' \neq n). \quad \text{(VI)}$$

In the limiting case $(n' = n)$, we begin with formulas 2.854-2 and 6.482, noting that in our notation $J_m(\varkappa) J'_m(\varkappa) = 0$,

$$\int_0^1 J_m(\varkappa x) J'_m(\varkappa x)\, x^2\, dx = \frac{1}{2}\frac{\partial}{\partial \varkappa}\int_0^1 J_m(\varkappa x)^2\, x\, dx =$$
$$= -\frac{1}{4}\frac{\partial}{\partial \varkappa}\left[J_{m-1}(\varkappa) J_{m+1}(\varkappa)\right] = \frac{1}{4}\frac{\partial}{\partial \varkappa}\left[J'_m(\varkappa)^2 - m^2\varkappa^{-2} J_m(\varkappa)^2\right] =$$
$$= \frac{1}{2}\left[J'_m(\varkappa) J''_m(\varkappa) + m^2\varkappa^{-3} J_m(\varkappa)^2\right].$$

Hence it follows that

$$\int_0^1 J_m(\mu_n^m x) J'_m(\mu_n^m x)\, x^2\, dx = \frac{1}{2} m^2 (\mu_n^m)^{-3} J_m(\mu_n^m)^2, \quad \text{(VII)}$$

and via formula 6.482 we obtain

$$\int_0^1 J_m(\nu_n^m x) J'_m(\nu_n^m x)\, x^2\, dx = -\frac{1}{2}(\nu_n^m)^{-1} J_{m-1}(\nu_n^m)^2. \quad \text{(VIII)}$$

d) Using formula 2.854, we see that for $\beta \neq \alpha$

$$\int_0^1 J'_m(\alpha x)\,J'_m(\beta x)\,x^3\,dx=\frac{\partial^2}{\partial\alpha\,\partial\beta}\int_0^1 J_m(\alpha x)\,J_m(\beta x)\,x\,dx=$$
$$=2\beta(\alpha^2-\beta^2)^{-2}[\beta J'_m(\alpha)J'_m(\beta)-J'_m(\alpha)J_m(\beta)-\alpha J''_m(\alpha)J_m(\beta)]-$$
$$-8\alpha\beta(\alpha^2-\beta^2)^{-3}[\beta J_m(\alpha)J'_m(\beta)-\alpha J'_m(\alpha)J_m(\beta)]+$$
$$+(\alpha^2-\beta^2)^{-1}[\beta J'_m(\alpha)J''_m(\beta)-\alpha J''_m(\alpha)J'_m(\beta)]-$$
$$-2\alpha(\alpha^2-\beta^2)^{-2}[J_m(\alpha)J'_m(\beta)+\beta J_m(\alpha)J''_m(\beta)-\alpha J'_m(\alpha)J'_m(\beta)].$$

Thus, again using formulas 6.481 and 6.482, we obtain

$$\int_0^1 J'_m(\nu_n^m x)\,J'_m(\nu_{n'}^m x)\,x^3\,dx=$$
$$=2\left[(\nu_n^m)^2+(\nu_{n'}^m)^2\right]\left[(\nu_n^m)^2-(\nu_{n'}^m)^2\right]^{-2}J_{m-1}(\nu_n^m)\,J_{m-1}(\nu_{n'}^m)\qquad (n'\neq n),\qquad \text{(IX)}$$

$$\int_0^1 J'_m(\mu_n^m x)\,J'_m(\mu_{n'}^m x)\,x^3\,dx=$$
$$=2\left[(\mu_n^m)^2-(\mu_{n'}^m)^2\right]^{-2}\left[2\mu_n^m\mu_{n'}^m-m^2\left(\frac{\mu_{n'}^m}{\mu_n^m}+\frac{\mu_n^m}{\mu_{n'}^m}\right)\right]J_m(\mu_n^m)\,J_m(\mu_{n'}^m)\quad (n'\neq n).\qquad \text{(X)}$$

§6. PARAMETRIC RESONANCE IN WAVE-PROPAGATION PROBLEMS IN PERIODIC STRUCTURES. ELECTROMAGNETIC WAVEGUIDES

The methods developed in Chapter V for systems with finitely many degrees of freedom may be applied to certain problems concerning wave propagation in continuous media, provided these problems are approximable by systems with finitely many degrees of freedom. We shall be concerned with wave propagation in periodic electromagnetic waveguides. A qualitative discussion may be found in /167a/ (see also the literature cited there). The other approach worked out in §5 for acoustic waveguides is also applicable to electromagnetic waveguides loaded by a beam of particles /160b, c, d; 162c, d/.

6.1. Statement of the problem

Let W be a three-dimensional region containing a segment of the z-axis ($z_0\leqslant z\leqslant z_1$), whose boundary is a perfectly conducting (metallic) surface Π and the regions $S(z_0)$ and $S(z_1)$ formed by the planes $z=z_0$ and $z=z_1$. Assume that all cross-sections $S(z)$ of W by planes $z=\text{const}$ ($z_0\leqslant z\leqslant z_1$) are simply-connected and bounded (see Figure 35).

Assume that the components of the field of an electromagnetic wave vary harmonically in time ($\sim\exp(-i\omega t)$). To determine the field at frequency ω in the region W we must determine the complex vector amplitudes $\mathcal{H}(\mathbf{r})$ and $\mathcal{E}(\mathbf{r})$ of the magnetic and electric fields of the wave; in the absence of sources, these satisfy Maxwell's equations

$$\text{rot}\,\mathcal{H} = -ik\varepsilon\mathcal{E}, \tag{6.1}$$
$$\text{rot}\,\mathcal{E} = ik\mu\mathcal{H}. \tag{6.2}$$

The permittivity ε and magnetic permeability μ of the medium are generally complex-valued and depend on the space coordinates $\mathbf{r}\{x, y, z\}$; $k=\omega/c$ is the wave number. The equations $\text{div}\,(\varepsilon\mathcal{E})=0$ and $\text{div}\,(\mu\mathcal{H})=0$ hold automatically if $k\neq 0$, because of (6.1) and (6.2). Assuming that Π is the surface of a real conductor, we use the boundary condition of Leontovich:

$$\{(\text{rot}\,\mathcal{H})_\tau + ikw\,[\varepsilon\mathcal{H}\times\mathbf{n}]\}_\Pi = 0 \qquad \left(w=\frac{1}{2}(1+i)\sqrt{\frac{\omega\mu_m}{2\pi\sigma_m}}\right), \tag{6.3}$$

where $\mathbf{n}$ is the unit normal to the surface Π, μ_m and σ_m are the magnetic permeability and permeance of the metal, and τ denotes projection onto the plane tangent to Π.

From (6.1) we obtain

$$\mathcal{E} = i\,(k\varepsilon)^{-1}\,\text{rot}\,\mathcal{H}, \tag{6.4}$$

which, when inserted into (6.2), yields a second-order partial differential equation for $\mathcal{H}$:

$$\text{rot}\,(\varepsilon^{-1}\,\text{rot}\,\mathcal{H}) - k^2\mu\mathcal{H} = 0. \tag{6.5}$$

Thus, combining equation (6.5) with the boundary condition (6.3), we obtain a boundary-value problem for $\mathcal{H}(\mathbf{r})$. Once $\mathcal{H}(\mathbf{r})$ is known, we can use (6.4) to find $\mathcal{E}(\mathbf{r})$. However, this problem need not have a unique solution, for example, since the surface Π on which the boundary conditions are given is not closed.

We shall formulate problem (6.5)—(6.3) in an equivalent variational form, demanding that the boundary conditions be unique for the functional in question.

Consider the bilinear functional*

$$J(\mathcal{H}, \mathcal{H}_0) = \iiint\limits_W [\varepsilon^{-1}(\text{rot}\,\mathcal{H}, \text{rot}\,\mathcal{H}_0) - k^2\mu(\mathcal{H}, \mathcal{H}_0)]\,dV - ik\iint\limits_\Pi w(\mathcal{H}, \mathcal{H}_{0\tau})\,d\sigma. \tag{6.6}$$

It is easy to see that the condition $\delta\mathcal{H}_0\,J(\mathcal{H}, \mathcal{H}_0)=0$, where the variation extends over the class of all sufficiently smooth three-dimensional vector-functions $\mathcal{H}_0$ assuming constant values on $S(z_0)$ and $S(z_1)$, is equivalent to the boundary-value problem (6.5)—(6.3) for the vector $\mathcal{H}$.

6.2. Reduction of boundary-value problem to system of ordinary differential equations

Let us express the unknown $\mathcal{H}$ as a finite sum

$$\mathcal{H} = \sum_{n=1}^{N}\sum_{\alpha=1}^{3} f_n^\alpha(z)\,\mathcal{H}_n^\alpha(\mathbf{r}), \tag{6.7}$$

* As always, the scalar product is understood here in the sense of § I.1.1.

where $\{\mathcal{H}_n^\alpha(\mathbf{r})\}$ is some given system of vector-functions and the scalar functions $f_n^\alpha(z)$ are unknown. Substituting (6.7) into (6.6) and noting that

$$\operatorname{rot}\mathcal{H}=\sum_{n=1}^{N}\sum_{\alpha=1}^{3}\{f_n^\alpha(z)\operatorname{rot}\mathcal{H}_n^\alpha+[\operatorname{grad}f_n^\alpha\times\mathcal{H}_n^\alpha]\},$$

with analogous expressions for $\mathcal{H}_0$ and $\operatorname{rot}\mathcal{H}_0$, we can write our functional as

$$J(\mathcal{H},\mathcal{H}_0)=\int_{z_0}^{z_1}J(\mathcal{H},\mathcal{H}_0,z)\,dz, \tag{6.8}$$

where

$$\left.\begin{aligned}J(\mathcal{H},\mathcal{H}_0,z)=\sum_{n,\alpha}\sum_{s,\beta}\iint_{S(z)}\Big[\varepsilon^{-1}\Big(\Big\{f_s^\beta\operatorname{rot}\mathcal{H}_s^\beta+\frac{df_s^\beta}{dz}[\mathbf{e}_z\times\mathcal{H}_s^\beta]\Big\},\\ \Big\{g_n^\alpha\operatorname{rot}\mathcal{H}_n^\alpha+\frac{dg_n^\alpha}{dz}[\mathbf{e}_z\times\mathcal{H}_n^\alpha]\Big\}\Big)-k^2\mu f_s^\beta\bar{g}_n^\alpha(\mathcal{H}_s^\beta,\mathcal{H}_n^\alpha)\Big]dS-\\ -ik\sum_{n,\alpha}\sum_{s,\beta}\oint_{L(z)}\frac{w}{c_z}f_s^\beta\bar{g}_n^\alpha(\mathcal{H}_s^\beta,\mathcal{H}_n^\alpha)\,dL.\end{aligned}\right\} \tag{6.9}$$

Here $L(z)$ is the contour of the cross-section $S(z)$, $c_z=|\cos(\boldsymbol{\tau}_z,\mathbf{e}_z)|$, where $\boldsymbol{\tau}_z$ is a unit vector tangent to Π in the plane passing through the z-axis and the relevant point of $L(z)$. The system $\{g_n^\alpha\}$ defines a representation

$$\mathcal{H}_0=\sum_{n=1}^{N}\sum_{\alpha=1}^{3}g_n^\alpha(z)\,\mathcal{H}_n^\alpha(r);$$

finally, the summation in (6.9) is from 1 to N with respect to n and s, and from 1 to 3 with respect to α and β.

Varying (6.8) over the functions $g_n^\alpha(z)$ $(\alpha=1,2,3;\ n=1,2,\dots,N)$, we obtain a system of ordinary differential equations for the unknown functions $f_s^\beta(z)$ $(\beta=1,2,3;\ s=1,2,\dots,N)$:

$$\sum_{s,\beta}\Big\{\frac{d}{dz}\Big[G_{ns}^{\alpha\beta}(z)\frac{df_s^\beta}{dz}+Q_{ns}^{\alpha\beta}(z)f_s^\beta\Big]-R_{ns}^{\alpha\beta}(z)\frac{df_s^\beta}{dz}+P_{ns}^{\alpha\beta}(z)f_s^\beta\Big\}=0. \tag{6.10}$$

The coefficients of system (6.10) are determined by

$$\left.\begin{aligned}G_{ns}^{\alpha\beta}(z)&=\iint_{S(z)}\varepsilon^{-1}([\mathbf{e}_z\times\mathcal{H}_s^\beta],\ [\mathbf{e}_z\times\mathcal{H}_n^\alpha])\,dS,\\ Q_{ns}^{\alpha\beta}(z)&=\iint_{S(z)}\varepsilon^{-1}(\operatorname{rot}\mathcal{H}_s^\beta,[\mathbf{e}_z\times\mathcal{H}_n^\alpha])\,dS,\\ R_{ns}^{\alpha\beta}(z)&=\iint_{S(z)}\varepsilon^{-1}([\mathbf{e}_z\times\mathcal{H}_s^\beta],\ \operatorname{rot}\mathcal{H}_n^\alpha)\,dS,\\ P_{ns}^{\alpha\beta}(z)&=\iint_{S(z)}[\mu k^2(\mathcal{H}_s^\beta,\mathcal{H}_n^\alpha)-\varepsilon^{-1}(\operatorname{rot}\mathcal{H}_s^\beta,\operatorname{rot}\mathcal{H}_n^\alpha)]dS+\\ &\quad+ik\oint_{L(z)}\frac{w}{c_z}(\mathcal{H}_{s\tau}^\beta,\mathcal{H}_{n\tau}^\alpha)\,dL,\end{aligned}\right\} \tag{6.11}$$

where $\mathcal{H}^{\beta}_{s\tau}$ and $\mathcal{H}^{\alpha}_{n\tau}$ are the projections of the vectors $\mathcal{H}^{\beta}_{s}$ and $\mathcal{H}^{\alpha}_{n}$ on the tangent plane to Π. In matrix notation, we can rewrite (6.10) as three vector equations

$$\frac{d}{dz}\sum_{\beta=1}^{3}\left[\mathbf{G}^{\alpha\beta}(z)\frac{d\mathbf{f}^{\beta}}{dz}+\mathbf{Q}^{\alpha\beta}(z)\,\mathbf{f}^{\beta}\right]- \\ -\sum_{\beta=1}^{3}\mathbf{R}^{\alpha\beta}(z)\frac{d\mathbf{f}^{\beta}}{dz}+\sum_{\beta=1}^{3}\mathbf{P}^{\alpha\beta}(z)\,\mathbf{f}^{\beta}=0 \qquad (\alpha=1,2,3), \tag{6.12}$$

where

$$\mathbf{G}^{\alpha\beta}=\|G^{\alpha\beta}_{ns}\|^{N}_{n,\,s=1},\quad \mathbf{Q}^{\alpha\beta}=\|Q^{\alpha\beta}_{ns}\|^{N}_{1},\quad \mathbf{R}^{\alpha\beta}=\|R^{\alpha\beta}_{ns}\|^{n}_{1},$$
$$\mathbf{P}^{\alpha\beta}=\|P^{\alpha\beta}_{ns}\|^{N}_{1},\quad \mathbf{f}^{\beta}=(f^{\beta}_{1},f^{\beta}_{2},\ldots,f^{\beta}_{N})^{\tau}\qquad (\alpha,\beta=1,2,3).$$

By (6.11), each matrix-function $\mathbf{R}^{\beta\alpha}$ is the hermitian conjugate of $\mathbf{Q}^{\alpha\beta}$ $(\alpha,\beta=1,2,3)$ provided that $\varepsilon(\mathbf{r})$ is real. System (6.12) is degenerate (the determinant of the matrix of highest-order derivatives vanishes), so that there must be a linear dependence between the three components of $\mathcal{H}(\mathbf{r})$: $\operatorname{div}\mu\mathcal{H}=0$. We remove the degeneracy as follows. We shall assume that in (6.7) the index $\alpha=3$ corresponds to the longitudinal part of the field, so that

$$\mathcal{H}^{3}_{n}(\mathbf{r})=\eta_{n}(\mathbf{r})\,\mathbf{e}_{z},$$

and $\mathcal{H}^{1}_{n}(\mathbf{r})$ and $\mathcal{H}^{2}_{n}(\mathbf{r})$ have no components along the z-axis. Any basis satisfying these conditions will be called f u n d a m e n t a l. Hence, relative to a fundamental basis, we have

$$\mathbf{G}^{\alpha3}(z)=\mathbf{G}^{3\beta}(z)=\mathbf{Q}^{3\beta}(z)=\mathbf{R}^{\alpha3}(z)\equiv0\qquad(\alpha,\beta=1,2,3),$$

and the third equation $(\alpha=3)$ of system (6.12) enables us to express $\mathbf{f}^{3}(z)$ in terms of $\mathbf{f}^{1}(z)$ and $\mathbf{f}^{2}(z)$:

$$\mathbf{f}^{3}(z)=(\mathbf{P}^{33})^{-1}\left[\mathbf{R}^{31}\frac{d\mathbf{f}^{1}}{dz}-\mathbf{P}^{31}\mathbf{f}^{1}+\mathbf{R}^{32}\frac{d\mathbf{f}^{2}}{dz}-\mathbf{P}^{32}\mathbf{f}^{2}\right]. \tag{6.13}$$

Using (6.13), we eliminate $\mathbf{f}^{3}(z)$ from the first two $(\alpha=1,2)$ equations of (6.12) and write the resulting system of two vector equations for $\mathbf{f}^{1}(z)$ and $\mathbf{f}^{2}(z)$ as

$$\frac{d}{dz}\left[\mathbf{G}(z)\frac{d\mathbf{f}}{dz}+\mathbf{Q}(z)\,\mathbf{f}\right]-\mathbf{R}(z)\frac{d\mathbf{f}}{dz}+\mathbf{P}(z)\,\mathbf{f}=0. \tag{6.14}$$

Here $\mathbf{f}=\begin{pmatrix}\mathbf{f}^{1}\\ \mathbf{f}^{2}\end{pmatrix}$,

$$\left.\begin{aligned}
\mathbf{G}(z)&=\begin{pmatrix}\mathbf{G}^{11}+\mathbf{Q}^{13}(\mathbf{P}^{33})^{-1}\mathbf{R}^{31} & \mathbf{G}^{12}+\mathbf{Q}^{13}(\mathbf{P}^{33})^{-1}\mathbf{R}^{32}\\ \mathbf{G}^{21}+\mathbf{Q}^{23}(\mathbf{P}^{33})^{-1}\mathbf{R}^{31} & \mathbf{G}^{22}+\mathbf{Q}^{23}(\mathbf{P}^{33})^{-1}\mathbf{R}^{32}\end{pmatrix},\\
\mathbf{Q}(z)&=\begin{pmatrix}\mathbf{Q}^{11}-\mathbf{Q}^{13}(\mathbf{P}^{33})^{-1}\mathbf{P}^{31} & \mathbf{Q}^{12}-\mathbf{Q}^{13}(\mathbf{P}^{33})^{-1}\mathbf{P}^{32}\\ \mathbf{Q}^{21}-\mathbf{Q}^{23}(\mathbf{P}^{33})^{-1}\mathbf{P}^{31} & \mathbf{Q}^{22}-\mathbf{Q}^{23}(\mathbf{P}^{33})^{-1}\mathbf{P}^{32}\end{pmatrix},\\
\mathbf{R}(z)&=\begin{pmatrix}\mathbf{R}^{11}-\mathbf{P}^{13}(\mathbf{P}^{33})^{-1}\mathbf{R}^{31} & \mathbf{R}^{12}-\mathbf{P}^{13}(\mathbf{P}^{33})^{-1}\mathbf{R}^{32}\\ \mathbf{R}^{21}-\mathbf{P}^{23}(\mathbf{P}^{33})^{-1}\mathbf{R}^{31} & \mathbf{R}^{22}-\mathbf{P}^{23}(\mathbf{P}^{33})^{-1}\mathbf{R}^{32}\end{pmatrix},\\
\mathbf{P}(z)&=\begin{pmatrix}\mathbf{P}^{11}-\mathbf{P}^{13}(\mathbf{P}^{33})^{-1}\mathbf{P}^{31} & \mathbf{P}^{12}-\mathbf{P}^{13}(\mathbf{P}^{33})^{-1}\mathbf{P}^{32}\\ \mathbf{P}^{21}-\mathbf{P}^{23}(\mathbf{P}^{33})^{-1}\mathbf{P}^{31} & \mathbf{P}^{22}-\mathbf{P}^{23}(\mathbf{P}^{33})^{-1}\mathbf{P}^{32}\end{pmatrix}.
\end{aligned}\right\}\quad(6.15)$$

Setting

$$\mathbf{g}=\mathbf{G}(z)\frac{d\mathbf{f}}{dz}+\mathbf{Q}(z)\mathbf{f},\quad \mathbf{h}=\begin{pmatrix}\mathbf{f}\\ \mathbf{g}\end{pmatrix},$$

we convert equation (6.14) to the form

$$\mathbf{J}\frac{d\mathbf{h}}{dz}=\mathbf{H}(z)\mathbf{h},\quad(6.16)$$

where

$$\mathbf{J}=\begin{pmatrix}0 & -\mathbf{I}_{2N}\\ \mathbf{I}_{2N} & 0\end{pmatrix},\quad \mathbf{H}(z)=\begin{pmatrix}\mathbf{P}+\mathbf{R}\mathbf{G}^{-1}\mathbf{Q} & -\mathbf{R}\mathbf{G}^{-1}\\ -\mathbf{G}^{-1}\mathbf{Q} & \mathbf{G}^{-1}\end{pmatrix}.\quad(6.17)$$

6.3. Analysis of periodic waveguides

To simplify matters, we confine attention to the azimuthally symmetric case, in which there exist independent solutions $\sim\exp(im\varphi)$ $(m=0,\ \pm1,\ \pm2,\ \ldots)$ with a preassigned number m of azimuthal nodes. The boundary Π, permittivity ε and permeability μ of the medium, which are assumed to be D-periodic functions of z, are defined by

$$\left.\begin{aligned}
&\Pi(z)=a\,[1+\varkappa g_\Pi(s)],\quad \varepsilon(z)=\varepsilon_0\,[1+\varkappa g_\varepsilon(s)],\\
&\mu(z)=\mu_0\,[1+\varkappa g_\mu(s)],\quad g_{\Pi,\varepsilon,\mu}(s+2\pi)\equiv g_{\Pi,\varepsilon,\mu}(s),
\end{aligned}\right\}\quad(6.18)$$

where $\varkappa$ is a real "smallness" parameter; the new independent variable s is defined by

$$s=\vartheta z,\quad \vartheta=\frac{2\pi}{D};\quad(6.19)$$

the functions $g_{\Pi,\varepsilon,\mu}(s)$ are assumed to satisfy the equalities

$$\int_0^{2\pi} g_\Pi\,ds=0,\quad \int_0^{2\pi}\operatorname{Re} g_{\varepsilon,\mu}\,ds=0,$$

by virtue of suitable choice of the constants a, ε_0, μ_0, where $\operatorname{Im}\varepsilon_0=\operatorname{Im}\mu_0=0$ (and since $\Pi(z)$ is real, it is also true that $\operatorname{Im}a=0$).

The parameter $\varkappa$ is also introduced into the numerical values of the wave resistance w of the metal and the wave number k:

$$w = \varkappa w_1, \quad k = k_0 + \varkappa k_1. \tag{6.20}$$

We define a fundamental basis $\{\mathcal{H}_n^\alpha(\mathbf{r})\}$ (see (6.7)) in cylindrical coordinates r, φ, z by

$$\left.\begin{aligned} \mathcal{H}_n^{1,2}(\mathbf{r}) &= \frac{1}{\sqrt{2\pi}} e^{im\varphi} [\rho_{nm}^{1,2}(x)\,\mathbf{e}_r + \varphi_{nm}^{1,2}(x)\,\mathbf{e}_\varphi], \\ \mathcal{H}_n^3(\mathbf{r}) &= \frac{1}{\sqrt{2\pi}} e^{im\varphi} z_{nm}(x)\,\mathbf{e}_z, \quad x = \frac{r}{\Pi(z)}. \end{aligned}\right\} \tag{6.21}$$

Once we have determined the specific form of the functions $\rho_{nm}^{1,2}(x)$, $\varphi_{nm}^{1,2}(x)$ and $z_{nm}(x)$ in (6.21), we can use (6.11) and (6.15) to evaluate the coefficients of system (6.14), which are D-periodic matrix-functions of z. The actual solution of the system may be effected on a computer by using Hill determinants. We shall only consider the analytical aspects of the problem. To this end we define the various functions by

$$\left.\begin{aligned} \rho_{nm}^1(x) &= \frac{\sqrt{2}\,m}{\nu_n^m x} \frac{J_m(\nu_n^m x)}{J_{m-1}(\nu_n^m)}, & \varphi_{nm}^1(x) &= \frac{i\sqrt{2}}{J_{m-1}(\nu_n^m)} J_m'(\nu_n^m x), \\ \rho_{nm}^2(x) &= \frac{\sqrt{2}\,\mu_n^m}{\sqrt{(\mu_n^m)^2 - m^2}} \frac{J_m'(\mu_n^m x)}{J_m(\mu_n^m)}, & \varphi_{nm}^2(x) &= \frac{i\sqrt{2}\,m}{x\sqrt{(\mu_n^m)^2 - m^2}} \frac{J_m(\mu_n^m x)}{J_m(\mu_n^m)}, \\ z_{nm}(x) &= \frac{\sqrt{2}}{\sqrt{(\mu_n^m)^2 - m^2}} \frac{J_m(\mu_n^m x)}{J_n^m(\mu_n^m)}, \end{aligned}\right\} \tag{6.22}$$

where $J_m(x)$ are the cylindrical harmonics of the first kind and m-th order, ν_n^m and μ_n^m the n-th positive roots of the equations $J_m(x) = 0$ and $J_m'(x) = 0$. *

With a view to calculating the characteristic exponents of system (6.16), we transform the latter as follows. We shall consider values of ϑ on the ray

$$\vartheta = \vartheta_0 + \varkappa\vartheta_1, \tag{6.23}$$

and use the expansion

$$\frac{1}{\vartheta} = \frac{1}{\vartheta_0 + \varkappa\vartheta_1} = \frac{1}{\vartheta_0}\left[1 - \frac{\vartheta_1}{\vartheta_0}\varkappa + O(\varkappa^2)\right] \qquad (0 < \vartheta < 2\vartheta_0).$$

Transforming to the independent variable s as in (6.19) and expanding the matrix-function $\mathbf{H}(s, \varkappa)$ in powers of $\varkappa$,

$$\mathbf{H}(s, \varkappa) = \mathbf{H}_0 + \varkappa\mathbf{H}_1(s) + O(\varkappa^2),$$

we write system (6.16) as

$$\mathbf{J}\frac{d\mathbf{h}}{ds} = \left[\frac{1}{\vartheta_0}\mathbf{H}_0 + \varkappa\left(\frac{1}{\vartheta_0}\mathbf{H}_1 - \frac{\vartheta_1}{\vartheta_0^2}\mathbf{H}_0\right) + O(\varkappa^2)\right]\mathbf{h}. \tag{6.24}$$

* Note that ν_n^m and μ_n^m in this section are not precisely the same as in § 5 (the subscripts n differ by 1).

The calculations yield the $4N \times 4N$ matrix $\mathbf{H}_0$ as a direct sum of four diagonal blocks:

$$\mathbf{H}_0 = [(k_0 a)^2 \mu_0 \mathbf{I}_N - \varepsilon_0^{-1} \operatorname*{diag}_{n=1,\dots,N} (\nu_n^m)^2] \dotplus (k_0 a)^2 \mu_0 \mathbf{I}_N \dotplus$$
$$\dotplus \varepsilon_0 a^{-2} \mathbf{I}_N \dotplus [\varepsilon_0 a^{-2} \mathbf{I}_N - (k_0 a)^{-2} \mu_0^{-1} \operatorname*{diag}_{n=1,\dots,N} (\mu_n^m / a)^2]. \quad (6.25)$$

Express the matrix $\mathbf{H}_1(s)$ as

$$\mathbf{H}_1(s) = \{\mathbf{A}^{ij}(s)\}_{i,\, j=1}^4, \quad (6.26)$$

where $\mathbf{A}^{ij}(s)$ are $N \times N$ matrix-functions; carrying out the necessary computations, we obtain

$$A_{ns}^{11} = (k_0 a)^2 \mu_0 \left[2g_\Pi + g_\mu + 2\frac{k_1}{k_0}\right] \delta_{ns} + \varepsilon_0^{-1} g_\varepsilon (\nu_n^m)^2 \delta_{ns} + 2i w_1 k_0 a,$$

$$A_{ns}^{12} = \frac{2imw_1 k_0 a}{\sqrt{(\mu_s^m)^2 - m^2}}, \quad A_{ns}^{13} = \vartheta_0 \frac{dg_\Pi}{ds} \begin{cases} 2(\nu_n^m)^2 [(\nu_n^m)^2 - (\nu_s^m)^2]^{-1} & (n \neq s), \\ 0 & (n = s), \end{cases}$$

$$A_{ns}^{14} = 2m\vartheta_0 [(\mu_s^m)^2 - m^2]^{-1/2} \frac{dg_\Pi}{ds}, \quad A_{ns}^{21} = A_{sn}^{12} = \frac{2imw_1 k_0 a}{\sqrt{(\mu_n^m)^2 - m^2}},$$

$$A_{ns}^{22} = \mu_0 (k_0 a)^2 \left[2g_\Pi + g_\mu + 2\frac{k_1}{k_0}\right] \delta_{ns} + 2iw_1 \frac{m^2 k_0 a}{\sqrt{(\mu_n^m)^2 - m^2}\sqrt{(\mu_s^m)^2 - m^2}},$$

$$A_{ns}^{23} = 0, \quad A_{ns}^{24} = \vartheta_0 \frac{dg_\Pi}{ds} \begin{cases} \dfrac{2(\mu_s^m)^2}{(\mu_n^m)^2 - (\mu_s^m)^2} \sqrt{\dfrac{(\mu_n^m)^2 - m^2}{(\mu_s^m)^2 - m^2}} & (n \neq s), \\ [2m^2 - (\mu_s^m)^2]\,[(\mu_s^m)^2 - m^2]^{-1} & (n = s), \end{cases}$$

$$A_{ns}^{31} = A_{sn}^{13}, \quad A_{ns}^{32} = 0, \quad A_{ns}^{33} = \varepsilon_0 a^{-2} (g_\varepsilon - 2g_\Pi) \delta_{ns}, \quad A_{ns}^{34} = 0,$$

$$A_{ns}^{41} = A_{sn}^{14}, \quad A_{ns}^{42} = A_{sn}^{24}, \quad A_{ns}^{43} = 0, \quad A_{ns}^{44} = \frac{\varepsilon_0}{a^2} (g_\varepsilon - 2g_\Pi) \delta_{ns} +$$
$$+ \mu_0^{-1} (k_0 a)^{-2} \left(\frac{\mu_n^m}{a}\right)^2 \left[g_\mu + 4g_\Pi + 2\frac{k_1}{k_0}\right] \delta_{ns} + \frac{2iw_1 (k_0 a)^{-3} (\mu_0 a)^{-2} (\mu_n^m \mu_s^m)^2}{\sqrt{(\mu_n^m)^2 - m^2}\sqrt{(\mu_s^m)^2 - m^2}}.$$

The characteristic exponents $\alpha_j(\varkappa)$ of system (6.24) will be calculated, following § IV. 3. 2, in the form

$$\alpha_j(\varkappa) = \alpha_j^0 + i\nu_j \varkappa + O(\varkappa^2), \quad (6.27)$$

where $\alpha_j^0 = i\omega_j / \vartheta_0$. In order to determine the coefficients ν_j, we must find the eigenvalues λ_p and eigenvectors $\mathbf{c}_p$ of the matrix $\vartheta_0^{-1} \mathbf{J}^{-1} \mathbf{H}_0$:

$$\frac{1}{\vartheta_0} \mathbf{J}^{-1} \mathbf{H}_0 \mathbf{c}_p = \lambda_p \mathbf{c}_p.$$

The calculations give

$$\lambda_p = i\frac{\omega_p}{\vartheta_0}, \quad \omega_p = \operatorname{sign} p \begin{cases} \sqrt{k_0^2 \varepsilon_0 \mu_0 - (\nu_{|p|}^m / a)^2} & (1 \leqslant |p| \leqslant N) \\ \sqrt{k_0^2 \varepsilon_0 \mu_0 - (\mu_{|p|-N}^m / a)^2} & (N + 1 \leqslant |p| \leqslant 2N) \end{cases} \quad (6.28)$$

and the eigenvectors, normalized so that

$$i(\mathbf{J}\mathbf{c}_p,\ \mathbf{c}_q)=\gamma_{pq}=\begin{cases}1, & p=q>0,\\ -1, & p=q<0,\\ 0, & p\neq q\end{cases}$$

are the following $4N$-vectors:

$$\mathbf{c}_p=\begin{cases}\dfrac{\sqrt{\varepsilon_0}}{a\sqrt{2\omega_{|p|}}}(\mathbf{e}_{|p|}\dot{+}ia^2\varepsilon_0^{-1}\omega_p\mathbf{e}_{|p|}) & (1\leqslant|p|\leqslant N),\\ \dfrac{\sqrt{\omega_{|p|}}}{k_0a\sqrt{2\mu_0}}(\mathbf{e}_{|p|}\dot{+}i\mu_0(k_0a)^2\omega_p^{-1}\mathbf{e}_{|p|}) & (N+1\leqslant|p|\leqslant 2N),\end{cases} \tag{6.29}$$

where $\mathbf{e}_1, \ldots, \mathbf{e}_{2N}$ are unit $2N$-vectors.

Setting the perturbation frequency in (6.23) equal to the critical frequency (V. 1.25)

$$\vartheta_0=\frac{1}{l}(\omega_j+\omega_h)\qquad (j,\ h=1,\ \ldots,\ 2N;\ l=1,\ 2,\ \ldots), \tag{6.30}$$

we deduce the following formula from (IV. 3.10):

$$\nu_j=\frac{1}{2}\left[\sigma_{-h-h}-\sigma_{jj}\mp\sqrt{(\sigma_{jj}+\sigma_{-h-h})^2-4\sigma_{j-h}\sigma_{-hj}}\right]. \tag{6.31}$$

The numbers $\sigma_{\alpha\beta}$ are defined by (IV. 3.9):

$$\sigma_{\alpha\beta}=-\left(\left[\frac{1}{\vartheta_0}\mathbf{H}_1-\frac{\vartheta_1}{\vartheta_0^2}\mathbf{H}_0\right]^{(l)}\mathbf{c}_\alpha,\ \mathbf{c}_\beta\right)=-\frac{1}{\vartheta_0}(\mathbf{H}_1^{(l)}\mathbf{c}_\alpha,\ \mathbf{c}_\beta)+\frac{\vartheta_1}{\vartheta_0^2}\omega_\alpha\gamma_{\alpha\beta},\qquad l=m_\beta-m_\alpha.$$

Here and below the superscript is the number of the harmonic in the Fourier expansion of $\exp(ils)$.

After the calculations, we finally obtain the values of $\sigma_{\alpha\beta}$ for the three main cases:

1) Electrical resonances ($|\alpha|,\ |\beta|=1,\ 2,\ \ldots,\ N$):

$$\left.\begin{aligned}
&\text{a)}\quad \sigma_{\alpha\alpha}=\frac{\vartheta_1}{\vartheta_0^2}\omega_{|\alpha|}-\frac{\varepsilon_0}{a^2}\frac{k_0a}{2\omega_{|\alpha|}\vartheta_0}\left[k_0a\mu_0\left(g_\varepsilon^{(0)}+g_\mu^{(0)}+2\frac{k_1}{k_0}\right)+2iw_1\right],\\
&\text{b)}\quad \sigma_{\alpha\beta}=lg_{\text{II}}^{(l)}(\omega_{|\alpha|}\omega_{|\beta|})^{-\frac{1}{2}}\left[(\nu_{|\beta|}^m)^2-(\nu_{|\alpha|}^m)^2\right]^{-1}\left[\omega_\alpha(\nu_{|\beta|}^m)^2+\right.\\
&\qquad\qquad \left.+\,\omega_\beta(\nu_{|\alpha|}^m)^2\right]\qquad (\beta\neq\pm\alpha),\\
&\text{c)}\quad \sigma_{\alpha-\alpha}=\frac{1}{2a^2\omega_{|\alpha|}\vartheta_0}\{g_\varepsilon^{(l)}\left[a^2\omega_\alpha^2-(\nu_{|\alpha|}^m)^2\right]-\\
&\qquad\qquad -2g_{\text{II}}^{(l)}a^2(\omega_\alpha^2+k_0^2\varepsilon_0\mu_0)-g_\mu^{(l)}\varepsilon_0\mu_0(k_0a)^2\}.
\end{aligned}\right\} \tag{6.32}$$

2) Magnetic resonances ($|\alpha|,\ |\beta|=N+1,\ \ldots,\ 2N$):

$$\begin{aligned}
&\text{a)}\quad \sigma_{\alpha\alpha}=\frac{\vartheta_1}{\vartheta_0^2}\omega_{|\alpha|}-\frac{(k_0a)^2\varepsilon_0\mu_0}{2a^2\omega_{|\alpha|}\vartheta_0}\left(g_\varepsilon^{(0)}+g_\mu^{(0)}+2\frac{k_1}{k_0}\right)-\\
&\qquad\qquad -\frac{iw_1\left[(\mu_{|\alpha|-N}^m)^4+m^2a^2\omega_\alpha^2\right]}{\vartheta_0\mu_0k_0a^3\omega_{|\alpha|}\left[(\mu_{|\alpha|-N}^m)^2-m^2\right]},
\end{aligned} \tag{6.33}$$

$$\left.\begin{aligned}
&\text{b)}\quad \sigma_{\alpha\beta}=lg_{\Pi}^{(l)}\frac{\sqrt{\omega_{|\alpha|}\omega_{|\beta|}}}{(\mu^{m}_{|\beta|-N})^2-(\mu^{m}_{|\alpha|-N})^2}\left[\frac{(\mu^{m}_{|\beta|-N})^2}{\omega_\beta}\sqrt{\frac{(\mu^{m}_{|\alpha|-N})^2-m^2}{(\mu^{m}_{|\beta|-N})^2-m^2}}+\right.\\
&\qquad\left.+\frac{(\mu^{m}_{|\alpha|-N})^2}{\omega_\alpha}\sqrt{\frac{(\mu^{m}_{|\beta|-N})^2-m^2}{(\mu^{m}_{|\alpha|-N})^2-m^2}}\right]\qquad(\beta\neq\pm\alpha),\\
&\text{c)}\quad \sigma_{\alpha-\alpha}=\frac{\omega_{|\alpha|}}{2\vartheta_0}\left\{\frac{k_0^2}{\omega_\alpha^2}\mu_0\varepsilon_0 g_\varepsilon^{(l)}+\left[\left(\frac{\mu^{m}_{|\alpha|-N}}{a}\right)^2\frac{1}{\omega_\alpha^2}-1\right]g_{(\mu)}^{(l)}-\right.\\
&\qquad\left.-2\left[3-\frac{\varepsilon_0\mu_0 k_0^2}{\omega_\alpha^2}-\frac{l\vartheta_0}{\omega_\alpha}\frac{2m^2-(\mu^{m}_{|\alpha|-N})^2}{(\mu^{m}_{|\alpha|-N})^2-m^2}\right]g_{\Pi}^{(l)}\right\}.
\end{aligned}\right\}\qquad(6.33)$$

3) Hybrid resonances:

$$\sigma_{\alpha\beta}=\begin{cases}-lg_{\Pi}^{(l)}(\mu_0\varepsilon_0)^{1/2}\dfrac{k_0}{\omega_\beta}\left(\dfrac{\omega_{|\beta|}}{\omega_{|\alpha|}}\right)^{1/2}m\left[(\mu^{m}_{|\beta|-N})^2-m^2\right]^{-1/2}\\ \qquad\qquad (|\alpha|=1,\ \ldots,\ N;\ |\beta|=N+1,\ \ldots,\ 2N),\\ lg_{\Pi}^{(l)}(\mu_0\varepsilon_0)^{1/2}\dfrac{k_0}{\omega_\alpha}\left(\dfrac{\omega_{|\alpha|}}{\omega_{|\beta|}}\right)^{1/2}m\left[(\mu^{m}_{|\alpha|-N})^2-m^2\right]^{-1/2}\\ \qquad\qquad (|\alpha|=N+1,\ \ldots,\ 2N;\ |\beta|=1,\ \ldots,\ N).\end{cases}\qquad(6.34)$$

The formulas for $\sigma_{\beta\beta}$ in resonance of the first two types are obtained by replacing α by β in $\sigma_{\alpha\alpha}$ in formulas (6.32) and (6.33). In addition, when analyzing hybrid resonances one uses, along with (6.34), the formulas (6.32) and (6.33) for $\sigma_{\alpha\alpha}$ and $\sigma_{\beta\beta}$, respectively, with α replaced by β in the formula for $\sigma_{\alpha\alpha}$. For resonances of all three types we have

$$\sigma_{\beta\alpha}=\overline{\sigma_{\alpha\beta}}.\qquad(6.35)$$

The reader should also note that $\omega_{-\alpha}=-\omega_\alpha$.

Provided with the above formulas, one can trace the relationship between the dispersive properties of waves in a homogeneous circular cylinder and their properties in moderating systems with periodically modulated boundary and filling. As $\varkappa\to 0$ the periodic waveguide becomes a circular cylinder with homogeneous filling and perfectly conducting boundary, and substitution of $\varkappa=0$ in (6.27) yields the well-known values of the propagation constants for a cylinder.

Using the characteristic exponent $\alpha_j(\varkappa)$, which is determined up to $O(\varkappa^2)$ by substituting ν_j from (6.31) into (6.27), we can determine the group velocities of the waves $\mathcal{E}_{mj}$ $(j=1,\ \ldots,\ N)$ or $\mathcal{H}_{m,\,j-N}$ $(j=N+1,\ \ldots,\ 2N)$ and the phase velocities of the space harmonics in the pass-bands (when $w_1=0,\ g_\mu^{(0)}=g_\varepsilon^{(0)}=0$) by the formulas

$$\frac{1}{c}v_j^{\text{gr}}=\left(\vartheta\frac{d}{dk}\operatorname{Im}\alpha_j\right)^{-1},\quad \frac{1}{c}v_{jn}^{\text{ph}}=\frac{k}{\vartheta(\operatorname{Im}\alpha_j+n)}\qquad(n=0,\ \pm1,\ \pm2\ \ldots).\qquad(6.36)$$

In the pass-bands, which in our case always exist for real ε, μ and $w_1=0$, the values of $\alpha_j(\varkappa)$ are pure imaginary. The boundaries of the pass-bands are determined by the vanishing of the radicand in (6.31) and are readily found for resonances of all three types using the above formulas. In the interest of brevity, we omit the derivation here.

When ε or μ has a small imaginary part (lossy medium) or when $w_1 \neq 0$, it follows from our formulas that the values of $\alpha_j(\varkappa)$ are always complex, so that no pass-band exists in the strict sense of the term. The parameter $\mathrm{Re}\,\alpha_j$ which determines the attenuation of waves due to losses in the medium and at the walls is also readily calculated for small $\varkappa$ using the formulas.

Note that if $g_{\text{II}}(s) \equiv 0$ then $\sigma_{\alpha\beta} = 0$ for $\alpha \neq \beta$, and therefore a stop-band is formed only for principal resonances $(h=j)$. This case of a periodic medium differs from that of a periodic boundary, for in the latter there may be a stop-band for $h \neq j$ (combination resonances) as well. In combination resonances, there is a change in the sign of the dispersion as one goes from one branch of the dispersion curve to another. Using formulas (6.31), (6.32) with $m=0$, $g_\varepsilon = g_\mu = 0$, one can derive the formulas of /160d/ as special cases.

§7. APPLICATION OF THE BUBNOV-GALERKIN METHOD IN PARAMETRIC RESONANCE PROBLEMS FOR MECHANICAL SYSTEMS

In this chapter we have considered several problems of parametric resonance in mechanical systems, all described by partial differential equations, using the Bubnov-Galerkin method to approximate these systems by systems of n ordinary differential equations (below the method will be presented as applied to systems of a general form).

Strictly speaking, we have thereby replaced each problem by a different one. It is natural to expect that for sufficiency large n the solution of the second problem will be in some sense "close" to that of the original one. Whether this in fact occurs is far from simple to ascertain and the problem has not been studied in all generality. In many cases of practical importance, however, it has been shown that the solution of the second problem approximates that of the first in an appropriate sense. This applies in particular to the exponents of exponential growth and to the wide dynamic instability domains. In this section we shall establish in rigorous mathematical terms the sense in which the approximation is valid and formulate the necessary assumptions. The results are due to Derguzov and Fomin.

7.1. Initial system

Consider a system described by the partial differential matrix equation

$$\frac{\partial^2}{\partial t^2}\left[\mathbf{M}(z) - \frac{\partial}{\partial z}\mathbf{N}(z)\frac{\partial}{\partial z}\right]\mathbf{y} - \frac{\partial}{\partial z}\mathbf{R}(z)\frac{\partial}{\partial z}\mathbf{y} + \frac{\partial^2}{\partial z^2}\mathbf{P}(z)\frac{\partial^2}{\partial z^2}\mathbf{y} + \\ + \left[\frac{\partial}{\partial z}\mathbf{K}(z, \vartheta t)\frac{\partial}{\partial z}\mathbf{L}(z, \vartheta t) + \mathbf{L}(z, \vartheta t)^{*}\frac{\partial}{\partial z}\mathbf{K}(z, \vartheta t)^{*}\frac{\partial}{\partial z}\right]\mathbf{y} + \mathbf{Q}(z)\,\mathbf{y} = \mathbf{0}, \quad (7.1)$$

where z is a real variable, $0 \leqslant z \leqslant l$, t is time, $\mathbf{K}(z, \vartheta t)$, $\mathbf{L}(z, \vartheta t)$, $\mathbf{M}(z)$, $\mathbf{N}(z)$, $\mathbf{P}(z)$, $\mathbf{Q}(z)$, $\mathbf{R}(z)$ are $m \times m$ matrices whose elements are real functions of z. The products in (7.1) are products of operators:

$$\frac{\partial^2}{\partial z^2}\mathbf{P}(z)\frac{\partial^2}{\partial z^2}\mathbf{y}=\frac{\partial^2}{\partial z^2}\left[\mathbf{P}(z)\frac{\partial^2\mathbf{y}}{\partial z^2}\right],$$

$$\frac{\partial}{\partial z}\left[\mathbf{K}(z,\,\vartheta t)\frac{\partial}{\partial z}\mathbf{L}(z,\,\vartheta t)+\ldots\right]\mathbf{y}=\frac{\partial}{\partial z}\left\{\mathbf{K}(z,\,\vartheta t)\left[\frac{\partial}{\partial z}[\mathbf{L}(z,\,\vartheta t)\mathbf{y}]\right]\right\}+\ldots$$

and so on. The solution $\mathbf{y}=\mathbf{y}(z,\ t)$ is an m-vector. It is assumed that the elements of the matrices $\mathbf{K}(z,\ \vartheta t)$, $\mathbf{L}(z,\ \vartheta t)$, $\mathbf{M}(z)$, $\mathbf{N}(z)$, $\mathbf{P}(z)$, $\mathbf{Q}(z)$, $\mathbf{R}(z)$ have as many continuous derivatives with respect to z as required by equation (7.1) ($\mathbf{L}(z,\ \vartheta t)$ has two continuous derivatives, $\mathbf{K}(z,\ \vartheta t)$ one, $\mathbf{P}(z)$ two, and so on). It is also assumed that $\mathbf{P}(z)$, $\mathbf{Q}(z)$, $\mathbf{M}(z)$, $\mathbf{N}(z)$, $\mathbf{R}(z)$ are symmetric matrices and

$$\mathbf{M}(z)>0,\quad \mathbf{P}(z)>0,\quad \mathbf{N}(z)\geqslant 0,\quad \mathbf{R}(z)\geqslant 0 \tag{7.2}$$

for $0\leqslant z\leqslant l$, i. e., $\mathbf{M}(z)$ and $\mathbf{P}(z)$ are positive definite matrices, $\mathbf{N}(z)$ and $\mathbf{R}(z)$ positive semidefinite matrices, for all $0\leqslant z\leqslant l$. Recall that inequalities (7.2) are equivalent to the appropriate inequalities for quadratic forms. The matrices $\mathbf{K}(z,\ \tau)$, $\mathbf{L}(z,\ \tau)$ are 2π-periodic functions of $\tau=\vartheta t$ (not necessarily symmetric). In addition, we shall sometimes assume below (in §7.3) that these matrices are "small":

$$\mathbf{K}(z,\ \vartheta t)=\varepsilon\mathbf{K}_0(z,\ \vartheta t,\ \varepsilon),\quad \mathbf{L}(z,\ \vartheta t)=\varepsilon\mathbf{L}_0(z,\ \vartheta t,\ \varepsilon),$$

where $\mathbf{K}_0$, $\mathbf{L}_0$ are analytic functions of ε at $\varepsilon=0$. In such cases we shall be interested, as usual, in small values of ε; as for $\mathbf{Q}(z)$, we shall distinguish two cases. We shall assume that $\mathbf{Q}(z)$ is either a "small" matrix in the above sense, so that $\mathbf{Q}(z)=\varepsilon\mathbf{Q}_0(z,\ \varepsilon)$, where $\mathbf{Q}_0(z,\ \varepsilon)$ is an analytic function of ε, or a not necessarily "small" matrix satisfying the additional condition

$$\mathbf{Q}(z)>0 \quad \text{for} \quad 0\leqslant z\leqslant l. \tag{7.3}$$

Inequality (7.3) is also understood in the sense of the corresponding quadratic form.

Thus the terms of (7.1) containing $\mathbf{K}$ and $\mathbf{L}$, and sometimes also $\mathbf{Q}$, represent parametric excitation. The unperturbed equation is obtained from (7.1) by omitting all these terms. If $\mathbf{Q}(z)$ is not "small," so that (7.3) holds, the unperturbed equation is

$$\frac{\partial^2}{\partial t^2}\left[\mathbf{M}(z)-\frac{\partial}{\partial z}\mathbf{N}(z)\frac{\partial}{\partial z}\right]\mathbf{y}-\frac{\partial}{\partial z}\mathbf{R}(z)\frac{\partial}{\partial z}\mathbf{y}+\frac{\partial^2}{\partial z^2}\mathbf{P}(z)\frac{\partial^2}{\partial z^2}\mathbf{y}+\mathbf{Q}(z)\mathbf{y}=0. \tag{7.4}$$

If $\mathbf{Q}(z)$ is "small" (and then (7.3) need not be true), the last term on the left in (7.4) should be dropped (for then the term $\mathbf{Q}(z)\mathbf{y}$ becomes an "excitation").

Equations (7.1), (7.4) will be considered in conjunction with suitable homogeneous boundary conditions. In the case of a bar, the boundary conditions are determined by how the ends of the rod are attached (built in, hinged, etc.). All conditions of this type are selfadjoint, i. e., they have certain properties which we now proceed to describe. In the general case, the homogeneous boundary conditions are

$$\sum_{k=1}^{4} (\mathbf{A}_{ik}\mathbf{y}^{(k-1)}(0)+\mathbf{A}_{ik+4}\mathbf{y}^{(k-1)}(l))=0 \qquad (i=1,2,3,4), \tag{7.5}$$

where $\mathbf{A}_{ik}$ are real constant $m\times m$ matrices.

Let $\mathfrak{Y}$ denote the set of complex-valued vector-functions, four times continuously differentiable with respect to z on $[0, l]$, that satisfy conditions (7.5).

Recall that the parentheses $(\mathbf{y}_1, \mathbf{y}_2)$ denote the scalar product of two complex m-vectors: $(\mathbf{y}_1, \mathbf{y}_2)=\mathbf{y}_2^*\mathbf{y}_1$. Concerning the boundary conditions (7.5), we shall assume (selfadjoint property) that for any real $\mathbf{y}_1(z)\in\mathfrak{Y}$, $\mathbf{y}_2(z)\in\mathfrak{Y}$,

$$\left\{\left(\frac{\partial}{\partial z}\left[\mathbf{P}(z)\frac{\partial^2\mathbf{y}_1}{\partial z^2}\right], \mathbf{y}_2\right)-\left(\mathbf{P}(z)\frac{\partial^2\mathbf{y}_1}{\partial z^2}, \frac{\partial\mathbf{y}_2}{\partial z}\right)-\left(\mathbf{y}_1, \left[\frac{\partial}{\partial z}\mathbf{P}\frac{\partial^2\mathbf{y}_2}{\partial z^2}\right]\right)+\right.$$
$$\left.+\left(\frac{\partial\mathbf{y}_1}{\partial z}, \mathbf{P}(z)\frac{\partial^2\mathbf{y}_2}{\partial z^2}\right)-\left(\mathbf{R}\frac{\partial\mathbf{y}_1}{\partial z}, \mathbf{y}_2\right)+\left(\mathbf{y}_1, \mathbf{R}\frac{\partial\mathbf{y}_2}{\partial z}\right)\right\}\Big|_0^l=0, \tag{7.6}$$

$$\left\{\left(\mathbf{N}\frac{\partial\mathbf{y}_1}{\partial z}, \mathbf{y}_2\right)-\left(\mathbf{y}_1, \mathbf{N}\frac{\partial\mathbf{y}_2}{\partial z}\right)\right\}\Big|_0^l=0, \tag{7.7}$$

$$\left\{\left(\mathbf{K}\frac{\partial}{\partial z}\mathbf{L}\mathbf{y}_1, \mathbf{y}_2\right)+\left(\frac{\partial\mathbf{y}_1}{\partial z}, \mathbf{K}\mathbf{L}\mathbf{y}_2\right)-\left(\mathbf{y}_1\ \mathbf{K}\frac{\partial}{\partial z}\mathbf{L}\mathbf{y}_2\right)-\left(\mathbf{K}\mathbf{L}\mathbf{y}_1, \frac{\partial\mathbf{y}_2}{\partial z}\right)\right\}\Big|_0^l=0. \tag{7.8}$$

The symbol $\Big|_0^l$ has its usual meaning; for example, $\mathbf{y}(z)\Big|_0^l$ denotes $\mathbf{y}(l)-\mathbf{y}(0)$. We shall now explain how these rather cumbersome conditions arise.

If $\mathbf{Q}(z)$ is a "small" matrix, we consider the following operators on $\mathfrak{Y}$:

$$\mathfrak{M}=\mathbf{M}(z)-\frac{\partial}{\partial z}\mathbf{N}(z)\frac{\partial}{\partial z}, \tag{7.9}$$

$$\mathfrak{P}=\frac{\partial^2}{\partial z^2}\mathbf{P}(z)\frac{\partial^2}{\partial z^2}-\frac{\partial}{\partial z}\mathbf{R}(z)\frac{\partial}{\partial z}, \tag{7.10}$$

$$\mathfrak{R}(\vartheta t)=\frac{\partial}{\partial z}\mathbf{K}(z, \vartheta t)\frac{\partial}{\partial z}\mathbf{L}(z, \vartheta t)+\mathbf{L}(z, \vartheta t)^*\frac{\partial}{\partial z}\mathbf{K}(z, \vartheta t)^*\frac{\partial}{\partial z}+\mathbf{Q}(z), \tag{7.11}$$

but if $\mathbf{Q}(z)$ is not "small" we consider the operators (7.9) together with

$$\mathfrak{P}=\frac{\partial^2}{\partial z^2}\mathbf{P}(z)\frac{\partial^2}{\partial z^2}+\mathbf{Q}(z)-\frac{\partial}{\partial z}\mathbf{R}(z)\frac{\partial}{\partial z}, \tag{7.12}$$
$$\mathfrak{R}(\vartheta t)=\frac{\partial}{\partial z}\mathbf{K}(z, \vartheta t)\frac{\partial}{\partial z}\mathbf{L}(z, \vartheta t)+\mathbf{L}(z, \vartheta t)^*\frac{\partial}{\partial z}\mathbf{K}(z, \vartheta t)^*\frac{\partial}{\partial z}. \tag{7.13}$$

(We have used the same notation for the operators (7.10), (7.12) and (7.11), (7.13) in order to permit a unified treatment of both cases.)
As before, the products in (7.9) through (7.13) are products of operators.

With these conventions, equations (7.1), (7.3) take on a by now familiar form in both cases (see §§ II. 3.2, III. 4.1, IV. 3.3, V. 1.10):

$$\frac{\partial^2}{\partial t^2}\mathfrak{M}\mathbf{y}+[\mathfrak{P}+\mathfrak{R}(\vartheta t)]\,\mathbf{y}=0, \tag{7.14}$$

$$\frac{\partial^2}{\partial t^2}\mathfrak{M}\mathbf{y}+\mathfrak{P}\mathbf{y}=0. \tag{7.15}$$

Here (7.15) is the unperturbed equation and (7.14) the perturbed equation. Note that in contrast to the cases considered in previous chapters $\mathfrak{M}$, $\mathfrak{P}$, $\mathfrak{R}$ are operators in the infinite-dimensional space $\mathfrak{Y}$, not in a finite-dimensional space.

We define a scalar product for functions $y\in\mathfrak{Y}$ by

$$((\mathbf{y}_1,\ \mathbf{y}_2))=\int_0^l(\mathbf{y}_1,\ \mathbf{y}_2)\,dz. \tag{7.16}$$

Integrating by parts, using the symmetry properties of the matrices **M, N, P, Q, R** and conditions (7.6)–(7.8), we readily see that the operators $\mathfrak{M}$, $\mathfrak{P}$, $\mathfrak{R}$ are symmetric. This means that for any $\mathbf{y}_1\in\mathfrak{Y}$, $\mathbf{y}_2\in\mathfrak{Y}$

$$\left.\begin{aligned}((\mathfrak{M}\mathbf{y}_1,\ \mathbf{y}_2))=((\mathbf{y}_1,\ \mathfrak{M}\mathbf{y}_2)),\quad ((\mathfrak{P}\mathbf{y}_1,\ \mathbf{y}_2))=((\mathbf{y}_1,\ \mathfrak{P}\mathbf{y}_2)),\\ ((\mathfrak{R}\mathbf{y}_1,\ \mathbf{y}_2))=((\mathbf{y}_1,\ \mathfrak{R}\mathbf{y}_2)).\end{aligned}\right\} \tag{7.17}$$

For example, in the case of $\mathfrak{M}$, integration by parts gives

$$\begin{aligned}((\mathfrak{M}\mathbf{y}_1,\ \mathbf{y}_2))&=((\mathbf{M}\mathbf{y}_1,\ \mathbf{y}_2))-\left(\left(\frac{\partial}{\partial z}\left[\mathbf{N}\frac{\partial}{\partial z}\mathbf{y}_1\right],\ \mathbf{y}_2\right)\right)=\\ &=((\mathbf{y}_1,\ \mathbf{M}\mathbf{y}_2))-\left(\left(\mathbf{y}_1,\ \frac{\partial}{\partial z}\left[\mathbf{N}\frac{\partial}{\partial z}\mathbf{y}_2\right]\right)\right)+\left\{\left(\mathbf{N}\frac{\partial\mathbf{y}_1}{\partial z},\ \mathbf{y}_2\right)\Big|_0^l-\left(\mathbf{y}_1,\ \mathbf{N}\frac{\partial\mathbf{y}_2}{\partial z}\right)\Big|_0^l\right\}.\end{aligned}$$

By (7.7), the expression in curly brackets vanishes, and this gives the first equality of (7.17). The others are proved similarly.

In actual fact, conditions (7.6)–(7.8) express the validity of (7.17): they simply state that the appropriate "integrated" terms vanish.

Again integrating by parts, we obtain

$$\left.\begin{aligned}((\mathfrak{M}\mathbf{y},\ \mathbf{y}))&=((\mathbf{M}\mathbf{y},\ \mathbf{y}))+\left(\left(\mathbf{N}\frac{\partial\mathbf{y}}{\partial z},\ \frac{\partial\mathbf{y}}{\partial z}\right)\right),\\ ((\mathfrak{P}\mathbf{y},\ \mathbf{y}))&=((\mathbf{Q}\mathbf{y},\ \mathbf{y}))+\left(\left(\mathbf{P}\frac{\partial^2\mathbf{y}}{\partial z^2},\ \frac{\partial^2\mathbf{y}}{\partial z^2}\right)\right)\end{aligned}\right\} \tag{7.18}$$

in case **Q** is not "small." If **Q** is "small," we drop the first term on the right of (7.18). From (7.18) and (7.2), (7.3) it is evident that

$$((\mathfrak{M}\mathbf{y},\ \mathbf{y}))\geqslant\delta_1((\mathbf{y},\ \mathbf{y})),\quad ((\mathfrak{P}\mathbf{y},\ \mathbf{y}))\geqslant\delta_2\left(\left(\frac{\partial^2\mathbf{y}}{\partial z^2},\ \frac{\partial^2\mathbf{y}}{\partial z^2}\right)\right), \tag{7.19}$$

where $\delta_1>0$, $\delta_2>0$ are lower bounds for the eigenvalues of the matrices $\mathbf{M}(z)$ and $\mathbf{P}(z)$, respectively. It is readily seen that the second relation may be written

$$((\mathfrak{P}\mathbf{y},\ \mathbf{y}))\geqslant\delta((\mathbf{y},\ \mathbf{y})) \tag{7.19'}$$

for some $\delta > 0$, if we demand that there are no constant vectors $\mathbf{c}_1$, $\mathbf{c}_2$, at least one of which is not zero, such that $\mathbf{y}(z) = \mathbf{c}_1 + \mathbf{c}_2 z$ lies in $\mathfrak{Y}$. This condition guarantees that zero is not an eigenvalue of $\mathfrak{P}$ and will be assumed true throughout the sequel.*

We have assumed that the boundary conditions (7.5) are selfadjoint, satisfying the relations (7.6)—(7.8). This assumption may be formulated explicitly in terms of the matrices $\mathbf{A}_{ik}$ figuring in (7.5). To do this, it is convenient to write (7.5) in a different form. Let $\mathbf{w}$ denote the $8m$-vector whose components are, in order: the m components of $\mathbf{y}(0)$, then the m components of $\mathbf{y}'(0)$, and then those of $\mathbf{y}''(0)$, $\mathbf{y}'''(0)$, $\mathbf{y}(l)$, $\mathbf{y}'(l)$, $\mathbf{y}''(l)$, $\mathbf{y}'''(l)$. The boundary conditions (7.5) are now

$$\mathbf{A}\mathbf{w} = 0, \tag{7.5'}$$

where $\mathbf{A}$ is the $8m \times 8m$ matrix

$$\mathbf{A} = \begin{pmatrix} \mathbf{A}_{11} & \dots & \mathbf{A}_{18} \\ \cdot & \cdot\ \cdot\ \cdot & \cdot \\ \mathbf{A}_{41} & \dots & \mathbf{A}_{48} \\ 0 & \dots & 0 \\ \cdot & \cdot\ \cdot\ \cdot & \cdot \\ 0 & \dots & 0 \end{pmatrix}.$$

Let $\mathbf{S}$ denote the matrix of orthogonal projection onto the null-space of $\mathbf{A}$. Condition (7.5) is then equivalent to

$$\mathbf{S}\mathbf{w} = \mathbf{w}. \tag{7.5''}$$

We now consider (7.6)—(7.8). Let $\mathbf{w}_1$ and $\mathbf{w}_2$ denote vectors defined in terms of $\mathbf{y}_1(z)$ and $\mathbf{y}_2(z)$ in the same way as $\mathbf{w}$ was defined in terms of $\mathbf{y}(z)$. We may then write (7.6), (7.7), (7.8) as follows:

$$\mathbf{w}_2^* \mathbf{C}_1 \mathbf{w}_1 = 0, \tag{7.6'}$$

$$\mathbf{w}_2^* \mathbf{C}_2 \mathbf{w}_1 = 0, \tag{7.7'}$$

$$\mathbf{w}_2^* \mathbf{C}_3(\tau) \mathbf{w}_1 = 0, \tag{7.8'}$$

where $\mathbf{C}_1$, $\mathbf{C}_2$ are constant $8m \times 8m$ matrices and $\mathbf{C}_3(\tau)$ is a periodic matrix-function of τ; these matrices are readily expressed in terms of the various quantities entering into (7.6)—(7.8):

$$\mathbf{C}_1 = \left(\begin{array}{cccc|cccc} 0 & -\mathbf{R}(0) & \mathbf{P}'(0) & \mathbf{P}(0) & & & & \\ \mathbf{R}(0) & 0 & -\mathbf{P}(0) & 0 & & \mathbf{0} & & \\ -\mathbf{P}'(0) & \mathbf{P}(0) & 0 & 0 & & & & \\ -\mathbf{P}(0) & 0 & 0 & 0 & & & & \\ \hline & & & & 0 & \mathbf{R}(l) & -\mathbf{P}'(l) & -\mathbf{P}(l) \\ & \mathbf{0} & & & -\mathbf{R}(l) & 0 & \mathbf{P}(l) & 0 \\ & & & & \mathbf{P}'(l) & -\mathbf{P}(l) & 0 & 0 \\ & & & & \mathbf{P}(l) & 0 & 0 & 0 \end{array}\right),$$

* Inequality (7.19') is certainly true if $\mathbf{Q}(z) > 0$ for $0 \leq z \leq l$.

$$\mathbf{C}_2=\begin{pmatrix} \begin{matrix} 0 & \mathbf{N}(0) & 0 & 0\\ -\mathbf{N}(0) & 0 & 0 & 0\\ 0 & 0 & 0 & 0\\ 0 & 0 & 0 & 0 \end{matrix} & \mathbf{0}\\ \mathbf{0} & \begin{matrix} 0 & -\mathbf{N}(l) & 0 & 0\\ \mathbf{N}(l) & 0 & 0 & 0\\ 0 & 0 & 0 & 0\\ 0 & 0 & 0 & 0 \end{matrix} \end{pmatrix},$$

$$\mathbf{C}_3(\tau)=\begin{pmatrix} \begin{matrix} \mathbf{U}(0,\tau) & \mathbf{V}(0,\tau) & 0 & 0\\ -\mathbf{V}(0,\tau) & 0 & 0 & 0\\ 0 & 0 & 0 & 0\\ 0 & 0 & 0 & 0 \end{matrix} & \mathbf{0}\\ \mathbf{0} & \begin{matrix} -\mathbf{U}(l,\tau) & -\mathbf{V}(l,\tau) & 0 & 0\\ \mathbf{V}(l,\tau) & 0 & 0 & 0\\ 0 & 0 & 0 & 0\\ 0 & 0 & 0 & 0 \end{matrix} \end{pmatrix},$$

where

$$\mathbf{U}(z,\tau)=-\mathbf{K}(z,\tau)\mathbf{L}'(z,\tau)+\mathbf{L}'(z,\tau)^*\mathbf{K}(z,\tau)^*,$$
$$\mathbf{V}(z,\tau)=\mathbf{K}(z,\tau)\mathbf{L}(z,\tau)+\mathbf{L}(z,\tau)^*\mathbf{K}(z,\tau)^*,$$

and the prime denotes differentiation with respect to z. The symmetry properties of the operators $\mathfrak{P}$, $\mathfrak{M}$ and $\mathfrak{N}(\vartheta t)$ now become

$$\mathbf{SC}_1\mathbf{S}=\mathbf{0}, \tag{7.6''}$$
$$\mathbf{SC}_2\mathbf{S}=\mathbf{0}, \tag{7.7''}$$
$$\mathbf{SC}_3(\tau)\mathbf{S}\equiv\mathbf{0}. \tag{7.8''}$$

Conditions (7.6'') – (7.8'') are the desired symmetry conditions for $\mathfrak{P}$, $\mathfrak{M}$, $\mathfrak{N}(\vartheta t)$ in "explicit" form, in terms of the coefficients $\mathbf{A}_{ik}$ in (7.5).

For the sequel we shall require that the operator $\mathfrak{P}$ be *essentially selfadjoint*.* This generally imposes additional restrictions on the coefficients $\mathbf{A}_{ik}$.

The operator $\mathfrak{P}$ is essentially selfadjoint if and only if any of the following three conditions holds (we omit the equivalence proofs):

(S) the equality $((\mathfrak{P}\mathbf{y}_1,\mathbf{y}_2))=((\mathbf{y}_1,\mathfrak{P}\mathbf{y}_2))$ can hold for any function $\mathbf{y}_1(z)\in\mathfrak{Y}$ and some smooth vector-function $\mathbf{y}_2(z)$ only if $\mathbf{y}_2(z)$ satisfies the boundary condition (7.5).

(S') If $\mathbf{SC}_1\mathbf{w}=\mathbf{0}$ for some $8m$-vector $\mathbf{w}$ with components as specified above, then either (7.5) or the equivalent (7.5') is satisfied.

(S'') There exists $\tau\geqslant 0$ such that

$$\tau\mathbf{C}_1^*\mathbf{SC}_1\geqslant\mathbf{A}^*\mathbf{A}.$$

* That is to say (see, e.g., Akhiezer, N.I. and I.M. Glazman, Theory of Linear Operators in Hilbert Space, Moscow, Nauka, 1966), the closure of the symmetric operator $\mathfrak{P}$ is selfadjoint.

(This last inequality is understood, as usual, as the inequality for the appropriate quadratic forms.) Condition (S'') is readily converted into "explicit" relationships for the coefficients.

7.2. Bubnov-Galerkin truncation of the original equation – derivation of finite-dimensional system

We first consider the unperturbed equation (7.15) (or (7.4)). Let us look for particular solutions of the type

$$\mathbf{y}=e^{i\omega_k t}\mathbf{y}_k(z) \qquad (k=1,\ 2,\ \ldots), \tag{7.20}$$

where $\mathbf{y}_k(z)\in\mathfrak{Y}$. As follows from (7.14), the functions $\mathbf{y}_k(z)$ must satisfy the equations

$$\omega_k^2\mathfrak{M}\mathbf{y}_k=\mathfrak{P}\mathbf{y}_k \qquad (k=1,\ 2,\ \ldots). \tag{7.21}$$

It can be shown that under the above assumptions there exist a complete system of eigenfunctions $\mathbf{y}_k(z)\in\mathfrak{Y}$ and a system of "eigenvalues" $\omega_k^2\geqslant 0$, $\omega_k^2\longrightarrow\infty$ (some of which may coincide) satisfying (7.21). The system $\{\mathbf{y}_k\}$ is complete in the space L_2 of m-dimensional vector-functions $\mathbf{y}(z)$ with the scalar product (7.16), and these functions may be assumed to be orthogonal:

$$((\mathfrak{M}\mathbf{y}_j,\ \mathbf{y}_k))=\delta_{jk}. \tag{7.22}$$

These statements follow readily from standard theorems of operator theory.* The functions $\mathbf{y}_k(z)$ and the number ω_k are usually called the normal modes and natural frequencies of the unperturbed system. Note that in the case $\omega_j^2\neq\omega_k^2$ the orthogonality relations (7.22) are obvious. Indeed, using (7.21) and the symmetry of the operators $\mathfrak{P}$, $\mathfrak{M}$, we have $\omega_j^2((\mathfrak{M}\mathbf{y}_j,\ \mathbf{y}_k))=((\mathfrak{P}\mathbf{y}_j,\ \mathbf{y}_k))=((\mathbf{y}_j,\ \mathfrak{P}\mathbf{y}_k))=$ $=\omega_k^2((\mathbf{y}_j,\ \mathfrak{M}\mathbf{y}_k))=\omega_k^2((\mathfrak{M}\mathbf{y}_j,\ \mathbf{y}_k))$. Thus, if $\omega_j^2\neq\omega_k^2$ we have $((\mathfrak{M}\mathbf{y}_j,\ \mathbf{y}_k))=0$.

It can be shown (though we shall not need this) that any solution of the unperturbed equation (7.15) may be approximated to within arbitrary accuracy by an expression

$$\mathbf{y}^{(n)}(t,z)=\sum_{k=1}^{n}c_k e^{i\omega_k t}\mathbf{y}_k(z), \tag{7.23}$$

where c_k are suitably chosen constants; the accuracy increases with n.

* Using standard theorems of the theory of differential operators (see, e.g., Akhiezer, N.I. and I.M. Glazman, op. cit., and also /80/, Chapters VI, VIII), one readily proves that the Friedrichs extensions of the operators $\mathfrak{M}$ and $\mathfrak{P}$ to L_2 are selfadjoint, and their spectra consist of isolated eigenvalues. Rewrite equation (7.21) as $\mathfrak{P}^{-1/2}\mathfrak{M}\mathfrak{P}^{-1/2}\mathbf{u}_k=\lambda_k\mathbf{u}_k$, where $\mathbf{u}_k=\mathfrak{P}^{1/2}\mathbf{y}_k$, $\lambda_k=\omega_k^{-2}$. It can be shown that the operator $\mathfrak{P}^{-1/2}\mathfrak{M}\mathfrak{P}^{-1/2}$ is selfadjoint, positive and compact; hence it has a complete orthogonal system of eigenfunctions $\mathbf{u}_k$ and a system of eigenvalues $\lambda_k>0$. Setting $\mathbf{y}_k=\mathfrak{P}^{-1/2}\mathbf{u}_k$, $\omega_k^2=\lambda_k^{-1}>0$, we obtain a system of functions $\mathbf{y}_k$ and numbers ω_k^2 satisfying (7.21) and (7.22).

In fact,

$$\int_0^l |\mathbf{y}^{(n)}(t, z)-\mathbf{y}(t, z)|^2 dz \underset{n\to\infty}{\longrightarrow} 0,$$

and the convergence is uniform on any finite time interval.

Applied to our parametric resonance problems, the Bubnov-Galerkin procedure is as follows.

We wish to determine an approximate solution of the original ("unperturbed") equation (7.1) (or, equivalently, (7.14)) of type (7.23):

$$\mathbf{y}^{(n)}(t, z)=\sum_{k=1}^{n}\varphi_k(t)\,\mathbf{y}_k(z), \tag{7.24}$$

where $\varphi_k(t)$ are complex-valued functions. Insert (7.24) into the left-hand side of equation (7.14). On the right we get, instead of zero, a certain "residual." We stipulate that this residual be orthogonal to the functions $\mathbf{y}_1(z), \ldots, \mathbf{y}_k(z)$:

$$\left(\left(\frac{\partial^2}{\partial t^2}\mathfrak{M}\mathbf{y}^{(n)}+[\mathfrak{P}+\mathfrak{R}(\vartheta t)]\,\mathbf{y}^{(n)},\ \mathbf{y}_j\right)\right)=0 \qquad (j=1, \ldots, n). \tag{7.25}$$

This will be the system for determination of the functions $\varphi_k(t)$ $(k=1,\ldots, n)$. Thus the Bubnov-Galerkin method amounts to replacing the original equation (7.1) by a finite-dimensional system (7.25). In our case system (7.25) is substantially simplified owing to the orthogonality of the functions $\mathbf{y}_j=\mathbf{y}_j(z)$. In fact, substituting (7.24) into (7.25) and using (7.21), (7.22), we convert (7.25) to the form

$$\frac{d^2\varphi_j}{dt^2}+\omega_j^2\varphi_j+\sum_{k=1}^{n}\gamma_{kj}(\vartheta t)\,\varphi_k=0 \qquad (j=1, \ldots, n), \tag{7.26}$$

where

$$\gamma_{kj}(\vartheta t)=((\mathfrak{R}(\vartheta t)\,\mathbf{y}_k,\ \mathbf{y}_j)). \tag{7.27}$$

It follows from the last equality of (7.17) that $\gamma_{kj}=\bar{\gamma}_{jk}$. Thus (7.26) is a Hamiltonian system, which may be written as a single vector equation (V. 1.48). Using (7.16) and (7.11), we obtain for "small" $\mathbf{Q}(z)$

$$\gamma_{kj}(\vartheta t)=\int_0^l\Big[\Big(\frac{\partial}{\partial z}\mathbf{K}(z, \vartheta t)\frac{\partial}{\partial z}\mathbf{L}(z, \vartheta t)\,\mathbf{y}_k,\ \mathbf{y}_j\Big)+ \\ +\Big(\mathbf{L}(z, \vartheta t)^*\frac{\partial}{\partial z}\mathbf{K}(z, \vartheta t)^*\frac{\partial}{\partial z}\mathbf{y}_k,\ \mathbf{y}_j\Big)+(\mathbf{Q}(z)\,\mathbf{y}_k, \mathbf{y}_j)\Big]\,dz. \tag{7.28}$$

If $\mathbf{Q}(z)$ is not "small," the term $(\mathbf{Q}(z)\mathbf{y}_k,\ \mathbf{y}_j)$ in (7.28) is dropped and ω_j^2 and $\mathbf{y}_j(z)$ are of course different.

Thus, instead of a partial differential equation (7.1) we have to investigate a system of ordinary differential equations (7.26). If the matrices $\mathbf{K}(z, \vartheta t)=\varepsilon\mathbf{K}_0(z, \vartheta t, \varepsilon)$, $\mathbf{L}(z, \vartheta t)=\varepsilon\mathbf{L}_0(z, \vartheta t, \varepsilon)$ (and possibly also $\mathbf{Q}(z)=\varepsilon\mathbf{Q}_0(z, \varepsilon)$) are analytic functions of the "small" parameter ε, the coefficients $\gamma_{kj}(\vartheta t) = \varepsilon\gamma^0_{kj}(\vartheta t, \varepsilon)$ in (7.26) will also be analytic in ε at $\varepsilon=0$. We can now construct for system (7.26) "wide" and "narrow" dynamic instability domains, estimate the characteristic exponents, and so on (see § V. 1). The obvious question is now whether the various statements about stability, instability, order of growth of solutions of system (7.26) and so on yield analogous statements (in some sense) for the original equation (7.1) as $n\to\infty$. In other words, is there a rigorous justification of the Bubnov-Galerkin method as applied to parametric resonance problems, and if there is, under what assumptions?

7.3. Justification of the Bubnov-Galerkin method in parametric resonance problems. Derguzov-Fomin theorem

Let $\mathbf{y}(z, t)$ be a solution of equation (7.1). We define the norm $\|\mathbf{y}(z, t)\|$ by

$$\|\mathbf{y}(z, t)\|^2 = \int_0^l (\mathfrak{P}\mathbf{y}, \mathbf{y})\, dz + \int_0^l \left(\mathfrak{M}\frac{\partial \mathbf{y}}{\partial t}, \frac{\partial \mathbf{y}}{\partial t}\right) dz. \tag{7.29}$$

This formula applies whether $\mathbf{Q}(z)$ is "small" or not. It follows from (7.19) and (7.19') that the right-hand side of (7.29) is indeed positive if $\mathbf{y}(z) \not\equiv \mathbf{0}$.

The exponent of exponential growth of solutions of equations (7.1) is defined to be

$$\alpha = \sup_{\mathbf{y}} \overline{\lim_{t\to\infty}} \frac{1}{t} \ln \|\mathbf{y}(z, t)\|, \tag{7.30}$$

where the supremum is taken over all solutions of equation (7.1). It follows from (7.30) (this may be regarded as an equivalent definition of α) that for any $\delta > 0$ there exist a constant $c_\delta > 0$ and a solution $\mathbf{y}^0(z, t)$ such that, first,

$$\|\mathbf{y}^0(z, t)\| > c_\delta e^{(\alpha-\delta)t} \|\mathbf{y}^0(z, 0)\|$$

and, second, for any solution $\mathbf{y}(z, t)$ of equation (7.1),

$$\|\mathbf{y}(z, t)\| < c_\delta e^{(\alpha+\delta)t} \|\mathbf{y}(z, 0)\|.$$

Similarly one defines the exponent of exponential growth α_n of the solutions of system (7.26): let $\tilde{\mathbf{y}}(t)$ denote a vector with components $\varphi_1(t), \ldots, \varphi_n(t)$ and consider the standard norm $\|\tilde{\mathbf{y}}\| = \sqrt{(\tilde{\mathbf{y}}, \tilde{\mathbf{y}})}$; then

$$\alpha_n = \sup_{\tilde{\mathbf{y}}} \lim_{t\to\infty} \frac{1}{t} \ln \|\tilde{\mathbf{y}}(t)\|,$$

where the supremum is taken over all solutions of system (7.26). Instead, one can consider n linearly independent solutions $\tilde{\mathbf{y}}^{(j)}$ of the system, and then

$$\alpha_n = \max_{1\leqslant j\leqslant n} \overline{\lim_{t\to\infty}} \frac{1}{t} \ln \|\tilde{\mathbf{y}}^{(j)}\|.$$

Let $\mathbf{K}(z, \vartheta t) = \varepsilon \mathbf{K}_0(z, \vartheta t, \varepsilon)$, $\mathbf{L}(z, \vartheta t) = \varepsilon \mathbf{L}_0(z, \vartheta t, \varepsilon)$ and possibly also $\mathbf{Q}(z) = \varepsilon \mathbf{Q}_0(z, \vartheta t, \varepsilon)$ be analytic in ε at $\varepsilon = 0$. (Recall that we are considering both cases together.) Then we can construct dynamic instability domains in the (ε, ϑ)-plane for system (7.26). The cusps of these wedge-shaped domains will be at points $(0, \vartheta_{jh}^{(k)})$, where

$$\vartheta_{jh}^{(k)} = \frac{\omega_j + \omega_h}{k} \qquad (1 \leqslant j \leqslant h \leqslant n;\ k = 1, 2, \ldots) \tag{7.31}$$

are the critical frequencies of system (7.26). Note that as n increases the set of critical frequencies (7.31) expands. Since $\omega_j \to \infty$ as $j \to \infty$, it follows from (7.31) that the frequencies $\vartheta_{jh}^{(m)}$ fill the half-axis $\vartheta > 0$ densely as $n \to \infty$.* Some of the critical frequencies (7.31) will have the property that the instability domains attached at $(0, \vartheta_{jh}^{(m)})$ are wide (see § V.1). We shall call such frequencies $\vartheta_{jh}^{(m)}$ *dangerous* and denote them by $\tilde{\vartheta}_{jh}^{(m)}$. Recall (§ V. 1) that a frequency $\vartheta_{jh}^{(k)}$ is dangerous if the k-th Fourier coefficient of the function $\gamma_{jh}(t)$ does not vanish:

$$\int_0^{2\pi} \gamma_{jh}(t) e^{-ikt} dt \neq 0.$$

The dangerous frequencies need not be dense in $\vartheta > 0$ as $n \to \infty$.

For example, if $\gamma_{kj}(t)$ are trigonometric polynomials (as is often the case in applied problems), the only point of accumulation of the dangerous frequencies $\tilde{\vartheta}_{jh}^{(m)}$ is $\vartheta = 0$.

Let μ be an arbitrary real number and consider the ray $\vartheta = \vartheta_0 + \mu\varepsilon$, $\varepsilon > 0$, through the point $(0, \vartheta_0)$ in the $\varepsilon\vartheta$-plane. Set

$$\beta(\vartheta_0, \mu) = \lim_{\varepsilon \to 0} \frac{\alpha(\varepsilon, \vartheta_0 + \varepsilon\mu)}{\varepsilon}, \tag{7.32}$$

where $\alpha(\varepsilon, \vartheta)$ is the exponent of exponential growth (7.30) for equation (7.1). The quantity $\beta(\vartheta_0, \mu)$ characterizes the degree of exponential increase of the solutions of equation (7.1) with parameters $\vartheta = \vartheta_0 + \varepsilon\mu$, $\varepsilon > 0$, for small $\varepsilon > 0$. If $\beta(\vartheta_0, \mu) = 0$, the exponent is $o(\varepsilon)$ as $\varepsilon \to 0$. If $\beta(\vartheta_0, \mu) > 0$, it is of the same order of magnitude as ε itself.

The following theorems lay a rigorous foundation for the Bubnov-Galerkin method in parametric resonance problems, justifying the replacement of equation (7.1) by system (7.26) for sufficiently large n. They demonstrate the sense in which system (7.26) provides a substitute for equation (7.1) and state the conditions under which the procedure is legitimate

Theorem I (Derguzov and Fomin). Under the assumptions formulated in subsection 1, suppose the system (7.26) has been constructed as described. Let α and α_n be the exponents of exponential growth of the solutions

* In fact if $\alpha_j \to \infty$ as $j \to \infty$, then the numbers α_j/k where k is a natural number) are dense in the positive half-axis. Indeed, the distance between consecutive numbers in the j-th sequence $\{k/\alpha_j\}$, where $k = 1, 2, \ldots$ and j is fixed, may be made arbitrarily small if j is large enough. It follows that for any $\vartheta > 0$, $\delta > 0$ there are numbers k/α_j in the interval $(\vartheta - \delta, \vartheta + \delta)$. Consequently, these numbers, and hence also α_j/k, are dense in the positive half-axis.

of equation (7.1) and system (7.26), respectively. Suppose the matrix $\mathbf{N}(z)$ *in (7.1) is positive definite for any* $z \in [0, l]$. *Then*

$$\lim_{n\to\infty} \alpha_n = \alpha \tag{7.33}$$

and the limit is uniform in any interval $0 < \vartheta_1 \leqslant \vartheta \leqslant \vartheta_2$.

Theorem II (Derguzov and Fomin). Under all the assumptions of Theorem I, except the one stipulating that $\mathbf{N}(z)$ *is positive definite for all* $0 \leqslant z \leqslant l$, * *let the eigenvalues* ω_k *satisfy the condition*

$$\omega_{k+1} - \omega_k \longrightarrow \infty \quad as \quad k \longrightarrow \infty. \tag{7.34}$$

Then the conclusions of Theorem I are valid.

Theorem III (Fomin). Let the following conditions hold for equation (7.1):

1) *The time-dependent matrix-coefficients are trigonometric polynomials of degree at most* m, *e.g.,*

$$\mathbf{K}(z, \tau) = \sum_{|p|<m} \mathbf{K}_p(z) e^{ip\tau}.$$

2) *The "fundamental" case holds: there exist a unique pair of natural frequencies* ω_j, ω_h *(see (7.21)) and a unique natural number* $k < m$ *such that*

$$\vartheta_0 = \frac{\omega_j + \omega_h}{k}. \tag{7.35}$$

3) *For all natural frequencies* ω_s *congruent to* $\pm\omega_j$ *or* $\pm\omega_h$ $\mathrm{mod}\,\vartheta_0$, **

$$|\omega_s \pm \omega_{r_0}| > m\vartheta_0, \quad |\omega_s \mp \omega_{s_0}| > m\vartheta_0.$$

Then

$$\beta(\vartheta_0, \mu) > 0 \tag{7.36}$$

(see (7.32)) if and only if

$$\mu_1 < \mu < \mu_2,$$

where

$$\mu_{1,2} = \frac{1}{k}\left(\frac{\gamma_{jj}^{(0)}}{2\omega_j} + \frac{\gamma_{hh}^{(0)}}{2\omega_h} \mp \frac{|\gamma_{jh}^{(k)}|}{\sqrt{\omega_j\omega_h}}\right). \tag{7.37}$$

Here $\gamma_{jh}^{(k)}$ *are the Fourier coefficients of the function* $\gamma_{jh}(\tau)$ *(see (7.28)):*

$$\gamma_{jh}(\tau) \sim \sum_k \gamma_{jh}^{(k)} e^{ik\tau}.$$

* Recall that according to (7.19) it is still true that $\mathbf{N}(z) \geqslant 0$ $(0 \leqslant z \leqslant l)$.

** ω_s is congruent to ω_j mod ϑ_0 if $(\omega_s - \omega_j)/\vartheta_0$ is an integer.

The proofs of these theorems are fairly complicated and will not be given here. We add a few words of explanation. The relation (7.33) shows that by taking n sufficiently large one can make the exponents of exponential growth of the solutions of (7.1) arbitrarily near those of (7.26). Theorem III shows that the wide instability domains (and dangerous critical frequencies) of the finite-dimensional system (7.26) yield the most dangerous instability domains of the original system (7.1). (In these regions the solutions increase at least exponentially with exponent $\alpha=\varepsilon\,[\beta(\vartheta_0,\ \mu)+o(1)]$ as $\varepsilon\to 0$.) There is no such correspondence for stability domains. This is natural: as $n\to\infty$ the dynamic instability domains of system (7.26) fill out the $(\varepsilon,\ \vartheta)$-plane ever more densely. (As mentioned, the critical frequencies fill out the half-axis $\vartheta>0$ densely as $n\to\infty$.) Examples can be given of equations (7.1) whose stability domains on the $(\varepsilon,\ \vartheta)$-plane contain no interior points near the ϑ-axis, i. e., in any, arbitrarily small disk on the $(\varepsilon,\ \vartheta)$-plane there are points for which equation (7.1) has solutions $\mathbf{y}(z,\ t)$ such that $\|\mathbf{y}(z,\ t)\|\to\infty$ as $t\to\infty$ (the norm is defined by (7.29)). Despite this, Derguzov has shown that for a certain specific system there may be sets of measure zero on the $(\varepsilon,\ \vartheta)$-plane (intervals parallel to the ε-axis) for all points of which equation (7.1) is stable (all solutions bounded uniformly in t in some natural metric).

A characteristic property of the wide instability domains of the finite-dimensional systems (7.26) $(n\to\infty)$ is that outside them (more precisely, on the ray $\vartheta=\vartheta_0+\mu\varepsilon,\ \varepsilon>0$, for sufficiently small $\varepsilon>0$ — for small ε this ray is in none of the instability domains) the norm $\|\mathbf{y}(z,\ t)\|$, though it increases, does so "weakly": the exponent of exponential growth is of order less than that of ε.

It is in this sense that some of the critical frequencies $\tilde{\vartheta}_{jh}^{(k)}$ (i. e., the frequencies (7.31) for which the k-th Fourier coefficient of $\gamma_{jh}(\vartheta t)$ does not vanish) are "dangerous." (The functions $\gamma_{jh}(\vartheta t)$ are defined by (7.27).) In the "fundamental" case the corresponding dangerous instability domains for equation (7.1) are defined by

$$\vartheta_{jh}^{(k)}+\mu_1\varepsilon+\ldots<\vartheta<\vartheta_{jh}^{(k)}+\mu_2\varepsilon+\ldots, \tag{7.38}$$

where (see (V. 1.29), (V. 1.30)) $\mu_{1,2}$ are defined by (7.37). The exponent of exponential growth of two solutions corresponding to the two multipliers which tend to

$$\rho=\exp\frac{2\pi i\omega_j}{\vartheta_{jh}^{(k)}}=\exp\frac{2\pi i\omega_h}{\vartheta_{jh}^{(k)}} \tag{7.39}$$

at $\varepsilon=0$ are determined by the formula of § V. 1.5; they are

$$\alpha_{jh}^{(k)}=\beta_{jh}^{(k)}\varepsilon+O(\varepsilon^{1+\delta}),\quad \delta>0, \tag{7.40}$$

where

$$\beta_{jh}^{(k)}=\frac{1}{\vartheta_h^{(k)}}\left\{\frac{1}{2}\,[\chi_{hh}(\mu)-\chi_{-j-j}(\mu)]-\sqrt{\Delta(\mu)}\right\}. \tag{7.41}$$

(We are considering points $(\varepsilon,\ \vartheta)$ on the ray $\vartheta=\vartheta_{jh}^{(k)}+\mu\varepsilon,\ \varepsilon>0,\ \varepsilon\to 0$.)

If the parametric excitation in (7.1) contains only finitely many Fourier harmonics, then each bounded strip $(0<\vartheta'\leqslant\vartheta<\vartheta'',\ 0<\varepsilon\leqslant\text{const})$ may contain only finitely many wide instability domains. For this case the quantity $\beta(\vartheta_0,\ \mu)$ in the Derguzov-Fomin theorems is

$$\beta(\vartheta_0,\ \mu)=\max\beta_{jh}^{(k)}, \tag{7.42}$$

where the maximum is taken over all $j,\ h,\ k$ (the number of triples is finite) for which $\vartheta_{jh}^{(k)}=\vartheta_0$.

It follows from the above account that the first-approximation formulas of § V.1 and the formulas for the dangerous critical frequencies also remain valid if we apply them not to equation (7.1) but to the formal infinite system

$$\frac{d^2\varphi_j}{dt^2}+\omega_j^2\varphi_j+\sum_{k=1}^{\infty}\gamma_{kj}(\vartheta t)\,\varphi_k=0 \qquad (j=1,\ 2,\ \ldots), \tag{7.43}$$

obtained from (7.26) by letting $n\longrightarrow\infty$. Here we have used the fact that the coefficients $\gamma_{kj}(\vartheta t)$ are independent of n. System (7.43) may also be obtained by substituting the formal series

$$\mathbf{y}(z,\ t)=\sum_{1}^{\infty}\varphi_k(t)\mathbf{y}_k(z)$$

into equation (7.1) and comparing the coefficients of the functions $\mathbf{y}_k(z)$. This procedure is often used in applications (as we ourselves did in §3).

These operations may be given a rigorous meaning (treating system (7.43) in the space l_2 of square-summable number sequencies), but we shall not dwell on this here.

7.4. Verification of the conditions of Theorems I – III for actual systems

a) The system of differential equations for flexural-torsional vibrations of a suspension bridge (§1, equations (1.1)) lead (at $q_y=0$) to equation (7.1) with

$$\left.\begin{aligned}
&\mathbf{y}=\begin{pmatrix}\eta\\ \Theta\end{pmatrix},\quad \mathbf{P}(z)=\begin{Vmatrix}EFr_1^2 & 0\\ 0 & EJ_\omega\end{Vmatrix},\quad \mathbf{K}=\mathbf{I}_2,\quad \mathbf{R}(z)=\mathbf{0},\\
&\mathbf{L}(z,\ \tau)=\begin{Vmatrix}0 & M_y(z,\ \tau)\\ 0 & 0\end{Vmatrix},\quad \mathbf{M}(z)=\begin{Vmatrix}\rho F & 0\\ 0 & \rho Fr^2\end{Vmatrix},\\
&\mathbf{N}(z)=\begin{Vmatrix}\rho Fr^2 & 0\\ 0 & \rho J_\omega\end{Vmatrix},\quad \mathbf{Q}(z)=\begin{Vmatrix}0 & 0\\ 0 & \frac{1}{2}qb\end{Vmatrix}.
\end{aligned}\right\} \tag{7.44}$$

The boundary conditions are

$$\mathbf{y}(0,\ t)=\mathbf{y}''(0,\ t)=\mathbf{y}(l,\ t)=\mathbf{y}''(l,\ t)=\mathbf{0},$$

so that we can write them in the form of (7.5) with $\mathbf{A}_{11}=\mathbf{I}_2$, $\mathbf{A}_{23}=\mathbf{I}_2$, $\mathbf{A}_{35}=\mathbf{I}_2$, $\mathbf{A}_{47}=\mathbf{I}_2$ and all other $\mathbf{A}_{ij}$'s are 2×2 zero matrices. Thus the matrix $\mathbf{A}$ of (7.5') is

$$\mathbf{A}=\begin{pmatrix} \mathbf{I}_2 & 0 & 0 & 0 & 0 & 0 & 0 & 0 \\ 0 & 0 & \mathbf{I}_2 & 0 & 0 & 0 & 0 & 0 \\ 0 & 0 & 0 & 0 & \mathbf{I}_2 & 0 & 0 & 0 \\ 0 & 0 & 0 & 0 & 0 & 0 & \mathbf{I}_2 & 0 \\ 0 & 0 & 0 & 0 & 0 & 0 & 0 & 0 \\ 0 & 0 & 0 & 0 & 0 & 0 & 0 & 0 \\ 0 & 0 & 0 & 0 & 0 & 0 & 0 & 0 \\ 0 & 0 & 0 & 0 & 0 & 0 & 0 & 0 \end{pmatrix}. \tag{7.45}$$

To check the symmetry conditions (7.6") – (7.8") and to verify that the relevant operator is essentially selfadjoint, we must construct a projection matrix $\mathbf{S}$ whose range is the null-space of the matrix (7.46). The matrix $\mathbf{S}$ is uniquely determined by the following conditions:

1) the equation $\mathbf{Aw}=\mathbf{0}$, where $\mathbf{w}$ is an $8m$-vector, is equivalent to the equation $\mathbf{Sw}=\mathbf{w}$,

2) $\mathbf{S}^*=\mathbf{S}$,

3) $\mathbf{S}^2=\mathbf{S}$.

Let $\mathbf{w}_1\ldots,\mathbf{w}_8$ be 2-vectors ($m=2$) and

$$\mathbf{w}=\mathbf{w}_1\dot{+}\mathbf{w}_2\dot{+}\ldots\dot{+}\mathbf{w}_8=\begin{pmatrix}\mathbf{w}_1\\ \cdots\\ \mathbf{w}_8\end{pmatrix}$$

an $8m$-vector (16-vector in this case). It follows from (7.45) that the equation $\mathbf{Aw}=0$ is equivalent to the equations

$$\mathbf{w}_1=0,\ \mathbf{w}_3=0,\ \mathbf{w}_5=0,\ \mathbf{w}_7=0.$$

Thus condition 1 holds for the block-diagonal matrix

$$\mathbf{S}=\begin{pmatrix} 0 & & & & & & & \\ & \mathbf{I}_2 & & & & & \mathbf{0} & \\ & & 0 & & & & & \\ & & & \mathbf{I}_2 & & & & \\ & & & & 0 & & & \\ & \mathbf{0} & & & & \mathbf{I}_2 & & \\ & & & & & & 0 & \\ & & & & & & & \mathbf{I}_2 \end{pmatrix}. \tag{7.46}$$

This matrix clearly satisfies the other conditions, so that it is the required projection matrix.

For the symmetry conditions (7.6")–(7.8") there is no need to determine the exact form of the matrices $\mathbf{C}_1$, $\mathbf{C}_2$, $\mathbf{C}_3(\tau)$. It will suffice to use the expressions derived for these matrices on pp. 523 – 524, interpreting the component block-matrices as the matrices (7.44). It turns out that (7.6")–(7.8") are satisfied, since the block matrices at even positions on the diagonals of $\mathbf{C}_1, \mathbf{C}_2, \mathbf{C}_3(\tau)$ are zero matrices. Thus the operators $\mathfrak{P}$, $\mathfrak{M}$ and $\mathfrak{R}(\tau)$ for our example are indeed symmetric.

We now show that $\mathfrak{P}$ is essentially selfadjoint, using condition (S'). The matrix $\mathbf{SC}_1$ is here

$$\mathbf{SC}_1=\left(\begin{array}{cc}\begin{matrix}0 & 0 & 0 & 0\\ 0 & 0 & -\mathbf{P}(0) & 0\\ 0 & 0 & 0 & 0\\ -\mathbf{P}(0) & 0 & 0 & 0\end{matrix} & \mathbf{0}\\ \mathbf{0} & \begin{matrix}0 & 0 & 0 & 0\\ 0 & 0 & \mathbf{P}(l) & 0\\ 0 & 0 & 0 & 0\\ \mathbf{P}(l) & 0 & 0 & 0\end{matrix}\end{array}\right). \tag{7.47}$$

Let $\mathbf{w}$ be a solution of the equation

$$\mathbf{SC}_1\mathbf{w}=\mathbf{0}. \tag{7.48}$$

Let $\mathbf{q}_1, \mathbf{q}_2, \ldots, \mathbf{q}_8$ be the 2-vectors making up the vector $\mathbf{w}$. Using (7.47), we deduce that equation (7.48) is equivalent to the system of equations

$$\mathbf{P}(0)\,\mathbf{q}_3=\mathbf{0},\quad \mathbf{P}(0)\,\mathbf{q}_1=\mathbf{0},\quad \mathbf{P}(l)\,\mathbf{q}_7=\mathbf{0},\quad \mathbf{P}(l)\,\mathbf{q}_5=\mathbf{0},$$

or, since the matrix $\mathbf{P}(z)$ is positive definite, to the system

$$\mathbf{q}_1=\mathbf{0},\quad \mathbf{q}_3=\mathbf{0},\quad \mathbf{q}_5=\mathbf{0},\quad \mathbf{q}_7=\mathbf{0}. \tag{7.49}$$

Hence it follows trivially that $\mathbf{Aw}=\mathbf{0}$, so that condition (S') is satisfied.

It is even easier to check condition (S''). In this case,

$$\begin{aligned}\mathbf{C}_1^*\mathbf{SC}_1 &= \operatorname{diag}(\mathbf{P}(0)^2,\ 0,\ \mathbf{P}(0)^2,\ 0,\ \mathbf{P}(l)^2,\ 0,\ \mathbf{P}(l)^2,\ 0),\\ \mathbf{A}^*\mathbf{A} &= \operatorname{diag}\ (\mathbf{I}_2,\ 0,\ \mathbf{I}_2,\ 0,\ \mathbf{I}_2,\ 0,\ \mathbf{I}_2,\ 0).\end{aligned}$$

Since the matrices $\mathbf{P}(0)$ and $\mathbf{P}(l)$ are positive definite, the necessary value of τ clearly exists and condition (S'') is satisfied.

We have thus shown that $\mathfrak{P}$ is essentially selfadjoint.

In this case, then, all the conditions of Theorem I are satisfied, and this justifies use of the Bubnov-Galerkin procedure.

b) The equations for dynamic stability of a thin-walled bar (3.3) may be written in the form of (7.1) with

$$\mathbf{y}=\begin{pmatrix}\xi\\ \eta\\ \Theta\end{pmatrix},\quad \mathbf{P}(z)=\begin{Vmatrix}EJ_y & 0 & 0\\ 0 & EJ_x & 0\\ 0 & 0 & EJ_\omega\end{Vmatrix},\quad \mathbf{K}=\mathbf{I}_3,$$

$$\mathbf{L}=\begin{Vmatrix}\frac{1}{2}p(\vartheta t) & 0 & -p(\vartheta t)(e_y-a_y)\\ 0 & \frac{1}{2}p(\vartheta t) & p(\vartheta t)(e_x-a_x)\\ 0 & 0 & -\frac{1}{2}GJ_d+p(\vartheta t)\left[\frac{1}{2}r^2+\beta_x e_x+\beta_y e_y\right]\end{Vmatrix},$$

$$\mathbf{M}(z)=\gamma g^{-1}F\begin{Vmatrix}1 & 0 & a_y\\ 0 & 1 & -a_x\\ a_y & -a_x & r^2\end{Vmatrix},\qquad \mathbf{N}(z)=\gamma g^{-1}\begin{Vmatrix}J_y & 0 & 0\\ 0 & J_x & 0\\ 0 & 0 & J_\omega\end{Vmatrix},\qquad \mathbf{R}(z)=\mathbf{0}.$$

In this case $\mathbf{Q}(z)\equiv\mathbf{0}$.

Suppose that the bar is hinged at both ends. Then the boundary conditions are

$$\mathbf{y}(0)=\mathbf{y}(l)=\mathbf{y}''(0)=\mathbf{y}''(l)=\mathbf{0}.$$

These conditions may also be expressed in the form of (7.5). In this example, the symmetry conditions (7.6'')–(7.8'') are readily verified, as is the fact that $\mathfrak{P}$ is essentially selfadjoint. The requisite arguments and operations are similar to those used in the preceding example. Since $\mathbf{N}(z)>\mathbf{0}$, we again have all the assumptions of Theorem I.

The verification is similar for a bar both of whose ends are built in (clamped) or one end is hinged and the other clamped, etc.

c) It is known /13, 18/ that small transverse plane vibrations of a bar due to a longitudinal force are described by the equation

$$EJ\frac{\partial^4 u}{\partial x^4}+\Phi(\vartheta t)\frac{\partial^2 u}{\partial x^2}+m\frac{\partial^2 u}{\partial t^2}=0, \tag{7.50}$$

where EJ is the flexural stiffness of the bar, m its linear density, $\Phi(t)$ a periodic applied longitudinal force. In addition, one has boundary conditions defining how the ends of the bar are attached. For example: hinged ($u=u''=0$ at the appropriate end), built in ($u=u'=0$) and various combinations of these. Equation (7.50) is clearly in the form of (7.1). There is no difficulty in checking the validity of conditions (7.6)–(7.8) for these types of attachment. In this case (7.50) has the form of equation (7.1) with $\mathbf{N}(z)\equiv\mathbf{0}$, so that Theorem I is inapplicable. It is readily seen, however, that the natural frequencies ω_k of the unperturbed system (equations (7.51) with $\Phi(\tau)\equiv 0$) display the asymptotic behavior $\omega_k\sim\alpha k^2$, where α is a scalar factor, so that the assumptions of Theorem II hold. If the Fourier series of the force $\Phi(\tau)$ in equation (7.50) includes only finitely many harmonics,

$$\Phi(\tau)=\sum_{|p|<m}\Phi^{(p)}e^{ip\tau}$$

and $\sum|\Phi^{(p)}|\equiv\varepsilon$ is small, one can also use Theorem III, whose conditions 2 and 3 are not overly restrictive for the system in question (in fact, they are almost always satisfied).

d) The problem of a vibrating thin-walled bar with two symmetry axes, under a transverse load $q(z,\vartheta t)$ of fixed direction, leads to a system of two differential equations for the lateral displacement $u(z,t)$ and the angle of torsion $\varphi(z,t)$:

$$\left.\begin{aligned}EJ_y\frac{\partial^4 u}{\partial z^4}+\frac{\partial^2}{\partial z^2}(M\varphi)+m\frac{\partial^2 u}{\partial t^2}=0,\\ M\frac{\partial^2 u}{\partial z^2}+EJ_\omega\frac{\partial^4\varphi}{\partial z^4}-\frac{\partial}{\partial z}\left(GJ_d\frac{\partial}{\partial z}\varphi\right)+q(z,\vartheta t)\,l\varphi+m\rho^2\frac{\partial^2\varphi}{\partial t^2}=0,\end{aligned}\right\} \tag{7.51}$$

where EJ_y, EJ_ω, EJ_d are the flexural, sectorial and torsional cross-sectional stiffnesses, $M=M(\vartheta t, z)$ is the bending moment due to the impressed load $q(\vartheta t, z)$, m is the linear mass of the pipe, ρ the polar radius of inertia, l the distance from the line of application of the transverse load to the centroid line.

The conditions at the ends may be:

1) hinged: $u=\varphi=\frac{\partial^2 u}{\partial z^2}=\frac{\partial^2 \varphi}{\partial z^2}=0$,

2) built in: $u=\varphi=\frac{\partial u}{\partial z}=\frac{\partial \varphi}{\partial z}=0$.

The boundary conditions may be any combination of these conditions for the two ends. System (7.51) has the form of (7.1) with

$$\mathbf{y}=\begin{pmatrix}u\\ \varphi\end{pmatrix},\quad \mathbf{M}(z)=\begin{Vmatrix}m & 0\\ 0 & m\rho^2\end{Vmatrix},\quad \mathbf{N}(z)\equiv 0,\quad \mathbf{P}(z)=\begin{Vmatrix}EJ_y & 0\\ 0 & EJ_\omega\end{Vmatrix},$$
$$\mathbf{R}(z)=\begin{Vmatrix}0 & 0\\ 0 & EJ_d\end{Vmatrix},\quad \mathbf{K}=\begin{Vmatrix}0 & 1\\ 1 & 0\end{Vmatrix},\quad \mathbf{L}=\begin{Vmatrix}0 & M\\ 0 & 0\end{Vmatrix},\quad \mathbf{Q}(z)=\begin{Vmatrix}0 & 0\\ 0 & ql\end{Vmatrix}.$$

The boundary conditions (7.6)—(7.8) are clearly satisfied for all combinations.

It is not hard to show that in this case the natural frequencies ω_k "diverge' (i. e., satisfy (7.34)), so that all the assumptions of Theorem II are valid. If $q(z, \tau)$ is a trigonometric polynomial in τ, of small absolute value (so that the small parameter ε may be chosen as $\max\limits_{\substack{0\leqslant z\leqslant l\\ 0\leqslant\tau\leqslant 2\pi}} |q(z, \tau)|$), one can use Theorem III. To this end one must of course verify also conditions 2 and 3 of that theorem, but these are generally valid for systems of this type.

Chapter VII

LYAPUNOV'S METHOD FOR ESTIMATING THE CHARACTERISTIC CONSTANT

Introduction

In this chapter we shall develop effective sufficient conditions for the stability and instability of the trivial solution of Hill's equation (§2), based on Lyapunov's method for estimating the characteristic constant. Lyapunov /180g/ evolved a technique for analyzing the stability of the trivial solution of Hill's equation

$$\ddot{y}+p(t)y=0 \qquad (p(t+T)=p(t)\geqslant 0), \tag{0.1}$$

where $p(t)$ is a real continuous T-periodic function, assumed to be non-negative. Lyapunov gave a brief account of the gist of the method in /180f/. The basic paper /180g/ was preceded by brief notes /180d, e/ presenting without proof certain propositions concerning the roots of the equation $A^2-1=0$; in /180e/ the function $p(t)$ was allowed to be of variable sign. (For the definition of the function $A=A(\lambda)$, see below.)

We begin with a brief outline of the method. Following Lyapunov, we let $f(t)$ and $\varphi(t)$ denote the real solutions of equation (0.1) satisfying initial conditions

$$f(0)=1, \quad \dot{f}(0)=0, \quad \varphi(0)=0, \quad \dot{\varphi}(0)=1.$$

Then the characteristic equation (II. 2. 3) is

$$\det[\mathbf{X}(T)-\rho\mathbf{I}_2]=\begin{vmatrix} f(T)-\rho & \varphi(T) \\ \dot{f}(T) & \dot{\varphi}(T)-\rho \end{vmatrix}=0$$

or, in expanded form,

$$\rho^2-2A\rho+1=0, \tag{0.2}$$

where the free term of the equation is 1 by Liouville's formula (II. 1.11). The constant

$$A=\frac{1}{2}[f(T)+\dot{\varphi}(T)] \tag{0.3}$$

was called by Lyapunov the characteristic constant of equation (0.1). The roots of equation (0.2) (the multipliers of equation (0.1)) are

$$\rho_{1,2}=A\mp\sqrt{A^2-1}.$$

If $A^2<1$, then ρ_1 and ρ_2 are complex conjugates and lie on the unit circle. If $A^2>1$, they are real, one of them is >1 in absolute value and the other is <1. If $A^2=1$, then $\rho_1=\rho_2=A=\pm 1$, and by the identity (II. 2.5) one of the solutions of equation (0.1) is periodic ($A=1$) or antiperiodic ($A=-1$). Is the second independent solution in each of these limiting cases ($A=\pm 1$) also periodic (antiperiodic) or does it increase with t? The answer to this question depends on the structure of the matrix $\mathbf{X}(T)-\rho\mathbf{I}_2$ (§ II. 2.6).

Thus, the condition $A^2<1$ is necessary (not counting the limiting cases $A=\pm 1$) and sufficient for the trivial solution of equation (0.1) to be stable. However, this condition in itself is not effective, for it assumes that a fundamental set of solutions is available.

To determine upper and lower bounds for the characteristic constant, Lyapunov proposed to represent the constant A by a series, obtained by considering the equation

$$\ddot{x}+\lambda p(t)x=0$$

instead of (0.1) and expanding the characteristic constant $A(\lambda)$ of this equation in powers of λ:

$$A(\lambda)=1-A_1\lambda+A_2\lambda^2-\dots$$

Setting $\lambda=1$, we obtain an infinite series for the characteristic constant A of equation (0.1):

$$A=1-A_1+A_2-\dots+(-1)^n A_n+\dots$$

The method of successive approximations yields explicit formulas for the terms A_n of this series, but they become considerably more complicated as n increases.

The following propositions (established in /180g/ for the case $p(t)\geqslant 0$) are fundamental for Lyapunov's method.

1) For any natural number ν, if

$$1-A_1+A_2-\dots+A_{2\nu}\leqslant -1 \tag{0.4}$$

then

$$A<-1,$$

and if

$$1-A_1+A_2-\dots+A_{2\nu-1}\geqslant 1 \tag{0.5}$$

then

$$A>1.$$

2) For any natural numbers ν and μ, if

$$1-A_1+A_2-\ldots+A_{2\nu}\leqslant 1, \tag{0.6}$$
$$1-A_1+A_2-\ldots-A_{2\mu-1}\geqslant -1 \tag{0.7}$$

then, respectively,

$$A<1, \quad A>-1.$$

Thus each of inequalities (0.4) or (0.5) for some ν is a sufficient condition for instability.

Any pair of inequalities (0.6) and (0.7), for suitable natural ν and μ, is a sufficient condition for the trivial solution of equation (0.1) to be stable.

Unlike the conditions $A^2<1$ and $A^2>1$, these new conditions are effective, since the coefficients $A_1, A_2, \ldots$ may be determined recursively. Even more: for any periodic function $p(t)$ (not counting the exponential case $A=\pm 1$) these conditions for instability (and stability) will necessarily hold for sufficiently large ν (and μ).

Lyapunov proved these two propositions by an extremely tedious analysis of the explicit formulas for the coefficients A_n (/180g/, pp. 415 – 439). We are going to adopt a different approach, based on certain theorems of complex function theory and properties of singular boundary-value problems (§III. 4). Not only shall we be able to prove Lyapunov's results more elegantly: the method will enable us to extend his ideas to a broader class of equations and systems with periodic coefficients.

Lyapunov's method will be developed here only for a real system of two equations. It carries over without essential modification to Hamiltonian systems of two equations. For the extension to canonical (and Hamiltonian) systems of arbitrary order, see /224k, 204g, h/.

§1. PROPERTIES OF THE CHARACTERISTIC FUNCTION $A(\lambda)$ OF A SECOND-ORDER CANONICAL SYSTEM OF LINEAR DIFFERENTIAL EQUATIONS WITH PERIODIC COEFFICIENTS

1.1. Definition of the characteristic function $A(\lambda)$; conditions on the coefficient matrix

According to (II. 3.1), the expanded form of a canonical system for $k=1$ is

$$\left.\begin{aligned}\frac{d\xi_1}{dt}&=h_{12}\xi_1+h_{22}\xi_2,\\ \frac{d\xi_2}{dt}&=-h_{11}\xi_1-h_{12}\xi_2.\end{aligned}\right\} \tag{1.1}$$

Suppose that the coefficients are linear functions of a real parameter λ:

$$h_{ij}=h_{ij}(t,\lambda)=h_{ij}^{(0)}(t)+\lambda h_{ij}^{(1)}(t),$$

where $h_{ij}^{(0)}(t)$ and $h_{ij}^{(1)}(t)$ $(i,\ j=1,\ 2)$ are real integrable T-periodic piecewise-continuous functions of t. Write system (1.1) as a vector equation

$$\mathbf{J}\frac{d\mathbf{x}}{dt}=\mathbf{H}(t;\ \lambda)\,\mathbf{x}, \tag{1.2}$$

where

$$\mathbf{x}=\begin{pmatrix}\xi_1\\ \xi_2\end{pmatrix},\quad \mathbf{J}=\begin{Vmatrix}0 & -1\\ 1 & 0\end{Vmatrix},$$

and $\mathbf{H}(t;\ \lambda)$ is a symmetric matrix:

$$\mathbf{H}(t;\ \lambda)=\mathbf{H}_0(t)+\lambda\mathbf{H}_1(t)\qquad (\mathbf{H}_k(t)=\|h_{ij}^{(k)}(t)\|_1^2;\ k=0,1). \tag{1.3}$$

Let $\mathbf{X}(t,\ \lambda)$ be the matrizant (§II. 1. 2) and $\mathbf{X}(T,\ \lambda)$ the monodromy matrix of equation (1.2). Since the trace of the matrix of system (1.1) vanishes $(\mathrm{Tr}\ \mathbf{J}^{-1}\mathbf{H}(t,\ \lambda)\equiv h_{12}+(-h_{12})=0)$, it follows from Liouville's formula (II. 1.11) that $\det\mathbf{X}(t,\ \lambda)\equiv 1$. Hence the characteristic equation (II. 2.3) is

$$\det[\mathbf{X}(T;\ \lambda)-\rho\mathbf{I}_2]=\rho^2-2A(\lambda)\rho+1=0, \tag{1.4}$$

where

$$A(\lambda)=\frac{1}{2}\mathrm{Tr}\,\mathbf{X}(T;\ \lambda).$$

The function $A(\lambda)$ will be called the *characteristic function* of equation (1.2).

We claim that

$$|A(\lambda)|\leqslant c_1 e^{c_2|\lambda|}, \tag{1.5}$$

where c_1 and c_2 are positive constants.

The matrizant $\mathbf{X}(t,\ \lambda)$ satisfies the equation

$$\frac{d\mathbf{X}}{dt}=\mathbf{J}^{-1}\mathbf{H}(t,\ \lambda)\,\mathbf{X}$$

and the condition $\mathbf{X}(0,\lambda)=\mathbf{I}_2$. Using the usual estimate for the norm $|\mathbf{X}(t,\ \lambda)|$ (§I. 1. 2), we obtain

$$|\mathbf{X}(t,\ \lambda)|\leqslant|\mathbf{I}_2|\exp\int_0^t|\mathbf{J}^{-1}\mathbf{H}(t_1,\ \lambda)|\,dt_1.$$

Using (1.3) and setting $t=T$, we get

$$|\mathbf{X}(T,\ \lambda)|\leqslant|\mathbf{I}_2|\exp\left[|\mathbf{J}|\int_0^T|\mathbf{H}_0(t)|\,dt+|\mathbf{J}|\cdot|\lambda|\int_0^T|\mathbf{H}_1(t)|\,dt\right],$$

and by property 4 of the norm (§I. 1.2) this implies (1.5).

We have as yet imposed no restrictions on the real periodic symmetric matrices $\mathbf{H}_0(t)$ and $\mathbf{H}_1(t)$, other than integrability in the interval $(0, T)$. Henceforth we shall assume that the following conditions hold:

(A) For any real 2-vector $\mathbf{y}=\|\eta_j\|\neq\mathbf{0}$ and any $t\in[0, T]$ which is not a discontinuity point of any element of $\mathbf{H}_1(t)$,

$$(\mathbf{H}_1(t)\mathbf{y}, \mathbf{y})=h_{11}^{(1)}(t)\eta_1^2+2h_{12}^{(1)}(t)\eta_1\eta_2+h_{22}^{(1)}(t)\eta_2^2\geqslant 0.$$

Condition (A), which we shall write briefly as $\mathbf{H}_1(t)\geqslant 0$, means that for the relevant values t

$$h_{11}^{(1)}(t)\geqslant 0,\quad h_{22}^{(1)}(t)\geqslant 0,\quad h_{11}^{(1)}(t)h_{22}^{(1)}(t)-[h_{12}^{(1)}(t)]^2\geqslant 0.$$

(B) Let $\mathbf{x}_0(t)$ be a vector-function such that, throughout the interval $(0, T)$ (except at discontinuity points of $\mathbf{H}_0(t)$ and $\mathbf{H}_1(t)$),

$$\mathbf{H}_1(t)\mathbf{x}_0(t)\equiv\mathbf{0},\quad \mathbf{J}\frac{d\mathbf{x}_0}{dt}=\mathbf{H}_0(t)\mathbf{x}_0,\quad \mathbf{x}_0(T)=\rho\mathbf{x}_0(0), \tag{1.6}$$

where ρ is some complex number, $|\rho|=1$; then $\mathbf{x}_0(t)\equiv\mathbf{0}$.

Condition (B) will hold, in particular, if condition (A) is strengthened as follows: the quadratic form $(\mathbf{H}_1(t)\mathbf{y}, \mathbf{y})$ is positive definite for all $t\in[0, T]$, except discontinuity points of the elements of $\mathbf{H}_1(t)$. The number ρ in condition (B) may of course be $+1$ or -1.

If condition (A) holds, condition (B) means that if $\mathbf{x}$ is a solution of the boundary-value problem

$$\mathbf{J}\frac{d\mathbf{x}}{dt}=[\mathbf{H}_0(t)+\lambda\mathbf{H}_1(t)]\mathbf{x},\qquad \mathbf{x}(T)=\rho\mathbf{x}(0)\qquad (|\rho|=1) \tag{1.7}$$

such that

$$\int_0^T(\mathbf{H}_1(t)\mathbf{x}(t), \mathbf{x}(t))\,dt=0, \tag{1.8}$$

then $\mathbf{x}(t)\equiv\mathbf{0}$. Indeed, it follows from conditions (A) and (1.8) that $\mathbf{H}_1(t)\mathbf{x}(t)\equiv\mathbf{0}$, and then conditions (1.7) and (B) imply that $\mathbf{x}\equiv\mathbf{0}$.

If condition (B) fails to hold, the spectrum of the selfadjoint boundary-value problem (1.7) coincides with the entire complex plane. In fact, any vector $\mathbf{x}_0(t)\not\equiv\mathbf{0}$ satisfying (1.6) is a solution of (1.7) for any λ.

Equation (1.2) is the special case of equation (III. 4.1) with $2k=2$, in which $i^{-1}\mathbf{G}=\mathbf{J}$, $\mathbf{H}(t)=\mathbf{H}_0(t)$, $\mathbf{Q}(t)=\mathbf{H}_1(t)$. The boundary-value problem (1.7) is the same as problem (III. 4.1), (III. 4.5). As shown in §III.4.1, we have (III.4.7), i. e., $((\mathbf{Lu}, \mathbf{v}))=((\mathbf{u}, \mathbf{Lv}))$, where $\mathbf{u}, \mathbf{v}$ are arbitrary vector-functions satisfying (III. 4.5), and

$$\mathbf{Lx}\equiv\mathbf{J}\frac{d\mathbf{x}}{dt}-\mathbf{H}_0(t)\mathbf{x},\qquad ((\mathbf{u}, \mathbf{v}))=\int_0^T(\mathbf{u}(t), \mathbf{v}(t))\,dt.$$

It follows from (A) and (B) that the selfadjointness conditions (I), (II) of § III. 4. 1 are satisfied. By Theorem 1 of § III. 4.1, the set of values λ for which there is a nontrivial solution $(\mathbf{x}(t) \not\equiv \mathbf{0})$ to the boundary-value problem (1.7) (for some fixed ρ, $|\rho|=1$) is at most countable and its only point of accumulation is $\lambda=\infty$. The eigenvalues of this boundary-value problem (if they exist) are real. These statements are summarized in the following

Lemma. If conditions (A) and (B) are satisfied, the spectrum of the boundary-value problem (1.7) is real.

1.2. Properties of the zeros of the entire function $A(\lambda)-\alpha$

Property I. All zeros of the entire function $A(\lambda)-\alpha$, where α is a real number of absolute value at most 1, are real.

Proof. Let λ_0 be a zero of $A(\lambda)-\alpha$:

$$A(\lambda_0)=\alpha \qquad (-1 \leqslant \alpha \leqslant 1).$$

Then the roots of the characteristic equation (1.4) (multipliers of system (1.2) for $\lambda=\lambda_0$) are

$$\rho_{1,2}(\lambda_0)=\alpha \pm i\sqrt{1-\alpha^2} \tag{1.9}$$

and

$$|\rho_1(\lambda_0)|=|\rho_2(\lambda_0)|=1.$$

Consider the boundary-value problem (1.7) for $\rho=\rho_1(\lambda_0)$. (One can also take $\rho=\rho_2(\lambda_0)$.) The number λ_0 is an eigenvalue of this problem. Indeed, inserting

$$\mathbf{x}(t;\ \lambda)=\mathbf{X}(t;\ \lambda)\,\mathbf{x}(0;\ \lambda)$$

into the boundary conditions, we get

$$[\mathbf{X}(T,\ \lambda)-\rho_1\mathbf{I}_2]\,\mathbf{x}(0;\ \lambda)=\mathbf{0}.$$

Thus the problem has a nontrivial solution for numbers satisfying the characteristic equation (1.4)

$$\det\,[\mathbf{X}(T;\ \lambda)-\rho_1\mathbf{I}_2]=\rho_1^2-2A(\lambda)\rho_1+1=0$$

and only for these numbers. Substituting (1.9) into this equation, we obtain an equation for the eigenvalues: $A(\lambda)=\alpha$, so that λ_0 is an eigenvalue of the boundary-value problem, as claimed. It follows from the lemma of §1.1 that λ_0 is a real number, and Property I is proved.

To prove the other properties of the function $A(\lambda)-\alpha$, we shall need the following simple

Lemma. Let λ *be a real number and* $\mathbf{a}$ *an eigenvector of the monodromy matrix* $\mathbf{X}_0=\mathbf{X}(T, \lambda)$ *of system (1.2), belonging to a nonreal root of the characteristic equation (1.4) of absolute value* 1:

$$\mathbf{X}_0\mathbf{a}=\rho\mathbf{a}, \quad |\rho|=1, \quad \rho\neq\pm 1.$$

Then $(\mathbf{Ja}, \mathbf{a})\neq 0$.

This was proved at the end of § III. 2.4 for an arbitrary $2k\times 2k$ symplectic matrix $\mathbf{X}_0$. We present a direct proof. Taking complex conjugates and noting that $\bar{\mathbf{X}}_0=\mathbf{X}_0$, we have

$$\mathbf{X}_0\bar{\mathbf{a}}=\bar{\rho}\bar{\mathbf{a}} \qquad (\bar{\rho}\neq\rho).$$

We now calculate the matrix

$$\|\mathbf{a}, \bar{\mathbf{a}}\|^*\mathbf{J}\|\mathbf{a}, \bar{\mathbf{a}}\|=\begin{pmatrix}\bar{\mathbf{a}}^\tau\\ \mathbf{a}^\tau\end{pmatrix}\cdot\|\mathbf{Ja}, \mathbf{J}\bar{\mathbf{a}}\|=\begin{Vmatrix}\bar{\mathbf{a}}^\tau\mathbf{Ja} & \bar{\mathbf{a}}^\tau\mathbf{J}\bar{\mathbf{a}}\\ \mathbf{a}^\tau\mathbf{Ja} & \mathbf{a}^\tau\mathbf{J}\bar{\mathbf{a}}\end{Vmatrix}.$$

(Recall that the star denotes transposition plus complex conjugation.)

For any vectors $\mathbf{a}$ and $\mathbf{b}$,

$$\mathbf{a}^*\mathbf{Jb}=(\mathbf{Jb}, \mathbf{a}).$$

A direct check shows that $(\mathbf{Ja}, \bar{\mathbf{a}})=0$ for any vector. Therefore

$$\|\mathbf{a}, \bar{\mathbf{a}}\|^*\mathbf{J}\|\mathbf{a}, \bar{\mathbf{a}}\|=\begin{Vmatrix}(\mathbf{Ja}, \mathbf{a}) & (\mathbf{J}\bar{\mathbf{a}}, \mathbf{a})\\ (\mathbf{Ja}, \bar{\mathbf{a}}) & (\mathbf{J}\bar{\mathbf{a}}, \bar{\mathbf{a}})\end{Vmatrix}=\begin{Vmatrix}(\mathbf{Ja}, \mathbf{a}) & 0\\ 0 & \overline{(\mathbf{Ja}, \mathbf{a})}\end{Vmatrix}. \qquad (1.10)$$

Since $\operatorname{Im}\rho\neq 0$, it follows that $\mathbf{a}$ and $\bar{\mathbf{a}}$ are eigenvectors of $\mathbf{X}(T, \lambda)$ belonging to distinct eigenvalues ρ and $\bar{\rho}$. Consequently (§ I. 1.8), $\mathbf{a}$ and $\bar{\mathbf{a}}$ are linearly independent, so that $\det\|\mathbf{a}, \bar{\mathbf{a}}\|\neq 0$. Thus the determinant of the left-hand side of (1.10) does not vanish. But the determinant of the right-hand side of (1.10) is precisely $|(\mathbf{Ja}, \mathbf{a})|^2$. Consequently, $(\mathbf{Ja}, \mathbf{a})\neq 0$, Q. E. D.

We now write the eigenvector $\mathbf{a}$ as

$$\mathbf{a}=\begin{pmatrix}\alpha_1\\ \alpha_2\end{pmatrix}.$$

Obviously,

$$i(\mathbf{Ja}, \mathbf{a})=-i(\alpha_2\bar{\alpha}_1-\alpha_1\bar{\alpha}_2)=2\operatorname{Im}\alpha_2\bar{\alpha}_1.$$

By the lemma, this real number is not zero. Recall that by the definition of § III. 1. 2, a nonreal multiplier ρ of system (1.2) (nonreal root of the characteristic equation (1.4)) is said to be of the first kind if

$$i(\mathbf{Ja}, \mathbf{a})>0,$$

and of the second kind if

$$i(\mathbf{Ja}, \mathbf{a})<0.$$

Property II. All the zeros of the entire function $A(\lambda)-\alpha$, $-1<\alpha<1$ (which are real by Property I), are simple. In other words, $dA/d\lambda \neq 0$ for $-1<A(\lambda)<+1$.

Proof. Let λ_* be some zero of $A(\lambda)-\alpha$,

$$A(\lambda_*)=\alpha \qquad (-1<\alpha<1).$$

By Property I, λ_* is a real number.

The roots of equation (1.4) are

$$\rho(\lambda)=A(\lambda)\pm\sqrt{A^2(\lambda)-1}.$$

We have

$$\rho_{1,2}(\lambda_*)=\alpha\pm\sqrt{\alpha^2-1},$$

so that $\rho_1(\lambda_*)$ and $\rho_2(\lambda_*)$ are real and of absolute value 1.

The number $\rho_0=\rho_1(\lambda_*)$ is a simple and (by the lemma of §1.1) definite multiplier of equation (1.2).

By the theorem of §III. 1.5, $\rho_1(\lambda)$ is an analytic function of λ at $\lambda=\lambda_*$; the derivative at λ_* is

$$\left(\frac{d\rho_1}{d\lambda}\right)_{\lambda=\lambda_*}=\delta_1\rho_0, \tag{1.11}$$

where δ_1 is defined by (III. 1.17). We claim that $\delta_1\neq 0$. It follows from (1.3) that $\frac{\partial \mathbf{H}}{\partial\lambda}=\mathbf{H}_1(t)$. Therefore

$$\delta_1=\frac{i}{\langle \mathbf{a}_1,\mathbf{a}_1\rangle}\int_0^T(\mathbf{H}_1\mathbf{x}_1,\mathbf{x}_1)\,dt, \tag{1.12}$$

where $\mathbf{a}_1\neq\mathbf{0}$ is an eigenvector of the monodromy matrix:

$$\mathbf{X}(T,\lambda_*)\,\mathbf{a}_1=\rho_0\mathbf{a}_1,$$

and $\mathbf{x}_1(t)=\mathbf{X}(t,\lambda_*)\,\mathbf{a}_1$ is the corresponding solution of equation (1.2). By the lemma proved above, $\langle \mathbf{a}_1,\mathbf{a}_1\rangle\equiv i(\mathbf{J}\mathbf{a}_1,\mathbf{a}_1)\neq 0$. It was shown previously that conditions (B) and (1.8) imply that $\mathbf{x}_1(t)\equiv\mathbf{0}$, so that $\mathbf{a}_1=\mathbf{0}$. Hence $\delta_1\neq 0$.

We could have avoided using the theorem of §III. 1.5. Indeed, since $A(\lambda)$ is an entire function, the only singular points of the function $\rho(\lambda)$ are branch points, at which $A^2(\lambda)=1$. Since $A^2(\lambda_*)=\alpha^2\neq 1$, the function $\rho_1(\lambda)$ (like $\rho_2(\lambda)$) is analytic in a sufficiently small neighborhood of λ_*. The formula for $d\rho_1(\lambda)/d\lambda$ is established by the simple arguments in the proof of the theorem in §III. 1.5.

Setting $\rho(\lambda_*)=e^{i\varphi}$, we see that

$$\frac{\rho'(\lambda_*)}{\rho(\lambda_*)}=i\frac{d\varphi}{d\lambda}.$$

Formulas (1.11) and (1.2) may be rewritten

$$\left.\frac{d\arg\rho(\lambda)}{d\lambda}\right|_{\lambda=\lambda^*}=\frac{\int_0^T(\mathbf{H}_1(t)\,\mathbf{x}(t,\lambda_*),\,\mathbf{x}(t,\lambda_*))\,dt}{i(\mathbf{J}\mathbf{a}(\lambda_*),\,\mathbf{a}(\lambda_*))}. \tag{1.13}$$

Note that in formulas (1.12) and (1.13) $|\rho(\lambda_*)|=1$, $\rho(\lambda_*)\neq\pm 1$. Thanks to conditions (A) and (B), the numerator of (1.12) is positive and $\rho'(\lambda_*)\neq 0$.

We have thus shown that if $|A(\lambda)|<1$ then

$$\rho'_{1,2}(\lambda)\neq 0.$$

Since

$$2A(\lambda)=\rho_1+\rho_2=\rho_1+\rho_1^{-1},$$

it follows that

$$\frac{dA}{d\lambda}=\frac{1}{2}\left(1-\frac{1}{\rho_1^2}\right)\frac{d\rho_1}{d\lambda}.$$

Thus, if $|A(\lambda)|<1$ we have

$$\frac{dA}{d\lambda}\neq 0,$$

so that the curve $A(\lambda)$ (λ real) has no stationary points in the strip $-1<A(\lambda)<1$. And this means that the zeros of the function $A(\lambda)-\alpha$ $(-1<\alpha<1)$ are simple, proving Property II.

Property III. Let λ_* be a real stationary point of the function $A(\lambda)$ (i.e., $A'(\lambda_*)=0$). Then

$$A(\lambda_*)A''(\lambda_*)<0. \tag{1.14}$$

(Note that by Property II $|A(\lambda_*)|\geqslant 1$.)

Proof. By (1.5), the entire function $A(\lambda)$ is of order at most 1. By Hadamard's theorem (see, e.g., /91/), we can express $A(\lambda)$ as an infinite product

$$A(\lambda)=\lambda^m e^{\gamma\lambda+\beta}\prod_{j=1}^{\infty}\left(1-\frac{\lambda}{\lambda_j}\right)e^{\frac{\lambda}{\lambda_j}},$$

where $m=0$ or $m=1$, and λ_j $(j=1, 2, \ldots)$ are all the zeros of $A(\lambda)$ (which we know are real and simple). Then

$$[\ln A(\lambda)]''=\frac{A''(\lambda)A(\lambda)-[A'(\lambda)]^2}{A^2(\lambda)}=-\frac{m}{\lambda^2}-\sum_{j=1}^{\infty}\frac{1}{(\lambda-\lambda_j)^2}<0.$$

At $\lambda=\lambda_*$ we have $A'(\lambda_*)=0$, and so (1.14) holds. Q. E. D.

We combine these properties of $A(\lambda)$ into a theorem, first proved for system (1.2) by Krein /166g/.

Theorem. Let system (1.2) satisfy conditions (A) and (B) of §1.1. Then:

a) *The function $A(\lambda)-\alpha$ $(-1\leqslant\alpha\leqslant 1)$ has only real simple zeros if $-1<\alpha<1$.*

b) *If $\lambda=\lambda_*$ is a stationary point of $A(\lambda)$ on the real axis, then (1.14) holds. By (a), it is also true that $|A(\lambda_*)|\geqslant 1$.*

Corollary. The zeros of the functions $A(\lambda)-1$ and $A(\lambda)+1$ (which are real) are at most of multiplicity 2 (otherwise we would have a contradiction to part (b) of the theorem).

Interpreted geometrically, the theorem states that any stationary point $\lambda=\lambda_*$ of the function $A(\lambda)$ is an extremum point: a maximum if $A(\lambda_*)\geqslant 1$ and a minimum if $A(\lambda_*)\leqslant -1$. Inside the strip $-1<A(\lambda)<+1$ the function $A(\lambda)$ has no stationary points or, consequently, extrema. Thus the curve of $A(\lambda)$ has the shape shown in Figure 37.

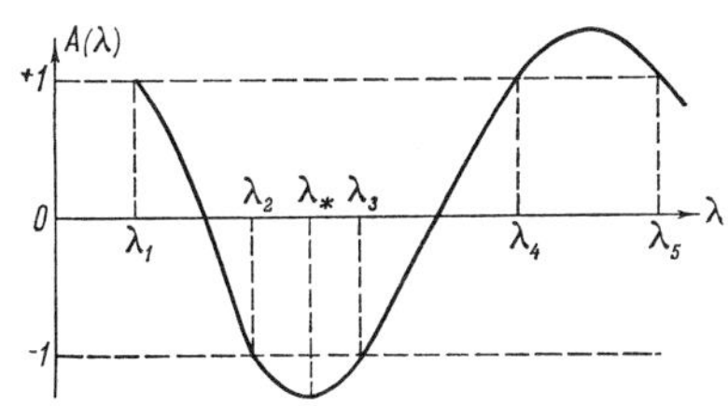

FIGURE 37.

Remark. If we replace condition (A) by

$$\text{(A}'\text{)} \qquad \mathbf{H}_1(t)\leqslant 0,$$

the theorem remains valid — simply substitute $\lambda'=-\lambda$, and everything is reduced to the previous case.

1.3. Behavior of multipliers with increasing parameter λ

For any real λ satisfying the inequality $|A(\lambda)|<1$, the multipliers $\rho_1(\lambda)$, $\rho_2(\lambda)=\overline{\rho_1(\lambda)}$ are distinct and lie on the unit circle (see proof of Property II). Of course, they are both definite. As follows from § III. 4.3., if λ is increased the multipliers move along the unit circle in a well-defined direction. We now wish to examine this motion more closely. Property III will enable us to draw certain additional conclusions (not following from the general discussion of § III. 4.3) concerning the behavior of the multipliers when they lie outside the unit circle.

Recall that the multipliers $\rho_1(\lambda)$, $\rho_2(\lambda)$ of system (1.2) are given by

$$\rho_{1,2}(\lambda)=A(\lambda)\pm\sqrt{A^2(\lambda)-1}. \tag{1.15}$$

We distinguish two cases: $|A(\lambda)|<1$ and $|A(\lambda)|>1$.

1) If $-1<A(\lambda)<1$, it follows from (1.15) that the roots are nonreal and lie on the unit circle:

$$\rho_1(\lambda)=e^{i\varphi(\lambda)}, \quad \rho_2(\lambda)=e^{-i\varphi(\lambda)} \qquad (0<\varphi(\lambda)<\pi).$$

By (1.13),

$$\varphi'(\lambda)=\frac{\int_0^T (\mathbf{H}_1(t)\,\mathbf{x}(t),\ \mathbf{x}(t))dt}{i(\mathbf{J}\,\mathbf{a}(\lambda),\ \mathbf{a}(\lambda))},\tag{1.16}$$

where $\mathbf{a}(\lambda)$ is an eigenvector of the matrix $\mathbf{X}(T,\ \lambda)-\rho_1(\lambda)\,\mathbf{I}_2$. Because of conditions (A) and (B), the numerator of (1.16) is positive, and so the sign of $\varphi'(\lambda)$ is determined by that of the real denominator, which is not zero (see lemma in §1.2). To be precise: $\varphi'(\lambda)>0$ for multipliers of the first kind, $\varphi'(\lambda)<0$ for multipliers of the second kind.

Now suppose that as λ increases from λ_1 to λ_2 the function $A(\lambda)$ varies from $+1$ to -1 (Figure 37). Then $\rho_1(\lambda)$ and $\rho_2(\lambda)$ move along the unit circle from $\rho=1$ to $\rho=-1$ (Figure 38). The multiplier of the first kind, $\rho_1(\lambda)$, traverses the upper semicircle and $\arg\rho_1(\lambda)$ increases monotonically, while the multiplier of the second kind, $\rho_2(\lambda)=\overline{\rho_1(\lambda)}$, traverses the lower semicircle and $\arg\rho_2(\lambda)$ decreases monotonically.

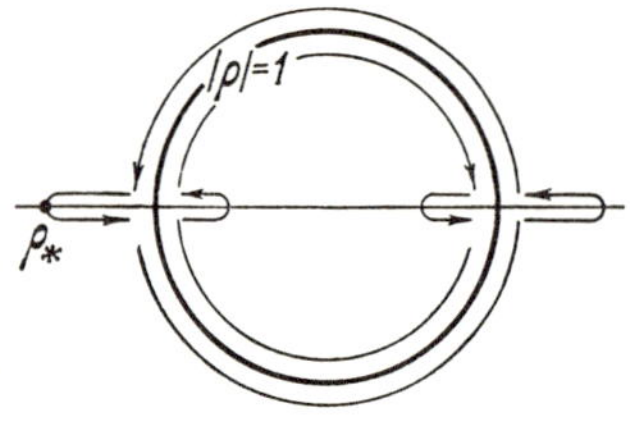

FIGURE 38.

2) If $A(\lambda)<-1$ or $A(\lambda)>1$, the multipliers $\rho_{1,2}(\lambda)$ are real and of absolute value $\neq 1$.

We have

$$\frac{d\rho_1}{dA}=1+\frac{A}{\sqrt{A^2-1}},\quad \frac{d\rho_2}{dA}=1-\frac{A}{\sqrt{A^2-1}},$$

whence

$$\operatorname{sign}\frac{d\rho_1}{dA}=-\operatorname{sign}\frac{d\rho_2}{dA}=\operatorname{sign}A.\tag{1.17}$$

Together with Property III, this readily implies the following assertion.

As λ increases in an interval (λ_2,λ_3) in which $A(\lambda)<-1$ (Figure 37), the multiplier $\rho_2(\lambda)$ moves monotonically from $\rho=-1$ to the left along the real axis, reaching a limiting position ρ_* (corresponding to $\lambda=\lambda_*$), and then returns, moving monotonically back to $\rho=-1$ (Figure 38).

Indeed, it follows from

$$\frac{d\rho_{1,2}}{d\lambda}=\frac{d\rho_{1,2}}{dA}\frac{dA}{d\lambda}$$

and from (1.17) that in the interval (λ_2, λ_*) the function $\rho_2(\lambda)$ is real and decreasing ($\rho_1(\lambda)$ is real and increasing), whereas in (λ_*, λ_3), $\rho_2(\lambda)$ is increasing ($\rho_1(\lambda)$ is decreasing).

It follows from the aforesaid that as λ increases in an interval on which $|A(\lambda)|<1$ the multipliers $\rho_1(\lambda), \rho_2(\lambda)$ move monotonically along the unit circle, but if λ increases in intervals where $|A(\lambda)|>1$ the multipliers $\rho_1(\lambda)$, $\rho_2(\lambda)$ move monotonically from $\rho=1$ (from $\rho=-1$) to certain limiting positions and then back, again monotonically, to $\rho=1$ (to $\rho=-1$). This behavior is illustrated schematically in Figure 38.

1.4. Hill's equation with homogeneously occurring parameter

Consider the equation

$$\frac{d^2\eta}{dt^2}+\lambda p(t)\eta=0 \qquad (p(t+T)=p(t)\not\equiv 0), \tag{1.18}$$

where $p(t)$ is a real piecewise-continuous function, integrable on $(0, T)$, not necessarily of fixed sign, and λ is a real parameter. Denote

$$\zeta=\frac{1}{\lambda}\frac{d\eta}{dt} \qquad (\lambda\neq 0)$$

and write equation (1.18) as a system

$$\left.\begin{aligned} \frac{d\eta}{dt}&=\lambda\zeta, \\ \frac{d\zeta}{dt}&=-p(t)\eta. \end{aligned}\right\} \tag{1.19}$$

This system is of the form of (1.2) with

$$\mathbf{x}=\begin{pmatrix}\eta\\ \zeta\end{pmatrix}, \quad \mathbf{H}_0(t)=\begin{Vmatrix} p(t) & 0\\ 0 & 0\end{Vmatrix}, \quad \mathbf{H}_1=\begin{Vmatrix} 0 & 0\\ 0 & 1\end{Vmatrix}.$$

Clearly, the matrix $\mathbf{H}_1$ satisfies condition (A) of §1.1. We verify condition (B). If $\mathbf{H}_1\mathbf{x}_0\equiv\mathbf{0}$, then

$$\begin{Vmatrix} 0 & 0\\ 0 & 1\end{Vmatrix}\cdot\begin{pmatrix}\eta\\ \zeta_0\end{pmatrix}=\begin{pmatrix}0\\ \zeta_0\end{pmatrix}\equiv 0, \text{ i. e., } \zeta_0\equiv 0,$$

and so

$$\mathbf{x}_0=\begin{pmatrix}\eta_0\\ 0\end{pmatrix}.$$

The vector $\mathbf{x}_0$ must be a solution of the boundary-value problem (1.6), so

$$\frac{d\eta_0}{dt}=0, \quad 0=\frac{d\zeta_0}{dt}=-p(t)\eta_0.$$

It follows that $\eta_0=\text{const}$, and since $p(t)\equiv 0$, we must have $\eta_0=0$. Thus $\mathbf{x}_0=\mathbf{0}$ and condition (B) is indeed satisfied.

If $\eta(t;\ \lambda)$ is a solution of equation (1.18) such that $\eta(t+T;\lambda)=\rho(\lambda)\,\eta(t;\ \lambda)$, it is clear that the corresponding vector $\mathbf{x}=\begin{pmatrix}\eta\\ \zeta\end{pmatrix}$ has the property

$$\mathbf{x}(t+T;\ \lambda)=\rho(\lambda)\,\mathbf{x}(t;\ \lambda),$$

and conversely. Hence equation (1.18) has the same multipliers $\rho_{1,2}(\lambda)$ and thus also the same characteristic function

$$A(\lambda)=\frac{1}{2}[\rho_1(\lambda)+\rho_2(\lambda)]$$

as system (1.19). By the theorem of §1.3, the characteristic function $A(\lambda)$ of equation (1.18) has properties (a) and (b). This result was first obtained by Lyapunov and presented (without proof) in /180d, e/.

Later we shall need the following simple proposition, also due to Lyapunov /60/.

Lemma. If the coefficient $p(t)\not\equiv 0$ in Hill's equation is nonnegative and

$$A(\lambda)=1-A_1\lambda+A_2\lambda^2-\ldots \qquad (1.20)$$

is its characteristic equation, then

$$A_n>0 \qquad (n=1,\ 2,\ \ldots). \qquad (1.21)$$

Setting $\lambda=-1$ and $p=-p_0(t)$, we get

Corollary. If the coefficient $p(t)\not\equiv 0$ of Hill's equation

$$\frac{d^2\eta}{dt^2}+p_0(t)\,\eta=0$$

is nonpositive, then its characteristic constant A satisfies the inequality $A>1$.

Proof of lemma. The characteristic function $A(\lambda)$ of equation (1.18) is defined by (0.3) as

$$A(\lambda)=\frac{1}{2}[f(T;\ \lambda)+\dot{\varphi}(T;\ \lambda)], \qquad (1.22)$$

where $f(t;\ \lambda)$, $\varphi(t;\ \lambda)$ are the solutions of equation (1.18) with initial values

$$\left\|\begin{matrix} f, & \varphi\\ \dot{f}, & \dot{\varphi}\end{matrix}\right\|_{t=0}=\left\|\begin{matrix}1 & 0\\ 0 & 1\end{matrix}\right\|.$$

The functions $f(t;\ \lambda)$, $\varphi(t;\ \lambda)$ are entire functions of λ. Substituting

$$\eta=f(t;\ \lambda)=f_0(t)-\lambda f_1(t)+\lambda^2 f_2(t)-\ldots$$

into equation (1.18) and comparing coefficients of like powers of λ, we see that

$$\ddot{f}_0(t)=0,$$
$$\ddot{f}_n(t)-p(t)f_{n-1}(t)=0 \qquad (n=1, 2, \ldots).$$

Using the initial conditions $f_0(0)=1,\ \dot{f}_0(0)=0,\ f_n(0)=\dot{f}_n(0)\,(n=1,2,\ldots)$, we find that

$$f_0(t)=1, \quad f_n(t)=\int_0^t (t-t_1)\,p(t_1)\,f_{n-1}(t_1)\,dt_1 \qquad (n=1, 2, \ldots). \tag{1.23}$$

Similarly,

$$\varphi(t;\lambda)=\varphi_0(t)-\lambda\varphi_1(t)+\lambda^2\varphi_2(t)-\ldots,$$

where

$$\varphi_n(t)=\int_0^t (t-t_1)\,p(t_1)\,\varphi_{n-1}(t_1)\,dt_1 \qquad (n=1, 2, \ldots). \tag{1.24}$$

It follows from (1.22) that

$$A(\lambda)=1-A_1\lambda+A_2\lambda^2-\ldots,$$

where

$$A_n=\frac{1}{2}\,[f_n(T)+\dot{\varphi}_n(T)].$$

Note moreover that $A(0)=1$ and

$$-A'(0)=A_1=\frac{1}{2}T\int_0^T p(t)\,dt=\frac{1}{2}T^2p_{\text{av}}. \tag{1.25}$$

Since $p(t)\geqslant 0$ and $p(t)\not\equiv 0$, it follows via formulas (1.23) and (1.24) that for $t>0$

$$f_n(t)>0, \quad \dot{\varphi}_n(t)=\int_0^t p(t_1)\,\varphi_{n-1}(t_1)\,dt_1>0 \qquad (n=1, 2, \ldots).$$

This proves (1.21).

We again consider equation (1.18) with $p(t)\geqslant 0$. It follows from (1.20), (1.21) that if $\lambda<0$ then $A(\lambda)>1$ and $A(\lambda)$ tends monotonically to infinity as $\lambda\to-\infty$.

Later (§2.2) we shall show that the entire function $A(\lambda)$ for equation (1.18) is of order at most $\frac{1}{2}$. Hence one easily deduces that $A(\lambda)$ has infinitely many zeros. Indeed, by Hadamard's theorem /91/ we have an expansion

$$A(\lambda)=\prod_{j=1}^{\infty}\left(1-\frac{\lambda}{\lambda_j}\right).$$

If $A(\lambda)$ had only finitely many zeros, this product would be a polynomial, and then $A_n=0$ for sufficiently large n contradicting (1.21). Hence it follows that each of the functions $A(\lambda)-1$, $A(\lambda)+1$ also has infinitely many zeros, for otherwise we would have a contradiction to the theorem of §1.3. The curve $A(\lambda)$ for $p(t)\geqslant 0$ has the shape shown in Figure 39a.

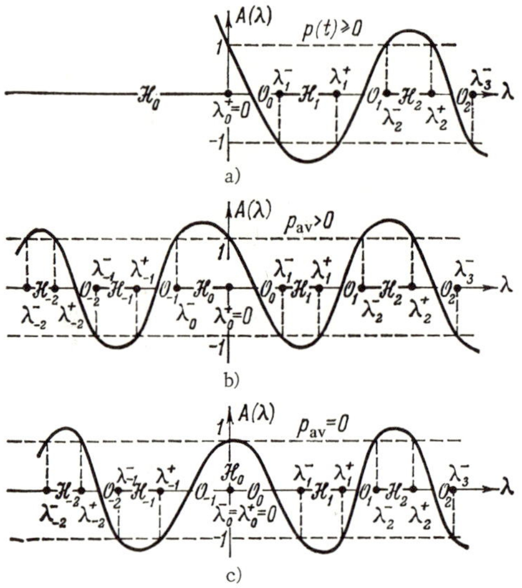

FIGURE 39.

We now allow the function $p(t)$ to change sign. Then the coefficients A_n $(n=1, 2, \ldots)$ will also generally have different signs. We shall show later (§ VIII. 6.2) that in this case too each of the functions $A(\lambda)-1$, $A(\lambda)+1$ has infinitely many zeros, extending to infinity along the λ-axis in both directions. The curve of $A(\lambda)$ has the shape shown in Figure 39b, c.

1.5. Stability and instability zones for Hill's equation with homogeneously occurring parameter

Figure 39 outlines the shapes of the characteristic function $A(\lambda)$ of equation (1.18) for nonnegative $p(t)$ and for a function $p(t)$ of variable sign with $p_{av}\geqslant 0$. * The curves reflect the properties of $A(\lambda)$ established above. If $p_{av}=0$ (Figure 39c), then $A'(0)=0$ by (1.25) and so, by the theorem of §1.2, $A(0)=1$ is a maximum of $A(\lambda)$. Note that if $\lambda\neq 0$ the values of the maxima or minima may be $+1$ or -1, respectively — this is not shown in the figure.

* If $p_{av}<0$, we need only replace λ by $-\lambda$ and $p(t)$ by $-p(t)$ in equation (1.18).

If $p(t)$ is not of fixed sign, the roots $\lambda_{2n}^{\pm}$ of the equation $A(\lambda)=1$ and $\lambda_{2n+1}^{\pm}$ of the equation $A(\lambda)=-1$ may be arranged as follows:

$$\ldots<\lambda_{-2}^{-}\leqslant\lambda_{-2}^{+}<\lambda_{-1}^{-}\leqslant\lambda_{-1}^{+}<\lambda_{0}^{-}=0<\lambda_{1}^{-}\leqslant\lambda_{1}^{+}<\lambda_{2}^{-}\leqslant\lambda_{2}^{+}<\ldots \quad (1.26)$$

Recalling that if $A(\lambda)=\pm 1$ then $\rho_{1,2}(\lambda)=A(\lambda)$ and referring back to §1.2, we note that the numbers $\lambda_{2j}^{\pm}$ $(j=0, \pm 1, \pm 2, \ldots)$ are the eigenvalues of the boundary-value problem

$$\ddot{\eta}+\lambda p(t)\eta=0, \quad \eta(T)=\eta(0), \quad \dot{\eta}(T)=\dot{\eta}(0), \quad (1.27)$$

and $\lambda_{2j+1}^{\pm}$ $(j=0, \pm 1, \pm 2, \ldots)$ are the eigenvalues of the boundary-value problem

$$\ddot{\eta}+\lambda p(t)\eta=0, \quad \eta(T)=-\eta(0), \quad \dot{\eta}(T)=-\dot{\eta}(0). \quad (1.28)$$

If the sequence (1.26) contains terms such that $\lambda_{2i}^{-}=\lambda_{2i}^{+}$ (or $\lambda_{2i+1}^{-}=\lambda_{2i+1}^{+}$) $(i=0, \pm 1, \pm 2, \ldots)$, then by the corollary to the theorem of §1.2 these are twofold roots of the equation $A(\lambda)=+1$ (or $A(\lambda)=-1$). The other numbers $\lambda_j^{\pm}$ are simple roots. In all cases we have

$$\lambda_i^{+}<\lambda_{i+1}^{-} \qquad (i=0, \pm 1, \pm 2, \ldots),$$

as is readily verified from Figure 39.

If $p(t)\geqslant 0$, then $A(\lambda)>1$ for $\lambda<0$, as shown at the end of §1.4. The sequence (1.26) then begins with the term $\lambda_0^{+}=0$ and extends to infinity only in one direction. Thus, if p is nonnegative, all the eigenvalues are nonnegative.

We now turn to the general case, in which $p(t)$ need not be of fixed sign. The whole real λ-axis is partitioned into open intervals $\mathcal{O}_i$:

$$(\lambda_i^{+}, \lambda_{i+1}^{-}) \qquad (i=0, \pm 1, \pm 2, \ldots)$$

in which $-1<A(\lambda)<1$ (numbered according to the index of the left endpoint) and a complementary set $\mathcal{H}$ which generally consists of points and closed intervals (but if $p(t)\geqslant 0$ also contains the negative λ-axis).

Let $\lambda\in\mathcal{O}_i$. Then $|A(\lambda)|<1$ and so all solutions of equation (1.18) are bounded for $-\infty<t<+\infty$ (the trivial solution is stable). The intervals $(\lambda_j^{+}, \lambda_{j+1}^{-})$ are therefore known as *stability zones*. The closed intervals $\mathcal{H}_i$

$$[\lambda_i^{-}, \lambda_i^{+}] \qquad (i=0, \pm 1, \pm 2, \ldots)$$

are also numbered according to their left endpoints. If $\lambda\in\mathcal{H}_j$ and $\lambda_j^{-}\neq\lambda_j^{+}$ then $|A(\lambda)|>1$ and so equation (1.18) has an exponentially increasing solution (the trivial solution is unstable). For this reason the intervals $\mathcal{H}_i$ are called *instability zones*.

It remains to characterize the endpoints λ of the intervals $\mathcal{H}_i$ $(i=0, \pm 1, \pm 2, \ldots)$. Assume, therefore, that λ is one of the roots of the equation

$A(\lambda)=1$ (or $A(\lambda)=-1$). If this root is twofold and nonzero, it follows from the general theory of boundary-value problems that problem (1.27) or (1.28) has two linearly independent solutions. This means that all solutions of equation (1.18) are either periodic or semiperiodic (i. e., satisfy the inequality $\eta(t+T)=\pm\eta(t)$, the actual sign being the same as that of A). On the other hand, if the root is simple, the boundary-value problem has only one linearly independent solution. Hence the matrix $\mathbf{X}(T,\lambda)$ has a multiple elementary divisor. In this case, then, one of the independent solutions will satisfy the preceding equality, while the other will be unbounded, increasing as t. Thus, if λ is an endpoint of one of the intervals $\mathcal{H}_i$, the trivial solution of equation (1.18) is unstable unless $\mathcal{H}_i$ contracts to a point, in which case it is stable.

1.6. Stability and instability zones for Hill's equation with inhomogeneously occurring parameter

Consider the equation

$$\frac{d^2\eta}{dt^2}+[\lambda q(t)-r(t)]\,\eta=0, \tag{1.29}$$

where

$$q(t+T)=q(t)\geqslant 0, \quad r(t+T)=r(t) \tag{1.30}$$

are piecewise-continuous real functions, integrable on $(0,T)$. Equation (1.29) may be written in the form of (1.2) with

$$\mathbf{x}=\begin{pmatrix}\eta\\ \frac{d\eta}{dt}\end{pmatrix},\quad \mathbf{H}(t,\lambda)=\mathbf{H}_0(t)+\lambda\mathbf{H}_1(t),\quad \mathbf{H}_0=\begin{Vmatrix}-r & 0\\ 0 & 1\end{Vmatrix},\quad \mathbf{H}_1=\begin{Vmatrix}q & 0\\ 0 & 0\end{Vmatrix}. \tag{1.31}$$

Condition (A) is obviously valid. If $\mathbf{H}_1\mathbf{x}\equiv\mathbf{0}$, then $q(t)\eta(t)\equiv 0$. Since there is an interval on which $q(t)>0$, it follows that $\eta(t)=0$ on this interval, and therefore everywhere. Thus condition (B) also holds. By the theorem of §1.2, the function $A(\lambda)$ has properties (a), (b) and so the corresponding curve is as shown in Figure 37.

Suppose now that the functions $r(t)$ and $q(t)$ are bounded below: $r(t)\geqslant r_0$ and $q(t)\geqslant q_0>0$. Then $\lambda q(t)-r(t)<0$ for sufficiently large $-\lambda$. By the corollary to the lemma of §1.4, $A(\lambda)>1$ for these values λ. Consequently, if λ is negative and of sufficiently large absolute value, the trivial solution of equation (1.29) is unstable (all solutions are nonoscillating).

As before, we consider the boundary-value problems obtained by taking equation (1.29) along with boundary conditions

$$\eta(T)=\eta(0), \quad \dot\eta(T)=\dot\eta(0) \tag{1.32}$$

and

$$\eta(T) = -\eta(0), \quad \dot{\eta}(T) = -\dot{\eta}(0). \tag{1.33}$$

As before, we see that the eigenvalues of the problem with conditions (1.32) are the roots $\lambda_{2j+1}^{\pm}$ of the equation $A(\lambda) = 1$, while those of the problem with (1.33) are the roots $\lambda_{2j}^{\pm}$ of the equation $A(\lambda) = -1$. It follows from properties (a), (b) of $A(\lambda)$ (theorem, §1.2) that under the above assumptions $(q(t) \geqslant q_0 > 0,\ r(t) \geqslant r_0)$ the eigenvalues of these problems are bounded below and, suitably indexed, may be arranged in a sequence

$$\lambda_0^+ < \lambda_1^- \leqslant \lambda_1^+ < \lambda_2^- \leqslant \lambda_2^+ < \dots \tag{1.34}$$

The intervals $(\lambda_j^+, \lambda_{j+1}^-)$ are the stability zones (here $|A(\lambda)| < 1$), while $(\lambda_i^-, \lambda_i^+)$ (if not empty) are the instability zones (where $|A(\lambda)| > 1$).

In § VIII. 6.2 we shall establish some additional facts:

1) If $\lambda = \lambda_n^{\pm}$, equation (1.29) has one periodic solution, which has exactly n zeros in the half-open interval $0 \leqslant t < T$.

2) If $\lambda < \lambda_0^+$, the solutions of equation (1.29) are nonoscillating (each solution has either no zeros or one).

3) If $\lambda > \lambda_0^+$, the solutions of equation (1.29) are oscillating (each solution has infinitely many zeros).

1.7. Qualitative pattern of stability and instability domains in parameter space for Hill's equation

Consider the instance of Hill's equation

$$\frac{d^2y}{ds^2} + c^2\,[1 + \varepsilon\varphi(\theta s)]\,y = 0,$$

where $\varphi(\theta s)$ is an integrable piecewise-continuous periodic function of s with period $2\pi/\theta$. Setting $\theta s = t$, we write the equation as

$$\frac{d^2y}{dt^2} + [\mu + \nu\varphi(t)]\,y = 0, \tag{1.35}$$

where

$$\mu = \frac{c^2}{\theta^2}, \qquad \nu = \frac{c^2}{\theta^2}\varepsilon, \quad \varphi(t + 2\pi) = \varphi(t)$$

(in this case $\mu \geqslant 0$, but for the sake of generality we shall also consider the case $\mu < 0$).

The characteristic equation of equation (1.35) is (0.2)

$$\rho^2 - 2A\rho + 1 = 0,$$

where the characteristic constant A (see (0.3)) is an entire function of μ and γ: $A = A(\mu, \nu)$.

This subsection is devoted to a qualitative characterization of the stability and instability domains in the plane of the real parameters μ, ν.

A point (μ_0, ν_0) in the (μ, ν)-plane will be called a stability (instability) point if the trivial solution of equation (1.35) is stable (unstable) at $\mu=\mu_0$, $\nu=\nu_0$. A point (μ_1, ν_1) will be called a strong stability (instability) point if all points in some neighborhood are stability (instability) points. In other words, (μ_1, ν_1) is a strong stability point if all solutions of equation (1.35) are bounded both at the point itself $(\mu=\mu_1, \nu=\nu_1)$ and in some neighborhood thereof. A point (μ_2, ν_2) is a strong instability point if equation (1.35) has unbounded solutions both at $\mu=\mu_2$, $\nu=\nu_2$ and at any point in some neighborhood thereof.

Thus, the point (μ, ν) will be strongly stable if

$$|A(\mu, \nu)|<1,$$

and strongly unstable if

$$|A(\mu, \nu)|>1.$$

The set of boundary curves $|A(\mu, \nu)|=1$ contains both stability points and instability points, and "as a rule" (but not always!) contains no strong stability (instability) points. The sets of strong stability points and strong instability points are clearly open.

Note that if $|A(\mu_0, \nu_0)|<1$, the property of stability is preserved not only by small variations of μ_0, ν_0 (which is obvious) but also by any small deformations of the function $\varphi(t)$ in equation (1.35) — "small" in the sense of the norm

$$\|\varphi\|=\sqrt{\int_0^T |\varphi(t)|\,dt}.$$

Indeed, if $|A(\mu_0, \nu_0)|<1$ the multipliers of system (1.2) are distinct and therefore definite. Hence, by the Krein-Gel'fand-Lidskii theorem (§ III. 3.2), system (1.2) is strongly stable, and this implies our assertion.*

Consider the axis $\nu=0$, i. e., the equation

$$\frac{d^2y}{dt^2}+\mu y=0.$$

Determining a fundamental matrix for this equation, we deduce directly that the characteristic constant for the period 2π is $A(\mu, 0)=\cos(2\pi\sqrt{\mu})$ (Figure 40). Points $(\mu, 0)$ for which $|\cos(2\pi\sqrt{\mu})|<1$ are strong stability points, while points $(\mu, 0)$ for which $|\cos(2\pi\sqrt{\mu})|=1$ $(\mu>0)$ are only stability points, and they may be on the common boundary of stability and instability domains. In the latter case,

$$\mu=\mu_n=\frac{n^2}{4} \qquad (n=1, 2, \ldots).$$

* The assertion also follows from the theorem stating that the matrizant $\mathbf{X}(T)$ (and therefore also $A=\frac{1}{2}\operatorname{Tr}\mathbf{X}(T)$) is a continuous function of the coefficients.

Thus points $(\mu, 0)$ on the positive μ-axis which lie on the common boundary of stability and instability domains in the (μ, ν)-plane must have the form $(\mu_n, 0)$, $n=0, 1, 2, \ldots$ All other points $(\mu, 0)$ on the positive μ-axis $(\mu \neq \mu_n)$ lie inside the stability domains.

This means that for small ν there can be instability points (μ, ν) only in neighborhoods of the points $(\mu_n, 0)$. Now let $\nu^* \neq 0$ be an arbitrary number. The curve $A(\mu, \nu^*)$ has the shape shown in Figure 41 (§1.6) (to fix ideas, the figure refers to the case $\varphi_{av} > 0$). We showed previously (see (1.34)) that the endpoints of the stability and instability zones on the μ-axis

$$\ldots < \mu_1^-(\nu^*) \leqslant \mu_1^+(\nu^*) < \mu_2^-(\nu^*) \leqslant \mu_2^+(\nu^*) < \ldots \tag{1.36}$$

are the eigenvalues of the periodic and antiperiodic boundary-value problems. As $\nu^* \to 0$, the pattern of Figure 41 must tend to that of Figure 40, i. e.,

$$\mu_n^{\pm}(0) = \frac{n^2}{4} \qquad (n=0, 1, 2, \ldots).$$

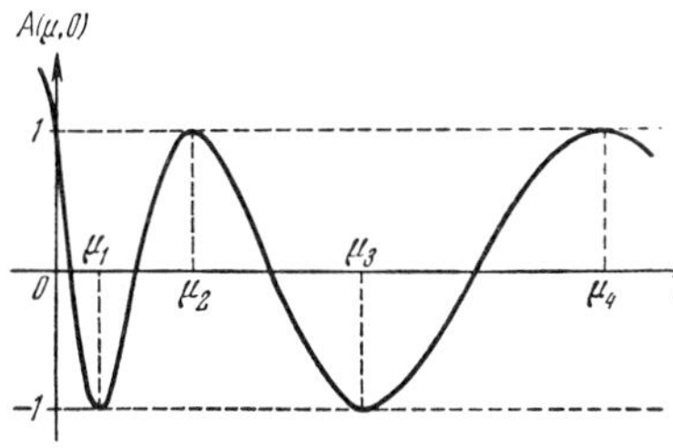

FIGURE 40.

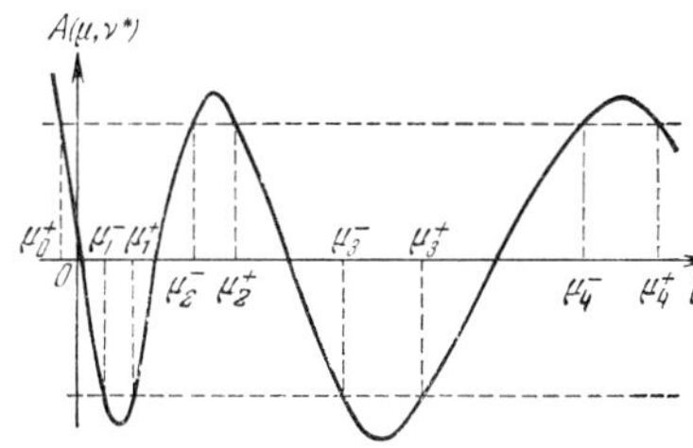

FIGURE 41.

Thus the curves $\mu = \mu_n^{\pm}(\nu)$ on the (μ, ν)-plane are the boundaries of the stability and instability domains (Figure 42). The following characteristics of the relative positions of these curves and of the stability and instability domains are readily noted.

1. The functions $\mu_n^{\pm}(\nu)$ are continuous and single-valued: any straight line $\nu = \text{const}$ cuts each of the curves in exactly one point. Indeed, the eigenvalues are numbered in increasing order: the n-th root of the equation $A(\mu,\nu) = +1$ (or $A(\mu, \nu) = -1$) is a continuous function of ν. Thus the functions $\mu = \mu_n^{\pm}(\nu)$ are continuous and (by definition) single-valued functions of ν.

Along any curve $\mu = \mu_n^{\pm}(\nu)$, an eigenfunction has exactly n zeros on $[0, T]$. Thus, knowing a solution of equation (1.35) for $\mu = \mu_n^{\pm}(\nu)$ (but not knowing the number n), one can determine n from the behavior of the solutions.

2. Inequalities (1.36) hold, so that curves $\mu = \mu_n^{\pm}(\nu)$ and $\mu = \mu_{n'}^{\pm}(\nu)$ $(n \neq n')$ cannot intersect or even touch one another. On the other hand, the curves $\mu = \mu_n^-(\nu)$ and $\mu = \mu_n^+(\nu)$ may touch; this happens when equality occurs somewhere in (1.36).

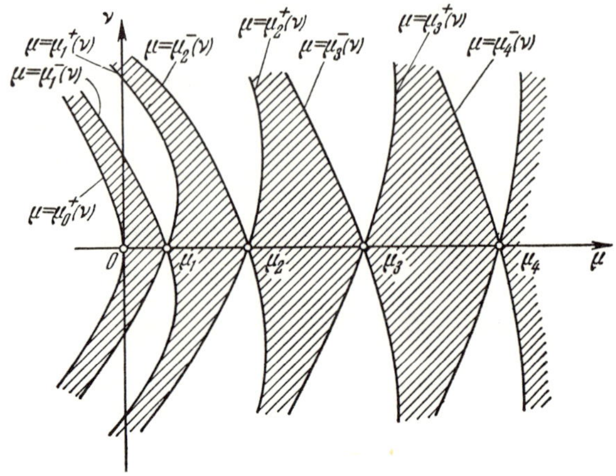

FIGURE 42.

The general pattern of the stability and instability domains is shown in Figure 42. In particular, the set of strong stability points is the union of several domains (stability domains); the n-th stability domain intersects the μ-axis in the n-th stability zone (μ_n^-, μ_n^+). The set of strong instability points is also a union of domains.

3. The stability and instability domains are directionally convex (see definition in §III. 6.3). This means that if (μ_1, ν_1) and (μ_2, ν_2) are points in the same stability (or instability) domain, then the latter will also contain points (μ, ν) for which

$$\mu_1+\nu_1\varphi(t)\leqslant\mu+\nu\varphi(t)\leqslant\mu_2+\nu_2\varphi(t) \quad \text{for all} \quad t\in[0, T].$$

As far as the stability domains are concerned, this assertion follows from the theorem of § III. 6. 3. The general case will be proved in § VIII.1.

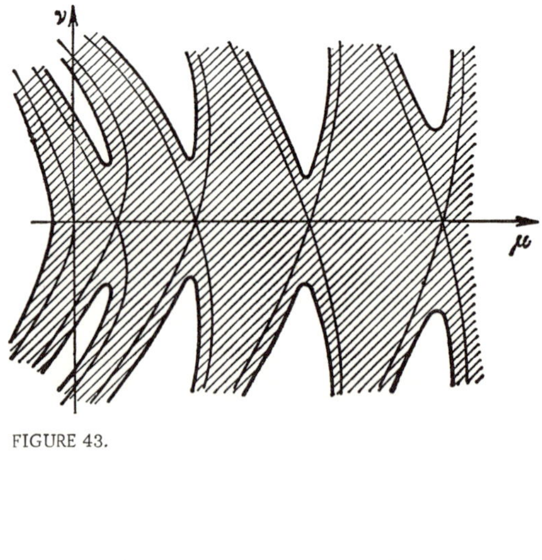

FIGURE 43.

This in turn implies the following property of the common boundary curves of stability and instability domains: if (μ_0, ν_0) is a one-sided tangent vector of such a curve, the function $\mu_0+\nu_0\varphi(t)$ necessarily changes sign.

Figure 43 shows the relative positions of the domains in the (μ, ν)-plane for which $|\rho_j|\leqslant e^{\varkappa}$ $(\varkappa>0,\ j=1,\ 2)$, where ρ_1, ρ_2 are the roots of the characteristic equation (1.36). As $\varkappa\to 0$, the curves $|\rho|=e^{\varkappa}$ (or, equivalently, the curves $A=c$, where c is a real number of absolute value greater than 1, $|c|\to 1$) approach the boundaries of the stability domains. By continuity considerations, therefore, the general shape of these curves for small $\varkappa>0$ is as shown by the thick lines in Figure 43.

§2. ESTIMATE FOR THE CHARACTERISTIC CONSTANT OF HILL'S EQUATION

2.1. Power series representation of characteristic function of auxiliary equation

Consider the equation

$$\frac{d^2x}{dt^2}+p(t)\,x=0 \tag{2.1}$$

and assume that the real piecewise-continuous T-periodic function $p(t)$, integrable on $(0, T)$, satisfies the inequality

$$p(t)\geqslant -a^2 \tag{2.2}$$

everywhere except at discontinuity points, where a is a nonnegative number. Consider the auxiliary equation

$$\frac{d^2x}{dt^2}-a^2x+\lambda\,[p(t)+a^2]\,x=0. \tag{2.3}$$

Equation (2.1) is equation (2.3) with $\lambda=1$. Any solution x of equation (2.3) is an entire function of λ (see §1.2):

$$x=x(t;\ \lambda)=x_0(t)+\lambda x_1(t)+\ldots+\lambda^n x_n(t)+\ldots$$

Substituting this into (2.3) and comparing coefficients of like powers of λ, we obtain a sequence of differential equations

$$\frac{d^2x}{dt^2}-a^2x_0=0,\qquad \frac{d^2x_n}{dt^2}-a^2x_n=-\,[p(t)+a^2]\,x_{n-1}\qquad (n=1,\ 2,\ \ldots).$$

If $x(0)$ and $\dot{x}(0)$ are independent of λ, then

$$x_0(0)=x(0),\quad \dot{x}_0(0)=\dot{x}(0);\quad x_n(0)=\dot{x}_n(0)=0\qquad (n=1,\ 2,\ \ldots)$$

and

$$x_0(t) = x(0)\operatorname{ch} at + \frac{1}{a}\dot{x}(0)\operatorname{sh} at,$$

$$x_n(t) = -\frac{1}{a}\int_0^t \operatorname{sh} a(t-t_1)\,[p(t_1)+a^2]\,x_{n-1}(t_1)\,dt_1 \qquad (n=1,\ 2,\ \ldots).$$

In particular, the solutions $f(t;\ \lambda)$ and $\varphi(t;\ \lambda)$ of equation (2.3) with initial values

$$f(0;\ \lambda)=1,\quad f'_t(0;\ \lambda)=0;\quad \varphi(0;\ \lambda)=0,\quad \varphi'_t(0;\ \lambda)=1$$

may be represented by alternating series

$$f(t;\ \lambda)=f_0(t)-\lambda f_1(t)+\lambda^2 f_2(t)-\ldots,$$
$$\varphi(t;\ \lambda)=\varphi_0(t)-\lambda\varphi_1(t)+\lambda^2\varphi_2(t)-\ldots,$$

where

$$\left.\begin{aligned} f_0(t)&=\operatorname{ch} at,\ f_n(t)=\frac{1}{a}\int_0^t \operatorname{sh} a(t-t_1)\,[p(t_1)+a^2]\,f_{n-1}(t_1)\,dt_1,\\ \varphi_0(t)&=\frac{1}{a}\operatorname{sh} at,\ \varphi_n(t)=\frac{1}{a}\int_0^t \operatorname{sh} a(t-t_1)\,[p(t_1)+a^2]\,\varphi_{n-1}(t_1)\,dt_1.\end{aligned}\right\} \qquad (2.4)$$

It follows from (2.2) and (2.4) that the functions $f_n,\ \varphi_n$ are nonnegative for $t\geqslant 0$.

The characteristic function $A(\lambda)$ of equation (2.3) is

$$A(\lambda)=\frac{1}{2}[f(T;\ \lambda)+\varphi'_t(T;\ \lambda)], \qquad (2.5)$$

whose series representation is

$$A(\lambda)=A_0-A_1\lambda+A_2\lambda^2+\ldots+(-1)^n A_n\lambda^n+\ldots, \qquad (2.6)$$

where

$$A_k=\frac{1}{2}[f_k(T)+\dot{\varphi}_k(T)] \qquad (k=0,\ 1,\ 2,\ \ldots). \qquad (2.7)$$

Since

$$\dot{\varphi}_n(t)=\int_0^t \operatorname{ch} a(t-t_1)\,[p(t_1)+a^2]\,\varphi_{n-1}(t_1)\,dt_1\geqslant 0,$$

it follows that $A_n>0$ $(n=0,\ 1,\ 2,\ \ldots)$ and the series (2.6) is alternating. It may happen that $A_n=0$ only if $p(t)+a^2\equiv 0$; this follows from (2.4) and (2.7). We shall of course assume that $p(t)+a^2\not\equiv 0$ $(0\leqslant t\leqslant T)$.

The first coefficients of the series (2.6) are

$$\left.\begin{aligned} A_0 &= \operatorname{ch} aT, \\ A_1 &= \frac{1}{2a}\int_0^T \operatorname{sh} a(T-t_1)\,[p(t_1)+a^2]\operatorname{ch} at_1\,dt_1 + \\ &+\frac{1}{2a}\int_0^T \operatorname{ch} a(T-t_1)\,[p(t_1)+a^2]\operatorname{sh} at_1\,dt_1 = \frac{1}{2a}\operatorname{sh} aT\int_0^T [p(t)+a^2]\,dt. \end{aligned}\right\} \quad (2.8)$$

Similarly, we obtain A_2 and, in general, A_n:

$$\left.\begin{aligned} A_2 &= \frac{1}{2a^2}\int_0^T dt_1\int_0^{t_1} \operatorname{sh} a(T-t_1+t_2)\operatorname{sh} a(t_1-t_2)\,r_1r_2\,dt_2, \\ A_n &= \frac{1}{2a^n}\int_0^T dt_1\int_0^{t_1} dt_2 \ldots \int_0^{t_{n-1}} \operatorname{sh} a(T-t_1+t_n)\operatorname{sh} a(t_1-t_2)\ldots \\ &\qquad \ldots \operatorname{sh} a(t_{n-1}-t_n)\,r_1\ldots r_n\,dt_n \qquad (n=3,\ \ldots), \end{aligned}\right\} \quad (2.9)$$

where we have used the abbreviations $r(t)=p(t)+a^2$, $r_i=r(t_i)$.

Note that all the previous calculations and formulas remain valid for $a=0$, provided we set

$$\operatorname{ch} a\tau\,|_{a=0} = 1, \quad \frac{1}{a}\operatorname{sh} a\tau\,\Big|_{a=0} = \tau.$$

When $a=0$, formulas (2.8) and (2.9) coincide with Lyapunov's formula (10) in /180g/:

$$A_0 = 1, \quad A_1 = \frac{T}{2}\int_0^T p(t)\,dt, \quad A_2 = \frac{1}{2}\int_0^T dt_1\int_0^{t_1}(T-t_1+t_2)(t_1-t_2)\,p_1p_2\,dt_2,$$

$$A_n = \frac{1}{2}\int_0^T dt_1\int_0^{t_1} dt_2 \ldots \int_0^{t_{n-1}} (T-t_1+t_n)(t_1-t_2)\ldots(t_{n-1}-t_n)\,p_1p_2\ldots p_n\,dt_n.$$

Lyapunov also derived formulas for A_2 and A_3 as simple integrals in the case $a=0$ (the case of a nonnegative periodic function p) /180g/. To this end, express p as

$$p = c + \frac{d^2\varphi(t)}{dt^2},$$

where c is some constant and $\varphi(t)$ a periodic function such that

$$\int_0^T \varphi(t)\,dt = 0.$$

Then

$$A_1 = \frac{1}{2}cT^2, \quad A_2 = \frac{1}{24}c^2T^2 - \frac{T}{2}\int_0^T \left(\frac{d\varphi}{dt}\right)^2 dt,$$

$$A_3 = \frac{1}{120}c^3T^6 - \frac{1}{12}cT^3\int_0^T\left(\frac{d\varphi}{dt}\right)^2 dt + 2cT\int_0^T \varphi^2\,dt - T\int_0^T \varphi\left(\frac{d\varphi}{dt}\right)^2 dt.$$

We reiterate that these formulas are valid only for $a=0$.

2.2. Proof of fundamental inequality

Lemma (Krein). If the entire function

$$F(\lambda) = a_0 - a_1\lambda + a_2\lambda^2 - \dots$$

admits a representation

$$F(\lambda) = a_0\prod_{j=1}^{\infty}\left(1-\frac{\lambda}{\lambda_j}\right) \qquad (0<\lambda_1\leqslant\lambda_2\leqslant\dots) \tag{2.10}$$

which is uniformly convergent in any disk $|\lambda|\leqslant r$, *then*

$$\frac{a_{j+1}}{a_j}\leqslant\frac{j}{j+1}\frac{a_j}{a_{j-1}} \qquad (j=1,\ 2,\ \dots). \tag{2.11}$$

Proof. Let

$$\Pi_k(\lambda) = a_0\prod_{j=1}^{k}\left(1-\frac{\lambda}{\lambda_j}\right) = a_0 - a_1^{(k)}\lambda + a_2^{(k)}\lambda^2 - \dots + (-1)^k a_k^{(k)}\lambda^k,$$

where the positive numbers $\lambda_1, \dots, \lambda_k$ are those figuring in (2.10). This function converges uniformly to $F(\lambda)$ as $k\to\infty$, if λ is constrained to vary in any compact subset of the complex plane. By Weierstrass' theorem, * the derivatives of $\Pi_k(\lambda)$ converge uniformly to the corresponding derivatives of $F(\lambda)$. Therefore $a_j^{(k)}\to a_j$ as $k\to\infty$ $(j=1,\ 2,\ \dots)$. Since all the roots of the polynomial $\Pi_k(\lambda)$ are positive, it follows from Newton's theorem /99/ that

$$\frac{a_{j+1}^{(k)}}{a_j^{(k)}} < \frac{j\,(k-j)}{(j+1)\,(k-j+1)}\frac{a_j^{(k)}}{a_{j-1}^{(k)}} \qquad (j=1,\ \dots,\ k-1).$$

Letting $k\to\infty$, we obtain (2.11), proving the lemma.

We now show that the entire function $A(\lambda)$ is of order less than unity. Equation (2.3) may be written as a system of two equations

$$\frac{d\mathbf{u}}{dt} = \mathbf{A}(t)\,\mathbf{u}, \tag{2.12}$$

* See, e.g., Privalov, I.I., Introduction to the Theory of Functions of a Complex Variable, p.196. – Moscow, Gostekhizdat, 1948. (Russian)

where $\lambda > 0$ and

$$\mathbf{u} = \begin{pmatrix} x \\ \frac{1}{\sqrt{\lambda}} \frac{dx}{dt} \end{pmatrix}, \quad \mathbf{A}(t) = \begin{Vmatrix} 0 & \sqrt{\lambda} \\ \frac{1}{\sqrt{\lambda}} a^2 - \sqrt{\lambda}\,[p(t) + a^2] & 0 \end{Vmatrix}.$$

The matrizant $\mathbf{U}(t;\ \lambda)$ of this system satisfies the equation

$$\mathbf{U}(t;\ \lambda) = \mathbf{I}_2 + \int_0^t \mathbf{A}(t_1)\,\mathbf{U}(t_1, \lambda)\,dt_1,$$

since $\mathbf{U}(0;\ \lambda) = \mathbf{I}_2$. We estimate the norm in the usual way (see §I. 1.2):

$$|\mathbf{U}(T;\ \lambda)| \leqslant |\mathbf{I}_2|\, e^{\int_0^T |\mathbf{A}(t)|\,dt}.$$

Since the norm of a matrix is bounded above by the sum of absolute values of its elements, it follows that as $|\lambda| \to \infty$

$$|\mathbf{U}(T;\ \lambda)| < c_1 \exp\left(c_2 \frac{1}{\sqrt{|\lambda|}} + c_3 \sqrt{|\lambda|}\right), \tag{2.13}$$

where c_1, c_2, c_3 are positive constants. If $x(t;\ \lambda)$ is a solution of equation (2.3) such that

$$x(t + T;\ \lambda) = \rho(\lambda)\, x(t;\ \lambda),$$

then it is clear that the vector

$$\mathbf{u} = \begin{pmatrix} x \\ \frac{1}{\sqrt{\lambda}} \frac{dx}{dt} \end{pmatrix}$$

will have the property

$$\mathbf{u}(t + T;\ \lambda) = \rho(\lambda)\, \mathbf{u}(t;\ \lambda)$$

and conversely. Hence equation (2.3) will have the same multipliers $\rho_{1,2}(\lambda)$ and thus also the same characteristic function

$$A(\lambda) = \frac{1}{2}[\rho_1(\lambda) + \rho_2(\lambda)]$$

as system (2.12). Since the characteristic function of the latter is

$$A(\lambda) = \frac{1}{2} \mathrm{Tr}\, \mathbf{U}(T;\ \lambda),$$

it follows that $A(\lambda)$ will satisfy the estimate (2.13), possibly with different positive constants. Consequently, for any $\varepsilon > 0$ there exists $k(\varepsilon)$ such that

$$A(\lambda) < \exp\left[k(\varepsilon)\lambda^{\frac{1}{2}+\varepsilon}\right] \quad \text{as} \quad \lambda \to +\infty.$$

This means that $A(\lambda)$ is of order at most 1/2, as claimed.

Hence, by Hadamard's theorem /91/, $A(\lambda)$ admits a representation

$$A(\lambda) = A_0 \prod_{n=1}^{\infty}\left(1 - \frac{\lambda}{\lambda_n}\right), \tag{2.14}$$

where $\lambda_1, \lambda_2, \ldots$ are the zeros of $A(\lambda)$.

Before we can apply Krein's lemma to $A(\lambda)$, we must show that the λ_n's are real and positive. It follows from (2.6) that $A(\lambda)$ cannot have negative real zeros, so it will suffice to show that the λ_n's are real.

It was shown in §1.6 that equation (2.3) (which is equation (1.29) with $q(t) = p(t) + a^2 \geqslant 0$, $r(t) = a^2$) satisfies conditions (A), (B) (§1.1). Thus by the theorem of §1.2 the zeros of $A(\lambda)$ are real.

We have thus shown that the numbers λ_n in (2.14) are real and positive. The "fundamental inequality" for the terms of (2.6) now follows from Krein' lemma and formula (2.14):

$$\frac{A_{n+1}}{A_n} \leqslant \frac{n}{n+1}\frac{A_n}{A_{n-1}} \qquad (n = 1, 2, \ldots). \tag{2.15}$$

2.3. Lyapunov's upper and lower bounds for the characteristic constant

Theorem. Define constants $A^{(n)}$ *by*

$$A^{(n)} = A_0 - A_1 + A_2 - \ldots + (-1)^n A_n \qquad (n = 1, 2, \ldots), \tag{2.16}$$

where $A_0, A_1, A_2, \ldots$ *are given by (2.8) and (2.9). Then each of the following conditions is sufficient for stability or instability of the trivial solution of equation (2.1):*

$$\begin{array}{llll} & \textit{Stability} & & \textit{Instability} \\ -1 \leqslant A^{(1)} & \text{and} & A^{(2)} \leqslant 1 & A^{(1)} \geqslant 1, \\ -1 \leqslant A^{(3)} & \text{and} & A^{(2)} \leqslant 1 & A^{(2)} \leqslant -1, \\ -1 \leqslant A^{(3)} & \text{and} & A^{(4)} \leqslant 1 & A^{(3)} \geqslant 1, \\ -1 \leqslant A^{(5)} & \text{and} & A^{(4)} \leqslant 1 & A^{(4)} \leqslant -1, \\ \ldots & \ldots & \ldots & \ldots \end{array} \tag{2.17}$$

If equation (2.1) has no T-periodic or T-antiperiodic solutions, these conditions will all hold for sufficiently large n.*

Proof. It follows from (2.15) that the coefficients A_n have only one maximum: as n increases from $n=0$ to some $n=n_0$ the coefficients A_n

* Recall that a solution $x(t)$ is said to be antiperiodic if $x(t+T) = \pm x(t)$. Note that the boundary points of the stability and instability domains in the function space $\{p(t)\}$ are functions for which equation (2.1) has a multiplier $\rho = \pm 1$, that is, T-periodic or T-antiperiodic solutions.

increase, and for $n > n_0$ they decrease, tending to zero. (It may happen that $n_0 = 0$, so that A_n decreases with increasing n, beginning from $n = 0$.) Indeed, let n_0 indicate the first maximum: $A_{n_0+1} \leqslant A_{n_0}$ and $A_{n+1} > A_n$ for $n < n_0$. This number certainly exists, for since the series (2.6) is convergent for $\lambda = 1$ we have $A_n \to 0$ as $n \to \infty$. Since $A_{n_0+1} \leqslant A_{n_0}$ by the choice of n_0, it follows from (2.15) (with $n = n_0 + 1$) that $A_{n_0+2} < A_{n_0+1}$, and then from the same inequality with $n = n_0 + 2$ that $A_{n_0+3} < A_{n_0+2}$, etc. Thus there is only one maximum.

It follows from (2.6) that the characteristic constant of equation (2.1) is determined by the series

$$A = A_0 - A_1 + A_2 - \ldots + (-1)^n A_n + \ldots \tag{2.18}$$

For each natural number n, let

$$A^{(n)} = A_0 - A_1 + \ldots + (-1)^n A_n \tag{2.19}$$

so that

$$A = A^{(n)} - (A_{n+1} - A_{n+2}) - (A_{n+3} - A_{n+4}) - \ldots \quad \text{for even } n,$$
$$A = A^{(n)} + (A_{n+1} - A_{n+2}) + (A_{n+3} - A_{n+4}) + \ldots \quad \text{for odd } n.$$

Therefore, if $n + 1 \geqslant n_0$, then

$$A < A^{(n)} \quad \text{for even } n, \tag{2.20}$$
$$A > A^{(n)} \quad \text{for odd } n, \tag{2.21}$$

and so

$$\text{if } A^{(n)} \leqslant 1 \ (n \text{ even}), \text{ then } A < 1, \tag{2.22}$$
$$\text{if } A^{(n)} \leqslant -1 \ (n \text{ even}), \text{ then } A < -1, \tag{2.23}$$
$$\text{if } A^{(n)} \geqslant 1 \ (n \text{ odd}), \text{ then } A > 1, \tag{2.24}$$
$$\text{if } A^{(n)} \geqslant -1 \ (n \text{ odd}), \text{ then } A > -1. \tag{2.25}$$

This is true on the assumption that $n_0 \leqslant n + 1$. We now show that, conversely, each of the inequalities for $A^{(n)}$ in (2.22)–(2.25) implies that $n_0 \leqslant n + 1$; in fact, we prove a stronger assertion: If $A^{(1)} \geqslant -1$, then $n_0 \leqslant 1$; in all other cases, each inequality for $A^{(n)}$ in (2.22)–(2.25) implies $n_0 \leqslant n - 1$. First let

$$A^{(1)} = A_0 - A_1 \geqslant -1 \quad \text{or} \quad A_1 \leqslant 1 + A_0 = 1 + \operatorname{ch} aT.$$

Using (2.15), we obtain for $n = 1$

$$A_2 \leqslant \frac{A_1}{2 \operatorname{ch} aT} A_1 \leqslant \frac{1 + \operatorname{ch} aT}{2 \operatorname{ch} aT} A_1 \leqslant A_1, \quad \text{i.e.,} \quad n_0 \leqslant 1.$$

If $A^{(1)} = A_0 - A_1 \geqslant 1$, then

$$A_1 \leqslant A_0 - 1 < A_0, \quad \text{i. e.,} \quad n_0 = 0.$$

Let $n > 1$ be odd and suppose that (2.24), (2.25) hold, i. e.,

$$A^{(n)} - \alpha = (A_0 - A_1 - \alpha) + (A_2 - A_3) + \dots + (A_{n-1} - A_n) \geqslant 0, \quad (2.25a)$$

where $\alpha = 1$ or $\alpha = -1$. Suppose that $n_0 > n - 1$, so that $n_0 \geqslant n$. Since $A_n > A_{n-1}$ for $n \leqslant n_0$, all parenthetical terms in (2.25a), from the second on, are negative. In addition, $A_0 - A_1 - \alpha < 0$, since $A_0 - A_1 - \alpha \geqslant 0$ would imply $A^{(1)} = A_0 - A_1 \geqslant \alpha \geqslant -1$ and, as shown previously, $n_0 \leqslant 1$, i. e., $n \leqslant n_0 \leqslant 1$, contrary to assumption. Thus (2.25a) cannot hold if $n_0 > n - 1$. Therefore $n_0 \leqslant n - 1$.

Let $n \geqslant 2$ be even and suppose (2.22), (2.23) are satisfied, so that $A^{(n)} - \alpha = (A_0 - \alpha) + (-A_1 + A_2) + (-A_3 + A_4) + \dots + (-A_{n-1} + A_n) \leqslant 0$, where $\alpha = 1$ or $\alpha = -1$. If $n \leqslant n_0$ the first parenthetical term in this inequality is nonnegative while the others are positive. Therefore $n_0 \leqslant n - 1$.

This proves the sufficiency of conditions (2.17).

The last assertion of the theorem follows from the fact that $A^{(n)} \to A$ as $n \to \infty$, and if $-1 < A < 1$ then from some n onward we have $1 < A^{(n)} < 1$. Similarly, $A < -1$ implies $A^{(n)} < -1$ and $A > 1$ implies $A^{(n)} > 1$, for sufficiently large n. This completes the proof.

Finally, then, excluding the exceptional case $|A| = 1$ we have a finite procedure for determining whether the trivial solution of equation (2.1) is stable or unstable: determine the coefficients A_n using the formulas of §2.1, then evaluate $A^{(n)}$ by (2.16); it can then be decided in finitely many steps which of conditions (2.17) is satisfied. This procedure for Hill's equation (2.1) is known as Lyapunov's method.

As shown above, the inequality $A_{n_0+1} \leqslant A_{n_0}$ (which necessarily holds for some n_0) implies the validity of the two-sided estimates (2.20), (2.21) for the characteristic exponent $n \geqslant n_0 - 1$. In the unstable case, these inequalities also provide an estimate for the order of increase of the solutions of equation (2.1). We now proceed to a more detailed study of this question.

2.4. Estimates for the characteristic exponents of Hill's equation

The characteristic exponents μ_1 and μ_2 of equation (2.1) are real or pure imaginary and $\mu_1 + \mu_2 = 0$ (see §II. 3. 5). Let $\mu = \mathrm{Re}\,\mu_1 \geqslant 0$. Suppose we wish to determine whether $\mu < \mu_0$ or $\mu > \mu_0$, where $\mu_0 > 0$ is some given number. Since

$$\pm\mu = \frac{1}{T} \ln |A \mp \sqrt{A^2 - 1}| = \frac{1}{T} \ln |\rho|^{\pm 1}$$

(where ρ, ρ^{-1} are the multipliers of equation (2.1)) and $A = (\rho + \rho^{-1})/2$, it follows that $\mu < \mu_0$ (or $\mu > \mu_0$) is true if and only if

$$|A| < \operatorname{ch} \mu_0 T \qquad (\text{or } |A| > \operatorname{ch} \mu_0 T).$$

To determine which of these alternatives holds, we calculate the coefficients A_n until we obtain the inequality $A_{n_0+1} \leqslant A_{n_0}$. By (2.20) and (2.21), for $n \geqslant n_0 - 1$:

$$\begin{array}{ll}
\text{if } A^{(n)} \leqslant \operatorname{ch}\mu_0 T \ (n \text{ even}), & \text{then } A < \operatorname{ch}\mu_0 T, \\
\text{if } A^{(n)} \leqslant -\operatorname{ch}\mu_0 T \ (n \text{ even}), & \text{then } A < -\operatorname{ch}\mu_0 T, \\
\text{if } A^{(n)} \geqslant \operatorname{ch}\mu_0 T \ (n \text{ odd}), & \text{then } A > \operatorname{ch}\mu_0 T, \\
\text{if } A^{(n)} \geqslant -\operatorname{ch}\mu_0 T \ (n \text{ odd}), & \text{then } A > -\operatorname{ch}\mu_0 T.
\end{array}$$

Thus, provided that $\mu \neq \mu_0$, we can always determine in finitely many steps which of the inequalities $\mu > \mu_0$ or $\mu < \mu_0$ is true. If $\mu = \mu_0$, then finitely many operations will suffice to establish the inequality $\mu_0 - \varepsilon < \mu < \mu_0 + \varepsilon$ for any $\varepsilon > 0$.

As in §2.3, it is easy to list a sequence of sufficient conditions for the inequalities $\mu > \mu_0$ and $\mu < \mu_0$:

$$\left.\begin{array}{ll}
\mu < \mu_0: & \mu > \mu_0: \\
 & A^{(1)} \geqslant \operatorname{ch}\mu_0 T, \\
-\operatorname{ch}\mu_0 T \leqslant A^{(1)} < A^{(2)} \leqslant \operatorname{ch}\mu_0 T, & A^{(2)} \leqslant -\operatorname{ch}\mu_0 T, \\
-\operatorname{ch}\mu_0 T \leqslant A^{(3)} < A^{(2)} \leqslant \operatorname{ch}\mu_0 T, & A^{(3)} \geqslant \operatorname{ch}\mu_0 T, \\
-\operatorname{ch}\mu_0 T \leqslant A^{(3)} < A^{(4)} \leqslant \operatorname{ch}\mu_0 T, & A^{(4)} \leqslant -\operatorname{ch}\mu_0 T, \\
\dots\dots\dots\dots & \dots\dots
\end{array}\right\} \qquad (2.26)$$

In applied problems, the coefficient $p(t)$ usually depends on parameters. Each of inequalities (2.26) (together with the appropriate inequality $A_{n_0+1} \leqslant A_{n_0}$) defines the domain in the parameter space in which $\mu < \mu_0$ $(\mu > \mu_0)$

2.5. Estimates for the necessary number of coefficients A_n

We showed in §1.6 that the roots $\lambda_0^+, \lambda_2^-, \lambda_2^+, \lambda_4^-, \lambda_4^+, \dots$ of the equation $A(\lambda) = 1$ and the roots $\lambda_1^-, \lambda_1^+, \lambda_3^-, \lambda_3^+, \dots$ of the equation $A(\lambda) = -1$ may be arranged in a sequence (1.34).

The intervals $(-\infty, \lambda_0^+)$, $(\lambda_\nu^-, \lambda_\nu^+)$ are the instability zones, the (nonempty) intervals $(\lambda_{\nu-1}^+, \lambda_\nu^-)$ the stability zones $(\nu = 1, 2, \dots)$ (see §1.6). We shall call the interval $(-\infty, \lambda_0^+)$ the zeroth instability zone $\mathscr{H}_0$, the intervals $(\lambda_n^-, \lambda_n^+)$ and $(\lambda_{n-1}^+, \lambda_n^-)$ the n-th instability zone $\mathscr{H}_n$ and the $(n-1)$-th stability zone $\mathscr{O}_{n-1}$, respectively $(n = 1, 2, \dots)$.

We shall say that equation (2.1) (or the function $p(t)$) belongs to the n-th stability (instability) domain if $\lambda = 1$ is in the n-th stability (instability) zone for equation (2.3) with the same function $p(t)$. *

Let us determine λ_n^-, λ_n^+ for the function $p(t) = n^2\pi^2 T^{-2}$; equation (2.3) is

$$\frac{d^2x}{dt^2} + \lambda \frac{n^2\pi^2}{T^2} x = 0 \qquad (\lambda > 0).$$

Having found a fundamental matrix, one verifies directly that the characteristic constant over the period T is $A(\lambda) = \cos n\pi\sqrt{\lambda}$. It follows from the geometric properties of this function that the instability zones are empty — they

* The justification for this definition will follow from the results of Chapter VIII, where we shall establish the existence of stability and instability domains in the Hamiltonian space (§ VIII.1.8) and prove the assertion cited here as a definition (§ VIII.6.2).

shrink to the points $n\pi\sqrt{\lambda}=k\pi$ $(k=1, 2,\ldots)$, which are the endpoints of the $(k-1)$-th and k-th stability zones. For the boundary of the $(n-1)$-th and n-th stability zone, we have $n\sqrt{\lambda}=n$, so that $\lambda=1$ and so $\lambda_n^-=\lambda_n^+=1$.

By the comparison theorem /87/, the numbers λ_n^-, λ_n^+ decrease with increasing $p(t)$. Thus, if $p(t)\geqslant n^2\pi^2T^{-2}$ on $[0, T]$, then λ_n^-, $\lambda_n^+<1$, so that $p(t)$ may belong only to stability domains with subscript less than n and to instability domains with subscript larger than n.

On the other hand, if $p(t)\leqslant n^2\pi^2T^{-2}$ $(0\leqslant t\leqslant T)$, then λ_n^-, $\lambda_n^+>1$, so that $p(t)$ may belong only to stability or instability domains with subscript less than n.

Theorem. *Let $p(t)$ belong to the m-th stability domain. In order to establish this fact by Lyapunov's method, one must evaluate at least $m+2$ of the coefficients A_n.*

Proof. Denote

$$R_n(\lambda)=A_{n+1}\lambda^{n+1}-A_{n+2}\lambda^{n+2}+\ldots$$

and write

$$A(\lambda)=A^{(n)}(\lambda)+(-1)^{n+1}R_n(\lambda).$$

We claim that if $n\geqslant n_0$ and $0<\lambda\leqslant 1$, the function $R_n(\lambda)$ is positive and increasing. That it is positive follows from the definition of n_0, since $A_{n_0+1}\leqslant A_{n_0}$, $A_{n+1}<A_n$ for $n>n_0$. It remains to show that $dR_n/d\lambda>0$ for $0<\lambda<1$. We have

$$\frac{dR_n}{d\lambda}=\sum_{k=n+1,\, n+3,\,\ldots}[kA_k\lambda^k-(k+1)A_{k+1}\lambda^{k+1}].$$

The bracketed expression is positive for $k>n_0$ and $0<\lambda<1$ by (2.15), so that $dR_n/d\lambda>0$ for $0<\lambda<1$.

Assume first that m is even. By Lyapunov's method, we obtain $A^{(n)}(1)\geqslant -1$ for some odd n and $A^{(n)}(1)\leqslant 1$ for some even n. For the latter,

$$-1\leqslant A(1)\leqslant A^{(n)}(1)\leqslant 1$$

and so $0<R_n(1)<2$. Hence also

$$0\leqslant R_n(\lambda)<2 \qquad (0\leqslant\lambda\leqslant 1).$$

For positive λ not exceeding 1, lying in odd-numbered instability zones, we have $A(\lambda)<-1$, so that

$$A^{(n)}(\lambda)=A(\lambda)+R_n(\lambda)<1.$$

For positive λ not exceeding 1 and belonging to even-numbered instability zones, we have $A(\lambda)>1$, so that

$$A^{(n)}(\lambda)=A(\lambda)+R_n(\lambda)>1.$$

Since $\lambda=1\in\mathfrak{G}_m$, it follows that there are $m+1$ instability zones on $(-\infty, 1)$. In addition, $A^{(n)}\leqslant 1$. Consequently, the n-th degree polynomial $A^{(n)}(\lambda)$ takes the value 1 at least $m+1$ times (Figure 44). Therefore $n\geqslant m+1$, and since n is even this means that $n\geqslant m+2$.

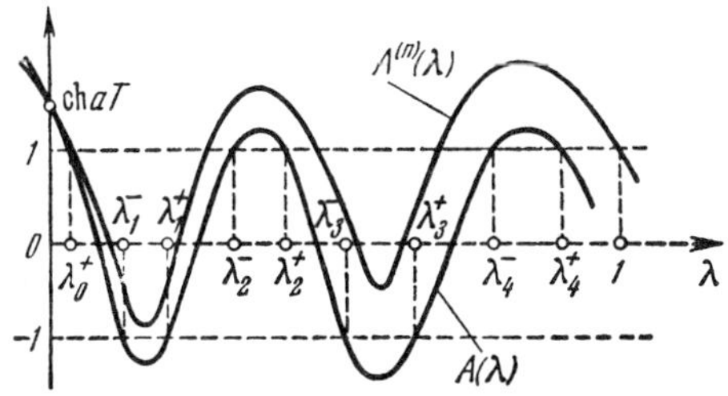

FIGURE 44.

If m is odd, the same reasoning applied to the inequality $A^{(n)}\geqslant -1$ with odd n will again yield $n\geqslant m+2$. Thus, in either case one must evaluate at least $m+2$ of the coefficients A_j. Q. E. D.

In particular, any function which remains "close to" c^2, where

$$\frac{n^2\pi^2}{T^2}\leqslant c^2\leqslant\frac{(n+1)^2\pi^2}{T^2},$$

will belong to the n-th stability domain. Thus the amount of work involved in Lyapunov's method increases to the degree that $p(t)$ exceeds the quantity π^2T^{-2}.

2.6. Simple stability and instability tests for the trivial solution of Hill's equation

In the tests presented below, a function $p(t)$ will belong (or not) to the n-th stability (instability) domain in the sense of the definitions in §2.5.

Test I. If

$$p(t)\geqslant -a^2 \tag{2.27}$$

and

$$\int_0^T [p(t)+a^2]\,dt\leqslant 2a\,\mathrm{th}\left(\frac{aT}{2}\right), \tag{2.28}$$

then $p(t)$ is in the zeroth instability domain.

Test II. If (2.27) holds and

$$\int_0^T p(t)\,dt \geqslant 0, \quad \int_0^T [p(t)+a^2]\,dt \leqslant 2a \operatorname{cth} \frac{aT}{2}, \tag{2.29}$$

then $p(t)$ is in the zeroth stability domain.

Proof of Test II. The second inequality in (2.29) is $A^{(1)} \geqslant -1$. We shall show that this inequality implies not only that $A > -1$ (as follows from (2.27)), but also that $\lambda_1^- > 1$.

Suppose the contrary: $\lambda_1^- \leqslant 1$. Then $\lambda_1^+ < 1$, since $\lambda_1^- \leqslant 1 \leqslant \lambda_1^+$ would imply that $A \leqslant -1$. Consider the function

$$p_\mu(t) = (1-\mu)\,p(t) - \mu a^2 \qquad (0 \leqslant \mu \leqslant 1).$$

Let A_μ denote the corresponding characteristic constant. Since

$$p_\mu(t) \geqslant -a^2, \quad \int_0^T [p_\mu(t)+a^2]\,dt \leqslant 2a \operatorname{cth} \frac{aT}{2},$$

it follows that $A_\mu > -1$ for all $\mu \in [0, 1]$. Let $A_\mu(\lambda)$ and $\lambda_n^-(\mu)$, $\lambda_n^+(\mu)$ denote the characteristic function $A(\lambda)$ and numbers λ_n^-, λ_n^+ for equation (2.3) with the function $p_\mu(t)$. Since $A_\mu(\lambda) \to \operatorname{ch} aT$ as $\mu \to 1$, it follows that $\lambda_n^\pm(\mu) \to \infty$. Since by assumption $\lambda_1^-(0) \leqslant 1$, there exists $\mu \in (0, 1)$ such that

$$\lambda_1^-(\mu) \leqslant 1 \leqslant \lambda_1^+(\mu).$$

But then $A_\mu \leqslant -1$, which is impossible. Consequently, $\lambda_1^- > 1$.

The rest of the proof of Test II is based on the following

Lemma. If

$$\int_0^T p(t)\,dt \geqslant 0,$$

then all solutions of equation (2.1) are oscillating /72/.

Indeed, if the solutions were nonoscillating, equation (2.1) would have a solution

$$x(t) = e^{kt} u(t),$$

where k, $u(t)$ are real, $u(t+T) = u(t)$ and $u(t) \neq 0$ for all t. Then

$$\int_0^T p\,dt = -\int_0^T \frac{\ddot{x}}{x}\,dt = -\int_0^T \frac{k^2 u + 2k\dot{u} + \ddot{u}}{u}\,dt = -k^2 T - \int_0^T \frac{\dot{u}^2}{u^2}\,dt < 0,$$

proving the lemma.

Going back to Test II, we thus have $\lambda_0^+ < 1$, so that $\lambda_0^+ < 1 < \lambda_1^+$ and so $p(t)$ belongs to the zeroth stability domain, completing the proof.

The proof of Test I is entirely analogous. Inequality (2.28) is simply the explicit form of the inequality $A^{(1)} \geqslant 1$ (see (2.8)). Instability follows from the theorem of §2.3 as applied to the first instability test of (2.17).

It can be shown that condition (2.28) is best possible. In other words, for any $\varepsilon > 0$ we can find a T-periodic function $p_0(t)$ satisfying condition (2.27) and the condition

$$\int_0^T [p_0(t) + a^2]\, dt \leqslant 2a \operatorname{th} \frac{aT}{2} + \varepsilon, \tag{2.30}$$

but the trivial solution of equation (2.1) is stable.

Reviewing the proof of conditions (2.29), we see that they are also best possible: for any $\varepsilon > 0$ there exist a function $p_0(t)$ in the zeroth instability domain such that

$$\int_0^T p_0(t)\, dt \geqslant -\varepsilon, \quad \int_0^T [p_0(t) + a^2]\, dt \leqslant 2a \operatorname{cth} \frac{aT}{2},$$

and a function $p_1(t)$ in the first instability domain such that

$$\int_0^T p_1(t)\, dt \geqslant 0, \quad \int_0^T [p_1(t) + a^2]\, dt \leqslant 2a \operatorname{cth} \frac{aT}{2} + \varepsilon.$$

Thus, if we have only a lower bound for the function $p(t)$ and $p_{\text{av}} = \frac{1}{T} \int_0^T p(t)\, dt$, then neither of inequalities (2.28), (2.29) can be improved.

§3. GENERALIZATION OF LYAPUNOV'S METHOD TO SYSTEMS OF TWO LINEAR FIRST-ORDER DIFFERENTIAL EQUATIONS WITH PERIODIC COEFFICIENTS

3.1. Transformation of initial system and evaluation of characteristic constant

Suppose that the first-approximation equations for the perturbed motion of a dynamic system are

$$\frac{dx_1}{dt} = a_{11}(t)\, x_1 + a_{12}(t)\, x_2, \quad \frac{dx_2}{dt} = a_{21}(t)\, x_1 + a_{22}(t)\, x_2, \tag{3.1}$$

where $a_{ij}(t)$ $(i, j = 1, 2)$ are real integrable piecewise-continuous periodic functions with period $T > 0$. We write (3.1) as a vector equation:

$$\frac{d\mathbf{u}}{dt} = \mathbf{A}(t)\, \mathbf{u} \qquad \left(\mathbf{u} = \begin{pmatrix} x_1 \\ x_2 \end{pmatrix}, \ \mathbf{A}(t) = \| a_{ij}(t) \|_1^2 \right). \tag{3.2}$$

For the sequel it is important to observe that any system (3.1) with bounded coefficients $a_{ij}(t)$ $(i, j = 1, 2)$ may be reduced by an orthogonal transformation with periodic coefficients to a form in which the functions on the second diagonal of the coefficient matrix are of fixed signs. Indeed, consider the substitution

$$\mathbf{w}=\exp\left(\frac{2k\pi}{T}t\mathbf{J}\right)\mathbf{u}\left(\mathbf{w}=\begin{pmatrix}z_1\\ z_2\end{pmatrix},\quad \mathbf{J}=\begin{Vmatrix}0 & -1\\ 1 & 0\end{Vmatrix},\quad k=\pm 1,\ \pm 2,\ \ldots\right). \qquad (3.3)$$

In terms of our new variables, system (3.1) becomes

$$\frac{d\mathbf{w}}{dt}=\mathbf{D}(t)\mathbf{w},\quad \mathbf{D}(t)=\|d_{ij}(t)\|_1^2=\frac{2k\pi}{T}\mathbf{J}+\exp\left(\frac{2k\pi}{T}t\mathbf{J}\right)\mathbf{A}(t)\exp\left(-\frac{2k\pi}{T}t\mathbf{J}\right). \qquad (3.4)$$

The matrix-function $\mathbf{D}(t)$ is periodic and if k is sufficiently large the functions d_{12} and d_{21} will be of fixed sign.

After these remarks, we return to system (3.1), assuming that the functions a_{12} and a_{21} are of fixed sign. Writing

$$\int_0^T a_{11}(t)\,dt=\alpha,\quad \int_0^T a_{22}(t)\,dt=\beta,$$

we may assume without loss of generality that $\alpha\geqslant\beta$. If

$$\int_0^T \operatorname{Tr}\mathrm{A}(t)\,dt\equiv\alpha+\beta>0,$$

then $\det\mathbf{U}(t)\longrightarrow\infty$ as $t\longrightarrow\infty$ by virtue of Liouville's formula (II.1.11) ($\mathbf{U}(t)$ is the matrizant of system (3.1)). Thus, if $\alpha+\beta>0$ the trivial solution of system (3.1) and, a fortiori, the unperturbed motion of the dynamic system described in the first approximation by system (3.1), are unstable. For the sequel we assume that

$$\alpha+\beta\leqslant 0,$$

or, since $\alpha\geqslant\beta$,

$$\beta\leqslant\alpha\leqslant-\beta. \qquad (3.5)$$

It follows from (3.5) that $\beta\leqslant 0$, and if $\beta=0$ then $\alpha=0$ as well. We must thus investigate the domain

$$\beta\leqslant\alpha\leqslant-\beta,\quad \beta\leqslant 0$$

in the (α,β)-plane, together with the rays forming its boundary.

The substitutions

$$x_1=x\exp\int_0^t a_{11}(t_1)\,dt_1,\quad x_2=y\exp\int_0^t\left[a_{22}(t_1)+\frac{\alpha-\beta}{T}\right]dt_1 \qquad (3.6)$$

bring system (3.1) to the form

$$\frac{dx}{dt}=r(t)y,\quad \frac{dy}{dt}=q(t)x-ay\qquad\left(a=\frac{\alpha-\beta}{T}\geqslant 0\right), \qquad (3.7)$$

where r and q denote T-periodic functions of fixed sign:

$$\left.\begin{aligned} r(t) &= a_{12}(t)\exp\int_0^t [a_{22}(t_1)-a_{11}(t_1)+a]\,dt_1, \\ q(t) &= a_{21}(t)\exp\int_0^t [a_{11}(t_1)-a_{22}(t_1)-a]\,dt_1. \end{aligned}\right\} \tag{3.8}$$

We can write system (3.7) as

$$\frac{d\mathbf{v}}{dt} = \mathbf{B}(t)\,\mathbf{v} \qquad \left(\mathbf{v} = \begin{pmatrix} x \\ y \end{pmatrix},\quad \mathbf{B}(t) = \begin{Vmatrix} 0 & r(t) \\ q(t) & -a \end{Vmatrix}\right). \tag{3.9}$$

To determine the characteristic constant A of system (3.7), we consider instead the system

$$\frac{dx}{dt} = r(t)\,y, \quad \frac{dy}{dt} = -\varepsilon q(t)\,x - ay. \tag{3.10}$$

Let $x^{(1)}(t;\ \varepsilon)$, $y^{(1)}(t,\ \varepsilon)$ denote the solution of system (3.10) satisfying initial conditions $x^{(1)}(0;\ \varepsilon) = 1$, $y^{(1)}(0;\ \varepsilon) = 0$, and $x^{(2)}(t;\ \varepsilon)$, $y^{(2)}(t;\ \varepsilon)$ the solution defined by the initial conditions $x^{(2)}(0;\ \varepsilon) = 0$, $y^{(2)}(0;\ \varepsilon) = 1$. Then the matrizant $\mathbf{V}(t;\ \varepsilon)$ of system (3.10),

$$\mathbf{V}(t;\ \varepsilon) = \begin{Vmatrix} x^{(1)}(t;\ \varepsilon) & x^{(2)}(t;\ \varepsilon) \\ y^{(1)}(t;\ \varepsilon) & y^{(2)}(t;\ \varepsilon) \end{Vmatrix},$$

and the characteristic function $A(\varepsilon)$ of the system,

$$A(\varepsilon) = \frac{1}{2}\operatorname{Tr}\mathbf{V}(T;\ \varepsilon) = \frac{1}{2}\left[x^{(1)}(T;\ \varepsilon) + y^{(2)}(T;\ \varepsilon)\right], \tag{3.11}$$

are analytic functions of ε (§ II. 1.3).

We write the solution of system (3.10) as

$$x(t;\ \varepsilon) = x_0(t) + x_1(t)\,\varepsilon + x_2(t)\,\varepsilon^2 + \ldots, \quad y(t;\ \varepsilon) = y_0(t) + y_1(t)\,\varepsilon + y_2(t)\,\varepsilon^2 + \ldots,$$

and insert it into (3.10); comparison of the coefficients of like powers of ε gives

$$\frac{dx_0}{dt} = r(t)\,y_0, \quad \frac{dy_0}{dt} = -ay_0, \tag{3.12}$$

$$\frac{dx_n}{dt} = r(t)\,y_n, \quad \frac{dy_n}{dt} = -q(t)\,x_{n-1} - ay_n \qquad (n = 1,\ 2,\ \ldots). \tag{3.13}$$

Integrating equations (3.13) with initial conditions $x_n(0) = y_n(0) = 0$, we get

$$x_n(t) = \int_0^t r_1 y_n(t_1)\,dt_1, \quad y_n(t) = -e^{-at}\int_0^t e^{at_1} q_1 x_{n-1}(t_1)\,dt_1 \quad (n = 1,\ 2,\ \ldots), \tag{3.14}$$

where $r_i=r(t_i)$, $q_i=q(t_i)$. Integrating (3.12) with initial conditions $x_0^{(1)}(0)=1$, $y_0^{(1)}(0)=0$, on the one hand, and $x_0^{(2)}(0)=0$, $y_0^{(2)}(0)=1$, on the other, we have

$$x_0^{(1)}(t)=1,\quad y_0^{(1)}(t)=0,\quad x_0^{(2)}(t)=\int_0^t e^{-at_1}r_1\,dt_1,\quad y_0^{(2)}=e^{-at}.$$

Thus the desired expressions for $x_n^{(1)}(T)$ and $y_n^{(2)}(T)$ are

$$x_n^{(1)}(T)=(-1)^n\int_0^T e^{-at_1}r_1\,dt_1\int_0^{t_1} e^{at_2}q_2\,dt_2\,\ldots\int_0^{t_{2n-2}} e^{-at_{2n-1}}r_{2n-1}\,dt_{2n-1}\int_0^{t} e^{at_{2n}}q_{2n}\,dt_{2n},$$

$$y_n^{(2)}(T)=$$
$$=(-1)^n e^{-aT}\int_0^T e^{at_1}q_1\,dt_1\int_0^{t_1} e^{-at_2}r_2\,dt_2\,\ldots\int_0^{t_{2n-2}} e^{at_{2n-1}}q_{2n-1}\,dt_{2n-1}\int_0^{t_{2n-1}} e^{-at_{2n}}r_{2n}\,dt_{2n}.$$

From (3.11) we obtain a series representation of the characteristic function $A(\varepsilon)$:

$$A(\varepsilon)=A_0+A_1\varepsilon+A_2\varepsilon^2+\ldots,$$
$$A_0=\frac{1}{2}(1+e^{-aT}),\quad A_n=\frac{1}{2}[x_n^{(1)}(T)+y_n^{(2)}(T)]\qquad (n=1,\,2,\,\ldots). \tag{3.15}$$

At $\varepsilon=-1$ system (3.10) coincides with system (3.7). Thus the characteristic constant* A of system (3.7) is

$$A=A(-1)=A_0-A_1+A_2-\ldots+(-1)^n A_n+\ldots \tag{3.16}$$

Estimates for A_n may be derived from upper and lower bounds for the functions r and q; these estimates provide some information on the rate of convergence of the series (3.16). We shall not do this, however, since we are interested in an upper bound for the quotient of two consecutive terms of the series — this will be the fundamental inequality. We first bring the expressions for A_n to a more convenient form.

We begin with

$$A_1=-\frac{1}{2}\left(\int_0^T e^{-at_1}r_1\,dt_1\int_0^{t_1} e^{at_2}q_2\,dt_2+e^{-aT}\int_0^T e^{at_1}q_1\,dt_1\int_0^{t_1} e^{-at_2}r_2\,dt_2\right). \tag{3.17}$$

Changing the order of integration in the first integral and setting

$$\psi(t)=\int_0^t e^{-at_1}r_1\,dt_1\qquad (\psi_i=\psi(t_i)), \tag{3.18}$$

we write (3.17) as

$$A_1=-\frac{1}{2}\int_0^T e^{at_1}[\psi(T)-(1-e^{-aT})\psi_1]\,q_1\,dt_1. \tag{3.19}$$

* In the notation of /204c/, A denoted a constant differing from our characteristic constant by a factor $2/(1+e^{-aT})$.

We now deal with $x_n^{(1)}(T)$ and $y_n^{(2)}(T)$. Assuming that $t_1, \ldots, t_{2n}$ are variables such that $T > t_1 > \ldots > t_{2n} > 0$, we integrate in the expression for $x_n^{(1)}(T)$ with respect to $t_1, t_3, \ldots, t_{2n-1}$, and in that for $y_n^{(2)}(T)$ with respect to $t_2, t_4, \ldots, t_{2n}$. Modifying the notation so that the remaining variables are $t_1, \ldots, t_n$, write the results as

$$x_n^{(1)}(T) = (-1)^n \int_0^T dt_1 \int_0^{t_1} dt_2 \ldots \int_0^{t_{n-1}} e^{at_1}[\psi(T) - \psi_1] e^{at_2}(\psi_1 - \psi_2) \ldots \\ \ldots e^{at_n}(\psi_{n-1} - \psi_n) q_1, \ldots q_n \, dt_n,$$

$$y_n^{(2)}(T) = (-1)^n \int_0^T dt_1 \int_0^{t_1} dt_2 \ldots \int_0^{t_{n-1}} e^{-aT} e^{at_1} e^{at_2}(\psi_1 - \psi_2) e^{at_3}(\psi_2 - \psi_3) \ldots \\ \ldots e^{at_n}(\psi_{n-1} - \psi_n) \psi_n q_1 \ldots q_n \, dt_n.$$

Consequently,

$$A_n = (-1)^n \frac{1}{2} \int_0^T dt_1 \int_0^{t_1} dt_2 \ldots \int_0^{t_{n-1}} e^{at_1}(\psi_1 - \psi_2) \ldots \\ \ldots e^{at_{n-1}}(\psi_{n-1} - \psi_n) e^{at_n}[\psi(T) - \psi_1 + \psi_n e^{-aT}] q_1 \ldots q_n \, dt_n. \quad (3.20)$$

Following Lyapunov (/180g/, subsections 5, 6) and retaining his notation, we consider $n+1$ variables $t_1, \ldots, t_{n+1}$ and, letting $j_1, \ldots, j_k$ denote some sequence of numbers $1, \ldots, n+1$, set

$$\exp(at_{j_1})(\psi_{j_1} - \psi_{j_2}) q_{j_1} \ldots \exp(at_{j_{k-1}})(\psi_{j_{k-1}} - \psi_{j_k}) q_{j_{k-1}} \times \\ \times \exp(at_{j_k})[\psi(T) - \psi_{j_1} + \psi_{j_k} e^{-aT}] q_{j_k} = [j_1, \ldots, j_k].$$

Formulas (3.19) and (3.20) now become

$$\left. \begin{aligned} A_1 &= -\frac{1}{2} \int_0^T [n+1] \, dt_{n+1}, \\ A_n &= (-1)^n \frac{1}{2} \int_0^T dt_1 \int_0^{t_1} dt_2 \ldots \int_0^{t_{n-1}} [1, \ldots, n] \, dt_n. \end{aligned} \right\} \quad (3.21)$$

Thus the product $(-1)^{n+1} 4 A_n A_1$ is given by

$$\int \ldots \int [1, \ldots, n][n+1] \, dt_1 \ldots dt_{n+1}.$$

We now extend the $(n+1)$-tuple integral to all values of $t_1, \ldots, t_{n+1}$ such that $T > t_1 > \ldots > t_n > 0$; $T > t_{n+1} > 0$. This integral is the sum of integrals of the same form obtained if we let t_{n+1} vary in turn in each of the intervals $(0, t_n)$, (t_n, t_{n-1}), $\ldots$, (t_2, t_1), (t_1, T), i.e.,

$$(-1)^n 4 A_n A_1 = \int_0^T dt_1 \int_0^{t_1} dt_2 \ldots \int_0^{t_n} [1, \ldots, n][n+1] \, dt_{n+1} + \\ + \sum_{\nu=1}^{n} \int_0^T dt_1 \int_0^{t_1} dt_2 \ldots \int_0^{t_{n-1}} dt_n \int_{t_\nu}^{t_{\nu-1}} [1, \ldots, n][n+1] \, dt_{n+1} \qquad (t_0 \equiv T).$$

Considering one of the integrals in the sum, we apply a circular permutation to the variables $t_\nu, \ldots, t_{n+1}$, thus obtaining

$$\int_0^T dt_1 \int_0^{t_1} dt_2 \ldots \int_0^{t_{\nu-2}} dt_{\nu-1} \int_0^{t_{\nu-1}} dt_{\nu+1} \int_0^{t_{\nu+1}} dt_{\nu+2} \ldots \int_0^{t_n} dt_n \times$$
$$\times \int_{t_{\nu+1}}^{t_{\nu-1}} [1, \ldots, \nu-1, \nu+1, \nu+2, \ldots, n+1][\nu]\, dt_{n+1}$$

and, changing the order of integration, we write this as

$$\int_0^T dt_1 \int_0^{t_1} dt_2 \ldots \int_0^{t_n} [1, \ldots, \nu-1, \nu+1, \nu+2, \ldots, n+1][\nu]\, dt_{n+1}.$$

Thus the product in question is

$$(-1)^{n+1} 4A_n A_1 = \int_0^T dt_1 \int_0^{t_1} dt_2 \ldots \int_0^{t_n} \sum_{\nu=1}^{n+1} [j_1, \ldots, j_n][\nu]\, dt_{n+1},$$

where $j_1, \ldots, j_n$ are the numbers $1, \ldots, n+1$, excluding ν, arranged in increasing order. With a view to further treatment of the sum figuring in this integral and formal simplification of the arguments, we divide throughout by $[1, \ldots, n+1]$, thus considering the sum

$$\sum_{\nu=1}^{n+1} \frac{[j_1, \ldots, j_n][\nu]}{[1, \ldots, n+1]} = S.$$

Hence

$$(-1)^{n+1} 4A_n A_1 = \int_0^T dt_1 \int_0^{t_1} dt_2 \ldots \int_0^{t_n} [1, \ldots, n+1]\, S\, dt_{n+1} \qquad (3.22)$$

where the detailed expression for S is

$$S = \frac{\psi(T) - \psi_2 + \psi_{n+1} e^{-aT}}{\psi(T) - \psi_1 + \psi_{n+1} e^{-aT}} \frac{\psi(T) - \psi_1 (1 - e^{-aT})}{\psi_1 - \psi_2} + \sum_{i=2}^{n} \frac{\psi_{i-1} - \psi_{i+1}}{\psi_{i-1} - \psi_i} \frac{\psi(T) - \psi_i (1 - e^{-aT})}{\psi_i - \psi_{i+1}} +$$
$$+ \frac{\psi(T) - \psi_{n+1} (1 - e^{-aT})}{\psi_n - \psi_{n+1}} \frac{\psi(T) - \psi_1 + \psi_n e^{-aT}}{\psi(T) - \psi_1 + \psi_{n+1} e^{-aT}}. \qquad (3.23)$$

The fundamental inequality for the terms of (3.16) will be established by estimating (3.23). We first consider a special case in which the estimate turns out to be sharper than in the general case.

3.2. Proof of fundamental inequality

Fundamental inequality in the case $r(t) \equiv 1$. We first assume that $r(t) \equiv 1$. This is the case for a three-term second-order linear equation (see §3.7). We have

$$\psi(t)=\frac{1-e^{-at}}{a},$$

and (3.23) is

$$S=(1-e^{-aT})\sum_{i=1}^{n}\frac{1-e^{-(\xi_{i-1}+\xi_i)}}{(1-e^{-\xi_{i-1}})(1-e^{-\xi_i})}$$

where

$$a(t_1-t_2)=\xi_1,\ \ldots,\ a(t_n-t_{n+1})=\xi_n,\quad a(T-t_1+t_{n+1})=\xi_{n+1}\qquad(\xi_0\equiv\xi_{n+1}).$$

Note that

$$\xi_j>0,\quad \xi_1+\ldots+\xi_{n+1}=aT. \tag{3.24}$$

To find the minimum of $S/(1-e^{-aT})$, we consider the function

$$\Phi(\xi_1,\ \ldots,\ \xi_{n+1})=\frac{S}{1-e^{-aT}}+\lambda(\xi_1+\ldots+\xi_{n+1}-aT)$$

and evaluate

$$\frac{\partial\Phi}{\partial\xi_\nu}=-\frac{2e^{-\xi_\nu}}{(1-e^{-\xi_\nu})^2}+\lambda\qquad(\nu=1,\ \ldots,\ n+1).$$

Equating these partial derivatives to zero and noting that $\lambda>0,\ e^{-\xi_\nu}<1$, we get

$$e^{-\xi_\nu}=1+\frac{1}{\lambda}-\sqrt{\left(1+\frac{1}{\lambda}\right)^2-1}.$$

Thus $\xi_1=\xi_2=\ldots=\xi_{n+1}$ and conditions (3.24) yield a unique stationary point

$$\xi_\nu=\frac{aT}{n+1}.$$

It remains to show that these values determine the smallest value of $S/(1-e^{-aT})$. We have

$$\frac{S}{1-e^{-aT}}=-(n+1)+2\sum_{\nu=1}^{n+1}\frac{1}{1-e^{-\xi_\nu}}.$$

Let us determine the minimum value of this expression in the domain (3.24). The boundary of the domain is defined by the $(n-1)$-dimensional hyperplanes

$$\left.\begin{aligned}\xi_1+\ldots+\xi_{n+1}&=aT,\\ \xi_\nu&=0\end{aligned}\right\}\quad(\nu=1,\ \ldots,\ n+1).$$

As one approaches the boundary, the expression $S/(1-e^{-aT})$ can only increase (and moreover without bound). This implies that the above values indeed yield the minimum value:

$$\inf\left[\frac{S}{1-e^{-aT}}\right]=(n+1)\frac{1-e^{-\frac{2aT}{n+1}}}{\left(1-e^{-\frac{aT}{n+1}}\right)^2}=(n+1)\operatorname{cth}\frac{aT}{2(n+1)}.$$

Referring to (3.22) and noting that by (3.21)

$$A_{n+1}=(-1)^{n+1}\frac{1}{2}\int_0^T dt_1\int_0^{t_1} dt_2\ldots\int_0^{t_n}[1,\ \ldots,\ n+1]\,dt_{n+1},$$

we obtain the inequality

$$2|A_n||A_1|>(n+1)(1-e^{-aT})|A_{n+1}|\operatorname{cth}\frac{aT}{2(n+1)}.$$

Replacing $n+1$ by n, we write this inequality

$$\left|\frac{A_n}{A_{n-1}}\right|<2(1-e^{-aT})^{-1}\operatorname{th}\frac{aT}{2n}\frac{|A_1|}{n} \qquad (n\geqslant 2). \tag{3.25}$$

This is the fundamental inequality for $r(t)\equiv 1$.

Note that if $0\leqslant a<\infty$

$$\frac{1}{n}\leqslant 2(1-e^{-aT})^{-1}\operatorname{th}\frac{aT}{2n}<2.$$

In the limiting case $a=0$, inequality (3.25) coincides with Lyapunov's inequality (26) /180g/.

F u n d a m e n t a l i n e q u a l i t y i n t h e g e n e r a l c a s e. In the general case, we first observe that the signs of ψ and $\dot{\psi}$ are the same as that of $r(t)$, hence as that of $a_{12}(t)$. Since the values of the independent variable decrease with increasing index, each of the fractions in (3.23) is greater than unity; therefore $S>n+1$. Referring to (3.22), we have

$$2|A_n||A_1|>(n+1)|A_{n+1}|.$$

Replacing $n+1$ by n, we write this inequality as

$$\left|\frac{A_n}{A_{n-1}}\right|<2\frac{|A_1|}{n} \qquad (n\geqslant 2). \tag{3.26}$$

One can prove an inequality which is sometimes sharper. Observe that

$$\frac{\psi(T)-\psi_2+\psi_{n+1}e^{-aT}}{\psi(T)-\psi_1+\psi_{n+1}e^{-aT}}\geqslant\frac{\psi(T)-\psi_2+\psi_{n+1}}{\psi(T)-\psi_1+\psi_{n+1}},$$

$$\frac{\psi(T)-\psi_1+\psi_n e^{-aT}}{\psi(T)-\psi_1+\psi_{n+1}e^{-aT}}[\psi(T)-\psi_{n+1}(1-e^{-aT})]\geqslant\frac{\psi(T)-\psi_1+\psi_n}{\psi(T)-\psi_1+\psi_{n+1}}\psi(T)e^{-aT},$$

$$\psi(T)-\psi_i(1-e^{-aT})\geqslant\psi(T)e^{-aT} \qquad (i=1,\ \ldots,\ n+1),$$

where equality can hold only if $a=0$. Denote

$$\psi_1-\psi_2=\eta_1, \ldots, \psi_n-\psi_{n+1}=\eta_n, \quad \psi(T)-\psi_1+\psi_{n+1}=\eta_{n+1}$$

and note that

$$\eta_1+\ldots+\eta_{n+1}=\psi(T). \tag{3.27}$$

Referring to (3.23), we obtain

$$S \geqslant \psi(T)e^{-aT}\sum_{i=1}^{n+1}\frac{\eta_{i-1}+\eta_i}{\eta_{i-1}\eta_i}.$$

To determine the minimum of the right-hand side, we consider the function

$$F(\eta_1, \ldots, \eta_{n+1})=\psi(T)e^{-aT}\sum_{i=1}^{n+1}\frac{\eta_{i-1}+\eta_i}{\eta_{i-1}\eta_i}+\lambda[\eta_1+\ldots+\eta_{n+1}-\varphi(T)]$$

and evaluate

$$\frac{\partial F}{\partial \eta_\nu}=-\frac{2\psi(T)e^{-aT}}{\eta_\nu^2}+\lambda \qquad (\nu=1, \ldots, n+1).$$

Equating the partial derivatives to zero and remembering that $\lambda>0$, $\eta_\nu>0$, we obtain

$$\eta_\nu=\sqrt{\frac{2\psi(T)e^{-aT}}{\lambda}} \qquad (\nu=1, \ldots, n+1).$$

Thus $\eta_1=\eta_2=\ldots=\eta_{n+1}$ and condition (3.27) yields a unique stationary point

$$\eta_\nu=\frac{1}{n+1}\psi(T).$$

The expression in which we are interested may be written as

$$2\psi(T)e^{-aT}\sum_{\nu=1}^{n+1}\frac{1}{\eta_\nu}.$$

Proceeding as in the treatment of the special case $r(t)\equiv 1$, we deduce from this that the stationary point indeed determines a minimum:

$$S \geqslant 2(n+1)^2 e^{-aT}.$$

By (3.22),

$$|A_n||A_1|>(n+1)^2 e^{-aT}|A_{n+1}|.$$

Replacing $n+1$ by n, we write this inequality as

$$\left|\frac{A_n}{A_{n-1}}\right| < e^{aT}\frac{|A_1|}{n^2} \qquad (n \geqslant 2). \tag{3.28}$$

If $a=0$ inequality (3.28) coincides with Lyapunov's inequality (26) /180g/. It is stronger than (3.26) for $n > \frac{1}{2}e^{aT}$, which is always the case if $aT < \ln 4$.

Inequalities (3.26) and (3.28) (and inequality (3.25) for $r(t) \equiv 1$) show that the upper bound for the quotient $|A_n/A_{n-1}|$ decreases monotonically with increasing n. However, as we saw in §2, when using Lyapunov's method to estimate the characteristic constant it is of major importance to have the fundamental inequality in the form of (2.15), for then one knows that the quotient itself (and not merely its upper bound!) decreases monotonically with increasing n. An inequality of this kind will be called a *strong fundamental inequality*.

To establish a strong fundamental inequality, we shall have to reduce the initial system of equations to canonical form.

3.3. Reduction of system to canonical form. Characteristic function of canonical system

We may assume that system (3.2) has already been brought (via the substitutions (3.6)) to the form of (3.9) with a triangular coefficient matrix. The substitution

$$\mathbf{v} = e^{-\frac{1}{2}at}\mathbf{x} \qquad \left(\mathbf{x} = \begin{pmatrix}\xi_1\\ \xi_2\end{pmatrix}\right) \tag{3.29}$$

brings system (3.9) to the form

$$\frac{d\mathbf{x}}{dt} = \left[\mathbf{B}(t) + \frac{1}{2}a\,\mathbf{I}_2\right]\mathbf{x}.$$

This system may be written as

$$\mathbf{J}\frac{d\mathbf{x}}{dt} = \mathbf{H}(t)\,\mathbf{x} \qquad \left(\mathbf{J} = \begin{Vmatrix} 0 & -1\\ 1 & 0\end{Vmatrix},\quad \mathbf{H}(t) = \begin{Vmatrix} -q(t) & \frac{1}{2}a\\ \frac{1}{2}a & r(t)\end{Vmatrix}\right), \tag{3.30}$$

so that it is a second-order canonical system of linear differential equations with periodic coefficients.

Of course, we could have derived system (3.30) directly from the initial system (3.2), via the transformation

$$x_1 = \xi_1 \exp\int_0^t \left[a_{11}(t_1) - \frac{1}{2}a\right]dt_1, \quad x_2 = \xi_2 \exp\int_0^t \left[a_{22}(t_1) + \frac{1}{2}a\right]dt_1,$$

which is simply the product of the transformations (3.6) and (3.29).

Proceeding as in §§1 and 2, we "provide" system (3.30) with a parameter λ, considering the canonical system

$$\left.\begin{aligned} &\mathbf{J}\frac{d\mathbf{x}}{dt}=[\mathbf{H}_0(t)+\lambda\mathbf{H}_1(t)]\,\mathbf{x},\\ &\mathbf{H}_0(t)=\begin{Vmatrix} 0 & \frac{1}{2}a \\ \frac{1}{2}a & r(t)\end{Vmatrix},\quad \mathbf{H}_1(t)=\begin{Vmatrix} -q(t) & 0 \\ 0 & 0\end{Vmatrix}.\end{aligned}\right\} \tag{3.31}$$

System (3.30) is system (3.31) with $\lambda=1$. Let $A^{(k)}(\lambda)$ denote the characteristic function of system (3.31):

$$A^{(k)}(\lambda)=\frac{1}{2}\mathrm{Tr}\,\mathbf{X}(T;\lambda)\qquad\left(\mathbf{J}\frac{d\mathbf{X}}{dt}=[\mathbf{H}_0+\lambda\mathbf{H}_1]\,\mathbf{X},\ \mathbf{X}(0;\lambda)=\mathbf{I}_2\right).$$

It is clear (see §1.1) that $A^{(k)}(\lambda)$ is an entire function of λ. We shall show that it is of order less than 1. System (3.31) may be written

$$\frac{d\mathbf{z}}{dt}=\mathbf{F}(t;\lambda)\,\mathbf{z}\qquad\left(\mathbf{z}=\begin{pmatrix}\sqrt{\lambda}\,\xi_1\\ \xi_2\end{pmatrix},\ \mathbf{F}(t;\lambda)=\begin{Vmatrix}\frac{1}{2}a & \sqrt{\lambda}\,r(t)\\ \sqrt{\lambda}\,q(t) & -\frac{1}{2}a\end{Vmatrix},\ \lambda>0\right). \tag{3.32}$$

The matrizant of system (3.32) is

$$\mathbf{Z}(t;\lambda)=\mathbf{I}_2+\int_0^t\mathbf{F}(t_1;\lambda)\,\mathbf{Z}(t_1;\lambda)\,dt_1,$$

since $\mathbf{Z}(0;\lambda)=\mathbf{I}_2$. The usual estimate of the norm (see §I. 1.2) gives

$$|\mathbf{Z}(T;\lambda)|\leqslant|\mathbf{I}_2|\,e^{\int_0^T|\mathbf{F}(t;\lambda)|\,dt}.$$

The norm of a matrix is bounded above by the sum of absolute values of its elements, and so, for $\lambda>0$,

$$|\mathbf{Z}(T;\lambda)|<c_1e^{c_2+c_3\sqrt{\lambda}},$$

where c_1, c_2, c_3 are positive constants. The matrizants $\mathbf{X}(T;\lambda)$ and $\mathbf{Z}(T;\lambda)$ satisfy the identity

$$\mathbf{X}(T;\lambda)=\begin{Vmatrix}\frac{1}{\sqrt{\lambda}} & 0\\ 0 & 1\end{Vmatrix}\mathbf{Z}(T;\lambda)\qquad(\lambda>0).$$

Consequently, for any $\varepsilon>0$ there exists $c(\varepsilon)$ such that

$$|A^{(k)}(\lambda)|<\exp\left[c(\varepsilon)\,\lambda^{\frac{1}{2}+\varepsilon}\right]\quad\text{as}\quad\lambda\to+\infty.$$

Hence the order of $A^{(k)}(\lambda)$ is at most 1/2. By Hadamard's theorem /91/, we have the representation

$$A^{(k)}(\lambda)=A_0^{(k)}\prod_{n=1}^{\infty}\left(1-\frac{\lambda}{\lambda_n}\right), \tag{3.33}$$

where $\lambda_1, \lambda_2, \ldots$ are the zeros of $A^{(k)}(\lambda)$.

We claim that if the functions $q(t)$ and $r(t)$ are of fixed sign the numbers λ_n are real and positive. If conditions (A) and (B) (§1.1) are satisfied, the fact that the zeros of $A^{(k)}(\lambda)$ are real will follow from part (a) of the theorem in §1.2.

We first offer a remark concerning condition (A). The proof of part (a) of the theorem of §1.2 goes through if the matrix $\mathbf{H}_1(t)$ is positive in the sense of its quadratic form; one assumes throughout that condition (B) is satisfied (see remark to the theorem of §1.2). In this section we shall understand condition (A) in an extended sense:

$$\mathbf{H}_1(t)\geqslant 0 \quad \text{or} \quad \mathbf{H}_1(t)\leqslant 0 \qquad (0\leqslant t\leqslant T).$$

The explicit form of the matrices $\mathbf{H}_0(t)$ and $\mathbf{H}_1(t)$ is given in (3.31). Thus the extended condition (A) means (see (3.8)) that

$$a_{21}(t)\leqslant 0 \quad \text{or} \quad a_{21}(t)\geqslant 0 \qquad (0\leqslant t\leqslant T). \tag{3.34}$$

We now consider condition (B). Let us solve the boundary-value problem

$$\mathbf{J}\frac{d\mathbf{x}_0}{dt}=\mathbf{H}_0(t)\,\mathbf{x}_0, \quad \mathbf{x}_0(T)=\rho\mathbf{x}_0(0) \qquad (|\rho|=1). \tag{3.35}$$

The general solution of the system $\mathbf{J}\,d\mathbf{x}/dt=\mathbf{H}_0(t)\,\mathbf{x}$, or, in expanded form,

$$\frac{d\xi_1}{dt}=\frac{1}{2}a\xi_1+r(t)\,\xi_2, \quad \frac{d\xi_2}{dt}=-\frac{1}{2}a\xi_2,$$

is

$$\xi_1=c_1e^{\frac{1}{2}at}+c_2e^{\frac{1}{2}at}\int_0^t e^{-at_1}r(t_1)\,dt_1, \quad \xi_2=c_2e^{-\frac{1}{2}at}.$$

The boundary conditions dictate that for ξ_2

$$c_2e^{-\frac{1}{2}aT}=\rho c_2 \qquad (|\rho|=1) \tag{3.36}$$

and if $a\neq 0$, as we assume for the moment, then $c_2=0$. The boundary condition for ξ_1 then implies that $c_1=0$ as well. Thus, if $a\neq 0$ problem (3.35) has no nontrivial solutions and thus condition (B) holds.

Now let $a=0$. Condition (3.36) implies that either

$$\text{a) } c_2\neq 0,\ \rho=1, \text{ or} \qquad \text{b) } c_2=0.$$

In case (a) the boundary condition in (3.35) for ξ_1 gives

$$\int_0^T r(t)\,dt = 0.$$

Henceforth we shall be interested only in the case that $a_{12}(t)$ (and hence also $r(t)$) is of fixed sign for $t \in [0, T]$. Then the above inequality implies that $r(t) \equiv 0$ for $t \in [0, T]$. This case is of no interest, for if $a = 0$ and $r(t) \equiv 0$, then system (3.30) is integrable.

In case (b) the boundary condition for ξ_1 in (3.35) can hold only if $\rho = 1$, and we have

$$\mathbf{x}_0 = \begin{pmatrix} c_1 \\ 0 \end{pmatrix}.$$

Now consider

$$\mathbf{H}_1(t)\,\mathbf{x}_0 = \begin{Vmatrix} -q(t) & 0 \\ 0 & 0 \end{Vmatrix} \begin{pmatrix} c_1 \\ 0 \end{pmatrix} = \begin{pmatrix} -q(t)\,c_1 \\ 0 \end{pmatrix}.$$

If $\mathbf{H}_1(t)\,\mathbf{x}_0 \equiv \mathbf{0}$, then either $q(t) \equiv 0$ for $t \in [0, T]$ or $c_1 = 0$. The first case is uninteresting, for if $a = 0$ and $q(t) \equiv 0$ system (3.30) is integrable. In the second case $\mathbf{x}_0 \equiv \mathbf{0}$ and so condition (B) is satisfied. Thus, if

$$q(t) \not\equiv 0 \qquad (0 \leqslant t \leqslant T),$$

condition (B) holds in case (b).

To summarize: if the functions $a_{12}(t)$ and $a_{21}(t)$ (and hence also $r(t)$ and $q(t)$) are of fixed sign and do not vanish identically for $0 \leqslant t \leqslant T$, then conditions (A) and (B) of §1.1 are satisfied. By virtue of (3.3), the assumption that the functions on the second diagonal of the coefficient matrix of system (3.1) are of fixed sign involves no loss of generality.

According to the theorem of §1.2, then all the zeros of the entire function $A^{(k)}(\lambda)$ are real.

3.4. Proof of strong fundamental inequality

We now wish to see, under the assumptions made at the end of the preceding subsection, when it can be stated that all the zeros of $A^{(k)}(\lambda)$ are positive. If we apply the substitution (3.29) to system (3.31), the result is system (3.10) with λ in place of $(-\varepsilon)$. The matrizants $\mathbf{V}(t;\ \varepsilon)$ and $\mathbf{X}(t;\ \lambda)$ of systems (3.10) and (3.31) satisfy the identity (see (3.29))

$$\mathbf{V}(t;\ \lambda) = e^{-\frac{1}{2}at}\,\mathbf{X}(t;\ -\lambda).$$

Thus, for the characteristic functions of these systems,

$$A(\lambda) = \frac{1}{2}\operatorname{Tr}\mathbf{V}(T;\ \lambda) = \frac{1}{2}e^{-\frac{1}{2}aT}\operatorname{Tr}\mathbf{X}(T;\ -\lambda) = e^{-\frac{1}{2}aT}A^{(k)}(-\lambda).$$

Formula (3.15) yields the following representation of $A^{(k)}(\lambda)$:

$$A^{(k)}(\lambda)=e^{\frac{1}{2}aT}[A_0-A_1\lambda+A_2\lambda^2-\ldots+(-1)^nA_n\lambda^n+\ldots]. \qquad (3.37)$$

What can be said of the sign of $A_\nu (\nu=1, 2, \ldots)$? It follows from (3.18) that $\psi(t)$ is a monotone nondecreasing function if $r(t)\geqslant 0$ and monotone nonincreasing if $r(t)\leqslant 0$ for $t\in[0, T]$. Referring to (3.20), we see that the expressions $(\psi_1-\psi_2), \ldots, (\psi_{n-1}-\psi_n), [\psi(T)-\psi_1+\psi_n e^{-aT}]$ are nonnegative if $r(t)\geqslant 0$ (nonpositive if $r(t)\leqslant 0$) and the integrand vanishes identically for all $0<t_n<t_{n-1}<\ldots<t_1<T$ only if $r(t)q(t)\equiv 0$ $(0\leqslant t\leqslant T)$. This is excluded, since then the system can be integrated interval by interval.

1) If $a_{12}(t)$ and $a_{21}(t)$ (and hence also $r(t)$ and $q(t)$) are of fixed sign on $[0, T]$, and their signs are different, it follows from (3.20) that

$$\operatorname{sign} A_\nu=+1 \qquad (\nu=0, 1, 2, \ldots). \qquad (3.38)$$

In that case neither $A^{(k)}(\lambda)$ nor, consequently, $A(\lambda)$ can have negative zeros, since when $\lambda<0$ the right-hand side of (3.37) is positive. Consequently, the numbers $\lambda_1, \lambda_2, \ldots$ in (3.33) are real and positive, so that by Krein's lemma (§2.2)

$$\frac{A_{n+1}}{A_n}\leqslant\frac{n}{n+1}\frac{A_n}{A_{n-1}} \qquad (n=1, 2, \ldots). \qquad (3.39)$$

Inequality (3.39) shows that the quotient of two consecutive terms of the alternating series (3.16) decreases monotonically with increasing n — this is what we have called the **strong fundamental inequality**. We reiterate that the proof is valid only provided the functions $a_{12}(t)$ and $a_{21}(t)$ have fixed but opposite signs.

2) If $a_{12}(t)$ and $a_{21}(t)$ are of the same fixed sign on $[0, T]$, then

$$\operatorname{sign} A_\nu=(-1)^\nu \qquad (\nu=0, 1, 2, \ldots)$$

and the terms of the series (3.16) are positive:

$$A(\lambda)=A_0+|A_1|\lambda+A_2\lambda^2+\ldots+|A_{2n-1}|\lambda^{2n-1}+A_{2n}\lambda^{2n}+\ldots$$

We then consider the function

$$A^*(\lambda)=A(-\lambda)=A_0-|A_1|\lambda+A_2\lambda^2-\ldots-|A_{2n-1}|\lambda^{2n-1}+A_{2n}\lambda^{2n}-\ldots$$

All the reasoning of the previous case is valid for the function $A^*(\lambda)$, since it cannot have negative zeros, and so we obtain the **strong fundamental inequality** in this case as

$$\left|\frac{A_{n+1}}{A_n}\right|\leqslant\frac{n}{n+1}\left|\frac{A_n}{A_{n-1}}\right| \qquad (n=1, 2, \ldots). \qquad (3.40)$$

3.5. Stability analysis

Since the exponents figuring in the substitution (3.6) satisfy the assumptions of Lemma VII in subsection 6 of /60/, the characteristic numbers of the solutions of system (3.1) and those of the transformed system (3.7) satisfy the identities

$$\chi\{x_1, x_2\}=\chi\{x, y\}-\frac{\alpha}{T}.$$

Since a system of equations with periodic coefficients is regular, it follows that if $\chi\{x, y\}>\alpha/T$ the unperturbed motion is stable, while if $\chi\{x, y\}<\alpha/T$ it is unstable (/60/, subsection 13).

The characteristic equation of system (3.7) is

$$\det[\mathbf{V}(T;\ -1)-\rho\mathbf{I}_2]=\rho^2-2A\rho+e^{\beta-\alpha}=0,$$

since $\mathrm{Tr}\,\mathbf{B}(t)=-a=(\beta-\alpha)/T$. Throughout the sequel we shall assume that the characteristic constant is given by the series (3.16). The roots of the characteristic equation are

$$\rho_{1,2}=A\pm\sqrt{A^2-e^{\beta-\alpha}}. \qquad (3.41)$$

If $A^2<e^{\beta-\alpha}$ the roots are complex and

$$|\rho_1|=|\rho_2|=e^{\frac{1}{2}(\beta-\alpha)};$$

if $A^2\geqslant e^{\beta-\alpha}$ they are real and the root of maximum absolute value is

$$|\rho|_{\max}=|A|+\sqrt{A^2-e^{\beta-\alpha}}.$$

The characteristic number of the solutions of system (3.7) is (/60/, subsection 46)

$$\chi\{x, y\}=-\frac{1}{T}\operatorname{Re}\ln|\rho|_{\max}.$$

Thus, if $\operatorname{Re}\ln|\rho|_{\max}<-\alpha$, the unperturbed motion of the dynamic system whose perturbed motion is described in the first approximation by (3.1) is stable, and if $\operatorname{Re}\ln|\rho|_{\max}>-\alpha$ it is unstable.

Recall that, throughout,

$$\beta\leqslant\alpha\leqslant-\beta \qquad (\beta\leqslant 0), \qquad (3.5)$$

as shown in §3.1.

(1) If $|A|\leqslant e^{\frac{1}{2}(\beta-\alpha)}$, then $\ln|\rho|_{\max}=\frac{1}{2}(\beta-\alpha)$ and if

$$\frac{1}{2}(\beta-\alpha)<-\alpha,$$

i. e., $\alpha+\beta<0$, the unperturbed motion is stable (and unstable for $\alpha+\beta>0$; see also §3.1). The case $\alpha+\beta=0$ requires special consideration (see §3.11).

(2) If $|A|>e^{\frac{1}{2}(\beta-\alpha)}$, then if

$$\operatorname{Re}\ln[|A|+\sqrt{A^2-e^{\beta-\alpha}}]<-\alpha$$

the unperturbed motion is stable. This may also be written as

$$\sqrt{A^2-e^{\beta-\alpha}}<e^{-\alpha}-|A| \tag{3.42}$$

and a necessary condition for this inequality to hold is

$$|A|<e^{-\alpha}.$$

This condition is consistent with our assumptions only provided $\alpha+\beta<0$ and this in turn is possible (by (3.5)) only if $\beta<0$. Under these conditions inequality (3.42) may be squared, and, solving for $|A|$, we get

$$|A|<\frac{1}{2}(e^{-\alpha}+e^{\beta}).$$

The right-hand side of this inequality is bounded above by $e^{-\alpha}$. Thus, if

$$|A|<\frac{1}{2}(e^{-\alpha}+e^{\beta}) \qquad (\alpha+\beta<0), \tag{3.43}$$

the unperturbed motion is stable.

If

$$\operatorname{Re}\ln\left[|A|+\sqrt{A^2-e^{\beta-\alpha}}\right]>-\alpha,$$

or, equivalently,

$$\sqrt{A^2-e^{\beta-\alpha}}>e^{-\alpha}-|A|, \tag{3.44}$$

the unperturbed motion is unstable. This inequality is certainly true if

$$|A|\geqslant e^{-\alpha},$$

and this in turn is always true in the present case if $\alpha+\beta\geqslant 0$.

Assuming now that $\alpha+\beta<0$ and $|A|<e^{-\alpha}$, let us square inequality (3.44); solving for $|A|$, we obtain

$$|A|>\frac{1}{2}(e^{-\alpha}+e^{\beta}).$$

Thus, if either

$$\alpha+\beta\geqslant 0,\ |A|>e^{\frac{1}{2}(\beta-\alpha)}, \quad \text{or} \quad \alpha+\beta<0,\ |A|>\frac{1}{2}(e^{-\alpha}+e^{\beta}),$$

the unperturbed motion is unstable. As for the lower bound $\frac{1}{2}(e^{-\alpha}+e^{\beta})$, if $\alpha+\beta<0$ it is at least

$$A_0=\frac{1}{2}(1+e^{\beta-\alpha})$$

for $\alpha\leqslant 0$ and less than A_0 if $\alpha>0$.

We can summarize the above results as follows.

Lemma 1. If $\alpha+\beta>0$, the unperturbed motion is unstable if $|A|>\frac{1}{2}(e^{-\alpha}+e^{\beta})$; the interval $|A|<\frac{1}{2}(e^{-\alpha}+e^{\beta})$ requires special consideration (see §3.11).

If $\alpha+\beta<0$, the unperturbed motion is stable if $|A|<\frac{1}{2}(e^{-\alpha}+e^{\beta})$ and unstable if $|A|>\frac{1}{2}(e^{-\alpha}+e^{\beta})$.

We now wish to ascertain whether Lyapunov's method can be used to estimate the characteristic constant A of system (3.7).

We first consider the case when the functions $a_{12}(t)$ and $a_{21}(t)$ have different signs. As shown in §3.4, we then have $A_n>0$ $(n=0, 1, 2, \ldots)$ and (3.16) is an alternating series. Proceeding as in §2.3 we deduce from (3.39) that the terms A_n of (3.16) have only one maximum: as n increases from $n=0$ to some n_0 $(n_0=0, 1, 2, \ldots)$, A_n increases, and for $n>n_0$ it tends monotonically to zero.

For each natural number n, set

$$S_n=A_0-A_1+\ldots+(-1)^nA_n$$

so that

$$A=S_n-(A_{n+1}-A_{n+2})-(A_{n+3}-A_{n+4})-\ldots \quad \text{for even } n.$$
$$A=S_n+(A_{n+1}-A_{n+2})+(A_{n+3}-A_{n+4})+\ldots \quad \text{for odd } n.$$

Therefore, if $n+1\geqslant n_0$ it follows from the choice of n_0 that

$$A<S_n \quad \text{for even } n, \quad A>S_n \quad \text{for odd } n$$

and thus

$$\text{if } S_n\leqslant\frac{1}{2}(e^{-\alpha}+e^{\beta}) \quad (n \text{ even}), \text{ then } A<\frac{1}{2}(e^{-\alpha}+e^{\beta}), \tag{3.45}$$
$$\text{if } S_n\leqslant-\frac{1}{2}(e^{-\alpha}+e^{\beta}) \quad (n \text{ even}), \text{ then } A<-\frac{1}{2}(e^{-\alpha}+e^{\beta}), \tag{3.46}$$
$$\text{if } S_n\geqslant\frac{1}{2}(e^{-\alpha}+e^{\beta}) \quad (n \text{ odd}), \text{ then } A>\frac{1}{2}(e^{-\alpha}+e^{\beta}), \tag{3.47}$$
$$\text{if } S_n\geqslant-\frac{1}{2}(e^{-\alpha}+e^{\beta}) \quad (n \text{ odd}), \text{ then } A>-\frac{1}{2}(e^{-\alpha}+e^{\beta}). \tag{3.48}$$

As in §2.3, these inequalities follow from the assumption $n_0\leqslant n+1$. We now ask: in what cases does one of inequalities (3.45)–(3.48) imply the inequality $n_0\leqslant n+1$?

Lemma 2. Inequalities (3.46) and (3.47) are valid for any natural number n, irrespective of whether $n_0 \leqslant n+1$ or not. Inequalities (3.45) and (3.48) are valid for any natural number n if

$$0 \leqslant \alpha \leqslant -\beta \qquad (\beta \leqslant 0),$$

again irrespective of whether $n_0 \leqslant n+1$ or not.

Proof. (1) First let $n=1$ and

$$S_1 = A_0 - A_1 \geqslant \frac{1}{2}(e^{-\alpha}+e^{\beta}), \text{ i.e. }, A_1 \leqslant A_0 - \frac{1}{2}(e^{-\alpha}+e^{\beta}) < A_0.$$

The inequality $A_1 < A_0$ means that $n_0 = 0$. Thus condition (3.47) ($n=1$) implies that $n_0 = 0$, so that it is certainly true that $n_0 \leqslant n+1 = 2$.

(2) Again let $n=1$ and

$$S_1 = A_0 - A_1 \geqslant -\frac{1}{2}(e^{-\alpha}+e^{\beta}), \quad \text{i. e.,} \quad A_1 \leqslant A_0 + \frac{1}{2}(e^{-\alpha}+e^{\beta}) = \\ = \frac{1}{2}(e^{-\alpha}+e^{\beta}+e^{\beta-\alpha}+1). \quad (3.49)$$

Hence it follows from (3.39) for $n=1$ that

$$A_2 \leqslant \frac{1}{2}\frac{A_1}{A_0}A_1 \leqslant \frac{1}{2}(e^{-\alpha}+e^{\beta}+e^{\beta-\alpha}+1)(1+e^{\beta-\alpha})^{-1}A_1. \quad (3.50)$$

We wish to determine in what cases the coefficient of A_1 on the right is at most unity, i. e., when it is true that

$$(1-e^{-\alpha})(1-e^{\beta}) \geqslant 0. \quad (3.51)$$

If $\beta = 0$ this is true, but if $\beta < 0$ the number α must be nonnegative. Since we may assume without loss of generality that α lies in the interval (3.5), we see that if

$$0 \leqslant \alpha \leqslant -\beta \qquad (\beta \leqslant 0) \quad (3.52)$$

then it follows from (3.50) that $A_2 \leqslant A_1$; hence $n_0 \leqslant 1$, so that $n_0 < n+1$.

(3) Now let n be an odd number greater than 1, and suppose that $S_n \geqslant \frac{1}{2}(e^{-\alpha}+e^{\beta})$, i. e.,

$$\left(A_0 - A_1 - \frac{1}{2}e^{-\alpha} - \frac{1}{2}e^{\beta}\right) + (A_2 - A_3) + \ldots + (A_{n-1} - A_n) \geqslant 0. \quad (3.53)$$

Let $n_0 \geqslant n$. Then, by the choice of n_0, all the parenthetical terms in (3.53) are negative. Thus inequality (3.53) is impossible if $n_0 \geqslant n$, and so $n_0 \leqslant n-1$.

(4) Let n be an odd number greater than 1 and suppose that $S_n \geqslant -\frac{1}{2}(e^{-\alpha}+e^{\beta})$, i.e.,

$$\left(A_0 - A_1 + \frac{1}{2}e^{-\alpha} + \frac{1}{2}e^{\beta}\right) + (A_2 - A_3) + \ldots + (A_{n-1} - A_n) \geqslant 0. \quad (3.54)$$

Suppose that $n_0 \geqslant n$. Then again all the terms in parentheses, starting from the second, are negative. If inequality (3.52) is true, the first term is also negative, for otherwise

$$A_0 - A_1 + \frac{1}{2}(e^{-\alpha} + e^{\beta}) \geqslant 0,$$

and it would follow as in (2) above that $n_0 \leqslant 1$. In this case inequality (3.54) is impossible if $n_0 \geqslant n$, and so $n_0 \leqslant n-1$.

(5) Now let n be an even number and $S_n \leqslant -\frac{1}{2}(e^{-\alpha}+e^{\beta})$, i. e.,

$$\left(A_0 + \frac{1}{2}e^{-\alpha} + \frac{1}{2}e^{\beta}\right) + (-A_1 + A_2) + \ldots + (-A_{n-1} + A_n) \leqslant 0.$$

If we assume that $n_0 \geqslant n$, we at once obtain a contradiction, for then all parenthetical terms on the left of this inequality are positive. Thus $n_0 \leqslant n-1$.

(6) Finally, let n be even and assume that $S_n \leqslant \frac{1}{2}(e^{-\alpha}+e^{\beta})$, i. e.,

$$\left(A_0 - \frac{1}{2}e^{-\alpha} - \frac{1}{2}e^{\beta}\right) + (-A_1 + A_2) + \ldots + (-A_{n-1} + A_n) \leqslant 0. \qquad (3.54a)$$

Suppose that $n_0 \geqslant n$. Then all terms on the left, starting from the second, are positive. The first term is equal to

$$\frac{1}{2}(1-e^{-\alpha})(1-e^{\beta})$$

which is nonnegative because of (3.52). Thus, if

$$0 \leqslant \alpha \leqslant -\beta \qquad (\beta \leqslant 0)$$

then inequality (3.54a) is impossible with $n_0 \geqslant n$; hence $n_0 \leqslant n-1$. This completes the proof of the lemma.

Stability theorem. Let equations (3.1) describe, in the first approximation, the perturbed motion of a dynamic system, and assume that its coefficients $a_{12}(t)$, $a_{21}(t)$ are of different but fixed signs, and $a_{12}(t)a_{21}(t) \not\equiv 0$ for $0 \leqslant t \leqslant T$. Let*

$$S_n = A_0 - A_1 + \ldots + (-1)^n A_n,$$

where A_0, A_1, A_2, ... are given by formulas (3.15), (3.19), (3.20), (3.18), (3.8), and

$$\alpha = \int_0^T a_{11}(t)\,dt, \quad \beta = \int_0^T a_{22}(t)\,dt.$$

1. *If either of the conditions*

* If $a_{12}(t)a_{21}(t) \equiv 0$, system (3.1) is integrable in closed form over each interval in which either $a_{12}(t) \equiv 0$ or $a_{21}(t) \equiv 0$.

$$S_{2\nu}\leqslant-\frac{1}{2}(e^{-\alpha}+e^{\beta}), \tag{3.46}$$

$$S_{2\nu-1}\geqslant\frac{1}{2}(e^{-\alpha}+e^{\beta}) \qquad (\nu=1, 2, \ldots) \tag{3.47}$$

holds, the unperturbed motion is unstable.

2. *If* $0\leqslant\alpha<-\beta$ $(\beta<0)$ *and both inequalities*

$$S_{2\nu}\leqslant\frac{1}{2}(e^{-\alpha}+e^{\beta}),\ S_{2\mu-1}\geqslant-\frac{1}{2}(e^{-\alpha}+e^{\beta}) \tag{3.55}$$

(where ν *and* μ *are arbitrary natural numbers) hold, then the unperturbed motion is stable.*

This theorem follows in an obvious way from Lemmas 1 and 2.

3.6. Sufficient conditions for stability and instability when $a_{12}(t)$ and $a_{21}(t)$ have different signs

Relying on the theorem of §3.5, we now derive the sufficient conditions for stability and instability that follow from Lyapunov's method; these conditions require knowledge of the first three terms in the series (3.16) (not counting the "zeroth").

(1) Let $\nu=1$ in inequality (3.47). Then: *if*

$$A_1\leqslant\frac{1}{2}(1-e^{-\alpha})(1-e^{\beta}), \tag{3.56}$$

the unperturbed motion of the dynamic system whose perturbed motion is described in the first approximation by system (3.1) is unstable.

(2) Setting $\nu=1$ in inequality (3.46), we obtain: *if*

$$A_2\leqslant A_1-\frac{1}{2}(1+e^{-\alpha})(1+e^{\beta}), \tag{3.57}$$

the unperturbed motion is unstable.

(3) Finally, setting $\nu=2$ in inequality (3.47) we obtain: *if*

$$A_2\geqslant\frac{1}{2}(e^{-\alpha}-1)(1-e^{\beta})+A_1+A_3, \tag{3.58}$$

the unperturbed motion is unstable.

It is worth noting that the numbers α and β are not subjected to any restrictions other than the inequalities

$$\beta\leqslant\alpha\leqslant-\beta \qquad (\beta\leqslant 0),$$

which, as we ascertained in §3.1, involve no loss of generality.

We now proceed to sufficient conditions for stability. Recall that we are now assuming

$$\beta \leqslant \alpha < -\beta \qquad (\beta < 0),$$

postponing the subcase $\alpha+\beta=0$ to §3.11.

(4) Setting $\nu=1, \mu=1$ in inequalities (3.55), we obtain: *if*

$$A_2 - \frac{1}{2}(e^{-\alpha}-1)(1-e^{\beta}) \leqslant A_1 \leqslant \frac{1}{2}(1+e^{-\alpha})(1+e^{\beta}), \qquad 0 \leqslant \alpha < -\beta \quad (3.59)$$

the unperturbed motion is stable.

(5) Finally, setting $\nu=1, \mu=2$ in (3.55), we find: *if*

$$A_3 + A_1 - \frac{1}{2}(1+e^{-\alpha})(1+e^{\beta}) \leqslant A_2 \leqslant A_1 + \frac{1}{2}(e^{-\alpha}-1)(1-e^{\beta}), \qquad 0 \leqslant \alpha < -\beta \quad (3.60)$$

the unperturbed motion is stable.

We now discuss the case not covered by inequalities (3.55): $\alpha < 0$.

(6) We first demand that $A_0 \leqslant \frac{1}{2}(e^{-\alpha}+e^{\beta})$, i. e.,

$$(1-e^{-\alpha})(1-e^{\beta}) \leqslant 0,$$

which may hold if $\alpha \leqslant 0$. It follows from the equality

$$A = A_0 - (A_1 - A_2) - (A_3 - A_4) - \ldots$$

that if $A_2 \leqslant A_1$ then $A < \frac{1}{2}(e^{-\alpha}+e^{\beta})$. Inequalities (3.26) and (3.28) ($n=2$) give

$$\frac{A_2}{A_1} < A_1 \quad \text{and} \quad \frac{A_2}{A_1} < \frac{1}{4}e^{\alpha-\beta} A_1,$$

so that if $A_1 \leqslant 1$ or $A_1 \leqslant 4e^{\beta-\alpha}$, then $A_2 < A_1$. We now demand that $S_1 \geqslant -\frac{1}{2}(e^{-\alpha}+e^{\beta})$, i. e.,

$$A_1 \leqslant \frac{1}{2}(1+e^{-\alpha})(1+e^{\beta}).$$

It follows from the equality

$$A = S_1 + (A_2 - A_3) + (A_3 - A_4) + \ldots,$$

that if $A_2 \leqslant A_1$ then $A > -\frac{1}{2}(e^{-\alpha}+e^{\beta})$. Combining all these findings, we obtain: *if*

$$\alpha \leqslant 0, \quad A_1 \leqslant \min\left[\max(1;\ 4e^{\beta-\alpha});\ \frac{1}{2}(1+e^{-\alpha})(1+e^{\beta})\right], \qquad (3.61)$$

the unperturbed motion is stable.

The upper bound for A_1 must be chosen as large as possible. Therefore, if

$$8e^{\beta-\alpha} \geqslant (1+e^{-\alpha})(1+e^{\beta})$$

then the last inequality of (3.61) may be written

$$A_1 \leqslant \frac{1}{2}(1+e^{-\alpha})(1+e^{\beta}),$$

and if

$$8e^{\beta-\alpha} \leqslant (1+e^{-\alpha})(1+e^{\beta})$$

the inequality becomes

$$A_1 \leqslant \begin{cases} 4e^{\beta-\alpha} & \text{if} \quad \alpha-\beta \leqslant \ln 4, \\ 1 & \text{if} \quad \alpha-\beta \geqslant \ln 4. \end{cases}$$

We again consider the case $\beta \leqslant \alpha < 0$.

(7) Setting $n=3$ in inequalities (3.26) and (3.28), we see that if

$$A_1 \leqslant \frac{3}{2} \quad \text{or} \quad A_1 \leqslant 9e^{\beta-\alpha},$$

then necessarily $A_2 > A_3 > A_4 > \ldots$

If $A_2 \leqslant \frac{3}{2} A_1$, it follows from (3.39) ($n=2$) that $A_2 \geqslant A_3 > A_4 > \ldots$

Not wishing for the moment to use the coefficient A_3 in estimating A, we observe that in each of these cases

$$A_0 - A_1 < A < A_0 - A_1 + A_2.$$

Therefore, if $A_0 - A_1 \geqslant -\frac{1}{2}(e^{-\alpha}+e^{\beta})$, i. e.,

$$A_1 \leqslant \frac{1}{2}(1+e^{-\alpha})(1+e^{\beta}),$$

then also $A > -\frac{1}{2}(e^{-\alpha}+e^{\beta})$. Similarly if $A_0 - A_1 + A_2 \leqslant \frac{1}{2}(e^{-\alpha}+e^{\beta})$, i. e.,

$$A_2 - \frac{1}{2}(e^{-\alpha}-1)(1-e^{\beta}) \leqslant A_1,$$

then also $A < \frac{1}{2}(e^{-\alpha}+e^{\beta})$.

Thus: *if one of the inequalities*

$$A_1 \leqslant \frac{3}{2}, \quad or \quad A_1 \leqslant 9e^{\beta-\alpha}, \quad or \quad A_2 \leqslant \frac{3}{2} A_1 \tag{3.62}$$

is satisfied, and also

$$A_2 - \frac{1}{2}(e^{-\alpha}-1)(1-e^{\beta}) \leqslant A_1 \leqslant \frac{1}{2}(1+e^{-\alpha})(1+e^{\beta}), \tag{3.63}$$

then the unperturbed motion is stable.

(8) Setting $n=4$ in (3.26) and (3.28), we find that if

$$A_1 \leqslant 2 \quad \text{or} \quad A_1 \leqslant 16e^{\beta-\alpha},$$

then necessarily $A_3 > A_4 > A_5 > \ldots$

If $A_3 \leqslant \frac{4}{3} A_2$, it follows from inequality (3.39) ($n=3$) that $A_3 \geqslant A_4 > A_5 > \ldots$

In either of these cases,

$$A_0 - A_1 + A_2 - A_3 < A < A_0 - A_1 + A_2.$$

Thus: *if one of the inequalities*

$$A_1 \leqslant 2, \ \textit{or} \quad A_1 \leqslant 16e^{\beta-\alpha}, \quad \textit{or} \quad A_3 \leqslant \frac{4}{3} A_2$$

is satisfied, and also

$$A_1 + A_3 - \frac{1}{2}(1+e^{-\alpha})(1+e^{\beta}) \leqslant A_2 \leqslant A_1 + \frac{1}{2}(e^{-\alpha}-1)(1-e^{\beta}),$$

then the unperturbed motion is stable.

We remind the reader here that we are referring throughout to the unperturbed motion of a dynamic system whose perturbed motion is described in the first approximation by system (3.1).

Example. Consider a system whose perturbed motion is governed in the first approximation by equations

$$\frac{dx_1}{dt} = x_2, \quad \frac{dx_2}{dt} = -\gamma x_1 - a(1+\delta \cos t)\, x_2 \qquad (\gamma > 0, \ a > 0, \ \delta > 0).$$

We have (see §3.1)

$$\alpha = 0, \quad \beta = -2\pi a, \quad r(t) = e^{-\delta \sin t}, \quad q(t) = -\gamma e^{\delta \sin t}.$$

We expand the functions r and q in Fourier series:

$$r(t) = \sum_{k=0}^{\infty} (-1)^k [a_{2k} \cos 2kt - a_{2k+1} \sin (2k+1)\, t],$$

$$q(t) = -\gamma \sum_{k=0}^{\infty} (-1)^k [a_{2k} \cos 2kt + a_{2k+1} \sin (2k+1)\, t]$$

where the coefficients a_k are defined by power series (see below, (3.87)) in δ. We determine A_1 from (3.17):

$$A_1 = \frac{1}{2} \pi a \sigma \gamma (1 - e^{-2\pi a}) \qquad \left(\sigma = 2\frac{a_0^2}{a^2} - \sum_{k=1}^{\infty} \left[\frac{a_{2k-1}^2}{a^2+(2k-1)^2} - \frac{a_{2k}^2}{a^2+4k^2}\right]\right).$$

The stability domain of the unperturbed motion in $a\gamma\delta$-space is defined by the inequalities obtained from (3.61):

$$\gamma \leqslant \frac{2}{\pi a\sigma \operatorname{th} \pi a} \quad \text{for} \quad a \leqslant \frac{1}{2\pi} \ln 3,$$
$$\gamma \leqslant \frac{8}{\pi a\sigma (1+e^{2\pi a}) \operatorname{th} \pi a} \quad \text{for} \quad \frac{1}{2\pi} \ln 3 \leqslant a \leqslant \frac{1}{2\pi} \ln 4,$$
$$\gamma \leqslant \frac{2}{\pi a\sigma (1+e^{-2\pi a}) \operatorname{th} \pi a} \quad \text{for} \quad a \geqslant \frac{1}{2\pi} \ln 4.$$

Note that the sufficient condition (14) of /204a/ applied to this example gives $\delta < 1$.

3.7. Mechanical system with one degree of freedom, constant dissipation of energy and periodic restoring force

The perturbed motion of a system of the type described in the heading is given by an equation of type (3.7) with $r(t) \equiv 1$:

$$\ddot{x} + a\dot{x} + p(t)x = 0 \qquad (p(t+T) = p(t) = -q(t) \geqslant 0,\ a > 0). \tag{3.64}$$

We introduce dimensionless time:

$$t = \frac{T}{2\pi}\tau$$

and, to bring out the role of the parameters, write equation (3.64) as

$$\frac{d^2x}{d\tau^2} + \alpha\frac{dx}{d\tau} + [\mu + \nu\varphi(\tau)]\,x = 0, \tag{3.65}$$

where

$$\alpha = \frac{T}{2\pi}a, \quad \mu + \nu\varphi(\tau) = \frac{T^2}{4\pi^2}p\left(\frac{T}{2\pi}\tau\right), \quad \varphi(\tau + 2\pi) = \varphi(\tau).$$

The characteristic equation is

$$\rho^2 - 2A\rho + e^{-2\pi a} = 0, \tag{3.66}$$

where the free term is determined using Liouville's formula (II. 1.11), and the characteristic constant A is an entire function of the parameters α, μ, ν. This follows from the fact that

$$2A = \operatorname{Tr}\mathbf{X}(2\pi),$$

where the monodromy matrix $\mathbf{X}(2\pi)$ (the value of the matrix $\mathbf{X}(\tau)$ at $\tau = 2\pi$) is an entire function of α, μ, ν (see § 1.1). The roots of equation (3.66) are

$$\rho_{1,2} = A \mp \sqrt{A^2 - e^{-2\pi a}}.$$

If

$$|A| > \frac{1}{2}(1+e^{-2\pi a}),$$

then one of the roots is greater than 1 in absolute value and the unperturbed motion is unstable; but if

$$|A| < \frac{1}{2}(1+e^{-2\pi a}),$$

then both roots are of absolute value at most 1 and the unperturbed motion is asymptotically stable. Since $\rho_1\rho_2 = e^{-2\pi a}$ and the coefficients of the characteristic equation (3.66) are real, it follows that the roots are symmetric with respect to both the real axis and the circle $|\rho| = e^{-\pi a}$. Thus the roots are either on the real axis and symmetric with respect to the circle $|\rho| = e^{-\pi a}$, or on the circle itself, symmetric with respect to the real axis (Figure 45).

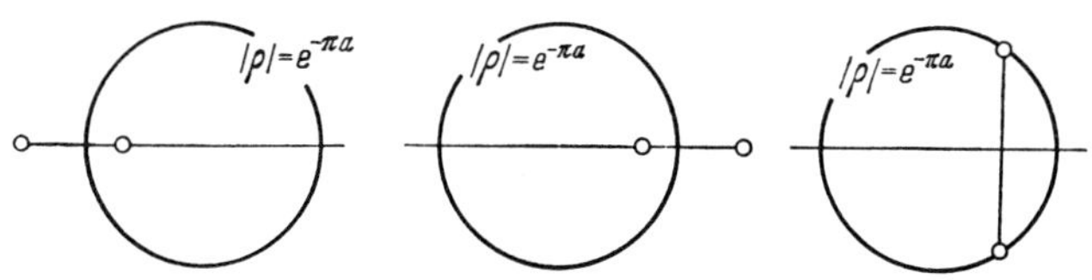

FIGURE 45.

In the limiting case for stability, we have $|A| = \frac{1}{2}(1+e^{-2\pi a})$. Then $\rho_1 = \pm 1$, $\rho_2 = \pm e^{-2\pi a}$. But this means (see § II. 2.7) that in the limiting case (remember that $a > 0$!) one solution is periodic or antiperiodic and the other decays.

Consider the parameter space (α, μ, ν). We shall call (α_0, μ_0, ν_0) a *stability point* if all solutions of the equation

$$\frac{d^2x}{d\tau^2} + \alpha_0\frac{dx}{d\tau} + [\mu_0 + \nu_0\varphi(\tau)]\,x = 0 \tag{3.67}$$

tend to zero as $\tau \to +\infty$(the trivial solution of equation (3.67) is asymptotically stable), and an *instability point* if at least one solution of the equation is unbounded as $\tau \to +\infty$ (the trivial solution is unstable). If the inequality

$$|A(\alpha, \mu, \nu)| < \frac{1}{2}(1+e^{-2\pi a}) \tag{3.68}$$

is false, then (α, μ, ν) is a stability point, and if the opposite inequality is true (α, μ, ν) is an instability point.

The common boundary $|A(\alpha, \mu, \nu)| = \frac{1}{2}(1+e^{-2\pi a})$ of these sets is the set of all points $(\alpha_0^*, \mu_0^*, \nu_0^*)$ for which one solution of equation (3.67) is periodic or antiperiodic while the other solution tends to zero as $\tau \to +\infty$.

If $a=0$, equation (3.65) reduces to (2.1). Recall that at $a=0$ the trivial solution of equation (2.1) cannot be asymptotically stable.

The familiar substitution

$$x = ye^{-\frac{1}{2}\alpha\tau}$$

reduces equation (3.65) to Hill's equation

$$\frac{d^2y}{d\tau^2} + [\mu^* + \nu\varphi(\tau)]\, y = 0 \qquad \left(\mu^* = \mu - \frac{1}{4}\alpha^2\right). \tag{3.69}$$

If the trivial solution of equation (3.64) is to be asymptotically stable, all solutions of equation (3.69) must satisfy the inequality

$$|y| < \text{const}\, e^{\left(\frac{1}{2}\alpha - \delta\right)\tau},$$

where $\delta > 0$. Hence, by the discussion in § II. 2.7, the roots of the characteristic equation of equation (3.69) (the multipliers of the latter) must satisfy the inequality

$$|\rho| < e^{\frac{1}{2}\alpha T}.$$

Consider the (μ^*, ν)-plane. At $\alpha=0$ equation (3.69) coincides with (2.1). The boundaries of the stability domains for this case are indicated by dashed lines in Figure 46. The full lines indicate the boundaries of the stability domains for some small fixed $\alpha > 0$, and the domains themselves are hatched.

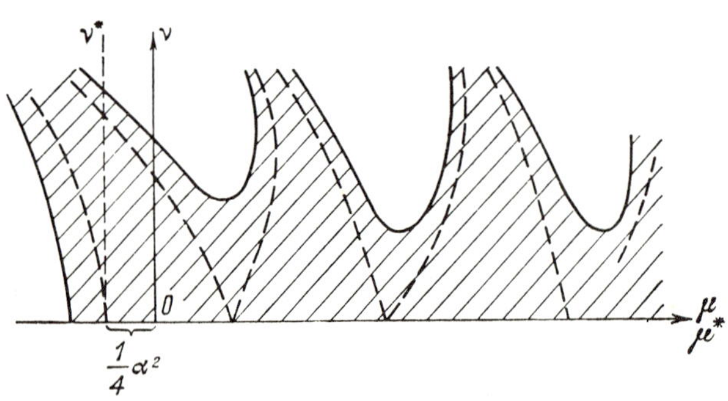

FIGURE 46.

3.8. Mechanical system with one degree of freedom, constant dissipation of energy and periodic restoring force. Sufficient conditions for stability and instability. Two examples

We begin with the equations of perturbed motion for a system of the above type, in the form of (2.64):

$$\ddot{x}+a\dot{x}+p(t)\,x=0 \qquad (p(t+T)=p(t)\geqslant 0,\ a>0). \tag{3.70}$$

The roots of the characteristic equation (3.41) are less than unity in absolute value only if $|A|<{}^1/_2(1+e^{-aT})$. Under this assumption, the unperturbed motion of the mechanical system is asymptotically stable (/60/, Theorem II, §55); but if $|A|>{}^1/_2(1+e^{-aT})$, one of the roots is greater than unity in absolute value and the unperturbed motion is unstable (/102/, §72).

In this case, it is convenient to base the successive evaluation of $A_1,\ A_2,\ A_3,\ \ldots$ not on formulas (3.20) but on (3.14); integration of the latter by parts gives

$$x_n(t)=\frac{1}{a}\int_0^t p_1x_{n-1}(t_1)\,dt_1-\frac{1}{a}e^{-at}\int_0^t e^{at_1}p_1x_{n-1}(t_1)\,dt_1,$$

$$y_n(t)=e^{-at}\int_0^t e^{at_1}p_1\,dt_1\int_0^{t_1} y_{n-1}(t_2)\,dt_2.$$

Hence, by (3.15),

$$A_n=\frac{1}{2a}\left[\int_0^T p_1x_{n-1}(t_1)\,dt_1-e^{-aT}\int_0^T e^{at_1}p_1x_{n-1}(t_1)\,dt_1+\right.$$
$$\left.+\,ae^{-aT}\int_0^T e^{at_1}p_1\,dt_1\int_0^{t_1} y_{n-1}(t_2)\,dt_2\right].$$

The first three terms of the series (3.16) are

$$\left.\begin{aligned}
A_1&=\frac{1-e^{-aT}}{2a}\int_0^T p\,dt,\\
A_2&=\frac{1+e^{-aT}}{4a^2}\left(\int_0^T p\,dt\right)^2-\frac{e^{-aT}}{2a^2}\int_0^T e^{-at}p\,dt\int_0^T e^{at}p\,dt-\\
&\qquad-\frac{1-e^{-aT}}{2a^2}\int_0^T e^{-at}p\,dt\int_0^t e^{at_1}p_1\,dt_1,\\
A_3&=\frac{1-e^{-aT}}{2a^3}\int_0^T p\,dt\int_0^t p_1\,dt_1\int_0^{t_1} p_2\,dt_2+\frac{1}{2a^3}[-K_1(p;\ T,\ a)+\\
&+e^{-aT}K_1(p;\ T,\ -a)+K_2(p;\ T,\ a)-e^{-aT}K_2(p;\ T,\ -a)-\\
&\qquad-K_3(p;\ T,\ a)+e^{-aT}K_3(p;\ T,\ -a)],
\end{aligned}\right\} \tag{3.71}$$

where

$$K_1(p;\ T,\ a)=\int_0^T p\,dt\int_0^t e^{-at_1}p_1\,dt_1\int_0^{t_1} e^{at_2}p_2\,dt_2,$$

$$K_2(p;\ T,\ a)=\int_0^T e^{-at}p\,dt\int_0^t p_1\,dt_1\int_0^{t_1} e^{at_2}p_2\,dt_2,$$

$$K_3(p;\ T,\ a)=\int_0^T e^{-at}p\,dt\int_0^t e^{at_1}p_1\,dt_1\int_0^{t_1} p_2\,dt_2.$$

We replace equation (3.70) by a system of type (3.7):

$$\frac{dx}{dt}=y,\quad \frac{dy}{dt}=-p(t)\,x-ay.$$

Note that, in the notation of §3.1, $\alpha=0$, $\beta=-aT$ and the functions $q(t)=-p(t)$ and $r(t)\equiv 1$ have different signs. Thus the quantity $^1/_2(e^{-\alpha}+e^{\beta})$ is equal to $^1/_2(1+e^{-aT})$ and so we can apply the results of §3.6 to analyze stability.

We first show that use of inequality (3.25) improves the result that follows from (3.61). For $n=2$, inequality (3.25) is

$$\frac{A_2}{A_1}<\frac{1}{1-e^{-aT}}A_1\,\mathrm{th}\frac{aT}{4}.$$

Since (3.25) is true for any n, it follows that if

$$A_1\leqslant(1-e^{-aT})\,\mathrm{cth}\frac{aT}{4}, \tag{3.72}$$

then necessarily $A_1>A_2>A_3>\ldots$, and so $A_0-A_1<A<A_0$. Since $A_0={}^1/_2(1+e^{-aT})$, the condition $A<{}^1/_2(1+e^{-aT})$ is satisfied, while the condition $A>-{}^1/_2(1+e^{-aT})$ will hold if

$$A_0-A_1\geqslant-\frac{1}{2}(1+e^{-aT}),\ \text{ i.e.,}\ \ A_1\leqslant 1+e^{-aT}. \tag{3.73}$$

But

$$1+e^{-aT}<(1-e^{-aT})\,\mathrm{cth}\frac{aT}{4},$$

and so inequality (3.72) is a trivial consequence of (3.73). Using the first formula in (3.71), we write (3.73) as

$$\mathrm{th}\frac{aT}{2}\int_0^T p(t)\,dt\leqslant 2a. \tag{3.74}$$

We have thus shown that *if condition (3.74) is satisfied the unperturbed motion of our mechanical system is stable.* As far as we know, this was first proved by Einaudi /246/. In Lyapunov's method for estimating the characteristic constant, this result is only the first step.

The instability condition (3.56) cannot hold for our system, since the right-hand side of this inequality vanishes at $\alpha=0$.

The instability conditions (3.57) and (3.58) imply that if either

$$A_2 \leqslant A_1-(1+e^{-aT}), \tag{3.75}$$

or

$$A_2 \geqslant A_1+A_3, \tag{3.76}$$

the unperturbed motion of the system is unstable.

The stability conditions (3.59) are a consequence of Einaudi's condition (3.74).

Finally, the stability conditions (3.60) imply that if

$$A_1+A_3-(1+e^{-aT}) \leqslant A_2 \leqslant A_1, \tag{3.77}$$

the unperturbed motion of the system is stable.

Further stability analysis using Lyapunov's method is possible only if one incorporates the terms following A_3 in the series (3.16).

Example 1. Consider the equation

$$\ddot{x}+\alpha\dot{x}+\left(1+\lambda\cos\frac{2\pi}{T}t\right)x=0 \qquad (\alpha>0,\ 0<\lambda<1). \tag{*}$$

This equation was studied in connection with parametric resonance and the theory of electrical machines with parametric excitation /2/. It also lies at the basis of various problems in the dynamic stability of elastic systems with friction proportional to velocity /7/.

Introducing dimensionless time $\tau=2\pi t/T$, we write equation (*) as

$$\frac{d^2y}{d\tau^2}+a\frac{dy}{d\tau}+c(1+\lambda\cos\tau)y=0 \qquad \left(a=\frac{T}{2\pi}\alpha,\ c=\frac{T^2}{4\pi^2}\right).$$

The first three terms of (3.16), according to (3.71), are

$$A_1=\frac{\pi c}{a}(1-e^{-2\pi a}), \qquad A_2=\frac{\pi^2c^2}{a^2}(1+e^{-2\pi a})-\frac{\pi c^2}{2a}\left(\frac{2}{a^2}+\frac{\lambda^2}{1+a^2}\right)(1-e^{-2\pi a}),$$

$$A_3=-\frac{\pi^2c^3}{a^2}\left(\frac{2}{a^2}+\frac{\lambda^2}{1+a^2}\right)(1+e^{-2\pi a})+$$
$$+\frac{\pi c^3}{a}\left[\frac{2}{a^4}+\frac{2\pi^2}{3a^2}+\frac{2\lambda^2}{a^2(1+a^2)}-\frac{\lambda^2}{a^2(1+a^2)^2}+\frac{\lambda^2}{(1+a^2)^2}\right](1-e^{-2\pi a}).$$

In the first approximation,* the stability domain in the parameter space $ac\lambda$ is defined by inequality (3.74):

$$\pi c \leqslant a \operatorname{cth} \pi a. \tag{3.78}$$

* The approximations will be indexed in accordance with the subscript of the last coefficient A_n actually calculated. The inequalities developed here guarantee asymptotic stability or instability (as the case may be) of the trivial solution of equation (*).

In the third approximation, the stability domain is defined by (3.77):

$$\frac{\pi c}{a}-\left[1+\left(\frac{2}{a^2}+\frac{\lambda^2}{1+a^2}\right)\frac{c}{2}\right]\operatorname{th}\pi\alpha\leqslant 0, \tag{3.79}$$

$$\left\{1+\left(\frac{2}{a^2}+\frac{\lambda^2}{1+a^2}\right)\frac{c}{2}+\left[\frac{2}{a^4}+\frac{2\pi^2}{3a^2}+\frac{2\lambda^2}{a^2(1+a^2)}-\frac{\lambda^2}{a^2(1+a^2)^2}+\right.\right.$$
$$\left.\left.+\frac{\lambda^2}{(1+a^2)^2}\right]c^2\right\}\operatorname{th}\pi a-\frac{a}{\pi c}-\frac{\pi c}{a}-\frac{\pi c^2}{a}\left(\frac{2}{a^2}+\frac{\lambda^2}{1+a^2}\right)\leqslant 0. \tag{3.80}$$

It can be shown that conditions (3.75) and (3.76) are impossible in our case; in other words, if $A<-{}^1/_2(1+e^{-aT})$ or $A>{}^1/_2(1+e^{-aT})$, this can be established only on the basis of terms occurring later than A_3 in the series (3.16).

Two sufficient conditions for stability of a somewhat more general equation than (3.70) were established in /204a/. As applied to equation (*), these conditions are (see (15) and (17) in /204a/):

$$\alpha\geqslant\sqrt{1+\lambda}-\sqrt{1-\lambda}, \tag{3.81}$$

$$\alpha\geqslant\frac{\pi\lambda}{T\sqrt{1-\lambda^2}}. \tag{3.82}$$

Figure 47 illustrates the curves defined by inequalities (3.78)–(3.82) for $\lambda=0.5$ on the (T, α)-plane. The resulting stability domain is hatched. Conditions for "unbounded" stability in this example were discussed in § II. 4.4.

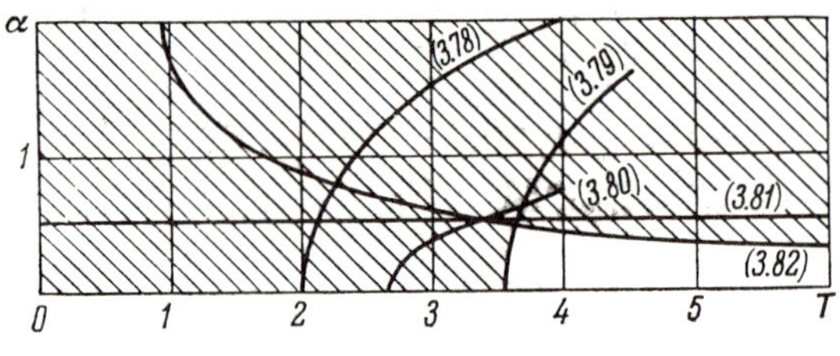

FIGURE 47.

Example 2. We now apply the results of this subsection to the equation

$$\ddot{x}+a\dot{x}+c(1+\lambda\cos t+\mu\cos 2t)x=0 \qquad (a\geqslant 0,\ c>0). \tag{3.83}$$

The special case $a=0$ of this equation was studied by Lyapunov (/180g/, subsections 20–24). Lyapunov proved that in this case the conditions for $p(t)$ to be nonnegative may be expressed as follows (see (48) in /180g/):

$$-1\leqslant\mu<1,\quad \lambda^2\leqslant N,\quad N=\begin{cases}(1+\mu)^2 & \left(\mu\leqslant\frac{1}{3}\right),\\ 8\mu(1-\mu) & \left(\mu\geqslant\frac{1}{3}\right).\end{cases}$$

Assuming these inequalities fulfilled, we investigate the conditions derived from the first two terms of the series (3.16). By formulas (3.71) these are

$$A_1=\frac{\pi c}{a}(1-e^{-2\pi a}),$$
$$A_2=\frac{\pi^2c^2}{a^2}(1+e^{-2\pi a})-\frac{\pi c^2}{2a}\left(\frac{2}{a^2}+\frac{\lambda^2}{1+a^2}+\frac{\mu^2}{4+a^2}\right)(1-e^{-2\pi a}).$$

The boundary of the stability domain in the (c, a)-plane is defined in the first approximation by the curve $A_1=1+e^{-2\pi a}$, or $c=a\,\mathrm{cth}\,(\pi a)/\pi$. This curve is shown in Figure 48 as $c=f(a)$. The instability domain is defined in the second approximation by inequality (3.75); in this case:

$$\pi^2c^2+a^2\leqslant\pi ca\left[1+\frac{1}{2}c\left(\frac{2}{a^2}+\frac{\lambda^2}{1+a^2}+\frac{\mu^2}{4+a^2}\right)\right]\mathrm{th}\,\pi a. \tag{3.84}$$

Setting $a=0$, we obtain Lyapunov's condition (49) /180g/:

$$\frac{2}{3}\pi^4c^2-2\pi^2c+2<\pi^2\left(\lambda^2+\frac{1}{4}\mu^2\right)c^2,$$

which, as Lyapunov showed, may hold only if

$$0.275\ldots\leqslant\mu\leqslant 0.748\ldots,\quad N\geqslant\lambda^2\geqslant\frac{1}{6}\pi^2-\frac{1}{4}\mu^2.$$

To illustrate, we take $\lambda=4/3$, $\mu=1/2$, and Lyapunov's conditions give $0.16901\leqslant c\leqslant 0.25299$. Inequality (3.84) is satisfied in the domain demarcated in Figure 48 by a dashed line; this domain intersects the c-axis in the interval (0.169, 0. 253], and it is the instability domain in the first approximation.

We now apply the stability conditions (15) and (17) of /204a/. The first of these may be written $a>\sqrt{M}-\sqrt{m}$, where M and m are upper and lower bounds, respectively, for $p(t)$. At $\lambda=4/3$, $\mu=1/2$, we have $M=\frac{17}{6}c$, $m={}^1/_{18}c$, and the first condition becomes $a\geqslant 1.4475\sqrt{c}$. The second condition is $a\geqslant-{}^1/_2\,\dot{p}/p$; at $\lambda=4/3$, $\mu=1/2$, we have

$$-\frac{\dot{p}(t)}{2p(t)}=\frac{4\sin t+3\sin 2t}{6+8\cos t+3\cos 2t}.$$

The extrema of this function are determined from the equation

$$12\cos^3 t+36\cos^2 t+30\cos t+7=0.$$

This gives $(\cos t)_1=-0.39885$ and $(\cos t)_2=-0.82206$, and

$$\sup\left[-\frac{\dot{p}(t)}{2p(t)}\right]=1.92935.$$

The curve $a=1.4475\sqrt{c}$ is shown in Figure 48.

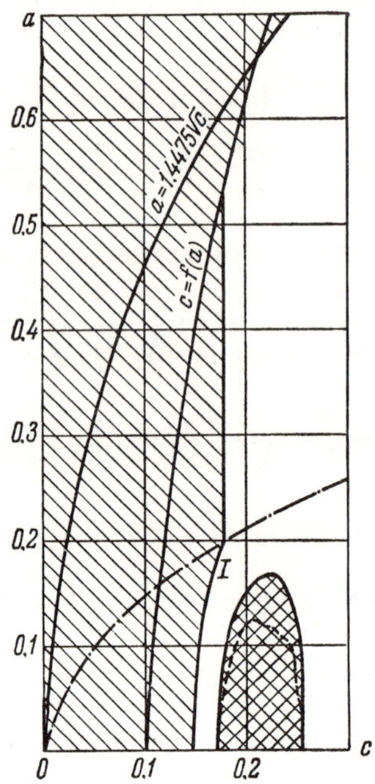

FIGURE 48.

We add a remark. The familiar substitution

$$x = ze^{-\frac{1}{2}at}$$

brings our equation to the form

$$\ddot{z} + p^*(t)\,z = 0$$

$$\left(p^*(t) = -\frac{1}{4}a^2 + c\left(1 + \frac{4}{3}\cos t + \frac{1}{2}\cos 2t\right)\right).$$

Since the characteristic numbers satisfy the equality

$$\chi\{x\} = \chi\{z\} + \frac{1}{2}a,$$

it follows that if $\chi\{z\} > -{}^1/_2\,a$ the trivial solution of the original equation is stable, and if $\chi\{z\} < -{}^1/_2\,a$ it is unstable. Hence we have stability for $|A^*| > \operatorname{ch}\pi a$ and instability for $|A^*| < \operatorname{ch}\pi a$ (the characteristic constant A^* is that of the last equation above). To determine the stability and instability domains directly by Lyapunov's method /180g/, one must allow for the displacement of the boundary in the case of A^* and add the condition that $p^*(t)$ is nonnegative. In this example the last-named condition gives $c \geqslant 4.5a^2$. The curve $c = 4.5a^2$ is shown by the dash-dot curve.

The boundary of the instability domain in the second approximation is shown in Figure 48 by the full line cutting the c-axis in the interval [0.169, 0.253]; the improved instability domain thus obtained is cross-hatched in the figure.

The boundary of the stability domain in the third approximation is the curve I intersecting the curve $c=4.5a^2$. The resulting stability domain is hatched.

3.9. Special case $a=0$. Example

If $\alpha=\beta$, the results of §3.5 imply the following conclusions:

If $\alpha>0$, the unperturbed motion is unstable.

If $\alpha=0$ the unperturbed motion is unstable if $|A|>\operatorname{ch}\alpha$; the interval $|A|<\operatorname{ch}\alpha$ requires separate consideration (see §3.11).

If $\alpha<0$ the unperturbed motion is stable if $|A|<\operatorname{ch}\alpha$ and unstable if $|A|>\operatorname{ch}\alpha$.

If $a=0$, we can apply Lyapunov's results /180g/. We introduce the new independent variable

$$\tau=\int_0^t r(t_1)dt_1.$$

System (3.7) becomes

$$\frac{d^2x}{d\tau^2}+p(\tau)x=0 \qquad \left(p(\tau)=-\left[\frac{q(t)}{r(t)}\right]_{t=t(\tau)}\right). \tag{3.85}$$

Here $p(\tau)$ is of fixed sign and (as can be shown) integrable over the period. All of Lyapunov's arguments for system (3.7) (/180g/, subsections 3–19) may be repeated for this case $(a=0)$.

However, our reduction of the original problem for $a=0$ to Hill's equation (3.85) with a periodic function of fixed sign does not imply that Lyapunov's method for estimating the characteristic constant can be applied without modification. The point is that the boundary values of A, at which stability goes over into instability (or conversely), are no longer ± 1, as in Lyapunov's treatment, but $\pm\operatorname{ch}\alpha$. This is true whether $p(\tau)\geqslant 0$ or $p(\tau)\leqslant 0$. In the latter case it is clear that the unperturbed motion need not be unstable at all.

Thus stability analysis of the initial system (3.1) reduces to estimating the characteristic exponents of Hill's equation (3.85). The final tests for stability and instability will follow from the results of §2.4. In calculations it is convenient to use the following formulas of Lyapunov (/180g/, subsection 19) for A_1, A_2, A_3. Express the function $p(\tau)$ as

$$p(\tau)=c+\vartheta''(\tau),$$

where c is some constant and $\vartheta''(\tau)$ is the second derivative of a periodic function $\vartheta(\tau)$ such that

$$\int_0^\omega \vartheta(\tau)\,d\tau = 0$$

(ω is the period of $p(\tau)$). Then

$$A_1 = \frac{1}{2}c\omega^2, \quad A_2 = \frac{1}{24}c^2\omega^2 - \frac{1}{2}\omega\int_0^\omega \vartheta'^2\,d\tau,$$

$$A_3 = \frac{1}{720}c^3\omega^6 - \frac{1}{12}c\omega^3\int_0^\omega \vartheta'^2\,d\tau + 2c\omega\int_0^\omega \vartheta^2\,d\tau - \omega\int_0^\omega \vartheta\vartheta'^2\,d\tau$$

where the prime denotes differentiation with respect to τ.

Example. Suppose that the perturbed motion is described in the first approximation by a single second-order equation

$$\ddot{x} + \delta\sin t\,\dot{x} + \gamma x = 0 \qquad (\delta > 0,\ \gamma > 0). \tag{3.86}$$

We first form a qualitative idea of the relative positions of stability and instability domains on the (γ, δ)-plane. Consider the averaged equation

$$\ddot{x} + \gamma x = 0.$$

The weak stability points are

$$\gamma = \frac{n^2\pi^2}{T^2} = \frac{n^2\pi^2}{(2\pi)^2} = \frac{n^2}{4} \qquad (n = 1,\ 2,\ \ldots).$$

Thus the stability domains in the first quadrant must contain the intervals $(0,\ 1/4)$, $(1/4,\ 1)$, $(1,\ 9/4)$, ... on the γ-axis. At $\gamma = 0$ the trivial solution of equation (3.86) is unstable. Thus the axis $\gamma = 0$ belongs to the zeroth instability domain. The qualitative pattern of stability and instability domains is as in Figure 42 (p. 556), where the approximate stability domains are hatched. We now use Lyapunov's method to estimate the characteristic constant.

Transforming to the independent variable

$$\tau = \int_0^t e^{\delta(\cos t_1 - 1)}\,dt_1,$$

we write the equation as

$$x'' + p(\tau)\,x = 0 \qquad \left(p(\tau) = \gamma e^{-2\delta[\cos t(\tau) - 1]}\right).$$

In our example,

$$c = \gamma e^{2\delta}, \quad \vartheta'' = \gamma e^{2\delta}\left(^{-2\delta\cos t(\tau)} - 1\right).$$

Expand the function $e^{\delta\cos t}$ in Fourier series:

$$e^{\delta \cos t} = \sum_{k=0}^{\infty} a_k \cos kt,$$

where the coefficients a_k are defined by the following power series in δ:

$$\left.\begin{aligned}
&a_0 = 1 + \frac{1}{4}\delta^2 + \frac{1}{64}\delta^4 + \frac{1}{2304}\delta^6 + \frac{1}{147\,456}\delta^8 + \ldots,\\
&a_1 = \delta + \frac{1}{8}\delta^3 + \frac{1}{192}\delta^5 + \frac{1}{9216}\delta^7 + \ldots,\\
&a_2 = \frac{1}{4}\delta^2 + \frac{1}{48}\delta^4 + \frac{1}{1536}\delta^6 + \frac{1}{92\,160}\delta^8 + \ldots,\\
&a_3 = \frac{1}{24}\delta^3 + \frac{1}{384}\delta^5 + \frac{1}{15\,360}\delta^7 + \ldots,\\
&a_4 = \frac{1}{192}\delta^4 + \frac{1}{3840}\delta^6 + \frac{1}{184\,320}\delta^8 + \ldots,\ a_5 = \frac{1}{1920}\delta^5 + \frac{1}{46\,080}\delta^7 + \ldots,\\
&a_6 = \frac{1}{23\,040}\delta^6 + \frac{1}{645\,120}\delta^8 + \ldots,\ a_7 = \frac{1}{322\,560}\delta^7 + \ldots,\ a_8 = \frac{1}{5\,160\,960}\delta^8 + \ldots
\end{aligned}\right\} \qquad (3.87)$$

Then

$$\vartheta' = \int_0^{\tau} \vartheta''(\tau)\, d\tau = \int_0^{t} \vartheta''[\tau(t)] \frac{d\tau}{dt}\, dt = -2\gamma e^{\delta} \sum_{\varkappa=1,3,\ldots} \frac{a_\varkappa}{\varkappa} \sin \varkappa t.$$

Similarly, calculation of the function ϑ gives

$$\vartheta = \gamma \sum_{\varkappa=1,3,\ldots} \sum_{k=0}^{\infty} \frac{a_\varkappa a_k}{\varkappa(\varkappa+k)} [\cos(\varkappa+k)t - 1] - \gamma \sum_{\varkappa=1,3,\ldots} \sum_{k=0}^{\infty} \frac{a_\varkappa a_k}{\varkappa(\varkappa-k)} [\cos(\varkappa-k)t - 1].$$

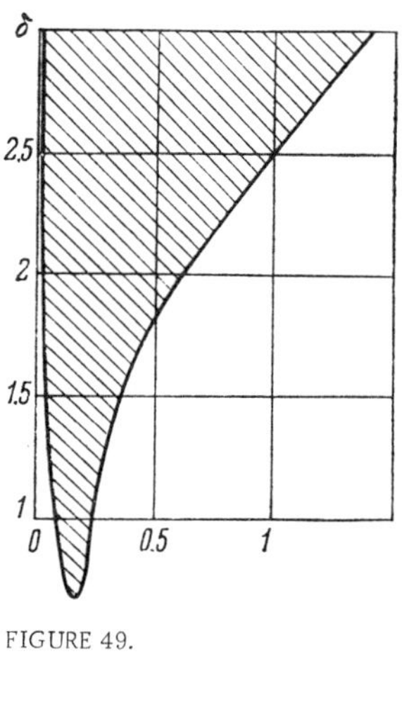

FIGURE 49.

The expressions for A_1 and A_2 are

$$A_1 = 2\pi^2 a_0^2 \gamma,$$

$$A_2 = \frac{2}{3}\pi^4 a_0^4 \gamma^2 - 4\pi^2 a_0 \gamma^2 \left[\sum_{\varkappa=1,3,\ldots} \frac{a_\varkappa^2}{\varkappa^2}\left(a_0 - \frac{1}{2}a_{2\varkappa}\right) + \sum_{\substack{\lambda=3,5,\ldots \\ \lambda>\varkappa}} \sum_{\varkappa=1,3,\ldots} \frac{a_\varkappa a_\lambda}{\varkappa\lambda}(a_{\lambda-\varkappa} - a_{\lambda+\varkappa}) \right].$$

The instability domain in the (γ, δ)-plane (hatched region in Figure 49) is defined by inequality (3.57); all its points are in fact instability points. An approximate calculation of the first two instability domains may be found in Valeev /129a/.

3.10. Sufficient conditions for stability and instability when $a_{12}(t)$ and $a_{21}(t)$ are of the same sign

In this case all the terms of the series

$$A = A_0 - A_1 + A_2 - A_3 + \ldots \tag{3.16}$$

are positive, since sign $A_\nu = (-1)^\nu$ $(\nu = 0, 1, 2, \ldots)$. Clearly, $A > A_0$, and by the theorem of §3.5 the unperturbed motion of the dynamic system whose perturbed motion is described in the first approximation by equation (3.1) is unstable if

$$A_0 = \frac{1}{2}(1 + e^{\beta-\alpha}) \geqslant \frac{1}{2}(e^{-\alpha} + e^{\beta}), \text{ i. e., } (1 - e^{-\alpha})(1 - e^{\beta}) \geqslant 0.$$

Recall that without loss of generality (except in the case $(\alpha + \beta > 0)$ when the motion is always unstable), we may assume that

$$\beta \leqslant \alpha \leqslant -\beta \qquad (\beta \leqslant 0).$$

Consequently, if $\alpha \geqslant 0$ then $A > A_0 \geqslant {}^1/_2(e^{-\alpha} + e^{\beta})$ and the unperturbed motion is unstable. Thus we have proved

Theorem I. Assume that the functions $a_{12}(t)$ and $a_{21}(t)$ in equation (3.1) are of the same fixed sign (i.e., maintaining the same sign, they may vanish, provided their product does not vanish identically on $[0, T]$). If the mean value of either of the functions $a_{11}(t)$ or $a_{22}(t)$ is nonnegative, the unperturbed motion is unstable.

Corollary. If the periodic coefficient $p(t)$ of the equation

$$\ddot{x} + s(t)\dot{x} + p(t)x = 0 \quad (s(t+T) = s(t),\ p(t+T) = p(t)) \tag{3.88}$$

is nonpositive for all $t\,(0 \leqslant t \leqslant T)$, the trivial solution of equation (3.88) is unstable, regardless of the behavior of the second periodic coefficient $s(t)$.

To prove this for $p(t) \not\equiv 0$, it suffices to write equation (3.88) as a system

$$\frac{dx}{dt} = \dot{x},$$

$$\frac{d\dot{x}}{dt} = -p(t)x - s(t)\dot{x}$$

and to apply the theorem. If $p(t)\equiv 0$, instability is an immediate consequence of the form of the general solution of the equation

$$\ddot{x}+s(t)\dot{x}=0.$$

Returning to system (3.1), we observe that we have still to examine the case

$$\beta\leqslant\alpha<0.$$

The unperturbed motion will be stable if

$$A<\frac{1}{2}(e^{-\alpha}+e^{\beta}),$$

and unstable if

$$A>\frac{1}{2}(e^{-\alpha}+e^{\beta}).$$

By inequality (3.26) for $n=2, 3, \ldots,$ we have

$$A_2<A_1^2,\quad |A_3|<\frac{2}{3}|A_1|A_2<\frac{2}{3}|A_1|^3,\ \ldots,\ |A_n|<\frac{2^{n-1}}{n!}|A_1|^n,$$

and so

$$A<A_0+|A_1|+\frac{1}{2}\sum_{n=2}^{\infty}\frac{2^n}{n!}|A_1|^n=\frac{1}{2}e^{\beta-\alpha}+\frac{1}{2}e^{2|A_1|}.$$

If

$$\frac{1}{2}e^{\beta-\alpha}+\frac{1}{2}e^{2|A_1|}\leqslant\frac{1}{2}(e^{-\alpha}+e^{\beta}),$$

or, equivalently,

$$|A_1|\leqslant\frac{1}{2}\ln(e^{-\alpha}+e^{\beta}-e^{\beta-\alpha}), \tag{3.89}$$

the unperturbed motion is stable.

We now apply inequality (3.28):

$$A_2<\frac{1}{4}e^{\alpha-\beta}A_1^2,\ |A_3|<\frac{1}{9}e^{\alpha-\beta}|A_1|A_2<\frac{1}{36}e^{2(\alpha-\beta)}|A_1|^3,\ \ldots$$

$$\ldots,|A_n|<\frac{1}{(n!)^2}e^{(n-1)(\alpha-\beta)}|A_1|^n.$$

Thus

$$A<A_0+A_1+\sum_{n=2}^{\infty}\frac{1}{(n!)^2}e^{(n-1)(\alpha-\beta)}|A_1|^n=\frac{1}{2}(1-e^{\beta-\alpha})+e^{\beta-\alpha}I_0(2\sqrt{e^{\alpha-\beta}|A_1|}),$$

where $I_0(z)$ is the Bessel function of an imaginary argument.

If

$$\frac{1}{2}(1-e^{\beta-\alpha})+e^{\beta-\alpha}I_0(2\sqrt{e^{\alpha-\beta}|A_1|})\leqslant\frac{1}{2}(e^{-\alpha}+e^{\beta}),$$

that is,

$$I_0(2\sqrt{e^{\alpha-\beta}|A_1|})\leqslant\frac{1}{2}(1+e^{\alpha}+e^{-\beta}-e^{\alpha-\beta}), \tag{3.90}$$

then the unperturbed motion is stable. Inequality (3.90) is inconsistent with (3.89) if $\alpha-\beta<\ln 4$.

If we begin the estimation of the terms in (3.16) from $|A_3|$, we see similarly that if either

$$A_2\leqslant\frac{1}{e^{2|A_1|}-1-2|A_1|}A_1^2[(e^{-\alpha}-1)(1-e^{\beta})-2|A_1|] \tag{3.91}$$

or

$$A_2\leqslant\frac{[(e^{-\alpha}-1)(1-e^{\beta})-2|A_1|]\,e^{2(\alpha-\beta)}A_1^2}{8[I_0(2\sqrt{e^{\alpha-\beta}|A_1|})-1-e^{\alpha-\beta}|A_1|]}, \tag{3.92}$$

then the unperturbed motion is stable.

Finally, we use inequality (3.40) to estimate the terms of (3.16):

$$\frac{A_2}{|A_1|}\leqslant\frac{1}{2}\frac{|A_1|}{A_0},$$
$$\frac{|A_3|}{A_2}\leqslant\frac{2}{3}\frac{A_2}{|A_1|},$$
$$\frac{A_4}{|A_3|}\leqslant\frac{3}{4}\frac{|A_3|}{A_2},\ \ldots$$

Summing the series, we have

$$\begin{aligned}A&\leqslant A_0+|A_1|+\frac{1}{2}\frac{A_1^2}{A_0}+\frac{1}{3}\frac{|A_1|A_2}{A_0}+\frac{1}{4}\frac{A_1A_3}{A_0}+\ldots=\\&=A_0+\frac{2}{3}|A_1|+\frac{1}{6}\frac{A_1^2}{A_0}+\frac{|A_1|}{A_0}\left[\frac{1}{3}A_0+\frac{1}{3}|A_1|+\frac{1}{3}A_2+\frac{1}{4}|A_3|+\ldots\right].\end{aligned} \tag{3.93}$$

Since the sum of the bracketed series is less than $^1/_3 A$, it follows that

$$A\left(1-\frac{1}{3}\frac{|A_1|}{A_0}\right)<A_0+\frac{2}{3}|A_1|+\frac{1}{6}\frac{A_1^2}{A_0}.$$

If $|A_1|\geqslant 3A_0$, this inequality is clearly true. But if $|A_1|<3A_0$, then

$$A<\frac{6A_0^2+4|A_1|A_0+A_1^2}{6A_0-2|A_1|},$$

and if

$$\frac{6A_0^2+4|A_1|A_0+A_1^2}{3A_0-|A_1|}\leqslant e^{-\alpha}+e^{\beta}, \tag{3.94}$$

the unperturbed motion is stable.

If we write inequality (3.93) as

$$A\leqslant A_0+\frac{3}{4}|A_1|+\frac{1}{4}\frac{A_1^2}{A_0}+\frac{1}{12}\frac{|A_1|A_2}{A_0}+ \\ +\frac{|A_1|}{A_0}\left[\frac{1}{4}A_0+\frac{1}{4}|A_1|+\frac{1}{4}A_2+\frac{1}{4}|A_3|+\dots\right],$$

we find that

$$A\left(1-\frac{1}{4}\frac{|A_1|}{A_0}\right)<A_0+\frac{3}{4}|A_1|+\frac{1}{4}\frac{A_1^2}{A_0}+\frac{1}{12}\frac{|A_1|A_2}{A_0},$$

and this is certainly true if $|A_1|\geqslant 4A_0$. But if $|A_1|<4A_0$, then the condition

$$\frac{12A_0^2+9|A_1|A_0+3A_1^2+|A_1|A_2}{4A_0-|A_1|}<\frac{3}{2}(e^{-\alpha}+e^{\beta}) \tag{3.95}$$

will ensure that the unperturbed motion is stable.

It is much simpler to derive sufficient conditions for instability. Indeed, $A>A_0+|A_1|$, and therefore if

$$A_0+|A_1|\geqslant\frac{1}{2}(e^{-\alpha}+e^{\beta})$$

then also $A>{}^1/_2(e^{-\alpha}+e^{\beta})$. Thus, if

$$|A_1|\geqslant\frac{1}{2}(e^{-\alpha}-1)(1-e^{\beta}) \tag{3.96}$$

the unperturbed motion is unstable.

Beginning with the inequality $A>A_0+|A_1|+A_2$, we see that if

$$A_2\geqslant\frac{1}{2}(e^{-\alpha}-1)(1-e^{\beta})-|A_1|, \tag{3.97}$$

then the unperturbed motion is unstable.

The derivation of further stability and instability conditions by this technique is obvious. The final result is as follows.

Theorem II. Suppose that the coefficients $a_{12}(t)$ and $a_{21}(t)$ of system (3.1), representing the first approximation to the perturbed motion of the dynamic system, are functions of the same fixed sign and that $a_{12}(t)a_{21}(t)\not\equiv 0$ for $0\leqslant t\leqslant T$. Set

$$\alpha=\int_0^T a_{11}(t)\,dt,\quad \beta=\int_0^T a_{22}(t)\,dt.$$

Let $\beta\leqslant\alpha<0$,* *define* A_0, A_1, A_2 *by formulas* (3.15), (3.19), (3.20), (3.18), (3.8).

Any one of the inequalities

$$|A_1|\leqslant\frac{1}{2}\ln(e^{-\alpha}+e^{\beta}-e^{\beta-\alpha}),$$

$$I_0(2\sqrt{e^{\alpha-\beta}|A_1|})\leqslant\frac{1}{2}(1+e^{\alpha}+e^{-\beta}-e^{\alpha-\beta}),$$

$$A_2\leqslant\frac{1}{e^{2|A_1|}-1-2|A_1|}A_1^2[(e^{-\alpha}-1)(1-e^{\beta})-2|A_1|],$$

$$A_2\leqslant\frac{[(e^{-\alpha}-1)(1-e^{\beta})-2|A_1|]e^{2(\alpha-\beta)}A_1^2}{8[I_0(2\sqrt{e^{\alpha-\beta}|A_1|})-1-e^{\alpha-\beta}|A_1|]},$$

$$\frac{6A_0^2-4|A_1|A_0+A_1^2}{3A_0-|A_1|}\leqslant e^{-\alpha}+e^{\beta}\qquad(|A_1|<3A_0),$$

$$\frac{12A_0^3+9|A_1|A_0+3A_1^2+|A_1|A_2}{4A_0-|A_1|}<\frac{3}{2}(e^{-\alpha}+e^{\beta})\qquad(|A_1|<4A_0)$$

is a sufficient condition for the unperturbed motion to be stable.

Either of the inequalities

$$|A_1|\geqslant\frac{1}{2}(e^{-\alpha}-1)(1-e^{\beta}),\quad A_2\geqslant\frac{1}{2}(e^{-\alpha}-1)(1-e^{\beta})-|A_1|$$

is a sufficient condition for the unperturbed motion to be unstable.

3.11. Sufficient conditions for stability of the trivial solution of system (3.1) when $\alpha+\beta=0$

Our treatment of the stability of a dynamic system whose perturbed motion is governed in the first approximation by system (3.1) has provided no conditions for stability in the case $\alpha+\beta=0$. The difficulty is that when $\alpha+\beta=0$ (a case which includes systems (3.1) of canonical form, $a_{22}(t)\equiv-a_{11}(t)$), the stability of the trivial solution cannot be asymptotic; for this reason, we cannot guarantee stability of the unperturbed motion.

In this subsection we prove that the sufficient conditions for stability of the unperturbed motion established in §3.6 for the case $\alpha+\beta<0$ guarantee stability of the trivial solution of system (3.1) when $\alpha+\beta=0$.

Let $\mathbf{U}(t)$ denote the fundamental matrix of system (3.1) normalized by the initial condition $\mathbf{U}(0)=\mathbf{I}_2$ (the matrizant of the system). Then the characteristic equation of system (3.1) with $\alpha+\beta=0$ is

$$\rho^{*2}-2A^*\rho^*+1=0\qquad(2A^*=\operatorname{Tr}\mathbf{U}(T)),$$

for if $\alpha+\beta=0$ Liouville's formula (II. 1.11) gives $\det\mathbf{U}(T)=1$. If $|A^*|<1$ the roots of this equation are complex and distinct and lie on the unit circle, so that the trivial solution of system (3.1) is stable. In applying Lyapunov's method, however, it is more convenient to calculate the characteristic constant A of system (3.7) than A^*. There is a simple relationship between these two constants. It follows from (3.6) that if $\alpha+\beta=0$, then

* For the case in which these inequalities are not satisfied, see §3.1 and Theorem I of this subsection.

$$\mathbf{U}(T) = \left\| \begin{matrix} \exp \int_0^T p_{11}(t)\,dt & 0 \\ 0 & \exp \int_0^T \left[p_{22}(t) + 2\frac{\alpha}{T} \right] dt \end{matrix} \right\| \mathbf{V}(T) = e^{\alpha}\mathbf{V}(T),$$

where $\mathbf{V}(T)$ is the matrizant of system (3.7). Recalling (3.11) and the fact that $\alpha = \beta$ implies $aT = 2\alpha$, we see that

$$2A^* = \mathrm{Tr}\,\mathbf{U}(T) = e^{\alpha}\mathrm{Tr}\,\mathbf{V}(T) = 2e^{\alpha}A, \quad \text{i.e.,} \quad A^* = e^{\alpha}A.$$

The condition $|A^*| < 1$ is equivalent to $|A| < e^{-\alpha}$, and so, if

$$\alpha + \beta = 0, \quad |A| < e^{-\alpha},$$

the trivial solution of system (3.1) is stable. We now see that the sufficient conditions obtained in §3.6 carry over to the case $\alpha + \beta = 0$, except that they no longer ensure stability of the unperturbed motion, but only of the trivial solution of system (3.1).

3.12. Direct application of Lyapunov's method by reducing system (3.1) to Hill's equation

We now show that a system of two linear first-order differential equations with periodic coefficients may be reduced to Hill's equation with a nonnegative periodic coefficient. The latter may be treated directly by Lyapunov's technique of stability analysis /180g/ (see also §2 for the case $a = 0$).

As shown in §3.1, system (3.1) may be reduced by an orthogonal transformation (3.3) with periodic coefficients to the form

$$\frac{d\mathbf{w}}{dt} = \mathbf{D}(t)\,\mathbf{w} \qquad (\mathbf{D}(t) = \| d_{ij}(t) \|_1^2). \tag{3.4}$$

The matrix-function $\mathbf{D}(t)$ is T-periodic, and for sufficiently large k we have $d_{12}(t) < 0$ and $d_{21}(t) > 0$ in (3.3); moreover, $\inf d_{12}(t)$ and $\inf |d_{21}(t)|$ may be made arbitrarily large, but $d_{11}(t)$ and $d_{22}(t)$ remain bounded. The trivial solution of system (3.4) is stable if and only if that of system (3.1) is stable.

Let us represent system (3.4) by a single equation, say in z_1, assuming that the functions $d_{11}(t)$ and $d_{12}(t)$ have piecewise-continuous derivatives. We obtain

$$\ddot{z}_1 - \left(d_{11} + d_{22} + \frac{1}{d_{12}} \dot{d}_{12} \right) \dot{z}_1 + \left(d_{11}d_{22} - d_{12}d_{21} - \dot{d}_{11} + \frac{d_{11}\dot{d}_{12}}{d_{12}} \right) z_1 = 0. \tag{3.98}$$

The familiar substitution

$$z_1 = z \exp\left[\frac{1}{2} \int \left(d_{11} + d_{22} + \frac{1}{d_{12}} \dot{d}_{12} \right) dt \right] \tag{3.99}$$

brings equation (3.98) to the form of Hill's equation

$$\ddot{z}+p(t)z=0,$$

where

$$p(t)=-d_{12}d_{21}+\frac{1}{4}\Big[-(d_{11}-d_{22})^2-2\dot{d}_{11}+ +2\dot{d}_{22}+2\frac{d_{11}-d_{22}}{d_{12}}\dot{d}_{12}+2\frac{\ddot{d}_{12}}{d_{12}}-3\left(\frac{\dot{d}_{12}}{d_{12}}\right)^2\Big] \quad (3.100)$$

(we are assuming that $\dot{d}_{22}$ and $\ddot{d}_{12}$ are piecewise-continuous).

For sufficiently large k, the bracketed expression is of the order of $k\|A\|$, where $\|A\|$ is the norm of the matrix $A(t)$; the sign of $p(t)$ is determined by the positive number $-d_{12}d_{21}$, whose infimum is of the order of k^2 as follows from our initial assumption. In order to use Lyapunov's method /180g/, we must note that the limits within which the characteristic constant $\tilde{A}$ of equation (3.100) varies may be displaced. Namely, since the characteristic numbers of the solutions of equations (3.98) and (3.100) satisfy the identity

$$\chi\{z_1\}=\chi\{z\}-\frac{1}{2}(d_{11}+d_{22})_{\text{av}},$$

it follows that when $(d_{11}+d_{22})_{\text{av}}>0$ the trivial solution of equation (3.98) (and hence also of the initial system (3.1)) is unstable; when $(d_{11}+d_{22})_{\text{av}}\leqslant 0$ it is stable, provided

$$|\tilde{A}|<\operatorname{ch}\frac{1}{2}T(d_{11}+d_{22})_{\text{av}},$$

and unstable if the sign of this inequality is reversed. It is easy to see that

$$d_{11}(t)+d_{22}(t)\equiv a_{11}(t)+a_{22}(t).$$

We end the chapter with a few observations.

1. Lyapunov's method of stability analysis, which applies to the case $p(t)\geqslant 0$, may also be carried over to the case of a periodic coefficient of variable sign; the procedure is based on representing the equation as a system (3.1).

2. Assume that the initial system (3.1) is canonical. Then (3.4) is also a canonical system. The mean value of the integrand in the substitution (3.99) is zero. Thus the trivial solution of equation (3.100) is stable if and only if the same is true for system (3.1). However, if the Hamiltonian of the original system belongs to the n_0-th stability domain, that of the new system (3.4) (and hence also that of the final Hill's equation (3.100)) belongs to the (n_0+k)-th stability domain (see § VIII. 1.10). The number k should be chosen so large in absolute value that the function $p(t)$ in (3.100) is nonnegative. If the original system was easy to analyze by Lyapunov's method, the theorem

of §2.5 tells us that investigation of the new (two-term) equation requires considerably more work. This justifies the efforts we have been making to extend Lyapunov's method directly to systems of two equations.

3. If the initial system (3.1) is not canonical, it can be shown that the conclusion remains unchanged: if the system is amenable to analysis by Lyapunov's method, the two-term equation (for $\boldsymbol{k}$ of large absolute value) may involve a prohibitive amount of calculation.

Chapter VIII

CANONICAL SYSTEM OF TWO LINEAR DIFFERENTIAL EQUATIONS WITH PERIODIC COEFFICIENTS. HILL'S EQUATION

In this chapter we consider a very simple type of system: canonical systems of two equations. Hill's equation, which is of great importance in applications, is one of the many systems reducible to this form. Thanks to the simplicity of the subject, the methods of investigation will be more elementary than those of Chapter III. On the other hand, we shall be able to go into much greater detail than in the general case.

Many of the results established in this chapter carry over directly to canonical systems in general. For example, we showed in Chapter III, §6, that any stability test for canonical second-order systems yields a stability test for canonical systems of arbitrary order. The same is true of Hill's equation and systems of Hill's equations. Considerable attention will therefore be devoted in this chapter to stability tests.

Certain aspects of the structure of the Hamiltonian space, too sophisticated to fall within the scope of this book in the general case, admit an elementary treatment for canonical systems of second order. These will be examined in §1. In §1 we shall construct a three-dimensional "model" of the function space of Hamiltonians which reflects the structure of the latter quite faithfully (in a sense to be clarified below). The model enables one to visualize the structure of various sets in the Hamiltonian space, corresponding to different restrictions or properties of the solutions. This will be of great assistance in regard to our final goal: effective (i. e., dependent on the coefficients) conditions for stability and instability and estimates for the characteristic exponents (§§ 2, 4).

Most of the stability tests and estimates for characteristic exponents in this chapter will be sharp, i. e., "best possible" if no other information is forthcoming. A few words on the significance of these tests are in order. With the growing sophistication of computer techniques, it is now quite a simple matter to verify whether a *given equation* is stable or unstable. However, many problems of present-day technology involve equations whose coefficients are not known with full accuracy. For example, suppose that we are studying the equation

$$\frac{d^2y}{dt^2}+\alpha\frac{dy}{dt}+q(t)\,y=0 \qquad (q(t+T)\equiv q(t)) \tag{0.1}$$

knowing only α, the period T, and the numbers $\max q(t)$, $\min q(t)$, $q_{\text{av}}=\frac{1}{T}\int_0^T q(t)dt$. It is required to determine whether the trivial solution of this equation is stable, or to estimate its characteristic exponents. Problems of this kind form the main subject matter of this chapter. A "sharp" condition for, say, asymptotic stability would be a necessary and sufficient condition, in terms of the above five parameters, guaranteeing asymptotic stability of the trivial solution of any equation (0.1) for which these parameters have the prescribed values.

Of the other results, we shall show in §§1, 2 that the main properties of interest here (stability, instability, etc.), as well as the characteristic exponent in the unstable case, may be expressed in terms of two numbers — the maximum (M) and minimum (m) angles through which the solution-vectors turn in time T. With bounds for these numbers, one can establish effective stability and instability conditions.

It will be shown in §6 that the geometric approach developed in §1 and the estimates for the angles of rotation of the solutions (§2) yield several results concerning boundary-value problems which involve λ nonlinearly; among other things, one can establish several deep properties of the characteristic function $A(\lambda)$, stated without proof in Chapter VII.

The reader concerned with practical aspects (specific stability and instability tests, estimates for characteristic exponents) is advised to read §§1.1, 1.2 and, omitting the proofs, familiarize himself with the main results of §§1.4—1.10 (especially the definition of the stability and instability domains $\mathcal{O}_n$ and $\mathcal{H}_n$, and their boundary sets Π_n; see §1.8), then proceeding to §§2, 3, 4. The main results are concentrated in the theorems of §§ 1—6, which are stated for the most part in such a way that they may be read independently of the remaining text.

We append one more remark concerning the notation. Recall that in a canonical equation (see (II. 3.4))

$$\tilde{\mathbf{J}}\frac{d\mathbf{z}}{dt}=\tilde{\mathbf{H}}(t)\cdot\mathbf{z}$$

the matrix $\tilde{\mathbf{J}}$ is an arbitrary real skew-symmetric matrix. In Chapters II through VII we specialized $\tilde{\mathbf{J}}$ to be

$$\mathbf{J}_{2k}=\begin{Vmatrix}0 & -\mathbf{I}_k\\ \mathbf{I}_k & 0\end{Vmatrix}.$$

In this chapter, however, it will be more convenient to take

$$\tilde{\mathbf{J}}=\begin{Vmatrix}0 & 1\\ -1 & 0\end{Vmatrix},$$

i. e., $\tilde{\mathbf{J}}=-\mathbf{J}_2$.

§1. GENERAL REMARKS. STRUCTURE OF HAMILTONIAN SPACE

1.1. Reduction to canonical form

In this chapter we study systems of two scalar equations

$$\frac{d\zeta_1}{dt}=a_{11}(t)\zeta_1+a_{12}(t)\zeta_2,$$
$$\frac{d\zeta_2}{dt}=a_{21}(t)\zeta_1+a_{22}(t)\zeta_2,$$

or, in vector notation, the equation

$$\frac{d\mathbf{z}}{dt}=\mathbf{A}(t)\mathbf{z}\quad\left(\mathbf{z}=\begin{pmatrix}\zeta_1\\ \zeta_2\end{pmatrix},\quad \mathbf{A}(t)=\|a_{ij}(t)\|_1^2\right). \tag{1.1}$$

As usual, we assume that the functions $a_{ij}(t)$ $(i, j=1, 2)$ are real, piecewise-continuous, integrable on $(0, T)$, * and periodic with period T.

Hill's equation

$$\frac{d^2y}{dt^2}+p(t)y=0\qquad(p(t+T)\equiv p(t)),$$

which is frequently encountered in applications, may be reduced to (1.1) by the substitution $\zeta_1=\frac{dy}{dt}$, $\zeta_2=y$; one then has

$$\mathbf{A}(t)=\begin{Vmatrix}0 & -p(t)\\ 1 & 0\end{Vmatrix}.$$

The substitution

$$\mathbf{z}=e^{\frac{1}{2}\int_0^t \operatorname{Tr}\mathbf{A}(t_1)\,dt_1}\mathbf{x}$$

(recall that $\operatorname{Tr}\mathbf{A}(t)=a_{11}(t)+a_{22}(t)$) reduces system (1.1) to a canonical system

$$\mathbf{J}\frac{d\mathbf{x}}{dt}=\mathbf{H}(t)\mathbf{x}, \tag{1.2}$$

in which

$$\mathbf{J}=\begin{Vmatrix}0 & 1\\ -1 & 0\end{Vmatrix},\quad \mathbf{H}(t)=\begin{Vmatrix}\alpha(t) & \beta(t)\\ \beta(t) & \gamma(t)\end{Vmatrix} \tag{1.3}$$

and

* For the sake of generality and/or convenience, we shall sometimes assume below that these functions are summable (Lebesgue-integrable) on $(0, T)$.

$$\alpha(t)=a_{21}(t),\quad \beta(t)=\frac{1}{2}[a_{22}(t)-a_{11}(t)],\quad \gamma(t)=-a_{12}(t).$$

If $\operatorname{Tr}\mathbf{A}(t)\equiv 0$, then $\mathbf{z}\equiv\mathbf{x}$. Thus $\operatorname{Tr}\mathbf{A}(t)\equiv 0$ is a necessary and sufficient condition for system (1.1) to be canonical. In particular, Hill's equation may be reduced directly to canonical form (1.2) with the Hamiltonian

$$\mathbf{H}(t)=\begin{Vmatrix}1 & 0\\ 0 & p(t)\end{Vmatrix}.$$

Let

$$\frac{1}{2T}\int_0^T \operatorname{Tr}\mathbf{A}(t)dt=\delta.$$

Then

$$\mathbf{z}(t)=e^{\delta t+\varphi(t)}\mathbf{x}(t),$$

where $\varphi(t)$ is a T-periodic function. Thus the derivation of conditions for asymptotic stability of the trivial solution of system (1.1) reduces to estimation of the characteristic exponents of system (1.2). The canonical system (1.2) in itself is of considerable practical importance. In view of the above reduction of arbitrary systems to canonical form, we shall discuss not only stability analysis but also estimates for the characteristic exponents of system (1.2).

By Liouville's formula (II. 1.11), the matrizant $\mathbf{Z}(t)$ of system (1.1) satisfies the equality

$$\det\mathbf{Z}(t)=\exp\int_0^t \operatorname{Tr}\mathbf{A}(t_1)\,dt_1=\Delta_0(t)\,e^{\delta t},$$

where $\Delta_0(t)$ is a T-periodic function. Thus, if $\delta>0$ system (1.1) has solutions which are unbounded as $t\to+\infty$. If $\delta<0$, one can use Lyapunov V-functions to establish sufficient conditions for asymptotic stability (see /60, 102, 54/).

The case $\delta=0$ requires more delicate arguments. Here the solutions of system (1.1) are bounded if and only if those of system (1.2) are bounded, and the stability (if it holds) cannot be asymptotic. Indeed, if $\delta=0$ the function $\det\mathbf{Z}(t)=\Delta_0(t)$ is periodic. In case of asymptotic stability, $\det\mathbf{Z}(t)\to 0$ as $t\to+\infty$, so that $\Delta_0(t)$ could not be periodic.

In this chapter, then, we are going to study mainly system (1.2).

1.2. Characteristic equation and Floquet-Lyapunov theorem for canonical systems

Since $\operatorname{Tr}\tilde{\mathbf{J}}^{-1}\mathbf{H}(t)\equiv 0$, it follows that

$$\det\mathbf{X}(t)\equiv 1,$$

where $\mathbf{X}(t)$ is the matrizant of equation (1.2). Thus the characteristic equation (see (II. 2.3))

$$\det [\mathbf{X}(T) - \rho \mathbf{I}_2] = 0$$

of equation (1.2) is

$$\rho^2 - 2A\rho + 1 = 0, \tag{1.4}$$

where

$$A = \frac{1}{2} \operatorname{Tr} \mathbf{X}(T). \tag{1.5}$$

The number A is known as the Lyapunov characteristic constant of equation (1.2) (see (VII. 0.3)).

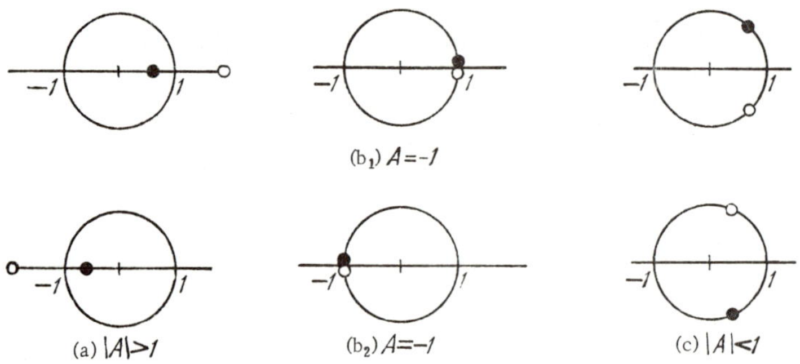

FIGURE 50.

Depending on the value of the characteristic constant, we distinguish three cases (Figure 50):

(a) $|A| > 1$; the multipliers (roots of the characteristic equation) $\rho_1, \rho_2 = \rho_1^{-1}$ are either both real and positive or both real and negative, and different from ± 1.

(b) $A = \pm 1$; then $\rho_1 = \rho_2 = A = \pm 1$.

(c) $|A| < 1$; the multipliers lie on the unit circle:

$$\rho_1 = e^{i\varphi}, \quad \rho_2 = e^{-i\varphi} \qquad (0 < \varphi < \pi).$$

In case (c) one of the multipliers is of the first kind and the other of the second (see §III. 1.4). Both cases are illustrated in Figure 50c.

With regard to stability, the situation in these three cases is as follows:

(a) Instability; there is one solution which increases exponentially as $t \to \infty$.

(b) There is at least one T-periodic solution (if $\rho_1 = \rho_2 = A = 1$) or T-antiperiodic solution (if $\rho_1 = \rho_2 = A = -1$).

(c) Stability; any solution-vector is either an almost-periodic function or a linear combination of two linearly independent solutions

$$\mathbf{x}_1(t)=e^{i\frac{\varphi}{T}t}\mathbf{u}(t), \quad \mathbf{x}_2(t)=e^{-i\frac{\varphi}{T}t}\overline{\mathbf{u}}(t),$$

where $\mathbf{u}(t)$ is a complex T-periodic vector-function.

A more detailed characterization of cases (a) and (b) will follow later from the Floquet-Lyapunov theorem.

For the present case, we have an improved version of the Floquet-Lyapunov theorem: (cf. § II. 2.2):

Theorem. The matrizant $\mathbf{X}(t)$ *of equation (1.2) may be expressed as*

$$\mathbf{X}(t)=\mathbf{F}(t)e^{t\mathbf{K}}, \tag{1.6}$$

where:

(α) $\mathbf{F}(t)$ *is a* T*-periodic* $(\mathbf{F}(t+T)\equiv\mathbf{F}(t))$ *or* T*-antiperiodic* $(\mathbf{F}(t+T)\equiv-\mathbf{F}(t))$ *real matrix-function;* $d\mathbf{F}/dt$ *is an integrable piecewise-continuous matrix-function;*

(β) $\det\mathbf{F}(t)\equiv 1$;

(γ) $\mathbf{K}$ *is a real matrix,* $\operatorname{Tr}\mathbf{K}=0$.

Conversely, any matrices $\mathbf{F}(t)$ *and* $\mathbf{K}$ *satisfying conditions* (α), (β) *and* (γ) *define (via (1.6)) the matrizant of an equation of type (1.3) with coefficient matrix*

$$\tilde{\mathbf{J}}^{-1}\mathbf{H}(t)=\frac{d\mathbf{X}}{dt}\mathbf{X}^{-1},$$

whose elements are periodic, integrable, piecewise-continuous functions; moreover, $\mathbf{H}(t)^{\tau}\equiv\mathbf{H}(t)$.

Proof. Let ρ_1 and ρ_2 be the eigenvalues of the matrix $\mathbf{X}(T)$ (the roots of equation (1.4)). Since A is a real number, ρ_1 and ρ_2 are complex conjugates $(\rho_2=\overline{\rho}_1)$ if they are not real. Thus ρ_1 and ρ_2 are either (1) both real and positive, (2) both real and negative, or (3) complex numbers on the unit circle.

In the first case, there exists a real matrix $\ln\mathbf{X}(T)$, in the second a real matrix $\ln[-\mathbf{X}(T)]$, and in the third real matrices $\ln\mathbf{X}(T)$ and $\ln[-\mathbf{X}(T)]$. This follows from § I. 2.7, but it will be proved independently in §1.4 below. Define a matrix

$$\mathbf{K}=\frac{1}{T}\ln[\pm\mathbf{X}(T)],$$

where the sign is chosen so that $\mathbf{K}$ is real. Define $\mathbf{F}(t)$ by (1.6). Inserting (1.6) into the identity (II. 2.2)

$$\mathbf{X}(t+T)=\mathbf{X}(t)\mathbf{X}(T),$$

we see that $\mathbf{F}(t+T)=\pm\mathbf{F}(t)$, where the sign is the same as that in the definition of $\mathbf{K}$. Thus part (α) of the theorem is proved. It follows from (1.6) via (II. 1.25) that $\det\mathbf{F}(t)=\det[\mathbf{X}(t)e^{-t\mathbf{K}}]=e^{-t\operatorname{Tr}\mathbf{K}}$. Since $\det\mathbf{F}(t)$ is a periodic function and $\operatorname{Tr}\mathbf{K}$ is real, it follows that $\operatorname{Tr}\mathbf{K}=0$, and so $\det\mathbf{F}(t)=1$. This proves parts (β) and (γ).

We now prove the converse. Denote

$$\mathbf{A}(t) = \frac{d\mathbf{X}}{dt}\mathbf{X}^{-1} = \frac{d\mathbf{F}}{dt}\mathbf{F}^{-1} + \mathbf{F}\mathbf{K}\mathbf{F}^{-1}.$$

Using (α), (β), we see that $\mathbf{A}(t)$ is integrable and piecewise-continuous and $\mathbf{A}(t+T) \equiv \mathbf{A}(t)$. Since

$$\det \mathbf{X}(t) \equiv 1, \quad \det \mathbf{X}(t) = e^{\int_0^t \operatorname{Tr} \mathbf{A}(t_1)\, dt_1},$$

it follows that

$$\int_0^t \operatorname{Tr} \mathbf{A}(t_1)\, dt_1 \equiv 0.$$

Hence $\operatorname{Tr} \mathbf{A}(t) \equiv 0$, so that the equation in question is indeed canonical. This completes the proof.

It follows from the above theorem that for each matrix $\mathbf{A}(t)$ we have a (not necessarily unique) pair $\{\mathbf{F}(t), \mathbf{K}\}$. Below we shall ascertain whether conditions can be added so as to make this correspondence single-valued. To this end we shall have to study the function $\mathbf{K} = \ln \mathbf{X}$ in more detail. Before doing so, we describe a simple special case in which the canonical equation (1.2) is integrable. *

1.3. An integrable case. Example

We wish to establish necessary and sufficient conditions for equation (1.2) to be orthogonally reducible (i.e., by an orthogonal transformation) to an equation with constant matrix. In the notation of §1.2, we have $\mathbf{X}(t) = \mathbf{F}(t)e^{t\mathbf{K}}$, and we are assuming that $\mathbf{F}(t)$ is an orthogonal matrix. Any 2×2 orthogonal matrix (see § I. 1.4) has the form

$$\begin{Vmatrix} \xi_1 & \eta_1 \\ \xi_2 & \eta_2 \end{Vmatrix},$$

where $\xi_1^2 + \xi_2^2 = 1$, $\eta_1^2 + \eta_2^2 = 1$, $\xi_1\eta_1 + \xi_2\eta_2 = 0$. Let $\xi_1 = \cos\vartheta$, $\xi_2 = \sin\vartheta$. Then $\eta_1 = -\sin\vartheta$, $\eta_2 = \cos\vartheta$. Hence

$$\mathbf{F}(t) = e^{-\vartheta(t)\tilde{\mathbf{J}}} = \begin{Vmatrix} \cos\vartheta(t) & -\sin\vartheta(t) \\ \sin\vartheta(t) & \cos\vartheta(t) \end{Vmatrix}.$$

As before, denoting $\mathbf{A}(t) = \tilde{\mathbf{J}}^{-1}\mathbf{H}(t)$ and $\mathbf{K} = \tilde{\mathbf{J}}^{-1}\mathbf{L}$, we obtain

$$\tilde{\mathbf{J}}^{-1}\mathbf{H}(t) = \mathbf{A}(t) = \frac{d\mathbf{X}}{dt}\mathbf{X}^{-1} = \frac{d\mathbf{F}}{dt}\mathbf{F}^{-1} + \mathbf{F}\mathbf{K}\mathbf{F}^{-1}.$$

* § 1.3 will not be used below, so the reader may proceed directly to § 1.4.

Hence

$$\mathbf{H}(t)=\frac{d\vartheta}{dt}\mathbf{I}_2+e^{-\vartheta\mathbf{J}}\mathbf{L}e^{\vartheta\mathbf{J}}\ . \tag{1.7}$$

This is the general formula of the Hamiltonian of an equation (1.2) of the desired type. Setting

$$\mathbf{H}(t)=\begin{Vmatrix} z+y & x \\ x & z-y \end{Vmatrix}, \quad \mathbf{L}=\begin{Vmatrix} c+b & a \\ a & c-b \end{Vmatrix}, \tag{1.8}$$

and performing some manipulations, we get

$$x=a\cos 2\vartheta+b\sin 2\vartheta, \quad y=-a\sin 2\vartheta+b\cos 2\vartheta, \quad z=\frac{d\vartheta}{dt}+c. \tag{1.9}$$

If $\frac{d\vartheta}{dt}=\text{const}$, then $\vartheta(t)=(\pi mt/T)+\vartheta_0$, and we obtain

$$\left.\begin{aligned} x&=a\cos 2\vartheta(t)+b\sin 2\vartheta(t),\\ y&=-a\sin 2\vartheta(t)+b\cos 2\vartheta(t),\\ z&=\frac{\pi m}{T}+c,\ \vartheta(t)=\frac{\pi mt}{T}+\vartheta_0. \end{aligned}\right\} \tag{1.10}$$

These formulas describe uniform motion along a circle in a plane perpendicular to the z-axis, with center on the z-axis. Conversely, any motion of this kind may be expressed as (1.10), with the Hamiltonian (1.8) falling into the desired class.

We can summarize our result as follows. Equation (1.2) is orthogonally reducible to an equation with constant matrix if and only if, with the elements of $\mathbf{H}(t)$ notated as in (1.8), the functions x, y, z may be expressed as (1.9). In particular, if the functions $x=x(t)$, $y=y(t)$, $z=z(t)$ describe uniform motion on a circle

$$x^2+y^2=\text{const}, \quad z=\text{const},$$

then equation (1.2) is orthogonally reducible to an equation with constant matrix. The Hamiltonian is defined by formulas (1.8), where a, b, c are given by (1.9) or (1.10), and the matrizant is

$$\mathbf{X}(t)=e^{-\vartheta\mathbf{J}}e^{-t\mathbf{JL}}=(\mathbf{I}_2\cos\vartheta-\mathbf{J}\sin\vartheta)e^{-t\mathbf{JL}}. \tag{1.11}$$

E x a m p l e . The equations for free vibrations of a gyropendulum in a two-gyroscope compass, for a circular course, simplified by disregarding nutation, are

$$\left.\begin{aligned} \dot{\xi}&=\mu\sin\omega t\cos(\omega t)\,\xi-[\nu+\mu\cos^2\omega t]\,\eta,\\ \dot{\eta}&=[\nu+\mu\sin^2\omega t]\xi-\mu\sin\omega t\cos(\omega t)\,\eta, \end{aligned}\right\} \tag{1.12}$$

where $\mu=m_1gl_1/H$, $\nu=m_2gl_2/H$ are constants; $\vartheta=\omega t$ is the course angle, $\omega=\text{const}$. System (1.12) is of canonical form, since the trace of the coefficient matrix is identically zero. In our case the coefficient matrix is

$$\tilde{J}^{-1}\mathbf{H}=\left\|\begin{matrix}\mu\sin\omega t\cos\omega t & -\nu-\mu\cos^2\omega t\\ \nu+\mu\sin^2\omega t & -\mu\sin\omega t\cos\omega t\end{matrix}\right\|.$$

Hence

$$\mathbf{H}=\left\|\begin{matrix}\nu+\mu\sin^2\omega t & -\mu\sin\omega t\cos\omega t\\ -\mu\sin\omega t\cos\omega t & \nu+\mu\cos^2\omega t\end{matrix}\right\|.$$

In the notation of (1.8), we have

$$\left.\begin{aligned} x&=-\mu\sin\omega t\cos\omega t=-\frac{1}{2}\mu\sin 2\omega t,\\ y&=\frac{1}{2}\mu(\sin^2\omega t-\cos^2\omega t)=-\frac{1}{2}\mu\cos 2\omega t,\\ z&=\mu+\nu.\end{aligned}\right\}\qquad(1.13)$$

We have thus shown that formulas (1.13) describe uniform motion on a circle $x^2+y^2=\text{const}$, $z=\text{const}$. Consequently, system (1.12) is reducible by an orthogonal matrix, that is to say, integrable in closed form. Comparing (1.9) and (1.13), we obtain

$$a=0,\quad b=-\frac{1}{2}\mu,\quad \vartheta=\omega t,\quad c=\mu+\nu-\omega.$$

Therefore,

$$\mathbf{L}=\left\|\begin{matrix}\frac{1}{2}\mu+\nu-\omega & 0\\ 0 & \frac{3}{2}\mu+\nu-\omega\end{matrix}\right\|,\quad \mathbf{K}=\tilde{J}^{-1}\mathbf{L}=\left\|\begin{matrix}0 & -\frac{3}{2}\mu-\nu+\omega\\ \frac{1}{2}\mu+\nu-\omega & 0\end{matrix}\right\|.$$

Denote the eigenvalues of $\mathbf{K}$ by $\pm\lambda$. Clearly,

$$\lambda=+\sqrt{\left(\frac{1}{2}\mu+\nu-\omega\right)\left(-\frac{3}{2}\mu-\nu+\omega\right)}.$$

The matrizant of system (1.12) is

$$\left\|\begin{matrix}\xi_1 & \xi_2\\ \eta_1 & \eta_2\end{matrix}\right\|=e^{-\vartheta\tilde{J}}e^{-t\tilde{J}\mathbf{L}}=$$

$$=\left\|\begin{matrix}\cos\omega t & -\sin\omega t\\ \sin\omega t & \cos\omega t\end{matrix}\right\|\cdot\left\|\begin{matrix}\operatorname{ch}\lambda t & \frac{1}{\lambda}\left(\omega-\nu-\frac{3}{2}\mu\right)\operatorname{sh}\lambda t\\ \frac{1}{\lambda}\left(\frac{1}{2}\mu+\nu-\omega\right)\operatorname{sh}\lambda t & \operatorname{ch}\lambda t\end{matrix}\right\|.$$

It is evident from the form of the matrix $\mathbf{K}$ that the motion is unstable for $^1/_2\,\mu+\nu<\omega<{}^3/_2\,\mu+\nu$ and stable for $\omega<{}^1/_2\,\mu+\nu$ and $\omega>{}^3/_2\,\mu+\nu$.

1.4. Lemma on real logarithms of real 2 × 2 matrices. Geometric interpretation

Let $\mathfrak{R}^3=\{\mathbf{K}\}$ denote the set of all real matrices $\mathbf{K}$ with trace zero. Writing $\mathbf{K}$ as

$$\mathbf{K}=\left\|\begin{matrix}-x & y-z\\ y+z & x\end{matrix}\right\|,$$

we see that $\tilde{\mathfrak{R}}^3$ is [isomorphic to] ordinary three-dimensional space, each point (x, y, z) corresponding to the above matrix $\mathbf{K}$. The characteristic equation $\det(\mathbf{K}-\lambda\mathbf{I}_2)=0$ is

$$\lambda^2=x^2+y^2-z^2.$$

We introduce the following notation: the cone $x^2+y^2=z^2$ will be denoted by $\tilde{\Pi}$; the exterior $x^2+y^2>z^2$ and interior $x^2+y^2<z^2$ of $\tilde{\Pi}$ will be denoted by $\tilde{\mathcal{H}}$ and $\tilde{\mathcal{O}}$, respectively (Figure 51). We shall show below that the sets $\tilde{\Pi}$, $\tilde{\mathcal{H}}$ and $\tilde{\mathcal{O}}$ correspond to periodic (antiperiodic), unbounded and bounded solutions, respectively. It is clear that the cone $\tilde{\Pi}$ is the locus of singular matrices: $\det\mathbf{K}=0$, $\lambda=0$. In the domain $\tilde{\mathcal{H}}$ we have $\det\mathbf{K}<0$ and $\lambda_{1,2}$ are real. In $\tilde{\mathcal{O}}$, $\det\mathbf{K}>0$ and $\lambda_{1,2}$ are pure imaginary. The locus of matrices with the same eigenvalues λ is the hyperboloid $\tilde{\Gamma}=\{x^2+y^2-z^2=\lambda^2\}$, of two sheets if the λ's are imaginary and of one sheet if they are real. The upper sheet of the cone $\tilde{\Pi}$ (without the vertex) is denoted by $\tilde{\Pi}_+$, the lower sheet (without vertex) by $\tilde{\Pi}_-$, the domain interior to the upper sheet by $\tilde{\mathcal{O}}_+$ and that interior to the lower sheet by $\tilde{\mathcal{O}}_-$. The upper and lower sheets of a hyperboloid $\tilde{\Gamma}$ of two sheets will be denoted by $\tilde{\Gamma}_+$ and $\tilde{\Gamma}_-$ (Figure 51).

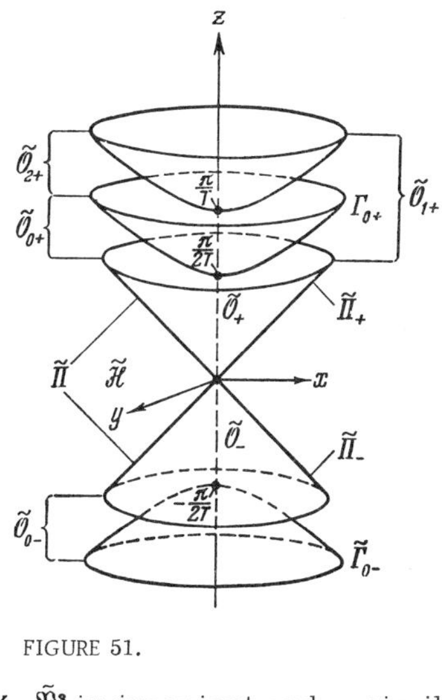

FIGURE 51.

Since $\mathrm{Tr}(\mathbf{S}^{-1}\mathbf{KS})=\mathrm{Tr}\,\mathbf{K}$, $\tilde{\mathfrak{R}}^3$ is invariant under similarity transformations. Since eigenvalues are invariants of similarity transformations, it follows that as $\mathbf{S}$ varies the matrix $\mathbf{K}_S=\mathbf{S}^{-1}\mathbf{KS}$ varies over the same one-sheeted or two-sheeted hyperboloid (cone) as $\mathbf{K}$.

If $\mathbf{K}$ is on a two-sheeted hyperboloid $\tilde{\Gamma}$, then $\mathbf{K}_S$ is on the same sheet as $\mathbf{K}$ if $\det\mathbf{S}>0$ and on the other sheet if $\det\mathbf{S}<0$. Indeed, we may assume without loss of generality that $\det\mathbf{S}=\pm 1$. Let $\det\mathbf{S}=1$. We shall show below

(§1.5) that the set of real matrices $\mathbf{S}$ with determinant $+1$ is connected. Thus there is a curve joining $\mathbf{S}$ continuously to the identity matrix $\mathbf{I}_2$, so that we can join $\mathbf{K}_S$ to $\mathbf{K}$. Thus $\mathbf{K}_S$ must lie on the same sheet as $\mathbf{K}$. If $\det \mathbf{S}=-1$, then $\mathbf{S}=\mathbf{S}_0\mathbf{S}_1$, where

$$\mathbf{S}_0=\begin{Vmatrix}1 & 0\\ 0 & -1\end{Vmatrix}, \quad \det \mathbf{S}_1=1.$$

There is a curve joining $\mathbf{S}_1$ to $\mathbf{I}_2$, so we can join $\mathbf{K}_S$ to $\mathbf{K}_{S_0}$. If $\mathbf{K}\sim(x, y, z)$, then $\mathbf{K}_{S_0}\sim(x, -y, -z)$. Clearly $\mathbf{K}_{S_0}$ lies on the opposite sheet to $\mathbf{K}$, and so $\mathbf{K}_S$ is also on the opposite sheet.

The converse is also valid. If $\mathbf{K}_1$ and $\mathbf{K}_2$ lie on the same hyperboloid $\lambda^2=x^2+y^2-z^2$, then they are similar: $\mathbf{K}_2=\mathbf{S}^{-1}\mathbf{K}_1\mathbf{S}$.

Indeed, since $\mathbf{K}_1$ and $\mathbf{K}_2$ are not on the cone $\tilde{\Pi}$, their eigenvalues are distinct. Since $\det\mathbf{K}_1=\det\mathbf{K}_2$, they have the same eigenvalues, and so $\mathbf{K}_1$ and are similar.

It follows from general theorems of linear algebra (§I. 1.11) that any matrix $\mathbf{K}\in\tilde{\mathfrak{R}}^3$ may be brought by a similarity transformation to one of the following forms:

(a) $\mathbf{K}=\mathbf{S}\begin{Vmatrix}\lambda & 0\\ 0 & -\lambda\end{Vmatrix}\mathbf{S}^{-1}$ (matrices in $\tilde{\mathcal{H}}$),

(b) $\mathbf{K}=\mathbf{S}\begin{Vmatrix}0 & \varepsilon\\ 0 & 0\end{Vmatrix}\mathbf{S}^{-1}$ (matrices on the cone $\tilde{\Pi}$),

(c) $\mathbf{K}=\mathbf{S}\begin{Vmatrix}0 & -\psi\\ \psi & 0\end{Vmatrix}\mathbf{S}^{-1}=-\psi\mathbf{S}\tilde{\mathbf{J}}\mathbf{S}^{-1}$ (matrices in $\tilde{\mathfrak{G}}$).

Here λ, ε, ψ are real numbers and $\mathbf{S}$ a real matrix. Since

$$\mathbf{S}^{-1}e^{t\mathbf{K}}\mathbf{S}=\exp(t\mathbf{S}^{-1}\mathbf{K}\mathbf{S}),$$

we obtain

$$e^{t\mathbf{K}}=\mathbf{S}\begin{Vmatrix}e^{\lambda t} & 0\\ 0 & e^{-\lambda t}\end{Vmatrix}\mathbf{S}^{-1} \quad \text{if} \quad \mathbf{K}\in\tilde{\mathcal{H}},$$
$$e^{t\mathbf{K}}=\mathbf{S}\begin{Vmatrix}1 & \varepsilon t\\ 0 & 1\end{Vmatrix}\mathbf{S}^{-1} \quad \text{if} \quad \mathbf{K}\in\tilde{\Pi},$$
$$e^{t\mathbf{K}}=\mathbf{S}e^{-\psi t\tilde{\mathbf{J}}}\mathbf{S}^{-1}=\mathbf{S}(\mathbf{I}_2\cos\psi t-\tilde{\mathbf{J}}\sin\psi t)\mathbf{S}^{-1} \quad \text{if} \quad \mathbf{K}\in\tilde{\mathfrak{G}}.$$

These formulas, together with the Floquet-Lyapunov theorem of §1.2 (see (1.6)), imply: the solutions of system (1.2) are unbounded if $\mathbf{K}\in\tilde{\mathcal{H}}$; if $\mathbf{K}\in\tilde{\Pi}$ but $\mathbf{K}\neq\mathbf{0}$, there is one periodic or antiperiodic solution (with initial value $\mathbf{x}(0)=\mathbf{S}\begin{pmatrix}1\\0\end{pmatrix}$); if $\mathbf{K}\in\tilde{\mathfrak{G}}$ all the solutions are bounded (the trivial solution is stable). Recalling that the multipliers $\rho_{1,2}$ are related to the characteristic exponents $\lambda_{1,2}=\pm\sqrt{-\det\mathbf{K}}$ by $\rho_{1,2}=\exp(\lambda_{1,2}T)$, we see that cases (a), (b), (c) of §1.2 are characterized by the following types of solution:

(a′) The multipliers $\rho_1=\rho_2^{-1}$ are real, distinct from ± 1, $\mathbf{K}\in\tilde{\mathcal{H}}$. There are two linearly independent solutions

$$\mathbf{x}_1(t)=e^{\lambda t}\mathbf{u}_1(t), \quad \mathbf{x}_2(t)=e^{-\lambda t}\mathbf{u}_2(t),$$

where $e^{\pm\lambda T}=|\rho_{1,2}|$, $\lambda>0$ and the vector-functions $\mathbf{u}_1(t)$, $\mathbf{u}_2(t)$ are T-periodic if $\rho_1=\rho_2^{-1}>0$ and T-antiperiodic if $\rho_1=\rho_2^{-1}<0$. The trivial solution of equation (1.2) is unstable.

(b′) The multipliers are $\rho_1=\rho_2=\pm 1$. Depending on whether $\mathbf{X}(T)=\pm\mathbf{I}_2$ or not, we have two subcases:

(b'_1) $\mathbf{X}(T)\neq\pm\mathbf{I}_2$, and so $\varepsilon\neq 0$, $\mathbf{K}\in\tilde{\Pi}_+\cup\tilde{\Pi}_-$. There are two linearly independent solutions $\mathbf{x}_1(t)=\mathbf{u}_1(t)$, $\mathbf{x}_2(t)=t\mathbf{u}_2(t)$, where $\mathbf{u}_1(t)$, $\mathbf{u}_2(t)$ are T-periodic if $\rho_1=\rho_2=1$ and T-antiperiodic if $\rho_1=\rho_2=-1$. The trivial solution of equation (1.2) is unstable.

(b'_2) $\mathbf{X}(T)=\pm\mathbf{I}_2$, $\mathbf{K}=\mathbf{0}$ or $\mathbf{K}$ lies on the two-sheeted hyperboloid $\tilde{\Gamma}_n=\{x^2+y^2-z^2=-n^2\pi^2/T^2\}$. There are two T-periodic (if $\mathbf{X}(T)=\mathbf{I}_2$) or T-antiperiodic (if $\mathbf{X}(T)=-\mathbf{I}_2$) linearly independent solutions $\mathbf{x}_1(t)$, $\mathbf{x}_2(t)$. The trivial solution is stable.

(c) The multipliers $\rho_1=e^{i\psi}$, $\rho_2=e^{-i\psi}$ lie on the unit circle and are distinct from ± 1, $\mathbf{K}\in\tilde{\mathfrak{G}}$, $\mathbf{K}\neq 0$, $\mathbf{K}\notin\tilde{\Gamma}_n$. There are two linearly independent solutions

$$\mathbf{x}_1(t)=e^{i\frac{\psi}{T}}\mathbf{w}(t),\qquad \mathbf{x}_2(t)=e^{-i\frac{\psi}{T}}\bar{\mathbf{w}}(t),$$

where $\mathbf{w}(t)$ is a complex T-periodic vector-function.

Throughout this section the reader should remember this classification of solutions for different matrices $\mathbf{K}\in\tilde{\mathfrak{K}}^3$.

We now wish to consider whether the representation (1.6) uniquely determines the matrix $\mathbf{K}$ (and hence also $\mathbf{F}(t)$). Denote $\mathbf{B}=\mathbf{X}(T)$. Recall (see §1.2) that the matrix $T\mathbf{K}$ is a solution of the equation

$$e^{\mathbf{Y}}=\pm\mathbf{B}, \tag{1.14}$$

the sign being chosen so that the solution $\mathbf{Y}$ is real.

We know (§I. 3.4) that, by a similarity transformation with a real matrix and suitable choice of sign, the matrix $\pm\mathbf{B}$ may be reduced to one of the forms:

(a″) $$\begin{Vmatrix}\mu & 0\\ 0 & \mu^{-1}\end{Vmatrix},\quad \mu>0,\quad \mu\neq 1,$$

(b″) $$\mathbf{I}_2+\begin{Vmatrix}0 & 1\\ 0 & 0\end{Vmatrix},$$

(c″) $$\begin{Vmatrix}\alpha & \beta\\ -\beta & \alpha\end{Vmatrix},\quad \alpha^2+\beta^2=1.$$

Case (a″) will hold if the eigenvalues ρ_1, $\rho_2=\rho_1^{-1}$ of $\mathbf{B}$ are real and $\rho_{1,2}\neq\pm 1$, case (b″) if $\rho_1=\rho_2=\pm 1$ and $\mathbf{B}\neq\pm\mathbf{I}_2$, and case (c″) if either $\rho_{1,2}=e^{\pm i\psi}$, $0<\psi<\pi$, or $\mathbf{B}=\pm\mathbf{I}_2$ (and so $\rho_1=\rho_2=\pm 1$).

These cases exhaust all possibilities.

We shall need the following proposition concerning the logarithm of a real matrix with determinant 1. *

* It is easy to see that the determination of all real solutions of equation (1.14) in the general case may be reduced to the case $\det\mathbf{B}=1$. All references to equation (1.14) mean an equation for the matrix $\mathbf{Y}$ with a definite sign on the right, i.e., the equation $e^{\mathbf{Y}}=\zeta\mathbf{B}$ where $\zeta=\pm 1$.

Lemma. All real solutions of equation (1.14) are determined as follows.

1) If $\pm\mathbf{B}$ is reduced by suitable choice of sign and a similarity transformation to case (a″) or (b″), there is a solution only if the eigenvalues of $\pm\mathbf{B}$ are positive. This condition determines the sign in (1.14). The solution is unique and is defined as follows:

$$\text{in case (a''): } \ln(\pm\mathbf{B}) = \pm\frac{2\mu\ln\mu}{\mu^2-1}\mathbf{B} - \frac{\mu^2+1}{\mu^2-1}\ln\mu\,\mathbf{I}_2, \tag{1.15}$$

$$\text{in case (b''): } \ln(\pm\mathbf{B}) = \pm\mathbf{B} - \mathbf{I}_2. \tag{1.16}$$

2) If $\mathbf{B} = \pm\mathbf{I}_2$, there are real solutions for any choice of sign in (1.14) and they are defined by

$$\ln(\pm\mathbf{I}_2) = n\pi\mathbf{S}^{-1}\tilde{\mathbf{J}}\mathbf{S} \qquad (n = 0, \pm 1, \pm 2, \ldots), \tag{1.17}$$

where n is even if the plus sign is chosen and odd otherwise, $\mathbf{S}$ is any nonsingular real matrix.

3) If $\pm\mathbf{B}$ may be brought to case (c″) and $\mathbf{B}\neq\pm\mathbf{I}_2$, there is a real solution for any choice of sign in (1.14) and all values of $\ln(\pm\mathbf{B})$ are defined by

$$\ln(\pm\mathbf{B}) = \frac{\psi+n\pi}{\sin\psi}\mathbf{B} - (\psi+n\pi)\operatorname{ctg}\psi\,\mathbf{I}_2 \qquad (n = 0, \pm 1, \pm 2, \ldots), \tag{1.18}$$

where $\exp(\pm i\psi)$ are the eigenvalues of $\mathbf{B}$ $(0<\psi<\pi)$ and n is even if the plus sign is chosen, odd otherwise.

Formulas (1.15)–(1.18) follow at once from standard formulas for 2×2 matrices (see §I. 2.3). The only point requiring proof is that all real solutions of equation (1.14) are given by these formulas. We shall nevertheless present an independent proof of (1.15)–(1.18).

Proof. Since $\mathbf{S}^{-1}\mathbf{B}\mathbf{S} = e^{\mathbf{S}^{-1}\mathbf{Y}\mathbf{S}}$, we may assume from the start that $\pm\mathbf{B}$ has one of the forms (a″), (b″), (c″). The eigenvalues ξ_i of the matrix $\mathbf{B}$ are related to the eigenvalues η_i of $\mathbf{Y}$ by $e^{\eta_i} = \pm\xi_i$ (the sign being the same as in (1.14)), and moreover $\eta_1+\eta_2 = 0$ since $\det\mathbf{B} = e^{\eta_1+\eta_2} = 1$.

In case (a″) the numbers η_i are real. In case (a″), therefore, the sign in (1.14) must be chosen so that the eigenvalues of $\pm\mathbf{B}$ are positive. Then the eigenvalues of $\mathbf{Y}$ are uniquely determined — they are $\pm\ln\mu$. Consequently,

$$\mathbf{Y} = \mathbf{S}^{-1}\begin{Vmatrix}\ln\mu & 0\\ 0 & -\ln\mu\end{Vmatrix}\mathbf{S}.$$

Inserting this into (1.14) with the matrix $\pm\mathbf{B}$ of type (a″), we see that $\mathbf{B}$ commutes with any diagonal matrix. Therefore

$$\mathbf{Y} = \begin{Vmatrix}\ln\mu & 0\\ 0 & -\ln\mu\end{Vmatrix},$$

so that the real matrix $\mathbf{Y}$ is uniquely defined. Recalling that $\pm\mathbf{B}$ has the form (a''), we express this solution in the form of (1.15). Applying a suitable similarity transformation, we see that the solution has the same form for the original matrix $\mathbf{B}$.

In case (b'') the matrix $\mathbf{Y}$ (if it exists) has a multiple elementary divisor (reduces to an elementary Jordan matrix), since otherwise both $\mathbf{B}$ and $\mathbf{Y}$ would be reducible to diagonal form. Therefore

$$\mathbf{Y}=\mathbf{S}^{-1}\begin{Vmatrix}0 & 1\\ 0 & 0\end{Vmatrix}\mathbf{S},$$

and, inserting in (1.14), we find that $\mathbf{Y}=\begin{Vmatrix}0 & 1\\ 0 & 0\end{Vmatrix}$. Thus the sign in (1.14) must again be chosen so that the eigenvalues of $\pm\mathbf{B}$ are positive. We thus have $\mathbf{Y}=(\pm\mathbf{B})-\mathbf{I}_2$. Returning via similarity to the original matrices $\mathbf{Y}$ and $\mathbf{B}$, we obtain (1.16). Thus, in the unbounded case ($\mathbf{X}(T)$ reducible to the form (a'') or (b'')) the matrix $\mathbf{K}$ is unique, and therefore so is $\mathbf{F}(t)$.

In case (c'') the matrix $\mathbf{Y}$ (if it exists) is reducible to the form (c), since if it were reducible to either of forms (a) or (b) ($\varepsilon\neq 0$) it would follow that $\pm\mathbf{B}$ is reducible to form (a'') or (b''), respectively, which is not the case. We first find the matrices $\mathbf{Y}$ of the form $\beta\tilde{\mathbf{J}}$ (all others are similar to these). With the complex number $\zeta=\alpha+i\beta$ we associate the matrix

$$\mathbf{K}_\zeta=\begin{Vmatrix}\alpha & \beta\\ -\beta & \alpha\end{Vmatrix}=\alpha\mathbf{I}_2+\beta\tilde{\mathbf{J}}.$$

This correspondence is an isomorphism, i. e., $\mathbf{K}_\zeta$ is merely another notation for the complex number ζ. Therefore all solutions of equation (1.14) in case (c'') are determined by the formula

$$\mathbf{Y}=\ln(\pm\mathbf{B})\sim\ln(\pm\zeta)=\ln|\zeta|+i(\arg\zeta+n\pi)=i(\arcsin\beta+n\pi),$$

i. e.,

$$\mathbf{Y}=(\arcsin\beta+n\pi)\,\tilde{\mathbf{J}}.$$

The number n is even if the plus sign is chosen in (1. 14), odd otherwise. Thus the solution $\mathbf{Y}$ must have the form

$$\mathbf{Y}=(\arcsin\beta+n\pi)\,\mathbf{S}^{-1}\tilde{\mathbf{J}}\mathbf{S}.$$

Inserting this in (1.14), we see that

$$(-1)^n\mathbf{S}^{-1}(\alpha\mathbf{I}_2+\beta\tilde{\mathbf{J}})\,\mathbf{S}=\pm(\alpha\mathbf{I}_2+\beta\tilde{\mathbf{J}})$$

or

$$\beta\mathbf{S}^{-1}\tilde{\mathbf{J}}\mathbf{S}=\beta\tilde{\mathbf{J}}.$$

If $\beta\neq 0$, then $\mathbf{S}^{-1}\tilde{\mathbf{J}}\mathbf{S}=\tilde{\mathbf{J}}$, and all solutions $\mathbf{Y}$ are given by

$$\mathbf{Y} = (\arcsin\beta + n\pi)\,\tilde{\mathbf{J}}.$$

Noting that $\mathbf{B} = \alpha\mathbf{I}_2 + \beta\tilde{\mathbf{J}}$, $\alpha^2 + \beta^2 = 1$, we transform this formula to the form of (1.18). If $\beta = 0$, then $\mathbf{Y} = n\pi\mathbf{S}^{-1}\tilde{\mathbf{J}}\mathbf{S}$. In this case $\mathbf{B} = \pm\mathbf{I}_2$, and the general formula for $\mathbf{Y}$ is given by (1.17), with n even if the sign is plus and odd otherwise, and $\mathbf{S}$ any nonsingular matrix. This completes the proof of the lemma.

The geometric interpretation of this lemma will be important in what follows. By (1.14),

$$\det e^{\mathbf{Y}} = e^{\operatorname{Tr}\mathbf{Y}} = 1, \quad \operatorname{Tr}\mathbf{Y} = 0,$$

so that all the real solutions $\mathbf{Y}$ of equation (1.14) are in $\tilde{\mathfrak{R}}^3$.

It is easy to see that the only possible combinations of matrices $\mathbf{Y}$ and $\pm\mathbf{B}$ are: cases (a) and (a''), cases (b) and (b''), or cases (c) and (c''). We are assuming here that in case (b) $\varepsilon \neq 0$ and the matrix $\mathbf{Y} = 0$ reduces to case (c) with $\psi = 0$.

Thus, in the "unbounded case" ($\mathbf{X}(T) = \mathbf{B}$ reduces to the form (a') or (b')) the matrix $\mathbf{K} = T^{-1}\ln[\pm\mathbf{X}(T)]$ is uniquely defined and $\mathbf{K} \in \tilde{\mathscr{H}} \cup \tilde{\Pi}_+ \cup \tilde{\Pi}_-$. In the "bounded case" ($\mathbf{X}(T)$ reducible to form (c'); $\mathbf{K} \in \tilde{\mathfrak{G}}$ or $\mathbf{K} = \mathbf{0}$) the matrix $\mathbf{K}$ is not uniquely defined. We discuss this case in greater detail.

Let $\mathbf{K} \in \tilde{\mathfrak{G}}$. Consider the straight line in $\tilde{\mathfrak{R}}^3$ through the origin and the point $\mathbf{K}$. Throughout the sequel, $\mathbf{J}_K$ will denote the point in $\tilde{\mathfrak{R}}^3$ at the intersection of this line and the upper sheet of the hyperboloid $x^2 + y^2 - z^2 = -1$. In other words, $\mathbf{J}_\mathbf{K} = \pm\dfrac{1}{\sqrt{\det\mathbf{K}}}\mathbf{K}$, with the plus sign if $\mathbf{K} \in \tilde{\mathfrak{G}}_+$ and minus if $\mathbf{K} \in \tilde{\mathfrak{G}}_-$. Since the matrices $\mathbf{J}_\mathbf{K}$ and $(-\tilde{\mathbf{J}})$ are on the same sheet of the hyperboloid $x^2 + y^2 - z^2 = -1$, it follows by what we proved above that

$$\mathbf{J}_\mathbf{K} = -\mathbf{S}^{-1}\tilde{\mathbf{J}}\mathbf{S}, \quad \det\mathbf{S} > 0.$$

It follows readily from (1.18) that all solutions of equation (1.14) may be derived from one such solution, using the formula

$$\mathbf{Y} = \mathbf{Y}_0 + n\pi\mathbf{J}_{\mathbf{Y}_0} \qquad (n = 0, \pm 1, \pm 2, \ldots),$$

so that all the solutions (1.18) lie on one straight line in $\tilde{\mathfrak{R}}^3$. Similarly, $\ln(-\mathbf{B}) = \ln\mathbf{B} + \pi\mathbf{J}_{\mathbf{Y}_0}$ (here ln denotes a many-valued function). Thus the matrices $\mathbf{J}_\mathbf{K}$ play the role of the imaginary unit. Formula (1.17) may be written as $\ln(\pm\mathbf{I}_2) = n\pi\mathbf{J}_\mathbf{K}$, where $\mathbf{K}$ is any matrix in $\tilde{\mathfrak{G}}$. The locus of the values of $\ln(\pm\mathbf{I}_2)$ is the union of the hyperboloids $x^2 + y^2 - z^2 = -n^2\pi^2$. The domains between adjacent sheets of the hyperboloids $x^2 + y^2 - z^2 = -n^2\pi^2$ are clearly unicity domains for the function $e^\mathbf{K}$ ($\mathbf{K} \in \tilde{\mathfrak{R}}^3$). Moreover, if $\mathbf{K}_1$ and $\mathbf{K}_2$ lie in one of these domains, then

$$e^{\mathbf{K}_1} \neq e^{\mathbf{K}_2}, \quad e^{\mathbf{K}_1} \neq -e^{\mathbf{K}_2}.$$

We shall use these geometric corollaries of the lemma quite frequently.

We now turn to the question of the uniqueness of $\mathbf{X}(t)$ as defined by (1.6). Let $\tilde{\Gamma}_{0+}$ and $\tilde{\Gamma}_{0-}$ denote the upper $(z>0)$ and lower $(z<0)$ sheets of the hyperboloid $x^2+y^2-z^2=-\pi^2/(4T^2)$, $\tilde{\mathfrak{G}}_{0+}$ and $\tilde{\mathfrak{G}}_{0-}$ the upper $(z>0)$ and lower $(z<0)$ domains between the cone $x^2+y^2-z^2=0$ $(\tilde{\Pi})$ and this hyperboloid, $\tilde{\mathfrak{G}}_{1+}$ the domain between the upper sheet $\tilde{\Pi}_+$ of the cone $x^2+y^2-z^2=0$ and the upper sheet $\tilde{\Gamma}_{0+}$ of the hyperboloid $x^2+y^2-z^2=-\frac{\pi^2}{T^2}$ (see Figure 51). It follows from the above discussion that the matrix $\mathbf{K}=T^{-1}\ln[\pm\mathbf{X}(T)]$ is uniquely defined if we take $\mathbf{K}$ in the domain $\tilde{\mathfrak{R}}_0=\tilde{\mathcal{H}}\cup\tilde{\Pi}\cup\tilde{\mathfrak{G}}_{0+}\cup\tilde{\mathfrak{G}}_{0-}\cup\tilde{\Gamma}_{0+}$. Instead of $\tilde{\mathfrak{R}}_0$, we can also take the domain $\tilde{\mathfrak{R}}_1=\tilde{\mathcal{H}}\cup\tilde{\Pi}\cup\tilde{\mathfrak{G}}_{1+}$. Recalling that the matrix-function $\mathbf{F}(t)$ is uniquely defined by $\mathbf{K}$, we obtain the following useful proposition.

Theorem (complement to theorem of § 1.2). The matrix $\mathbf{K}$ *in (1.6) may always be chosen in the domain* $\tilde{\mathfrak{R}}_0$, *and when this is done the matrices* $\mathbf{K}$ *and* $\mathbf{F}(t)$ *in (1.6) are uniquely defined. Instead of* $\tilde{\mathfrak{R}}_0$, *one can also take the domain* $\tilde{\mathfrak{R}}_1$.

It is clear that $\tilde{\mathfrak{G}}_{0+}\cup\tilde{\mathfrak{G}}_{0-}\cup\tilde{\Gamma}_{0+}$ may be replaced by any "half-open" domain between the sheets of two-sheeted hyperboloids distant π/T from each other along the z-axis, lying on different sides of the origin. The same arguments go through if $\tilde{\Gamma}_{0+}$ is replaced by $\tilde{\Gamma}_{0-}$.

1.5. Structure of the group of 2 × 2 symplectic matrices

Let $\mathfrak{G}=\{\mathbf{X}\}$ denote the set of real 2×2 matrices with determinant unity. The matrizant $\mathbf{X}(t)$ of equation (1.2) is a trajectory in $\mathfrak{G}$. Our next problem is to determine the "structure" of the set $\mathfrak{G}$, in particular, to ascertain the "relative positions" in this set of matrices $\mathbf{X}$ with various prescribed eigenvalues; we shall be especially interested in the case of monodromy matrices $\mathbf{X}(T)$ of equations (1.2) with bounded and unbounded solutions.

We first examine the structure of $\mathfrak{G}$ "in the large." Namely, we shall show that, endowed with a natural metric, $\mathfrak{G}$ is homeomorphic to the interior of a torus. Let $\mathbf{X}=\|\mathbf{a},\mathbf{b}\|$, so that $\mathbf{a}$ and $\mathbf{b}$ denote the columns of $\mathbf{X}$. Let α be the angle between the vectors $\mathbf{a}$ and $\mathbf{b}$. Since $\det\mathbf{X}=|\mathbf{a}|\cdot|\mathbf{b}|\sin\alpha=1$, it follows that $0<\alpha<\pi$ and the matrix $\mathbf{X}$ is uniquely determined by three numbers: the vector $\mathbf{a}$ (or $\mathbf{b}$) and the angle α. Consider the real three-dimensional space $\{(\xi,\ \eta,\ \zeta)\}$ and set $\mathbf{a}=\begin{pmatrix}\xi\\ \eta\end{pmatrix}$, $\alpha=\zeta$. The corresponding points will fill out an open doubly-connected domain $0<\zeta<\pi$, $\xi^2+\eta^2\neq 0$, which is clearly homeomorphic to the interior of a torus, as claimed.

An important feature of this model is the following. Let us assume that the domain $0<\zeta<\pi$, $\xi^2+\eta^2\neq 0$, is deformed into the interior of a torus in such a way that points on planes passing through the ζ-axis move in these planes throughout. Recalling that the matrizant $\mathbf{X}(t)$, $0\leqslant t\leqslant T$, is a curve in $\mathfrak{G}$, we see that the angle φ between the meridional planes through the matrices $\mathbf{I}_2$ and $\mathbf{X}(T)$ is the angle formed by the column $\mathbf{a}$ of $\mathbf{X}(T)$ and the vector $\mathbf{e}_1=\begin{pmatrix}1\\ 0\end{pmatrix}$ (the corresponding column of $\mathbf{I}_2$). A column of $\mathbf{X}(t)$ is a solution of equation (1.3). Thus the angle φ between these planes is the angle through which the solution-vector $\mathbf{x}(t)$ $(\mathbf{x}(0)=\mathbf{e}_1)$ rotates in time T.

We now wish to determine the relative positions in $\mathfrak{G}$ of matrices with a specific canonical structure — matrices reducible to one of the forms (a''), (b''), (c'') of §1.4. Since the analogous problem for real matrices with trace zero was solved in §1.4, we shall try to reduce the present problem to the study of matrices $\mathbf{K}\in\tilde{\mathfrak{R}}^3$.

Let $\tilde{\mathfrak{R}}^+$ and $\tilde{\mathfrak{R}}^-$ be two copies of the space $\tilde{\mathfrak{R}}^3$ defined above. All the sets defined in §1.4 will be denoted by the same symbols with additional superscripts plus (for $\tilde{\mathfrak{R}}^+$) or minus (for $\tilde{\mathfrak{R}}^-$). Recall that $\tilde{\mathfrak{R}}_0$ was defined as the subset of $\tilde{\mathfrak{R}}^3$ between the sheets $\tilde{\Gamma}_{0-}$ and $\tilde{\Gamma}_{0+}$ of the hyperboloid $x^2+y^2-z^2=-\pi^2/(4T^2)$, "half-open" in the sense that $\tilde{\Gamma}_{0-}$ is not a subset of $\tilde{\mathfrak{R}}_0$ but $\tilde{\Gamma}_{0+}$ is; thus $\tilde{\mathfrak{R}}_0=\tilde{\mathscr{H}}\cup\tilde{\Pi}\cup\tilde{\mathscr{O}}_{0+}\cup\tilde{\mathscr{O}}_{0-}\cup\tilde{\Gamma}_{0+}$ (see §1.4). The corresponding sets in $\tilde{\mathfrak{R}}^+$ and $\tilde{\mathfrak{R}}^-$ will be

$$\tilde{\mathfrak{R}}_0^+=\tilde{\mathscr{H}}^+\cup\tilde{\Pi}^+\cup\tilde{\mathscr{O}}_{0+}^+\cup\tilde{\mathscr{O}}_{0-}^+\cup\tilde{\Gamma}_{0+}^+,\quad \tilde{\mathfrak{R}}_0^-=\tilde{\mathscr{H}}^-\cup\tilde{\Pi}^-\cup\tilde{\mathscr{O}}_{0+}^-\cup\tilde{\mathscr{O}}_{0-}^-\cup\tilde{\Gamma}_{0+}^-.$$

The positions of these sets in $\tilde{\mathfrak{R}}_0^+, \tilde{\mathfrak{R}}_0^-$ are shown schematically in Figure 52; the reader is advised to familiarize himself with the situation.

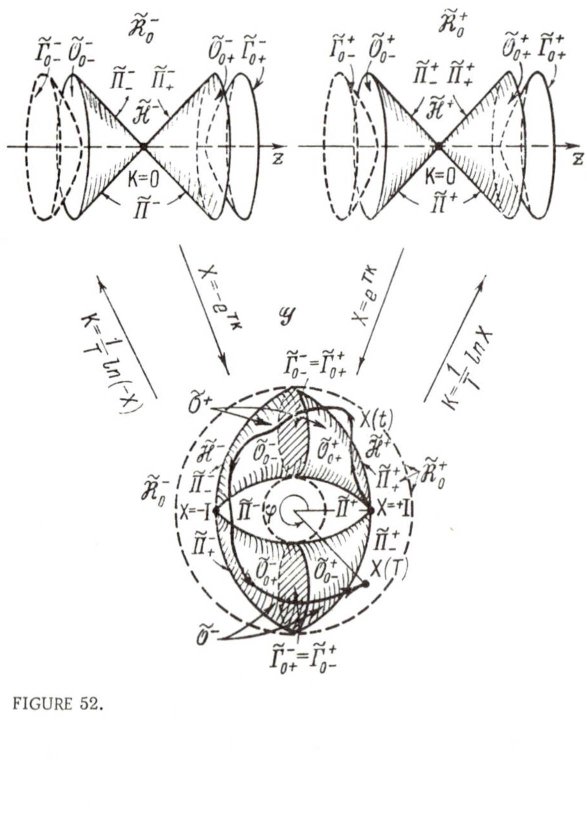

FIGURE 52.

We now introduce mappings*

$$\tilde{\mathfrak{R}}_0^+ \xrightarrow{X=e^{TK}} \mathfrak{G}, \quad \tilde{\mathfrak{R}}_0^- \xrightarrow{X=-e^{TK}} \mathfrak{G}$$

of the sets $\tilde{\mathfrak{R}}_0^+$ and $\tilde{\mathfrak{R}}_0^-$ into $\mathfrak{G}$. The first mapping is defined by $\mathbf{X}=e^{T\mathbf{K}}$ and the second by $\mathbf{X}=e^{-T\mathbf{K}}$. Since $\det e^{T\mathbf{K}}=e^{T\operatorname{Tr}\mathbf{K}}$ and the matrices $\mathbf{K}\in\mathfrak{R}^+$, $\mathbf{K}\in\mathfrak{R}^-$ have trace zero, it follows that in either case $\det\mathbf{X}=1$, so that the ranges of these mappings are indeed in $\mathfrak{G}$. It is by no means trivial to show that, first, the image of $\tilde{\mathfrak{R}}_0^+\cup\tilde{\mathfrak{R}}_0^-$ in $\mathfrak{G}$ exhausts all of $\mathfrak{G}$, and, second, the mapping is one-to-one. However, the proof is implicit in the reasoning at the end of §1.4. Indeed, it was shown there that for any matrix $\mathbf{X}\in\mathfrak{G}$ we have either $\mathbf{K}=T^{-1}\ln\mathbf{X}\in\tilde{\mathfrak{R}}_0^+$ or $\mathbf{K}=T^{-1}\ln(-\mathbf{X})\in\tilde{\mathfrak{R}}_0^-$. We also showed that only one value of the many-valued function $T^{-1}\ln(\pm\mathbf{X})$ lies in $\tilde{\mathfrak{R}}_0$. This means that for any two matrices $\mathbf{K}_1\in\tilde{\mathfrak{R}}_0^+$, $\mathbf{K}_2\in\tilde{\mathfrak{R}}_0^-$ we have $e^{T\mathbf{K}_1}=-e^{T\mathbf{K}_2}$, so that the mapping of $\tilde{\mathfrak{R}}_0^+\cup\tilde{\mathfrak{R}}_0^-$ onto $\mathfrak{G}$ is indeed one-to-one.

We claim that the image of the set $\tilde{\Gamma}_{0-}^-$ under the mapping $\mathbf{X}=-e^{T\mathbf{K}}$ coincides with the image of $\tilde{\Gamma}_{0+}^+$ under the mapping $\mathbf{X}=e^{T\mathbf{K}}$. (Recall that $\tilde{\Gamma}_{0-}^-$ is not a subset of $\tilde{\mathfrak{R}}_0^-$.) Indeed, it follows from $\mathbf{K}_1\in\tilde{\Gamma}_{0+}^+$, $\mathbf{K}_2\in\tilde{\Gamma}_{0-}^-$ that**

$$\mathbf{K}_1=\frac{\pi}{2T}\tilde{\mathbf{J}}_1, \quad \mathbf{K}_2=-\frac{\pi}{2T}\tilde{\mathbf{J}}_2,$$

where $\tilde{\mathbf{J}}_1=\mathbf{S}_1^{-1}\tilde{\mathbf{J}}\mathbf{S}_1$, $\tilde{\mathbf{J}}_2=\mathbf{S}_2^{-1}\tilde{\mathbf{J}}\mathbf{S}_2$, $\det\mathbf{S}_1=1$, $\det\mathbf{S}_2=1$.

When $\mathbf{S}_1$, $\mathbf{S}_2$ run through $\mathfrak{G}$, the matrices and $\mathbf{K}_2$ run through $\tilde{\Gamma}_{0+}^+$, $\tilde{\Gamma}_{0-}^-$, respectively. If $\mathbf{S}_1=\mathbf{S}_2$, then $e^{T\mathbf{K}_1}=e^{-T\mathbf{K}_2}=-e^{T\mathbf{K}_2}$, as claimed. Similarly, if $\mathbf{K}_1\in\tilde{\Gamma}_{0+}^-$, $\mathbf{K}_2\in\tilde{\Gamma}_{0-}^+$, then $e^{T\mathbf{K}_1}=-e^{T\mathbf{K}_2}$, so that the images of the sets $\tilde{\Gamma}_{0+}^-$ and $\tilde{\Gamma}_{0-}^+$ coincide.

To summarize: the set $\tilde{\mathfrak{R}}_0^+\cup\tilde{\mathfrak{R}}_0^-$ is mapped in one-to-one fashion into $\mathfrak{G}$ and the images of the sets $\tilde{\Gamma}_{0-}^-\subset\tilde{\mathfrak{R}}^-$ and $\tilde{\Gamma}_{0+}^+\subset\tilde{\mathfrak{R}}^+$ coincide, as do the images of $\tilde{\Gamma}_{0+}^-\subset\tilde{\mathfrak{R}}^-$ and $\tilde{\Gamma}_{0-}^+\subset\tilde{\mathfrak{R}}^+$. This means that $\mathfrak{G}$ may be thought of as the set obtained via the above mapping by "gluing together" the sets $\tilde{\Gamma}_{0+}^+$ and $\tilde{\Gamma}_{0-}^-$, on the one hand, and the sets $\tilde{\Gamma}_{0+}^-$ and $\tilde{\Gamma}_{0-}^+$, on the other. The matrices $\mathbf{K}=\mathbf{0}\in\tilde{\mathfrak{R}}_0^+$ and $\mathbf{K}=\mathbf{0}\in\tilde{\mathfrak{R}}_0^-$ are mapped respectively onto $\mathbf{X}=\mathbf{I}_2\in\mathfrak{G}$ and $\mathbf{X}=-\mathbf{I}_2\in\mathfrak{G}$. The cones $\tilde{\Pi}_+$ and $\tilde{\Pi}_-$ are deformed, occupying the positions shown in Figure 52. In general, denoting the images in $\mathfrak{G}$ of the sets $\tilde{\mathscr{H}}^-$, $\tilde{\Pi}^-$, ..., $\tilde{\mathscr{H}}_-\tilde{\mathfrak{R}}_0^+$, $\tilde{\mathfrak{R}}_0^-$ by the same symbols, we see that all these sets lie in $\mathfrak{G}$ as shown in Figure 52. (Recall that in the figure $\mathfrak{G}$ is the interior of a torus; the torus itself, shown by the dashed curves, is not in $\mathfrak{G}$.)

We now consider the properties of matrices $\mathbf{X}$ in these sets in greater detail.

Matrices $\mathbf{K}\in\mathscr{H}^+\subset\mathfrak{R}^+$ are reducible to the form (a) of §1.4. To these correspond matrices $\mathbf{X}=e^{T\mathbf{K}}\in\tilde{\mathscr{H}}^+\subset\mathfrak{G}$ which may be reduced to the form (a'') of §1.4. Consequently, the characterizing property of matrices $\mathbf{X}\in\tilde{\mathscr{H}}^+\subset\mathfrak{G}$ is that they have distinct positive eigenvalues. Since the matrix $\mathbf{K}_s=\mathbf{S}^{-1}\mathbf{K}\mathbf{S}\in\tilde{\mathfrak{R}}^+$ describes a one-sheeted hyperboloid $\lambda^2=x^2+y^2-z^2$ in $\tilde{\mathfrak{R}}_0^+$ as $\mathbf{S}$ varies, it follows that $\mathbf{X}_s=\mathbf{S}^{-1}\mathbf{X}\mathbf{S}$ describes a surface in $\mathfrak{G}$, topologically equivalent to the hyperboloid, obtained by the deformation $\mathbf{X}=e^{T\mathbf{K}}$; the

* Recall that T is the period of the coefficients in (1.3). We may assume without loss of generality that $T=1$.

** In the expressions for $\tilde{\mathbf{J}}_1$ we have $\tilde{\mathbf{J}}\in\tilde{\mathfrak{R}}_0^+$, and in that for $\tilde{\mathbf{J}}_2$ we have $\tilde{\mathbf{J}}\in\tilde{\mathfrak{R}}_0^-$.

relative position of this surface in $\mathfrak{G}$ is readily pictured. As $\mu = e^{\lambda} \to 1$, these "hyperboloids" in $\mathfrak{G}$ contract about their "middle" and finally become the "cone" $\tilde{\Pi}_+ \subset \mathfrak{G}$.

The cone $\tilde{\Pi}$ in $\tilde{\mathfrak{R}}^3$ is the union of its upper sheet $\tilde{\Pi}_+$, its lower sheet $\tilde{\Pi}_-$, and the vertex $\mathbf{K} = \mathbf{0}$:

$$\tilde{\Pi} = \tilde{\Pi}_+ \cup \tilde{\Pi}_- \cup \{0\}.$$

For the sets $\tilde{\mathfrak{R}}^+$, $\tilde{\mathfrak{R}}^-$, we have

$$\tilde{\Pi}^+ = \tilde{\Pi}^+_+ \cup \tilde{\Pi}^+_- \cup \{0^+\}, \quad \tilde{\Pi}^- = \tilde{\Pi}^-_+ \cup \tilde{\Pi}^-_- \{0^-\}.$$

Therefore the sets $\tilde{\Pi}^+$, $\tilde{\Pi}^-$ in $\mathfrak{G}$ are the following unions:

$$\tilde{\Pi}^+ = \tilde{\Pi}^+_+ \cup \tilde{\Pi}^+_- \cup \{\mathbf{I}\}, \quad \tilde{\Pi}^- = \tilde{\Pi}^-_+ \cup \tilde{\Pi}^-_- \cup \{-\mathbf{I}\},$$

where $\mathbf{I}$ is the identity matrix.

For matrices $\mathbf{K} \in \tilde{\Pi} \subset \tilde{\mathfrak{R}}^3$, we have the representation (b) of §1.4, with $\varepsilon > 0$ for $\mathbf{K} \in \tilde{\Pi}_+$, $\varepsilon < 0$ for $\mathbf{K} \in \tilde{\Pi}_-$, and clearly $\varepsilon = 0$ if $\mathbf{K} = \mathbf{0}$. Consequently, matrices $\mathbf{X} \in \tilde{\Pi}^+ \subset \mathfrak{G}$ are similar to the matrix

$$\begin{Vmatrix} 1 & \varepsilon \\ 0 & 1 \end{Vmatrix},$$

with $\varepsilon > 0$ for $\mathbf{X} \in \tilde{\Pi}^+_+$, $\varepsilon < 0$ for $\mathbf{X} \in \tilde{\Pi}^+_-$, and $\varepsilon = 0$ if $\mathbf{X} = \mathbf{I}$. An analogous assertion is true for matrices $\mathbf{X} \in \tilde{\Pi}^- \subset \mathfrak{G}$, with $\mathbf{X}$ replaced by $-\mathbf{X}$.

Similarly, we see that $\tilde{\mathscr{H}} \subset \mathfrak{G}$ is the set of all matrices $\mathbf{X}$ with distinct negative eigenvalues.

Now consider the set $\tilde{\mathcal{O}}^+_0 \subset \tilde{\mathfrak{R}}^+$. This is the set of matrices $\mathbf{K}$ of type (c) in §1.4, where $0 < \psi < \pi/(2T)$ and $\det \mathbf{S} > 0$. The number $i\psi$ is an eigenvalue of the first kind for this matrix. Consequently, $\tilde{\mathcal{O}}^+_{0+} \subset \mathfrak{G}$ contains the matrices $\mathbf{X} = \exp(T\mathbf{K})$ similar to the matrix (c″) of §1.4, with an eigenvalue $e^{i\psi_0}$ of the first kind,* where $\psi_0 = T\psi$ $\left(0 < \psi_0 < \frac{\pi}{2}\right)$. Similarly, we deduce that matrices $\mathbf{X} = \exp(T\mathbf{K})$ in the sets $\tilde{\Gamma}^+_{0+}$, $\tilde{\mathcal{O}}^-_{0-}$, $\tilde{\mathcal{O}}^+_{0-}$, $\tilde{\Gamma}^+_{0-}$, $\tilde{\mathcal{O}}^-_{0+}$ reduce to the form (c″) of §1.4, with an eigenvalue $e^{i\psi_0}$ of the first kind, where ψ_0 takes the following respective values:

$$\psi_0 = \frac{\pi}{2}, \quad \frac{\pi}{2} < \psi_0 < \pi, \quad -\pi < \psi_0 < -\frac{\pi}{2}, \quad \psi_0 = -\frac{\pi}{2}, \quad -\frac{\pi}{2} < \psi_0 < 0.$$

Denote

$$\tilde{\mathcal{O}}^+ = \tilde{\mathcal{O}}^+_{0+} \cup \tilde{\Gamma}^+_{0+} \cup \tilde{\mathcal{O}}^+_{0-} \subset \mathfrak{G}, \quad \tilde{\mathcal{O}}^- = \tilde{\mathcal{O}}^-_{0+} \cup \tilde{\Gamma}^-_{0+} \cup \tilde{\mathcal{O}}^-_{0-} \subset \mathfrak{G}$$

(see Figure 52). It follows from our discussion that $\tilde{\mathcal{O}}^+$ and $\tilde{\mathcal{O}}^-$ are simply-connected domains in $\mathfrak{G}$. The sets $\tilde{\mathcal{O}}^+$, $\tilde{\mathcal{O}}^-$ consist of the matrices $\mathbf{X}$

* For the definition of eigenvalues of the first (second) kind, see § III.1.2. In our case the indefinite scalar product is $\langle x, y \rangle = i(\tilde{J}x, y)$.

with eigenvalues $e^{i\psi_0}$ (of the first kind) and $e^{-i\psi_0}$ (of the second kind), where $0 < \psi_0 < \pi$ if $\mathbf{X} \in \tilde{\mathcal{O}}^+$ and $-\pi < \psi_0 < 0$ if $\mathbf{X} \in \tilde{\mathcal{O}}^-$.

If $\mathbf{X} \in \tilde{\mathcal{O}}^+ \cup \tilde{\mathcal{O}}^-$ is a fixed matrix and $\mathbf{S} \in \mathfrak{G}$ runs through the whole set $\mathfrak{G}$, then $\mathbf{X}_S = \mathbf{S}^{-1}\mathbf{X}\mathbf{S}$ describes the connected and simply-connected set obtained as above by deformation of the appropriate sheet of a two-sheeted hyperboloid in $\mathfrak{R}^+$ or $\mathfrak{R}^-$.

We augment our notation as follows:

$$\tilde{\mathcal{O}} = \tilde{\mathcal{O}}^+ \cup \tilde{\mathcal{O}}^-, \quad \tilde{\mathcal{H}} = \tilde{\mathcal{H}}^+ \cup \tilde{\mathcal{H}}^-, \quad \tilde{\Pi} = \tilde{\Pi}^+ \cup \tilde{\Pi}^-.$$

In other words, $\tilde{\mathcal{O}}$ is the set of matrices $\mathbf{X}$ whose eigenvalues are distinct and lie on the unit circle, $\tilde{\mathcal{H}}$ the set of matrices $\mathbf{X}$ with distinct real eigenvalues, and $\tilde{\Pi}$ the set of matrices $\mathbf{X}$ with a twofold eigenvalue $\rho = 1$ ($\tilde{\Pi}^+$) or $\rho = -1$ ($\tilde{\Pi}^-$). The spectra of these matrices for the different subsets of $\mathfrak{G}$ are shown in Figure 53.

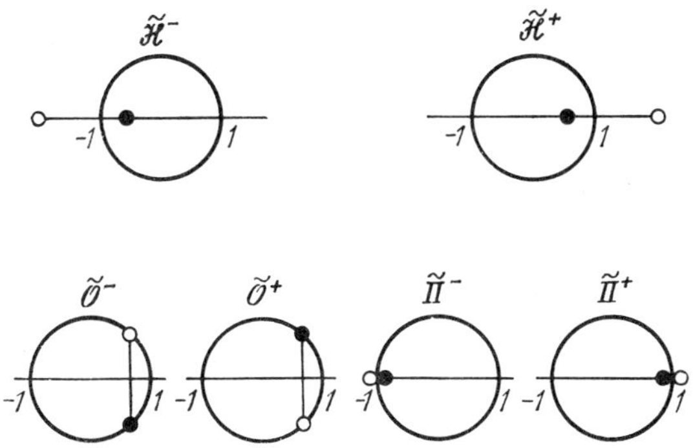

FIGURE 53.

The above discussion shows that the mappings we have constructed provide a lucid visualization of the relative positions of matrices $\mathbf{X}$ in $\mathfrak{G}$ for different types of canonical structure.

1.6. Structure of Hamiltonian space $\mathfrak{H}^3$ and three-dimensional model of $\mathfrak{H}^3$

The set of all Hamiltonians $\mathbf{H}(t)$ in (1.3) or, alternatively, the set of triples of T-periodic functions $\alpha(t)$, $\beta(t)$, $\gamma(t)$ in (1.3) will be denoted by $\mathfrak{H}^3$. $\mathfrak{H}^3$ is a linear space. We define a norm in $\mathfrak{H}^3$ by*

* From here until the end of the section, we shall assume that the functions $\alpha(t)$ $\beta(t)$, $\gamma(t)$ are Lebesgue-integrable on $[0, T]$, so that $\mathfrak{H}^3$ is a complete normed space /46/. In this case the matrizant $\mathbf{X}(t)$ is absolutely continuous and the analog of the theorem in § II.1.7 is valid (the proof is word for word the same as that given there). The reader unacquainted with the theory of functions of a real variable may assume, as before, that $\alpha(t)$, $\beta(t)$, $\gamma(t)$ are Riemann-integrable piecewise-continuous functions, although in this case $\mathfrak{H}^3$ is not a complete space, so that various assertions below as to certain subsets of $\mathfrak{H}^3$ being closed will then fail to hold (as their proofs utilize the completeness of $\mathfrak{H}^3$).

$$\|\mathbf{H}(t)\| = \int_0^T |\alpha|\,dt + \int_0^T |\beta|\,dt + \int_0^T |\gamma|\,dt.$$

In this subsection we shall construct a three-dimensional model of $\mathfrak{L}^3$ which will provide a clear picture of the relative positions in $\mathfrak{L}^3$ of Hamiltonians $\mathbf{H}(t)$ for which the appropriate equations (1.2) have solutions with various asymptotic properties. We proceed directly to this construction.

We showed in § II. 1.7 that $\mathfrak{L}^3$ can be mapped homeomorphically (i. e., in one-to-one and bicontinuous fashion) onto the set of matrizants $\mathbf{X}(t)$, $0 \leqslant t \leqslant T$, normed by the distance function

$$\rho(\mathbf{X}_1(t)\,\mathbf{X}_2(t)) = \max_{0 \leqslant t \leqslant T} |\mathbf{X}_1(t) - \mathbf{X}_2(t)| + \int_0^T \left|\frac{d\mathbf{X}_1}{dt} - \frac{d\mathbf{X}_2}{dt}\right| dt.$$

In the present case, $\det \mathbf{X}(t) \equiv 1$ (see §1.2). Thus $\mathbf{X}(t) \in \mathfrak{G}$, $0 \leqslant t \leqslant T$, so that the matrizant $\mathbf{X}(t)$ is a curve in $\mathfrak{G}$, beginning at the point $\mathbf{X}(0) = \mathbf{I}$ (Figure 52). If we agree to identify topologically equivalent (i. e., homeomorphic) sets, we may assume that $\mathfrak{L}^3$ is the set of these curves in $\mathfrak{G}$. The asymptotic properties of solutions of equation (1.2) are determined by the monodromy matrix $\mathbf{X}(T)$, which is the "end" of the curve $\mathbf{X}(t)$, $0 \leqslant t \leqslant T$. Using the results of §1.5 concerning the relative positions in $\mathfrak{G}$ of matrices $\mathbf{X}$ of different canonical structure, we shall be able to examine the asymptotic behavior of solutions of equation (1.2) on the basis of the position of the appropriate endpoint of the curve $\mathbf{X}(t)$ in $\mathfrak{G}$. Thus, case (a') of §1.4 will hold if $\mathbf{X}(T) \in \tilde{\mathcal{H}}^+$ or $\mathbf{X}(T) \in \tilde{\mathcal{H}}^-$; the functions $\mathbf{u}_1(t)$, $\mathbf{u}_2(t)$ will be T-periodic if $\mathbf{X}(T) \in \tilde{\mathcal{H}}^+$ and T-antiperiodic if $\mathbf{X}(T) \in \tilde{\mathcal{H}}^-$. Case (b') will hold if $\mathbf{X}(T) \in \tilde{\Pi}^-$ or $\mathbf{X}(T) \in \tilde{\Pi}^+$. The functions $\mathbf{u}_1(t)$, $\mathbf{u}_2(t)$ will be T-periodic if $\mathbf{X}(T) \in \Pi^-$ and T-antiperiodic if $\mathbf{X}(T) \in \tilde{\Pi}^+$. Case ($\mathrm{b}_1'$) will hold if $\mathbf{X}(T) = \mathbf{I}$ (all solutions T-periodic) or if $\mathbf{X}(T) = -\mathbf{I}$ (all solutions T-antiperiodic). Case (c') will hold if $\mathbf{X}(T) \in \tilde{\mathcal{O}}^+$ or $\mathbf{X}(T) \in \tilde{\mathcal{O}}^-$. A Hamiltonian $\mathbf{H}(t)$ for which $\mathbf{X}(T) \in \tilde{\mathcal{O}} = \tilde{\mathcal{O}}^+ \cup \tilde{\mathcal{O}}^-$ is strongly stable, a Hamiltonian $\mathbf{H}(t)$ for which $\mathbf{X}(T) \in \tilde{\mathcal{H}} = \tilde{\mathcal{H}}^- \cup \tilde{\mathcal{H}}^+$ is strongly unstable (for the definitions, see § III. 5.1). (Recall that $\tilde{\mathcal{H}}^-$, $\tilde{\mathcal{H}}^+$, $\tilde{\mathcal{O}}^-$, $\tilde{\mathcal{O}}^+$ are open domains in $\mathfrak{G}$.)

We introduce further notation as follows. The set of curves $\mathbf{X}(t)$, $(0 \leqslant t \leqslant T,\ \mathbf{X}(0) = \mathbf{I})$ whose ends belong to one of the subsets

$$\tilde{\mathcal{O}},\ \tilde{\mathcal{O}}^\pm;\quad \tilde{\mathcal{H}},\ \tilde{\mathcal{H}}^\pm;\quad \tilde{\Pi}_i^\pm,\ \tilde{\Pi} = \tilde{\Pi}^+ \cup \tilde{\Pi}^- \text{ in } \mathfrak{G}$$

will be denoted by the same symbol without the tilde $\sim$. For example, the following inclusion relations are equivalent:

$$\mathbf{X}(t) \in \mathcal{O}^+ \text{ and } \mathbf{X}(T) \in \tilde{\mathcal{O}}^+,\quad \mathbf{X}(t) \in \mathcal{O}^- \text{ and } \mathbf{X}(T) \in \tilde{\mathcal{O}}^-,$$
$$\mathbf{X}(t) \in \mathcal{O} \text{ and } \mathbf{X}(T) \in \tilde{\mathcal{O}}\ (= \tilde{\mathcal{O}}^+ \cup \tilde{\mathcal{O}}^-),\quad \mathbf{X}(t) \in \mathcal{H} \text{ and } \mathbf{X}(T) \in \tilde{\mathcal{H}}\ (= \tilde{\mathcal{H}}^+ \cup \tilde{\mathcal{H}}^-),$$

etc.

The set of corresponding Hamiltonians in $\mathfrak{L}^3$ will be denoted by the same letter as the set of curves in $\mathfrak{G}$. Thus

$$\mathfrak{L}^3 = \mathcal{O} \cup \Pi \cup \mathcal{H},\quad \mathcal{O} = \mathcal{O}^+ \cup \mathcal{O}^-,$$
$$\Pi = \Pi^+ \cup \Pi^-,\quad \mathcal{H} = \mathcal{H}^+ \cup \mathcal{H}^-.$$

It follows from our discussion that $\mathcal{G}\subset\mathfrak{L}^3$ is the set of strongly stable Hamiltonians and $\mathcal{H}\subset\mathfrak{L}^3$ the set of strongly unstable Hamiltonians.

If a Hamiltonian $\mathbf{H}(t)\in\mathfrak{L}^3$ is deformed continuously (in the sense of the above norm), the curve $\mathbf{X}(t)$ is also deformed continuously (in the sense of the metric $\rho[\mathbf{X}_1(t), \mathbf{X}_2(t)]$ introduced above). In particular, the endpoint of $\mathbf{X}(t)$ will move continuously. It follows at once that none of the sets $\mathcal{G}$, Π, $\mathcal{H}$, $\mathcal{G}^+$, $\mathcal{G}^-$, Π^+, Π^-, $\mathcal{H}^+$, $\mathcal{H}^-$ is connected in $\mathfrak{L}^3$. As an example, let us consider the set $\mathcal{G}$. Let $\mathbf{H}_1(t)\in\mathcal{G}$, $\mathbf{H}_2(t)\in\mathcal{G}$ and suppose there exists a continuous deformation $\mathbf{H}(t, s)$ of $\mathbf{H}_1(t)$ into $\mathbf{H}_2(t)$ which never leaves $\mathcal{G}$: $\mathbf{H}(t, 0)=\mathbf{H}_1(t)$, $\mathbf{H}(t, 1)=\mathbf{H}_2(t)$, $\mathbf{H}(t, s)\in\mathcal{G}$ for $0\leqslant s\leqslant 1$. Let $\mathbf{X}_1(t)\sim\mathbf{H}_1(t)$, $\mathbf{X}_2(t)\sim\mathbf{H}_2(t)$ $\mathbf{X}(t, s)\sim\mathbf{H}(t, s)$ be the corresponding matrizants. Since $\mathbf{X}(t, s)\in\mathcal{G}$ for $0\leqslant s\leqslant 1$, it follows that $\mathbf{X}(T, s)\in\tilde{\mathcal{G}}^+=\tilde{\mathcal{G}}^+\cup\tilde{\mathcal{G}}^-$ for $0\leqslant s\leqslant 1$, so that for any $0\leqslant s\leqslant 1$ the matrizant $\mathbf{X}(T,s)$ belongs to one of the sets $\tilde{\mathcal{G}}^+$ or $\tilde{\mathcal{G}}^-$. Since $\mathbf{X}(T, s)$ depends continuously on s, $0\leqslant s\leqslant 1$, and the open sets $\tilde{\mathcal{G}}^+$ and $\tilde{\mathcal{G}}^-$ are disjoint, it follows that the ends of the curves, $\mathbf{X}(T, 0)=\mathbf{X}_1(T)$ and $\mathbf{X}(T, 1)=\mathbf{X}_2(T)$, must be either both in $\tilde{\mathcal{G}}^+$ or both in $\tilde{\mathcal{G}}^-$. Hence curves $\mathbf{X}_1(t)$, $\mathbf{X}_2(t)\in\mathcal{G}$ with ends $\mathbf{X}_1(T)\in\tilde{\mathcal{G}}^+$, $\mathbf{X}_2(T)\in\tilde{\mathcal{G}}^-$ cannot be deformed continuously into one another, remaining throughout in $\mathcal{G}$. Thus the set $\mathcal{G}\subset\mathfrak{L}^3$ is not connected.

It follows, then, that a necessary condition for one of the sets $\mathcal{G}, \Pi, \ldots, \mathcal{H}^-$ in $\mathfrak{L}^3$ to be connected is that the corresponding set $\tilde{\mathcal{G}}, \tilde{\Pi}, \ldots, \tilde{\mathcal{H}}^-$ in $\mathfrak{G}$ be connected. This condition is however not sufficient. For example, consider the set $\mathcal{G}^+$; the corresponding set $\tilde{\mathcal{G}}^+$ of monodromy matrices $\mathbf{X}(T)$ is connected. It is easy to see that $\mathcal{G}^+$ is not connected. Indeed, let $\mathbf{H}(t)\in\mathcal{G}^+$, let $\mathbf{X}(t)$ be the corresponding matrizant, and $\mathbf{X}(T)\in\tilde{\mathcal{G}}^+$ (Figure 52). Let φ be the angle between the meridional planes through the matrices $\mathbf{X}=\mathbf{I}$ and $\mathbf{X}=\mathbf{X}(T)$, as defined in §1.6 (see Figure 52). Let $\mathbf{X}'(t)$ be some other curve in $\mathfrak{G}$ such that $\mathbf{X}'(0)=\mathbf{I}$, $\mathbf{X}'(T)=\mathbf{X}(T)$, but for which $\varphi'=\varphi+2\pi$. (This curve "winds" once more in $\mathfrak{G}$ than $\mathbf{X}(t)$.) Let $\mathbf{H}'(t)$ be the corresponding Hamiltonian. It is obvious that $\mathbf{H}'(t)\in\mathcal{G}^+$. We shall show that $\mathbf{H}(t)$ cannot be continuously deformed into $\mathbf{H}'(t)$, with all intermediate Hamiltonians remaining in $\mathcal{G}^+$. (Recall that during the deformation we can "glue together" ends of curves, i. e., there is no need to maintain $\mathbf{X}(T)=\mathbf{X}'(T)$; but the end of the curve must remain throughout in the set $\tilde{\mathcal{G}}^+$.) Let $\mathbf{X}(t, s)$ be the deformation:

$$\mathbf{X}(t, 0)=\mathbf{X}(t), \quad \mathbf{X}(t, 1)=\mathbf{X}'(t), \quad \mathbf{X}(T, s)\in\tilde{\mathcal{G}}^+$$
$$(0\leqslant t\leqslant T, \quad 0\leqslant s\leqslant 1).$$

Let $\varphi(s)$ be the angle of rotation of the curve $\mathbf{X}(t, s)$. It is clear that $\varphi(s)$ is continuous and $\varphi(0)=\varphi$, $\varphi(1)=\varphi'=\varphi+2\pi$. Consequently, $\mathbf{X}(T, s)$ is a closed curve in $\tilde{\mathcal{G}}^+$ (since $\mathbf{X}(T, 0)=\mathbf{X}(T, 1)=\mathbf{X}(T)$) with angle of rotation $\varphi(1)-\varphi(0)=2\pi$. However, there are no such curves in $\tilde{\mathcal{G}}^+$: it is evident from the form of $\tilde{\mathcal{G}}^+$ that any closed curve in $\tilde{\mathcal{G}}^+$ may be contracted to a point, so that for any such curve $\varphi(1)-\varphi(0)=0$. Thus $\mathcal{G}^+$ is not a connected set.

A set $\tilde{\mathfrak{M}}\in\mathfrak{G}$ is said to be *simply-connected* in $\mathfrak{G}$ if any closed curve in $\tilde{\mathfrak{M}}$ may be contracted to a point by a continuous deformation (which may produce "intermediate" curves not in $\tilde{\mathfrak{M}}$, but naturally still in $\mathfrak{G}$). It follows from the foregoing that if $\tilde{\mathfrak{M}}$ is connected and simply-connected in $\mathfrak{G}$, then the set $\mathfrak{M}$ of curves $\mathbf{X}(t)$ $(0\leqslant t\leqslant T,\ \mathbf{X}(0)=\mathbf{I})$ with ends $\mathbf{X}(T)\in\tilde{\mathfrak{M}}$, and hence also the set $\mathfrak{M}\subset\mathfrak{L}^3$ of corresponding Hamiltoians, is not connected.

It is easy to see that each of the sets $\bar{\mathscr{H}}^-$, $\bar{\mathscr{H}}^+$, $\bar{\Pi}^-$, $\bar{\Pi}^+$ is simply-connected in $\mathfrak{G}$. Consequently, none of the sets $\mathscr{H}^-$, $\mathscr{H}^+$, Π^-, Π^+ in $\mathfrak{L}^3$ is connected. We shall show later that each of these sets, as well as the sets $\mathcal{O}^+$, $\mathcal{O}^-$, is the union of countably many connected components in $\mathfrak{L}^3$.

This assertion and many other similar propositions will follow at once from our model of the space $\mathfrak{L}^3$, which we now proceed to construct.

We first construct the so-called universal covering space $\mathfrak{R}^3$ of $\mathfrak{G}$. * Recall that $\mathfrak{G}$ may be obtained as described above by gluing together domains $\tilde{\mathfrak{R}}_0^+$ and $\tilde{\mathfrak{R}}_0^-$ (see Figure 52). To construct $\mathfrak{R}^3$, we first deform $\tilde{\mathfrak{R}}_0^-$ and $\tilde{\mathfrak{R}}_0^+$ in such a way that the hyperboloids bounding them are converted into parallel planes, and the domains $\tilde{\mathfrak{R}}_0^-$, $\tilde{\mathfrak{R}}_0^+$ themselves become the region between the planes (see Figure 54). Each of the domains $\tilde{\mathfrak{R}}_0^-$, $\tilde{\mathfrak{R}}_0^+$ shown in Figure 54 is to be treated as a domain in the three-dimensional space obtained by rotating the plane of the figure about the horizontal dash-dot line. Now take a countable set of copies of $\tilde{\mathfrak{R}}_0^-$, which we denote by $\tilde{\mathfrak{R}}_n$ $(n=\pm 1, \pm 3, \ldots)$ and a countable set of copies of $\tilde{\mathfrak{R}}_0^+$, which we denote by $\tilde{\mathfrak{R}}_n$ $(n=0, \pm 2, \pm 4)$. "Glue" these domains together in sequence, as indicated in Figure 55. The various sets introduced previously are assigned symbols as shown in Figure 55. The three-dimensional space resulting from this construction, illustrated in Figure 56, is the required space $\mathfrak{R}^3$.

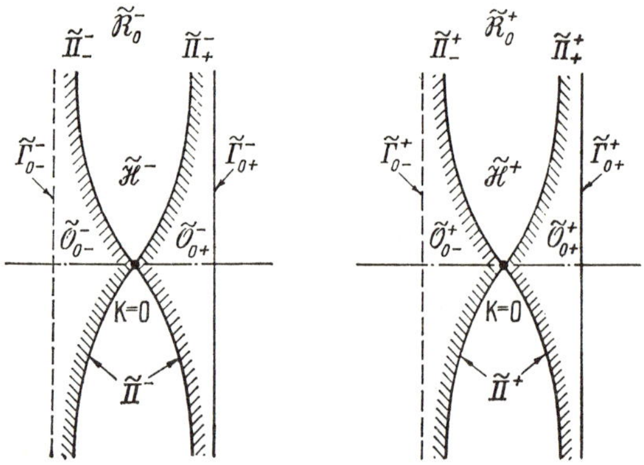

FIGURE 54.

The important point in this construction of the space $\mathfrak{R}^3$ from $\tilde{\mathfrak{R}}_0^+$, $\tilde{\mathfrak{R}}_0^-$ is that with each point $\mathbf{C}\in\mathfrak{R}^3$ one associates a matrix $\mathbf{K}\in\tilde{\mathfrak{R}}_0^-\cup\tilde{\mathfrak{R}}_0^+$ or, because of the one-to-one correspondence $\tilde{\mathfrak{R}}_0^+\cup\tilde{\mathfrak{R}}_0^-\sim\mathfrak{G}$, a matrix $\mathbf{X}\in\mathfrak{G}$. Thus, under the mapping $\mathfrak{R}^3\to\tilde{\mathfrak{R}}_0^+\cup\tilde{\mathfrak{R}}_0^-$ a set $\tilde{\mathfrak{R}}_n$ goes into $\tilde{\mathfrak{R}}_0^-$ if n is odd and into $\tilde{\mathfrak{R}}_0^+$ if n is even. The mapping $\mathfrak{R}^3\to\mathfrak{G}$ is obtained by composition with the mapping $\tilde{\mathfrak{R}}_0^+\cup\tilde{\mathfrak{R}}_0^-\to\mathfrak{G}$. These mappings of $\mathfrak{R}^3$ into $\tilde{\mathfrak{R}}_0^+\cup\tilde{\mathfrak{R}}_0^-$ and into $\mathfrak{G}$ will be called the natural mappings.

* For the definition of the universal covering space, see, e.g., /76a/. The geometric constructions given below are intuitive and not rigorous, but the appropriate rigorous reasoning is easily reconstructed.

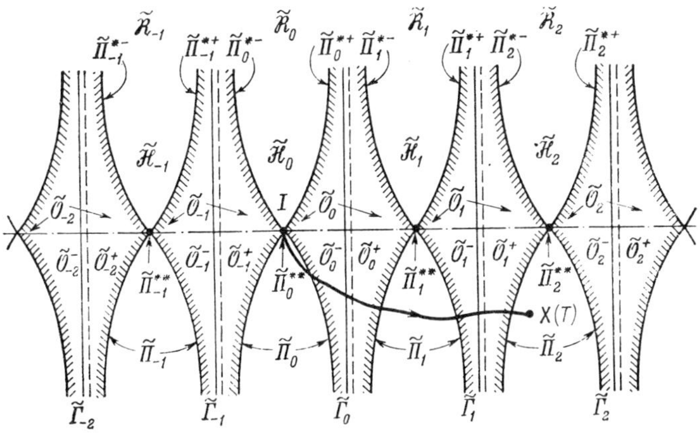

FIGURE 55.

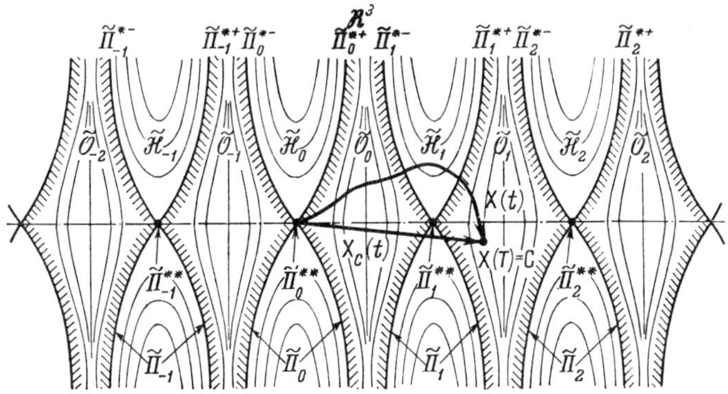

FIGURE 56.

The entire space $\mathfrak{R}^3$ is partitioned into sets $\tilde{\mathcal{O}}_n$, $\tilde{\mathscr{H}}_n$, $\tilde{\Pi}_n^+$, $\tilde{\Pi}_n^-$, $\tilde{\Pi}_n^{**}$ as shown in Figure 55. These sets are obtained from the corresponding subsets $\tilde{\mathcal{O}}_{0\pm}^{\pm}$, $\tilde{\Pi}_{\pm}^{\pm}$, $\tilde{\mathscr{H}}^{\pm}$ of $\tilde{\mathfrak{R}}_0^+$, $\tilde{\mathfrak{R}}_0^-$ by "gluing" operations which should be clear from an examination of Figures 52, 54, 55 and 56. In the process, the set $\tilde{\Gamma}_{0\mp}^{-}$ in $\tilde{\mathfrak{R}}_{2n-1}$ is "glued" to the set $\tilde{\Gamma}_{0-}^{\pm}$ in $\tilde{\mathfrak{R}}_{2n}$ and the set $\tilde{\Gamma}_{0+}^{+}\subset\tilde{\mathfrak{R}}_{2n}$ to the set $\tilde{\Gamma}_{0}^{-}\subset\tilde{\mathfrak{R}}_{2n+1}$. The points $\mathbf{K}=\mathbf{0}$ in $\tilde{\mathfrak{R}}_{2n+1}$ and $\tilde{\mathfrak{R}}_{2n}$ are denoted in $\mathfrak{R}^3$ by Π_n^{**}, the sheets of the cones $\tilde{\Pi}_{\pm}^{\pm}$ in $\tilde{\mathfrak{R}}_{2n\pm1}$ and $\tilde{\mathfrak{R}}_{2n}$ (without vertices) are denoted in $\mathfrak{R}^3$ by $\tilde{\Pi}_n^{*\pm}$, the cones $\tilde{\Pi}^+\subset\tilde{\mathfrak{R}}_{2n}$, $\tilde{\Pi}^-\subset\tilde{\mathfrak{R}}_{2n+1}$ are denoted by Π_n. The rest of the notation is clear from the following table, which also brings out the correspondences between the sets in each row of the table in the spaces $\mathfrak{R}^3$, $\mathfrak{G}$ and $\tilde{\mathfrak{R}}^+\cup\tilde{\mathfrak{R}}^-$ (the symbols in the first column will be explained below).

Recall that $\tilde{\Pi}_{\pm}^{-}$, $\tilde{\Pi}_{\pm}^{+}$ are the sheets of the cones (without vertices) in $\tilde{\mathfrak{R}}_0^-\cup\tilde{\mathfrak{R}}_0^+$ and $\tilde{\mathfrak{G}}$, obtained by the deformation described above from the sheets of the cone $\tilde{\Pi}$ (without vertices) in $\tilde{\mathfrak{R}}^3$.

TABLE

$\mathfrak{L}^3$	$\mathfrak{R}^3$	$\mathfrak{S}$	$\tilde{\mathfrak{R}}_0^+ \cup \tilde{\mathfrak{R}}_0^-$
$\mathscr{H}_{2n+1}$	$\tilde{\mathscr{H}}_{2n+1}$	$\tilde{\mathscr{H}}^-$	$\tilde{\mathscr{H}}^-$
$\mathscr{H}_{2n}$	$\tilde{\mathscr{H}}_{2n}$	$\tilde{\mathscr{H}}^+$	$\tilde{\mathscr{H}}^+$
$\mathfrak{O}_{2n+1}$	$\tilde{\mathfrak{O}}_{2n+1}$	$\tilde{\mathfrak{O}}^-$	$\tilde{\mathfrak{O}}^- = \tilde{\mathfrak{O}}_{0+}^- \cup \tilde{\Gamma}_{0+}^- \cup \tilde{\mathfrak{O}}_{0-}^- \subset \tilde{\mathfrak{R}}_0^-$
$\mathfrak{O}_{2n}$	$\tilde{\mathfrak{O}}_{2n}$	$\tilde{\mathfrak{O}}^+$	$\tilde{\mathfrak{O}}^+ = \tilde{\mathfrak{O}}_{0+}^+ \cup \tilde{\Gamma}_{0+}^+ \cup \tilde{\mathfrak{O}}_{0-}^+ \subset \tilde{\mathfrak{R}}_0^+$
Π_{2n+1}^{**}	$\tilde{\Pi}_{2n+1}^{**}$	$\mathbf{X} = -\mathbf{I}$	$\mathsf{K} = 0 \in \tilde{\mathfrak{R}}_0^-$
Π_{2n}^{**}	$\tilde{\Pi}_{2n}^{**}$	$\mathbf{X} = \mathbf{I}$	$\mathsf{K} = 0 \in \tilde{\mathfrak{R}}_0^+$
$\Pi_{2n+1}^{*\pm}$	$\tilde{\Pi}_{2n+1}^{*\pm}$	$\tilde{\Pi}_{\pm}^-$	$\tilde{\Pi}_{\pm}^- \subset \tilde{\mathfrak{R}}_0^-$
$\Pi_{2n}^{*\pm}$	$\tilde{\Pi}_{2n}^{*\pm}$	$\tilde{\Pi}_{\pm}^+$	$\tilde{\Pi}_{\pm}^+ \subset \tilde{\mathfrak{R}}_0^+$
Π_{2n+1}	$\tilde{\Pi}_{2n+1}$	$\tilde{\Pi}^- = \tilde{\Pi}_+^- \cup \{-\mathbf{I}\} \cup \tilde{\Pi}_-^-$	$\tilde{\Pi}^- = \tilde{\Pi}_+^- \cup \{0\} \cup \tilde{\Pi}_-^- \subset \tilde{\mathfrak{R}}_0^-$
Π_{2n}	$\tilde{\Pi}_{2n}$	$\tilde{\Pi}^+ = \tilde{\Pi}_+^+ \cup \{\mathbf{I}\} \cup \tilde{\Pi}_-^+$	$\tilde{\Pi}^+ = \tilde{\Pi}_+^+ \cup \{0\} \cup \tilde{\Pi}_-^+ \subset \tilde{\mathfrak{R}}_0^+$

We shall now show how the three-dimensional space $\mathfrak{R}^3$ serves as a faithful model (in the sense to be explained below) for the function space $\mathfrak{L}^3$. With each Hamiltonian $\mathbf{H}(t) \in \mathfrak{L}^3$ we have associated a curve $\mathbf{X}(t) \in \mathfrak{S}$, $\mathbf{X}(0) = \mathbf{I}$ $(0 \leqslant t \leqslant T)$. We first consider some curve in $\mathfrak{R}^3$ with initial point $\tilde{\Pi}_0^{**}$ (see Figure 56). The natural mapping of $\mathfrak{R}^3$ into $\mathfrak{S}$ takes any such curve into a curve $\mathbf{X}(t) \in \mathfrak{S}$, $\mathbf{X}(0) = \mathbf{I}$. It is important to observe that the converse is also valid: any curve $\mathbf{X}(t)$ in $\mathfrak{S}$ with initial point $\mathbf{X}(0) = \mathbf{I}$ is uniquely determined by a curve in $\mathfrak{R}^3$ with initial point $\tilde{\Pi}_0^{**}$. For example, the curve $\mathbf{X}(t)$ illustrated in Figure 52 corresponds to the curve shown in Figure 55. Note that when the first curve crosses the boundary $\tilde{\Gamma}_{0-}^- = \tilde{\Gamma}_{0+}^+ \subset \mathfrak{S}$ between $\tilde{\mathfrak{R}}_0^+ \in \mathfrak{S}$ and $\tilde{\mathfrak{R}}_0^- \in \mathfrak{S}$ on this curve, the corresponding curve in $\mathfrak{R}^3$ (Figure 54) goes from $\tilde{\mathfrak{R}}_0$ to $\tilde{\mathfrak{R}}_1$. As the first curve crosses the boundary $\tilde{\Gamma}_{0+}^- = \tilde{\Gamma}_{0-}^+ \in \mathfrak{S}$, the curve in $\mathfrak{R}^3$ goes from $\tilde{\mathfrak{R}}_1$ to $\tilde{\mathfrak{R}}_2$. In general, the subscript of the domain $\tilde{\mathfrak{R}}_n$ to be considered is uniquely determined by the increment of the argument of the first column of the matrix $\mathbf{X}(t)$, which is equal to the angle φ in Figure 52. This subscript is the integer n in the formula $n\pi - \frac{\pi}{2} < \varphi \leqslant n\pi + \frac{\pi}{2}$. (We are using the model of $\mathfrak{S}$ described at the beginning of §1.5.)

Different curves in $\mathfrak{S}$ with common endpoints will generally correspond to curves in $\mathfrak{R}^3$ with different endpoints. *The endpoints of curves in $\mathfrak{R}^3$ will coincide if and only if the corresponding curves in $\mathfrak{S}$ can be deformed continuously into one another.*

Thus, to each Hamiltonian $\mathbf{H}(t) \in \mathfrak{L}^3$ corresponds a curve in $\mathfrak{R}^3$ with initial point $\tilde{\Pi}_0^{**}$, and the correspondence is one-to-one and bicontinuous (in the sense of the metrics introduced above). This curve ends at a point whose image under the natural mapping of $\mathfrak{R}^3$ into $\mathfrak{S}$ is $\mathbf{X}(T)$. Briefly, we might say that $\mathfrak{L}^3$ is identified with the set of curves $\mathbf{X}(t)$ in $\mathfrak{R}^3$ with initial point $\tilde{\Pi}_0^{**}$.

We now introduce the set of Hamiltonians $\mathbf{H}(t)$ such that the endpoints $\mathbf{X}(T)$ of the corresponding curves $\mathbf{X}(t) \in \mathfrak{R}^3$ are points of one of the sets in the second column of the above table; this set will be denoted by the same symbol, without the tilde $\sim$. Thus:

$$\mathbf{H}(t) \in \mathscr{H}_n \text{ if and only if } \mathbf{X}(T) \in \tilde{\mathscr{H}}_n \subset \mathfrak{R}^3,$$
$$\mathbf{H}(t) \in \mathscr{O}_n \text{ if and only if } \mathbf{X}(T) \in \tilde{\mathscr{O}}_n \subset \tilde{\mathfrak{R}}^3,$$
$$\mathbf{H}(t) \in \Pi_n^{\bullet -} \text{ if and only if } \mathbf{X}(T) \in \tilde{\Pi}_n^{\bullet -} \subset \tilde{\mathfrak{R}}^3,$$

and so on.

We have thus defined all the sets in the first column of the table.

We now show that

$$\mathfrak{L}^3 = \Omega_0 \times \mathfrak{R}^3, \tag{1.19}$$

where the symbol $\times$ denotes the cartesian product of sets and Ω_0 is the set of all absolutely continuous closed curves $\mathbf{F}(t)$ in $\mathfrak{G}$ such that $\mathbf{F}(0) = \mathbf{F}(t) = \mathbf{I}_2$. We define a metric in Ω_0 by

$$\rho(\mathbf{F}_1, \mathbf{F}_2) = \sup_{0 \leqslant t \leqslant T} |\mathbf{F}_1(t) - \mathbf{F}_2(t)| + \int_0^T \left| \frac{d\mathbf{F}_1(t)}{dt} - \frac{d\mathbf{F}_2(t)}{dt} \right| dt.$$

The meaning of (1.19) is that we can define a homeomorphism (one-to-one and bicontinuous mapping) taking each Hamiltonian $\mathbf{H}(t) \in \mathfrak{L}^3$ onto a pair $\{\mathbf{F}(t), \mathbf{C}\}$, where $\mathbf{F}(t) \in \Omega_0$, $\mathbf{C} \in \mathfrak{R}^3$. This we now proceed to do.

With each point $\mathbf{C} \in \mathfrak{R}^3$ we associate a special curve $\mathbf{X}_\mathbf{C}(t)$, defined by uniform motion along a straight line from the point $\tilde{\Pi}_0^{\bullet\bullet}$ to $\mathbf{C}$ in the time interval from $t = 0$ to $t = T$. The natural mapping of $\mathfrak{R}^3$ into $\mathfrak{G}$ takes this curve into a certain curve in $\mathfrak{G}$, which we again denote by $\mathbf{X}_\mathbf{C}(t)$.

Let $\mathbf{H}(t) \in \mathfrak{L}^3$, let $\mathbf{X}(t) \in \mathfrak{G}$ be the corresponding matrizant and $\mathbf{X}(t) \in \mathfrak{R}^3$ the corresponding curve in $\mathfrak{R}^3$. Let $\mathbf{C} = \mathbf{X}(T) \in \mathfrak{R}^3$. Let $\mathbf{X}_\mathbf{C}(t) \in \mathfrak{G}$ be the corresponding curve in $\mathfrak{G}$, and set

$$\mathbf{F}(t) = \mathbf{X}(t)\, \mathbf{X}_c(t)^{-1} \qquad (\in \mathfrak{G}).$$

Since $\mathfrak{R}^3$ is the usual three-dimensional space and the curves $\mathbf{X}(t) \in \mathfrak{R}^3$ and $\mathbf{X}_\mathbf{C}(t) \in \mathfrak{R}^3$ have common endpoints (see Figure 56), it follows that the curve $\mathbf{X}(t) \in \mathfrak{R}^3$ may be deformed continuously into $\mathbf{X}_\mathbf{C}(t) \in \mathfrak{R}^3$. The same holds true for the corresponding curves $\mathbf{X}(t) \in \mathfrak{G}$ and $\mathbf{X}_\mathbf{C}(t) \in \mathfrak{G}$. Consequently, the curve $\mathbf{F}(t)$ may be contracted in $\mathfrak{G}$ to the point $\mathbf{X} = \mathbf{I}_2$. The set of all absolutely continuous curves $\mathbf{F}(t)$, $0 \leqslant t \leqslant T$, in $\mathfrak{G}$ such that $\mathbf{F}(0) = \mathbf{F}(T) = \mathbf{I}_2$ which can be contracted to a point is precisely Ω_0. The metric in Ω_0 was defined above.

Thus, for every Hamiltonian $\mathbf{H}(t) \in \mathfrak{L}^3$, hence also for the matrizant $\mathbf{X}(t)$ of the corresponding equation (1.3), we have a well-defined point $\mathbf{C} \in \mathfrak{R}^3$ and a curve $\mathbf{F}(t) \in \Omega_0$. Since the monodromy matrix $\mathbf{X}(T)$ is defined by the point $\mathbf{C}$ alone (it is the image of $\mathbf{C} \in \mathfrak{R}^3$ under the natural mapping $\mathfrak{R}^3 \to \mathfrak{G}$), it follows that the point $\mathbf{C} \in \mathfrak{R}^3$ itself will determine the behavior of the solutions as $t \to \infty$. The inverse correspondence is obvious: to each pair $\{\mathbf{F}(t), \mathbf{C}\}$ correspond the matrizant $\mathbf{X}(t) = \mathbf{F}(t)\, \mathbf{X}_\mathbf{C}(t)$ and the Hamiltonian $\mathbf{H}(t) = \tilde{\mathbf{J}}^{-1}(d\mathbf{X}/dt)\, \mathbf{X}^{-1}$. The one-to-one correspondence we have constructed, associating each Hamiltonian $\mathbf{H}(t) \in \mathfrak{L}^3$ with a pair $\mathbf{F}(t) \in \Omega_0$, $\mathbf{C} \in \mathfrak{R}^3$, is expressed in set-theoretic terms by (1.19).

We now show that this correspondence is indeed a homeomorphism. Under any continuous deformation of the Hamiltonian $\mathbf{H}(t) \in \mathfrak{L}^3$, the corresponding curve $\mathbf{X}(t) \in \mathfrak{R}^3$ undergoes a continuous deformation (in the metric $\rho[\mathbf{X}_1(t), \mathbf{X}_2(t)]$), and therefore so does its endpoint $\mathbf{C}$. It is clear that $\mathbf{X}_\mathbf{C}(t)$ and, therefore, also $\mathbf{F}(t)$ depend continuously on $\mathbf{C}$ (in the metric $\rho[\mathbf{X}_1(t), \mathbf{X}_2(t)]$). Thus the pair $\{\mathbf{F}(t), \mathbf{C}\}$ depends continuously on $\mathbf{H}(t) \in \mathfrak{L}^3$. The converse follows from the formula $\mathbf{X}(t) = \mathbf{F}(t)\mathbf{X}_\mathbf{C}(t)$ and the continuity theorem of § II. 1.7.

We have thus defined the correspondence (1.19) and proved that it is a homeomorphism.

In view of (1.19), we can represent the subsets $\mathscr{O}_n$, $\mathscr{H}_n$, Π_n, $\Pi_n^{\bullet -}$, $\Pi_n^{\bullet\bullet}$, $\Pi_n^{\bullet +}$ of $\mathfrak{L}^3$ defined above as follows:

$$\left.\begin{array}{ll} \mathscr{O}_n = \Omega_0 \times \tilde{\mathscr{O}}_n, & \mathscr{H}_n = \Omega_0 \times \tilde{\mathscr{H}}_n, \\ \Pi_n = \Omega_0 \times \tilde{\Pi}_n, & \Pi_n^{\bullet\bullet} = \Omega_0 \times \tilde{\Pi}_n^{\bullet\bullet}, \\ \Pi_n^{\bullet +} = \Omega_0 \times \tilde{\Pi}_n^{\bullet +}, & \Pi_n^{\bullet -} = \Omega_0 \times \tilde{\Pi}_n^{\bullet -}. \end{array}\right\} \quad (1.19a)$$

We shall show later (§1.7) that Ω_0 is connected. Thus, by (1.19), for each connected set $\tilde{\mathfrak{M}} \subset \mathfrak{R}^3$ we have a connected set

$$\mathfrak{M} = \Omega_0 \times \tilde{\mathfrak{M}}$$

of Hamiltonians in $\mathfrak{L}^3$. Since the sets $\tilde{\mathscr{O}}_n$, $\tilde{\mathscr{H}}_n$, $\tilde{\Pi}_n$, $\tilde{\Pi}_n^{\bullet -}$, $\tilde{\Pi}_n^{\bullet\bullet}$, $\tilde{\Pi}_n^{\bullet +}$ are clearly connected, each of the sets (1.19a) is connected. It also follows from (1.19a) that the sets $\mathscr{O}_n$, $\mathscr{H}_n$ are open and Π_n is closed.

We note that the sets listed in the first column of the table are taken by the natural mappings $\mathfrak{L}^3 \to \mathfrak{R}^3 \to \mathfrak{G}$ onto the corresponding sets in the second and third columns. Thus, for example, to Hamiltonians $\mathbf{H}(t)$ in $\mathscr{H}_{2n+1}$ correspond monodromy matrices in the set $\tilde{\mathscr{H}} \subset \mathfrak{G}$, and similar statements hold true for all other subsets of $\mathfrak{L}^3$ in the first column.

It should now be clear that the three-dimensional space $\mathfrak{R}^3$ provides a highly intuitive model of the function space $\mathfrak{L}^3$. Given a point $\mathbf{C} \in \mathfrak{R}^3$, we can reconstruct (via the natural mapping $\mathfrak{R}^3 \to \mathfrak{G}$) the monodromy matrix $\mathbf{X}(T)$. Different Hamiltonians "projecting" onto one point $\mathbf{C} \in \mathfrak{R}^3$ correspond to matrizants $\mathbf{X}_1(t)$ and $\mathbf{X}_2(t)$ which, first, have common endpoints and, second, may be continuously deformed into one another with their common endpoint $\mathbf{X}_1(T) = \mathbf{X}_2(T)$ held fixed. Thus the differences between Hamiltonians that "project" onto the same point $\mathbf{C} \in \mathfrak{R}^3$ are "immaterial" as far as we are concerned here.

As mentioned above, we associate with any connected set $\tilde{\mathfrak{M}} \subset \mathfrak{R}^3$ a connected set of Hamiltonians $\mathfrak{M} = \Omega_0 \times \tilde{\mathfrak{M}} \subset \mathfrak{L}^3$. Thus, the structure of the space $\mathfrak{R}^3$ as shown in Figure 56 at once implies that the set of strongly stable (strongly unstable) Hamiltonians is the union in $\mathfrak{L}^3$ of countably many open domains $\mathscr{O}_n$, $n = 0, \pm 1, \ldots$ ($\mathscr{H}_n$, $n = 0, \pm 1, \ldots$); the set of Hamiltonians for which equation (1.2) has a T-periodic (T-antiperiodic) solution is the union in $\mathfrak{L}^3$ of countably many closed sets Π_{2n} (Π_{2n+1}); and so on. The sets $\mathscr{O}_n$ ($\tilde{\mathscr{O}}_n$) will be called *stability domains* and the sets $\mathscr{H}_n$ ($\tilde{\mathscr{H}}_n$) *instability domains*. Equations with Hamiltonians in the various sets listed in the table have the labeled spectra illustrated in Figure 53. For example, if $\mathbf{H}(t) \in \mathscr{H}_{2n+1}$, then $\mathbf{X}(t) \in \tilde{\mathscr{H}}^-$, and so the spectrum of the equation will have the appropriate form, as shown in Figure 53 for $\tilde{\mathscr{H}}^-$.

To end this subsection, we review the three important properties of $\mathfrak{R}^3$ as a model for $\mathfrak{L}^3$.

I. $\mathfrak{R}^3$ is three-dimensional euclidean space; there is a mapping of $\mathfrak{L}^3$ onto $\mathfrak{R}^3$, i. e., for each Hamiltonian $\mathbf{H}(t) \in \mathfrak{L}^3$ we have a point $\mathbf{C} = \mathbf{C}[\mathbf{H}(t)] \in \mathfrak{R}^3$, and conversely, for each point $\mathbf{C} \in \mathfrak{R}^3$ we have a corresponding Hamiltonian. This mapping is continuous (in the sense of the metric in $\mathfrak{L}^3$).

II. The following mappings are defined on $\mathfrak{R}^3$: 1) into the space $\tilde{\mathfrak{R}}^3$ of the real matrices with trace zero; 2) onto the space $\mathfrak{G} = \{\mathbf{X}\}$ of real unimodular matrices; 3) onto the set $\Xi = \{\zeta\}$ of labeled spectra. The last two mappings are defined by the mappings of $\mathfrak{R}^3$ onto $\mathfrak{G}$ and of $\mathfrak{G}$ onto Ξ:

$$\mathfrak{R}^3 \longrightarrow \tilde{\mathfrak{R}}^3 = \{\mathbf{K}\}, \quad \mathfrak{R}^3 \longrightarrow \mathfrak{G} = \{\mathbf{X}\} \longrightarrow \Xi = \{\zeta\}.$$

All these mappings are continuous. The mapping $\mathfrak{R}^3 \longrightarrow \mathfrak{G}$ is a local homeomorphism, that is to say, it maps sufficiently small neighborhoods of points $\mathbf{C} \in \mathfrak{R}^3$ in one-to-one and bicontinuous fashion onto neighborhoods of the corresponding points $\mathbf{X} \in \mathfrak{G}$. Let $\mathbf{H}(t) \in \mathfrak{L}^3$, let $\mathbf{X}(t)$ be the matrizant of the corresponding equation (1.2) and $\mathbf{C} = \mathbf{C}[\mathbf{H}(t)] \in \mathfrak{R}^3$. Suppose that under the above mappings

$$\mathbf{C} \longrightarrow \mathbf{K}, \quad \mathbf{C} \longrightarrow \mathbf{X} \longrightarrow \zeta.$$

Then $\mathbf{K}$ is the exponent in the Floquet-Lyapunov representation (1.6), $\mathbf{X}$ is the monodromy matrix $\mathbf{X}(T)$, and ζ the labeled spectrum of $\mathbf{X}(T)$. In particular, all pre-images $\mathbf{H}(t)$ of a point $\mathbf{C} \in \mathfrak{R}^3$ correspond to the same monodromy matrix $\mathbf{X}(T)$.

III. A point $\mathbf{C} \in \mathfrak{R}^3$ is the "projection" under the mapping $\mathfrak{L}^3 \longrightarrow \mathfrak{R}^3$ of the set of all Hamiltonians which, first, have the same monodromy matrix and, second, have the property that the corresponding trajectories $\mathbf{X}(t)$ in $\mathfrak{G}$ may be deformed continuously into one another without displacing their endpoints

$$\mathbf{X}(0) = \mathbf{I}_2 \text{ and } \mathbf{X}(T).$$

Property III means that equations (1.2) with different Hamiltonians belonging to the same point $\mathbf{C} \in \mathfrak{R}^3$ define the same final "positions" of the vectors $\mathbf{x}(T)$ in the phase space $\{\mathbf{x}\}$ and that the corresponding motions $\mathbf{x}(t)$ may be deformed continuously into one another without altering this final situation.

In descriptive terms, property II means that the elements of $\mathfrak{R}^3$ may be "colored" in one of three types of "color". This "color" may be either a matrix $\mathbf{K} \in \tilde{\mathfrak{R}}^3$, a matrix $\mathbf{X} \in \mathfrak{G}$, or a labeled spectrum $\zeta \in \Xi$. The "color scheme" of a point $\mathbf{C} \in \mathfrak{R}^3$ characterizes the properties of the monodromy matrix of the corresponding equation (1.2).

Figure 56 may be viewed as being "colored" in this way by considering the mappings

$$\mathfrak{R}^3 \longrightarrow \mathfrak{G}, \quad \mathfrak{R}^3 \longrightarrow \tilde{\mathfrak{R}}^3, \quad \mathfrak{R}^3 \longrightarrow \Xi.$$

The model $\mathfrak{R}^3$ provides a "picture" of the position in $\mathfrak{L}^3$ of any set of Hamiltonians defined in terms of the properties of the monodromy matrices of the corresponding equations (1.2). For example, suppose we are studying the set $\mathfrak{M}_\rho$ of Hamiltonians such that equation (1.2) has prescribed multipliers ρ and ρ^{-1}. Since the multipliers are fixed, the corresponding matrices $\mathbf{K}$ in (1.6) have fixed eigenvalues. Thus these matrices fill out hyperboloids $\det(\mathrm{K}-\alpha \mathbf{I}_2)=0$, or $x^2+y^2-z^2=\alpha^2$. Under the mapping from $\mathfrak{R}^3$ to $\tilde{\mathfrak{R}}_0^-$ and $\tilde{\mathfrak{R}}_0^+$, and then to $\mathfrak{G}$ (see Figure 52) and to $\mathfrak{R}^3$ (Figures 54, 55), these hyperboloids become surfaces ("deformed hyperboloids") in the domains $\mathfrak{G}^{\pm}_{0\pm}$ (deformed two-sheeted hyperboloids, $|\rho|=1$, $\rho\neq\pm 1$) or in $\tilde{\mathcal{H}}^-$, $\tilde{\mathcal{H}}^+$ (deformed one-sheeted hyperboloids, $|\rho|\neq 1$, ρ real). The corresponding surfaces are shown in Figure 56 by thin curves. (As mentioned above, the partition of $\mathfrak{R}^3$ in Figure 56 should be envisaged as three-dimensional, rotating the plane of the figure about the dash-dot axis.) If $\rho=\rho^{-1}=1$ these surfaces become the cones $\tilde{\Pi}_n$. The domain $\tilde{\mathfrak{G}}^+\subset\mathfrak{G}$ contains the matrices whose multiplier of the first kind lies on the upper semicircle $|\rho|=1$, and the domain $\tilde{\mathfrak{G}}^-\subset\mathfrak{G}$ those whose multiplier of the first kind is on the lower semicircle $|\rho|=1$ (Figure 53).

Thus the set of Hamiltonians $\mathbf{H}(t)$ for which the multipliers are fixed and distributed as shown in Figure 53 ($\tilde{\mathfrak{G}}^+$) is the union in $\mathfrak{L}^3$ of countably many disjoint closed connected and simply-connected sets; in the model $\mathfrak{R}^3$ (Figure 56), these are the surfaces represented by thin curves (deformed sheets of one-sheeted hyperboloids) in the domains $\tilde{\mathfrak{G}}_{2n}$ ($n=0, \pm 1, \pm 2, \ldots$). For the spectrum marked $\tilde{\mathfrak{G}}^-$ in Figure 53, we have a similar proposition, except that now the surfaces lie in the domains $\tilde{\mathfrak{G}}_{2n+1}$ ($n=0, \pm 1, \pm 2, \ldots$). For the spectrum $\tilde{\mathcal{H}}^+$ in Figure 53, the corresponding set of Hamiltonians is the union of a countable set of disjoint closed connected (but not simply-connected!) subsets, represented in $\mathfrak{R}^3$ by deformed one-sheeted hyperboloids lying in the domains $\tilde{\mathcal{H}}_{2n}$ ($n=0, \pm 1, \pm 2, \ldots$). Finally, for the spectrum marked $\tilde{\mathcal{H}}^-$ in Figure 53 we have deformed one-sheeted hyperboloids in the domains $\tilde{\mathcal{H}}_{2n+1}$ ($n=0, \pm 1, \pm 2, \ldots$). All these surfaces are represented in Figure 56 by thin curves.

Similarly, one shows, for example, that the set of Hamiltonians $\mathbf{H}(t)$ in $\mathfrak{L}^3$ such that

$$|\mathbf{x}(t)|\leqslant \text{const}\, e^{\alpha t}$$

(with given $\alpha>0$) is an open set, which is the union in $\mathfrak{L}^3$ of countably many open, connected, but not simply-connected, domains — the "exteriors" of certain deformed two-sheeted hyperboloids lying in the domains $\tilde{\mathcal{H}}_n$ ($n=0, \pm 1, \pm 2, \ldots$).

Our three-dimensional model $\mathfrak{R}^3$ of the function space $\mathfrak{L}^3$ lends great clarity and elegance to our further investigations of equation (1.2), to be continued at the end of this section and in the succeeding sections. Despite this, the next two subsections will make no use of the results of this subsection. Independently of these results, we shall define (in a rather formal manner) the various sets in Figure 56. These definitions will rest directly on the representation (1.6). We shall also need some properties of the matrix-functions $\mathbf{F}(t)$ in (1.6), which we now proceed to study.

1.7. Structure of set $\{\mathbf{F}(t)\}$

Let Ω denote the set of absolutely continuous matrix-functions $\mathbf{F}(t)$ such that $\mathbf{F}(t+T) \equiv \mathbf{F}(t)$ or $\mathbf{F}(t+T) \equiv -\mathbf{F}(t)$, $\det \mathbf{F}(t) \equiv 1$, $\mathbf{F}(0) = \mathbf{I}_2$. Recall that the metric in Ω is

$$\rho(\mathbf{F}_1, \mathbf{F}_2) = \max_{0 \leqslant t \leqslant T} |\mathbf{F}_1(t) - \mathbf{F}_2(t)| + \int_0^T \left| \frac{d\mathbf{F}_1}{dt} - \frac{d\mathbf{F}_2}{dt} \right| dt,$$

where the matrix norm is defined as in § I. 1.2. The set Ω then becomes a metric space /46/.

Let $\mathbf{a} \neq \mathbf{0}$ be an arbitrary vector and $\mathbf{x} = \mathbf{F}(t)\mathbf{a}$. Let φ_x denote the *rotation* of the vector $\mathbf{x}$ in time T, that is, the angle through which it turns in that time. Since $\mathbf{F}(T) = \pm\mathbf{I}$, it follows that $\varphi_x = n\pi$, where n is even if $\mathbf{F}(t)$ is periodic and odd if $\mathbf{F}(t)$ is antiperiodic ($n = 0, \pm 1, \pm 2, \ldots$).

The following proposition is almost obvious: The number n is independent of the choice of the vector $\mathbf{a} \neq \mathbf{0}$, but depends only on $\mathbf{F}(t)$. Indeed, let $\mathbf{x} = \mathbf{F}(t)\mathbf{a}$, $\mathbf{y} = \mathbf{F}(t)\mathbf{b}$. We have

$$\det \| \mathbf{x}, \mathbf{y} \| = \det \mathbf{F}(t) \det \| \mathbf{a}, \mathbf{b} \| = \det \| \mathbf{a}, \mathbf{b} \|.$$

Thus the area of the parallelogram spanned by $\mathbf{x}$ and $\mathbf{y}$ is an invariant of the motion. Thus neither vector can "overtake" the other; neither can the two vectors form an angle $\geqslant \pi$. Since the rotations of $\mathbf{x}$ and $\mathbf{y}$ may differ only by integral multiples of π, they must coincide, as claimed.

Thus the number n is defined only by the matrix $\mathbf{F}(t)$; we denote it by $n_\mathbf{F}$.

Remark. The interval $(0, T)$ may be replaced in the definition of $n_\mathbf{F}$ by the interval $(t, t_0 + T)$ for arbitrary t_0. In other words, if $\vartheta(t)$ is the argument of the vector $\mathbf{F}(t)\mathbf{a}$, then $\vartheta(t_0 + T) - \vartheta(t_0) = n_\mathbf{F}\pi$. Indeed, $\mathbf{F}(t+T) = \pm\mathbf{F}(t)$. Therefore $\vartheta(t+T) - \vartheta(t) = n(\mathbf{a}, t)\pi$, where the number $n(\mathbf{a}, t)$ is independent of t since $n(\mathbf{a}, t)$ is an integer and a continuous function of t. But $n(\mathbf{a}, 0) = n_\mathbf{F}$. Consequently, $\vartheta(t+T) - \vartheta(t) = n_\mathbf{F}\pi$.

Let Ω_n denote the set of matrices $\mathbf{F}(t)$ such that $n_\mathbf{F} = n$. Then

$$\Omega = \bigcup_{n=-\infty}^{\infty} \Omega_n, \quad \Omega_n \cap \Omega_m = 0 \quad \text{if} \quad n \neq m.$$

Theorem I (structure of Ω).

I. *The set Ω_n is connected.*

II. *Points in Ω_n and Ω_m ($m \neq n$) cannot be joined by a continuous curve lying entirely in Ω.*

In other words, Ω is the union of the countable set of components Ω_n.

We first present a nonrigorous but intuitive proof. $\mathbf{F}(t)$ is a trajectory in $\mathfrak{G}$, beginning at the point $\mathbf{I}_2$ and ending either at $\mathbf{I}_2$ or at $-\mathbf{I}_2$. The angle φ (Figure 52), which is the rotation of the first column of $\mathbf{F}(t)$, i. e., of $\mathbf{F}(t)\begin{pmatrix}1\\0\end{pmatrix}$, in time t, is $n_\mathbf{F}\pi$. Thus, Ω_n is the set of all trajectories which turn in $\mathfrak{G}$ through an angle $n\pi$. It is obvious that two such curves with the same numbers n can be deformed continuously into one another, with their endpoints held fixed, whereas this is impossible for curves with different n's. We now proceed to the rigorous proof.

Proof. We first prove II, and then I. We begin by showing that $n_{\mathbf{F}}$ is a continuous function of $\mathbf{F}$ on Ω. Let

$$\mathbf{x}(t)=\mathbf{F}(t)\mathbf{a},\quad \mathbf{y}(t)=\mathbf{G}(t)\mathbf{a}\quad (\mathbf{F}(t)\in\Omega,\ \mathbf{G}(t)\in\Omega),$$
$$|\mathbf{x}(t)-\mathbf{y}(t)|<|\mathbf{F}(t)-\mathbf{G}(t)||\mathbf{a}|.$$

If $|\mathbf{F}(t)-\mathbf{G}(t)|<\varepsilon$, where ε is small, then $\mathbf{y}(t)$ revolves around 0 as many times as $\mathbf{x}(t)$ ($\mathbf{x}(t)$ never vanishes). Thus $n_{\mathbf{F}}$ is a continuous function of $\mathbf{F}$. Now let $\mathbf{F}_s$ be a continuous trajectory in Ω $(0\leqslant s\leqslant 1)$; $\mathbf{F}_0\in\Omega_n$, $\mathbf{F}_1\in\Omega_m$, $m\neq n$. Then $n_{\mathbf{F}_s}=n(s)$ is a continuous function of s, $n(0)=1$, $n(1)=m$; but this is absurd, because $n(s)$ takes only integer values.

We now prove I. The simplest transformation in Ω_n is a uniform rotation of the plane through $n\pi$ in the positive sense: $\exp(-n\pi t\tilde{\mathbf{J}}/T)$. Let $\mathbf{F}(t)\in\Omega_n$. It will suffice to show that we can join $\mathbf{F}(t)$ to $\exp(-n\pi t\tilde{\mathbf{J}}/T)$ by a continuous curve in Ω_n. Let $\mathbf{F}(t)=\|\mathbf{p},\mathbf{q}\|$, $\vartheta_{\mathbf{p}}=\arg\mathbf{p}$, $\vartheta_{\mathbf{q}}=\arg\mathbf{q}$, $\det\mathbf{F}(t)=|\mathbf{p}||\mathbf{q}|\sin(\vartheta-\vartheta_{\mathbf{p}})\equiv 1$. Then

$$0<\vartheta_{\mathbf{q}}(t)-\vartheta_{\mathbf{p}}(t)<\pi,$$

so that the vector $\mathbf{q}(t)$ always moves "ahead" of the vector $\mathbf{p}(t)$, preceding it by an angle $<\pi$. Let $\mathbf{p}'(t)=-\tilde{\mathbf{J}}\mathbf{p}(t)$ be a vector perpendicular to $\mathbf{p}(t)$ (see Figure 57) and

$$\mathbf{F}'(t)=\frac{1}{|\mathbf{p}|^2}\|\mathbf{p},\mathbf{p}'\|,\quad \det\mathbf{F}'(t)\equiv 1,$$
$$\mathbf{F}'(t+T)=(-1)^n\mathbf{F}'(t),\quad \text{i. e.,}\ \mathbf{F}'(t)\in\Omega.$$

Since $\mathbf{p}$ turns through the angle $n\pi$ in time T, it follows that $n_{\mathbf{F}'}=n$, so that $\mathbf{F}'(t)\in\Omega_n$. We join $\mathbf{F}(t)$ to $\mathbf{F}'(t)$ as follows. Join $\mathbf{q}(t)$ to $\mathbf{p}'(t)$ by a "straight-line segment" $\mathbf{p}_s(t)=(1-s)\mathbf{q}(t)+s\mathbf{p}'(t)$ $(0\leqslant s\leqslant 1)$, and set

$$\delta_s(t)=\det\|\mathbf{p},\ \mathbf{p}_s\|=(1-s)\det\|\mathbf{p},\ \mathbf{q}\|+s\det\|\mathbf{p},\mathbf{p}'\|=(1-s)+s|\mathbf{p}|^2>0$$
$$(\delta_s(t+T)=\delta_s(t)).$$

The matrix

$$\mathbf{G}_s(t)=\left\|\mathbf{p},\ \frac{1}{\delta_s(t)}\mathbf{p}_s\right\|$$

is periodic or antiperiodic, since the vectors $\mathbf{p}$ and $\mathbf{p}_s$ have this property. Moreover, $d\mathbf{G}_s/dt$ exists and is summable, and $\det\mathbf{G}_s\equiv 1$. Thus $\mathbf{G}_s(t)\in\Omega$ and it joins $\mathbf{F}$ and $\mathbf{F}'$ in Ω_n.

It follows from II that $\mathbf{G}_s(t)\in\Omega_n$. It remains to join $\mathbf{F}'(t)$ to $\exp(-n\pi\tilde{\mathbf{J}}t/T)$. Writing $\mathbf{F}'(t)$ as

$$\mathbf{F}'(t)=\left\|\frac{\mathbf{p}}{|\mathbf{p}|},\ \frac{\mathbf{p}'}{|\mathbf{p}'|}\right\|,$$

we see that $\mathbf{F}'(t)$ is an orthogonal matrix, so that

$$\mathbf{F}'(t) = e^{-\vartheta(t)\tilde{\mathbf{J}}},$$

where

$$\vartheta(t) = \operatorname{Arg}\mathbf{p} = \vartheta_p(t)$$

is a suitable continuous branch of the argument of the vector $\mathbf{p}=\mathbf{p}(t)$. It is obvious that the "curve" $\mathbf{F}_s(t) = \exp[-\vartheta_s(t)\tilde{\mathbf{J}}]$, where

$$\vartheta_s(t) = (1-s)\vartheta(t) + s\frac{n\pi}{T}t, \qquad 0 \leqslant s \leqslant 1,$$

joins $\mathbf{F}'(t)$ and $\exp\left(-\frac{n\pi}{T}t\tilde{\mathbf{J}}\right)$ in Ω_n, proving the theorem.

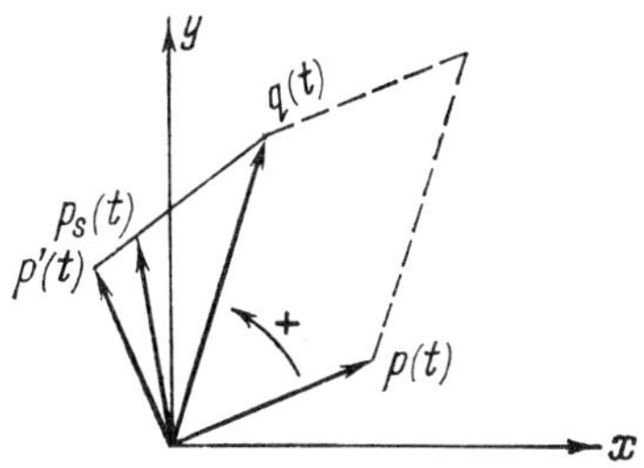

FIGURE 57.

Corollary 1. If $\mathbf{F}(t) \in \Omega_n$, *then* $\mathbf{F}^{\tau}(t) \in \Omega_{-n}$.

Proof. It is clear that $\exp(-n\pi t\tilde{\mathbf{J}}/T) \in \Omega_n$. Join $\mathbf{F}(t)$ continuously to $\exp(-n\pi t\tilde{\mathbf{J}}/T)$ by a curve $\mathbf{F}_s(t)$ $(0 \leqslant s \leqslant 1)$. Then $\mathbf{F}^{\tau}(t)$ is joined continuously by the curve $\mathbf{F}_s^{\tau}(t)$ to the matrix-function

$$[\exp(-n\pi t\tilde{\mathbf{J}}/T)]^{\tau} = \exp(-n\pi t\tilde{\mathbf{J}}^{\tau}/T) = \exp(n\pi t\tilde{\mathbf{J}}/T),$$

which is clearly in Ω_{-n}. Thus $\mathbf{F}^{\tau}(t) \in \Omega_{-n}$.

Corollary 2. If $\mathbf{F}(t) \in \Omega_n$ *and* $\mathbf{S}$ *is a constant nonsingular matrix, then* $\mathbf{SFS}^{-1} \in \Omega_n$ *if* $\det\mathbf{S} > 0$ *and* $\mathbf{SFS}^{-1} \in \Omega_{-n}$ *if* $\det\mathbf{S} < 0$.

Proof. If $\det\mathbf{S} > 0$, we may assume without loss of generality that $\det\mathbf{S} = 1$, so that $\mathbf{S} \in \mathfrak{G}$. Joining $\mathbf{S}$ continuously to $\mathbf{I}_2$, we thereby join $\mathbf{SFS}^{-1}$ to $\mathbf{F}$. Consequently, $\mathbf{SFS}^{-1} \in \Omega_n$. If $\det\mathbf{S} < 0$, we may assume without loss of generality that $\det\mathbf{S} = -1$. Then $\mathbf{S} = \mathbf{E}_0\mathbf{S}_0$, where $\det\mathbf{S}_0 = 1$ and $\mathbf{E}_0 = \begin{Vmatrix} 1 & 0 \\ 0 & -1 \end{Vmatrix}$. Now join $\mathbf{S}_0$ continuously to $\mathbf{I}_2$ and $\mathbf{F}(t)$ continuously to $\exp(-n\pi t\tilde{\mathbf{J}}/T)$. Then $\mathbf{SFS}^{-1}$ is thereby joined continuously to

$$\mathbf{E}_0\exp(-n\pi t\tilde{\mathbf{J}}/T)\mathbf{E}_0^{-1} = \exp(n\pi t\tilde{\mathbf{J}}/T),$$

since $\mathbf{E}_0\tilde{\mathbf{J}}\mathbf{E}_0^{-1} = -\tilde{\mathbf{J}}$. Therefore $\mathbf{SFS}^{-1} \in \Omega_{-n}$.

Corollary 3. $n_{F'F''} = n_{F'} + n_{F''}$.

Proof. Denote $n_{F'} = n_1$, $n_{F''} = n_2$. Join $\mathbf{F}'$ and $\mathbf{F}''$ by curves $\mathbf{F}'_s$ and $\mathbf{F}''_s$, respectively, to

$$\exp(-n_1 \pi t \tilde{\mathbf{J}}/T) \text{ and } \exp(-n_2 \pi t \tilde{\mathbf{J}}/T).$$

Then $\mathbf{F}'\mathbf{F}''$ is joined by the curve $\mathbf{F}'_s \mathbf{F}''_s$ to

$$\exp(-n_1 \pi t \tilde{\mathbf{J}}/T) \exp(-n_2 \pi t \tilde{\mathbf{J}}/T) = \exp[-(n_1 + n_2)\pi \tilde{\mathbf{J}}/T] \in \Omega_{n_1+n_2}.$$

Consequently, $\mathbf{F}' \cdot \mathbf{F}'' \in \Omega_{n_1+n_2}$.

Corollary 4. The domains Ω_n are homeomorphic to one another; in particular, each Ω_n is homeomorphic to Ω_0. The correspondence between $\mathbf{F}_n \in \Omega_n$ and $\mathbf{F}_0 \in \Omega_0$ may be defined by

$$\mathbf{F}_n = \mathbf{F}_0 \exp(-n \pi t \tilde{\mathbf{J}}/T).$$

Corollary 5. If $\mathbf{F}(t) \in \Omega_n$, then $\mathbf{F}(t)^m \in \Omega_{nm}$ $(m = 0, \pm 1, \pm 2, \ldots)$.

The proof follows the same lines as the preceding proofs.

Theorem II. Each of the sets Ω_n is both open and closed (in the sense of the above metric).

Proof. Let $\mathbf{F} \in \Omega_n$, $\mathbf{F}' \in \Omega$, $\rho(\mathbf{F}', \mathbf{F}) < \varepsilon$. For sufficiently small ε, the vectors $\mathbf{F}'(t)\mathbf{c}$ and $\mathbf{F}(t)\mathbf{c}$ may be brought arbitrarily close to each other in the uniform metric. In addition, they have the same initial and end points:

$$\mathbf{F}'(0)\mathbf{c} = \mathbf{F}(0)\mathbf{c} = \mathbf{c}, \qquad \mathbf{F}'(T)\mathbf{c} = \mathbf{F}(T)\mathbf{c} = (-1)^n \mathbf{c}.$$

Consequently, the rotations of these vectors coincide, and so $\mathbf{F}' \in \Omega_n$. Thus Ω_n is an open set.

Let $\mathbf{F}_k(t) \in \Omega_n$, $\rho(\mathbf{F}_k, \mathbf{F}_l) \to 0$ as $k, l \to \infty$. Set $\mathbf{H}_k = \tilde{\mathbf{J}}^{-1} \frac{d\mathbf{F}_k}{dt} \mathbf{F}_k^{-1}$. By the continuity theorem of § II. 1.7, $\rho'(\mathbf{H}_k, \mathbf{H}_l) \to 0$ as $k, l \to \infty$, where ρ' is the metric in $\mathfrak{L}^3$. Since $\mathfrak{L}^3$ is complete, there is a Hamiltonian $\mathbf{H} \in \mathfrak{L}^3$ such that $\rho'(\mathbf{H}_k, \mathbf{H}) \to 0$. Define $\mathbf{F}$ as a solution of the equation $\tilde{\mathbf{J}} \frac{d\mathbf{F}}{dt} = \mathbf{H}(t)\mathbf{F}$, $\mathbf{F}(0) = \mathbf{I}_2$. Since $\mathbf{F}_k(0) = \mathbf{I}_2$, it follows from the same theorem that $\rho(\mathbf{F}_k, \mathbf{F}) \to 0$. Consequently, $\mathbf{F}_k(T) \to \mathbf{F}(T)$ as $k \to \infty$, and since $\mathbf{F}_k(T) = (-1)^n \mathbf{I}_2$ it follows that $\mathbf{F}(T) = (-1)^n \mathbf{I}_2$. Thus $\mathbf{F}(t) \in \Omega$. Since $n_{\mathbf{F}}$ is a continuous function of $\mathbf{F}$ and $n_{\mathbf{F}_k} = n$, we have $n_{\mathbf{F}} = n$, so that $\mathbf{F}(t) \in \Omega_n$ and Ω_n is a closed set. Q. E. D.

1.8. Stability and instability domains in Hamiltonian space

Recall that $\mathfrak{L}^3$ is the set of all Hamiltonians $\mathbf{H}(t)$ in (1.3), normed by

$$\|\mathbf{H}(t)\| = \int_0^T |\alpha|\,dt + \int_0^T |\beta|\,dt + \int_0^T |\gamma|\,dt.$$

We are now going to define the sets $\mathscr{O}_n$, $\mathscr{H}_n$, $\Pi_n = \Pi_n^{*-} \cup \Pi_n^{**} \cup \Pi_n^{*+}$, independently of §§1.5, 1.6. This definition will often be used in the sequel. At the end of the subsection, we shall show that this new definition is equivalent to the old definitions of §1.6.

Each matrix $\mathbf{H}(t) \in \mathfrak{L}^3$ determines a unique fundamental matrix (matrizant) $\mathbf{X}(t)(\mathbf{X}(0) = \mathbf{I}_2)$ of system (1.2). Via formula (1.6), the matrizant $\mathbf{X}(t)$ determines a unique pair $\mathbf{F}(t)$, $\mathbf{K}$, where $\mathbf{F}(t) \in \Omega$, $\mathbf{K} \in \tilde{\mathfrak{R}}_1 = \tilde{\mathscr{H}} \cup \tilde{\Pi} \cup \tilde{\mathscr{O}}_{1+}$ (see theorem at end of §1.4). We write this correspondence as

$$\mathfrak{L}^3 = \Omega \times (\tilde{\mathscr{H}} \cup \tilde{\Pi} \cup \tilde{\mathscr{O}}_{1+}).$$

Let $\tilde{\mathbf{0}}$ denote the zero matrix ("origin") in $\tilde{\mathfrak{R}}^3$ (see §1.4). Then

$$\tilde{\Pi} = \tilde{\Pi}_+ \cup \tilde{\mathbf{0}} \cup \tilde{\Pi}_-. \tag{1.20}$$

We showed above (§1.7) that the set Ω is the union of countably many disjoint sets Ω_n. Therefore,

$$\mathfrak{L}^3 = \bigcup_n (\Omega_n \times \tilde{\mathscr{H}}) \cup (\Omega_n \times \tilde{\Pi}) \cup (\Omega_n \times \tilde{\mathscr{O}}_{1+})$$

and by (1.20)

$$\Omega_n \times \tilde{\Pi} = (\Omega_n \times \tilde{\Pi}_+) \cup (\Omega_n \times \tilde{\mathbf{0}}) \cup (\Omega_n \times \tilde{\Pi}_-).$$

We now define

$$\left.\begin{array}{ll} \mathscr{H}_n = \Omega_n \times \tilde{\mathscr{H}}, & \Pi_n^{*+} = \Omega_n \times \tilde{\Pi}_+, \\ \mathscr{O}_n = \Omega_n \times \tilde{\mathscr{O}}_{1+}, & \Pi_n^{*-} = \Omega_n \times \tilde{\Pi}_-, \\ \Pi_n = \Omega_n \times \tilde{\Pi}, & \Pi_n^{**} = \Omega_n \times \tilde{0}. \end{array}\right\} \tag{1.21}$$

These products of sets are understood throughout in the sense of the correspondence described above. Thus, $\mathbf{H}(t) \in \mathscr{H}_n$ means that $\mathbf{F}(t) \in \Omega_n$ and $\mathbf{K} \in \tilde{\mathscr{H}}$. Formulas (1.21) are the new definitions of the sets $\mathscr{H}_n$, $\mathscr{O}_n$, Π_n, Π_n^{*+}, Π_n^{**}, Π_n^{*-}. It follows from (1.20) and (1.21) that

$$\Pi_n = \Pi_n^{*+} \cup \Pi_n^{**} \cup \Pi_n^{*-}.$$

We now denote

$$\mathscr{H} = \bigcup_{n=-\infty}^{\infty} \mathscr{H}_n, \quad \mathscr{O} = \bigcup_{n=-\infty}^{\infty} \mathscr{O}_n, \quad \Pi = \bigcup_{n=-\infty}^{\infty} \Pi_n, \quad \Pi^{*+} = \bigcup_{n=-\infty}^{\infty} \Pi_n^{*+},$$

$$\Pi_n^{*-} = \bigcup_{n=-\infty}^{\infty} \Pi_n^{*-}, \quad \Pi^{**} = \bigcup_{n=-\infty}^{\infty} \Pi_n^{**}, \quad \Pi^* = \Pi^{*+} \cup \Pi^{*-}.$$

It is obvious that

$$\mathfrak{L}^3 = \mathscr{H} \cup \mathscr{O} \cup \Pi, \quad \Pi = \Pi^* \cup \Pi^{**},$$

$$\Pi_n = \Pi_n^{*+} \cup \Pi_n^{**} \cup \Pi_n^{*-}, \quad \mathscr{H} = \Omega \times \tilde{\mathscr{H}}, \quad \mathscr{O} = \Omega \times \tilde{\mathscr{O}}_{1+}, \text{ etc.}$$

As follows from §1.2, the matrizant $\mathbf{X}(t)$ will have one of the following forms:

$$\text{if } \mathbf{H}(t) \in \mathscr{H}: \quad \mathbf{X}(t) = \mathbf{F}(t)\,\mathbf{S}\begin{Vmatrix} e^{\lambda t} & 0 \\ 0 & e^{-\lambda t} \end{Vmatrix}\mathbf{S}^{-1} \qquad (\lambda > 0),$$

$$\text{if } \mathbf{H}(t) \in \mathscr{G}: \quad \mathbf{X}(t) = \mathbf{F}(t)\,\mathbf{S}\begin{Vmatrix} \cos\beta t & -\sin\beta t \\ \sin\beta t & \cos\beta t \end{Vmatrix}\mathbf{S}^{-1} \qquad \left(0 < \beta < \frac{\pi}{T}\right),$$

$$\text{if } \mathbf{H}(t) \in \Pi^{*\pm}: \quad \mathbf{X}(t) = \mathbf{F}(t)\,\mathbf{S}\begin{Vmatrix} 1 & \pm t \\ 0 & 1 \end{Vmatrix}\mathbf{S}^{-1},$$

$$\text{if } \mathbf{H}(t) \in \Pi^{**}: \quad \mathbf{X}(t) = \mathbf{F}(t).$$

Here $\mathbf{F}(t) \in \Omega$, $\mathbf{S}$ is a constant nonsingular real matrix.

Thus $\mathscr{H}$ is the set of all Hamiltonians $\mathbf{H}(t)$ for which equation (1.2) has an unbounded solution (the trivial solution is unstable) and no periodic or antiperiodic solutions of period T; $\mathscr{G}$ is the set of matrices $\mathbf{H}(t)$ for which all the solutions are bounded on $(-\infty, +\infty)$ and are almost-periodic functions, but there are no periodic or antiperiodic solutions of period T; Π^* is the set of $\mathbf{H}(t)$ for which only one solution is periodic or antiperiodic of period T; Π^{**} is the set of $\mathbf{H}(t)$ for which all solutions are periodic or antiperiodic of period T.

It follows from (1.21) that $\mathscr{H}_n$ and $\mathscr{G}_n$ are open connected subsets of $\mathfrak{L}^3$. Indeed, $\tilde{\mathscr{H}}$ and $\tilde{\mathscr{G}}_{1+}$ are open and connected subsets of $\tilde{\mathfrak{R}}^3$; by Theorems I, II of §1.7, Ω_n is an open connected subset of Ω. It follows from the first two equations in (1.21) that $\mathscr{H}_n$, $\mathscr{G}_n$ are open and connected subsets of $\mathfrak{L}^3$.

Similarly one shows that Π_n and Π_n^{**} are closed connected sets.

We have thus shown that $\mathscr{H}$ and $\mathscr{G}$ are unions in $\mathfrak{L}^3$ of countable sequences of domains $\mathscr{H}_n$ and $\mathscr{G}_n$, respectively, and Π is the union of a countable sequence of closed connected sets Π_n $(n = 0, \pm 1, \pm 2, \ldots)$.

As before we shall call $\mathscr{H}_n$ and $\mathscr{G}_n$ *instability* and *stability domains*, respectively.

We now prove that the sets introduced in §1.6 (see the table on p. 636) and those just defined (denoted by the same symbols) are in fact the same.*

By the definition of §1.6, every set in the first column of the table is the set of all Hamiltonians $\mathbf{H}(t) \in \mathfrak{L}^3$ for which the corresponding curve $\mathbf{X}(t)$ in $\mathfrak{R}^3$ (Figure 56) ends in the corresponding subset of $\mathfrak{R}^3$ in the second column of the table:

$$\mathbf{H}(t) \in \mathscr{H}_n \text{ if and only if } \mathbf{X}(T) \in \tilde{\mathscr{H}}_n \subset \mathfrak{R}^3$$

and so on. By the construction of §1.6, the matrix $\mathbf{K}$ in (1.6) is related to $\mathbf{X}(T) \in \mathfrak{G}$ by

$$\mathbf{X}(T) = e^{T\mathbf{K}},$$

if the end $\mathbf{X}(T)$ of the curve $\mathbf{X}(t)$ in $\mathfrak{R}^3$ is in a domain $\tilde{\mathfrak{R}}_n$ with even n, and by

$$\mathbf{X}(T) = -e^{T\mathbf{K}},$$

* The reader who has omitted §§ 1.5, 1.6 may go on directly to § 1.9.

if n is odd (see Figures 52, 54, 55). The matrix $\mathbf{F}(t)\in\Omega$ is defined by

$$\mathbf{F}(t)=\mathbf{X}(t)e^{-t\mathbf{K}} \qquad (\in\mathfrak{G}).$$

We claim that if $\mathbf{X}(T)\in\tilde{\mathfrak{R}}_n$, where $\mathbf{X}(t)$ is the corresponding curve in $\mathfrak{R}^3$, then $\mathbf{F}(t)\in\Omega_n$ with the same subscript n.

Consider the curve defined in $\mathfrak{G}$ by

$$\mathbf{F}_s(t)=\mathbf{X}(t)e^{(s-1)t\mathbf{K}}.$$

We have

$$\mathbf{F}_0(t)=\mathbf{F}(t),\quad \mathbf{F}_1(t)=\mathbf{X}(t),\quad \mathbf{F}_s(T)=(-1)^n e^{sT\mathbf{K}}.$$

Thus $\mathbf{F}_s(t)$ $(0\leqslant s\leqslant 1)$ is a deformation of the curve $\mathbf{F}(t)$ into the curve $\mathbf{X}(t)$. We shall show that $\mathbf{F}_s(T)$, which describes the end point of $\mathbf{F}_s(t)$ in the deformation process, lies in $\tilde{\mathfrak{R}}_0^+\subset\mathfrak{G}$ if n is even and in $\tilde{\mathfrak{R}}_0^-\subset\mathfrak{G}$ if n is odd. Indeed, recalling the construction of §1.6, we see that $\mathbf{K}\in\tilde{\mathfrak{R}}_0^+$ if $\mathbf{X}(T)\in\tilde{\mathfrak{R}}_n$ with even n, and $\mathbf{K}\in\tilde{\mathfrak{R}}_0^-$ if n is odd. The matrix $sT\mathbf{K}$, $0\leqslant s\leqslant 1$, describes a ray in $\tilde{\mathfrak{R}}_0^+$ (even n) or in $\tilde{\mathfrak{R}}_0^-$ (odd n).

According to the correspondence described in Figure 52, the matrix $(-1)^n e^{sT\mathbf{K}}$ describes a curve in $\tilde{\mathfrak{R}}_0^+$ (even n) or in $\tilde{\mathfrak{R}}_0^-$ (odd n). The curve in $\mathfrak{G}$ made up of the curves $\mathbf{X}(t)$ and $(-1)^n e^{sT\mathbf{K}}$ (s varying from $s=1$ to $s=0$) in succession has initial point $\mathbf{I}_2$ and endpoint $(-1)^n\mathbf{I}_2$. It is easy to see that the rotation of a column-vector of this curve is $n\pi$. (For example, for the curve $\mathbf{X}(t)$ shown in Figure 55, $n=2$; the corresponding curve $\mathbf{X}(t)$ in $\mathfrak{G}$ is shown in Figure 52; as s varies from $s=1$ to $s=0$, the curve $e^{sT\mathbf{K}}$ joins $\mathbf{X}(T)$ to $\mathbf{I}_2$, remaining throughout in $\tilde{\mathfrak{R}}_0^+$; the rotation for the composite curve is 2π.) On the other hand, the existence of a deformation $\mathbf{F}_s(t)$ shows that this composite curve may be deformed continuously into $\mathbf{F}(t)$. Consequently, the angle for the curve $\mathbf{F}(t)$ is also $n\pi$, so that $\mathbf{F}(t)\in\Omega_n$, as claimed.

We now consider the definitions of the sets in the first column of the table (§1.6). Let $\mathbf{H}(t)\in\mathscr{H}_n$. We have shown that in (1.6) we then have $\mathbf{F}(t)\in\Omega_n$, $\mathbf{K}\in\tilde{\mathscr{H}}$. This is equivalent to the first of formulas (1.21). Similarly, one proves the equivalence of the definitions §1.6 and (1.21) for the sets Π_n, Π_n^{*+}, Π_n^{**}, Π_n^{*-}.

Let $\mathbf{H}(t)\in\mathscr{G}_n$. Then, as shown previously, we have $\mathbf{K}\in\tilde{\mathscr{G}}_{0+}\cup\tilde{\Gamma}_{0+}\cup\tilde{\mathscr{G}}_{0-}$ in (1.6). Moreover (see Figures 54, 55), if $\mathbf{K}\in\tilde{\mathscr{G}}_{0+}\cup\tilde{\Gamma}_{0+}$, then $\mathbf{X}(T)\in\tilde{\mathfrak{R}}_n$ and $\mathbf{F}(t)\in\Omega_n$; if $\mathbf{K}\in\tilde{\mathscr{G}}_{0-}$, then $\mathbf{X}(T)\in\tilde{\mathfrak{R}}_{n+1}\subset\tilde{\mathfrak{R}}^3$ and $\mathbf{F}(t)\in\Omega_{n+1}$. It thus follows from the definition of $\mathscr{G}_n$ in §1.6 that

$$\mathscr{G}_n=(\Omega_n\times\tilde{\mathscr{G}}_{0+})\cup(\Omega_n\times\tilde{\Gamma}_{0+})\cup(\Omega_{n+1}\times\tilde{\mathscr{G}}_{0-}). \tag{1.22}$$

Now the second formula of (1.21) may be written

$$\mathscr{G}_n=(\Omega_n\times\tilde{\mathscr{G}}_{0+})\cup(\Omega_n\times\tilde{\Gamma}_{0+})\cup(\Omega_n\times\tilde{\mathscr{G}}_{2+}). \tag{1.22a}$$

Recall that $\tilde{\mathscr{G}}_{2+}$ is the open domain in $\mathfrak{R}^3$ between the upper sheets of the hyperboloids $x^2+y^2-z^2=-\pi^2/(4T^2)$ and $x^2+y^2-z^2=-\pi^2/T^2$ (Figure 51), and so

$$\tilde{\mathscr{G}}_{1+}=\tilde{\mathscr{G}}_{0+}\cup\tilde{\Gamma}_{0+}\cup\tilde{\mathscr{G}}_{2+}$$

(Figure 51). It will suffice to show that the last terms in (1.22) and (1.22a) are equal:

$$\Omega_n \times \tilde{\mathfrak{G}}_{2+} = \Omega_{n+1} \times \tilde{\mathfrak{G}}_{0-}. \tag{1.23}$$

Let $\mathbf{H}(t) \in \Omega_n \times \tilde{\mathfrak{G}}_{2+}$. This means that in (1.6) $\mathbf{F}(t) \in \Omega_n$, $\mathbf{K} \in \tilde{\mathfrak{G}}_{2+}$. Set*

$$\mathbf{F}_1(t) = \mathbf{F}(t) e^{\frac{t}{T} \mathbf{J}_k}, \quad \mathbf{K}_1 = \mathbf{K} - \frac{1}{T} \mathbf{J}_{\mathbf{K}}.$$

It is obvious that

$$\mathbf{K} \in \tilde{\mathfrak{G}}_{0-}.$$

Since $\exp\left(\frac{t}{T} \mathbf{J}_{\mathbf{K}}\right) \in \Omega_1$, it follows by Corollary 3 (§1.7) that $\mathbf{F}_1(t) \in \Omega_n$. The equality (1.23) now follows from

$$\mathbf{F}(t) e^{t\mathbf{K}} = \mathbf{F}_1(t) e^{t\mathbf{K}_1}$$

(both sides of which are equal to the matrizant $\mathbf{X}(t)$), and hence (1.22a) is also valid.

We have thus proved that the definitions of all the sets in the first column of the table are equivalent. The results of §1.6 also imply that (1.19a) and (1.21) are equivalent.

Remark. We have shown in passing that $\mathfrak{G}_n$ may be defined by (1.22). Reasoning in analogous fashion, one shows that $\mathfrak{G}_n$ may also be defined by the formula

$$\mathfrak{G}_n = (\Omega_n \times \tilde{\mathfrak{G}}_{0+}) \cup (\Omega_{n+1} \times \tilde{\Gamma}_{0-}) \cup (\Omega_{n+1} \times \mathfrak{G}_{0-}). \tag{1.24}$$

1.9. Qualitative behavior of solutions of equations with Hamiltonians in $\mathscr{H}_n$, $\mathfrak{G}_n$, Π_n. Rotation functions of solution-vectors

We shall now prove two simple propositions to be utilized below.**

Lemma I. Let $\mathbf{y}(t)$ *be a nonvanishing continuous vector-function,* $\mathbf{L}$ *a* 2×2 *matrix,* $\mathbf{z}(t) = \mathbf{L}\mathbf{y}(t)$, φ_y *and* φ_z *the rotations of the vectors* $\mathbf{y}(t)$ *and* $\mathbf{z}(t)$ *in time* T.

(a) *Let* $\det \mathbf{L} > 0$. *If* $\varphi_y = n\pi$, *then* $\varphi_z = n\pi$. *If* $n\pi < \varphi_y < (n+1)\pi$, *then* $n\pi < \varphi_z < (n+1)\pi$.

(b) *Let* $\det \mathbf{L} < 0$. *If* $\varphi_y = n\pi$, *then* $\varphi_z = -n\pi$. *If* $n\pi < \varphi_y < (n+1)\pi$, *then* $n\pi < -\varphi_z < (n+1)\pi$.

Proof. (a) Let $c = \sqrt{\det \mathbf{L}}$. Then $\mathbf{L} = c\mathbf{U}$, where $\det \mathbf{U} = 1$, so that $\mathbf{U} \in \mathfrak{G}$. Joining $\mathbf{U}$ continuously to $\mathbf{I}_2$ by a curve $\mathbf{U}_s \in \mathfrak{G}$ $(0 \leqslant s \leqslant 1)$, we see that the curve $\mathbf{L}_s = c^{1-s}\mathbf{U}_s$ joins $\mathbf{L}$ continuously to $\mathbf{I}_2$. Let $\mathbf{y}_s(t) = \mathbf{L}_s \mathbf{y}(t)$ and let $\varphi(s)$ be the corresponding rotation. It is clear that $\varphi(s)$ is continuous.

* Concerning $\mathbf{J}_{\mathbf{K}}$, see § 1.4.

** In this subsection, unless otherwise stated, all numbers will be real.

Let $\varphi_y = n\pi$. Then $\mathbf{y}(T) = \mathrm{const}\cdot\mathbf{y}(0)$. Applying the transformation $\mathbf{L}_s$, we obtain $\mathbf{y}_s(T) = \mathrm{const}\cdot\mathbf{y}_s(0)$, so that $\varphi(s) = k\pi$ for some integer k; $\varphi(0) = \varphi_y = n\pi$. Since $\varphi(s)$ is continuous, $\varphi_z = \varphi(1) = n\pi$.

Let $n\pi < \varphi_y < (n+1)\pi$. If this is false for $\varphi(1)$, then by continuity there exists s_0 such that $\varphi(s_0) = n\pi$ or $\varphi(s_0) = (n+1)\pi$. Then, as we have just proved, if $\mathbf{y}(t) = \mathbf{L}_{s_0}^{-1}\mathbf{y}_{s_0}(t)$ then $\varphi_y = n\pi$ or $\varphi_y = (n+1)\pi$, contrary to assumption.

b) $\mathbf{E}_0 = \begin{Vmatrix} 1 & 0 \\ 0 & -1 \end{Vmatrix}$ is the operator of reflection in the first axis. If $\mathbf{y}(t)$ turns through an angle φ, then the vector $\mathbf{y}_0(t) = \mathbf{E}_0\mathbf{y}(t)$ turns through $-\varphi$. We have

$$\mathbf{L} = (\mathbf{L}\mathbf{E}_0^{-1})\,\mathbf{E}_0 = \mathbf{L}_0\mathbf{E}_0, \quad \det\mathbf{L}_0 > 0, \quad \mathbf{z}(t) = \mathbf{L}\mathbf{y}(t) = \mathbf{L}_0\mathbf{y}_0(t).$$

If $\varphi_y = n\pi$, then $\varphi_{y_0} = -n\pi$ by (a), $\varphi_z = -n\pi$. If $n\pi < \varphi_y < (n+1)\pi$, then by (a), $n\pi < -\varphi_z < (n+1)\pi$. This proves Lemma I.

Note that if $\det\mathbf{L} = 0$ $(\mathbf{L} \neq 0)$, $\mathbf{z}(t)$ moves along a straight line through the origin (i. e., does not turn).

Lemma II. Let $\mathbf{y}(t)$ *be a nonvanishing continuous vector-function,* $\mathbf{F}(t) \in \Omega_n$, $\mathbf{z}(t) = \mathbf{F}(t)\,\mathbf{y}(t)$, φ_y *and* φ_z *the rotations of the appropriate vectors in the time interval from* $t = t_0$ *to* $t = t_0 + T$. *Then* $\varphi_z = n\pi + \varphi_y$ *at* $t_0 = 0$. *If* $t_0 \neq 0$, *then* $\varphi_z = (n+k)\pi$ *if* $\varphi_y = k\pi$, *and* $(n+k)\pi < \varphi_z < (n+k+1)\pi$ *if* $k\pi < \varphi_y < (k+1)\pi$. *

Proof. Set

$$\mathbf{F}_1(t) = \mathbf{F}(t + t_0)\,\mathbf{F}^{-1}(t_0).$$

Using the remark in §1.7, we deduce that the rotation of the vector $\mathbf{F}_1(t)\,\mathbf{c}$ from $t = 0$ to $t = T$ is $n\pi$. Therefore,

$$\mathbf{F}_1(t) \in \Omega_n, \quad \mathbf{z}(t) = \mathbf{F}_1(t - t_0)\,\mathbf{y}_1(t),$$

where $\mathbf{y}_1(t) = \mathbf{F}(t_0)\,\mathbf{y}(t)$. Using Theorem 1 of §1.7, we join $\mathbf{F}_1(t)$ continuously to an orthogonal transformation $\mathbf{G}(t) \in \Omega_n$ by a curve $\mathbf{G}_s(t)$. Let $\mathbf{z}_s(t) = \mathbf{G}_s(t - t_0)\,\mathbf{y}_1(t)$. Since $\mathbf{z}_s(t_0) = \mathbf{y}_1(t_0)$ and $\mathbf{z}_s(t_0 + T) = \pm\,\mathbf{y}_1(t_0 + T)$, it follows that the angle φ is a multiple of π and is therefore independent of s. Thus $\varphi_z = \varphi_{z_1}$, where $\mathbf{z}_1(t) = \mathbf{G}(t - t_0)\,\mathbf{y}_1(t)$. The latter motion may be represented as follows. The vector $\mathbf{y}_1(t)$ moves in a certain plane. This plane is rotated by the transformation $\mathbf{G}(t - t_0)$, as a rigid body, through an angle $n\pi$ in time T. The projection of $\mathbf{y}_1(t)$ on the fixed plane is the vector $\mathbf{z}_1(t)$. Clearly, $\varphi_{z_1} = n\pi + \varphi_{y_1}$. ** If $t_0 = 0$,

$$\mathbf{y}_1(t) = \mathbf{y}(t), \quad \mathbf{z}_1(t) = \mathbf{z}(t) \quad \text{and} \quad \varphi_z = n\pi + \varphi_y.$$

If $t_0 \neq 0$, then by Lemma I $\varphi_{y_1} = k\pi$ if $\varphi_y = k\pi$, and $k\pi < \varphi_{y_1} < (k+1)\pi$ if $k\pi < \varphi_{y_1} < (k+1)\pi$. This completes the proof of Lemma II.

We now examine the qualitative behavior of the solutions of equation (1.2) with Hamiltonians in the sets $\mathcal{H}_n$, $\mathcal{O}_n$, $\Pi_n^{\bullet+}$, $\Pi_n^{\bullet\bullet}$, $\Pi_n^{\bullet-}$.

By the theorem of §1.2, the solution of equation (1.2) is

$$\mathbf{x}(t) = \mathbf{F}(t)\,e^{t\mathbf{K}}\mathbf{x}(0).$$

* Recall that the distinctive feature of the time origin $t = 0$ is that $\mathbf{F}(0) = \mathbf{I}_2$.

** Of course, it is easy to prove this equality analytically.

Let $\mathbf{H}(t)\in\mathscr{H}_n$. Then $\mathbf{F}(t)\in\Omega_n$, $\mathbf{K}\in\tilde{\mathscr{H}}\subset\tilde{\mathfrak{R}}_3$ (see (1.21)). Let $\mathbf{a}_+$ and $\mathbf{a}_-$ be eigenvectors of $\mathbf{K}$,

$$\mathbf{K}\mathbf{a}_+=\lambda\mathbf{a}_+,\quad \mathbf{K}\mathbf{a}_-=-\lambda a_-,\quad e^{t\mathbf{K}}\mathbf{a}_+=e^{\lambda t}\mathbf{a}_+,\quad e^{t\mathbf{K}}\mathbf{a}_-=e^{-\lambda t}\mathbf{a}_-\qquad(\lambda>0).$$

Let $\mathbf{c}$ be an arbitrary vector, $\mathbf{c}=\alpha_1\mathbf{a}_++\alpha_2\mathbf{a}_-$. Denote

$$\mathbf{y}(t)=e^{t\mathbf{K}}\mathbf{c}=\alpha_1e^{\lambda t}\mathbf{a}_++\alpha_2e^{-\lambda t}\mathbf{a}_-.$$

All points on the plane $\{\mathbf{y}\}$ approach the straight line defined by the vector $\mathbf{a}_+$ and escape to infinity. The motion $\mathbf{y}=e^{t\mathbf{K}}\mathbf{c}$ on the plane $\{\mathbf{y}\}$ is a saddle point (Figure 58).

If $\mathbf{F}(t)$ is an orthogonal transformation then, as in Lemma II, the motion $\mathbf{x}(t)=\mathbf{F}(t)\mathbf{y}(t)$ is readily described intuitively as follows. Suppose that the motion $\{\mathbf{y}\}$ (saddle point) takes place on a "glass" plane $\mathbf{y}(t)$. This plane is rotated by $\mathbf{F}(t)$, as a rigid body, through an angle $n\pi$ in time T. Projecting the motion $\mathbf{y}(t)$ onto a fixed plane, we get the motion $\mathbf{x}(t)$.

Since the transformation $\mathbf{F}(t)$ need not be orthogonal, the "glass" plane $\{\mathbf{y}\}$ will generally not only rotate but will also be deformed. Since $\det F(t)\equiv 1$, the deformation is area-preserving. In time T, the plane, having turned through the angle $n\pi$, returns to its original position if n is even ($\mathbf{F}(T)=\mathbf{I}_2$) and to a position at an angle π to its original one if n is odd ($\mathbf{F}(T)=-\mathbf{I}_2$). The rotations of solutions $\mathbf{x}_1(t)=e^{\lambda t}\mathbf{F}(t)\mathbf{a}_+$ and $\mathbf{x}_2(t)=e^{-\lambda t}\mathbf{F}(t)\mathbf{a}_-$ in time T are $\varphi_{x_1}=\varphi_{x_2}=n\pi$. By Lemma II, for any other solution we have either $n\pi<\varphi_x<(n+1)\pi$ or $(n-1)\pi<\varphi_x<n\pi$. *

Let $\mathbf{H}(t)\in\mathfrak{G}_n$. The motion $\mathbf{x}=\mathbf{F}(t)\mathbf{y}$ may be interpreted as before, except that there will now be a center (node) on the plane $\{\mathbf{y}\}$ (motion in ellipses; see Figure 59).

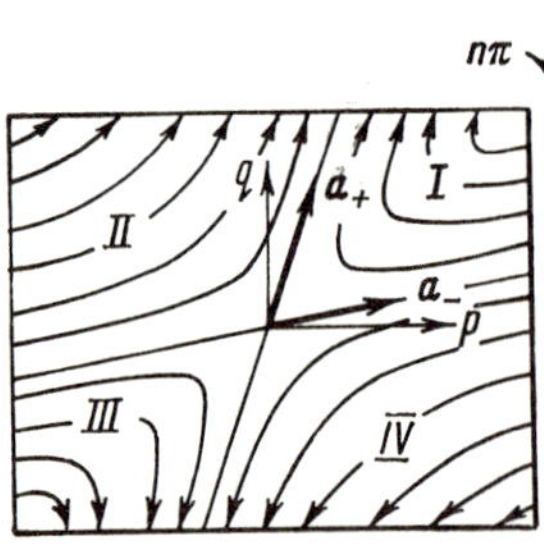

FIGURE 58.

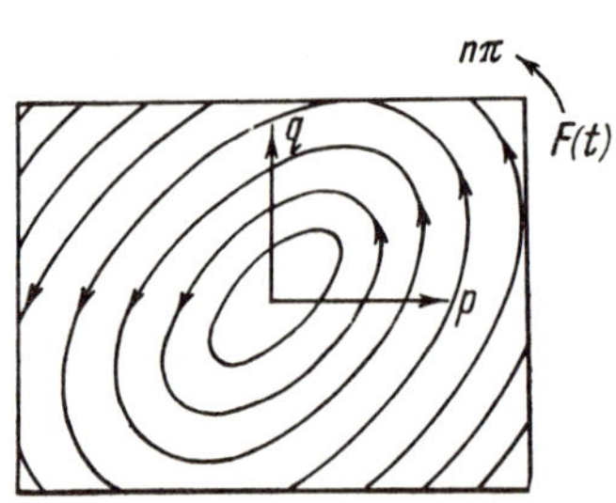

FIGURE 59.

We claim that in this case $n\pi<\varphi_x<(n+1)\pi$ for any solution $\mathbf{x}$. By Lemma II, it will suffice to prove that $0<\varphi_y<\pi$. By assumption, $\mathbf{K}\in\tilde{\mathfrak{G}}_1^+$. This means that $\mathbf{K}=-\vartheta\mathbf{S}^{-1}\tilde{\mathbf{J}}S$, $\det\mathbf{S}>0$, $0<\vartheta<\pi/T$. Multiplying the equality $\mathbf{y}(t)=e^{t\mathbf{K}}\mathbf{y}(0)$ on the left by $\mathbf{S}$, we see that $\mathbf{S}\mathbf{y}=\exp(-\vartheta t\tilde{\mathbf{J}})\mathbf{b}$, where $\mathbf{b}=\mathbf{S}\mathbf{y}(0)$. Thus the vector $\mathbf{S}\mathbf{y}(t)$ moves uniformly in a circle, describing an angle ϑT in time T. Since $0<\vartheta T<\pi$, Lemma I tells us that $0<\varphi_y<\pi$.

* Henceforth, unless otherwise stipulated, φ_x will always denote the rotation of the vector $\mathbf{x}$ in time T. We shall call φ_x simply the rotation, dropping the words "in time T."

It remains to consider the case $\mathbf{H}(t) \in \Pi_n^{*-} \cup \Pi_n^{*+} \cup \Pi_n^{**}$. First let $\mathbf{H}(t) \in \Pi_n^{*-}$. This means that $\mathbf{K} \in \tilde{\Pi}^-$, $\mathbf{F}(t) \in \Omega_n$. Set

$$\mathbf{E}_1 = \begin{Vmatrix} 0 & 1 \\ 0 & 0 \end{Vmatrix}.$$

Clearly, $\mathbf{E}_1 \in \Pi^-$. Therefore (§1.4) $\mathbf{K} = \mathbf{S}\mathbf{E}_1\mathbf{S}^{-1}$, $\det \mathbf{S} > 0$. Hence there exist two linearly independent vectors $\mathbf{a}_1$ and $\mathbf{a}_2$ such that $\mathbf{K}\mathbf{a}_1 = \mathbf{0}$, $\mathbf{K}\mathbf{a}_2 = \mathbf{a}_1$ ($\mathbf{a}_i = \mathbf{S}\mathbf{e}_i$, where $\mathbf{e}_i$ are the standard basis vectors, $i = 1, 2$). We have

$$e^{t\mathbf{K}}\mathbf{a}_1 = (\mathbf{I}_2 + t\mathbf{K})\,\mathbf{a}_1 = \mathbf{a}_1, \quad e^{t\mathbf{K}}\mathbf{a}_2 = (\mathbf{I}_2 + t\mathbf{K})\,\mathbf{a}_2 = \mathbf{a}_2 + t\mathbf{a}_1,$$

since $\mathbf{K}^2 = \mathbf{K}^3 = \ldots = \mathbf{0}$. If $\mathbf{c} = \alpha_1\mathbf{a}_1 + \alpha_2\mathbf{a}_2$ is an arbitrary vector, then $e^{t\mathbf{K}}\mathbf{c} = \mathbf{c} + \alpha_2 t\mathbf{a}_1$.

The general form of the motion is illustrated in Figure 60. The motion takes place along straight lines parallel to $\mathbf{a}_1$, with a velocity which increases with the distance of the line from $\mathbf{a}_1$. In addition, the whole plane is deformed (though areas are preserved), turning through an angle $n\pi$ in time T. For the rotations φ_x we again obtain $(n-1)\pi < \varphi_x \leqslant n\pi$, with equality holding for the solution $x(t)$ with $\mathbf{x}(0) = \mathrm{const}\cdot\mathbf{a}_1$, i. e., if and only if the solution is periodic (even n) or antiperiodic (odd n).

The situation is similar if $\mathbf{H}(t) \in \Pi_n^{*+}$. The motion on the plane $\{y\}$ is in the opposite direction and $n\pi \leqslant \varphi_x < (n+1)\pi$, with equality holding if and only if the solution is periodic or antiperiodic.

Finally, if $\mathbf{H}(t) \in \Pi_n^{**}$, it is clear that $\varphi_x = n\pi$ for any solution (all points on the plane $\{\mathbf{y}\}$ are fixed).

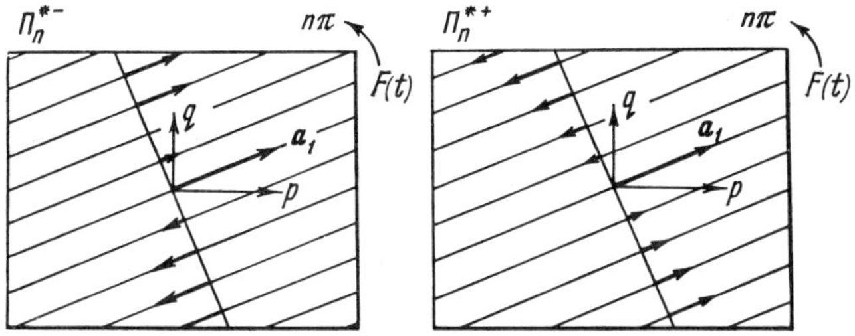

FIGURE 60.

Let $\mathbf{a}_\chi = \exp(-\chi\tilde{\mathbf{J}})\,\mathbf{a} = (\mathbf{I}_2 \cos\chi - \tilde{\mathbf{J}} \sin\chi)\,\mathbf{a}$, where $\mathbf{a}$ is some constant vector. It is clear that the vector $\mathbf{a}_\chi$ is obtained by rotating $\mathbf{a}$ in the positive sense through the angle χ. Let $\mathbf{x}_\chi(t) = \mathbf{X}(t)\,\mathbf{a}_\chi$ and let $\psi(\chi) = \varphi_{x_\chi}$ be the corresponding rotation.

We shall call $\psi(\chi)$ a *rotation function*. This function will play a major role in what follows.

Since $\mathbf{a}_{\chi+\pi} = -\mathbf{a}_\chi$, it follows that $\mathbf{x}_{\chi+\pi}(t) = -\mathbf{x}_\chi(t)$, and so $\psi(\chi+\pi) \equiv \psi(\chi)$.

The previous discussion shows that the behavior of the function $\psi(\chi)$ uniquely determines which of the sets $\mathcal{O}_n$, $\mathcal{H}_n$, $\Pi_n^{\bullet -}$, $\Pi_n^{\bullet +}$, $\Pi_n^{\bullet\bullet}$ contains $\mathbf{H}(t)$ (Figure 61). In order to determine this, given $\mathbf{H}(t)$, it suffices to have upper and lower bounds for the function $\psi(\chi)$, which is tantamount to knowing upper and lower bounds for the rotations of all solutions of the corresponding equation. Let us denote

$$m = \min_{0 \leqslant \chi \leqslant \pi} \psi(\chi), \quad M = \max_{0 \leqslant \chi \leqslant \pi} \psi(\chi).$$

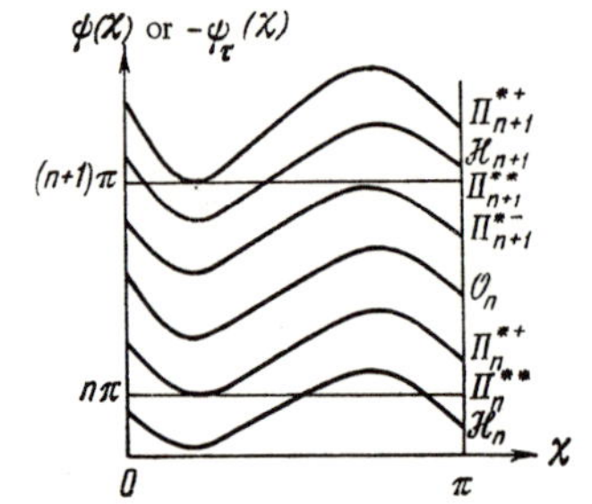

FIGURE 61.

The propositions proved above concerning the numbers m, M for each of the sets $\Pi_n^{\bullet -}$, $\Pi_n^{\bullet\bullet}$, $\Pi_n^{\bullet +}$, $\mathcal{H}_n$, $\mathcal{O}_n$ are combined in the following table.

Set	Numbers m, M	Set	Numbers m, M
$\Pi_n^{\bullet -}$	$(n-1)\pi < m < n\pi,\ M = n\pi$	$\Pi_n^{\bullet +}$	$m = n\pi,\ n\pi < M < (n+1)\pi$
$\mathcal{H}_n$	$m < n\pi < M$	$\mathcal{O}_n$	$n\pi < m \leqslant M < (n+1)\pi$
$\Pi_n^{\bullet\bullet}$	$m = M = n\pi$		

It follows from the discussion that the pattern of behavior of the integral curves in $\mathfrak{L}^3$ varies continuously, provided that they do not intersect the surfaces Π_n. Across Π_n, the pattern changes abruptly, so that these surfaces are bifurcation manifolds in $\mathfrak{L}^3$.

In the sequel we shall need the properties of rotations not only for solution-vectors $\mathbf{x} = \mathbf{X}(t)\mathbf{c}$ but also for vectors $\mathbf{x}_\tau = \mathbf{X}^\tau(t)\mathbf{c}$ (e.g., the rows of the matrix $\mathbf{X}(t)$)

Let $\psi_\tau(\chi)$ denote the rotation of the vector $\mathbf{X}^\tau(t)\mathbf{a}_\chi$ in time T, where $\mathbf{a}_\chi = \exp(\chi\tilde{\mathbf{J}})\mathbf{a}$ is some constant vector. We claim that in order to determine which of the sets $\mathcal{O}_n$, $\mathcal{H}_n$, $\Pi_n^{\bullet -}$, $\Pi_n^{\bullet +}$, $\Pi_n^{\bullet\bullet}$ contains the Hamiltonian $\mathbf{H}(t)$ we can use the function $\psi_\tau(\chi)$ instead of $\psi(\chi)$.

Let

$$\mathbf{E}_0 = \begin{Vmatrix} 1 & 0 \\ 0 & -1 \end{Vmatrix}$$

denote the operator of reflection in the horizontal axis.

The matrix $\mathbf{X}_1(t)=\mathbf{E}_0\mathbf{X}^{\tau}(t)\mathbf{E}_0^{-1}$ is a trajectory in the unimodular group $\mathfrak{G}$, with initial point $\mathbf{X}_1(0)=\mathbf{I}_2$, i. e., it is the matrizant of some equation (1.2).

Let σ denote the transformation converting an equation with fundamental matrix $\mathbf{X}(t)$ to an equation with fundamental matrix $\mathbf{X}_1(t)$:

$$\sigma\mathbf{X}=\mathbf{E}_0\mathbf{X}^{\tau}\mathbf{E}_0^{-1} \quad \text{or} \quad \sigma\mathbf{X}=\mathbf{E}_0(\mathbf{F}e^{t\mathbf{K}})^{\tau}\mathbf{E}_0^{-1}=\mathbf{F}_1e^{t\mathbf{K}_1},$$

where $\mathbf{F}_1=\sigma\mathbf{F}=\mathbf{E}_0e^{t\mathbf{K}^{\tau}}\mathbf{F}^{\tau}e^{-t\mathbf{K}^{\tau}}\mathbf{E}_0^{-1}$, $\mathbf{K}_1=\sigma\mathbf{K}=\mathbf{E}_0\mathbf{K}^{\tau}\mathbf{E}_0^{-1}$. Thus

$$\sigma\mathbf{K}=\begin{Vmatrix}1 & 0\\ 0 & -1\end{Vmatrix}\begin{Vmatrix}-x & y-z\\ y+z & x\end{Vmatrix}^{\tau}\begin{Vmatrix}1 & 0\\ 0 & -1\end{Vmatrix}=\begin{Vmatrix}-x & -y-z\\ -y+z & x\end{Vmatrix},$$

i. e., if $\mathbf{K}\sim(x,\ y,\ z)$ (see §1.4), then $\sigma\mathbf{K}\sim(x,\ -y,\ z)$. Therefore $\sigma\mathbf{K}$ lies in the same subset $\tilde{\mathcal{H}}$, $\tilde{\mathfrak{G}}$, $\tilde{\Pi}^+$ or $\tilde{\Pi}^-$ as $\mathbf{K}$.

Let $\mathbf{F}\in\Omega_n$. By Corollary 1 to the theorem of §1.7, $\mathbf{F}^{\tau}\in\Omega_{-n}$. Joining $\mathbf{K}$ continuously to zero, we thereby join $\exp(t\mathbf{K}^{\tau})\mathbf{F}^{\tau}\exp(-t\mathbf{K}^{\tau})$ continuously to $\mathbf{F}^{\tau}$. Hence, by the theorem of §1.7, $\exp(t\mathbf{K}^{\tau})\mathbf{F}^{\tau}\exp(-t\mathbf{K}^{\tau})\in\Omega_{-n}$. By Corollary 2 to that theorem, $\sigma\mathbf{F}\in\Omega_n$. It follows from (1.21) that the Hamiltonian $\sigma\mathbf{H}(t)$ lies in the same set $\mathcal{H}_n$, $\mathfrak{G}_n$, Π_n^{*+}, Π_n^{*-}, Π_n^{**} as $\mathbf{H}(t)$. Thus the function $\psi(\chi)$ in Figure 61 may be replaced by the rotation function $\sigma\psi(\chi)$ of the equation with Hamiltonian $\sigma\mathbf{H}(t)$.

It remains to show that the maximum and minimum of $\sigma\psi(\chi)$ coincide with those of $(-\psi_{\tau}(\chi))$.

Let $\mathbf{c}\neq\mathbf{0}$ be a vector, varying in such a way that arg $\mathbf{c}$ varies from 0 to π. The maximum and minimum of $\sigma\psi(\chi)$ coincide with the maximum and minimum rotations of the vectors $\mathbf{E}_0\mathbf{X}(t)^{\tau}\mathbf{E}_0^{-1}\mathbf{c}$ or, obviously, $\mathbf{E}_0\mathbf{X}(t)^{\tau}\mathbf{c}$. Since the vectors $\mathbf{E}_0\mathbf{X}^{\tau}\mathbf{c}$ and $\mathbf{X}^{\tau}\mathbf{c}$ are mutually symmetric with respect to the horizontal axis, their rotations differ only in sign:

$$\max\sigma\psi(\chi)=\max[-\psi_{\tau}(\chi)],\quad \min\sigma\psi(\chi)=\min[-\psi_{\tau}(\chi)],$$

Q. E. D.

We summarize the various propositions proved in this section as theorems.

Theorem I (qualitative behavior of solutions of system (1.2)). Any motion defined by system (1.2) may be represented as follows. The motion takes place on a "glass" plane $\{\mathbf{y}\}$, obeying an equation $\dot{\mathbf{y}}=\mathbf{K}\mathbf{y}$, where $\mathbf{K}$ is a constant real matrix with trace zero. Its trajectories are either hyperbolas (case (a) of §1.4 — Figure 58), ellipses (case (c) of §1.4, $0<\psi<\pi$ — Figure 59), or straight lines (case (b), §1.4, $\varepsilon\neq0$ — Figure 60), or else all points on the plane are fixed ($\mathbf{K}=0$). The whole plane $\{\mathbf{y}\}$ is subjected to an area-preserving deformation by a linear transformation $\mathbf{F}(t)$ and rotated through an angle $n\pi$ in time T. At the end of the time T the plane is either in its original position (even n) or at an angle π to its original position (odd n). The operator $\mathbf{F}(t)$ satisfies the conditions*

* That is to say, such that no change occurs in the area of the parallelogram spanned by any two radius-vectors, although their lengths and the angle between them change under the deformation.

$$\mathbf{F}(t+T)=(-1)^n\mathbf{F}(t),\quad \det\mathbf{F}(t)\not\equiv 1,\quad d\mathbf{F}/dt\in L(0,\ T).$$

The motion $\mathbf{x}(t)$ *is the projection of this motion on a fixed plane. In case (a),* $\mathbf{H}(t)\in\mathcal{H}_n$*; in case (b),* $\mathbf{H}(t)\in\Pi_n^{\bullet+}$ *if* $\varepsilon>0$ *and* $\mathbf{H}(t)\in\Pi_n^{\bullet-}$ *if* $\varepsilon<0$*; in case (c) (where* $0<\psi<\pi$*),* $\mathbf{H}(t)\in\mathfrak{S}_n$*; finally, if* $\mathbf{K}=\mathbf{0}$*, then* $\mathbf{H}(t)\in\Pi_n^{\bullet\bullet}$*. This exhausts all possibilities. Conversely, any motion of the type described is a motion* $\mathbf{x}(t)$ *for some equation (1.2).*

Theorem II *(characterization of stability domains* $\mathfrak{S}_n$*, instability domains* $\mathcal{H}_n$*, and "boundary" sets* $\Pi_n^{\bullet+}$, $\Pi_n^{\bullet-}$, $\Pi_n^{\bullet\bullet}$*, in terms of rotations).*

1) $\mathbf{H}(t)\in\mathfrak{S}_n$ *if and only if the rotation of any solution* $\mathbf{x}=\mathbf{x}(t)\neq\mathbf{0}$ *of equation (1.2) satisfies the inequality* $n\pi<\varphi_x<(n+1)\pi$.

$\mathbf{H}(t)\in\mathcal{H}_n$ *if and only if there exist two linearly independent solutions* $\mathbf{x}_1$ *and* $\mathbf{x}_2$ *with rotations* $\varphi_{x_1}=\varphi_{x_2}=n\pi$ *and there are also solutions with rotations both less than and greater than* $n\pi$*. In this case we have* $(n-1)\pi<\varphi_x<(n+1)\pi$ *for any solution* $\mathbf{x}\neq\mathbf{0}$*.*

$\mathbf{H}(t)\in\Pi_n^{\bullet+}$ *if and only if for any solution* $\mathbf{x}=\mathbf{0}$ *we have* $n\pi<\varphi_x\leqslant(n+1)\pi$ *and there is exactly one linearly independent solution with* $\varphi_x=(n+1)\pi$*.*

$\mathbf{H}(t)\in\Pi_n^{\bullet-}$ *if and only if* $(n-1)\pi<\varphi_x\leqslant n\pi$ *for any solution* $\mathbf{x}\neq\mathbf{0}$ *and there is exactly one linearly independent solution* $\mathbf{x}$ *such that* $\varphi_x=n\pi$*.*

$\mathbf{H}(t)\in\Pi_n^{\bullet\bullet}$ *if and only if* $\varphi_x=n\pi$ *for any solution* $\mathbf{x}\neq\mathbf{0}$*.*

Let $\psi(\chi)$ $(0\leqslant\chi\leqslant\pi)$ *be a rotation function, i.e., the rotation of a solution* $\mathbf{x}(t)\neq\mathbf{0}$ *for which* $\arg\mathbf{x}(0)=\chi$*. Let* ψ_1 *and* ψ_2 *be the minimum and maximum of* $\psi(\chi)$*. Necessary and sufficient conditions for a Hamiltonian* $\mathbf{H}(t)$ *to be in each of the sets* $\mathcal{H}_n$, $\mathfrak{S}_n$, $\Pi_n^{\bullet+}$, $\Pi_n^{\bullet-}$, $\Pi_n^{\bullet\bullet}$ *are listed in the table on p. 652.*

2) *Part 1 of the theorem is true with the rotations* φ_x *of solutions* $\mathbf{x}=\mathbf{X}(t)\mathbf{c}$ *replaced by* $(-\varphi_{x_\tau})$*, where* φ_{x_τ} *is the rotation of the vector* $\mathbf{x}_\tau=\mathbf{X}(t)^\tau\mathbf{c}$*, the word "solution" replaced by "vector* $\mathbf{c}\neq\mathbf{0}$*," and the function* $\psi(\chi)$ *replaced by* $[-\psi_\tau(\chi)]$*, where* $\psi_\tau(\chi)$ *is the rotation of the vector*

$$\mathbf{x}_\tau=\mathbf{X}(t)^\tau\mathbf{c},\quad \mathbf{c}\neq\mathbf{0},\quad \arg\mathbf{c}=\chi.$$

A graphical representation of the conclusions of Theorem II is shown in Figure 61.

Remark 1. The "rotations" in Theorem II are the angles through which the vectors $\mathbf{x}(t)$ or $\mathbf{x}_\tau(t)$ turn in the time from $t=0$ to $t=T$. Theorem II remains valid, however, if we define these angles in terms of the rotation of the vectors from t_0 to t_0+T, where t_0 is any real number (the same for different solutions $\mathbf{x}$ or vectors $\mathbf{x}_\tau$).

Indeed, rotations were originally defined for $t_0=0$. Let $\mathbf{x}=\mathbf{F}(t)\mathbf{y}(t)$, where $\mathbf{y}(t)=\exp(t\mathbf{K})\mathbf{c}$, $\mathbf{F}(t)\in\Omega_n$, t_0 is an arbitrary number, $\mathbf{x}_1=\mathbf{x}(t+t_0)$. Then $\mathbf{x}_1=\mathbf{F}(t+t_0)\mathbf{y}_1(t)$, where $\mathbf{y}_1(t)=\exp(t\mathbf{K})\mathbf{c}_1$, $\mathbf{c}_1=\exp(t_0\mathbf{K})\mathbf{c}$.

The rotation from $t=0$ to $t=t_0+T$ is given by $\varphi_{x_1}=n\pi+\varphi_{y_1}$. Since $\varphi_x=n\pi+\varphi_y$, it follows that φ_{x_1} is obtained from φ_x by replacing the initial vector $\mathbf{c}$ by $\mathbf{c}_1$. Thus the corresponding rotation functions $\psi(\chi)$ and $\psi_1(\chi)$ satisfy the identity $\psi_1(\chi)=\psi(\chi+\chi_1)$, where χ_1 is some number. We have thus shown that measuring the rotation from $t=t_0$ to $t=t_0+T$ is equivalent to replacing $\psi(\chi)$ by $\psi(\chi+\chi_1)$. This replacement does not alter the numbers M and m. This implies our assertion for solution-vectors $\mathbf{x}(t)$. The case of vectors $\mathbf{x}_\tau(t)$ is treated similarly.

Remark 2. If t_0 is an arbitrary number, the Hamiltonian $\mathbf{H}(t+t_0)$ is in the same set $\mathfrak{G}_n$, $\mathcal{H}_n$, $\Pi_n^{\bullet+}$, $\Pi_n^{\bullet-}$, $\Pi_n^{\bullet\bullet}$ as $\mathbf{H}(t)$.

Indeed, if $\mathbf{x}=\mathbf{x}(t)$ is a solution of the equation with Hamiltonian $\mathbf{H}(t)$, then $\mathbf{z}=\mathbf{x}(t+t_0)$ is a solution of the equation with Hamiltonian $\mathbf{H}(t+t_0)$. But we have already seen that

$$\min \varphi_x = \min \varphi_z, \quad \max \varphi_x = \max \varphi_z,$$

where the min and max range over all solutions. Which of the above sets contains a Hamiltonian depends solely on the numbers $m=\min \varphi_x$, $M=\max \varphi_x$. This implies our assertion.

The same conclusion follows from the Floquet-Lyapunov representation and the results of §1.8.

1.10. General form of Hamiltonians in $\mathfrak{G}_n$, $\mathcal{H}_n$, $\Pi_n^{\bullet+}$, $\Pi_n^{\bullet-}$, $\Pi_n^{\bullet\bullet}$. Equations with constant Hamiltonians

Theorem I.

(I) *The Hamiltonian* $\mathbf{H}(t)$ *in (1.2) may always be expressed as*

$$\mathbf{H}(t) = -\mathbf{P}(t)\,\mathbf{J}\frac{d}{dt}\mathbf{P}(t)^{\tau} + \mathbf{P}(t)\,\mathbf{H}_0\mathbf{P}(t)^{\tau}, \tag{1.25}$$

where $\mathbf{H}_0=\mathbf{H}_0^*$ *is a constant matrix and* $\mathbf{P}(t)\in\Omega_n$, *i.e.,* $\mathbf{P}(t)$ *is an absolutely continuous matrix-function such that (a)* $\det\mathbf{P}(t)\equiv 1$, *(b)* $\mathbf{P}(0)=\mathbf{I}_2$, *(c)* $\mathbf{P}(t+T)=(-1)^n\mathbf{P}(t)$, *and (d) is any constant vector* $\mathbf{c}\neq\mathbf{0}$ *the rotation of the vector* $\mathbf{P}(t)\mathbf{c}$ *in time* T *is* $n\pi$.

(II) *The Hamiltonian* $\mathbf{H}(t)$ *is in one of the sets* $\mathfrak{G}_n$, $\mathcal{H}_n$, $\Pi_n^{\bullet+}$, $\Pi_n^{\bullet-}$, $\Pi_n^{\bullet\bullet}$ *if and only if the matrix* $\mathbf{H}_0$ *in (1.25) satisfies the appropriate condition:*

$\mathbf{H}(t)\in\mathfrak{G}_n$	$0<\det\mathbf{H}_0<\frac{\pi^2}{T^2}$, $\operatorname{Tr}\mathbf{H}_0>0$, i.e., $0<\mathbf{H}_0<\frac{\pi}{T}\mathbf{C}$, where $\mathbf{C}^*=\mathbf{C}$, $\det\mathbf{C}=1$
$\mathbf{H}(t)\in\mathcal{H}_n$	$\det\mathbf{H}_0<0$
$\mathbf{H}(t)\in\Pi_n^{\bullet+}$	$\det\mathbf{H}_0=0$, $\operatorname{Tr}\mathbf{H}_0>0$
$\mathbf{H}(t)\in\Pi_n^{\bullet-}$	$\det\mathbf{H}_0=0$, $\operatorname{Tr}\mathbf{H}_0<0$
$\mathbf{H}(t)\in\Pi_n^{\bullet\bullet}$	$\mathbf{H}_0=\mathbf{0}$

(III) *The matrices* $\mathbf{P}(t)$ *and* $\mathbf{H}_0$ *satisfying the conditions of (I) are uniquely determined by the matrix* $\mathbf{H}(t)$.

(IV) *Part (I) remains valid if the normalizing condition* $\mathbf{P}(0)=\mathbf{I}_2$ *is dropped, but then the representation (1.25) ceases to be unique: for any two pairs* $\{\mathbf{P}'(t), \mathbf{H}_0'\}$, $\{\mathbf{P}''(t), \mathbf{H}_0''\}$ *satisfying (1.25) with the same matrix,*

$$\mathbf{P}''(t)=\mathbf{P}'(t)\,\mathbf{R}, \quad \mathbf{H}_0''=\mathbf{R}^{\tau}\mathbf{H}_0'\mathbf{R},$$

where $\mathbf{R}$ *is a constant matrix,* $\det\mathbf{R}=1$.

Proof. It follows from the representation of $\mathbf{K}=\tilde{\mathbf{J}}^{-1}\mathbf{H}_0$ in the form shown at the beginning of §1.4 that

$$\mathbf{H}_0=\begin{Vmatrix} z+y & x \\ x & z-y \end{Vmatrix}.$$

Using the definitions of the sets $\tilde{\mathfrak{G}}_{1+}$, $\tilde{\mathscr{H}}$, $\tilde{\Pi}^{\bullet+}$, $\tilde{\Pi}^{\bullet-}$, $\tilde{\Pi}^{\bullet\bullet}$ (see §1.4), we see that the two conditions in each of the following rows are equivalent:

a)	$\tilde{J}^{-1}\mathbf{H}_0\in\tilde{\mathfrak{G}}_{1+}$	and	$0<\det\mathbf{H}_0<\frac{\pi^2}{T^2}$,	$\operatorname{Tr}\mathbf{H}_0>0$,
b)	$\tilde{J}^{-1}\mathbf{H}_0\in\tilde{\mathscr{H}}$	and	$\det\mathbf{H}_0<0$,	
c)	$\tilde{J}^{-1}\mathbf{H}_0\in\tilde{\Pi}_n^{\bullet+}$	and	$\det\mathbf{H}_0=0$,	$\operatorname{Tr}\mathbf{H}_0>0$,
d)	$\tilde{J}^{-1}\mathbf{H}_0\in\tilde{\Pi}_n^{\bullet-}$	and	$\det\mathbf{H}_0=0$,	$\operatorname{Tr}\mathbf{H}_0<0$,
e)	$\tilde{J}^{-1}\mathbf{H}_0\in\tilde{\Pi}_n^{\bullet\bullet}$	and	$\mathbf{H}_0=0$.	

According to the definitions (1.21), the fact that $\mathbf{H}(t)$ is in one of the sets $\mathfrak{G}_n$, $\mathscr{H}_n$, $\Pi_n^{\bullet+}$, $\Pi_n^{\bullet-}$, $\Pi_n^{\bullet\bullet}$ implies that the matrizant $\mathbf{X}(t)$ may be expressed as

$$\mathbf{X}(t)=\mathbf{F}(t)\,e^{t\mathbf{K}},$$

where $\mathbf{F}(t)\in\Omega_n$ and $\mathbf{K}$ is in the appropriate set $\tilde{\mathfrak{G}}_{1+}$, $\tilde{\mathscr{H}}$, $\tilde{\Pi}^{\bullet+}$, $\tilde{\Pi}^{\bullet-}$, $\tilde{\Pi}^{\bullet\bullet}$. We have

$$\tilde{\mathbf{J}}^{-1}\mathbf{H}(t)=\dot{\mathbf{X}}\mathbf{X}^{-1}=\dot{\mathbf{F}}\mathbf{F}^{-1}+\mathbf{F}\mathbf{K}\mathbf{F}^{-1}.$$

Setting

$$\mathbf{K}=\tilde{\mathbf{J}}^{-1}\mathbf{H}_0,\quad \mathbf{F}=[\mathbf{P}^{\tau}]^{-1}\qquad (\mathbf{P}^{\tau}\tilde{\mathbf{J}}\mathbf{P}\equiv\tilde{\mathbf{J}}),$$

we obtain the formula (1.25) for $\mathbf{H}(t)$. By Corollaries 1 and 5 of §1.7 (for $m=-1$), $\mathbf{F}\in\Omega_n$ if and only if $\mathbf{P}\in\Omega_n$, whence we deduce parts (I) and (II) of Theorem I. Part (III) follows from the unicity of the correspondence

$$\mathbf{H}(t)\sim\{\mathbf{F}(t),\ \mathbf{K}\},\ \mathbf{F}(t)\in\Omega_n,\ \mathbf{K}\in\tilde{\mathscr{H}}\cup\tilde{\Pi}\cup\tilde{\mathfrak{G}}_{1+}$$

(see §1.8).

The proof of part (IV) is analogous, starting now from the representation $\mathbf{X}(t)=\mathbf{F}_1(t)\,e^{t\mathbf{K}_1}$, where $\mathbf{F}_1(t)=\mathbf{F}(t)\mathbf{C}$, $\mathbf{K}_1=\mathbf{C}^{-1}\mathbf{K}\mathbf{C}$, $\mathbf{C}$ being a constant matrix, $\det\mathbf{C}=1$.

Theorem II. Let $\mathbf{H}(t)$ and $\mathbf{H}_1(t)$ be Hamiltonians such that

$$\mathbf{H}_1(t)=-\mathbf{P}(t)\tilde{\mathbf{J}}\left(\frac{d\mathbf{P}(t)}{dt}\right)^{\tau}+\mathbf{P}(t)\,\mathbf{H}(t)\,\mathbf{P}(t)^{\tau},\qquad(1.26)$$

where $\mathbf{P}(t)$ is an absolutely continuous matrix-function satisfying conditions (a), (b), (c) of Theorem I. Then the set $\mathfrak{G}_n$, $\mathscr{H}_n$, $\Pi_n^{\bullet+}$, $\Pi_n^{\bullet-}$, $\Pi_n^{\bullet\bullet}$ containing $\mathbf{H}_1(t)$ is of the same type as that containing $\mathbf{H}(t)$, but with subscript increased by m:

if $\mathbf{H}(t) \in \mathcal{O}_n$, *then* $\mathbf{H}_1(t) \in \mathcal{O}_{n+m}$;
if $\mathbf{H}(t) \in \mathcal{H}_n$, *then* $\mathbf{H}_1(t) \in \mathcal{H}_{n+m}$

and so on.

Proof. Consider the equation $\tilde{\mathbf{J}}\dot{\mathbf{x}} = \mathbf{H}(t)\mathbf{x}$. Performing the substitution

$$\mathbf{x} = \mathbf{F}(t)\mathbf{y},$$

where $\mathbf{F}(t) = [\mathbf{P}(t)^{\tau}]^{-1}$, we obtain the equation $\tilde{\mathbf{J}}\dot{\mathbf{y}} = \mathbf{H}_1(t)\mathbf{y}$ for $\mathbf{y}$. Using Corollaries 1 and 5 of §1.7 (with $m=1$) and Lemmas I and II of §1.9, we see that if $\varphi_x = k\pi$ (where k is an integer), then $\varphi_y = (k+m)\pi$, and if

$$k\pi < \varphi_x < (k+1)\pi \text{ , then } \qquad (k+m)\pi < \varphi < (k+m+1)\pi.$$

Applying Theorem II of §1.9, we obtain the desired assertion.

The next theorem is similar to Theorem II but is sometimes more convenient to use.

Theorem III. Let $\mathbf{P}(t)$ *be an absolutely continuous matrix-function such that*

$$\mathbf{P}(t+T) = \pm\mathbf{P}(t), \quad \det \mathbf{P}(t) = \text{const} > 0.$$

Define an integer m *by*

$$\mathbf{P}(t)\mathbf{P}(0)^{-1} \in \Omega_m.$$

Change variables in the canonical equation

$$\tilde{\mathbf{J}}\frac{d\mathbf{x}}{dt} = \mathbf{H}(t)\mathbf{x} \qquad (\mathbf{H}(t+T) \equiv \mathbf{H}(t)) \tag{1.27}$$

by

$$\mathbf{x}_1 = \mathbf{P}(t)\mathbf{x}.$$

This gives a canonical equation for $\mathbf{x}_1$:

$$\tilde{\mathbf{J}}\frac{d\mathbf{x}_1}{dt} = \mathbf{H}_1(t)\mathbf{x}_1 \qquad (\mathbf{H}_1(t+T) \equiv \mathbf{H}_1(t)), \tag{1.27a}$$

where the Hamiltonian $\mathbf{H}_1(t)$ *is in a set of the partition*

$$\mathfrak{L}^3 = \bigcup_{k=-\infty}^{+\infty} (\mathcal{O}_k \cup \mathcal{H}_k \cup \Pi_k^{*-} \cup \Pi_k^{**} \cup \Pi_k^{*+})$$

of the same type as $\mathbf{H}(t)$, *but with subscript increased by* m:

if $\mathbf{H}(t) \in \mathcal{O}_n$, *then* $\mathbf{H}_1(t) \in \mathcal{O}_{n+m}$,
if $\mathbf{H}(t) \in \mathcal{H}_n$, *then* $\mathbf{H}(t) \in \mathcal{H}_{n+m}$,
if $\mathbf{H}(t) \in \Pi_n^{*-}$, *then* $\mathbf{H}_1(t) \in \Pi_{n+m}^{*-}$

and so on.

Proof. Let $\mathbf{X}(t)$ and $\mathbf{X}_1(t)$ be the matrizants of equations (1.27) and (1.27a), respectively. Then

$$\mathbf{X}_1(t)=\mathbf{P}(t)\mathbf{X}(t)\mathbf{P}_0^{-1}, \quad \text{where} \quad \mathbf{P}_0=\mathbf{P}(0),$$

or, in other words,

$$\mathbf{X}_1(t)=\mathbf{F}_0(t)\mathbf{U}(t), \quad \mathbf{F}_0(t)=\mathbf{P}(t)\mathbf{P}_0^{-1}, \quad \mathbf{U}(t)=\mathbf{P}_0\mathbf{X}(t)\mathbf{P}_0^{-1}.$$

Suppose that by the representation in (1.6) (see §1.8)

$$\mathbf{X}(t)=\mathbf{F}(t)e^{t\mathbf{K}}, \quad \text{where} \quad \mathbf{F}(t)\in\Omega_n, \quad \mathbf{K}\in\tilde{\mathscr{H}}\cup\tilde{\Pi}\cup\tilde{\mathscr{O}}_{1+}.$$

Then

$$\mathbf{X}_1(t)=\mathbf{F}_1(t)e^{t\mathbf{K}_1},$$

where

$$\mathbf{F}_1(t)=\mathbf{F}_0(t)[\mathbf{P}_0\mathbf{F}(t)\mathbf{P}_0^{-1}], \quad \mathbf{K}_1=\mathbf{P}_0\mathbf{K}\mathbf{P}_0^{-1}.$$

By Corollary 2 of §1.7, $\mathbf{P}_0\mathbf{F}(t)\mathbf{P}_0^{-1}\in\Omega_n$. Since $\mathbf{F}_0(t)\in\Omega_m$ by assumption, we have from Corollary 3 of §1.7 that $\mathbf{F}_1(t)\in\Omega_{n+m}$.

By the definitions (§§1.4, 1.8) of the sets $\tilde{\mathscr{H}}$, $\tilde{\Pi}_+$, $\tilde{\Pi}_-$, $\tilde{\mathscr{O}}_{1+}$ each of these sets is invariant under the transformation $\mathbf{K}_1=\mathbf{P}_0^{-1}\mathbf{K}\mathbf{P}_0$ if $\det\mathbf{P}_0>0$. Consequently, the matrix $\mathbf{K}_1$ is in the same set

$$\tilde{\mathscr{H}},\ \tilde{\Pi}_+,\ \tilde{\Pi}_-,\ \{0\},\ \tilde{\mathscr{O}}_{1+}$$

as $\mathbf{K}$. The conclusion of the theorem now follows at once from the definitions of $\mathscr{H}_n$, $\Pi_n^{\bullet+}$, $\Pi_n^{\bullet-}$, $\Pi_n^{\bullet\bullet}$, $\mathscr{O}_n$ in §1.8.

Remark. If $\det\mathbf{P}(0)<0$, we have an analogous but more cumbersomely formulated theorem. (In fact, now the type of set in the partition may also change; for example, if $m=0$ and $\mathbf{H}(t)\in\Pi_0^{\bullet+}$, then $\mathbf{H}_1(t)\in\Pi_0^{\bullet-}$.)

We now consider the equation

$$\tilde{\mathbf{J}}\frac{d\mathbf{x}}{dt}=\mathbf{H}_0\mathbf{x}$$

with constant Hamiltonian $\mathbf{H}_0$, treated as periodic with period T. We shall show that the specific set $\mathscr{O}_n$, $\mathscr{H}_n$, $\Pi_n^{\bullet+}$, $\Pi_n^{\bullet-}$, $\Pi_n^{\bullet\bullet}$ containing $\mathbf{H}_0$ is determined by the following table.

I	$\det\mathbf{H}_0<0$	$\mathbf{H}_0\in\mathscr{H}_0$
II	$\det\mathbf{H}_0=0,\ \ \mathrm{Tr}\,\mathbf{H}_0>0$	$\mathbf{H}_0\in\Pi_0^{\bullet+}$
III	$\det\mathbf{H}_0=0,\ \ \mathrm{Tr}\,\mathbf{H}_0<0$	$\mathbf{H}_0\in\Pi_0^{\bullet-}$
IV	$\mathbf{H}_0=0$	$\mathbf{H}_0=0\in\Pi_0^{\bullet\bullet}$
V	$\det\mathbf{H}_0=\dfrac{n^2\pi^2}{T^2},\ \ \mathrm{Tr}\,\mathbf{H}_0>0$	$\mathbf{H}_0\in\Pi_n^{\bullet\bullet}\quad (n=1,2,\ldots)$

VI	$\det \mathbf{H}_0 = \frac{n^2\pi^2}{T^2}$, $\operatorname{Tr} \mathbf{H}_0 < 0$	$\mathbf{H}_0 \in \Pi_{-n}^{\bullet\bullet}$	$(n=1, 2, \ldots)$
VII	$\frac{n^2\pi^2}{T^2} < \det \mathbf{H}_0 < \frac{(n+1)^2\pi^2}{T^2}$, $\operatorname{Tr} \mathbf{H}_0 > 0$	$\mathbf{H}_0 \in \mathfrak{S}_n$	$(n=1, 2, \ldots)$
VIII	$\frac{n^2\pi^2}{T^2} < \det \mathbf{H}_0 < \frac{(n+1)^2\pi^2}{T^2}$, $\operatorname{Tr} \mathbf{H}_0 < 0$	$\mathbf{H}_0 \in \mathfrak{S}_{-n}$	$(n=1, 2, \ldots)$

Conditions I, II, III, IV follow from part (II) of Theorem I, by putting $\mathbf{P}(t) \equiv \mathbf{I}_2$. In cases V — VIII we have $\det \mathbf{H}_0 = \varkappa^2 > 0$. The characteristic equation $\det[\tilde{\mathbf{J}}^{-1}\mathbf{H}_0 - \lambda \mathbf{I}_2] = 0$ is $\lambda^2 - \varkappa^2 = 0$, so that (see §1.4)

$$\tilde{\mathbf{J}}^{-1}\mathbf{H}_0 = -\varkappa \mathbf{S}^{-1}\tilde{\mathbf{J}}\mathbf{S},$$

where $\mathbf{S}$ is a real 2×2 matrix and, without loss of generality, $\det \mathbf{S} = 1$. Using the identity

$$\mathbf{S}^{\tau}\tilde{\mathbf{J}}\mathbf{S} = \tilde{\mathbf{J}},$$

we deduce that

$$\mathbf{H}_0 = \varkappa \mathbf{S}^{\tau}\mathbf{S}$$

and that the numbers $\operatorname{Tr} \mathbf{H}_0$ and $\varkappa$ have the same sign.

If $\mathbf{S} = \mathbf{I}_2$, then

$$\mathbf{x}(t) = e^{-\varkappa t \tilde{J}} \mathbf{x}(0),$$

so that $\mathbf{x}(t)$ is obtained from $\mathbf{x}(0)$ by rotation through the angle $\varkappa t$. Thus $\varphi_{\mathbf{x}} \equiv \varkappa T$ for $\mathbf{S} = \mathbf{I}_2$. Conditions V — VIII follow in this case from Theorem II of §1.9. Applying Theorem II for $\mathbf{P}(t) \equiv \mathbf{S}$, we obtain the proof in the general case.

Note that the matrix $\mathbf{H}_0'$ may be expressed as $\mathbf{H}_0' = \mathbf{S}^{\tau}\mathbf{S}$, where $\det \mathbf{S} = 1$, if and only if

$$\mathbf{H}_0' > 0, \quad \det \mathbf{H}_0' = 1. \tag{1.28}$$

Indeed, the matrix $\mathbf{H}_0' = \mathbf{S}^{\tau}\mathbf{S}$ satisfies (1.28). Conversely, setting, say $\mathbf{S} = (\mathbf{H}_0')^{1/2}$, we deduce from (1.28) that $\mathbf{H}_0' = \mathbf{S}^{\tau}\mathbf{S}$ and $\det \mathbf{S} = 1$.

Setting $\mathbf{H}_0 = \varkappa \mathbf{H}_0'$ in cases (VII), (VIII) and choosing $\varkappa$ so that $\mathbf{H}_0'$ satisfies (1.28), we obtain

Corollary. A constant Hamiltonian $\mathfrak{S}_n$ is in the stability domain if and only if $\mathbf{H}_0$ may be expressed as $\mathbf{H}_0 = \varkappa \mathbf{H}_0'$, where $\mathbf{H}_0'$ satisfies conditions (1.28) and $\frac{n\pi}{T} < \varkappa < \frac{(n+1)\pi}{T}$.

Setting $\mathbf{H}(t) \equiv \mathbf{H}_0$ in Theorem II, replacing $\mathbf{H}_1(t)$ by $\mathbf{H}(t)$, and using the corollary, we deduce the following theorem from the corollary.

Theorem IV. $\mathbf{H}(t) \in \mathfrak{S}_n$ *if and only if*

$$\mathbf{H}(t) = -\mathbf{P}(t)\,\tilde{\mathbf{J}} \frac{d}{dt} \mathbf{P}(t)^{\tau} + \varkappa \mathbf{P}(t)\,\mathbf{H}_0'\mathbf{P}(t)^{\tau}, \tag{1.29}$$

where

(I) $\mathbf{P}(t)$ *satisfies conditions (a), (c) of Theorem I and the rotation of any vector* $\mathbf{P}(t)\,\mathbf{c}$ *in time* T *is zero;*

(II) *the constant matrix* $\mathbf{H}_0' = (\mathbf{H}_0')^\tau$ *satisfies conditions (1.28);*
(III)

$$\frac{n\pi}{T} < \varkappa < \frac{(n+1)\pi}{T}.$$

§2. STABILITY AND INSTABILITY TESTS, ESTIMATES FOR CHARACTERISTIC EXPONENTS OF A CANONICAL SYSTEM OF TWO LINEAR DIFFERENTIAL EQUATIONS WITH PERIODIC COEFFICIENTS

2.1. Stability and instability tests based on estimates for rotation functions of solution-vectors

According to Theorem II of §1.9, the rotations of solution-vectors $\mathbf{x}(t) = \mathbf{X}(t)\mathbf{c}$ and vectors $\mathbf{x}_\tau(t) = \mathbf{X}(t)^\tau \mathbf{c}$ play an important part in stability and instability analysis of equation (1.2). We shall derive various simple estimates for these angles, which will then yield stability and instability tests.

We first agree on some terminology. If the trivial solution of equation (1.2) is Lyapunov-stable (all solutions are bounded as $t \to +\infty$), we shall sometimes say briefly that equation (1.2) is stable. Otherwise we shall say that the equation is unstable.

Let $\mathbf{z}(t)$ be some vector on the plane such that $d\mathbf{z}/dt$ exists and $\mathbf{z}(t) \neq 0$. Let $\vartheta(t)$ denote a continuous branch of $\operatorname{Arg} \mathbf{z}(t)$.

If p and q are the components of $\mathbf{z}(t)$, then

$$\frac{d\vartheta}{dt} = \frac{d}{dt} \operatorname{arctg} \frac{q}{p} = \frac{1}{p^2+q^2} \begin{vmatrix} p & \dot{p} \\ q & \dot{q} \end{vmatrix}, \quad \frac{d\vartheta}{dt} = \frac{\det \| \mathbf{z}, \dot{\mathbf{z}} \|}{(\mathbf{z}, \mathbf{z})}. \tag{2.1}$$

Using this formula and the identity $\det \| \mathbf{a}, -\mathbf{J}\mathbf{b} \| = (\mathbf{a}, \mathbf{b})$, we see that if $\mathbf{z} = \mathbf{x}(t)$ is a solution of equation (1.2), then

$$\frac{d\vartheta}{dt} = \frac{(\mathbf{H}\mathbf{x}, \mathbf{x})}{(\mathbf{x}, \mathbf{x})}, \tag{2.2}$$

$$\varphi_x = \int_0^T \frac{(\mathbf{H}\mathbf{x}, \mathbf{x})}{(\mathbf{x}, \mathbf{x})} dt. \tag{2.3}$$

As before, $\varphi_x = \vartheta(T) - \vartheta(0)$ denotes the rotation of the vector $\mathbf{x}$ in time T.

Let $h_{\max}(t)$ and $h_{\min}(t)$ denote the maximum and minimum eigenvalues of

$$\mathbf{H}(t) = \begin{Vmatrix} \alpha & \beta \\ \beta & \gamma \end{Vmatrix}.$$

It is easy to see that

$$h_{\substack{\max\\ \min}}(t)=\frac{1}{2}(\alpha+\gamma)\pm\sqrt{\frac{1}{4}(\alpha+\gamma)^2+\beta^2}\,. \tag{2.4}$$

Since $h_{\min}(t)\leqslant\frac{(\mathbf{Hx},\ \mathbf{x})}{(\mathbf{x},\ \mathbf{x})}\leqslant h_{\max}(t)$ (see (§I.1.58)), it follows from (2.2) that

$$\int_0^T h_{\min}(t)\,dt\leqslant\varphi_x\leqslant\int_0^T h_{\max}(t)\,dt. \tag{2.5}$$

From part 1 of Theorem II, §1.9, we obtain the following proposition:

If

$$n\pi<\int_0^T h_{\min}(t)\,dt\leqslant\int_0^T h_{\max}(t)\,dt<(n+1)\pi \tag{2.6}$$

for some $n=0,\ \pm1,\ \pm2,\ \ldots$, *then equation (1.2) is stable and* $\mathbf{H}(t)\in\mathfrak{G}_n$.

Now let $\mathbf{H}(t)=h(t)\,\mathbf{I}_2$ in equation (1.2). Then

$$\mathbf{X}(T)=\exp\left[-\tilde{\mathbf{J}}\int_0^T h(t)\,dt\right].$$

Consequently, for any solution,

$$\varphi_x=\int_0^T h(t)\,dt.$$

By part 1 of Theorem II, §1.9, the condition

$$n\pi<\int_0^T h(t)\,dt<(n+1)\pi$$

is necessary and sufficient for the Hamiltonian $\mathbf{H}(t)=h(t)\,\mathbf{I}_2$ to be in $\mathfrak{G}_n$. Thus condition (2.5) indicates how close the Hamiltonian $\mathbf{H}(t)$ must be to a Hamiltonian $h(t)\,\mathbf{I}_2$ in $\mathfrak{G}_n$ for $\mathbf{H}(t)$ to belong to $\mathfrak{G}_n$ as well.

It is easily checked that not all constant matrices $\mathbf{H}(t)\equiv\mathbf{C}\in\mathfrak{G}_n$ satisfy condition (2.6). We shall improve the condition in order to include all constant matrices in $\mathfrak{G}_n$. To this end, we substitute $\mathbf{x}=\mathbf{S}\mathbf{x}_1$, $\det\mathbf{S}=1$, in equation (1.2). We obtain an equation with Hamiltonian $\mathbf{H}_1(t)=\tilde{\mathbf{J}}\mathbf{S}^{-1}\tilde{\mathbf{J}}^{-1}\mathbf{H}(t)\,\mathbf{S}$. The characteristic equation $\det[\mathbf{H}_1(t)-h\mathbf{I}_2]=0$ is equivalent to the equation

$$\det[\mathbf{H}(t)-h\mathbf{H}_0]=0, \tag{2.7}$$

where

$$\mathbf{H}_0=\tilde{\mathbf{J}}\mathbf{S}\tilde{\mathbf{J}}^{-1}\mathbf{S}^{-1}.$$

Using the identity $\mathbf{S}\tilde{\mathbf{J}}=\tilde{\mathbf{J}}\,[\mathbf{S}^{\tau}]^{-1}$, we see that

$$\mathbf{H}_0=[\mathbf{S}^{\tau}]^{-1}\,\mathbf{S}^{-1}, \tag{2.8}$$

so that

$$\mathbf{H}_0 > 0, \quad \det \mathbf{H}_0 = 1. \tag{2.9}$$

Conversely, it follows from (2.9) that the matrix $\mathbf{H}_0$ may be written as (2.8) with, say, $\mathbf{S}=\mathbf{H}_0^{1/2}$.

Given vectors $\mathbf{x}$ and $\mathbf{x}_1=\mathbf{S}^{-1}\mathbf{x}$, the inequalities

$$n\pi < \varphi_x < (n+1)\pi$$

must hold either for both or for neither (see Lemma I, 1.9); hence the systems with Hamiltonians $\mathbf{H}(t)$ and $\mathbf{H}_1(t)$ are either both or neither in $\mathfrak{G}_n$.

Applying this test to the equation with matrix $\mathbf{H}_1(t)$, we obtain:

Test 1. Let

$$\mathbf{H}_0 = \begin{Vmatrix} \alpha_0 & \beta_0 \\ \beta_0 & \gamma_0 \end{Vmatrix},$$

where α_0, β_0, γ_0 are arbitrary numbers such that $\alpha_0>0$, $\gamma_0>0$, $\alpha_0\gamma_0-\beta_0^2=1$, and let $h_{\min}(t)$, $h_{\max}(t)$ be the roots of equation (2.7). If inequalities (2.6) hold for some $n=0, \pm 1, \pm 2, \ldots$, then equation (1.2) is stable and $\mathbf{H}(t)\in\mathfrak{G}_n$.

We saw above (see corollary at the end of §1.10) that a constant Hamiltonian $\mathbf{C}=\mathbf{C}^*$ is in $\mathfrak{G}_n$ if and only if $\mathbf{C}=h_0\mathbf{H}_0$, where

$$\frac{n\pi}{T} < h_0 < \frac{(n+1)\pi}{T}$$

and $\mathbf{H}_0$ is a matrix satisfying the conditions of the test (which are equivalent to (1.28)). Thus Test 1 covers all constant matrices.

We now derive a formula for φ_y, where $\mathbf{y}=\mathbf{X}(t)^\tau\mathbf{c}$. We have

$$\frac{d\mathbf{y}}{dt} = \mathbf{X}^\tau\mathbf{H}\tilde{\mathbf{J}}\mathbf{c}, \ \det\begin{Vmatrix}\mathbf{y}, & \frac{d\mathbf{y}}{dt}\end{Vmatrix} = -\det\|\mathbf{X}^\tau\mathbf{c}, \ \mathbf{X}^\tau\mathbf{H}\tilde{\mathbf{J}}^\tau\mathbf{c}\| =$$
$$= -\det\mathbf{X}^\tau\det\|\mathbf{c}, \ \mathbf{H}\tilde{\mathbf{J}}^\tau\mathbf{c}\| = \det\|\mathbf{H}\tilde{\mathbf{J}}^\tau\mathbf{c}, \ \tilde{\mathbf{J}}\tilde{\mathbf{J}}^{-1}\mathbf{c}\|.$$

Since $\det\|\mathbf{a}, \ \tilde{\mathbf{J}}\mathbf{b}\| = -(\mathbf{a}, \mathbf{b})$, it follows that

$$\det\begin{Vmatrix}\mathbf{y}, & \frac{d\mathbf{y}}{dt}\end{Vmatrix} = -(\mathbf{H}\tilde{\mathbf{J}}\mathbf{c}, \ \tilde{\mathbf{J}}\mathbf{c}).$$

Hence, by (2.1),

$$\varphi_y = -\int_0^T \frac{(\mathbf{H}\tilde{\mathbf{J}}\mathbf{c}, \ \tilde{\mathbf{J}}\mathbf{c})}{(\mathbf{y}, \mathbf{y})}\, dt. \tag{2.10}$$

Let us suppose that we have some estimate $0<\Phi_1(t)\leqslant(\mathbf{y}, \mathbf{y})\leqslant\Phi_2(t)$. Then (2.10) at once yields an estimate for φ_y and hence, by Theorem II of §1.9, a certain conclusion as to the stability or instability of equation (1.2). We now show that any estimate

$$\Phi_1(t)\leqslant(\mathbf{x}, \mathbf{x})\leqslant\Phi_2(t), \tag{a}$$

valid for all $\mathbf{c}=\mathbf{x}(0)$, $|\mathbf{c}|=1$, implies the estimate

$$\Phi_1(t)\leqslant(\mathbf{y},\ \mathbf{y})\leqslant\Phi_2(t), \tag{b}$$

again for all $\mathbf{c}=\mathbf{y}(0)$, $|\mathbf{c}|=1$; and conversely, an estimate of type (b) implies (a). Indeed, estimates (a) and (b) are equivalent respectively to

$$\Phi_1(t)\leqslant(\mathbf{X}^\tau\mathbf{X}\mathbf{c},\ \mathbf{c})\leqslant\Phi_2(t) \tag{a'}$$

and

$$\Phi_1(t)\leqslant(\mathbf{X}\mathbf{X}^\tau\mathbf{c},\ \mathbf{c})\leqslant\Phi_2(t). \tag{b'}$$

Condition (a') means that the minimum and maximum roots of the equation

$$\det(\mathbf{X}^\tau\mathbf{X}-\mu\mathbf{I}_2)=0 \tag{a''}$$

lie between Φ_1 and Φ_2. Condition (b') has the same meaning for the equation

$$\det(\mathbf{X}\mathbf{X}^\tau-\mu\mathbf{I}_2)=0. \tag{b''}$$

But equations (a'') and (b'') are identical: they both have the form

$$\mu^2-\beta\mu+1=0,$$

where $\beta=\operatorname{Tr}\mathbf{X}^\tau\mathbf{X}$ for equation (a'') and $\beta=\operatorname{Tr}\mathbf{X}\mathbf{X}^\tau$ for (b''). Since $\operatorname{Tr}\mathbf{X}^\tau\mathbf{X}=\operatorname{Tr}\mathbf{X}\mathbf{X}^\tau$, this proves our assertion.

It seems to be more difficult to derive estimate (b) directly from equation (1.2) than to derive (a). It is easy, on the other hand, to deduce (a) from equation (1.2), using Lyapunov's techniques. Let $\xi=(\mathbf{x},\ \mathbf{x})$. Then

$$\dot{\xi}=(-\tilde{\mathbf{J}}\mathbf{H}\mathbf{x},\ \mathbf{x})+(\mathbf{x},\ -\tilde{\mathbf{J}}\mathbf{H}\mathbf{x})=([\mathbf{H}\tilde{\mathbf{J}}-\tilde{\mathbf{J}}\mathbf{H}]\,\mathbf{x},\ \mathbf{x}).$$

It is readily checked that the eigenvalues of the matrix $\mathbf{H}\tilde{\mathbf{J}}-\tilde{\mathbf{J}}\mathbf{H}$ are $\pm g(t)$, where

$$g(t)=\sqrt{(\alpha-\gamma)^2+4\beta^2}. \tag{2.11}$$

Here $\alpha=\alpha(t)$, $\beta=\beta(t)$, $\gamma=\gamma(t)$ are the elements of the matrix $\mathbf{H}=\mathbf{H}(t)$ in (1.3). Then

$$-g\leqslant\frac{\dot{\xi}}{\xi}\leqslant g \quad\text{and}\quad \xi_0\exp\left(-\int_0^t g\,dt\right)\leqslant\xi\leqslant\xi_0\exp\int_0^t g\,dt.$$

Thus the bounds $\Phi_1(t)$, $\Phi_2(t)$ may be chosen as

$$\Phi_1(t)=\exp\left(-\int_0^t g(s)\,ds\right),\quad \Phi_2(t)=\exp\int_0^t g(s)\,ds. \tag{2.12}$$

Now let $\Phi_1(t)$, $\Phi_2(t)$ be any positive functions satisfying (a) for all $\mathbf{c}=\mathbf{x}(0)$, $|\mathbf{c}|=1$, and so also (b). We shall establish an estimate for the rotation (2.10).

Let us first assume that $\mathbf{H}(t)$ is a positive semidefinite matrix for all t, i. e., $(\mathbf{H}(t)\mathbf{c},\ \mathbf{c})\geqslant 0$ for any $\mathbf{c}$. Assuming that $(\mathbf{c},\ \mathbf{c})=1$, we obtain from (2.10) and (b)

$$\int_0^T \Phi_2(t)^{-1}(\mathbf{H}(t)\tilde{\mathbf{J}}\mathbf{c},\ \tilde{\mathbf{J}}\mathbf{c})\,dt\leqslant\varphi_y\leqslant\int_0^T \Phi_1(t)^{-1}(\mathbf{H}(t)\tilde{\mathbf{J}}\mathbf{c},\ \tilde{\mathbf{J}}\mathbf{c})\,dt.$$

We now change the initial conditions: let $\tilde{\mathbf{J}}\mathbf{c}=\exp(-\chi\tilde{\mathbf{J}})\,\mathbf{a}$ $(0\leqslant\chi\leqslant\pi)$, where $\mathbf{a}$ is a fixed constant vector, $(\mathbf{a},\ \mathbf{a})=1$. Set $-\varphi_y=\psi(\chi)$ (this differs slightly from the notation of §1.9). We have shown that

$$\Psi^-(\chi)\leqslant\psi(\chi)\leqslant\Psi^+(\chi)\qquad(0\leqslant\chi\leqslant\pi),\tag{2.13}$$

where

$$\Psi^-(\chi)=\int_0^T \Phi_2(t)^{-1}(\mathbf{H}(t)\tilde{\mathbf{J}}\mathbf{c},\ \tilde{\mathbf{J}}\mathbf{c})\,dt,\quad \Psi^+(\chi)=\int_0^T \Phi_1(t)^{-1}(\mathbf{H}(t)\tilde{\mathbf{J}}\mathbf{c},\ \tilde{\mathbf{J}}\mathbf{c})\,dt.$$

Substituting

$$\mathbf{J}\mathbf{c}=\exp(-\chi\tilde{\mathbf{J}})\,\mathbf{a}=(\mathbf{I}_2\cos\chi-\tilde{\mathbf{J}}\sin\chi)\,\mathbf{a},\qquad \mathbf{a}=\mathbf{e}_1=\begin{pmatrix}1\\0\end{pmatrix}$$

and introducing the notation

$$\left.\begin{aligned}
A^-&=\int_0^T \Phi_2(t)^{-1}\alpha(t)\,dt, & A^+&=\int_0^T \Phi_1(t)^{-1}\alpha(t)\,dt,\\
B^-&=\int_0^T \Phi_2(t)^{-1}\beta(t)\,dt, & B^+&=\int_0^T \Phi_1(t)^{-1}\beta(t)\,dt,\\
C^-&=\int_0^T \Phi_2(t)^{-1}\gamma(t)\,dt, & C^+&=\int_0^T \Phi_1(t)^{-1}\gamma(t)\,dt,
\end{aligned}\right\}\tag{2.14}$$

we see that

$$\Psi^\pm(\chi)=\frac{1}{2}(A^\pm+C^\pm)+B^\pm\sin 2\chi+\frac{1}{2}(A^\pm-C^\pm)\cos 2\chi.\tag{2.15}$$

If $n\pi<\Psi^-(\chi)\leqslant\Psi^+(\chi)<(n+1)\pi$ for some integer n, part 2 of Theorem II, §1.9, tells us that stability holds and $\mathbf{H}(t)\in\mathfrak{S}_n$.

If $\Psi^+(\chi')<n\pi$, $\Psi^-(\chi'')>n\pi$ for some integer n and χ',χ'', we have instability and $\mathbf{H}(t)\in\mathcal{H}_n$.

It is clear that the choice of the vector $\mathbf{a}$ is immaterial, for a change in $\mathbf{a}$ is equivalent to a transformation

$$\Psi^\pm(\chi)\longrightarrow\Psi^\pm(\chi+\chi_0).$$

Let us denote

$$\max_{\chi} \Psi^{+}(\chi) = M^{+}, \quad \max_{\chi} \Psi^{-}(\chi) = M^{-},$$
$$\min_{\chi} \Psi^{+}(\chi) = m^{+}, \quad \min_{\chi} \Psi^{-}(\chi) = m^{-}.$$

Obviously,

$$\left.\begin{aligned} M^{\pm} &= \frac{1}{2}(A^{\pm}+C^{\pm}) + \sqrt{\frac{1}{4}(A^{\pm}-C^{\pm})^2+(B^{\pm})^2}, \\ m^{\pm} &= \frac{1}{2}(A^{\pm}+C^{\pm}) - \sqrt{\frac{1}{4}(A^{\pm}-C^{\pm})^2+(B^{\pm})^2}. \end{aligned}\right\} \tag{2.16}$$

The inequalities $n\pi < m^{-} \leqslant M^{+} < (n+1)\pi$ imply stability (Figure 62), and the inequalities $M^{-} > n\pi$, $m^{+} < n\pi$ give instability (Figure 63).

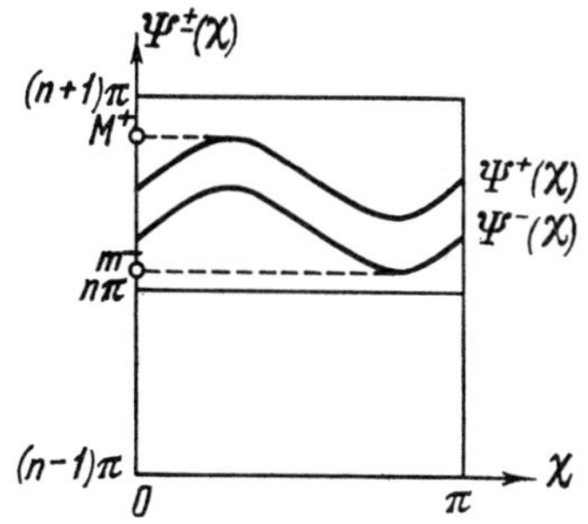

FIGURE 62.

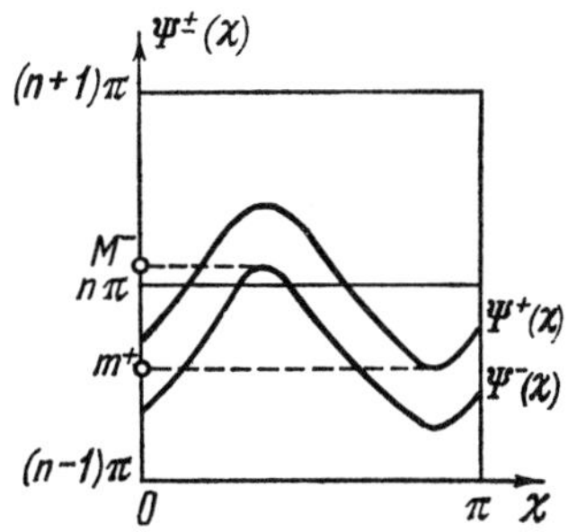

FIGURE 63.

Now let $(\mathbf{H}(t)\mathbf{c}, \mathbf{c}) \leqslant 0$. Then we must reverse the inequality signs in (2.13). If we retain the notation (2.14) and (2.15), we have $\Psi^{+}(\chi) \leqslant \psi(\chi) \leqslant \Psi^{-}(\chi)$; if the notation (2.16) is retained the inequalities $n\pi < m^{+} \leqslant M^{-} < (n+1)\pi$ imply stability and the inequalities $M^{+} > n\pi$, $m^{-} < n\pi$ instability.

The final result is as follows.

Let $\Phi_1(t)$, $\Phi_2(t)$ *be positive functions such that estimate (a) holds for all* $\mathbf{c} = \mathbf{x}(0)$, $|\mathbf{c}| = 1$ *(for example, the functions* $\Phi_{1,2}(t)$ *may be defined by (2.12), where* $\mathbf{g}(t)$ *is given by (2.11)).*

Let $(\mathbf{H}(t)\mathbf{c}, \mathbf{c}) \geqslant 0$ *for all* t *and* $\mathbf{c}$. *Define the numbers* M^{+}, M^{-}, m^{+}, m^{-} *by (2.14), (2.16). If* $n\pi < m^{-} \leqslant M^{+} < (n+1)\pi$, *then equation (1.2) is stable and* $\mathbf{H}(t) \in \mathcal{G}_n$. *But if* $M^{-} > n\pi$, $m^{+} < n\pi$, *equation (1.2) is unstable and* $\mathbf{H}(t) \in \mathcal{H}_n$.*

If $(\mathbf{H}(t)\mathbf{c}, \mathbf{c}) \leqslant 0$ *for all* t *and* $\mathbf{c}$, *the above assertions are valid, provided* $M^{\pm}$, $m^{\pm}$ *are replaced throughout by* $M^{\mp}$, $m^{\mp}$.

Now suppose that $(\mathbf{H}(t)\mathbf{c}, \mathbf{c}) > 0$ for some $\mathbf{c}$ and t, and $(\mathbf{H}(t)\mathbf{c}, \mathbf{c}) < 0$ for others. Fix a vector $\mathbf{c}$, and denote

$$E_c^{+} = E\{(\mathbf{H}(t)\mathbf{c}, \mathbf{c}) > 0\} \subset [0, T], \quad E_c^{-} = E\{(\mathbf{H}(t)\mathbf{c}, \mathbf{c}) < 0\} \subset [0, T].$$

* It can be shown that the conditions $M^{-} > n\pi$, $m^{+} < n\pi$ cannot be satisfied if $\Phi_{1,2}(t)$ are defined by (2.12).

Then (2.13) remains valid, provided we define

$$\Psi^{\pm}(\chi)=\int\limits_{E_c^+}\Phi_{1,2}(t)^{-1}(\mathbf{H}(t)\mathbf{c},\ \mathbf{c})\,dt+\int\limits_{E_c^-}\Phi_{2,1}(t)^{-1}(\mathbf{H}(t)\mathbf{c},\ \mathbf{c})\,dt.$$

Here $\mathbf{c}=\exp(-\chi\tilde{\mathbf{J}})\,\mathbf{a},\ |\mathbf{a}|=1$. Determining $\max\Psi^{\pm}(\chi)$ and $\min\Psi^{\pm}(\chi)$ in each specific case, we see as before that in the case illustrated in Figure 62 equation (1.2) is stable, but in that shown in Figure 63 the equation is unstable.

It is clear that if sharper estimates are available for $(\mathbf{y},\ \mathbf{y})$ (depending on χ), we obtain stronger stability and instability tests.

2.2. Arbitrarily sharp two-sided estimates for rotation functions; stability and instability tests

We return to the canonical equation (1.2). As before, let $\mathbf{x}(t)$ be a solution, $\operatorname{Arg}\mathbf{x}(0)=\chi\ (0\leqslant\chi\leqslant\pi)$, $\operatorname{Arg}\mathbf{x}(t)=\vartheta(t,\ \mathbf{x})$, and

$$\psi(\chi)=\vartheta(T,\ \mathbf{x})-\chi \tag{2.17}$$

the rotation of the solution. We showed above (§1.9) that the decision as to which of the sets Π_n^{**}, Π_n^+, Π_n^-, $\mathfrak{G}_n$, $\mathcal{H}_n$ contains the Hamiltonian $\mathbf{H}(t)$ depends on the relative positions on the rea axis of the numbers

$$m=\min_{\chi}\psi(\chi)\quad\text{and}\quad M=\max_{\chi}\psi(\chi)$$

(see table, §1.9). In this subsection we shall establish arbitrarily sharp two-sided estimates for m and M: we shall show how to calculate numbers $m_k^{\pm}$, $M_k^{\pm}$ $(k=0,\ 1,\ 2,\ \ldots)$ such that

$$m_k^-\leqslant m\leqslant m_k^+,\quad M_k^-\leqslant M\leqslant M_k^+ \tag{2.18}$$

and

$$m_k^{\pm}\longrightarrow m,\quad M_k^{\pm}\longrightarrow M\quad\text{as}\quad k\longrightarrow\infty. \tag{2.19}$$

These properties immediately yield stability and instability tests. Indeed, it follows from (2.18) and Theorem II of §1.9 (see also the table in §1.9) that the truth of the inequalities in the left-hand column of the following table, for some $k=0,\ 1,\ 2,\ \ldots$ and some integer n, is a sufficient condition for the condition in the right-hand column to be fulfilled:

$(n-1)\pi<m_k^-,\ M_k^+<n\pi$	$\mathbf{H}(t)\in\mathfrak{G}_n$
$m_k^+<n\pi<M_k^-$	$\mathbf{H}(t)\in\mathcal{H}_n$

The relations (2.19) and Theorem II of § 1.9 (as well as the table given there) show that for any Hamiltonian $\mathbf{H}(t)\in\mathfrak{G}\cup\mathcal{H}$ (i. e., not lying on the common boundary Π of the stability and instability domains) there is a natural number k such that one of the above inequalities holds. Thus these tests ($k=0, 1, 2, \ldots$; $n=0, \pm 1, \pm 2, \ldots$) exhaust the entire Hamiltonian space $\mathfrak{L}^3$ except for the "boundary" set Π.

Our goal is thus to find the numbers $m_k^{\pm}$, $M_k^{\pm}$.

It follows from (2.2) that the function $\vartheta=\vartheta(t, \chi)$ is the solution of the differential equation

$$\frac{d\vartheta}{dt}=\frac{\alpha+\gamma}{2}+\beta \sin 2\vartheta+\frac{\alpha-\gamma}{2}\cos 2\vartheta \tag{2.20}$$

with initial condition $\vartheta(0, \chi)=\chi$. Here $\alpha=\alpha(t)$, $\beta=\beta(t)$, $\gamma=\gamma(t)$ are the elements of the Hamiltonian (1.3).

We first consider the derivation of arbitrarily sharp two-sided estimates for the solution of equation (2.20). Substituting

$$\xi=\vartheta-\frac{1}{2}\int_0^t [\alpha(\tau)+\gamma(\tau)]\,d\tau, \tag{2.21}$$

we obtain a differential equation

$$\frac{d\xi}{dt}=\delta(t)\cos[2\xi+\sigma(t)]+\beta(t)\sin[2\xi+\sigma(t)], \tag{2.22}$$

where

$$\delta(t)=\frac{1}{2}[\alpha(t)-\gamma(t)], \quad \sigma(t)=\int_0^t [\alpha(\tau)+\gamma(\tau)]\,d\tau. \tag{2.23}$$

Equation (2.22) reduces to an integral equation. We shall study a more general integral equation

$$\xi(t)=\varkappa(t)+\int_0^t K[t, s, \xi(s)]\,ds. \tag{2.24}$$

In this equation $\xi(t)$ and $\varkappa(t)$ are real functions, continuous on $[0, T]$; $K(t, s, \xi)$ is jointly continuous in its variables and satisfies a Lipschitz condition with respect to ξ; in other words, for any $t, s \in [a, b]$ and any ξ

$$|K(t, s, \xi_1)-K(t, s, \xi_2)| \leqslant L|\xi_1-\xi_2|, \tag{2.25}$$

where L is a constant. (This condition may in fact be substantially weakened; it is sufficient to assume that the function $K(t, s, \xi)$ satisfies Carathéodory's conditions and is of bounded variation in ξ for almost all $(t, s) \in [0, T]\times[0, T]$.)

It is easily shown that equation (2.24) has a solution $\xi(t)$ on $[0, T]$, that $\xi(t)$ is continuous on $[0, T]$, and that the solution is unique in the class of functions summable on $[0, T]$. (This is proved in exactly the same way as the analogous assertion in Theorem I below.)

The function $K(t, s, \xi)$ may be expressed in infinitely many ways as the sum of increasing and decreasing functions of ξ. Let

$$K(t, s, \xi) = P^{in}(t, s, \xi) + P^{d}(t, s, \xi), \tag{2.26}$$
$$K(t, s, \xi) = Q^{in}(t, s, \xi) + Q^{d}(t, s, \xi) \tag{2.27}$$

be any two such representations, where P^{in}, P^{d}, Q^{in}, Q^{d} are jointly continuous functions, P^{in}, Q^{in} increasing and P^{d}, Q^{d} decreasing functions of ξ for fixed $t \in [0, T]$, $s \in [0, T]$:

$$P^{in}(t, s, \xi_1) \leqslant P^{in}(t, s, \xi_2), \quad P^{d}(t, s, \xi_1) \geqslant P^{d}(t, s, \xi_2) \quad \text{for} \quad \xi_1 \geqslant \xi_2$$

and analogous inequalities for Q^{in}, Q^{d}. We may assume without loss of generality that the functions P^{in}, P^{d}, Q^{in}, Q^{d} are Lipschitzian in ξ (it is readily seen that this is always possible).

We shall need the following theorem, due to Komlenko /159b/, which provides arbitrarily sharp upper and lower bound for the solution $\xi(t)$ of equation (2.24).

Theorem I. *Let the functions* K, P^{in}, P^{d}, Q^{in}, Q^{d} *satisfy the conditions stipulated above. Let* $\xi_0'(t)$, $\xi_0''(t)$ *be functions on* $[0, T]$ *satisfying the general inequalities*

$$\left.\begin{aligned} \xi_0'(t) &\leqslant \varkappa(t) + \int_0^t \{P^{in}[t, s, \xi_0'(s)] + P^{d}[t, s, \xi_0''(s)]\}\, ds, \\ \xi_0''(t) &\geqslant \varkappa(t) + \int_0^t \{Q^{d}[t, s, \xi_0'(s)] + Q^{in}[t, s, \xi_0''(s)]\}\, ds. \end{aligned}\right\} \tag{2.28}$$

Construct a sequence of functions $\xi_k'(t)$, $\xi_k''(t)$ $(k = 1, 2, \ldots)$ *by*

$$\left.\begin{aligned} \xi_{k+1}'(t) &= \varkappa(t) + \int_0^t \{P^{in}[t, s, \xi_k'(s)] + P^{d}[t, s, \xi_k''(s)]\}\, ds, \\ \xi_{k+1}''(t) &= \varkappa(t) + \int_0^t \{Q^{d}[t, s, \xi_k'(s)] + Q^{in}[t, s, \xi_k''(s)]\}\, ds. \end{aligned}\right\} \tag{2.29}$$

Then

$$\xi_k'(t) \leqslant \xi_{k+1}'(t) \leqslant \xi(t) \leqslant \xi_{k+1}''(t) \leqslant \xi_k''(t) \qquad (k = 0, 1, 2, \ldots) \tag{2.30}$$

and

$$\xi_k'(t) \to \xi(t), \quad \xi_k''(t) \to \xi(t) \quad \text{as} \quad k \to \infty, \tag{2.31}$$

uniformly on $[0, T]$, *where* $\xi(t)$ *is a solution of equation (2.24).*

Proof. Instead of (2.24), let us consider the system of equations

$$\left.\begin{aligned}\xi'(t)&=\varkappa(t)+\int_0^t\{P^{\text{in}}[t,s,\xi'(s)]+P^{\text{d}}[t,s,\xi''(s)]\}\,ds,\\ \xi''(t)&=\varkappa(t)+\int_0^t\{Q^{\text{d}}[t,s,\xi'(s)]+Q^{\text{in}}[t,s,\xi''(s)]\}\,ds,\end{aligned}\right\}\tag{2.32}$$

or, in vector notation,

$$\mathbf{z}(t)=\mathbf{h}(t)+\int_0^t\mathbf{r}[t,s,\mathbf{z}(s)]\,ds,\tag{2.33}$$

where

$$\mathbf{z}(t)=\begin{pmatrix}\xi'(t)\\ \xi''(t)\end{pmatrix},\quad \mathbf{h}(t)=\begin{pmatrix}\varkappa(t)\\ \varkappa(t)\end{pmatrix},\quad \mathbf{r}(t,s,\mathbf{z})=\begin{pmatrix}P^{\text{in}}(t,s,\xi')+P^{\text{d}}(t,s,\xi'')\\ Q^{\text{d}}(t,s,\xi')+Q^{\text{in}}(t,s,\xi'')\end{pmatrix}.\tag{2.34}$$

Formulas (2.29) are equivalent to the usual successive approximations to the solution of equation (2.33), for they may be written as

$$\mathbf{z}_{k+1}(t)=\mathbf{h}(t)+\int_0^t\mathbf{r}[t,s,\mathbf{z}_k(s)]\,ds,\quad\text{where}\quad \mathbf{z}_k(t)=\begin{pmatrix}\xi_k'(t)\\ \xi_k''(t)\end{pmatrix}.\tag{2.35}$$

We now show in the standard manner that the sequence of functions $\mathbf{z}_k(t)$ is uniformly convergent on $[0, T]$. Since the functions P^{in}, P^{d}, Q^{in}, Q^{d} are Lipschitzian in ξ', ξ'', it follows that $\mathbf{r}(t, s, \mathbf{z})$ is Lipschitzian in $\mathbf{z}$: there exists a constant L_0 such that

$$|\mathbf{r}(t,s,\mathbf{z}')-\mathbf{r}(t,s,\mathbf{z}'')|\leqslant L_0|\mathbf{z}'-\mathbf{z}''|$$

for any $t, s\in[0, T]$ and any $\mathbf{z}'$, $\mathbf{z}''$.

By (2.35),

$$|\mathbf{z}_{k+1}(t)-\mathbf{z}_k(t)|\leqslant\int_0^t L_0|\mathbf{z}_k(s)-\mathbf{z}_{k-1}(s)|\,ds\qquad(k=0,1,2,\ldots).\tag{2.36}$$

The same formula is true for $k=0$ if we define $\mathbf{z}_{-1}\equiv\mathbf{0}$ and increase the constant L_0 (if necessary) so that inequality (2.36) holds for $k=0$ as well. Using (2.36), we show inductively that

$$|\mathbf{z}_{k+1}(t)-\mathbf{z}_k(t)|\leqslant L_0^{k+1}\int_0^t\frac{(t-s)^k}{k!}|\mathbf{z}_0(s)|\,ds\leqslant\frac{ML_0^{k+1}T^{k+1}}{(k+1)!},$$

where $M=\sup\limits_{0\leqslant s\leqslant T}|\mathbf{z}_0(s)|$. Thus the series

$$\sum_{k=0}^{\infty}|\mathbf{z}_{k+1}(t)-\mathbf{z}_k(t)|$$

is uniformly convergent on $[0, T]$, and hence so is the sequence $\{\mathbf{z}_k(t)\}$.

Since the convergence is uniform, the limit function

$$\lim_{k\to\infty} \mathbf{z}_k(t)=\mathbf{z}(t) \tag{2.37}$$

is continuous on $[0, T]$ and satisfies equation (2.33).

We now show that equation (2.33) has a unique solution on $[0, T]$ (in the class of summable vector-functions on $[0, T]$). Let $\mathbf{z}'(t)$ and $\mathbf{z}''(t)$ be two solutions. Then by (2.33)

$$|\mathbf{z}'(t)-\mathbf{z}''(t)|\leqslant L_0\int_0^t|\mathbf{z}'(s)-\mathbf{z}''(s)|\,ds.$$

Denoting the right-hand side of this inequality by $\zeta(t)$, we rewrite it as $L_0^{-1}d\zeta/dt\leqslant\zeta(t)$. Let $\mathbf{z}'(t)\not\equiv\mathbf{z}''(t)$ (i. e., $\mathbf{z}'(t)\neq\mathbf{z}''(t)$ on a set of positive measure in $[0, T]$). Then there exists t_0 $(0\leqslant t_0<T)$ such that $\zeta(t)>0$ for $t_0<t\leqslant T$ and $\zeta(t)=0$ for $0\leqslant t\leqslant t_0$. If $t\geqslant t_1>t_0$, we have $\zeta(t)\leqslant\zeta(t_1)\exp L_0(t-t_1)$. Letting $t_1\to t_0$, we obtain $\zeta(t)\leqslant 0$. This contradiction shows that $\mathbf{z}'(t)\equiv\mathbf{z}''(t)$.

Proceeding in exactly the same way, one proves that equation (2.24) has a solution (which may be constructed in similar fashion by successive approximations) which is continuous and unique in the class of functions summable on $[0, T]$.

Let this solution be $\xi(t)$. Setting $\xi'=\xi''=\xi(t)$ in (2.32), we see by (2.26) and (2.27) that equation (2.32) is satisfied. Thus, since the solution of system (2.32) is unique, the solution (2.37) of system (2.32) has equal components $\xi(t)$, $\xi(t)$. Thus (2.37) and (2.31) are identical.

It remains only to show that inequalities (2.30) are satisfied.

Consider the cone $\xi'\geqslant 0$, $\xi''\leqslant 0$ on the (ξ',ξ'')-plane. We define a semi-order relation relative to this cone in the set of 2-vectors:

$$\mathbf{z}_1\geqslant\mathbf{z}_2, \quad\text{where}\quad \mathbf{z}_1=\begin{pmatrix}\xi_1'\\ \xi_1''\end{pmatrix},\quad \mathbf{z}_2=\begin{pmatrix}\xi_2'\\ \xi_2''\end{pmatrix},$$

if and only if

$$\xi_1'\geqslant\xi_2',\quad \xi_1''\leqslant\xi_2''.$$

Inequalities (2.30) may now be written

$$\mathbf{z}_k(t)\leqslant\mathbf{z}_{k+1}(t)\leqslant\mathbf{z}(t)\qquad(k=0, 1, 2, \ldots;\ 0\leqslant t\leqslant T). \tag{2.38}$$

It thus remains to prove (2.38).

The first inequality (2.38) for $k=0$ is satisfied. Indeed, the vector form of inequalities (2.28) (see (2.34)) is

$$\mathbf{z}_0(t)\leqslant\mathbf{h}(t)+\int_0^t\mathbf{r}[t, s, \mathbf{z}_0(s)]\,ds.$$

By (2.35), this means that $\mathbf{z}_0(t)\leqslant\mathbf{z}_1(t)$ $(0\leqslant t\leqslant T)$.

Now suppose that $z_{k-1}(t) \leqslant z_k(t)$ for some k and all $t \in [0, T]$; in other words, $\xi'_{k-1}(t) \leqslant \xi'_k(t)$, $\xi''_{k-1}(t) \geqslant \xi''_k(t)$. By the properties of P^{in}, P^d, we have

$$P^{in}[t, s, \xi'_{k-1}(s)] \leqslant P^{in}[t, s, \xi'_k(s)], \quad P^d[t, s, \xi''_{k-1}(s)] \leqslant P^d[t, s, \xi''_k(s)].$$

Therefore (cf. first formula of (2.29))

$$\xi'_k(t) = \varkappa(t) + \int_0^t \{P^{in}[t, s, \xi'_{k-1}(s)] + P^d[t, s, \xi''_{k-1}(s)]\} ds \leqslant$$
$$\leqslant \varkappa(t) + \int_0^t \{P^{in}[t, s, \xi'_k(s)] + P^d[t, s, \xi''_k(s)]\} ds = \xi'_{k+1}(t).$$

Similarly one shows that $\xi''_k(t) \geqslant \xi''_{k+1}(t)$. Thus $z_k(t) \leqslant z_{k+1}(t)$.

Thus $z_k(t) \leqslant z_{k+1}(t)$ for all k and all $t \in [0, T]$. It follows from (2.37) that $z_k(t) \leqslant z(t)$. This completes the proof of (2.38) and thus of the theorem.

In the rest of this subsection we shall use the following notation. For any real ζ, we set*

$$\zeta_+ = \frac{1}{2}(|\zeta| + \zeta), \quad \zeta_- = \frac{1}{2}(|\zeta| - \zeta). \tag{2.39}$$

Obviously,

$$\zeta = \zeta_+ - \zeta_-, \quad |\zeta| = \zeta_+ - \zeta_-, \quad \zeta_+ \geqslant 0, \quad \zeta_- \geqslant 0. \tag{2.40}$$

We now return to equation (2.22). Integrating, we write it as (2.24), where

$$\varkappa(t) = \chi, \quad K(t, s, \xi) = \delta(s) \cos[2\xi + \sigma(s)] + \beta(s) \sin[2\xi + \sigma(s)].$$

We define the functions P^{in}, P^d, Q^{in}, Q^d

$$P^{in}(t, s, \xi) = Q^{in}(t, s, \xi) = -2(|\delta(s)| + |\beta(s)|)\xi + \delta(s)_+ \cos[2\xi + \sigma(s)] + \beta(s)_+ \sin[2\xi + \sigma(s)],$$
$$P^d(t, s, \xi) = Q^d(t, s, \xi) = -2(|\delta(s)| + |\beta(s)|)\xi - \delta(s)_- \cos[2\xi + \sigma(s)] - \beta(s)_- \sin[2\xi + \sigma(s)].$$

These functions satisfy all the necessary conditions. Indeed, it is readily checked that they are jointly continuous, $\partial P^{in}/\partial\xi \geqslant 0$, $\partial P^d/\partial\xi \leqslant 0$, and by (2.40) we have (2.26) and (2.27).

Theorem I implies the following:

Theorem II. Set

$$\left.\begin{aligned} \xi'_0(t) &= \chi - \frac{1}{4}(e^{4\Delta(t)} - 1), \\ \xi''_0(t) &= \chi + \frac{1}{4}(e^{4\Delta(t)} - 1), \end{aligned}\right\} \tag{2.41}$$

where

$$\Delta(t) = \int_0^t (|\delta(\tau)| + |\beta(\tau)|) d\tau. \tag{2.42}$$

* In these notations the symbols $+$, $-$ are used as subscripts; when used as superscripts, $+$ and $-$ are simply indices, usually related to the supremum ($+$) and infimum ($-$) of some function.

Consider the two function sequences defined by

$$\left.\begin{aligned}
\xi'_{k+1}(t) &= \chi - 2\int_0^t (|\delta| + |\beta|)(\xi''_k - \xi'_k)\,ds + \\
&\quad + \int_0^t \delta_+ \cos(2\xi'_k + \sigma)\,ds + \int_0^t \beta_+ \sin(2\xi'_k + \sigma)\,ds - \\
&\qquad - \int_0^t \delta_- \cos(2\xi''_k + \sigma)\,ds - \int_0^t \beta_- \sin(2\xi''_k + \sigma)\,ds, \\
\xi''_{k+1}(t) &= \chi + 2\int_0^t (|\delta| + |\beta|)(\xi''_k - \xi'_k)\,ds - \\
&\quad - \int_0^t \delta_- \cos(2\xi'_k + \sigma)\,ds - \int_0^t \beta_- \sin(2\xi'_k + \sigma)\,ds + \\
&\qquad + \int_0^t \delta_+ \cos(2\xi''_k + \sigma)\,ds + \int_0^t \beta_+ \sin(2\xi''_k + \sigma)\,ds,
\end{aligned}\right\} \quad (2.43)$$

where $\sigma = \sigma(s)$, $\delta = \delta(s)$, $\beta = \beta(s)$, $\xi'_k = \xi'_k(s)$, $\xi''_k = \xi''_k(s)$ *and we have used the notation (2.23), (2.30). Then*

$$\xi'_k(t) \leqslant \xi'_{k+1}(t) \leqslant \xi(t) \leqslant \xi''_{k+1}(t) \leqslant \xi''_k(t) \qquad (k = 0, 1, 2, \ldots) \qquad (2.44)$$

and

$$\xi'_k(t) \to \xi(t), \quad \xi''_k(t) \to \xi(t) \quad \text{as} \quad k \to \infty, \qquad (2.45)$$

uniformly on $[0, T]$, *where* $\xi(t)$ *is a solution of equation (2.22),* $\xi(0) = \chi$.

To prove Theorem II we need only check that the functions $\xi'_0(t)$ and $\xi''_0(t)$ satisfy conditions (2.28). Using the above expressions for P^{in}, P^{d}, Q^{in}, Q^{d} and (2.40), we find that

$$\begin{aligned}
&\xi'_0(t) - \chi - \int_0^t P^{\text{in}}[t, s, \xi'_0(s)]\,ds - \int_0^t P^{\text{d}}[t, s, \xi''_0(s)]\,ds \geqslant \\
&\geqslant \xi'_0(t) - \chi - 2\int_0^t (|\delta(s)| + |\beta(s)|)[\xi''_0(s) - \xi'_0(s)]\,ds - \int_0^t (|\delta(s)| + |\beta(s)|)\,ds = 0, \\
&\xi''_0(t) - \chi - \int_0^t Q^{\text{d}}[t, s, \xi'_0(s)]\,ds - \int_0^t Q^{\text{in}}[t, s, \xi''_0(s)]\,ds \leqslant \\
&\leqslant \xi''_0(t) - \chi - 2\int_0^t (|\delta(s)| + |\beta(s)|)[\xi''_0(s) - \xi'_0(s)]\,ds + \int_0^t (|\delta(s)| + |\beta(s)|)\,ds = 0.
\end{aligned}$$

That the expressions on the right do indeed vanish follows from (2.41) (these are precisely equations defining the functions (2.41)).

We can now estimate the numbers m, M as promised at the beginning of this subsection. By (2.17), (2.21), (2.23),

$$\xi'_k(T) + \tfrac{1}{2}\sigma(T) - \chi \leqslant \psi(\chi) = \vartheta(T, \chi) - \chi \leqslant \xi''_k(T) + \tfrac{1}{2}\sigma(T)$$

and also

$$\xi'_k(T)+\frac{1}{2}\sigma(T)-\chi\to\psi(\chi),$$
$$\xi''_k(T)+\frac{1}{2}\sigma(T)-\chi\to\psi(\chi)$$
$$\text{as}\quad k\to\infty.$$

Consequently, conditions (2.18), (2.19) will hold if we put

$$\left.\begin{aligned} m_k^- &=\frac{1}{2}\sigma(T)+\min_{0\leqslant\chi\leqslant\pi}[\xi'_k(T)-\chi],\ m_k^+=\frac{1}{2}\sigma(T)+\min_{0\leqslant\chi\leqslant\pi}[\xi''_k(T)-\chi]\\ M_k^- &=\frac{1}{2}\sigma(T)+\max_{0\leqslant\chi\leqslant\pi}[\xi'_k(T)-\chi],\ M_k^+=\frac{1}{2}\sigma(T)+\max_{0\leqslant\chi\leqslant\pi}[\xi''_k(T)-\chi].\end{aligned}\right\}\tag{2.46}$$

(The values of $\xi'_k(T)$, $\xi''_k(T)$ clearly depend on χ but $\sigma(T)$ does not.) We summarize the results as follows.

Theorem III. Given the Hamiltonian of an equation (1.3), define a sequence of functions $\xi'_k(t)$, $\xi''_k(t)$ *by formulas (2.41), (2.43), where*

$$\delta(t)=\frac{1}{2}[\alpha(t)-\gamma(t)],\quad \sigma(t)=\int_0^t[\alpha(\tau)+\gamma(\tau)]\,d\tau,\quad \Delta(t)=\int_0^t(|\delta(\tau)|+|\beta(\tau)|)\,d\tau$$

and χ *is a real parameter* $(0\leqslant\chi\leqslant\pi)$. *Define the numbers* $m_k^\pm$, $M_k^\pm$, $(k=0,1,2,\ldots)$ *by formulas (2.46). If there exist* $k=0,1,2,\ldots$ *and* $n=0,\pm1,\pm2,\ldots$ *such that*

$$(n-1)\pi<m_k^-,\quad M_k^+<n\pi,\tag{2.47}$$

then equation (1.2) is stable and $\mathbf{H}(t)\in\mathfrak{G}_n$; *but if*

$$m_k^+<n\pi<M_k^-,\tag{2.48}$$

then equation (1.2) is unstable and $\mathbf{H}(t)\in\mathcal{H}_n$.

For any Hamiltonian $\mathbf{H}(t)$ *which is not on the common boundary* Π *of stability and instability domains (i.e., for which* $|A|\neq1$, *where* A *is the Lyapunov characteristic constant), there exist* k *and* n *such that one of these tests holds. Moreover, we then have (2.18) and (2.19).*

2.3. Estimates for characteristic exponents

As mentioned in §1.1, the estimation of characteristic exponents for the general equation (1.1),

$$\frac{d\mathbf{z}}{dt}=\mathbf{A}(t)\,\mathbf{z},$$

reduces to the same problem for a canonical equation (1.2),

$$\mathbf{J}\frac{d\mathbf{x}}{dt}=\mathbf{H}(t)\,\mathbf{x},$$

where

$$\tilde{\mathbf{J}} = \begin{Vmatrix} 0 & 1 \\ -1 & 0 \end{Vmatrix}, \quad \mathbf{H}(t) = \begin{Vmatrix} \alpha(t) & \beta(t) \\ \beta(t) & \gamma(t) \end{Vmatrix}.$$

The characteristic exponents (see § II. 2.5) λ_1 and λ_2 of equation (1.2), i. e., the eigenvalues of the matrix $\mathbf{K}$ in (1.6), are either real (cases (a), (b) of § 1.4) or pure imaginary (case (c)). Since $\operatorname{Tr}\mathbf{K}=0$, we have $\lambda_1+\lambda_2=0$. Let $\lambda=\operatorname{Re}\lambda_2\geqslant 0$. If $\lambda=0$, all solutions of system (1.2) are either bounded for $-\infty<t<+\infty$ (case (c)) or (in the exceptional case (b)) increase as t; if $\lambda>0$ there are exponentially increasing solutions (case (a)). The number $\lambda\geqslant 0$ will be called the exponent of exponential growth for solutions of equation (1.2).

As in § 1.9, we let φ_x denote the rotation of a solution-vector $\mathbf{x}(t)$ in the time from $t=0$ to $t=T$. Let the initial vector $\mathbf{x}(0)$ be

$$\mathbf{x}(0) = \begin{pmatrix} \cos\chi \\ \sin\chi \end{pmatrix}.$$

Let $\varphi_x=\psi(\chi)$ be the angle function introduced in § 1.9, and

$$m=\min\psi(\chi), \quad M=\max\psi(\chi) \qquad (0\leqslant\chi\leqslant\pi).$$

It was shown in § 1.9 that for a stability analysis we need only know the positions on the real axis of m and M relative to the numbers $n\pi$ $(n=0,\ \pm1,\ \pm2,\ \ldots)$ (see the table in § 1.9 and Theorem II, § 1.9). Theorem I of § 1.9, which described the qualitative behavior of the solutions, suggests that there should be some connection between the numbers m, M and the characteristic exponent λ. We shall show that this is indeed the case.

According to the Floquet-Lyapunov theorem (§ 1. 2), any solution of equation (1.2) may be written as

$$\mathbf{x}(t)=\mathbf{F}(t)\,\mathbf{y}(t), \quad \mathbf{y}(t)=e^{t\mathbf{K}}\mathbf{x}(0),$$

where $\mathbf{F}(t)\in\Omega_n$ for some integer n (for the definition, see § 1.6). By Theorem I of § 2.2,

$$\varphi_x=n\pi+\varphi_y, \tag{2.49}$$

where φ_y is the rotation of $\mathbf{y}(t)$ in the time from $t=0$ to $t=T$.

Let us assume that the equation is unstable. Then (see § 1.9) the number n in the above formula is the subscript of the instability domain containing the Hamiltonian $\mathbf{H}(t)$, and the matrix $\mathbf{K}$ has real eigenvalues λ and $-\lambda$. Let $\mathbf{a}_1$, $\mathbf{a}_2$ be corresponding normalized (and, of course, real and linearly independent) eigenvectors of $\mathbf{K}$:

$$\mathbf{K}\mathbf{a}_1=\lambda\mathbf{a}_1 \quad \mathbf{K}\mathbf{a}_2=-\lambda\mathbf{a}_2 \qquad (\lambda>0).$$

We may write $\mathbf{x}(0)$ as

$$\mathbf{x}(0)=\xi_1\mathbf{a}_1+\xi_2\mathbf{a}_2.$$

Then

$$\mathbf{y}(t)=e^{\lambda t}\xi_1\mathbf{a}_1+e^{-\lambda t}\xi_2\mathbf{a}_2$$

and by formula (2.1)

$$\varphi_y=2\lambda\xi_1\xi_2\sin\beta\int_0^T(\xi_1^2e^{-2\lambda t}+2\cos\beta\xi_1\xi_2+\xi_2^2e^{2\lambda t})^{-1}\,dt,$$

where

$$\cos\beta=(\mathbf{a}_1,\ \mathbf{a}_2),\quad \sin\beta=\det\|\,\mathbf{a}_1,\ \mathbf{a}_2\,\|,$$

i. e., β is the angle between the vectors $\mathbf{a}_1$, $\mathbf{a}_2$. We may assume without loss of generality that $0<\beta<\pi$. (If $-\pi<\beta<0$, we simply replace $\mathbf{a}_1$ by $-\mathbf{a}_1$.) To define the rotation function, we vary $\mathbf{x}(0)$, setting

$$\xi_1=\cos\chi,\quad \xi_2=\sin\chi\qquad(0\leqslant\chi\leqslant\pi).$$

Some elementary manipulations yield (for $\chi\neq\frac{1}{2}\pi$)

$$\varphi_y=\operatorname{arctg}\frac{e^{2\lambda T}\operatorname{tg}\chi+\cos\beta}{\sin\beta}-\operatorname{arctg}\frac{\operatorname{tg}\chi+\cos\beta}{\sin\beta}$$

and

$$\min_\chi\varphi_y=-\arccos\tau_1,\quad \max_\chi\varphi_y=\arccos\tau_2,$$

where

$$\tau_1=\frac{1-\operatorname{ch}\lambda T\cos\beta}{\operatorname{ch}\lambda T-\cos\beta},\quad \tau_2=\frac{1+\operatorname{ch}\lambda T\cos\beta}{\operatorname{ch}\lambda T+\cos\beta}\tag{2.50}$$

and the branches are so chosen that $0<\arccos\tau_{1,\,2}<\pi$.

Therefore

$$m=n\pi-\arccos\tau_1,\quad M=n\pi+\arccos\tau_2.\tag{2.51}$$

Formulas (2.50),(2.51) give us the desired relationship.

We now show that sufficiently sharp upper and lower bounds for the rotation function provide an estimate for the characteristic exponent λ. Suppose that

$$\tilde{m}\leqslant m\leqslant M\leqslant\tilde{M},\tag{2.52}$$

where the numbers $\tilde{m}$, $\tilde{M}$ are expressed explicitly in terms of $\mathbf{H}(t)$, so that they are readily available. If there is no number $n\pi$ between $\tilde{m}$ and $\tilde{M}$, where n is an integer (i. e., $(n-1)\pi < \tilde{m} \leqslant \tilde{M} < n\pi$ for some integer n), then by Theorem I of §1.9 $\lambda = 0$ and $\mathbf{H}(t) \in \mathcal{G}_n$. Suppose that for some integer n,

$$(n-1)\pi < \tilde{m} \leqslant n\pi \leqslant \tilde{M} < (n+1)\pi. \tag{2.53}$$

If $\lambda \neq 0$, then by Theorem II of §1.9 there is an integer n (a priori not the same as before) for which

$$(n-1)\pi < m < n\pi < M < (n+1)\pi. \tag{2.54}$$

Here n is the subscript of the instability domain $\mathcal{H}_n$ containing $\mathbf{H}(t)$; in other words, it is the integer in (2.51). By (2.52), the integers n in (2.53) and (2.54) must in fact coincide. Thus it follows from (2.53) that either $\lambda = 0$ or $\lambda > 0$ and $\mathbf{H}(t) \in \mathcal{H}_n$.

It follows from (2.50)—(2.54) that

$$\tau_1 \geqslant \cos(n\pi - \tilde{m}), \quad \tau_2 \geqslant \cos(\tilde{M} - n\pi). \tag{2.55}$$

These inequalities enable us to derive an estimate for the unknown λ. Consider the functions

$$\tau_1(\xi, \eta) = \frac{1+\xi\eta}{\xi+\eta}, \quad \tau_2(\xi, \eta) = \frac{1-\xi\eta}{\xi-\eta},$$

defined in the strip

$$\xi \geqslant 1, \quad -1 \leqslant \eta \leqslant 1.$$

The domains

$$\tau_1(\xi, \eta) \geqslant \sigma_1, \quad \tau_2(\xi, \eta) \geqslant \sigma_2$$

on the (ξ, η)-plane are as shown in Figure 64. Thus it follows from these inequalities that if $\sigma_1 + \sigma_2 > 0$ then

$$\xi \leqslant \xi_0,$$

where ξ_0 is a solution of the equations

$$\tau_1(\xi, \eta) = \sigma_1, \quad \tau_2(\xi, \eta) = \sigma_2.$$

Therefore, setting $\xi = \operatorname{ch} \lambda T$, $\eta = \cos\beta$ and carrying out a few calculations, we deduce from (2.55) and from the inequality

$$\cos(n\pi + \tilde{m}) + \cos(\tilde{M} - n\pi) > 0 \tag{2.56}$$

that

$$\operatorname{ch}\lambda T \leqslant \operatorname{ch}\lambda_0 T = \frac{1+\cos(\tilde{m}+\tilde{M})}{\cos(n\pi-\tilde{m})+\cos(\tilde{M}-n\pi)}. \tag{2.57}$$

Thus the bounds $\tilde{m}$, $\tilde{M}$ in (2.52), bounding the rotation function above and below, yield an estimate (2.57) for λ.

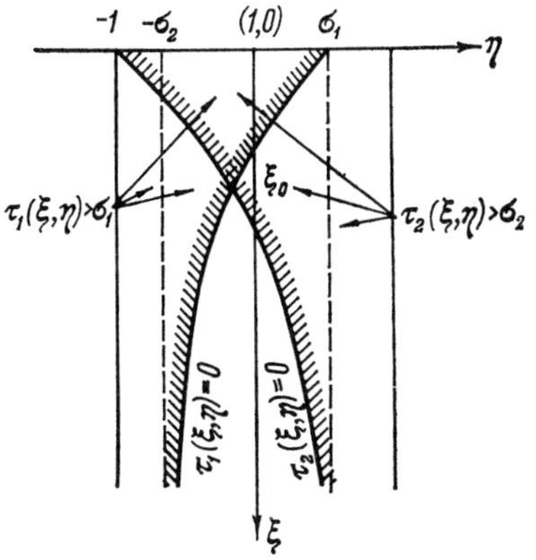

FIGURE 64.

Note that condition (2.56), which guarantees the validity of (2.57), is necessarily true if $\tilde{m}$, $\tilde{M}$ are sufficiently close to the actual values of m, M. Indeed, if we set $\tilde{m}=m$, $\tilde{M}=M$, the left-hand side of (2.56) is

$$\tau_1+\tau_2=\frac{2\operatorname{ch}\lambda T\cdot\sin^2\beta}{\operatorname{ch}^2\lambda T-\cos^2\beta}>0.$$

A simple estimate for the rotation function was established in §2.1. It has the form (2.4) and implies that we can take

$$\tilde{m}=\int_0^T h_{\min}(t)\,dt,\quad \tilde{M}=\int_0^T h_{\max}(t)\,dt, \tag{2.58}$$

where $h_{\min}\leqslant h_{\max}$ are the eigenvalues of the Hamiltonian $\mathbf{H}(t)$, i. e., the roots of the equation $\det[\mathbf{H}(t)-h\mathbf{I}_2]=0$. Instead of this equation, one can also consider equation (2.6). Indeed, this is equivalent to substituting $\mathbf{x}_1=\mathbf{S}\mathbf{x}$ in the original equation, and involves no change in the characteristic exponents (for more details, see §2.1). We formulate our result as

Test 1a. Let $\mathbf{H}_0$ be an arbitrary constant symmetric matrix such that

$$\mathbf{H}_0>0,\quad \det\mathbf{H}_0=1.$$

Define numbers $\tilde{m}$ and $\tilde{M}$ by formulas (2.58), where $h_{\min}(t)$, $h_{\max}(t)$ are the minimum and maximum roots of the equation

$$\det[\mathbf{H}(t)-h\mathbf{H}_0]=0.$$

Suppose that inequalities (2.53) and (2.56) hold for some $n=0, \pm 1, \pm 2, \ldots$ *Then the exponent* $\lambda \geqslant 0$ *of exponential growth for the solutions of equation (1.2) satisfies inequality (2.57).*

Let us explain the use of this test. Suppose we have a vector equation (1.2) with an "almost constant" Hamiltonian, and that the equation with the corresponding constant Hamiltonian has only bounded solutions. Any such Hamiltonian has the form $h_0\mathbf{H}_0$ (see §1.10), where h_0 is a number and $\mathbf{H}_0$ satisfies the assumptions of the test. We first assume that there is no "resonance," i. e., for some integer n

$$n\pi < h_0 < (n+1)\pi.$$

If $\mathbf{H}(t)$ is sufficiently close to $h_0\mathbf{H}_0$, then also

$$n\pi < \tilde{h}_1 \leqslant \tilde{h}_2 < (n+1)\pi. \tag{2.59}$$

By the test of §2.1, equation (1.2) is also stable (in fact, $\mathbf{H}(t) \in \mathfrak{G}_n$ and $h_0\mathbf{H}_0 \in \mathfrak{G}_n$). Essentially, inequality (2.59) is a measure of how close $\mathbf{H}(t)$ must be to $h_0\mathbf{H}_0$ in order not to lose the stability property.

Now consider the "resonant" case $h_0 = n\pi$ when inequalities (2.59) are not satisfied. If $\mathbf{H}(t)$ is sufficiently close to $h_0\mathbf{H}_0$, inequalities (2.53) and (2.56) will hold. Then we may have the phenomenon of parametric excitation, the solutions of equation (1.2) may be unbounded, but the exponent of exponential growth will always be bounded by (2.57).

Of course, Test 1a may also be applied in other situations, e. g., when $\mathbf{H}(t)$ is "substantially" different from any Hamiltonian $h_0\mathbf{H}_0$. If

$$\mathbf{H}_{\text{av}} = \frac{1}{T}\int_0^T \mathbf{H}(t)\,dt > 0,$$

we can take $\mathbf{H}_0 = (\det \mathbf{H}_{\text{av}})^{-1/2}\,\mathbf{H}_{\text{av}}$.

To elucidate the use of Test 1a in deriving stability conditions for the trivial solution of a noncanonical equation (1.1), we set

$$\mu_0 = \frac{1}{2T}\int_0^T \operatorname{Tr} A(t)\,dt.$$

If $\mu_0 > 0$, equation (1.1) has solutions which are unbounded as $t \to +\infty$.

Let $\mu_0 \leqslant 0$. The solutions of equations (1.1) and (1.2) obey the identity $\mathbf{z}(t) = \exp[\mu_0 t + \varphi(t)]\,\mathbf{x}(t)$, where $\varphi(t+T) = \varphi(t)$. The Hamiltonian $\mathbf{H}(t)$ of (1.2) is expressed in terms of $\mathbf{A}(t)$ as shown in §1.1.

It follows from the foregoing account that if there exists $n = 0, \pm 1, \pm 2, \ldots$ such that either $n\pi \leqslant \tilde{m} \leqslant \tilde{M} \leqslant (n+1)\pi$ or inequalities (2.53), (2.56) hold and $\operatorname{ch}\lambda_0 T < \operatorname{ch}\mu_0 T$, where $\operatorname{ch}\lambda_0 T$ is defined by (2.57), then any solution of equation (1.1) tends exponentially to zero as $t \to \infty$.

Before using the estimate (2.57), it is sometimes useful to substitute $\mathbf{x} = \mathbf{P}(t)\mathbf{x}_1$, where $\mathbf{P}(t)$ is a differentiable T-periodic matrix-function whose determinant is identically equal to unity. This substitution, applied to a canonical system (1.2), again yields a canonical system. It clearly does not alter the value of λ. Using one of the estimates for λ for the transformed system, we thereby obtain an estimate for the original system. It is important here that, as has been pointed out by Burdina /128a/, there always exists a matrix $\mathbf{P}(t)$ (even a triangular one) such that in the transformed system $\operatorname{ch}\lambda T=\operatorname{ch}\lambda_0 T$, where $\operatorname{ch}\lambda_0 T$ is defined by (2.39).

The estimates (2.57) and (II. 4.16) provide upper bounds for λ. We now proceed to derive two-sided estimates for λ; these will be more cumbersome.

As in §2.1, consider the vector function $\mathbf{y}=\mathbf{X}^{\tau}(t)\mathbf{c}$, where $\mathbf{c}=\begin{pmatrix}\cos\chi\\ \sin\chi\end{pmatrix}$ is a constant vector. Let $\varphi_y=\varphi_\tau(\chi)$ be the rotation of the vector $\mathbf{y}$ in time T. It was shown in §2.1 that

$$M=\max\varphi(\chi)=\max[-\varphi_y(\chi)],\quad m=\min\varphi(\chi)=\min[-\varphi_y(\chi)] \qquad (2.60)$$

(maximum and minimum over χ, $0\leqslant\chi\leqslant\pi$).

We shall use formulas (2.60) to derive two-sided estimates for the unknown numbers m and M.

In §2.1, on the assumption that $\mathbf{H}(t)\geqslant 0$ (i. e., $(\mathbf{H}(t)\mathbf{c},\mathbf{c})\geqslant 0$ for $0\leqslant t\leqslant T$ and all vectors $\mathbf{c}$), we established a two-sided estimate (2.13) for the rotation φ_y, in which the functions $\Psi^{\pm}(\chi)$ were defined by (2.15) and the numbers $A^{\pm}, B^{\pm}, C^{\pm}$ entering into (2.15) were defined by (2.14).

As in §2.1, let us denote

$$\max\Psi^{\pm}(\chi)=M^{\pm},\quad \min\Psi^{\pm}(\chi)=m^{\pm}.$$

Let $\mathbf{H}(t)\geqslant 0$. Then we have the formulas (2.16) of §2.1 for $M^{\pm}$, $m^{\pm}$. Moreover, by (2.60), (2.13), we have

$$m^-\leqslant m\leqslant m^+,\quad M^-\leqslant M\leqslant M^+.$$

Using these inequalities and making appropriate assumptions, we deduce from (2.50) and (2.51) two-sided estimates for $\operatorname{ch}\lambda T$.

The final result is as follows.

Theorem I. Let $\mathbf{H}(t)\geqslant 0$ *and let* $\Phi_1(t)$, $\Phi_2(t)$ *be positive functions satisfying inequalities (a) of §2.1 for all* $\mathbf{c}=\mathbf{x}(0)$. *(For example, these functions may be defined by (2.12), (2.11).) Define numbers* $m^{\pm}$, $M^{\pm}$ *by formulas (2.11), (2.14), (2.16).*

1) *If there exists* $n=0, 1, 2, \ldots$ *such that*

$$n\pi<m^-\leqslant M^-<(n+1)\pi,$$

then all the solutions of equations (1.2) are bounded for $-\infty<t<+\infty$ *and* $\mathbf{H}(t)\in\mathfrak{G}_n$.

2) *If*

$$\left.\begin{aligned}&(n-1)\pi<m^-\leqslant n\pi\leqslant M^+<(n+1)\pi,\\ &\cos(M^+-n\pi)+\cos(n\pi-m^-)>0,\end{aligned}\right\} \qquad (2.61)$$

then the exponent λ *of exponential growth for solutions of equation (1.2) satisfies the estimate*

$$\operatorname{ch}\lambda T \leqslant \frac{1+\cos(M^{+}-m^{-})}{\cos(M^{+}-n\pi)+\cos(n\pi-m^{-})}. \tag{2.62}$$

3) *If*

$$(n-1)\pi < m^{+} \leqslant n\pi \leqslant M^{-} < (n+1)\pi, \tag{2.63}$$

then system (1.2) has unbounded solutions $(\lambda > 0)$ *and*

$$\operatorname{ch}\lambda T \geqslant \frac{1+\cos(M^{-}+m^{+})}{\cos(M^{-}-n\pi)+\cos(n\pi-m^{+})}. \tag{2.64}$$

The condition $\cos(M^{-}-n\pi)+\cos(n\pi-m^{+})>0$ *must hold automatically. If condition (2.63) holds, then* $\mathbf{H}(t)\in\mathcal{H}_n$.

It is easy to state an analogous theorem for the case that $(\mathbf{H}(t)\mathbf{c},\ \mathbf{c})\leqslant 0$, $0\leqslant t\leqslant T$.

Remark. If the Hamiltonian $\mathbf{H}(t)$ in (1.2) is bounded but it is not true that $\mathbf{H}(t)\geqslant 0$, the equation may nevertheless be reduced to a similar equation with positive Hamiltonian. Indeed, substituting

$$\mathbf{x}=\exp(\vartheta\mathbf{J})\,\mathbf{x}_1, \quad \text{where} \quad \vartheta=\frac{p\pi}{T}t$$

and p is an integer, into equation (1.2), we obtain the equation $\mathbf{J}\dot{\mathbf{x}}_1=\mathbf{H}_1(t)\,\mathbf{x}_1$ with Hamiltonian

$$\mathbf{H}_1(t)=\frac{p\pi}{T}\mathbf{I}_2+e^{-\vartheta\mathbf{J}}\mathbf{H}(t)\,e^{\vartheta\mathbf{J}}.$$

Clearly, $\mathbf{H}_1(t)^*=\mathbf{H}_1(t)$ so that the new equation is also canonical. In addition, $\mathbf{H}_1(t+T)=\mathbf{H}_1(t)$, and if p is sufficiently large then

$$(\mathbf{H}_1(t)\mathbf{c},\ \mathbf{c})=\frac{p\pi}{T}+(\mathbf{H}(t)\mathbf{c},\ \mathbf{c})\geqslant 0$$

for any $0\leqslant t\leqslant T$ and any vector $\mathbf{c}$, $|\mathbf{c}|=1$.

Theorem III of §2.2 furnished arbitrarily sharp two-sided estimates for the numbers m and M. In fact, the main result of that theorem was a procedure for constructing numbers $m_k^{\pm}$, $M_k^{\pm}$ such that

$$m_k^{-}\leqslant m\leqslant m_k^{+},\quad M_k^{-}\leqslant M\leqslant M_k^{+},\quad m_k^{\pm}\to m,\quad M_k^{\pm}\to M \quad \text{as} \quad k\to\infty.$$

It follows from (2.50), (2.51) that

$$\operatorname{ch}\lambda T=\frac{1+\cos(M-m)}{\cos(M-n\pi)+\cos(n\pi-m)},$$

and as we have already stated the denominator in this expression is positive. Consequently, estimates (2.62), (2.64) with $M^{\pm}=M_k^{\pm}$, $m^{\pm}=m_k^{\pm}$ may be made arbitrarily sharp by increasing k.

We summarize the result as a theorem complementing Theorem III of §2.2.

Theorem II. Let $\mathbf{H}(t) \geqslant 0$ *and let the numbers* $m_k^{\pm}$, $M_k^{\pm}$ *be defined as in Theorem III of § 2.2. Then parts 1, 2 and 3 of Theorem I remain valid with* $m^{\pm}$, $M^{\pm}$ *replaced by* $m_k^{\pm}$, $M_k^{\pm}$ *for any* $k \geqslant 0$. *If* $\lambda > 0$, *then for sufficiently large* k *there exists* n *such that inequalities (2.61) – (2.64) hold, with* $m^{\pm}$, $M^{\pm}$ *replaced by* $m_k^{\pm}$, $M_k^{\pm}$. *In addition, the right-hand sides of the new inequalities (2.62), (2.63) tend to* $\operatorname{ch} \lambda T$ *as* $k \to \infty$.

In applications, one is often faced with the problem of calculating the exponent of exponential growth λ together with an estimate of the error made in the approximation. Theorem II provides the answer. The estimate is given by (2.62), (2.64) with $M^{\pm} = M_k^{\pm}$, $m^{\pm} = m_k^{\pm}$.

2.4. Comparison theorem. Directional convexity of stability and instability domains. Example

Consider equation (1.2) with a Hamiltonian $\mathbf{H}_\lambda(t)$ that depends on a parameter λ. We shall assume that $\partial \mathbf{H}_\lambda / \partial \lambda$ exists for almost all t and is summable on $[0, T]$.

Fix $\mathbf{x}(0) = \mathbf{c}$. The solution of the equation is $\mathbf{x}_\lambda(t) = \mathbf{X}(t)\mathbf{c}$. As before, let $\vartheta_\lambda(t) = \operatorname{Arg} \mathbf{x}_\lambda(t)$ be a continuous branch of the argument of the vector $\mathbf{x}_\lambda(t)$. We need a formula for $\partial \vartheta_\lambda / \partial \lambda$. We have

$$\vartheta_\lambda(t) = \operatorname{arctg} \frac{x_2}{x_1}, \quad \frac{\partial \vartheta_\lambda}{\partial \lambda} = \frac{\frac{\partial x_2}{\partial \lambda} x_1 - \frac{\partial x_1}{\partial \lambda} x_2}{x_1^2 + x_2^2} = \frac{\det \left\| \mathbf{x}_\lambda, \frac{\partial \mathbf{x}_\lambda}{\partial \lambda} \right\|}{(\mathbf{x}_\lambda, \mathbf{x}_\lambda)}.$$

Write equation (1.2) as

$$\frac{d\mathbf{x}_\lambda}{dt} = \mathbf{A}_\lambda(t) \mathbf{x}_\lambda,$$

where $\mathbf{A}_\lambda(t) = \mathbf{J}^{-1} \mathbf{H}_\lambda(t)$. Differentiating with respect to λ, we get

$$\frac{d}{dt} \frac{\partial \mathbf{x}_\lambda}{\partial \lambda} = \mathbf{A}_\lambda(t) \frac{\partial \mathbf{x}_\lambda}{\partial \lambda} + \frac{\partial \mathbf{A}_\lambda}{\partial \lambda} \mathbf{x}_\lambda,$$

whence, since $\partial \mathbf{x}_\lambda / \partial \lambda |_{t=0} = \mathbf{0}$, we obtain by formula (II. 1.17)

$$\frac{\partial \mathbf{x}_\lambda}{\partial \lambda} = \mathbf{X}(t) \int_0^t \mathbf{X}(s)^{-1} \frac{\partial \mathbf{A}_\lambda}{\partial \lambda} \mathbf{x}_\lambda(s)\, ds,$$

$$\Delta(t) = \det \left\| \mathbf{x}_\lambda, \frac{\partial \mathbf{x}_\lambda}{\partial \lambda} \right\| = \det \left[\mathbf{X}(t) \left\| \mathbf{c}, \int_0^t \mathbf{X}(s)^{-1} \frac{\partial \mathbf{A}_\lambda}{\partial \lambda} \mathbf{x}_\lambda(s)\, ds \right\| \right].$$

Since $\det \mathbf{X}(t) \equiv 1$,

$$\Delta(t) = \det\left\| \mathbf{c},\ \int_0^t \mathbf{X}(s)^{-1}\frac{\partial \mathbf{A}_\lambda}{\partial\lambda}\mathbf{x}_\lambda(s)\,ds \right\| = \int_0^t \det\left\| \mathbf{c},\ \mathbf{X}(s)^{-1}\frac{\partial \mathbf{A}_\lambda}{\partial\lambda}\mathbf{x}_\lambda(s) \right\| ds =$$

$$= \int_0^t \det\left\| \mathbf{X}(s)\,\mathbf{c},\ \frac{\partial \mathbf{A}_\lambda}{\partial\lambda}\mathbf{x}_\lambda(s) \right\| ds = \int_0^t \det\left\| \mathbf{x}_\lambda,\ \tilde{\mathbf{J}}^{-1}\frac{\partial \mathbf{H}_\lambda}{\partial\lambda}\mathbf{x}_\lambda \right\| ds.$$

It follows from the formula $\det\|\mathbf{a},\ \tilde{\mathbf{J}}^{-1}\mathbf{b}\| = (\mathbf{a},\ \mathbf{b})$ that

$$\Delta(t) = \int_0^t \left(\mathbf{x}_\lambda,\ \frac{\partial \mathbf{H}_\lambda}{\partial\lambda}\mathbf{x}_\lambda\right) ds. \tag{2.65}$$

Inserting $\Delta(t)$ into the expression for $\partial\vartheta_\lambda/\partial\lambda$, we obtain

$$\frac{\partial\vartheta_\lambda}{\partial\lambda} = \frac{1}{(\mathbf{x}_\lambda(t),\ \mathbf{x}_\lambda(t))}\int_0^t \left(\frac{\partial \mathbf{H}_\lambda}{\partial\lambda}\mathbf{x}_\lambda(s),\ \mathbf{x}_\lambda(s)\right) ds. \tag{2.66}$$

This formula is of great importance for the sequel.

Let $\mathbf{H}_1(t) \geqslant \mathbf{H}_2(t)$* be two Hamiltonians and $\mathbf{x}_1(t), \mathbf{x}_2(t)$ solutions of the equations with Hamiltonians $\mathbf{H}_1(t),\ \mathbf{H}_2(t)$,

$$\mathbf{x}_1(0) = \mathbf{x}_2(0), \quad \vartheta_j(t) = \operatorname{Arg}\mathbf{x}_j(t).$$

Let $\mathbf{H}_\lambda = \lambda\mathbf{H}_1 + (1-\lambda)\mathbf{H}_2$. Then $\partial\mathbf{H}_\lambda/\partial\lambda = \mathbf{H}_1 - \mathbf{H}_2 \geqslant 0$. It follows from formula (2.66) that $\partial\vartheta_\lambda/\partial\lambda \geqslant 0$, i. e., $\vartheta_1(t) \geqslant \vartheta_2(t)$. If there is a set of positive measure in the interval $[0,\ t_0]$ on which $\mathbf{H}_1(t) > \mathbf{H}_2(t)$ (if $\mathbf{H}_1(t)$ and $\mathbf{H}_2(t)$ are continuous, it suffices to demand that $\mathbf{H}_1(t_0) > \mathbf{H}_2(t_0)$), then (2.66) implies that $\vartheta_1(t) > \vartheta_2(t)$ for $t \geqslant t_0$.

We have proved the following

Theorem I (Comparison theorem). Let $\mathbf{x}_1(t)$ *and* $\mathbf{x}_2(t)$ *be solutions of equation (1.2) with Hamiltonians* $\mathbf{H}_1(t)$ *and* $\mathbf{H}_2(t)$*, respectively,* $\mathbf{x}_1(0) = \mathbf{x}_2(0) \neq 0$. *Let* $\operatorname{Arg}\mathbf{x}_j(t)$ *be a continuous branch of the argument of* $\mathbf{x}_j(t)$, $\operatorname{Arg}\mathbf{x}_1(0) = \operatorname{Arg}\mathbf{x}_2(0)$. *If* $\mathbf{H}_1(t) \geqslant \mathbf{H}_2(t)$*, then* $\mathbf{x}_1(t)$ *rotates "ahead" of* $\mathbf{x}_2(t)$*, i. e.,* $\operatorname{Arg}\mathbf{x}_1(t) \geqslant \operatorname{Arg}\mathbf{x}_2(t)$ *If there is a set of positive measure in the interval* $(0,\ t_0)$ *on which* $\mathbf{H}_1(t) > \mathbf{H}_2(t)$*, then* $\operatorname{Arg}\mathbf{x}_1(t) > \operatorname{Arg}\mathbf{x}_2(t)$ *for* $t \geqslant t_0$.

It is readily checked that as applied to the equation $\ddot{y} + p(t)y = 0$ Theorem I yields the familiar comparison theorem for zeros of solutions /87/.

The theorem carries over to arbitrary second-order systems. Namely, given two systems (1.1) with matrices $\mathbf{A}_1(t) = \|a_{ij}^{(1)}(t)\|$, $\mathbf{A}_2(t) = \|a_{ij}^{(2)}(t)\|$, the comparison theorem is valid for their solution-vectors, with the Hamiltonians

$$\mathbf{H}_k(t) = \left\| \begin{matrix} a_{21}^{(k)}(t) & \frac{1}{2}[a_{22}^{(k)}(t) - a_{11}^{(k)}(t)] \\ \frac{1}{2}[a_{22}^{(k)}(t) - a_{11}^{(k)}(t)] & -a_{12}^{(k)}(t) \end{matrix} \right\| \qquad (k = 1,\ 2)$$

(see reduction of system (1.1) to system (1.2) at the beginning of §1.1).

* We remind the reader that $\mathbf{H}_1 > \mathbf{H}_2$ is an abbreviated way of saying that $(\mathbf{H}_1\mathbf{c},\ \mathbf{c}) > (\mathbf{H}_2\mathbf{c},\ \mathbf{c})$ for all vectors $\mathbf{c}$, almost everywhere on $[0,\ T]$. The notation $\mathbf{H}_1 \geqslant \mathbf{H}_2$ means that $(\mathbf{H}_1\mathbf{c},\ \mathbf{c}) \geqslant (\mathbf{H}_2\mathbf{c},\ \mathbf{c})$ for all $\mathbf{c}$ and almost everywhere on $[0,\ T]$. We shall also write $\mathbf{H}_1 \geqq \mathbf{H}_2$ to mean that $(\mathbf{H}_1\mathbf{c},\ \mathbf{c}) \geqslant (\mathbf{H}_2\mathbf{c},\ \mathbf{c})$ for all $\mathbf{c}$ almost everywhere on $[0,\ T]$, but $(\mathbf{H}_1\mathbf{c}_0,\ \mathbf{c}_0) = (\mathbf{H}_2\mathbf{c}_0,\ \mathbf{c}_0)$ for some t and $\mathbf{c}_0 \neq 0$. Necessary and sufficient conditions for either of the inequalities $\mathbf{H}_1 > \mathbf{H}_2$, $\mathbf{H}_1 \geqslant \mathbf{H}_2$, $\mathbf{H}_1 \geqq \mathbf{H}_2$ are well known (see § I.1.14).

The comparison theorem does not assume that the coefficients are periodic.

Reasoning similar to that used for (2.66) yields the following formula for the vector $\mathbf{y}_\lambda(t) = \mathbf{X}^\tau(t)\mathbf{c}$:

$$\frac{\partial \operatorname{Arg} \mathbf{y}_\lambda(t)}{\partial \lambda} = -\frac{1}{(\mathbf{y}_\lambda(t), \mathbf{y}_\lambda(t))} \int_0^t \left(\frac{\partial \mathbf{H}_\lambda(s)}{\partial \lambda} \mathbf{z}_\lambda(s, t), \mathbf{z}_\lambda(s, t) \right) ds, \qquad (2.67)$$

where $\mathbf{z}_\lambda(s, t) = \mathbf{X}(s)\check{\mathbf{J}}\mathbf{X}(t)^\tau \mathbf{c}$. It follows from this formula that an analog of the comparison theorem is valid for the vectors $\mathbf{y}_\lambda = \mathbf{X}^\tau \mathbf{c}$. (In that case though, increase of the Hamiltonian implies a decrease in the rotation.)

The next theorem states that the stability domains $\mathfrak{G}_n$ and instability domains $\mathcal{H}_n$ are directionally convex (for the definition, see § III. 6.3).

Theorem II. Let $\mathbf{H}_1(t) \leqslant \mathbf{H}(t) \leqslant \mathbf{H}_2(t)$, where the Hamiltonians $\mathbf{H}_1(t)$ and $\mathbf{H}_2(t)$ belong to the same stability domain $\mathfrak{G}_n$ or instability domain $\mathcal{H}_n$. Then $\mathbf{H}(t)$ also belongs to the same stability (instability) domain $\mathfrak{G}_n$ ($\mathcal{H}_n$).

Proof. Let $\psi_1(\chi)$, $\psi(\chi)$, $\psi_2(\chi)$ be rotation functions for equations (1.3) with Hamiltonians $\mathbf{H}_1$, $\mathbf{H}$, $\mathbf{H}_2$, respectively (see §1.9). By the comparison theorem, $\psi_1(\chi) \leqslant \psi(\chi) \leqslant \psi_2(\chi)$.

Let $\mathbf{H}_1 \in \mathfrak{G}_n$, $\mathbf{H}_2 \in \mathfrak{G}_n$. By Theorem II of §1.9 (see also Figure 61), we have $n\pi < \psi_1(\chi) \leqslant \psi_2(\chi) < (n+1)\pi$. Therefore $n\pi < \psi(\chi) < (n+1)\pi$ and the same theorem implies that $\mathbf{H} \in \mathfrak{G}_n$.

Let $\mathbf{H}_1 \in \mathcal{H}_n$, $\mathbf{H}_2 \in \mathcal{H}_n$. By the same theorem, there exist χ', χ'' such that $\psi_1(\chi') > n\pi$, $\psi_2(\chi'') < n\pi$. Then $\psi(\chi') > n\pi$, $\psi(\chi'') < n\pi$ and by the same theorem $\mathbf{H} \in \mathcal{H}_n$. Q. E. D.

The converse of Theorem II is obvious: if $\mathbf{H}(t) \in \mathfrak{G}_n$ ($\mathbf{H}(t) \in \mathcal{H}_n$), then the domain $\mathfrak{G}_n$ ($\mathcal{H}_n$) contains infinitely many Hamiltonians $\mathbf{H}_1(t)$, $\mathbf{H}_2(t)$ such that $\mathbf{H}_1(t) < \mathbf{H}(t) < \mathbf{H}_2(t)$. This follows at once from the fact that $\mathfrak{G}_n$ and $\mathcal{H}_n$ are open sets.

Corollary. Each of the stability (instability) domains $\mathfrak{G}_n$ ($\mathcal{H}_n$) is a directionally wide set (for the definition, see § III. 6.1).

In other words, if equation (1.2) with a Hamiltonian $\mathbf{H}(t, s) = s\mathbf{H}_2(t) + (1-s)\mathbf{H}_1(t)$ $\mathbf{H}_2(t) \geqslant \mathbf{H}_1(t)$, is strongly stable (strongly unstable) for any $0 \leqslant s \leqslant 1$, then the equation is strongly stable (strongly unstable) with any Hamiltonian $\mathbf{H}(t)$ such that $\mathbf{H}_1(t) \leqslant \mathbf{H}(t) \leqslant \mathbf{H}_2(t)$.

Indeed, the equation with Hamiltonian $\mathbf{H}(t, s)$ $(0 \leqslant s \leqslant 1)$ is strongly stable if and only if the "segment" $\mathbf{H}(t, s)$ $(0 \leqslant s \leqslant 1)$ does not intersect the common boundary of $\mathfrak{G}$ and $\mathcal{H}$ Consequently, $\mathbf{H}(t, s)$ belongs to the same stability (instability) domain $\mathfrak{G}_n$ ($\mathcal{H}_n$) for all $0 \leqslant s \leqslant 1$. Thus the relevant domain also contains the "endpoints" of the segment — the Hamiltonians $\mathbf{H}_1(t)$ and $\mathbf{H}_2(t)$.

Recall that if an equation (1.2) with Hamiltonian $\mathbf{H}(t, s)$ is strongly stable, its multipliers $\rho(s)$ and $\rho(s)^{-1}$ are different from ± 1 and $|\rho(s)| = 1$ for $0 \leqslant s \leqslant 1$; strong instability implies that $|\rho(s)| > 1$ for $0 \leqslant s \leqslant 1$.

We now consider a geometric interpretation of the condition $\mathbf{H}_1 \leqslant \mathbf{H}_2$.

Consider the three-dimensional space $\mathfrak{R}^3$ of real matrices $\mathbf{K}$ with $\operatorname{Tr} \mathbf{K} = 0$, introduced in §1.4.

Let $\mathbf{H}=\mathbf{H}_1-\mathbf{H}_2$ and $\tilde{\mathbf{J}}^{-1}\mathbf{H}\in\tilde{\mathfrak{R}}^3$. We saw previously that the geometric meaning of the inequality $\mathbf{H}>0$ is $\tilde{\mathbf{J}}^{-1}\mathbf{H}\in\tilde{\mathscr{O}}^+$, while $\mathbf{H}<0$ means that $\tilde{\mathbf{J}}^{-1}\mathbf{H}\in\tilde{\mathscr{O}}^-$. Similarly, one deduces that $\mathbf{H}\geqq 0$ is equivalent to the condition $\tilde{\mathbf{J}}^{-1}\mathbf{H}\in\tilde{\Pi}^+$, and $\mathbf{H}\leqq 0$ is equivalent to $\tilde{\mathbf{J}}^{-1}\mathbf{H}\in\tilde{\Pi}^-$. (Recall that $\tilde{\Pi}^+,\tilde{\Pi}^-$ are the upper and lower sheets of the cone $\tilde{\Pi}=\{z^2=x^2+y^2\}\subset\tilde{\mathfrak{R}}^3$ without the vertex, and the symbol $\geqq$ means that $(\mathbf{Hc},\,\mathbf{c})\geqslant 0$ and $(\mathbf{Hc}_0,\,\mathbf{c}_0)=0$ for some $\mathbf{c}_0\neq\mathbf{0}$.) It follows that $\mathbf{H}>\mathbf{H}_0$ if and only if $\mathbf{J}^{-1}\mathbf{H}$ lies inside the upper sheet of the cone with center at $\mathbf{H}_0$, axis parallel to the z-axis, and generators inclined at 45° to the axis of the cone.

Similarly, one sees that $\mathbf{H}<\mathbf{H}_0$ if and only if $\tilde{\mathbf{J}}^{-1}\mathbf{H}$ lies inside the lower sheet of this cone, $\mathbf{H}\leqq\mathbf{H}_0$ if and only if $\tilde{\mathbf{J}}^{-1}\mathbf{H}$ lies on the lower sheet, and $\mathbf{H}\geqq\mathbf{H}_0$ if and only if $\tilde{\mathbf{J}}^{-1}\mathbf{H}$ lies on the upper sheet. Matrices $\mathbf{H}$ outside the cone are incomparable with $\mathbf{H}_0$ (Figure 65).

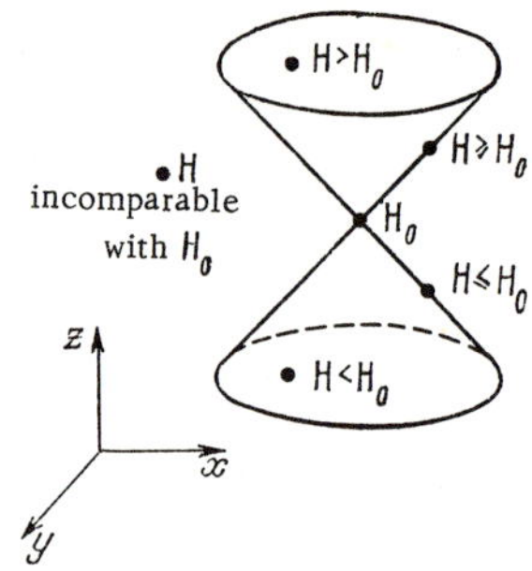

FIGURE 65.

Certain conclusions may be drawn by comparing $\mathbf{H}(t)$ with constant Hamiltonians $\mathbf{C}_1$ and $\mathbf{C}_2$ and using Theorem II. In fact, if $\tilde{\mathbf{J}}^{-1}\mathbf{H}(t)$ is a trajectory in $\tilde{\mathfrak{R}}^3$, in the region between the upper and lower sheets of cones whose vertices $(\tilde{\mathbf{J}}^{-1}\mathbf{C}_1)$ and $(\tilde{\mathbf{J}}^{-1}\mathbf{C}_2)$ lie in the instability domain $\tilde{\mathscr{H}}$ (that is to say, $\det\mathbf{C}_{1,2}<0$), then $\mathbf{H}(t)$ belongs to the zeroth instability domain $\mathscr{H}_0\subset\mathfrak{L}^3$. Indeed, $\tilde{\mathbf{J}}^{-1}\mathbf{C}\in\tilde{\mathscr{H}}_0$ implies that $\mathbf{C}$ belongs to the zeroth instability domain.

Similarly, if the vertices $\tilde{\mathbf{J}}^{-1}\mathbf{C}_1$ and $\tilde{\mathbf{J}}^{-1}\mathbf{C}_2$ lie in the domain $\tilde{\mathscr{O}}_n\subset\tilde{\mathfrak{R}}^3$ between neighboring sheets of the hyperboloids

$$z^2-x^2-y^2=\frac{n^2\pi^2}{T^2}\quad\text{and}\quad z^2-x^2-y^2=\frac{(n+1)^2\pi^2}{T^2}$$

(if $n=0$ or $n=-1$, we have a cone instead of a two-sheeted hyperboloid), and also $z>0$ if $n\geqslant 0$, $z<0$ if $n<0$, then $\mathbf{H}(t)\in\mathscr{O}_n\subset\mathfrak{L}^3$. To prove this it suffices to show that $\tilde{\mathbf{J}}^{-1}\mathbf{C}\in\tilde{\mathscr{O}}_n$ if and only if $\mathbf{C}\in\mathscr{O}_n\subset\mathfrak{L}^3$.

Indeed, $\tilde{\mathbf{J}}^{-1}\mathbf{C}\in\tilde{\mathscr{O}}_n$ implies that

$$\left.\begin{array}{lll}\text{if } n\geqslant 0, & \mathbf{C}>0, \text{ then} & \frac{n\pi}{T}<\sqrt{\det\mathbf{C}}<\frac{(n+1)\pi}{T},\\ \text{if } n<0, & \mathbf{C}<0, \text{ then} & -\frac{(n+1)\pi}{T}<-\sqrt{\det\mathbf{C}}<-\frac{n\pi}{T},\end{array}\right\}\qquad(2.68)$$

and this is the same as conditions VII, VIII of §1.10. Similarly, one sees that the upper sheet of the hyperboloid $z^2-x^2-y^2=\frac{n^2\pi^2}{T^2}$ in $\tilde{\mathfrak{R}}^3$ is "filled" by the matrices $\tilde{\mathbf{J}}^{-1}\mathbf{C}$ with Hamiltonians $\mathbf{C}\in\Pi_n^{\bullet\bullet}$, $n>0$, while its lower sheet is "filled" by the matrices $\tilde{\mathbf{J}}^{-1}\mathbf{C}$ with $\mathbf{C}\in\Pi_n^{\bullet\bullet}$, $n<0$. The upper sheet of the cone $z^2-x^2-y^2=0$ (without vertex) is filled (in the same sense) by the Hamiltonians $\mathbf{C}\in\Pi_0^{\bullet+}$, the lower sheet by the Hamiltonians $\mathbf{C}\in\Pi_0^{\bullet-}$. The zero matrix $\mathbf{C}=\mathbf{0}$ lies, of course, in $\Pi_0^{\bullet\bullet}$. Figure 66 shows the cone $x^2+y^2=z^2$ and hyperboloids $z^2-x^2-y^2=\frac{n^2\pi^2}{T^2}$ in $\tilde{\mathfrak{R}}^3$ and indicates the various domains.

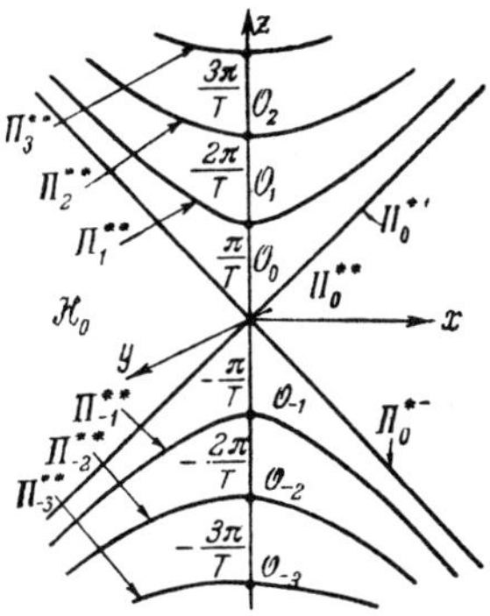

FIGURE 66.

The comparison of $\mathbf{H}(t)$ with constant matrices poses, however, quite difficult analytical problems. In order to facilitate the analytical treatment, we shall apply a certain transformation.

Let us denote

$$\begin{aligned}\beta_{av}&=\frac{1}{T}\int_0^T\beta(t)\,dt,\\ \delta(t)&=\int_0^t[\beta(t_1)-\beta_{av}]\,dt_1.\end{aligned} \tag{2.69}$$

$\delta(t)$ is clearly T-periodic. In equation (1.2), with $\mathbf{x}=\binom{p}{q}$, we substitute

$$p=e^{-\delta t}p_1,\quad q=e^{\delta t}q_1. \tag{2.70}$$

The result is the system

$$\dot{p}_1=-\beta_{av}p_1-\gamma_1(t)\,q_1,\quad \dot{q}_1=\alpha_1(t)\,p_1+\beta_{av}q_1, \tag{2.71}$$

where

$$\alpha_1(t) = \alpha(t)\, e^{-2\delta(t)}, \quad \gamma_1(t) = \gamma(t)\, e^{2\delta(t)}. \tag{2.72}$$

The new Hamiltonian $\mathbf{H}(t)$ is

$$\mathbf{H}_1(t) = \begin{Vmatrix} \alpha_1(t) & \beta_{av} \\ \beta_{av} & \gamma_1(t) \end{Vmatrix}. \tag{2.73}$$

This transformation does not affect the instability and stability domains, etc., since the vectors $\mathbf{x} = \begin{pmatrix} p \\ q \end{pmatrix}$ and $\mathbf{x}_1 = \begin{pmatrix} p_1 \\ q_1 \end{pmatrix}$ have the same rotations $\varphi_\mathbf{x}$ and $\varphi_{\mathbf{x}_1}$. Indeed, by (2.70), $\mathbf{x} = \mathbf{P}(t)\,\mathbf{x}_1$, where

$$\mathbf{P}(t) = \begin{Vmatrix} e^{\delta(t)} & 0 \\ 0 & e^{-\delta(t)} \end{Vmatrix}.$$

The curve

$$\mathbf{P}_s(t) = \begin{Vmatrix} e^{s\delta(t)} & 0 \\ 0 & e^{-s\delta(t)} \end{Vmatrix} \qquad (0 \leqslant s \leqslant 1)$$

obviously joins $\mathbf{P}(t)$ to $\mathbf{I}_2$. Therefore, by the theorem of §1.7, $\mathbf{P}(t) \in \Omega_0$, $\varphi_\mathbf{x} = \varphi_{\mathbf{x}_1}$. By Theorem III of §1.10, the Hamiltonians $\mathbf{H}(t)$ and $\mathbf{H}_1(t)$ belong to the same domain $\mathscr{O}_n$, $\mathscr{H}_n$, or Π_n. In particular, the original system is stable if and only if system (2.71) (with Hamiltonian (2.73)) is stable.

Let $\mathbf{C}^- \leqslant \mathbf{H}_1(t) \leqslant \mathbf{C}^+$. It is quite easy to determine $\mathbf{C}^\pm$:

$$\mathbf{C}^\pm = \begin{Vmatrix} \alpha_0^\pm & \beta_{av} \\ \beta_{av} & \gamma_0^\pm \end{Vmatrix}, \tag{2.74}$$

where

$$\alpha_0^+ = \max_t \alpha_1(t), \quad \gamma_0^+ = \max_t \gamma_1(t), \qquad \alpha_0^- = \min_t \alpha_1(t), \quad \gamma_0^- = \min_t \gamma_1(t). \tag{2.75}$$

By Theorem II, if $\mathbf{C}^\pm \in \mathscr{O}_n$, then $\mathbf{H}_1(t) \in \mathscr{O}_n$. As we have seen, this will be true if

$$\frac{n^2\pi^2}{T^2} < \det \mathbf{C}^\pm < \frac{(n+1)^2\pi^2}{T^2} \tag{2.76}$$

and $\alpha_0^\pm + \gamma_0^\pm > 0$ $(n > 0)$. If $\gamma_0^\pm + \alpha_0^\pm < 0$, then $\mathbf{C}^\pm \in \mathscr{O}_{-n-1}$ $(n > 0)$. If $\mathbf{H}_1(t) \not\equiv \text{const}$ (otherwise the system is integrable in closed form), the signs $<$ in (2.76) must be replaced by $\leqslant$ and instead of (2.76) we obtain

$$\frac{n^2\pi^2}{T^2} \leqslant \alpha_0^\pm \gamma_0^\pm - \beta_{av}^2 \leqslant \frac{(n+1)^2\pi^2}{T^2}. \tag{2.77}$$

If $\det \mathbf{C}^+ < 0$ and $\det \mathbf{C}^- < 0$, then $\mathbf{C}^\pm \in \mathscr{H}_0$, and so $\mathbf{H}_1(t) \in \mathscr{H}_0$ and $\mathbf{H}(t) \in \mathscr{H}_0$.

We have thus proved the following

Test 2. Define β_{av}, α_0^+, γ_0^+, α_0^-, γ_0^- *by formulas (2.69), (2.72) and (2.75). Assume that the functions* $\alpha_1(t)$ *and* $\gamma_1(t)$ *in (2.72) are not identically constant (otherwise the system is integrable in closed form), inequalities (2.77) hold and moreover* $\alpha_0^+ + \gamma_0^+ > 0$, $\alpha_0^- + \gamma_0^- > 0$. *Then all the solutions of equation (1.2) are bounded and* $\mathbf{H}(t) \in \mathfrak{G}_n\ (n > 0)$. *If inequalities (2.77) hold and* $\alpha_0^+ + \gamma_0^+ < 0$, $\alpha_0^- + \gamma_0^- < 0$, *then* $\mathbf{H}(t) \in \mathfrak{G}_{-n-1}\ (n > 0)$. *If* $\alpha_0^+\gamma_0^+ - \beta_{av}^2 < 0$ *and* $\alpha_0^-\gamma_0^- - \beta_{av}^2 < 0$, *then* $\mathbf{H}(t)$ *belongs to the zeroth instability domain.*

It is clear that this test accounts for all the constant Hamiltonians.

The test of §2.1 also follows from Theorem II. To verify this, we must let the matrices $\mathbf{H}_{1,2}(t)$ have the form $h(t)\mathbf{C}$. Now a matrix of this form may belong to only one instability domain — the zeroth. Thus we cannot derive a test analogous to that of §2.1 for instability domains. However, as follows from Theorem II, if we can find "unstable" constant Hamiltonians $\mathbf{C}_1$ and $\mathbf{C}_2$ (i. e., $\det \mathbf{C}_{1,2} < 0$)) such that $h_1(t)\mathbf{C}_1 \leqslant \mathbf{H}(t) \leqslant h_2(t)\mathbf{C}_2$ for some $h_1(t)$ and $h_2(t)$ such that $\int_0^T h_{1,2}(t)\,dt \neq 0$, then $\mathbf{H}(t)$ is in the zeroth instability domain.

We may clearly demand that $\det \mathbf{C}_{1,2} \leqslant 0$ without imposing the restriction $\int_0^T h_{1,2}(t)\,dt \neq 0$. It may then happen, however, that $\mathbf{H}(t) \in \Pi_0$. If we demand in addition that each of the inequalities $h_1(t)\mathbf{C}_1 \leqslant \mathbf{H}(t) \leqslant h_2(t)\mathbf{C}_2$ is strict on some set of positive measure, then $\mathbf{H}(t) \in \mathscr{H}_0$.

Theorem I is sometimes useful in conjunction with the results of §1.3. In fact, given a Hamiltonian $\mathbf{H}(t)$, suppose that we can construct Hamiltonians

$$\mathbf{H}_j(t) = \begin{Vmatrix} z_j(t) + y_j(t) & x_j(t) \\ x_j(t) & z_j(t) - y_j(t) \end{Vmatrix} \qquad (j = 1,\ 2),$$

where the functions $x_j(t)$, $y_j(t)$, $z_j(t)$ have the form (1.9) or (1.10), such that

$$\mathbf{H}_1(t) \leqslant \mathbf{H}(t) \leqslant \mathbf{H}_2(t),$$

and such that $\mathbf{H}_1(t)$, $\mathbf{H}_2(t)$ belong to the same instability or stability domain (this is readily checked, since the corresponding equations are immediately integrable; see §1.10); then $\mathbf{H}(t)$ will belong to the same instability or stability domain.

We now prove

Theorem III. $\mathbf{H}(t) \in \mathfrak{G}_n$ *if and only if there exist* $\mathbf{G}_1(t) \in \Pi_n^{**}$ *and* $\mathbf{G}_2(t) \in \Pi_{n+1}^{**}$ *such that* $\mathbf{G}_1(t) \leqslant \mathbf{H}(t) \leqslant \mathbf{G}_2(t)$, *and each of these inequalities is strict on some set of positive measure.*

Proof. Sufficiency. Let $\mathbf{x}(t)$, $\mathbf{x}'(t)$, $\mathbf{x}''(t)$ be solutions of equations (1.2) with Hamiltonians $\mathbf{H}(t)$, $\mathbf{G}_1(t)$, $\mathbf{G}_2(t)$, respectively, satisfying the same initial conditions: $\mathbf{x}(0) = \mathbf{x}'(0) = \mathbf{x}''(0) = \mathbf{a}$. By the comparison theorem,

$$\varphi_{x'} < \varphi_x < \varphi_{x''}.$$

Since $\mathbf{G}_1(t) \in \Pi_n^{**}$, $\mathbf{G}_2(t) \in \Pi_n^{**}$, it follows that $\varphi_{x'} = n\pi$, $\varphi_{x''} = (n+1)\pi$ (see Theorem II in §1.9), so that $n\pi < \varphi_x < (n+1)\pi$. By Theorem II of §1.9, $\mathbf{H}(t) \in \mathfrak{G}_n$.

Necessity. Let $\mathbf{H}(t) \in \mathfrak{G}_n$. By Theorem I of §1.10, we have a representation (1.25) with $\mathbf{P}(t) \in \Omega_n$, $0 < \mathbf{H}_0 < \frac{\pi}{T}\mathbf{C}_0$, $\det \mathbf{C}_0 = 1$. Replacing $\mathbf{H}_0$ in (1.25) first by zero and then by the matrix $\pi\mathbf{C}_0/T$, we get

$$\mathbf{G}_1(t) < \mathbf{H}(t) < \mathbf{G}_2(t),$$

where

$$\mathbf{G}_1(t) = -\mathbf{P}(t)\,\tilde{\mathbf{J}}\,\frac{d}{dt}\mathbf{P}(t)^\tau,$$
$$\mathbf{G}_2(t) = -\mathbf{P}(t)\,\tilde{\mathbf{J}}\,\frac{d}{dt}\mathbf{P}(t)^\tau + \mathbf{P}(t)\left[\frac{\pi}{T}\mathbf{C}_0\right]\mathbf{P}(t)^\tau.$$

Again by Theorem I of §1.10, $\mathbf{G}_1(t) \in \Pi_n^{**}$. The matrizant of the equation with Hamiltonian $\mathbf{G}_2(t)$ is

$$\mathbf{X}(t) = \mathbf{P}(t)\exp\left[\frac{\pi}{T}t\,\tilde{\mathbf{J}}^{-1}\mathbf{C}_0\right].$$

Since $\exp\left[\frac{\pi}{T}t\,\tilde{\mathbf{J}}^{-1}\mathbf{C}_0\right] \in \Omega_1$ (see §1.7), it follows by Corollary 3 in §1.7 that $n_{\mathbf{X}} = n_{\mathbf{P}} + 1 = n+1$, i. e., $\mathbf{X}(t) \in \Omega_{n+1}$. By Theorem II of §1.9, we have $\mathbf{G}_2(t) \in \Pi_{n+1}^{**}$. Q. E. D.

The general form of Hamiltonians $\mathbf{G}(t) \in \Pi_n^{**}$ was described in §1.10. The constant Hamiltonians $\mathbf{G} \in \Pi_n^{**}$ may be characterized as follows. The corresponding matrices $\tilde{\mathbf{J}}^{-1}\mathbf{G} \in \tilde{\mathfrak{R}}^3$ fill the sheets of the hyperboloids $z^2 - x^2 - y^2 = n^2\pi^2/T^2$ in $\tilde{\mathfrak{R}}^3$: the upper sheets for $n > 0$, the lower sheets for $n < 0$. Thus the form of a constant Hamiltonian $\mathbf{G} \in \Pi_n^{**}$ is

$$\mathbf{G} = \frac{n\pi}{T}\begin{Vmatrix} a_n & b_n \\ b_n & c_n \end{Vmatrix}, \text{ where } a_n > 0, \quad c_n > 0, \quad a_n c_n - b_n^2 = 1. \tag{2.78}$$

Another use of the transformation (2.70), converting the original equation to an equation with Hamiltonian (2.73), is to establish tests analogous to that furnished by Theorem III in §2.2. To this end, we construct a sequence of step functions $u_k^\pm(t)$, $v_k^\pm(t)$ $(0 \leqslant t \leqslant T)$ such that

$$u_k^-(t) \leqslant \alpha_1(t) \leqslant u_k^+(t), \quad v_k^-(t) \leqslant \gamma_1(t) \leqslant v_k^+(t)$$

almost everywhere on $[0, T]$, and

$$u_k^\pm(t) \to \alpha_1(t), \quad v_k^\pm(t) \to \gamma_1(t) \text{ as } k \to \infty.$$

(These limiting relations must hold in the metric of the space $L(0, T)$.)

Consider the equations with Hamiltonians

$$\mathbf{H}_k^\pm(t) = \begin{Vmatrix} u_k^\pm(t) & \beta_{av} \\ \beta_{av} & v_k^\pm(t) \end{Vmatrix}.$$

Let $\psi^+(\chi)$, $\psi^-(\chi)$ denote the corresponding rotation functions. Since the coefficients of the equations are step functions, the rotation functions are readily found explicitly. Let $\psi_1(\chi)$ be the rotation function for the equation with Hamiltonian $\mathbf{H}_1(t)$.

It follows from (2.73) that, almost everywhere on $[0, T]$,

$$\mathbf{H}_k^-(t) \leqslant \mathbf{H}_1(t) \leqslant \mathbf{H}_k^+(t).$$

By the comparison theorem,

$$\psi^-(\chi) \leqslant \psi_1(\chi) \leqslant \psi^+(\chi) \qquad (0 \leqslant \chi \leqslant \pi).$$

In addition, by the above limiting relations,

$$\psi_k^-(\chi) \to \psi_1(\chi), \quad \psi_k^+(\chi) \to \psi_2(\chi) \quad \text{as} \quad k \to \infty,$$

uniformly for $0 \leqslant \chi \leqslant \pi$.

If $\mathbf{H}(t)$ is not in the boundary set Π, then also $\mathbf{H}_1(t) \notin \Pi$. Thus, in order to determine to which domain $\mathfrak{G}_n$ or $\mathcal{H}_n$ the Hamiltonian $\mathbf{H}_1(t)$ belongs, it will suffice to consider infima and suprema of functions $\psi_k^\pm(\chi)$ for sufficiently large k (see table, §1.9). And this will settle the question for the original Hamiltonian $\mathbf{H}(t)$ as well.

The resulting stability/instability tests resemble those deduced from Theorem III of §2.2 (see (2.47), (2.48)), except that now the numbers $m_k^\pm$, $M_k^\pm$ are defined by

$$m_k^\pm = \min_{0 \leqslant \chi \leqslant \pi} \psi_k^\pm(\chi), \quad M_k^\pm = \max_{0 \leqslant \chi \leqslant \pi} \psi_k^\pm(\chi). \tag{2.79}$$

The following example illustrates the application of Test 2.

Example. Consider the system

$$\left.\begin{aligned} \dot{x}_1 &= \frac{1}{4} x_1 \cos t + \mu (2 - \sin t) x_2, \\ \dot{x}_2 &= -\mu x_1 - \frac{1}{4} x_2 \cos t. \end{aligned}\right\} \tag{2.80}$$

The characteristic equation of the averaged system

$$\lambda^2 + 2\mu^2 = 0,$$

has two pure imaginary roots, $\lambda = \pm i\sqrt{2}\mu$. This case is more difficult to investigate than that studied in the example of §II. 4.2.

The method of averaging is frequently used to investigate systems of this kind, having only one "small" parameter. One uses a nonsingular linear transformation with periodic coefficients to reduce system (2.80) to a system whose coefficients are constant up to quantities of order $O(\mu^p)$. The larger p, the more cumbersome is the necessary transformation (for an account of the method, see, e. g., Malkin /62/, §84). After dropping variable terms of the order of μ^p one obtains a "simplified system" with constant coefficients. If this system is asymptotically stable or exponentially unstable one can construct a Lyapunov function, which will also be a Lyapunov function for the transformed system, provided μ is sufficiently small (suitable estimates of μ values for which this situation occurs are established). For these μ values, the transformed system, and therefore also the original system, are stable or unstable, as the case may be.

In our case (as in general for any canonical system), this technique is useless in checking for stability. If the original system is stable, then, to whatever order we include terms in the simplified system, we always have the critical case (in Lyapunov's sense): the eigenvalues vanish. Indeed, otherwise it would follow that system (2.80) is asymptotically stable, but this is false since (2.80) is a canonical system (the sum of diagonal coefficients vanishes identically). Stability may hold for system (2.80) only in the non-asymptotic sense.

Let us apply Test 1 to system (2.80). In this case

$$\alpha(t)=\mu, \quad \beta(t)=\frac{1}{4}\cos t, \quad \gamma(t)=\mu(2-\sin t),$$
$$\beta_{av}=0, \quad \delta(t)=\frac{1}{4}\sin t, \quad \alpha_0^{\pm}=\mu e^{\pm 1/2}.$$

Let $\mu>0$. Elementary considerations show that

$$\gamma_0^+=\max\left[\gamma(t)\,e^{2\delta(t)}\right]=\mu\max\left[(2-\sin t)\,e^{\frac{1}{2}\sin t}\right]=2\mu,$$
$$\gamma_0^-=\min\left[\gamma(t)\,e^{2\delta(t)}\right]=e^{1/2}\mu.$$

The conditions $\alpha_0^++\gamma_0^+>0$, $\alpha_0^-+\gamma_0^->0$ are fulfilled. Inequality (2.77) means here that

$$n^2\leqslant 4\mu^2\leqslant 8\sqrt{e}\,\mu^2\leqslant(n+1)^2.$$

Thus system (2.80) is stable if

$$\frac{n}{2}<\mu<\frac{n+1}{\sqrt{8\sqrt{e}}}=0.275\,(n+1).$$

This inequality is consistent only if $n=0.1$. Hence, if

$$0<\mu<0.275\ldots,$$

then system (2.80) is stable (the Hamiltonian $\mathbf{H}(t)$ belongs to $\mathfrak{G}_0$). It is also stable if

$$0.5<\mu<0.552\ldots$$

(the Hamiltonian is then in $\mathfrak{G}_1$).

The case $\mu<0$ is treated similarly. By Test 1, if

$$-0.275<\mu<0,$$

then system (2.80) is stable and $\mathbf{H}(t)\in\mathfrak{G}_{-1}$. It is also stable if

$$-0.552<\mu<-0.5$$

and then $\mathbf{H}(t)\in\mathfrak{G}_{-2}$. Finally, it is obvious that if $\mu=0$ system (2.80) is again stable, $\mathbf{H}(t)\in\Pi_0^{**}$.

For large μ values, one obtains a sharper result from the test of §2.1, since for large μ the Hamiltonian of system (2.80)

$$\mathbf{H}(t)=\begin{Vmatrix} \mu & \frac{1}{4}\cos t \\ \frac{1}{4}\cos t & \mu(2-\sin t) \end{Vmatrix}$$

is "nearly" diagonal.

§3. APPLICATION OF THE RESULTS OF §§1, 2 TO HILL'S EQUATION

3.1. Position of Hamiltonians of Hill's equation in $\mathfrak{L}^3$

Throughout this section and the following ones we consider Hill's equation

$$\ddot{y}+p(t)y=0, \tag{3.1}$$

where $p(t)$ is a T-periodic function, Lebesgue-integrable on $[0, T]$. As shown in §1.1, this equation reduces to a canonical equation

$$\tilde{\mathbf{J}}\frac{d\mathbf{x}}{dt}=\mathbf{H}(t)\mathbf{x}, \tag{3.2}$$

where

$$\mathbf{x}=\begin{pmatrix}\dot{y}\\ y\end{pmatrix}, \quad \tilde{\mathbf{J}}=\begin{Vmatrix} 0 & 1 \\ -1 & 0 \end{Vmatrix}, \quad \mathbf{H}(t)=\begin{Vmatrix} 1 & 0 \\ 0 & p(t) \end{Vmatrix}. \tag{3.3}$$

The matrizant $\mathbf{X}(t)$ is

$$\mathbf{X}(t)=\begin{Vmatrix} \dot{y}_1 & \dot{y}_2 \\ y_1 & y_2 \end{Vmatrix}.$$

Recall that $\mathbf{X}(0)=\mathbf{I}_2$ and $\det\mathbf{X}(t)\equiv 1$.

Let $y(t)$ be any solution of equation (3.1). It is clear that the trajectory $\{\dot{y}(t), y(t)\}$ on the $(\dot{y},y)$-plane does not pass through the origin and cuts the $\dot{y}$-axis in such a way that its tangent in the positive direction (i. e., in the sense of increasing argument) at the intercept is vertical (see Figure 67).

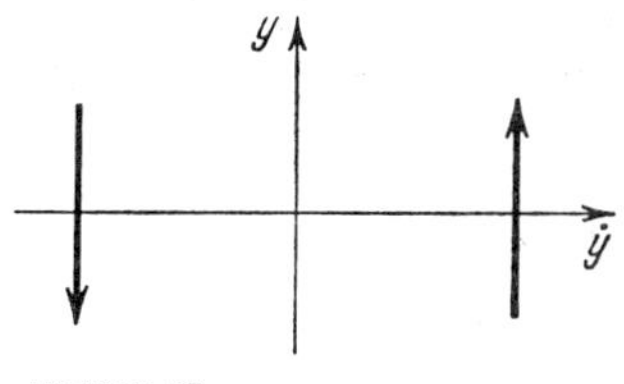

FIGURE 67.

Moreover, $y(t)$ is increasing if $\dot{y} > 0$ and decreasing if $\dot{y} < 0$. Thus, if the vector $\mathbf{x}$ turns through the angle π from time t_1 to time t_2, this rotation takes place in the positive direction and the function $y(t)$ has exactly one zero on $[t_1, t_2]$.

The same holds for the row-vector $\mathbf{y}^{\tau} = (y_1, y_2)$. Indeed, letting $\vartheta_{\mathbf{y}} = \operatorname{Arg} \mathbf{y}(t)$ denote (as usual) a continuous branch of the argument of $\mathbf{y}^{\tau}$, we have $\vartheta_{\mathbf{y}} = \operatorname{arctg} y_2/y_1$ and

$$\frac{d\vartheta_{\mathbf{y}}}{dt} = \frac{\dot{y}_2 y_1 - \dot{y}_1 y_2}{y_1^2 + y_2^2} = -\frac{1}{y_1^2 + y_2^2}.$$

Hence it is clear that $\mathbf{y}^{\tau}$ rotates monotonically in the negative sense.

The third formula of (3.3) defines a correspondence $p(t) \sim \mathbf{H}(t) \in \mathfrak{L}^3$. (Concerning $\mathfrak{L}^3$, see §1.6.) Henceforth we shall write simply $p(t) \in \mathfrak{L}^3$, and say that equation (3.1) belongs to the appropriate set in $\mathfrak{L}^3$. We shall now show that the position of the set of equations (3.1) (i. e., the set of Hamiltonians $\mathbf{H}(t)$ of type (3.3)) in $\mathfrak{L}^3$ is that shown schematically in Figure 68. In other words, we shall show that equation (3.1) cannot belong to instability or stability domains with negative subscripts, neither can it belong to Π_0^{**} or Π_0^{*-}. The first assertion is clear, for the vector $\mathbf{y}$ always rotates in the positive sense. For the same reason, equation (3.1) cannot be in Π_0^{*-}. Finally, if $p(t) \in \Pi_0^{**}$, then $y_1(t)$ and $y_2(t)$ are periodic nonvanishing functions. It follows from the standard theory of differential equations /87/ that

$$y_1(t) = y_2(t) \int_0^t \frac{dt_1}{y_2(t_1)^2}.$$

Hence

$$y_1(t+T) - y_1(t) = y_2(t) \int_0^T \frac{dt}{y_2(t)^2},$$

contradicting the periodicity of $y_1(t)$. (If $p(t) \in \Pi_k^{**}$, $k > 0$, the integral in this formula becomes improper and the above reasoning is meaningless.)

Thus the set of equations (3.1) must occupy (qualitatively speaking) the position shown in Figure 68 in the space $\mathfrak{L}^3$.

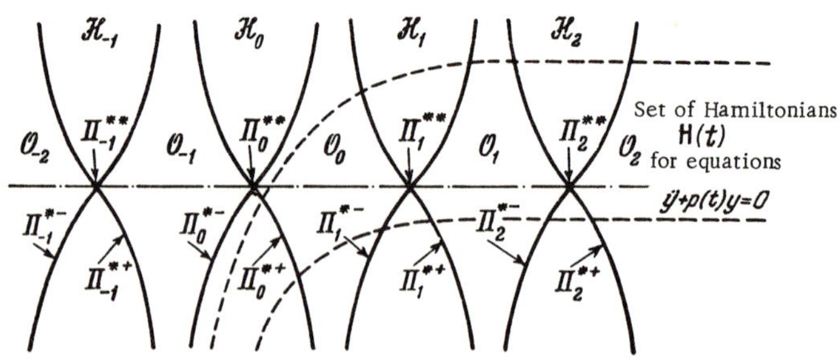

FIGURE 68.

The reader should note that we can equally well reduce equation (3.1) to a vector equation (3.2) in such a way that, say, the domains Π_k, $\mathcal{H}_k$, $\mathcal{O}_k$ with positive subscripts are empty.

3.2. Characterization of stability and instability domains in terms of zeros of solutions

Theorem I (Adamov). Let $t_0 < t_1 < t_2 < \ldots$ *be the consecutive zeros of a solution* $y(t)$ *of equation (3.1). Denote*

$$D_n = \sup(t_n - t_0), \quad d_n = \inf(t_n - t_0),$$

where the supremum and infimum extend over all solutions $y(t)$. *Then the two conditions in each of the following rows are equivalent (for* $n \geqslant 0$*):*

$$\begin{array}{lll} p(t) \in \mathcal{O}_n & \textit{and} & D_n < T < d_{n+1}, \\ p(t) \in \mathcal{H}_{n+1} & \textit{and} & d_{n+1} < T < D_{n+1}, \\ p(t) \in \Pi_{n+1}^{*-} & \textit{and} & d_{n+1} = T, \ D_{n+1} > T, \\ p(t) \in \Pi_{n+1}^{**} & \textit{and} & d_{n+1} = D_{n+1} = T, \\ p(t) \in \Pi_{n+1}^{*+} & \textit{and} & d_{n+1} < T, \ D_{n+1} = T. \end{array}$$

Proof. Let y and $\tilde{y}$ be any linearly independent solutions of equation (3.1), $\mathbf{y} = (y, \tilde{y})$. Since y and $\tilde{y}$ are linear combinations of the "normalized" solutions y_1 and y_2 appearing in the matrizant (see §3.1), we have $\mathbf{y} = (y_1, y_2)\mathbf{S}$, where $\mathbf{S}$ is some constant matrix. Since

$$\dot{\vartheta}_{\mathbf{y}} = -\frac{1}{y^2 + \tilde{y}^2} \begin{vmatrix} \dot{y} & \dot{\tilde{y}} \\ y & \tilde{y} \end{vmatrix} \quad \text{and} \quad \begin{vmatrix} \dot{y}_1 & \dot{y}_2 \\ y_1 & y_2 \end{vmatrix} = 1,$$

it follows that

$$\dot{\vartheta}_{\mathbf{y}} = -\frac{\det \mathbf{S}}{(\mathbf{y}, \mathbf{y})}.$$

Thus the vector $\mathbf{y}$ rotates monotonically. Let $y(t_0) = y(t_1) = \ldots = y(t_n) = 0$ and let $\varphi_{\mathbf{y}}$ be the rotation of $\mathbf{y}$ from $t = t_0$ to $t = t_0 + T$. We may assume without loss of generality that $\varphi_{\mathbf{y}} > 0$ (otherwise, we simply interchange y and $\tilde{y}$). In the time from t_i to t_{i+1}, the vector $\mathbf{y}$ clearly rotates through an angle π. Therefore the following assertions are equivalent:

$$\begin{array}{lll} \varphi_{\mathbf{y}} > n\pi & \text{and} & t_n - t_0 < T, \\ \varphi_{\mathbf{y}} = n\pi & \text{and} & t_n - t_0 = T, \\ \varphi_{\mathbf{y}} < n\pi & \text{and} & t_n - t_0 > T. \end{array}$$

Let $p(t) \in \mathcal{H}_n$ $(n \geqslant 1)$. Then there are solutions with $\varphi_{\mathbf{y}} > n\pi$ and $\varphi_{\mathbf{y}} < n\pi$ (see Figure 61), i.e., $d_n < T$, $D_n > T$.

If $p(t) \in \Pi_n^{\bullet -}$, then $\varphi_y < n\pi$ and there is one solution with $\varphi_y = n\pi$. Therefore $D_n > T$, $d_n = T$.

Similarly, if $p(t) \in \Pi_n^{\bullet +}$, then $d_n < T$, $D_n = T$.

If $p(t) \in \Pi_n^{\bullet\bullet}$, then clearly $d_n = D_n = T$.

Let $p(t) \in \mathcal{O}_n$. Then $n\pi < \varphi_y < (n+1)\pi$, and so $t_n - t_0 < T$. Hence also $D_n < T$. Similarly, the fact that $t_{n+1} - t_0 > T$ implies that $d_{n+1} > T$. This completes the proof.

This theorem was first proved (in a different terminological framework) by Adamov /114d/.

Theorem I says nothing about the sets $\mathcal{H}_0$ and $\Pi_0^{\bullet +}$. We claim that for functions $p(t)$ in these sets $d_1 = \infty$, $D_1 = \infty$, in other words, any solution $y(t)$ has at most one zero on any finite interval. Solutions with this property are said to be n o n o s c i l l a t i n g. Indeed, suppose that a solution $y(t)$ has at least two zeros in the interior of some interval (l_1T, l_2T), where l_1, l_2 are integers. Then the rotation of the solution-vector $\mathbf{x}(t)$ of equation (3.2) in the time from $t = l_1T$ to $t = l_2T$ is greater than π. On the other hand, $\mathbf{x}(t) = \mathbf{F}(t) \exp(t\mathbf{K})\mathbf{x}(0)$, where $\mathbf{F}(t) \in \Omega_0$ (see §1). Thus $\mathbf{F}(0) = \mathbf{F}(l_1T) = \mathbf{F}(l_2T) = \mathbf{I}_2$ and $\mathbf{F}(t)$ can be deformed continuously into $\mathbf{I}_2$. Consequently, the above rotation coincides with the rotation of the vector $\mathbf{z}(t) = \exp(t\mathbf{K})\mathbf{x}(0)$ from $t = l_1T$ to $t = l_2T$. The motion $\mathbf{z}(t) = \exp(t\mathbf{K})\mathbf{x}(0)$ is shown in Figure 58 for $p(t) \in \mathcal{H}_0$ and in Figure 60 for $p(t) \in \Pi_0^{\bullet +}$. In either case the rotation of $\mathbf{z}(t)$ over any time can never exceed π in absolute value. This contradiction shows that if $p(t) \in \mathcal{H}_0$ or $p(t) \in \Pi_0^{\bullet +}$ then any solution $y(t)$ is nonoscillating.

If $p(t)$ is in any other set ($\mathcal{O}_0$ or $\Pi_1 = \Pi_1^{\bullet -} \cup \Pi_1^{\bullet\bullet} \cup \Pi_1^{\bullet +}$ or $\mathcal{H}_1$ or $\mathcal{O}_1$, etc.), any solution of equation (3.1) is o s c i l l a t i n g, i.e., has infinitely many zeros on any infinite interval (t_0, ∞) Indeed, for any of these sets we have $\varphi_x \geqslant m > 0$ for any solution (see table in §1.9). Therefore the rotation of a solution $\mathbf{x}(t)$ of equation (3.2) in the time from $t = 0$ to $t = lT$ increases without bound as $l \to \infty$ and so (see Figure 66) the number of zeros $y(t)$ on $[0, lT]$ increases without bound as $l \to \infty$.

According to Sturm's theorem /87/, between any two zeros of one solution of equation (3.1) there lies at least one zero of another solution. This follows at once from simple geometric arguments. Indeed, let $y_1(t)$, $y_2(t)$ be two linearly independent solutions of equation (3.1) and $\mathbf{y}$ the vector with components $y_1(t), y_2(t)$. We showed above that $\mathbf{y}(t)$ rotates monotonically. Two consecutive zeros t_0 and t_1 of the function $y_1(t)$ correspond to rotation of this vector through π or $(-\pi)$. In either case, $y_2(t_*) = 0$ for some t_* $(t_0 < t_* < t_1)$, namely, the point at which the vector $\mathbf{y}(t)$ intersects the coordinate axis.

Thus the property of oscillation is inherent in the coefficient $p(t)$ of equation (3.1). We may therefore speak of equation (3.1) itself as being o s c i l l a t i n g or n o n o s c i l l a t i n g. We have thus shown that if $p(t) \in \mathcal{H}_0$ or $p(t) \in \Pi_0^{\bullet +}$, then equation (3.1) is nonoscillating, while in all other cases it is oscillating. Let $\vartheta_x(t)$ be the argument of the vector $\mathbf{x}(t)$ defined by the first formula of (3.3). We have proved that in the oscillating case

$$\vartheta_x(t) \to \infty \quad \text{as} \quad t \to \infty$$

whereas in the nonoscillating case $\vartheta_x(t)$ is a bounded function. If the function $p(t)$ is increased, the corresponding Hamiltonian (see third formula of (3.3)) is increased (or remains the same), and by the comparison theorem (§2.4) ϑ_x increases for any fixed t_0. Thus: *if equation (3.1) is oscillating, it remains oscillating if $p(t)$ is increased; if equation (3.1) is nonoscillating, it remains nonoscillating if $p(t)$ is decreased.*

If $p_1(t) \in \Pi_0^{*+}$ and $p_2(t) \underset{\not\equiv}{\leqslant} p_1(t)$, then $p_2(t) \in \mathscr{H}_0$. Indeed, by the comparison theorem, the corresponding rotation functions $\psi_1(\chi)$ and $\psi_2(\chi)$ satisfy the strict inequality $\psi_1(\chi) < \psi_2(\chi)$. The assertion now follows from Theorem II of §1.9.

As a corollary, we see that *if $p(t) \underset{\not\equiv}{\leqslant} 0$, then $p(t) \in \mathscr{H}_0$*. Recall that at the end of § VII.2 we established a simple criterion for oscillation:

If

$$p_{\text{av}} = \frac{1}{T}\int_0^T p(t)\,dt \geqslant 0, \qquad p(t) \not\equiv 0,$$

then equation (3.1) is oscillating.

We claim that any function $p(t)$ in $\mathscr{H}_0 \cup \Pi_0^{*+}$ may be expressed as

$$p(t) = -\dot{u}(t) - u(t)^2,$$

where $u(t)$ is an absolutely continuous T-periodic function such that

$$\frac{1}{T}\int_0^T u(t)\,dt = \lambda \geqslant 0,$$

where $\lambda > 0$ if $p \in \mathscr{H}_0$ and $\lambda = 0$ if $p(t) \in \Pi_0^{*+}$.

Indeed, if $p(t) \in \mathscr{H}_0 \cup \Pi_0^{*+}$, equation (3.1) has a solution $y = \exp(\lambda t)\, r(t)$, where $r(t+T) = r(t) > 0$ and $\lambda > 0$ if $p(t) \in \mathscr{H}_0$; $\lambda = 0$ if $p(t) \in \Pi_0^{*+}$. Setting $u(t) = d \ln r(t)/dt$, we get the desired representation.

It is easy to see that the converse is also valid: if $p(t)$ has the above form, then $p(t) \in \mathscr{H}_0$ if $\lambda > 0$ and $p(t) \in \Pi_0^{*+}$ if $\lambda = 0$. Indeed, setting

$$r(t) = \exp\int_0^t [u(t) - \lambda]\,dt,$$

we see that the function $y = \exp(\lambda t)\, r(t)$ is a solution of the corresponding equation (3.1). Consequently, $p(t) \in \mathscr{H}_0$ if $\lambda > 0$ and $p(t) \in \Pi_0^{*+}$ if $\lambda = 0$.

It follows from (3.3) that if $p_1(t) \geqslant p_2(t)$ the corresponding Hamiltonians stand in the same relation: $\mathbf{H}_1(t) \geqslant \mathbf{H}_2(t)$. Thus Theorem II of §2.4 implies

Theorem II. If $p_1(t) \leqslant p(t) \leqslant p_2(t)$ and the functions $p_1(t)$, $p_2(t)$ belong to the same stability domain $\mathfrak{G}_n$ or instability domain $\mathscr{H}_n$ then $p(t)$ is also in this domain.

The following interesting theorem, due to Adamov /114f/, is given without proof.

Theorem III (Adamov). The set $\mathscr{H}_0 \cup \Pi_0^{*+}$ (i.e., the set of functions $p(t)$ for which equation (3.1) is nonoscillating) is convex.

3.3. Zhukovskii's test and its generalization

Theorem III of §2.4 immediately implies a stability test for equation (3.1). To establish this test, we first determine the general form of functions $p(t) \in \Pi_n^{**}$ $(n=1, 2, \ldots)$.

Theorem. *Let* $r(t)$ *be an arbitrary function satisfying the following conditions:* $\ddot{r}(t)$ *exists and is Lebesgue-summable,*

$$r(t+T) = r(t) > 0, \quad \int_0^T \frac{dt}{r(t)^2} = 1. \tag{3.4}$$

Then the general form of a function $p(t) \in \Pi_n^{**}$ *is*

$$p(t) = \frac{n^2\pi^2}{r(t)^4} - \frac{\ddot{r}(t)}{r(t)} \qquad (n=1, 2, \ldots). \tag{3.5}$$

Proof. Let $p(t) \in \Pi_n^{**}$ $(n=1, 2, \ldots)$; let y_1, y_2 be linearly independent solutions of equation (3.1) satisfying initial conditions $y_1(0)=1$, $\dot{y}_1(0)=y_2(0)=0$, $\dot{y}_2(0)=1$. By part 2 of Theorem II, §1.9, the vector $\mathbf{y}$ with components y_1, y_2 rotates through an angle $(-n\pi)$ in time T and $\mathbf{y}(t+T)=(-1)^n \mathbf{y}(t)$. Let $\rho = \sqrt{y_1^2+y_2^2}$, $\vartheta = \arg \mathbf{y}$. Differentiating, we get

$$\frac{d\vartheta}{dt} = -\frac{1}{(\mathbf{y}, \mathbf{y})} = -\frac{1}{\rho^2}, \qquad \vartheta(t) - \vartheta(0) = -\int_0^t \frac{dt}{\rho^2}.$$

Since $\vartheta(T)-\vartheta(0) = -n\pi$, it follows that $\int_0^T \frac{dt}{\rho^2} = n\pi$. Let $r = \sqrt{n\pi}\,\rho$. It is clear that r satisfies conditions (3.4). A direct check shows that

$$p(t) = -\frac{\ddot{y}_1}{y_1} = -\frac{\frac{d^2}{dt^2}\left[\sqrt{n\pi}\, r \cos\left(n\pi \int_0^t \frac{dt}{r^2}\right)\right]}{\sqrt{n\pi}\, r \cos\left(n\pi \int_0^t \frac{dt}{r^2}\right)} = \frac{n^2\pi^2}{r^4} - \frac{\ddot{r}}{r}. \tag{3.6}$$

Conversely, let $r(t)$ be a function of the stipulated type, and define $p(t)$ by (3.5). Since the last equality in (3.6) holds for any function r, we see that the function

$$y_1 = \sqrt{n\pi}\, r \cos\left(n\pi \int_0^t \frac{dt}{r^2}\right)$$

is a solution of equation (3.1). Similarly,

$$y_2 = \sqrt{n\pi}\, r \sin\left(n\pi \int_0^t \frac{dt}{r^2}\right)$$

is a solution of equation (3.1). By (3.4), the vector $\mathbf{y}=(y_1, y_2)$ turns in time T through an angle $(-n\pi)$ and $\mathbf{y}(t+T)=(-1)^n\mathbf{y}(t)$. Hence it follows that the matrizant of equation (3.1) is in Ω_n, so that $p(t)\in\Pi_n^{**}$. Q. E. D.

Remark. If $n=0$, formula (3.5) clearly yields functions in Π_0^{*+}. The normalizing condition in (3.4) is then inessential.

Theorem III of §2.4 yields the following stability test.

*Test 1. Let $r_1(t)$ and $r_2(t)$ be arbitrary functions satisfying conditions (3.4). If**

$$\frac{n^2\pi^2}{r_1^4}-\frac{\ddot{r}_1}{r_1}\underset{\not\equiv}{\leqslant} p(t)\underset{\not\equiv}{\leqslant}\frac{(n+1)^2\pi^2}{r_2^4}-\frac{\ddot{r}_2}{r_2}, \tag{3.7}$$

for some $n=0, 1, 2, \ldots$, the trivial solution of equation (3.1) is stable and $p(t)\in\mathfrak{S}_n$.

The only case requiring verification is $n=0$. In that case, instead of Theorem III of §2.4, one uses the comparison theorem of §2.4 and Theorem II of §1.9 (Figure 61), and the proof becomes obvious.

Taking $r_1(t)=r_2(t)=\text{const}$ in Test 1, we obtain Zhukovskii's test /147/:

If $\frac{n^2\pi^2}{T^2}\underset{\not\equiv}{\leqslant} p(t)\underset{\not\equiv}{\leqslant}\frac{(n+1)^2\pi^2}{T^2}$, then $p(t)\in\mathfrak{S}_n$.

The geometric significance of Test 1 is evident: if the curve $p(t)$ lies between

$$p_n(t)=\frac{n^2\pi^2}{r_1^4}-\frac{\ddot{r}_1}{r_1} \quad \text{and} \quad p_{n+1}(t)=\frac{(n+1)^2\pi^2}{r_2^4}-\frac{\ddot{r}_2}{r_2},$$

then $p(t)\in\mathfrak{S}_n$. Figure 69 shows the curves $p_n(t)=p_n(t, \varepsilon)$ for

$$r_1(t)=r_2(t)=\frac{\sqrt{\pi}}{\sqrt{1+\varepsilon\sin 2t}} \qquad (T=\pi);\quad \varepsilon=0.08,\ 0.17,\ 0.25,\ 0.50;$$

$n=0, 1, 2, 3, 4, 5, 6$. The table on the following page gives the values of $p_n(t, \varepsilon)$ accurate up to four decimal digits, for $\varepsilon=0.08$, 0.17, 0.25, 0.50, 0.75, 0.95. Thus, *if the curve $p(t)$ lies between $p_n(t, \varepsilon_1)$ and $p_{n+1}(t, \varepsilon_2)$ for some $\varepsilon_1, \varepsilon_2$, $n=0, 1, \ldots, 6$, then all solutions of equation (3.1) $(T=\pi!)$ are bounded and $p(t)\in\mathfrak{S}_n$.* When using curves of this type, one should remember that all the curves may be displaced to the right or the left, since the transformation $p(t)\to p(t+c)$ preserves membership in a stability (or instability) domain (see Remark 2 in §1.9).

Test 1 implies the following

Corollary. Any stability domain $\mathfrak{S}_n$ $(n=0, 1, 2, \ldots)$ contains functions $p(t)$ with negative mean values of arbitrarily large absolute value.

Proof. By Test 1, it is sufficient to show that for any $n=0, 1, 2, \ldots$ there exists a function (3.5) with negative mean value p_{av} of arbitrarily large absolute value.

* The identity symbol in (3.7) and below stands for equality almost everywhere on $[0, T]$.

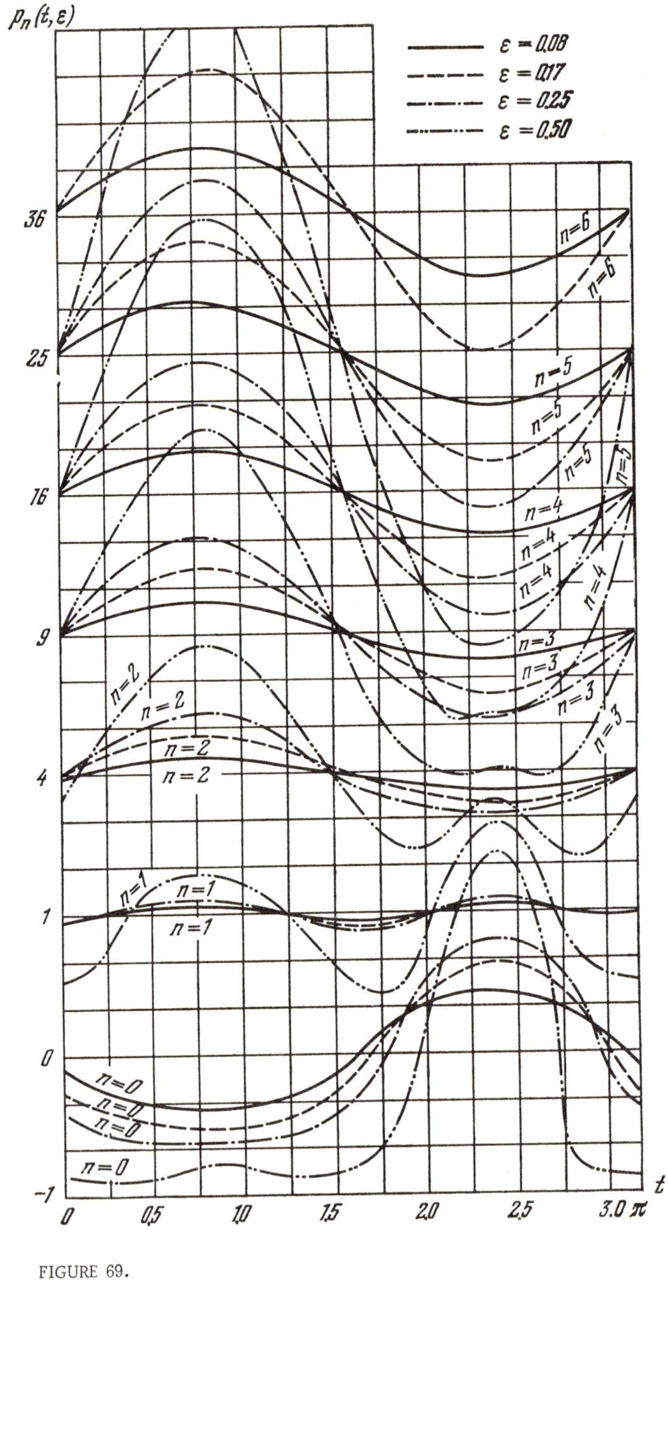

$p_n(t,\varepsilon)$
ε = 0.08
ε = 0.17
ε = 0.25
ε = 0.50
36
25
16
9
4
1
0
-1
n=6
n=5
n=4
n=3
n=2
n=1
n=0
0
0.5
1.0
1.5
2.0
2.5
3.0 π
t

FIGURE 69.

Table of functions $p_n(t, \varepsilon)$

t	$\varepsilon=0.08$	$\varepsilon=0.17$	$\varepsilon=0.25$
0.000	$n^2-0.0192$	$n^2-0.0867$	$n^2-0.1875$
0.349	$1.105\,n^2-0.1080$	$1.230\,n^2-0.2384$	$1.348\,n^2-0.3584$
0.698	$1.164\,n^2-0.1466$	$1.362\,n^2-0.2889$	$1.553\,n^2-0.3988$
1.047	$1.143\,n^2-0.1338$	$1.316\,n^2-0.2733$	$1.481\,n^2-0.3875$
1.396	$1.055n^2-0.0694$	$1.119\,n^2-0.1786$	$1.179\,n^2-0.2982$
1.745	$0.9461\,n^2+0.0382$	$0.8874\,n^2+0.0370$	$0.8366\,n^2-0.1131$
2.094	$0.8662\,n^2+0.1433$	$0.7273\,n^2+0.3154$	$0.6140\,n^2+0.4757$
2.443	$0.8486\,n^2+0.1704$	$0.6931\,n^2+0.3988$	$0.5681\,n^2+0.6436$
2.792	$0.8997\,n^2+0.0961$	$0.7932\,n^2+0.1817$	$0.7041\,n^2+0.2276$
3.142	$n^2-0.0192$	$n^2-0.0867$	$n^2-0.1875$

t	$\varepsilon=0.50$	$\varepsilon=0.75$	$\varepsilon=0.95$
0.000	$n^2-0.7500$	$n^2-1.6875$	$n^2-2.707$
0.349	$1.745\,n^2-0.7386$	$2.196\,n^2-1.102$	$2.595\,n^2-1.370$
0.698	$2.226\,n^2-0.6702$	$3.024\,n^2-0.8662$	$3.748\,n^2-0.988$
1.047	$2.053\,n^2-0.6957$	$2.722\,n^2-0.9421$	$3.323\,n^2-1.106$
1.396	$1.371\,n^2-0.7754$	$1.580\,n^2-1.351$	$1.756\,n^2-1.852$
1.745	$0.6877\,n^2-0.5517$	$0.5534\,n^2-2.006$	$0.4566\,n^2-4.279$
2.094	$0.3217\,n^2+0.9417$	$0.1231\,n^2+0.2631$	$0.0316\,n^2-12.24$
2.443	$0.2576\,n^2+1.854$	$0.0683\,n^2+4.917$	$0.0041\,n^2+9.566$
2.792	$0.4600\,n^2-0.0061$	$0.2676\,n^2-1.828$	$0.1510\,n^2-7.356$
3.142	$n^2-0.7500$	$n^2-1.6875$	$n^2-2.707$

Let

$$r=\frac{\sqrt{T}}{\sqrt{1+\frac{1}{2}\sin\frac{2m\pi}{T}t}}.$$

It is clear that for any integer m this function satisfies conditions (3.4). Integrating by parts, we obtain

$$\int_0^T p(t)\,dt=n^2\pi^2\int_0^T\frac{dt}{r^4}-\int_0^T\frac{\dot{r}^2}{r^2}\,dt=\frac{9n^2\pi^2}{8T}-\int_0^T\frac{\dot{r}^2}{r^2}\,dt.$$

We have

$$\frac{\dot{r}}{r}=-\frac{m\pi}{2T}\,\frac{\cos\frac{2m\pi}{T}}{1+\frac{1}{2}\sin\frac{2m\pi}{T}t},$$

whence

$$\left(\frac{\dot{r}}{r}\right)^2\geqslant\frac{m^2\pi^2}{9T^2}\cos^2\frac{2m\pi}{T}t,\qquad p_{\text{av}}=\frac{1}{T}\int_0^T p(t)\,dt\leqslant\left(\frac{9}{4}n^2-\frac{1}{9}m^2\right)\frac{\pi^2}{2T^2}.$$

For sufficiently large m, the number p_{av} will be negative and of arbitrarily large absolute value.

3.4. Results of §§2.1, 2.4 applied to Hill's equation

We now apply the test (2.5) of §2.1 to equation (3.1). Setting $\dot{y}=x^{(1)}$, $cy=x^{(2)}$, we obtain an equation (3.2) with Hamiltonian

$$\mathbf{H}(t)=\begin{Vmatrix} c & 0 \\ 0 & \frac{1}{c}p(t) \end{Vmatrix}. \tag{3.8}$$

Here $c>0$ is arbitrary. The smallest and largest eigenvalues of $\mathbf{H}(t)$ are

$$h_{\min}(t)=\frac{1}{c}p_c^-(t), \quad h_{\max}(t)=\frac{1}{c}p_c^+(t),$$

where

$$p_c^-(t)=\begin{cases} p(t), & \text{if } p(t)<c^2, \\ c^2, & \text{if } p(t)\geqslant c^2, \end{cases} \quad p_c^+(t)=\begin{cases} p(t), & \text{if } p(t)>c^2, \\ c^2, & \text{if } p(t)\leqslant c^2. \end{cases} \tag{3.9}$$

The numbers

$$\tilde{m}=\int_0^T h_{\min}(t)\,dt, \quad \tilde{M}=\int_0^T h_{\max}(t)\,dt, \tag{3.10}$$

that is to say, the upper and lower bounds for the rotation function $\psi(\chi)$, are in this case

$$\tilde{m}=\frac{1}{c}\int_0^T p_c^-(t)\,dt, \quad \tilde{M}=\frac{1}{c}\int_0^T p_c^+(t)\,dt. \tag{3.11}$$

Inequalities (2.5) may now be written

$$n\pi<\frac{1}{c}\int_0^T p_c^-(t)\,dt\leqslant\frac{1}{c}\int_0^T p_c^+(t)\,dt<(n+1)\pi. \tag{3.12}$$

It follows from these inequalities that the Hamiltonian (3.8) belongs to the stability domain $\mathfrak{G}_n$. Then the same holds for the Hamiltonian (3.3) (and, in our terminology, for the function $p(t)$). Indeed, we can write the substitution $\dot{y}=x^{(1)}$, $cy=x^{(2)}$ as

$$\mathbf{x}_1=\begin{Vmatrix} 1 & 0 \\ 0 & c \end{Vmatrix}\mathbf{x}, \quad \text{where} \quad \mathbf{x}_1=\begin{pmatrix} x^{(1)} \\ x^{(2)} \end{pmatrix}, \quad \mathbf{x}=\begin{pmatrix} \dot{y} \\ y \end{pmatrix}$$

and apply Theorem III of §1.10. By this theorem $\mathbf{H}(t)$ is in the same stability domain as $\mathbf{H}_1(t)$.

We claim that inequalities (3.12) may hold only if

$$\frac{n\pi}{T} < c < \frac{(n+1)\pi}{T}.$$

Let $c \leqslant n\pi/T$. Since $p_c^-(t)/c \leqslant c$, we have

$$\frac{1}{c}\int_0^T p_c^-(t)\,dt \leqslant Tc \leqslant n\pi,$$

and this contradicts (3.10). Therefore, $c > n\pi/T$. Similarly, one shows that

$$c < \frac{(n+1)\pi}{T}.$$

We have thus proved the following stability condition for Hill's equation (3.1):

Test 2. *Let c be an arbitrary number and $n \geqslant 0$ an integer such that*

$$\frac{n\pi}{T} < c < \frac{(n+1)\pi}{T}.$$

Define functions $p_c^-(t)$, $p_c^+(t)$ by (3.9). If inequalities (3.12) are satisfied, the trivial solution of equation (3.1) is stable and the equation is in the n-th stability domain $\mathfrak{S}_n$ $(n=0, 1, 2, \ldots)$.

Test 2 provides a measure of the proximity of $p(t)$ to the function

$$p(t) \equiv c^2, \quad \text{where} \quad \frac{n\pi}{T} < c < \frac{(n+1)\pi}{T},$$

which clearly belongs to the same stability domain $\mathfrak{S}_n$ as $p(t)$.

A convenient choice of the number c^2 in Test 2 is the mean value of $p(t)$.

Remark. The left-hand side of inequality (3.7) in Test 1 means that $\psi(\chi) > n\pi$, and the right-hand side that $\psi(\chi) < (n+1)\pi$ (see §2). A similar statement is true for inequalities (3.12) in Test 2. It follows that Tests 1 and 2 may be combined: Each of the conditions

$$\frac{n^2\pi^2}{r_1^4} - \frac{\ddot{r}_1}{r_1} \underset{\not\equiv}{\leqslant} p(t), \qquad \frac{1}{c}\int_0^T p_c^+(t)\,dt < (n+1)\pi \tag{3.13}$$

and

$$n\pi < \frac{1}{c}\int_0^T p_c^-(t)\,dt, \quad p(t) \leqslant \frac{(n+1)^2\pi^2}{r_2^4} - \frac{\ddot{r}_2}{r_2} \tag{3.14}$$

implies $p(t) \in \mathfrak{S}_n$. In (3.13) and (3.14), $r_1(t)$, $r_2(t)$ are arbitrary functions satisfying (3.4), and n is an integer. By the oscillation criterion at the end of §3.2, if $n=0$ the first inequalities in (3.13), (3.14) may be replaced by the conditions $p_{\text{av}} \geqslant 0$, $p(t) \not\equiv 0$.

Example. Consider the equation

$$\ddot{y}+(a+b\sin^{2k+1}2t)\,y=0,$$

where k is an integer. Assuming that $a>0$ and setting $c=\sqrt{a}$, we obtain the following sufficient condition for stability from (3.12):*

$$\left.\begin{array}{l} a>0, \quad n<\sqrt{a}<n+1, \\ |b|<N_k\pi\sqrt{a}(\sqrt{a}-n), \quad |b|<N_k\pi\sqrt{a}(n+1-\sqrt{a}), \end{array}\right\} \tag{3.15}$$

where

$$N_k=1 \quad \text{for} \quad k=0 \quad \text{and} \quad N_k=\frac{1\cdot 3\cdot 5\ldots(2k+1)}{2\cdot 4\ldots 2k} \quad \text{for } k>0.$$

Lyapunov's test (see § VII. 2.6)

$$p(t)\underset{\not\equiv}{\geqslant}0, \quad T\int_0^T p(t)\,dt\leqslant 4$$

gives the domain

$$|b|\leqslant a\leqslant\frac{\pi}{4},$$

while Zhukovskii's test (§3.3) gives

$$n^2\leqslant a\leqslant(n+1)^2, \quad |b|<a-n^2, \quad |b|<(n+1)^2-a.$$

It is readily checked that both these domains are subsets of (3.15).

The domains (3.15) may be enlarged by using the theorems of §2.4.

It is worth noting that as $k\longrightarrow\infty$ the function $p(t)=a+b\sin^{2k+1}2t$ approaches a constant in the integral sense. Thus the actual stability domains tend to fill out the entire half-plane $a>0$ as $k\longrightarrow\infty$. The same is true of the domains (3.15) as $k\longrightarrow\infty$. (By contrast, the domains given by the tests of Lyapunov and Zhukovskii are independent of k.)

We now proceed to estimate the exponent of exponential growth for solutions of equation (3.1) (see §2.3). The numbers $\tilde{m}$, $\tilde{M}$ figuring in the tests of §2.3 are defined by (2.58) or, in our case, (3.11). The test of §2.3 implies the following

Test 2a. Define numbers $\tilde{m}$, $\tilde{M}$ by (3.11), (3.9), where $c^2>0$ is arbitrary. Suppose that for some integer n

$$\left.\begin{array}{c} (n-1)\pi<\tilde{m}\leqslant n\pi<\tilde{M}<(n+1)\pi, \\ \cos(n\pi-\tilde{m})+\cos(\tilde{M}-n\pi)>0. \end{array}\right\} \tag{3.16}$$

Then the exponent $\lambda\geqslant 0$ of exponential growth for solutions of equation (3.1) satisfies the inequality

$$\operatorname{ch}\lambda T\leqslant\operatorname{ch}\lambda_0 T=\frac{1+\cos(\tilde{m}+\tilde{M})}{\cos(n\pi-\tilde{m})+\cos(\tilde{M}-n\pi)}. \tag{3.17}$$

* It is easy to see that if $n=0$ it is sufficient that $0<a<1$, $|b|<N_k\pi\sqrt{a}(1-\sqrt{a})$.

Let us return now to equation (3.2). We introduce a substitution similar to that performed above, but with a variable matrix:

$$\mathbf{x}_1 = \begin{Vmatrix} f(t)^{-1} & 0 \\ 0 & f(t) \end{Vmatrix} \mathbf{x}, \quad \mathbf{x} = \begin{pmatrix} \dot{y} \\ y \end{pmatrix},$$

where $f(t) > 0$ is absolutely continuous and T-periodic. We obtain an equation $\mathbf{J}\dot{\mathbf{x}}_1 = \mathbf{H}_1(t)\mathbf{x}_1$ with Hamiltonian

$$\mathbf{H}_1(t) = \begin{Vmatrix} f^2 & \frac{\dot{f}}{f} \\ \frac{\dot{f}}{f} & \frac{p}{f^2} \end{Vmatrix}. \tag{3.18}$$

It is clear that this equation is stable if and only if the original equation is stable. Even more: it follows from Theorem II of §1.9 that the Hamiltonians $\mathbf{H}(t)$ and $\mathbf{H}_1(t)$ belong to the same stability or instability domain. Let us assume that the function $p(t)$ in (3.1) is a positive, absolutely continuous function, and set

$$f(t) = \sqrt[4]{p(t)}.$$

Then the smallest and largest eigenvalues of the Hamiltonian (3.18) are

$$h_{\min}(t) = \sqrt{p} - \frac{1}{\sqrt[4]{p}} \left| \frac{d\sqrt[4]{p}}{dt} \right|, \quad h_{\max}(t) = \sqrt{p} + \frac{1}{\sqrt[4]{p}} \left| \frac{d\sqrt[4]{p}}{dt} \right|.$$

Observing that

$$\int_0^T \frac{1}{\sqrt[4]{p}} \left| \frac{d\sqrt[4]{p}}{dt} \right| dt = \frac{1}{2} \ln \frac{p(i_1)\ldots p(i_s)}{p(j_1)\ldots p(j_s)},$$

where $i_1, \ldots, i_s$ are the maximum points and $j_1, \ldots, j_s$ the minimum points of $p(t)$ on the half-open interval $0 \leqslant t < T$, we see that

$$\left.\begin{aligned} \tilde{m} &= \int_0^T \sqrt{p(t)}\,dt - \frac{1}{2} \ln \frac{p(i_1)\ldots p(i_s)}{p(j_1)\ldots p(j_s)}, \\ \tilde{M} &= \int_0^T \sqrt{p(t)}\,dt + \frac{1}{2} \ln \frac{p(i_1)\ldots p(i_s)}{p(j_1)\ldots p(j_s)}. \end{aligned}\right\} \tag{3.19}$$

Applying the test of §2.1, we obtain the following test, first established (using a different method) by Burdina /128b/.

T e s t 3 *(Burdina). Let the function* $p(t)$ *in (3.1) be a positive and absolutely continuous function and* $i_1, i_2, \ldots, i_s$ *its maximum points,* $j_1, j_2, \ldots, j_s$ *its minimum points in the interval* $0 \leqslant t < T$. *If*

$$n\pi < \tilde{m} \leqslant \tilde{M} < (n+1)\pi$$

for some integer $n = 0, 1, 2, \ldots,$ *where* $\tilde{m}, \tilde{M}$ *are defined by (3.19), then the trivial solution of equation (3.1) is stable and* $p(t) \in \mathfrak{G}_n$.

The test of §2.4 implies

Test 3a. Under the assumptions of Test 3, let the numbers $\tilde{m}$, $\tilde{M}$ be defined by (3.19) and suppose that (3.16) holds for some integer n. Then the exponent λ of exponential growth of solutions of equation (3.1) satisfies inequality (3.17).

Similar results are valid for any other choice of the function $f(t)$. The numbers $\tilde{m}$, $\tilde{M}$ are defined by formulas (3.10), where $h_{\min}(t) \leqslant h_{\max}(t)$ are the eigenvalues of the matrix (3.18). If (3.13) holds, we have stability, and $p(t) \in \mathfrak{G}_n$; if (3.14), (3.16) hold, we have the estimate (3.17).

3.5. Stability and instability domains on the plane (two parameters). Mathieu's equation

Consider the equation

$$\ddot{y} + [a + bq(t)]\, y = 0, \tag{3.20}$$

where $q(t)$ is a continuous function such that

$$q\left(t + \frac{\pi}{2}\right) = -q(t), \quad \max q(t) = -\min q(t) = 1. \tag{3.21}$$

The first condition implies $q(t+\pi) = q(t)$, so that $T = \pi$.

The Hamiltonian (3.3) for equation (3.20) is

$$\mathbf{H}(t) = \begin{Vmatrix} 1 & 0 \\ 0 & a + bq(t) \end{Vmatrix}. \tag{3.22}$$

Formula (3.22) defines a mapping taking the points (a, b) of the (a,b)-plane onto some two-dimensional manifold in the Hamiltonian space $\mathfrak{L}^3$. In this sense, the (a,b)-plane "cuts" the function space $\mathfrak{L}^3$. The sets $\mathfrak{G}_n$, $\mathcal{H}_n$, $\Pi_n^{\bullet +}$, $\Pi_n^{\bullet -}$, $\Pi_n^{\bullet\bullet}$ in $\mathfrak{L}^3$ define "projections" on the (a,b)-plane, which we shall continue to denote by the same letters. In other words, we shall say that a point (a, b) belongs to one of the sets in $\mathfrak{L}^3$ if the corresponding Hamiltonian (3.22) belongs to the set.

In § VII. 1.7 we established a number of qualitative properties characterizing the positions of the stability and instability domains in the (a,b)-plane. We shall reestablish some of these properties here, using only the results of this chapter, and in addition some new properties. (Some of the latter will follow specifically from the assumptions (3.21).) The main purpose of the following arguments is to illustrate the qualitative theorems of §1.2 and some of the simplest tests of this section and § VII. 2.

Consider the set of all points $(a, 0)$ on the (a,b)-plane. For $b=0$, $a<0$, the solutions of equation (3.20) are nonoscillating and there are exponentially increasing solutions. Therefore $(a, 0) \in \mathcal{H}_0$ for $a<0$. If $a = n^2$, $b=0$, where n is an integer, then all solutions of equation (3.20) are π-periodic (odd n) or π-antiperiodic (even n). Therefore $(n^2, 0) \in \Pi_n^{\bullet\bullet}$. If $n^2 < a < (n+1)^2$, $b=0$, all solutions of equation (3.20) are bounded functions. By Zhukovskii's test, the corresponding points $(a, 0)$ belong to $\mathcal{O}_n$. (This conclusion may also be reached by the use of rotation functions.) We have thus classified the points $(a, 0)$ as to membership in stability/instability domains.

We now consider points (a, b) for small $|b|$. Since $\mathcal{H}_0$, $\mathcal{O}_n$ are open sets, it follows that any point (a, b) with sufficiently small $|b|$ will lie in $\mathcal{H}_0$ if $a<0$ and in $\mathcal{O}_n$ if $n^2 < a < (n+1)^2$. An examination of the model $\mathfrak{R}^3$ (Figure 56) shows that in the neighborhood of the axis $b=0$ the domains $\mathcal{H}_n$ on the (a, b)-plane are necessarily wedge-shaped, with their "cusps" at the points $(n^2, 0)$.

The results of Chapter V (see §1.5) imply the following property:

1) *The set of points (a, b) in the neighborhood of a point $(n^2, 0)$, for which the trivial solution of equation (3.1) is exponentially unstable, has the shape of a wedge*

$$a_n^-(b) < a < a_n^+(b),$$

where

$$a_n^{\pm}(b) = n^2 \pm n\sqrt{2}\,|q^{(m)}|\cdot|b| + O(b^2) \tag{3.23}$$

and

$$q^{(m)} = \int_0^{\pi} e^{-2imt} q(t)\,dt. \tag{3.24}$$

(It follows from the first equality of (3.21) that the mean value of $q(t)$ is 0.) In the neighborhood of $(n^2, 0)$, the curves $a=a_n^{\pm}(b)$, $b \neq 0$, coincide with the sets $\Pi_n^{*\pm}$. Indeed, by the results of § VII. 1, equation (3.1) has one π-periodic (even n) or one π-antiperiodic (odd n) solution on these curves (provided $b \neq 0$), so that their points (a, b) are in the sets Π^{*-}, Π^{*+}. The fact that the curves $a=a_n^-(b)$ $(a=a_n^+(b))$ correspond to the set $\Pi_n^{*-}(\Pi_n^{*+})$ is verified, for example, by examination of the model $\mathfrak{R}^3$ (in fact, it was precisely in order to guarantee this correspondence that b was replaced by $|b|$ in (3.23)).

The points $(a, b) \in \Pi$ may fill out only curves on the (a, b)-plane. Indeed, for these points we have $A(a, b) = \pm 1$, where $A(a, b)$ is the Lyapunov characteristic constant, which is an analytic function of a and b.

Examination of the model $\mathfrak{R}^3$ immediately implies the following property:

2) *The boundaries of the domain $\mathcal{O}_n$ on the (a, b)-plane (which includes the segment $b=0$, $n^2 < a < (n+1)^2$) are disjoint.*

We note that the closures of the boundaries Π_n^{*-}, Π_n^{*+} (i.e., the continuations of the curves $a=a_n^{\pm}(b)$) need not be disjoint (their common points will be in Π_n^{**}).

By the first equality of (3.21), the substitution $t \to t+\frac{\pi}{2}$ converts an equation (3.20) with parameters (a, b) to an equation of the same form with parameters $(a, -b)$. But, as is true in general of any transformation $t \to t+t_0$, the new Hamiltonian must be in the same set $\mathcal{O}_n$, $\mathcal{H}_n$, Π_n^{*+}, Π_n^{*-}, Π_n^{**} as before (see Remark 2 to Theorem II, §1.9). Therefore:

3) *Each of the sets* $\mathfrak{S}_n$, $\mathcal{H}_n$, $\Pi_n^{\bullet +}$, $\Pi_n^{\bullet -}$, $\Pi_n^{\bullet\bullet}$ *in the* (a, b)*-plane is symmetric with respect to the* a*-axis.*

We now define a semi-ordering relation on the (a, b)-plane, setting

$$(a_1, b_1) \geqslant (a_2, b_2),$$

if and only if

$$a_1 + b_1 q(t) \geqslant a_2 + b_2 q(t)$$

for all $0 \leqslant t \leqslant \pi$; this is clearly equivalent to the inequality

$$a_1 - a_2 \geqslant |b_1 - b_2|.$$

Thus the points (a, b) for which

$$(a_1, b_1) \leqslant (a, b) \leqslant (a_2, b_2) \tag{3.25}$$

lie within and on the boundary of the rectangle with opposite vertices at the points (a_1, b_1) and (a_2, b_2) and sides parallel to the bisectors of the coordinate angles. The inequality $(a_1, b_1) > (a_2, b_2)$ is defined similarly (with $\geqslant$ replaced by $>$). Theorem II of 3.2 implies:

4) *The stability and instability domains on the* (a, b)*-plane are directionally convex (with respect to the above semi-ordering): if* (a_1, b_1) *and* (a_2, b_2) *are points in one stability (instability) domain, this domain contains the entire rectangle of points* (a, b) *defined by (3.25).*

This fundamental property has two corollaries:

5) *Any connected component of the stability set* $(\mathfrak{S})$ *or instability set* $(\mathcal{H})$ *is simply-connected.*

6) *The boundaries of the stability domains* $\mathfrak{S}_n$ *have "indefinite" directions: if* (a_0, b_0) *is a boundary point, then the points of the cones* $(a, b) > (a_0, b_0)$ *and* $(a, b) < (a_0, b_0)$ *in the neighborhood of this point cannot all lie on the boundary. If there is a tangent to the boundary at* (a_0, b_0), *the angle between the tangent and the* b*-axis is at most* $45°$ *in absolute value.*

Indeed, if there were a "hole" in a connected component of a stability (instability) domain, we could deduce, taking points (a_1, b) and (a_2, b) in the component on either side of the "hole," that there is a point (a_0, b), $a_1 < a_0 < a_2$, not in the domain. But $(a_1, b) \leqslant (a_0, b) \leqslant (a_2, b)$ and so by property 4 this is impossible. This proves property 5.

Now suppose that the boundary of a stability domain has a definite direction in the neighborhood of some point $(a_0, b_0) \in \Pi$. Then there are points $(a_1, b_1) \in \mathfrak{S}$, $(a_2, b_2) \in \mathfrak{S}$ in this neighborhood such that the interior of the rectangle (3.25) has nonempty intersection with the boundary of $\mathfrak{S}$. Thus the interior of the rectangle contains points $(a, b) \in \mathcal{H}$, which is impossible by virtue of property 4. This proves property 6.

It was shown in 3.2 that $p(t) \in \mathcal{H}_0$ if $p(t) \leqslant 0$, $p(t) \not\equiv 0$. Hence:

7) *Any point* (a, b) *such that* $a < 0$, $|b| \leqslant a$ *lies in* $\mathcal{H}_0$.

If $p_{av} \geqslant 0$, $p \not\equiv 0$, equation (3.1) is oscillating (see §3.2). Therefore:

8) *The domain* $\mathcal{H}_0$ *lies entirely in the half-plane* $a < 0$.

All these properties are illustrated by Figure 70, which shows the stability and instability domains for Mathieu's equation

$$\ddot{y}+(a+b\cos 2t)\,y=0. \tag{3.26}$$

(The boundaries Π_n^{*+}, Π_n^{*-} of the stability and instability domains are hatched parallel to the a-axis.) We know from § V.1.5 that the only wide instability domain for this equation is $\mathscr{H}_1$, which is "attached" at the point $a=1$, $b=0$. (Indeed, in this case, in the notation of (3.24) we have only $q^{(1)}\neq 0$; the other Fourier coefficients $q^{(m)}$ vanish.)

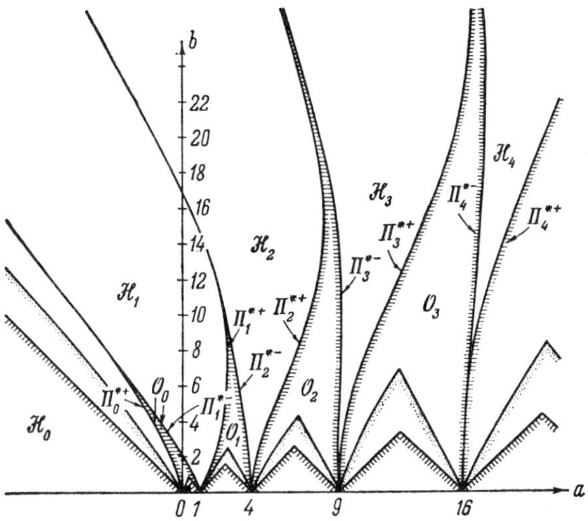

FIGURE 70.

An examination of Figure 70 suggests that for large $|b|$ all the stability domains tend to "bend backward" toward negative a values. This is indeed so. Even more: we can prove the following general property of equation (3.20):

9) *For any* $\varkappa$, $|\varkappa|>1$, *the ray* $b=\varkappa a$, $a<0$, *contains points* (a, b) *of* $\mathcal{O}_n$ *and points of* $\mathscr{H}_n$, *for any* $n>0$.

Thus any stability domain $\mathcal{O}_n$ and any instability domain $\mathscr{H}_n$ $(n>0)$ will contain points with large $|b|$ in an arbitrarily narrow sector $-a<b<(-1-\varepsilon)\,a$, $a<0$ (i.e., for arbitrarily small $\varepsilon>0$).

To prove this property, consider the straight line $b=\varkappa a$ on the (a,b)-plane, where $|\varkappa|>1$. The corresponding coefficient of equation (3.20), $p_\varkappa(t,a)=$ $=a[1+\varkappa q(t)]$, is a continuous function of variable sign. As shown in § VII.1.5 (and this will be shown independently again in §6 below), there is an infinite sequence of points

$$\ldots<\lambda_{-1}^-\leqslant\lambda_{-1}^+<\lambda_0^-\leqslant\lambda_0^+=0<\lambda_1^-\leqslant\lambda_1^+<\lambda_2^-\leqslant\lambda_2^+<\ldots$$

on the a-axis, going to infinity in both directions, such that $p_\varkappa(t, a) \in \mathfrak{S}_{|n|}$ if $\lambda_n^- < a < \lambda_{n+1}^+$ and $p_\varkappa(t, a) \in \mathcal{H}_{|n|}$ if $\lambda_n^- < a < \lambda_n^+$. The numbers $\lambda_n^\pm = \lambda_n^\pm(\varkappa)$, as roots of the equation $A(\lambda, \varkappa\lambda) = \pm 1$, are continuous functions of $\varkappa$. This implies the assertion.

The domains in Figure 70 hatched parallel to the bisectors of the coordinate angle are $|b| \leqslant -a$, $a < 0$, which is a subset of $\mathcal{H}_0$ by property 7, and also

$$n^2 \leqslant a - b < a + b \leqslant (n+1)^2 \qquad (n \geqslant 0),$$

which are subsets of the stability domains $\mathfrak{S}_n$ (with the same n) by Zhukovskii's test (§3.3). (Note that $b \geqslant 0$ everywhere.)

Somewhat larger domains (dotted boundaries in Figure 70) are obtained from the first step of the method of §2.4.

The small black triangle in the domain $\mathfrak{S}_0$ is

$$0 < b \leqslant a, \quad 0 \leqslant a \leqslant 0.406\ldots, \tag{3.27}$$

which is obtained by Lyapunov's test

$$p \underset{\not\equiv}{\geqslant} 0, \quad \pi \int_0^\pi p\,dt \leqslant 4. \tag{3.28}$$

Figure 71 illustrates $\mathfrak{S}_0$ on a larger scale. Also shown in the figure are the domains mentioned previously (the cross-hatched triangle is the domain (3.27) given by Lyapunov's test), the domain

$$|b| < \pi\sqrt{a} - a,$$

which follows from Test 1 of §3.4 with $c^2 = a + |b|$, $n = 0$, and the domain

$$a > 0, \quad a + |b| < 1,$$

obtained from the test (3.14) for $n = 0, r_1 \equiv \sqrt{\pi}, c^2 = a$. (Some of the tests in the next section yield even larger domains.)

We now consider the use of Lyapunov's method as described in § VII. 2.1. Following Lyapunov /180g/, we confine attention to the case $p(t) \geqslant 0$, which defines the domain $a \geqslant 0$, $|b| \leqslant a$. Using the formulas for the coefficients A_1, A_2, A_3 at the end of § VII. 2.1, we obtain

$$A_1 = \frac{1}{2}\pi^2 a, \quad A_2 = \frac{1}{24}\pi^4 a^2 - \frac{1}{16}\pi^2 b^2, \quad A_3 = \frac{1}{720}\pi^6 a^3 - \left(\frac{1}{96}\pi^4 - \frac{1}{16}\pi^2\right) ab^2.$$

Recall that

$$A^{(1)} = A_0 - A_1, \quad A^{(2)} = A_0 - A_1 + A_2, \quad A^{(3)} = A_0 - A_1 + A_2 - A_3.$$

The first stability condition (2.17) of § VII. 2.3,

$$-1 \leqslant A^{(1)} < A^{(2)} \leqslant 1,$$

again gives the domain (3.27) (cross-hatched triangle in Figure 71). The second stability condition (2.17) of § VII. 2.3,

$$-1 \leqslant A^{(3)} < A^{(2)} \leqslant 1 \tag{3.29}$$

defines the domain whose boundary is indicated by crosses in Figure 71.

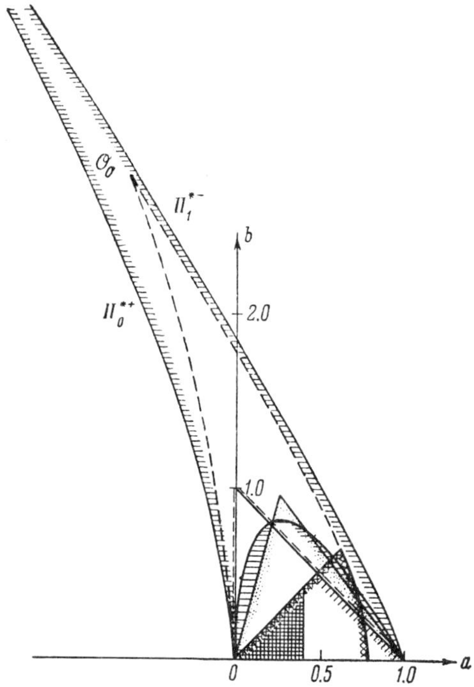

FIGURE 71.

We now consider the instability tests of § VII. 2.3. A few computations show that the domains $A^{(2)} \leqslant -1$ and $A^{(3)} \geqslant 1$ are both disjoint from the domain $|b| \leqslant a$. In this case, therefore, the coefficients $A^{(1)}$, $A^{(2)}$, $A^{(3)}$ alone are insufficient to determine instability domains.

All these arguments corroborate our conclusion in § VII. 2.5 that Lyapunov's method is best applied in cases when one has a priori information as to whether $p(t)$ belongs to a stability (or instability) domain with a small subscript.

The calculations also demonstrate how restrictive the condition $p(t) \geqslant 0$ (in our case: $|b| \leqslant a$) is in Lyapunov's method of /180g/. The domain (3.29) is demarcated in Figure 71 by a dashed line. It is clearly a proper subset of $\mathcal{G}_0$. All that we can say, however, is that $\mathcal{G}_0$ contains the intersection of (3.29) with $|b| \leqslant a$. It is reasonable to conjecture that (3.29) is also a sufficient condition for stability for functions $p(t)$ of variable sign (possibly only those such that $p\left(t+\frac{1}{2}\pi\right)=-p(t)\Big)$.

§4. STABILITY TESTS AND ESTIMATES FOR CHARACTERISTIC EXPONENTS OF HILL'S EQUATION — VARIATIONAL METHODS

4.1. Stability tests and the distance from functions $p(t) \equiv c^2 \in \mathfrak{G}_n$ to Π_n and Π_{n+1}

Denote

$$\| p(t) \| = \int_0^T | p(t) | \, dt. \tag{4.1}$$

A natural approach to the derivation of stability tests is as follows.

Consider a function $p(t) \equiv c^2$, where

$$\frac{n\pi}{T} < c < \frac{(n+1)\pi}{T}. \tag{4.2}$$

It follows, for example, from Zhukovskii's test (§3.3) that this function is in the n-th stability domain $\mathfrak{G}_n$. Denote

$$\Pi_n^{+} = \Pi_n^{\bullet\bullet} \cup \Pi_n^{\bullet +}, \quad \Pi_{n+1}^{-} = \Pi_{n+1}^{\bullet\bullet} \cup \Pi_{n+1}^{\bullet -}.$$

Each of these sets is connected, simply-connected and closed; their union is the boundary of the stability domain $\mathfrak{G}_n$ (see Figure 56 or 68 and §1.6). Set

$$\left. \begin{aligned} \rho_n'(c) &= \inf \| p(t) - c^2 \| \text{ over all } p(t) \in \Pi_n^{+}, \\ \rho_{n+1}''(c) &= \inf \| p(t) - c^2 \| \text{ over all } p(t) \in \Pi_{n+1}^{-}. \end{aligned} \right\} \tag{4.3}$$

Suppose that for some function $p(t)$

$$\| p(t) - c^2 \| < \min [\rho_n'(c), \ \rho_{n+1}''(c)]. \tag{4.4}$$

The geometric import of this condition is that the distance from $c^2 \in \mathfrak{G}_n$ to the function $p(t)$ is less than the distance from c^2 to the boundary of $\mathfrak{G}_n$. It follows from (4.4) that $p(t) \in \mathfrak{G}_n$. Indeed, if $p(t) \notin \mathfrak{G}_n$, the "segment" $p_\lambda(t) = \lambda p(t) + (1-\lambda) c^2$ $(0 \leqslant \lambda \leqslant 1)$ joining the functions $p(t)$ and $c^2 \in \mathfrak{G}_n$ must cut the boundary of $\mathfrak{G}_n$ in some function $p_{\lambda_0}(t)$ $(0 \leqslant \lambda_0 \leqslant 1)$. By (4.3),

$$\| p_{\lambda_0}(t) - c^2 \| > \min [\rho_n'(c), \ \rho_{n+1}''(c)].$$

But

$$\| p_{\lambda_0}(t) - c^2 \| = \lambda_0 \| p(t) - c^2 \| \leqslant \| p(t) - c^2 \| < \min [\rho_n'(c), \ \rho_{n+1}''(c)]$$

by (4.4). This contradiction shows that $p(t) \in \mathfrak{G}_n$.

Thus inequality (4.4) is a test for the stability of equation (3.1). This test will not be effective unless we can find explicit expressions for $\rho_n'(c)$ and $\rho_n''(c)$; we shall do this below.

Defining the norm $\|p\|$ differently, one derives other, analogous tests. Instead of c^2 one can take any function $p_0(t)\in\mathfrak{G}_n$ whose distance from the sets Π_n^+ and Π_{n+1}^- may be determined explicitly.

We claim that the sets Π_n^+, Π_{n+1}^- in (4.3) may be replaced by $\Pi_n=\Pi_n^+\cup\Pi_n^{*-}$ and $\Pi_{n+1}=\Pi_{n+1}^-\cup\Pi_{n+1}^{*+}$, respectively (this is more convenient when evaluating the numbers (4.3)). Indeed, it follows from the structure of the space $\mathfrak{R}^3$ (see Figure 56) that any curve in $\mathfrak{L}^3$ with endpoints at $c^2\in\mathfrak{G}_n$ and $p(t)\in\Pi_n^{*-}$ intersects the set $\Pi_n^+=\Pi_n^{*+}\cup\Pi_n^{**}$. In particular, this is true of the segment $p_\lambda(t)=\lambda p(t)+(1-\lambda)c^2$ $(0\leqslant\lambda\leqslant1)$. Consequently, for any function $p(t)\in\Pi_n^{*-}$ there exists a function $p_{\lambda_0}(t)\in\Pi_n^+$ $(0\leqslant\lambda_0\leqslant1)$ such that $\|p_{\lambda_0}(t)-c^2\|=\lambda_0\|p(t)-c^2\|\leqslant\|p(t)-c^2\|$. Thus the infimum of the norms $\|p(t)-c^2\|$ is the same over $p(t)\in\Pi_n=\Pi_n^{*+}\cup\Pi_n^{**}\cup\Pi_n^{*-}$ as over $p(t)\in\Pi_n^+=\Pi_n^{*+}\cup\Pi_n^{**}$. Similarly, the set Π_{n+1}^- in (4.3) may be replaced by Π_{n+1}. We may thus assume that in (4.4)

$$\left.\begin{aligned}\rho_n'(c)&=\inf\|p(t)-c^2\|\ \text{ over all }\ p(t)\in\Pi_n,\\ \rho_{n+1}''(c)&=\inf\|p(t)-c^2\|\ \text{ over all }\ p(t)\in\Pi_{n+1}.\end{aligned}\right\}\tag{4.5}$$

Note that in general $\rho_{n+1}'(c)\neq\rho_{n+1}''(c)$, since c and the integer n must satisfy (4.2).

We now show how to improve the test (4.4), using the structure of $\mathfrak{L}^3$ and the directional convexity of the domains $\mathfrak{G}_n$.

We first note that the "surfaces" $\tilde{\Pi}_n^+=\tilde{\Pi}_n^{**}$ in $\mathfrak{R}^3$ (see §1.6) are homeomorphic to a plane (see Figure 56); each of them divides $\mathfrak{R}^3\setminus\tilde{\Pi}_n^+$ (or $\mathfrak{R}^3\setminus\tilde{\Pi}_n^-$) into two open subsets, to the "left" and "right" of the surface. We shall accordingly say that $p(t)$ lies in $\mathfrak{L}^3$ to the "left" of $\Pi_n^\pm$, writing $p(t)<\Pi_n^\pm$, or to the "right" of $\Pi_n^\pm$, writing $p(t)>\Pi_n^\pm$, if the projection of the corresponding Hamiltonian in $\mathfrak{R}^3$ lies in the appropriate open subset. For example, $p(t)>\Pi_n^+$ if and only if $p(t)\in\mathfrak{G}_n\cup\Pi_{n+1}\cup\mathscr{H}_{n+1}\cup\ldots$ (see Figure 68).

If either $p(t)\in\Pi_n^+$ or $p(t)>\Pi_n^+$ and $p'(t)\geqslant p(t)$, then $p'(t)$ lies to the right of Π_n^+: $p'(t)>\Pi_n^+$. Indeed, by the third formula of (3.3), the corresponding Hamiltonians are such that $\mathbf{H}'(t)\geqslant\mathbf{H}(t)$. By the comparison theorem (§2.4), the rotation functions satisfy the inequality $\psi(\chi)'\geqslant\psi(\chi)$. By Theorem II of §1.9, $p'(t)$ lies to the right of Π_n^+.

Similarly one shows that if $p'(t)\leqslant p(t)$ and either $p(t)<\Pi_{n+1}^-$ or $p(t)\in\Pi_{n+1}^-$, then $p'(t)<\Pi_{n+1}^-$. We now show that if $\|p(t)-c^2\|<\rho_n'(c)$, where $c^2\geqslant n^2\pi^2/T^2$, then $p(t)>\Pi_n^+$. Indeed, otherwise, either $p(t)\in\Pi_n^+$ or $p(t)<\Pi_n^+$. Since the function $c^2\in\mathfrak{L}^3$ lies to the right of Π_n^+ (as $c^2>n^2\pi^2/T^2$), it follows that the segment $p_\lambda(t)=\lambda p(t)+(1-\lambda)c^2$ cuts Π_n^+, i.e., $p_{\lambda_0}(t)\in\Pi_n^+$ for some $0\leqslant\lambda_0<1$. (We are using the structure of $\mathfrak{L}^3$ as established in §1.6, illustrated in Figure 68, and also the representation (1.19).) Then by (4.3) we have $\|p_{\lambda_0}(t)-c^2\|\geqslant\rho_n'(c)$. On the other hand, $\|p_{\lambda_0}(t)-c^2\|=\lambda_0\|p(t)-c^2\|<\rho_n'(c)$, since by assumption $\|p(t)-c^2\|<\rho_n'(c)$.

Similarly one shows that if $\|p(t)-c^2\|<\rho_{n+1}''(c)$, where $c^2\leqslant(n+1)^2\pi^2/T^2$, then $p(t)<\Pi_{n+1}^-$.

Consider the functions

$$p_c^+(t)=\begin{cases}p(t), & \text{if}\quad p(t)\geqslant c^2,\\ c^2, & \text{if}\quad p(t)<c^2,\end{cases}\qquad p_c^-(t)=\begin{cases}c^2, & \text{if}\quad p(t)\geqslant c^2,\\ p(t), & \text{if}\quad p(t)<c^2.\end{cases}\tag{4.6}$$

Suppose that

$$\| p_c^-(t) - c_1^2 \| < \rho_n'(c_1), \quad \| p_c^+(t) - c_2^2 \| \leqslant \rho_{n+1}''(c_2), \tag{4.7}$$

where c^2 is an arbitrary number (possibly also negative) and

$$c_1^2 > \left(\frac{n\pi}{T}\right)^2, \quad c_2^2 < \left[\frac{(n+1)\pi}{T}\right]^2. \tag{4.8}$$

Here c_2^2 may also be negative. By the foregoing discussion, the first inequality of (4.7) means that $p_c^-(t) \succ \Pi_n^+$, and the second that $p_c^+(t) \prec \Pi_{n+1}^-$. Since $p_c^-(t) \leqslant p(t) \leqslant p_c^+(t)$, it follows by what we have proved that $p(t) \succ \Pi_n^+$, $p(t) \prec \Pi_{n+1}^-$, i. e., $p(t) \in \mathfrak{G}_n$.

Thus, if inequality (4.7) holds, Hill's equation is stable and $p(t) \in \mathfrak{G}_n$. Here $-\infty < c^2 < +\infty$, c_1, c_2 are arbitrary constants satisfying conditions (4.8), and $p_c^+(t)$, $p_c^-(t)$ are the functions (4.6).

The test (4.7) is superior (or, more precisely, not inferior) to (4.4): since

$$\| p_c^{\pm}(t) - c^2 \| \leqslant \| p(t) - c^2 \|,$$

it follows that if (4.4) holds then so does (4.7), with $c_1 = c_2 = c$.

As in (4.4), we may replace the norm (4.1) in (4.7) by any other norm with the usual properties. We may then use different norms in the two inequalities (4.7).

It follows from our discussion that the first (second) inequality of (4.7) may be replaced by any inequality guaranteeing that $p(t) \succ \Pi_n^+$ ($p(t) \prec \Pi_{n+1}^-$). Hence the conditions

$$p(t) \underset{\not\equiv}{\geqslant} \frac{n^2\pi^2}{T^2}, \quad \| p_c^+(t) - c_2^2 \| < \rho_{n+1}''(c_2), \tag{4.9}$$

where $c_2^2 < (n+1)^2\pi^2/T^2$, and also the conditions

$$p(t) \underset{\not\equiv}{\leqslant} \frac{(n+1)^2\pi^2}{T^2}, \quad \| p_c^-(t) - c_1^2 \| < \rho_n'(c_1), \tag{4.10}$$

where $c_1^2 > n^2\pi^2/T^2$, guarantee that $p(t) \in \mathfrak{G}_n$.

Indeed, since $c_2^2 < (n+1)^2\pi^2/T^2$, we have $c_2^2 \prec \Pi_{n+1}$. By the second inequality of (4.9), $p_c^+(t) \prec \Pi_{n+1}$. Since $p(t) \leqslant p_c^+(t)$, it is also true that $p(t) \prec \Pi_{n+1}$. By the first inequality of (4.9), $p(t) \succ \Pi_n$. Consequently, $\Pi_n \prec p(t) \prec \Pi_{n+1}$, so that $p(t) \in \mathfrak{G}_n$.

Conditions (4.10) are treated similarly. The constant c^2 in (4.9) and (4.10) is arbitrary (possibly negative). The value of c_2^2 may also be negative.

It follows from the theorem of §3.3 (characterization of functions $p(t) \in \Pi_n^{**}$) that the first inequalities in tests (4.9), (4.10) may be replaced by inequalities $p(t) \underset{\not\equiv}{\geqslant} p_n(t)$ (in (4.9)) or $p(t) \underset{\not\equiv}{\leqslant} p_{n+1}(t)$ (in (4.10)), where $p_n(t)$ is any function of the form (3.5) and $r(t)$ satisfies the assumptions of the theorem of §3.3.

To end this subsection, we note that the condition $p > \Pi_0^+$ holds if and only if (see §3.2) all solutions of equation (3.1) are oscillating. Thus the condition $\|p_c^-(t) - c_1^2\| < \rho_0'(c_1)$ in (4.9), (4.10), which guarantees that $p > \Pi_0^+$, may be replaced by

$$p(t) \not\equiv 0, \quad p_{\mathrm{av}} = \frac{1}{T}\int_0^T p(t)\,dt \geqslant 0.$$

Indeed, as shown in §3.2, this condition is also sufficient for all solutions of equation (3.1) to be oscillating.

4.2. Reduction of problem to two special variational problems. Borg's lemma

Let $p(t) \in \Pi_n$, $n \geqslant 1$, and let $y(t)$ be a periodic or antiperiodic solution of the corresponding equation (3.1). The function $y(t)$ has the following properties:

a) $y(t_0) = y(t_1) = \ldots = y(t_n) = 0 \quad (t_n = t_0 + T)$; $y(t) \neq 0$ for $t \neq t_i$;

b) y and $\dot{y}$ are absolutely continuous, $\ddot{y}/y$ is summable on $[t_0, t_0+T]$;

c) $\dot{y}(t_0+T) = (-1)^n \dot{y}(t_0)$.

Conversely, any function y with properties (a), (b), (c) defines a function $p(t) = -\ddot{y}/y \in \Pi_n$.

Obviously,

$$\int_0^T |p(t) - c^2|\,dt = \int_{t_0}^{t_0+T} |p(t) - c^2|\,dt = \int_{t_0}^{t_0+T} \left|\frac{\ddot{y}}{y} + c^2\right| dt.$$

The problem of determining $\rho_n'(c)$ in (4.5) leads to a variational problem: to find

$$\rho_n(c) = \inf \int_{t_0}^{t_0+T} \left|\frac{\ddot{y}}{y} + c^2\right| dt,$$

where the infimum extends over all functions y satisfying conditions (a), (b), (c).

The same problem is involved in determining $\rho_n''(c)$ in (4.5). The values of $\rho_n'(c)$, $\rho_n''(c)$ in (4.10), (4.9) are expressed in terms of $\rho_n(c)$ as follows:

$$\left.\begin{aligned} \rho_n'(c) &= \rho_n(c) \quad \text{if} \quad c^2 \geqslant \frac{n^2\pi^2}{T^2}, \\ \rho_n''(c) &= \rho_n(c) \quad \text{if} \quad c^2 < \frac{n^2\pi^2}{T^2}. \end{aligned}\right\} \qquad (4.11)$$

(In the second case, c^2 may be negative.) Now

$$\int_{t_0}^{t_0+T} \left|\frac{\ddot{y}}{y} + c^2\right| dt = \sum_{i=0}^{n-1} \int_{t_i}^{t_{i+1}} \left|\frac{\ddot{y}}{y} + c^2\right| dt$$

and $y(t)$ may be varied independently on each interval $[t_i, t_{i+1}]$, provided one observes the conditions $y(t_i)=y(t_{i+1})=0$ and leaves $\dot{y}(t_i)$, $\dot{y}(t_{i+1})$ unchanged; thus the determination of $\rho_n(c)$ amounts to finding

$$\inf\int_0^{\tau}\left|\frac{\ddot{y}}{y}+c^2\right|dt \tag{4.12}$$

where the infimum extends over all functions $y(t)$ satisfying the following conditions:

(a') $\qquad y(0)=y(\tau)=0, \quad y(t)>0 \text{ for } 0<t<\tau;$

(b') y and $\dot{y}$ are absolutely continuous; $\ddot{y}/y$ is Lebesgue-integrable on $[0, \tau]$, and also the conditions: $\dot{y}(0)=\varkappa_1$, $\dot{y}(\tau)=\varkappa_2$, where $\varkappa_1$ and $\varkappa_2$ are fixed.

We first consider the problem of determining (4.12) when the infimum is taken over all functions satisfying only conditions (a') and (b').

The following lemma of Borg /234a/ enables us to drop the absolute value symbol in the integral (4.12), thus simplifying the problem.

Borg's lemma. Let $y(t)$ satisfy conditions (a') and (b').

1. *If $c^2\tau^2>\pi^2$, there exists a function $h(t)$ (which we call the inner c^2-envelope of $y(t)$) satisfying conditions (a') and (b') and also:*

(I) $\qquad h(t)\leqslant y(t), \quad \ddot{h}+c^2h\geqslant 0,$

(II) *$h(t)=y(t)$ on some closed set E and $\ddot{h}+c^2h=0$ on the complement $[0, \tau]-E$.*

2. *If $c^2\tau^2<\pi^2$ (in particular, if c is pure imaginary), there exists a function $H(t)$ (the outer c^2-envelope of $y(t)$) satisfying conditions (a') and (b') and also:*

(I) $\qquad H(t)\geqslant y(t), \quad \ddot{H}+c^2H\leqslant 0,$

(II) *$H(t)=y(t)$ on some closed set E' and $\ddot{H}+c^2H=0$ on the complement $[0, \tau]-E'$.*

Proof. Let $c^2\tau^2>\pi^2$, so that the distance between any two consecutive zeros of the function $\varphi(t)=a\sin c\,(t-t')$ is less than τ. It is clear that for $0\leqslant t'<\tau-\pi/c$ and sufficiently small a we have $a\sin[c(t-t')]\leqslant y(t)$ (see Figure 72). We now increase a till the sine curve $a\sin c(t-t')$ touches the curve $y(t)$. Let E be the set of t values for which this can be done. Since the sine curve $a\sin c(t-t')$ touches $y(t)$ from below on the set E, it is readily seen that $\ddot{y}+c^2y\geqslant 0$ on E. Set $h(t)=y(t)$ on E. The set E is closed. Therefor its complement $[0, \tau]-E$ is a union of at most countably many intervals. On each of these intervals we set $h(t)$ equal to the specific sine function touching $y(t)$ at the endpoints of the interval. Then $\ddot{h}+c^2h=0$ on $[0,\tau]-E$. It is clear that the function $h(t)$ thus obtained satisfies all our requirements.

In case 2, $c^2\tau^2<\pi^2$, we consider the family of functions $\varphi(t)=a\sin c(t-t')$ if $c^2>0$, the functions $\varphi(t)=a(t-t')$ if $c=0$, and $\varphi(t)=a\,\mathrm{sh}\,c_1(t-t')$ if $c^2=-c_1^2<0$ $(-\infty<a<+\infty, 0\leqslant t'\leqslant\tau)$. The proof then follows the same lines as above, except that now the curves $\varphi(t)$ will lie above the curve $y(t)$.

We now go back to the problem of determining the infimum (4.12) over the class of functions satisfying conditions (a'), (b').

Let y be any function of this class and H its outer c^2-envelope. By conditions (I), (II),

$$\frac{\ddot{y}}{y}+c^2=\frac{\ddot{H}}{H}+c^2 \text{ on } E', \quad \left|\frac{\ddot{y}}{y}+c^2\right|\geqslant\frac{\ddot{H}}{H}+c^2=0 \text{ on } (0,\ \tau)-E'$$

(we are excluding from E' the points $t=0$, $t=\tau$, at which $y=H=0$). Therefore

$$\left|\frac{\ddot{y}}{y}+c^2\right|\geqslant\left|\frac{\ddot{H}}{H}+c^2\right|=-\left(\frac{\ddot{H}}{H}+c^2\right).$$

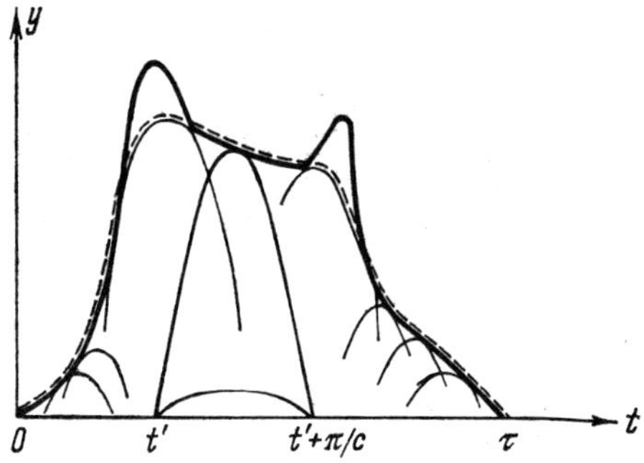

FIGURE 72.

Consequently, if $c^2\tau^2<\pi^2$ we may assume from the start that the problem is being solved for the outer c^2-envelope $H(t)$. Thus, if $c^2\tau^2<\pi^2$ the infimum (4.12) over functions y satisfying (a′), (b′) is equal to

$$\inf\int_0^\tau\left(-\frac{\ddot{y}}{y}-c^2\right)dt \tag{4.13}$$

over functions satisfying (a′), (b′) and also

(c$_1'$) $\qquad \ddot{y}+c^2y\leqslant 0$, i. e., $\ddot{y}/y\leqslant -c^2$.

Similarly, if $c^2\tau^2>\pi^2$, it follows from Borg's lemma that the infimum (4.12) over functions satisfying (a′), (b′) is equal to

$$\inf\int_0^\tau\left(\frac{\ddot{y}}{y}+c^2\right)dt \tag{4.14}$$

over functions satisfying (a′), (b′) and also

(c$_2'$) $\qquad \ddot{y}+c^2y\geqslant 0$, i. e., $\ddot{y}/y\geqslant -c^2$.

It is clear that problem (4.13) is equivalent to determining the supremum

$$\sup\int_0^\tau\left(\frac{\ddot{y}}{y}+c^2\right)dt.$$

In the next subsection we shall prove some theorems on the solution of these and some more general variational problems, which will be needed later.

4.3. Solution of variational problems of §4.2

To summarize, we are concerned with the following variational problems.

I) To find

$$\inf \int_0^\tau \frac{\ddot{y}}{y}\,dt \tag{4.15}$$

over all functions satisfying conditions (a'), (b') and the condition

$$-c^2 \leqslant \frac{\ddot{y}}{y} < +\infty.$$

(II) To find

$$\sup \int_0^\tau \frac{\ddot{y}}{y}\,dt \tag{4.16}$$

over all functions satisfying conditions (a'), (b') and the condition

$$-\infty \leqslant \frac{\ddot{y}}{y} < -c^2.$$

Fix $\dot{y}(0)$ and $\dot{y}(\tau)$. We may assume without loss of generality that $\dot{y}(0)=1$. Extending the problem, we set ourselves the goal of finding

$$\inf \int_0^\tau \frac{\ddot{y}}{y}\,dt \text{ and } \sup \int_0^\tau \frac{\ddot{y}}{y}\,dt$$

in the class of functions y satisfying the following conditions:

(a'') $y(0)=y(\tau)=0, \quad \dot{y}(0)=1, \quad \dot{y}(\tau)=-q<0, \quad y(t)>0$ for $0<t<\tau$.

(b'') The functions y and $\dot{y}$ are absolutely continuous; the function $\ddot{y}/y$ is Lebesgue-integrable on $[0, \tau]$.

(c'') We have $-a^2 \leqslant \ddot{y}/y \leqslant b^2$ almost everywhere on $(0, \tau)$, where a is a real number and b is either real or pure imaginary $(0 \leqslant a^2 \leqslant \infty, -a^2 \leqslant b^2 \leqslant \infty)$.

We shall need the solution to this more general problem in the sequel.

Note that the class of functions satisfying conditions (a''), (b''), (c'') is nonempty only if $-b^2 \leqslant \pi^2/\tau^2 \leqslant a^2$, as we shall indeed assume. In fact, setting $p(t)=-\ddot{y}/y$, we see from (c'') that $-b^2 \leqslant p(t) \leqslant a^2$. By the comparison theorem /87/, the distance τ between consecutive zeros of a solution $y(t)$ of the equation $\ddot{y}+p(t)y=0$ satisfies the inequality $\tau \geqslant \pi/a$ and, if $-b^2=b_1^2>0$, also $\tau \leqslant \pi/b_1$, as asserted. Moreover, by the comparison theorem, $p(t)=a^2$ almost everywhere if $\tau=\pi/a$ and $p(t)=b_1^2 \geqslant 0$ almost everywhere if $\tau=\pi/b_1$. Thus, if either $\tau=\pi/a$ or $\tau=\pi/b_1$ (the latter, in the case when $b_1^2=-b^2>0$), the above problems become meaningless, since the class of functions contains only one function

$$y=\frac{1}{2}\sin at \quad \text{or} \quad y=\frac{1}{b_1}\sin b_1 t$$

if $q=1$, and is empty if $q\neq 1$.

Let

$$J(y)=\int_0^\tau \frac{\ddot{y}}{y}\,dt.$$

It is easy to see that Euler's equation for $J(y)$ is in this case $\frac{d}{dt}\left(\frac{\dot{y}}{y}\right)=0$. The general solution is $y=Ce^{Dt}$, where C and D are arbitrary constants; this function is not in the class of admissible functions. Thus the standard methods of the calculus of variations are useless here.

Our variational problems will be solved by means of two theorems, cited below and proved in Appendices 1 and 2 at the end of the book. The proof in Appendix 1 will use no additional material; the (simpler) proof in Appendix 2 will be based on Pontryagin's maximum principle /76c/.

Theorem I. The infimum (4.15) for functions satisfying conditions (a''), (b''), (c'') is achieved at a function $y=y_0(t)$ defined as follows:

1) *Let $y_1(t)$ and $y_2(t)$ be two sine functions solving the equation $\ddot{y}+a^2y=0$ and satisfying the conditions*

$$y_2(\tau)=0, \quad \dot{y}_2(\tau)=-q; \quad y_1(0)=0, \quad \dot{y}_1(0)=1.$$

2) *Join the sine curves smoothly by a curve $y_3=Ce^{Dt}$ (solution of Euler's equation), i.e., determine numbers $\xi, \eta\ (0\leqslant\xi\leqslant\eta\leqslant\tau)$ and a function*

$$y_0(t)=\begin{cases} y_1(t) & \text{if } 0\leqslant t\leqslant\xi,\\ y_3(t) & \text{if } \xi<t\leqslant\eta,\\ y_2(t) & \text{if } \eta<t\leqslant\tau\end{cases}$$

in such a way that $y_0(t)$ and $\dot{y}_0(t)$ are continuous at $t=\xi$ and $t=\eta$ (Figure 73). The continuity conditions determine the numbers C and D. If the function thus defined is admissible ($D^2\leqslant b^2$), it solves the problem.

3) *If $D^2>b^2$, the sine curves are joined smoothly by a curve*

$$y=Ae^{bt}+Be^{-bt}.$$

The constants A, B are determined by the "joining" conditions – continuity of the final function and its derivatives at the points $t=\xi$, $t=\eta$.

The resulting function is the solution $y_0(t)$.

Simple calculations show (see Appendix 1) that

$$D=\frac{\ln q}{\tau-\frac{\pi}{a}}.$$

Thus, on the inner interval $\xi \leqslant t \leqslant \eta$

$$y_0 = Ce^{Dt} \quad \text{if} \quad b^2\left(\tau - \frac{\pi}{a}\right)^2 \geqslant (\ln q)^2,$$

$$y_0 = Ae^{bt} + Be^{-bt} \quad \text{if} \quad b^2\left(\tau - \frac{\pi}{a}\right)^2 < (\ln q)^2.$$

Theorem II. *If $b^2 = -b_1^2 \leqslant 0$, the supremum (4.16) in the class of functions satisfying conditions (a''), (b''), (c'') is achieved at a function $y = y_0(t)$ defined as follows. Join the sine curves**

$$y_1(t) = \frac{1}{b_1}\sin b_1 t, \quad y_2(t) = \frac{q}{b_1}\sin b_1(\tau - t) \tag{4.17}$$

smoothly by a third sine curve

$$y_3 = C_1 \sin at + C_2 \cos at,$$

i.e., determine numbers ξ, η $(0 \leqslant \xi \leqslant \eta \leqslant \tau)$ and a function

$$y_0(t) = \begin{cases} y_1(t) & \text{if} \quad 0 \leqslant t \leqslant \xi, \\ y_3(t) & \text{if} \quad \xi < t \leqslant \eta, \\ y_2(t) & \text{if} \quad \eta < t \leqslant \tau \end{cases}$$

such that $y_0(t)$ and $\dot{y}_0(t)$ are continuous at $t = \xi$ and $t = \eta$ (Figure 74). The continuity conditions determine the constants C_1, C_2. The function $y_0(t)$ solves the problem.

2) *If $b^2 \gtreqless 0$, $a^2 = \infty$, then*

$$\sup \int_0^\tau \frac{\ddot{y}}{y}\,dt$$

is not achieved in the above class of functions. Its numerical value is

$$b_2\tau + \frac{\dot{y}_2(\xi)}{y_2(\xi)} - \frac{\dot{y}_1(\xi)}{y_1(\xi)}, \tag{4.18}$$

where

$$y_1(t) = \frac{1}{b}\operatorname{sh} bt, \qquad y_2(t) = \frac{q}{b}\operatorname{sh} b(t - \tau), \tag{4.19}$$

and ξ, $0 < \xi < \tau_1$, is the abscissa of the point of intersection of the curves

$$y_1 = y_1(t), \quad y_2 = y_2(t)$$
$$(y_1(\xi) = y_2(\xi)).$$

If $b^2 = -b_1^2$, formulas (4.19) coincide with (4.17); in particular, if $b = 0$, then $y_1(t) = t$, $y_2(t) = (\tau - t)q$.

* If $b_1 = 0$, then $y_1(t) = t$, $y_2(t) = (\tau - t)q$.

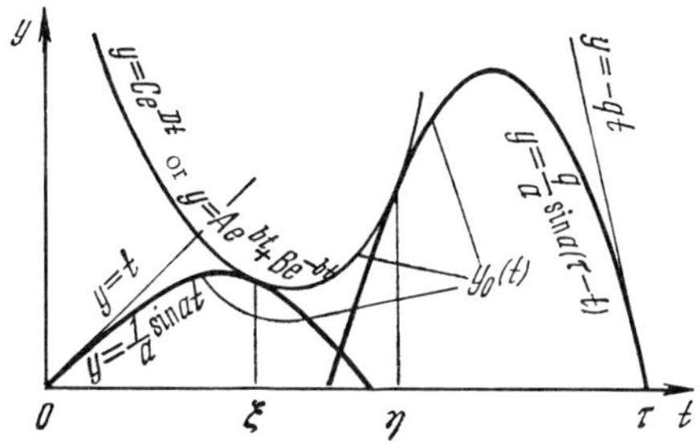

FIGURE 73.

Remark. In case 2 of Theorem II, let us set

$$y_0(t)=\begin{cases} y_1(t) & \text{if} \quad 0\leqslant t<\xi, \\ y_2(t) & \text{if} \quad \xi\leqslant t\leqslant \tau, \end{cases}$$

where ξ is determined, as before, from the conditions $y_1(\xi)=y_2(\xi)$ $(0<\xi<\tau)$. Since $\dot{y}_0(t)$ is discontinuous at $t=\xi$, $y_0(t)$ is not an admissible function. If we interpret the integral

$$J(y)=\int_0^\tau \frac{\ddot{y}}{y}\,dt$$

as a Stieltjes integral

$$J(y)=\int_0^\tau \frac{d\dot{y}}{y},$$

the latter is defined for $y=y_0(t)$ and $J(y_0)$ will be equal to (4.18). In this case $\sup J(y)$ is attained at a function $y_0(t)$ which is not in the admissible class.

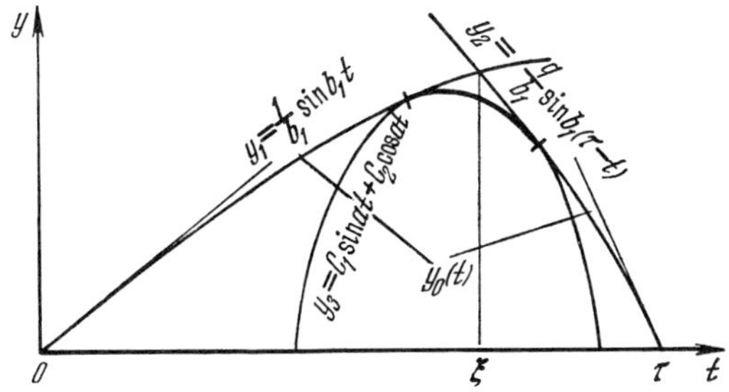

FIGURE 74.

A few words of explanation are in order. In problem (I) (Theorem I), the infimum (4.15) differs only in sign from

$$\sup \int_0^\tau p(t)\, dt,$$

where $p(t) = -\ddot{y}/y$ and, thanks to condition (c''), $-b^2 \leqslant p(t) \leqslant a^2$. Let us call any $p(t)$ satisfying these conditions an admissible function. Choose small numbers $t_1 > 0$, $t_2 > 0$. Given arbitrary admissible values of $p(t)$ on the intervals $(0, t_1)$, $(\tau - t_2, \tau)$, we can find an admissible function $p(t)$ on the inner interval $(t_1, \tau - t_2)$ such that conditions (a''), (b''), (c'') are satisfied. To this end, we must join the points $(t_1, y(t_1))$, $(t_2, y(t_2))$ by a curve $y = y(t)$ so as to maintain the conditions (a''), (b''), (c''), and then set $p(t) = -\ddot{y}/y$ for $t_1 \leqslant t \leqslant \tau - t_2$. (For this to be possible, the numbers $t_1 > 0$, $t_2 > 0$ must be sufficiently small.)

It is therefore natural to expect that the function $p(t)$ will assume the maximum possible values on the outer intervals $(0, t_1)$, $(\tau - t_2, \tau)$, i. e., $p(t) \equiv a^2$. This is precisely the conclusion of Theorem I.

Similarly, in problem (II), where we are looking for

$$\inf \int_0^\tau p(t)\, dt$$

it seems natural to assert (as we indeed do in Theorem II) that $p(t) = -b^2$ on the outer intervals $(0, t_1)$, $(\tau - t_2, \tau)$ for suitable $t_1 > 0$, $t_2 > 0$.

Theorem II states that in fact the numbers $t_1 > 0$, $t_2 > 0$ should be as large as possible, provided only that the points $(t_1, y_1(t_1))$, $(\tau - t_2, y_2(\tau - t_2))$ may be joined by an admissible curve $y = y(t)$. Theorem I states that a similar situation obtains in problem (I) if $D^2 > b^2$. But if $D^2 \leqslant b^2$, i.e., the sine curves $y = y_1(t)$, $y = y_2(t)$ can be joined smoothly by a solution of Euler's equation, then preference should naturally be given to an Euler curve.

In the sequel we shall need the following theorem, easily deduced from Theorems I and II.

*Theorem III.** *Let $Y_1^{(q)}$ denote the class of functions satisfying conditions (a''), (b'') and the condition*

$$(c_1) \qquad \frac{\ddot{y}}{y} + c^2 \geqslant 0,$$

where c^2 is any number $> \pi^2/\tau^2$. Then

$$\inf_{y \in Y_1^{(q)}} \int_0^\tau \left(\frac{\ddot{y}}{y} + c^2\right) dt = c\left[c\tau - \pi + \frac{(\ln q)^2}{c\tau - \pi}\right],$$

and the infimum is achieved at some function in $Y_1^{(q)}$.

2) *Let $Y_2^{(q)}$ denote the class of functions $y(t)$ satisfying conditions (a''), (b'') and the condition*

$$(c_2) \qquad -\frac{\ddot{y}}{y} - c^2 \geqslant 0,$$

* Part 3 of this theorem was first proved by Borg /234a/.

where c^2 is any number $< \pi^2/\tau^2$. Then

$$\inf_{y \in Y_2^{(q)}} \int_0^\tau \left(-\frac{\ddot{y}}{y} - c^2\right) dt = \begin{cases} c\,(q+q^{-1}+2\cos c\tau)\sin^{-1} c\tau & \text{if } 0 < c^2 < \dfrac{\pi^2}{\tau^2}, \\ (q+q^{-1}+2)\,\tau^{-1} & \text{if } c^2 = 0, \\ |c|\,(q+q^{-1}+2\,\mathrm{ch}\,|c|\tau)\,\mathrm{sh}^{-1}|c|\tau & \text{if } c^2 < 0, \end{cases}$$

but the infimum is not achieved in the class $Y_2^{(q)}$.

3) *Let Y_3 be the class of functions $y(t)$ satisfying condition (b″) and the condition*

$$\text{(a}_3\text{)} \qquad y(0) = y(\tau) = 0, \quad y(t) > 0 \ \text{ if } \ 0 < t < \tau.$$

Then

$$\inf_{y \in Y_3} \int_0^\tau \left|\frac{\ddot{y}}{y} + c^2\right| dt = f(c, \tau) = \begin{cases} |c|\,\mathrm{cth}\,\dfrac{c\tau}{2} & \text{if } c^2 < 0, \\ 4\tau & \text{if } c^2 = 0, \\ 2c\,\mathrm{ctg} < \dfrac{c\tau}{2} & \text{if } 0 < c^2\tau^2 < \pi^2, \\ c\,(c\tau - \pi) & \text{if } c^2\tau^2 > \pi^2. \end{cases}$$

If $c^2\tau^2 \geqslant \pi^2$, the infimum is achieved at a function $y \in Y_3$ such that $\dot{y}(\tau) = -\dot{y}(0)$. If $c^2\tau^2 < \pi^2$, the infimum is not achieved; there is a minimizing sequence of functions $y_k \in Y_3$ such that $\dot{y}_k(\tau) = -\dot{y}_k(0)$.

Condition (a$_3$) differs from condition (a″) in that the derivatives $\dot{y}$ are not specified at $t=0$, $t=\tau$.

Proof. We apply Theorem I with $a^2 = c^2$, $b^2 = \infty$. Case 1 holds, and

$$J = \inf_{y \in Y_1} \int_0^\tau \left(\frac{\ddot{y}}{y} + c^2\right) dt = \int_0^\tau \left(\frac{\ddot{y}_0}{y_0} + c^2\right) dt,$$

where $y_0(t)$ is equal to

$$\frac{1}{c}\sin ct \ \text{ if } \ 0 \leqslant t \leqslant \xi, \qquad Ae^{Bt} \ \text{ if } \ \xi \leqslant t \leqslant \eta, \quad \frac{q}{c}\sin c\,(t-\tau) \ \text{ if } \ \eta \leqslant t \leqslant \tau.$$

ξ and η are determined by the joining conditions:

$$\left.\begin{aligned} \frac{1}{c}\sin c\xi &= Ae^{B\xi}, & Ae^{B\eta} &= \frac{q}{c}\sin c\,(\tau-\eta), \\ \cos c\xi &= ABe^{B\xi}, & ABe^{B\eta} &= -q\cos c\,(\tau-\eta). \end{aligned}\right\} \tag{4.20}$$

We have

$$\int_0^\tau \left(\frac{\ddot{y}_0}{y_0} + c^2\right) dt = (B^2 + C^2)(\eta - \xi). \tag{4.21}$$

By (4.20),

$$A^2Be^{B(\xi+\eta)} = -\frac{q}{c}\sin c\xi \cdot \cos c\,(\tau-\eta) = \frac{q}{c}\sin c\,(\tau-\eta)\cos c\xi.$$

Hence

$$\sin c\,(\tau - \eta + \xi) = 0.$$

Since $0 < \tau - \eta < \frac{\pi}{c}$ and $0 < \xi < \frac{\pi}{c}$, it follows that

$$0 < c(\tau - \eta + \xi) < 2\pi.$$

Therefore $c(\tau - \eta + \xi) = \pi$, or

$$\eta - \xi = \tau - \frac{\pi}{c}. \tag{4.22}$$

It follows from (4.20) that

$$e^{B(\eta - \xi)} = \frac{Ae^{B\eta}}{Ae^{B\xi}} = \frac{\frac{q}{c} \sin c(\tau - \eta)}{\frac{1}{c} \sin c\xi} = q.$$

Therefore

$$B = \frac{\ln q}{\tau - \frac{\pi}{c}}. \tag{4.23}$$

Formulas (4.21)–(4.23) yield the infimum as required in problem (I).

2) By Theorem II,

$$J = \inf_{y \in Y_2} \int_0^\tau \left(-\frac{\ddot{y}}{y} - c^2 \right) dt = -\sup_{y \in Y_2} \int_0^\tau \left(\frac{\ddot{y}}{y} + c^2 \right) dt = \frac{\dot{y}_1(\xi)}{y_1(\xi)} - \frac{\dot{y}_2(\xi)}{y_2(\xi)}.$$

The functions $y_1(t)$ and $y_2(t)$ are defined by (4.17) with $-b^2 = b_1^2 = c^2$. To fix ideas, let $c^2 > 0$. Performing the substitutions, we obtain

$$J = c[\operatorname{ctg} c\xi + \operatorname{ctg} c(\tau - \xi)]. \tag{4.24}$$

ξ is determined from the equation $y_1(\xi) = y_2(\xi)$, or

$$\sin c\xi = q \sin c(\tau - \xi),$$

whence

$$q^{-1} = \frac{\sin c(\tau - \xi)}{\sin c\xi} = \sin c\tau \operatorname{ctg} c\xi - \cos c\tau, \quad \operatorname{ctg} c\xi = \frac{q^{-1} + \cos c\tau}{\sin c\tau}.$$

Similarly,

$$q = \sin c\tau \operatorname{ctg} c(\tau - \xi) - \cos c\tau,$$

whence

$$\operatorname{ctg} c(\tau - \xi) = \frac{q + \cos c\tau}{\sin c\tau}.$$

Inserting these expressions into (4.24), we obtain the required value for $c^2>0$. The case $c^2\leqslant 0$ is treated in similar fashion.

3) As shown at the end of §4.2, if $c^2\tau^2>\pi^2$,

$$\inf_{y\in Y_3}\int_0^\tau\left|\frac{\ddot y}{y}+c^2\right|dt=\inf\int_0^\tau\left(\frac{\ddot y}{y}+c^2\right)dt$$

over the functions of Y_3 for which $\ddot y/y+c^2\geqslant 0$. This set is the union of the sets $Y_1^{(q)}$ for all $q\geqslant 0$. Therefore, if $c^2\tau^2>\pi^2$,

$$\inf_{y\in Y_3}\int_0^\tau\left|\frac{\ddot y}{y}+c^2\right|dt=\inf_{q>0}\ \inf_{Y_1^{(q)}}\int_0^\tau\left(\frac{\ddot y}{y}+c^2\right)dt.$$

By part 1 of this theorem, the last expression is

$$\inf_{q>0} c\left[c\tau-\pi+\frac{(\ln q)^2}{c\tau-\pi}\right]=c\,(c\tau-\pi).$$

This infimum is achieved for $q=1$. Thus there exists a function $y_0\in Y_1^{(1)}$ (in particular, $\dot y(0)=-\dot y(\tau)$) such that

$$\inf_{y\in Y_3}\int_0^\tau\left|\frac{\ddot y}{y}+c^2\right|dt=\int_0^\tau\left|\frac{\ddot y_0}{y_0}+c^2\right|dt.$$

Similarly, if $c^2\tau^2<\pi^2$ we have (see end of §4.2)

$$\inf\int_0^\tau\left|\frac{\ddot y}{y}+c^2\right|dt=\inf\int_0^\tau\left(-\frac{\ddot y}{y}-c^2\right)dt$$

over the functions $y\in Y_3$ for which $-\ddot y/y-c^2\geqslant 0$. This set is the union of the sets $Y_2^{(q)}$ for all $q>0$. Consequently,

$$\inf_{y\in Y_3}\int_0^\tau\left|\frac{\ddot y}{y}+c^2\right|dt=\inf_{q>0}\ \inf_{Y_2^{(q)}}\int_0^\tau\left(-\frac{\ddot y}{y}-c^2\right)dt.$$

To fix ideas, let $c^2>0$. By part 2 of this theorem, the last infimum is

$$\inf_{q>0}\frac{c\,(q+q^{-1}+2\cos c\tau)}{\sin c\tau}=2c\cdot\operatorname{ctg}\frac{c\tau}{2}.$$

The infimum is achieved for $q=1$. Thus there exists a sequence of functions $y_k\in Y_2^{(1)}$ (in particular, $\dot y_k(0)=-\dot y_k(\tau)$) such that

$$\inf \int_0^\tau \left| \frac{\ddot{y}}{y} + c^2 \right| dt = \lim_{k\to\infty} \int_0^\tau \left| \frac{\ddot{y}_k}{y_k} + c^2 \right| dt.$$

This completes the proof of Theorem III.

Remark. According to what we have proved, in case 3 the minimizing function (if $c^2\tau^2 > \pi^2$) or minimizing sequence may be so chosen that $\dot{y}(0) = -\dot{y}(\tau) = 1$.

Using Theorems I, II for $q = 1$, we can immediately determine the explicit form of the minimizing function $y_0(t)$ $(c^2\tau^2 > \pi^2)$ and the minimizing sequence $y^{(k)}(t)$ $(c^2\tau^2 < \pi^2)$. In fact, if $c^2\tau^2 > \pi^2$, then

$$y_0(t) = \frac{1}{c}\sin ct \quad \text{for} \quad 0 \leqslant t < \frac{\pi}{2c}, \quad y_0(t) = 1 \quad \text{for} \quad \frac{\pi}{2c} \leqslant t < \tau - \frac{\pi}{2c},$$
$$y_0(t) = \frac{1}{c}\sin c(\tau - t) \quad \text{for} \quad \tau - \frac{\pi}{2c} \leqslant t < \tau.$$

It is clear that $y_0(t) \in Y_3$ and

$$\int_0^\tau \left| \frac{\ddot{y}_0}{y_0} + c^2 \right| dt = c(c\tau - \pi).$$

(By Theorem I, in this case $y = C\exp Dt$ on the inner interval; it follows from the joining conditions that $y = 1$.)

In the case $c^2\tau^2 < \pi^2$, the limit function $y_0(t)$ (which is not in Y_3) is

$$y_0(t) = \begin{cases} y_1(t) & \text{for} \quad 0 \leqslant t < \frac{1}{2}\tau, \\ y_2(t) & \text{for} \quad \frac{1}{2}\tau < t < \tau, \end{cases}$$

where $y_1(t) = \frac{1}{c}\sin ct$, $y_2(t) = \frac{1}{c}\sin c(\tau - t)$. A minimizing sequence $y^{(k)} \in Y_3$ is obtained by joining these sine curves smoothly by the curve $y(t, a_k) = C_1^{(k)}\sin a_k t + C_2^{(k)}\cos a_k t$:

$$y^{(k)} = \begin{cases} \frac{1}{c}\sin ct & \text{if} \quad 0 \leqslant t < \xi_k, \\ y(t, a_k) & \text{if} \quad \xi_k \leqslant t < \eta_k, \\ \frac{1}{c}\sin c(\tau - t) & \text{if} \quad \eta_k \leqslant t < \tau. \end{cases}$$

The numbers ξ_k, η_k are determined by the continuity of $y^{(k)}(t)$ and $\dot{y}^{(k)}(t)$ at the points ξ_k and η_k (note that by considerations of symmetry $\xi_k + \eta_k = \tau$), and $\{a_k\}$ is any sequence such that $a_k \to +\infty$ as $k \to \infty$. Then $\xi_k \to \frac{1}{2}\tau$, $\eta_k \to \frac{1}{2}\tau$ as $k \to \infty$, and

$$\int_0^\tau \left| \frac{\ddot{y}^{(k)}}{y^{(k)}} + c^2 \right| dt = \int_0^\tau \left(-\frac{\ddot{y}^{(k)}}{y^{(k)}} - c^2 \right) dt,$$

$$\frac{\dot{y}_1\left(\frac{1}{2}\tau\right)}{y_1\left(\frac{1}{2}\tau\right)} - \frac{\dot{y}_2\left(\frac{1}{2}\tau\right)}{y_2\left(\frac{1}{2}\tau\right)} = 2c\operatorname{ctg}\frac{c\tau}{2} \quad \text{as} \quad k \to \infty.$$

4.4. Stability tests for Hill's equation with estimate for $\int_0^T |p_c^{\pm}(t)-c^2|\,dt$

Our goal is to develop explicit expressions for $\rho_n'(c)$, $\rho_{n+1}''(c)$ in (4.7), (4.9) and (4.10).

As shown in §4.2, the quantities $\rho_n'(c)$, $\rho_{n+1}''(c)$ are easily expressed in terms of $\rho_n(c)$ (see (4.11)), where

$$\rho_n(c)=\inf\int_{t_0}^{t_0+T}\left|\frac{\ddot{y}}{y}+c^2\right|dt$$

and the infimum is taken over all functions y satisfying conditions (a), (b), (c) of §4.2. By part 3 of Theorem III in §4.3,

$$\int_{t_0}^{t_0+T}\left|\frac{\ddot{y}}{y}+c^2\right|dt=\sum_{i=1}^{n}\int_{t_{i-1}}^{t_i}\left|\frac{\ddot{y}}{y}+c^2\right|dt\geqslant\sum_{i=1}^{n}f(c,\tau_i),$$

where $\tau_i=t_i-t_{i-1}$; the value of $f(c,\tau)$ was specified in Theorem III. Define

$$\min\sum_i f(c,\tau_i)$$

subject to the constraints $\tau_i\geqslant 0$, $\sum\tau_i=T$. The function $f(c,\tau)$ is not differentiable with respect to τ at $\tau=\pi/c$. Smoothing the function in the neighborhood of this point, we deduce by elementary arguments that the minimum is attained at $\tau_1=\tau_2=\ldots=\tau_n=\frac{1}{n}T$. Therefore, if $n\geqslant 1$,

$$\rho_n(c)\geqslant\sum_{i=1}^{n}f\left(c,\frac{T}{n}\right)=nf\left(c,\frac{T}{n}\right).$$

In fact,

$$\rho_n(c)=nf\left(c,\frac{T}{n}\right)\qquad(n\geqslant 1).$$

Indeed, assume first that $\frac{T}{n}>\frac{\pi}{c}$. It follows from part 3 of Theorem III, §4.3, that for each interval $(t_i,t_{i+1})=\left(\frac{iT}{n},\frac{i+1}{n}T\right)$ we can construct a function $y_i(t)\in Y_3$ such that

$$f\left(c,\frac{T}{n}\right)=\int_{t_i}^{t_{i+1}}\left|\frac{\ddot{y}_i}{y_i}+c^2\right|dt,$$

and moreover $\dot{y}_i(t_i)=-\dot{y}_i(t_{i+1})$. We may assume without loss of generality that

$$(-1)^i y_i(t)>0,\quad \dot{y}_i(t_i)=(-1)^i.$$

Set

$$y(t)=y_1(t) \quad \text{if} \quad t_0 \leqslant t < t_1 = t_0+\frac{T}{n},$$

$$y(t)=y_2(t) \quad \text{if} \quad t_1 \leqslant t < t_2 = t_0+\frac{2T}{n},$$

$$\cdots\cdots\cdots\cdots\cdots\cdots\cdots$$

$$y(t)=y_n(t) \quad \text{if} \quad t_{n-1} \leqslant t \leqslant t_n = t_0+T.$$

It follows from the conditions $y_i(t_i)=0$, $\dot{y}_i(t_i)=(-1)^i$ that the functions $y(t)$, $\dot{y}(t)$ are continuous on $[0, T]$. Thus the function $y(t)$ satisfies condition (a), (b), (c) and

$$\int_0^T \left| \frac{\ddot{y}}{y} + c^2 \right| dt = nf\left(c, \frac{T}{n}\right).$$

Consequently, $\rho_n(c) = nf\left(c, \frac{T}{n}\right)$.

The case $\frac{T}{n} < \frac{\pi}{c}$ is treated similarly, except that instead of the functions $y_i(t)$ one considers suitable minimizing sequences.

By (4.11) and the expression for $f(c, \tau)$ in part 3 of Theorem III, §4.3, we have

$$\rho_n'(c) = c(Tc - n\pi) \qquad \left(c \geqslant \frac{n\pi}{T} > 0\right), \tag{4.25}$$

and in this case the infimum is achieved;

$$\rho_{n+1}''(c) = \begin{cases} 2c(n+1)\operatorname{ctg}\dfrac{Tc}{2(n+1)} & \text{if} \quad c^2 > 0, \\ \dfrac{4(n+1)^2}{T^2} & \text{if} \quad c^2 = 0, \\ 2|c|(n+1)\operatorname{cth}\dfrac{T|c|}{2(n+1)} & \text{if} \quad c^2 < 0 \end{cases} \tag{4.26}$$

$$\left(c^2 \leqslant \frac{(n+1)^2\pi^2}{T^2}\right),$$

and in this case the infimum is not achieved.

The above derivation is valid for the case when $c > n\pi/T$ in (4.25) and $c^2 < (n+1)^2\pi^2/T^2$ in (4.26). If $c = n\pi/T$ in (4.25) and $c^2 = (n+1)^2\pi^2/T^2$ in (4.26), the formulas are still valid. Indeed, it then follows from (4.25) and (4.26) that

$$\rho_n'\left(\frac{n\pi}{T}\right) = 0 \quad \text{and} \quad \rho_{n+1}''\left(\frac{(n+1)\pi}{T}\right) = 0.$$

It follows directly from the definition (4.5) of the functions $\rho_n'(c)$, $\rho_{n+1}''(c)$ (see §4.1) that these formulas are valid, for

$$\frac{n^2\pi^2}{T^2} \in \Pi_n^{**}, \quad \frac{(n+1)^2\pi^2}{T^2} \in \Pi_{n+1}^{**}.$$

Together with formulas (4.25), (4.26), our tests (4.7), (4.9), (4.10) now become effective. We state the tests in full, assuming for simplicity that $c_1 = c_2 = c$ and using the remark at the end of §4.1 for the case $n=0$.

Test 3. Let c be an arbitrary constant such that $c \neq k\pi/T$, $k = 0, 1, 2, \ldots$ and n the integer defined by

$$\frac{n\pi}{T} < c < \frac{(n+1)\pi}{T}.$$

Let $p_c^{\pm}(t)$ be defined by (4.6). If $n \geqslant 1$ and

$$\left.\begin{aligned} &\int_0^T [c^2 - p_c^-(t)]\,dt < c\,(Tc - n\pi), \\ &\int_0^T [p_c^+(t) - c^2]\,dt \leqslant 2c\,(n+1)\operatorname{ctg}\frac{Tc}{2(n+1)} \end{aligned}\right\} \tag{4.27}$$

or if $n = 0$ and

$$p(t) \not\equiv 0, \quad \int_0^T p(t)\,dt \geqslant 0$$

and

$$\int_0^T |p_c^+(t) - c^2|\,dt \leqslant 2c\operatorname{ctg}\frac{cT}{2}, \tag{4.28}$$

then all solutions of equation (3.1) are bounded on $(-\infty, +\infty)$ and the corresponding Hamiltonian (3.2) belongs to the n-th stability domain $\mathfrak{G}_n$.

In the second inequalities (4.27) and (4.28) we have replaced $<$ by $\leqslant$ in view of the fact that in these cases the infimum is not attained.

Test 4. Suppose that for some integer n

$$p(t) \geqslant \frac{n^2\pi^2}{T^2} \qquad (n \geqslant 1)$$

or

$$p(t) \not\equiv 0, \quad \int_0^T p(t)\,dt \geqslant 0 \qquad (n = 0).$$

Let c^2 be an arbitrary constant such that $c^2 < (n+1)^2\pi^2/T^2$; define $p_c^+(t)$ by (4.6). If

$$\int_0^T [p_c^+(t) - c^2]\,dt \leqslant \rho''_{n+1}(c), \tag{4.29}$$

where $\rho''_{n+1}(c)$ is defined by (4.26), then all solutions of equation (3.1) are bounded on $(-\infty, +\infty)$ and the Hamiltonian of the corresponding equation (3.2) belongs to the n-th stability domain $\mathfrak{G}_n$.

Inequality (4.29) has the sign $\leqslant$ instead of $<$ as in (4.9), since we have already proved that the infimum in the formula (4.3) for $\rho''_{n+1}(c)$ is not attained.

Test 5. Suppose that for some integer n

$$p(t) \underset{\not\equiv}{\leqslant} \frac{(n+1)^2\pi^2}{T^2}$$

and let $c>0$ *be an arbitrary constant such that* $c^2 \geqslant n^2\pi^2/T^2$. *Define the function* $p_c^-(t)$ *by (4.6). If*

$$\left.\begin{aligned} &\int_0^T [c^2 - p_c^-(t)]\,dt < c(Tc - n\pi) \quad \text{if} \quad n \geqslant 1, \\ &p(t) \not\equiv 0, \quad \int_0^T p(t)\,dt \geqslant 0 \quad \text{if} \quad n = 0, \end{aligned}\right\} \tag{4.30}$$

then all solutions of equation (3.1) are bounded on $(-\infty, \infty)$ *and the Hamiltonian of the corresponding equation (3.2) belongs to the* n*-th stability domain* $\mathfrak{G}_n$.

The following specialized tests are obtained by setting $c = n\pi/T$ in Test 4 and $c = (n+1)\pi/T$ in Test 5; they are particularly simple.

Test 4a. Suppose that for some integer n

$$T\int_0^T \left[p(t) - \frac{n^2\pi^2}{T^2}\right] dt \leqslant 2\pi n(n+1)\,\mathrm{tg}\,\frac{\pi}{2(n+1)} \tag{4.31}$$

and let

$$p(t) \underset{\not\equiv}{\geqslant} \frac{n^2\pi^2}{T^2}, \quad \text{if} \quad n \geqslant 1;$$

$$p(t) \not\equiv 0, \quad \int_0^T p(t)\,dt \geqslant 0 \quad \text{if} \quad n = 0.$$

Then $p(t) \in \mathfrak{G}_n$.

Test 5a. Suppose that for some integer $n \geqslant 1$

$$p(t) \underset{\not\equiv}{\leqslant} \frac{n^2\pi^2}{T^2} \quad \text{and} \quad T\int_0^T \left[\frac{n^2\pi^2}{T^2} - p(t)\right] dt < n\pi^2.$$

Then $p(t) \in \mathfrak{G}_{n-1}$.

Setting $n=0$ in Test 4a, we get Lyapunov's test

$$p(t) \underset{\not\equiv}{\geqslant} 0, \quad T\int_0^T p(t)\,dt \leqslant 4,$$

which "picks out" functions from the stability domain $\mathfrak{G}_0$. Thus Test 4a may be viewed as a generalization and extension of Lyapunov's test to the domains $\mathfrak{G}_n\,(n \geqslant 1)$.

Tests 4a and 5a also recall Zhukovskii's test (see §3.3).

In the interests of a unified treatment, our formulation of Test 5a for the case $n=0$ is somewhat cruder than the proposition deduced from Test 5 when $c=\pi/T$. For $n=0$, $c=\pi/T$, Test 5 is quite simple:

$$p(t)\not\equiv 0, \quad p_{\text{av}}=\frac{1}{T}\int_0^T p(t)\,dt\geqslant 0, \quad p(t)\underset{\not\equiv}{\leqslant}\frac{\pi^2}{T^2}.$$

The formulation of Test 4 for $n=0$, $c=0$ is as follows:

Test 6 (Krein /166d/, Borg /234b/). Let*

$$p(t)\not\equiv 0, \quad \int_0^T p(t)\,dt\geqslant 0,$$

and

$$T\int_0^T p_+(t)\,dt\leqslant 4, \tag{4.32}$$

where

$$p_+(t)=\begin{cases} p(t) & \textit{if} \quad p(t)\geqslant 0,\\ 0 & \textit{if} \quad p(t)<0.\end{cases}$$

Then $p(t)\in\mathfrak{G}_0$.

Remark. As shown at the end of §4.1, the condition

$$p(t)\geqslant\frac{n^2\pi^2}{T^2} \quad \text{or} \quad p(t)\underset{\not\equiv}{\leqslant}\frac{(n+1)^2\pi^2}{T^2}$$

in Tests 4, 4a, 5 and 5a may be replaced respectively by the conditions

$$p(t)\geqslant\frac{n^2\pi^2}{r(t)^4}-\frac{\ddot r(t)}{r(t)} \quad \text{or} \quad p(t)\leqslant\frac{(n+1)^2\pi^2}{r(t)^4}-\frac{\ddot r(t)}{r(t)},$$

where $r(t)$ is any function such that

$$r(t+T)=r(t)>0, \quad \int_0^T\frac{dt}{r(t)^2}=1.$$

We now examine the sharpness of the constants on the right of inequalities (4.27) – (4.32). Assume that $n^2\pi^2/T^2\leqslant c^2\leqslant(n+1)^2\pi^2/T^2$ (in Tests 4 and 5). It will be shown that under these assumptions the constants in question are best possible — no other values will yield stronger stability tests (i. e., enlarge the set of functions covered by the tests).

It is clearly sufficient to consider Tests 3, 4, 5.

* This version of Test 6 is due to Krein /166d/. Borg /234b/ proves a somewhat weaker test, obtained from (4.32) if $p_+(t)$ is replaced by $|p(t)|$.

Theorem. Suppose that some constant on the right of inequalities (4.27) – (4.30), (4.32) in Tests 3, 4, 5 has been replaced by a larger number; alternatively, replace the condition $p_{av}=\frac{1}{T}\int_0^T p\,dt \geqslant 0$ *in Tests 3, 4, 5* $(n=0)$ *by a condition* $p_{av} \geqslant -\varepsilon$, $\varepsilon > 0$. *Then the conclusion of the appropriate stability test fails to hold in each case; that is to say, there exists a function* $q(t) \in \mathfrak{L}^3$ *satisfying the modified test such that the equation* $\ddot{y}+q(t)y=0$ *has an exponentially increasing solution.*

Proof. We first consider Test 3. Suppose that the stipulated modification has been made in the first inequality of (4.27).

By (4.25) and (4.3), the right-hand side of this inequality is

$$\rho_n'(c) = \inf_{p \in \Pi_n} \| p - c^2 \|.$$

We have seen that the infimum in this formula is attained at some function $\tilde{p}=\tilde{p}(t) \in \Pi_n$ such that $\tilde{p}(t) \leqslant c^2$. It follows from (4.6) that $\tilde{p}_c^-(t)=\tilde{p}(t)$ and so $\rho_n'(c) = \|\tilde{p}-c^2\| = \|\tilde{p}_c^- - c^2\|$.

Since $\tilde{p}(t) \in \Pi_n$, and Π_n is the boundary of the instability domain $\mathcal{H}_n$, we see that for any $\varepsilon > 0$ there is a function $q(t) \in \mathcal{H}_n$ such that

$$\| q(t) - p(t) \| < \varepsilon.$$

Using (4.6), we obtain

$$\int_0^T |q_c^-(t) - c^2|\,dt = \int_{q<c^2} |q(t)-c^2|\,dt \leqslant$$

$$\leqslant \int_{q<c^2} |\tilde{p}-c^2|\,dt + \int_{q<c^2} |\tilde{p}-q|\,dt \leqslant \int_0^T |\tilde{p}-c^2|\,dt + \varepsilon.$$

Therefore

$$\int_0^T |q_c^-(t)-c^2|\,dt \leqslant \rho_n'(c)+\varepsilon. \tag{4.33}$$

On the other hand, again using (4.6), we have

$$\int_0^T |q_c^+(t)-c^2|\,dt = \int_{q>c^2} (q-c^2)\,dt = \int_{q>c^2} (\tilde{p}-c^2)\,dt + \int_{q>c^2} (q-\tilde{p})\,dt.$$

Since $\tilde{p} \leqslant c^2$, the last expression is bounded above by $\|q-\tilde{p}\| < \varepsilon$. Therefore

$$\int_0^T |q_c^+(t)-c^2|\,dt \leqslant \varepsilon. \tag{4.34}$$

It follows from (4.33) and (4.34) that if we replace the constant $\rho_n'(c)=c(Tc-n\pi)$ in the first inequality of (4.27) by any larger quantity, the assertion of Test 3 is false: there is a function $q(t)\in\mathcal{H}_n$ which satisfies the second inequality of (4.27) and, instead of the first inequality of (4.27), the new inequality (4.33).

We now modify the second inequality of (4.27), $\|p_c^+-c^2\|\leqslant\rho''_{n+1}(c)$, replacing it by an inequality

$$\|p_c^+-c^2\|<\rho''_{n+1}(c)+\varepsilon, \tag{4.35}$$

where $\varepsilon>0$. Since

$$\rho''_{n+1}(c)=\inf_{p\in\Pi_{n+1}}\|p-c^2\|,$$

there is a function $\tilde{p}\in\Pi_{n+1}$ such that

$$\|\tilde{p}-c^2\|<\rho''_{n+1}(c)+\frac{1}{2}\varepsilon. \tag{4.36}$$

In addition, as shown previously, $\tilde{p}(t)\geqslant c^2$, i. e., (see (4.6)) $\tilde{p}_c^+=\tilde{p}$. Since Π_{n+1} is the boundary of $\mathcal{H}_{n+1}$, there exists a function $q\in\mathcal{H}_{n+1}$ such that

$$\|q-\tilde{p}\|<\frac{1}{2}\varepsilon. \tag{4.37}$$

Using (4.6), we obtain

$$\|q_c^+-c^2\|=\int_{q>c^2}|q-c^2|\,dt\leqslant\int_{q>c^2}|\tilde{p}-c^2|\,dt+\int_{q>c^2}|q-\tilde{p}|\,dt\leqslant$$
$$\leqslant\|\tilde{p}-c^2\|+\|q-\tilde{p}\|.$$

It follows from (4.36), (4.37) that the function q satisfies inequality (4.35).

Further, we have

$$\|q_c^--c^2\|=\int_{q<c^2}(c^2-q)\,dt=\int_{q<c^2}(c^2-\tilde{p})\,dt+\int_{q<c^2}(\tilde{p}-q)\,dt.$$

Since $\tilde{p}\geqslant c^2$, this expression is at most

$$\int_{q<c^2}(\tilde{p}-q)\,dt\leqslant\|\tilde{p}-q\|<\frac{1}{2}\varepsilon.$$

Making ε smaller (if necessary), we can ensure that q satisfies the second inequality of (4.27). Consequently, there exists a function $q\in\mathcal{H}_{n+1}$ satisfying the assumptions of the modified Test 3, as required.

Now consider Test 4. We have proved that there exists a function $p\in\Pi_{n+1}$ such that (4.36) holds and $p(t)\geqslant c^2>n^2\pi^2/T^2$. Using the remark to Theorem III, §4.3, we see that this function may be defined as follows. Divide the interval $[0,\ T]$ by the points $t_j=jT/(n+1)$ $(j=0,1,\ldots,n)$ into $n+1$ subintervals. The required function $y(t)$ is defined on $[t_j,\ t_{j+1}]$ by joining the sine curves

$$(-1)^j c^{-1}\sin c(t-t_j),\quad (-1)^j c^{-1}\sin c(t_{j+1}-t)$$

smoothly by a sine curve $A_j \sin at + B_j \cos at$, where a is sufficiently large. Thus,

$$y(t)=\begin{cases}(-1)^j c^{-1}\sin c(t-t_j) & \text{if } t_j \leqslant t < \xi_j,\\ A_j \sin at + B_j \cos at & \text{if } \xi_j \leqslant t < \eta_j,\\ (-1)^j c^{-1}\sin c(t_{j+1}-t) & \text{if } \eta_j \leqslant t \leqslant t_{j+1},\end{cases}$$

$$\tilde{p}(t)=\begin{cases}c^2 & \text{if } t_j \leqslant t < \xi_j,\\ a^2 & \text{if } \xi_j \leqslant t < \eta_j,\\ c^2 & \text{if } \eta_j \leqslant t < t_{j+1}.\end{cases}$$

The numbers ξ_j, η_j, A_j, B_j are determined by the continuity of the functions $y(t)$, $\dot{y}(t)$ at ξ_j, η_j.

We may obviously modify the function $y(t)$ on the last interval $[t_n, t_{n+1}]$ in such a way that $\dot{y}(T)=(-1)^{n+1}\mu$, where $\mu > 1$ is sufficiently close to 1. Thus, $y(t)$ is not modified on any interval except the last, and there

$$y(t)=\begin{cases}(-1)^n c^{-1}\sin(t-t_n) & \text{if } t_n \leqslant t < \xi_n',\\ A_n' \sin at + B_n' \cos at & \text{if } \xi_n' \leqslant t < \eta_n',\\ \mu(-1)^n c^{-1}\sin c(t_{n+1}-t) & \text{if } \eta_n' \leqslant t < t_{n+1},\end{cases}$$

where the numbers ξ_n', η_n', A_n', B_n' are determined by the continuity conditions for $y(t)$ and $\dot{y}(t)$ at ξ_n', η_n'. The resulting function $q(t)=-\ddot{y}/y$ satisfies the condition $q(t) \geqslant c^2$. Since $q(t+T)=q(t)$, the solution $y(t)$ just constructed will be such that $y(t+T)=(-1)^{n+1}\mu y(t)$. This means that $y(t)$ is unbounded on $(0, \infty)$. The function $y(t)$ vanishes exactly $n+1$ times on $[0, T)$. Therefore $q(t) \in \mathcal{H}_{n+1}$. As before, we deduce from (4.37) that (4.35) is true with $p=q$. Thus the function q satisfies all the assumptions of the modified Test 4, but $q \in \mathcal{H}_{n+1}$.

Finally, we consider Test 5.

Using the remark to Theorem III at the end of §4.3, we see that a function $\tilde{p} \in \Pi_n$ with $\rho_n'(c)=\|c^2-\tilde{p}\|$ may be defined as follows in terms of a suitable solution $y(t)$. Divide the interval $[0, T]$ by points $t_j = jT/n$ $(j=0, 1, \ldots, n+1)$ into $n+1$ subintervals. Then divide each subinterval $[t_j, t_{j+1}]$ further by

$$\Delta_j' = \left[t_j, t_j+\frac{\pi}{2c}\right], \quad \Delta_j'' = \left[t_j+\frac{\pi}{2c}, t_{j+1}-\frac{\pi}{2c}\right], \quad \Delta_j''' = \left[t_j-\frac{\pi}{2c}, t_{j+1}\right],$$

and define $y(t)$ on these subintervals by

$$y(t)=\begin{cases}(-1)^j c^{-1}\sin c(t-t_j) & \text{if } t \in \Delta_j',\\ 1 & \text{if } t \in \Delta_j'',\\ (-1)^j c^{-1}\sin c(t_{j+1}-t) & \text{if } t \in \Delta_j'''.\end{cases}$$

(The functions $y(t)$ and $\dot{y}(t)$ are clearly continuous.) Thus,

$$\tilde{p}(t) = -\frac{\ddot{y}}{y} = \begin{cases}c^2 & \text{if } t \in \Delta_j',\\ 0 & \text{if } t \in \Delta_j'',\\ c^2 & \text{if } t \in \Delta_j'''.\end{cases}$$

Modifying $y(t)$ (always maintaining the continuity of $y(t)$, $\dot{y}(t)$) in such a way that $y(T)=0$, $\dot{y}(T)=(-1)^n\mu$, where $\mu>1$ is sufficiently close to 1, we can ensure that the new function $\tilde{p}(t)$, which we denote by $q(t)$, satisfies the inequality $q(t)\geqslant c^2$ and the number $\|q-p\|$ is sufficiently small. We need modify $y(t)$ only on the last subinterval, where we set

$$y(t)=\begin{cases}(-1)^{n-1}c^{-1}\sin c(t-t_{n-1}) & \text{if} \quad t_{n-1}\leqslant t<\xi,\\ A+Bt & \text{if} \quad \xi\leqslant t<\eta,\\ \mu(-1)^{n-1}c^{-1}\sin c(t_n-t) & \text{if} \quad \eta\leqslant t\leqslant t_{n+1}.\end{cases}$$

The numbers ξ, η, A, B are determined by the joining conditions (continuity of $y(t)$ and $\dot{y}(t)$ at ξ, η). Then the function $q(t)=-\ddot{y}/y$ satisfies the required conditions. Since y has exactly n zeros on $(0, T)$ and $\dot{y}(T)=(-1)^n\mu\dot{y}(0)$, it follows that $y(t+T)=(-1)^n\mu y(t)$, so that $q(t)\in\mathcal{H}_n$. As before, we see that (4.33) will hold provided μ is sufficiently close to 1. Thus $q(t)$ satisfies all the assumptions of Test 5 but $q(t)\in\mathcal{H}_n$.

It remains to discuss what happens if we replace the inequality $p_{\text{av}}\geqslant 0$ in Tests 3, 4, 5 by an inequality $p_{\text{av}}\geqslant-\varepsilon$, $\varepsilon>0$. But in this case the function $p_\varepsilon(t)\equiv-\varepsilon$ satisfies the conditions of the modified test, and it is clear that $p_\varepsilon(t)\in\mathcal{H}_0$. This completes the proof.

4.5. Borg's test

Let us return to Hill's equation (3.1). Set

$$p_{\text{av}}=\frac{1}{T}\int_0^T p(t)\,dt. \tag{4.38}$$

Test 7 (Borg /234a/). Suppose that for some integer n

$$\frac{n^2\pi^2}{T^2}<p_{\text{av}}<\frac{(n+1)^2\pi^2}{T^2},$$

$$\int_0^T|p(t)-p_{\text{av}}|\,dt<2\sqrt{p_{\text{av}}}(T\sqrt{p_{\text{av}}}-n\pi), \quad \text{if} \quad n\geqslant 1, \tag{4.39}$$

$$\int_0^T|p(t)-p_{\text{av}}|\,dt\leqslant 4\sqrt{p_{\text{av}}}(n+1)\operatorname{ctg}\frac{T\sqrt{p_{\text{av}}}}{2(n+1)}. \tag{4.40}$$

Then all solutions of equation (3.1) are bounded on $(-\infty, +\infty)$ and the corresponding Hamiltonian in (3.2) is in the stability domain $\mathfrak{G}_n$.

Proof. Let us apply Test 3 of §4.4 with $c=\sqrt{p_{\text{av}}}$. By (4.6),

$$\int_0^T|p_c^- -c^2|\,dt=\int_{p<c^2}(c^2-p)\,dt, \quad \int_0^T|p_c^+ -c^2|\,dt=\int_{p\geqslant c^2}(p-c^2)\,dt. \tag{4.41}$$

Since $c^2=p_{\text{av}}$, it follows from (4.38) that

$$\frac{1}{T}\int_0^T(c^2-p)\,dt=0.$$

Hence

$$\int_{p<c^2} (c^2-p)\,dt = \int_{p\geqslant c^2} (p-c^2)\,dt.$$

Since

$$\int_0^T |p-c^2|\,dt = \int_{p<c^2} (c^2-p)\,dt + \int_{p\geqslant c^2} (p-c^2)\,dt,$$

it follows from these two relations and from (4.41) that

$$\int_0^T |p_c^- - c^2|\,dt = \int_0^T |p_c^+ - c^2|\,dt = \frac{1}{2}\int_0^T |p-c^2|\,dt.$$

Inequalities (4.27) now coincide with (4.39), (4.40), completing the proof.

R e m a r k . It follows from the theorem of §4.4 that none of the numbers on the right of (4.39), (4.40) may be replaced by a larger number without affecting the validity of the test.

Employing some additional reasoning, we can give an alternative, stronger version of Borg's test.

Given a function $p \in L[0, T]$, define numbers

$$\alpha = p_{\text{av}} = \frac{1}{T}\int_0^T p\,dt, \quad \beta = \frac{1}{T}\int_0^T |p(t)-\alpha|\,dt. \tag{4.42}$$

The set of all $p(t) \in L$ with fixed values of (4.42) will be denoted by $\mathfrak{M}_{\alpha\beta}$.

D e f i n i t i o n . Call a point (α_0, β_0) on the (α, β)-plane a *s t a b i l i t y p o i n t* if there exists $\delta > 0$ such that all solutions of Hill's equation with

$$p(t) \in \mathfrak{M}_{\alpha\beta}, \quad \text{where} \quad |\alpha-\alpha_0| < \delta, \quad |\beta-\beta_0| < \delta,$$

are bounded on $(-\infty, +\infty)$. Otherwise we call (α_0, β_0) an *i n s t a b i l i t y p o i n t*.

It is evident from the definition that the set of stability points is open: it is therefore a union of countably many domains. We wish to determine these domains explicitly. The answer is furnished by the following

T h e o r e m (Borg /234a/). A point (α, β) is a stability point if and only if

$$\alpha > 0, \quad \alpha \neq \frac{k^2\pi^2}{T^2} \quad (k = 1, 2, \ldots), \tag{4.43}$$

$$\left.\begin{aligned} &\beta < \min(\Psi_n'(\alpha), \Psi_{n+1}''(\alpha)), && \text{if} \quad n \geqslant 1, \\ &\beta < \Psi_2''(\alpha), && \text{if} \quad n = 0, \end{aligned}\right\} \tag{4.44}$$

where n is the integer defined by

$$\frac{n^2\pi^2}{T^2} < \alpha < \frac{(n+1)^2\pi^2}{T^2}, \tag{4.45}$$

and the functions $\Psi_n'(\alpha)$ and $\Psi_{n+1}(\alpha)$ are defined by

$$\left.\begin{aligned}\Psi_n'(\alpha)&=2\alpha\left(1-\frac{n\pi}{T\sqrt{\alpha}}\right),\\ \Psi_{n+1}''(\alpha)&=4\,\frac{n+1}{T}\sqrt{\alpha}\operatorname{ctg}\left(\frac{1}{2}\frac{T\sqrt{2}}{n+1}\right).\end{aligned}\right\}\tag{4.46}$$

The fact that any point (α, β) satisfying these conditions is indeed a stability point follows from Test 7, since inequality (4.43) is equivalent to (4.39), (4.40). It follows from the theorem of §4.4 that points on the boundary of the domains (4.43), (4.44) (except, of course, the points $\beta=0$, $\alpha>0$, $\alpha\neq n^2\pi^2/T^2$) are instability points. Thus we need only prove that all points outside the domains (4.43), (4.44) are instability points. The proof is exceedingly cumbersome and will not be given here. In outline, one proceeds as follows. One first shows that for any such point (α, β) there is a function $p(t)\in\Pi$, $p\in\mathfrak{M}_{\alpha\beta}$. It may be assumed that this is a step function. Varying the function in a special way, one obtains a function $p\in\mathcal{H}$, $p\in\mathfrak{M}_{\alpha\beta}$. In so doing, one actually constructs (as in §4.4) not $p(t)$ itself but a suitable solution $y(t)$, joined together smoothly from various sine curves.

Test 7 was derived from Test 3. It can be shown that there are functions satisfying Test 3 but not Test 7. There are also examples of functions satisfying Test 4a but not Test 7, and conversely.

4.6. Stability tests for Hill's equation using $\max p(t)$, $\min p(t)$ and p_{av}

We consider Hill's equation (3.1) with a T-periodic piecewise-continuous integrable function $p(t)$ such that

$$0\leqslant b^2\leqslant p(t)\leqslant a^2. \tag{4.47}$$

Denote

$$M=\int_0^T p(t)\,dt. \tag{4.48}$$

What conditions (necessary and sufficient) must be satisfied by the numbers T, a^2, b^2, M in order that all solutions of Hill's equation (3.1) with coefficients satisfying (4.47) and (4.48) be bounded? Since we shall make use of the "additional" information available here about $p(t)$, the stability tests derived below will be more complicated than those of §§4.4 and 4.5.

Let us first see what stability domains may contain a function $p(t)$ satisfying inequalities (4.47). Define integers k_1 and $k_2>k_1$ by

$$\frac{k_1^2\pi^2}{T^2}\leqslant b^2<\frac{(k_1+1)^2\pi^2}{T^2},\quad \frac{k_2^2\pi^2}{T^2}<a^2\leqslant\frac{(k_2+1)^2\pi^2}{T^2}. \tag{4.48a}$$

By the comparison theorem and §1.8, $p(t)$ may belong only to stability domains $\mathfrak{G}_k$ with subscripts $k=k_1, k_1+1, \ldots, k_2$. Denote

$$A_k = \sup_{p\in\Pi_k,\ b^2\leqslant p(t)\leqslant a^2} \int_0^T p(t)\,dt, \quad B_k = \inf_{p\in\Pi_k,\ b^2\leqslant p(t)\leqslant a^2} \int_0^T p(t)\,dt. \tag{4.49}$$

We shall show that if $M > A_k$ the function $p(t)$ lies "to the right" of the surface Π_k in $\mathfrak{L}^3$ (see Figure 68):

$$p(t)\in\mathfrak{G}_k\cup\Pi_{k+1}\cup\mathcal{H}_{k+1}\cup\mathfrak{G}_{k+1}\cup\Pi_{k+2}\cup\mathcal{H}_{k+2}\cup\ldots$$

Suppose the contrary. We may enlarge $p(t)$, taking instead, say, $p_1(t)=p(t)+c$, without violating the conditions $M>A_k$, $p_1(t)\leqslant a^2$. For a suitable choice of c, we see that $p_1(t)=p(t)+c\in\Pi_k$, $b^2\leqslant p_1(t)\leqslant a^2$, and the inequality $M>A_k$ holds for $p_1(t)$. But this is impossible by virtue of the definition of A_k.

Similarly, the condition $M<B_k$ means that in the model $\tilde{\mathfrak{R}}^3$ the function $p(t)$ lies "to the left" of the surface Π_k. Therefore,

$$\left.\begin{array}{rl} M<B_{k_1+1} & \text{implies that } p(t)\in\mathfrak{G}_{k_1}, \\ A_{k_1+1}<M<B_{k_1+2} & \text{implies that } p(t)\in\mathfrak{G}_{k_1+1}, \\ \cdots\cdots\cdots & \cdots\cdots\cdots\cdots\cdots \\ A_{k_2-1}<M<B_{k_2} & \text{implies that } p(t)\in\mathfrak{G}_{k_2-1}, \\ A_{k_2}<M & \text{implies that } p(t)\in\mathfrak{G}_{k_2}. \end{array}\right\} \tag{4.50}$$

Let us find A_k. Clearly,

$$A_k = -\inf_{y\in\Pi_k,\ b^2\leqslant -\frac{\ddot{y}}{y}\leqslant a^2} \int_0^T \frac{\ddot{y}}{y}\,dt.$$

As in §4.3, one readily shows that the infimum is achieved at a function $y_0(t)$ with equidistant zeros. Denoting the (fixed) distance between the zeros by τ, we see that $\tau=T/k$,

$$A_k=-k\int_0^\tau \frac{\ddot{y}_0}{y_0}\,dt.$$

Assume first that $b^2\geqslant 0$. By Theorem I of §4.3, we have $(q=1)$

$$y_0=\begin{cases} y_1=\dfrac{1}{a}\sin at & \text{if } 0\leqslant t\leqslant\xi, \\ y_3=C\cos b\left(\dfrac{\tau}{2}-t\right) & \text{if } \xi\leqslant t\leqslant\tau-\xi, \\ y_2=\dfrac{1}{a}\sin a(\tau-t) & \text{if } \tau-\xi\leqslant t\leqslant\tau, \end{cases}$$

where C and ξ are determined by the joining conditions $y_1(\xi)=y_3(\xi)$, $\dot{y}_1(\xi)=\dot{y}_3(\xi)$. These conditions readily imply a transcendental equation for ξ:

$$\varphi(\xi)=\frac{a}{b}, \text{ where } \varphi(\xi)=\operatorname{tg} a\xi\cdot\operatorname{tg} b\left(\frac{1}{2}\tau-\xi\right). \tag{4.51}$$

The solution ξ must be such that

$$0 < a\xi < \frac{1}{2}\pi, \tag{4.52}$$

and also $0 < b(^1/_2\tau - \xi) < \pi/2$, which, as is readily seen, is a consequence of (4.52). It follows from the expression for y_0 that

$$A_k = b^2T + 2k\xi(a^2 - b^2) \tag{4.53}$$

(ξ depends on k). The condition $A_k < M$ is equivalent to

$$\xi < \frac{d}{2k}, \tag{4.54}$$

where

$$d = \frac{M - b^2T}{a^2 - b^2}. \tag{4.55}$$

The function $\varphi(\xi)$ increases monotonically in $(0, \pi/2a)$, from 0 to ∞. Thus condition (4.54) is equivalent to the conditions

$$\text{either } \frac{d}{2k} \geqslant \frac{\pi}{2a}, \quad \text{or } \frac{d}{2k} < \frac{\pi}{2a} \quad \text{and} \quad \varphi(\xi) = \frac{a}{b} < \varphi\left(\frac{d}{2k}\right). \tag{4.56}$$

Denoting

$$T_k = \varphi\left(\frac{d}{2k}\right) \tag{4.57}$$

we see that

$$T_k = \operatorname{tg}\frac{ad}{2k} \cdot \operatorname{tg}\frac{bD}{2k}, \quad \text{where } D = \frac{a^2T - M}{a^2 - b^2}. \tag{4.58}$$

It follows from (4.56) that $A_k < M$ if and only if

$$\text{either } d \geqslant \frac{\pi k}{a}, \quad \text{or } d < \frac{\pi k}{a} \quad \text{and} \quad T_k > \frac{a}{b}. \tag{4.59}$$

Now let $b^2 \leqslant 0$. Then by Theorem I of §4.3 we have $y_0 = y_3 = C$ on the inner interval. Therefore $\xi = \pi a/2$, $A_k = ka\pi$. The condition $A_k < M$ is equivalent to

$$ak\pi < M. \tag{4.60}$$

We now determine B_k. By (4.49),

$$B_k = -\sup_{y \in \Pi_k,\ b^2 \leqslant -\frac{\ddot{y}}{y} \leqslant a^2} \int_0^T \frac{\ddot{y}}{y}\, dt,$$

and the supremum is achieved at a function $y_0(t)$, the distance between whose zeros is a constant $\tau = T/k$. By Theorem II of §4.3,

$$y_0 = \begin{cases} y_1 = \frac{1}{b}\sin bt & \text{if} \quad 0 \leqslant t \leqslant \eta, \\ y_3 = C\cos a\left(t - \frac{1}{2}\tau\right) & \text{if} \quad \eta \leqslant t \leqslant \tau - \eta, \\ y_2 = \frac{1}{b}\sin b(\tau - t) & \text{if} \quad \tau - \eta \leqslant t \leqslant \tau, \end{cases}$$

where C and η are determined by the joining conditions $y_1(\eta) = y_3(\eta)$, $\dot{y}_1(\eta) = \dot{y}_3(\eta)$. These conditions lead to a transcendental equation for η:

$$\psi(\eta) = \frac{b}{a}, \quad \text{where} \quad \varphi(\eta) = \operatorname{tg} b\eta \cdot \operatorname{tg} a\left(\frac{1}{2}\tau - \eta\right). \tag{4.61}$$

The root η must satisfy the inequalities

$$0 < a\left(\frac{1}{2}\tau - \eta\right) < \frac{1}{2}\pi \tag{4.62}$$

and also the inequalities $0 < b\eta < \pi/2$, which follow from (4.62). Thus

$$B_k = a^2T - 2k\eta(a^2 - b^2). \tag{4.63}$$

The condition $B_k > M$ is equivalent to

$$\eta < \frac{D}{2k}. \tag{4.64}$$

Denoting $\tau/2 - \eta = \xi'$, we see that ξ' is a root of the equation $\varphi(\xi') = b/a$ $(0 < a\xi' < \pi/2)$. Condition (4.64) means that

$$\xi' > \frac{\tau}{2} - \frac{D}{2k} = \frac{d}{2k},$$

and this is equivalent to

$$\frac{d}{2k} < \frac{\pi}{2a} \quad \text{and} \quad \varphi\left(\frac{d}{2k}\right) < \varphi(\xi') = \frac{b}{a}.$$

Thus $B_k > M$ if and only if

$$d < \frac{\pi k}{a} \quad \text{and} \quad T_k < \frac{b}{a}. \tag{4.65}$$

Finally, if $b^2 > 0$, the inequalities in (4.50) are equivalent to the following:

$$\left.\begin{array}{ll}
d<\frac{\pi(k_1+1)}{a}, \quad T_{k_1+1}<\frac{b}{a} & (\mathfrak{G}_{k_1}),\\
\left.\begin{array}{l}\text{either } d<\frac{\pi(k_1+1)}{a}, \quad T_{k_1+1}>\frac{a}{b}, \quad T_{k_1+2}<\frac{b}{a},\\
\text{or } \frac{\pi(k_1+1)}{a}\leqslant d<\frac{\pi(k_1+2)}{a}, \quad T_{k_1+2}<\frac{b}{a},\end{array}\right\} & (\mathfrak{G}_{k_1+1}),\\
\cdots\cdots\cdots\cdots\cdots\cdots & \\
\left.\begin{array}{l}\text{either } d<\frac{\pi(k_2-1)}{a}, \quad T_{k_2-1}>\frac{a}{b}, \quad T_{k_2}<\frac{b}{a},\\
\text{or } \frac{\pi(k_2-1)}{a}\leqslant d<\frac{\pi k_2}{a}, \quad T_{k_2}<\frac{b}{a},\end{array}\right\} & (\mathfrak{G}_{k_2-1}),\\
\left.\begin{array}{l}\text{either } d\geqslant\frac{\pi k_2}{a},\\
\text{or } d<\frac{\pi k_2}{a}, \quad T_{k_2}>\frac{a}{b}\end{array}\right\} (\mathfrak{G}_{k_2}). &
\end{array}\right\} \tag{4.66}$$

We summarize the result in the following two theorems.

Theorem I. A sufficient condition for all solutions of Hill's equation to be bounded when $p(t)$ satisfies (4.47), (4.48) ($b^2>0$) is that conditions (4.66) hold. In that case, validity of the inequalities ($\mathfrak{G}_k$) implies $p(t)\in\mathfrak{G}_k$. The integers k_1, k_2 are defined by (4.48a), d and T_k by (4.55) and (4.58).

If $b^2=-b_1^2\leqslant 0$, inequalities (4.66) become:

$$\left.\begin{array}{lr}
d<\frac{\pi(k_1+1)}{a}, \quad S_{k_1+1}<\frac{1}{a} & (\mathfrak{G}_{k_1}),\\
d<\frac{\pi(k_1+2)}{a}, \quad S_{k_1+2}<\frac{1}{a}, \quad a(k_1+1)\pi<M & (\mathfrak{G}_{k_1+1}),\\
\cdots\cdots\cdots\cdots\cdots\cdots & \\
d<\frac{\pi k_2}{a}, \quad S_{k_2}<\frac{1}{a}, \quad a(k_2-1)\pi<M & (\mathfrak{G}_{k_2-1}),\\
ak_2\pi<M & (\mathfrak{G}_{k_2}).
\end{array}\right\} \tag{4.67}$$

Here

$$S_k=\frac{1}{b_1}\operatorname{th}\frac{b_1D}{2k}\cdot\operatorname{tg}\frac{ad}{2k} \quad \text{if} \quad b_1>0 \quad \text{and} \quad S_k=\frac{D}{2k}\operatorname{tg}\frac{ad}{2k} \quad \text{if} \quad b_1=0,$$

and the numbers d, D are defined by (4.55), (4.58).

Theorem II. A sufficient condition for all solutions of Hill's equation (3.1) to be bounded when $p(t)$ is any function satisfying (4.47), (4.48) ($b^2=-b_1^2\leqslant 0$) is that conditions (4.67) hold. In that case, validity of inequalities ($\mathfrak{G}_k$) implies $p(t)\in\mathfrak{G}_k$.

Using the fact that $(-A_k)$ and $(-B_k)$ are not merely lower and upper bounds but infima and suprema, one can show that conditions (4.66) and (4.67) are also necessary for the appropriate conclusions to hold in Theorems I, II.

4.7. Estimates of characteristic exponents and stability tests for equations $\ddot{x}+P(t)\dot{x}+Q(t)x=0$

Our next problem is to estimate the characteristic exponents of equation (3.1). We can at once extend the problem, since estimation of the characteristic exponents of equation (3.1) makes it possible to do the same for the more general equation

$$\ddot{x}+P(t)\dot{x}+Q(t)x=0, \tag{4.68}$$

where $P(t)$ and $Q(t)$ are T-periodic functions such that $P(t)$, $\dot{P}(t)$, $Q(t)$ are Lebesgue-integrable. Indeed, the usual substitution

$$x=ye^{-\frac{1}{2}\int_0^t P(t_1)\,dt_1} \tag{4.69}$$

reduces equation (4.68) to the form of (3.1) with

$$p(t)=Q(t)-\frac{1}{4}P^2(t)-\frac{1}{2}\dot{P}(t), \tag{4.70}$$

and under our assumptions this function is integrable and T-periodic. Let

$$P_{\text{av}}=\frac{1}{T}\int_0^T P(t)\,dt=2\mu_0. \tag{4.71}$$

Then $x=e^{-\mu_0 t}e^{R(t)}y$, where

$$R(t)=\frac{1}{2}\left[\frac{t}{T}\int_0^T P(t)\,dt-\int_0^t P(t_1)\,dt_1\right]$$

is a continuous periodic function. Thus, in order to estimate the characteristic exponents for solutions x of equation (4.68) we need only do the same for solutions y of equation (3.1).

Let $\mu>0$, and denote the set of Hamiltonians $\mathbf{H}(t)$ in (3.3) for which $y(t)=O(e^{\mu t})$ by $\mathfrak{L}_\mu$. To determine the position of $\mathfrak{L}_\mu$ in the function space $\mathfrak{L}^3$, we observe that the corresponding matrices $\mathbf{K}$ (§1.9) fill out the interior of a one-sheeted hyperboloid in $\mathfrak{R}^3$:

$$x^2+y^2-z^2=\mu^2.$$

Therefore (see end of §1.6) the set $\mathfrak{L}_\mu^3$ of all Hamiltonians $\mathbf{H}(t)$ in (3.3) for which the solution $\mathbf{x}(t)$ of equation (3.2) satisfies the estimate $\mathbf{x}(t)=O(e^{\mu t})$ occupies the position in $\mathfrak{L}^3$ shown by the dotted region in Figure 75. $\mathfrak{L}_\mu$ is the intersection of $\mathfrak{L}_\mu^3$ with the set of all Hamiltonians $\mathbf{H}(t)$ in (3.3), and so its structure is also shown in Figure 75 (except that, as shown above, it does not contain the sets $\mathfrak{G}_n$, $\mathcal{H}_n$, Π_n with negative subscripts, nor does it contain Π_0^{++} or Π_0^{+-}). Its boundary $\Gamma^{(\mu)}$ is the set of all $p(t)$ (more precisely, of all Hamiltonians $\mathbf{H}(t)$ in (3.3)) such that the solutions of equation (3.1) have characteristic exponents $\pm\mu$. $\Gamma^{(\mu)}$ is the union of countably many subsets $\Gamma_k^{(\mu)}$, each a subset of the corresponding instability domain $\mathcal{H}_k$. Thus, if $p(t) \in \Gamma_k^{(\mu)}$, the corresponding equation (3.1) has a solution

$$y = e^{\mu t} u(t),$$

where $u(t+T) = (-1)^k u(t)$ and $u(t)$ has exactly k zeros on $[t_0, t_0+T]$: $u(t_0) = u(t_1) = \ldots = u(t_{k-1}) = u(t_0+T) = 0$ $(t_0 < t_1 < t_2 < \ldots < t_{k-1} < t_0+T)$.

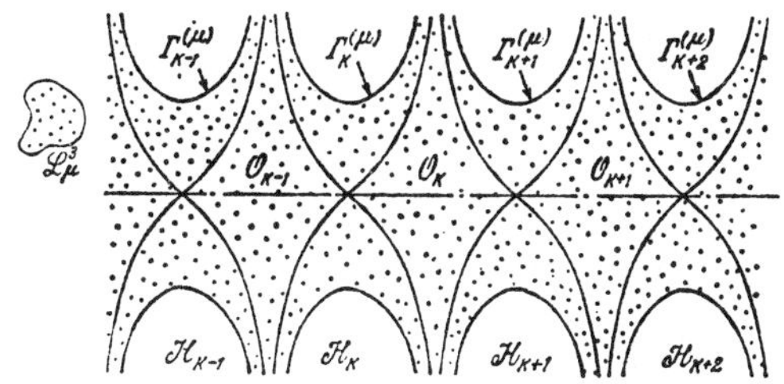

FIGURE 75.

Set $b^2 = \min p(t)$, $a^2 = \max p(t)$ (b^2 and a^2 may also be negative, i. e., b and a may be pure imaginary). We wish to find a criterion for $p(t)$ to lie in $\mathfrak{L}_\mu$, in the form

$$a^2 + \mu^2 \geqslant 0, \quad \int_0^T [a^2 - p(t)]\, dt \leqslant \Phi(a, \mu), \tag{I}$$

$$b^2 + \mu^2 \geqslant 0, \quad \int_0^T [p(t) - b^2]\, dt \leqslant \Psi(b, \mu). \tag{II}$$

It is easily seen that the condition $a^2 + \mu^2 \geqslant 0$ implies $p(t) \equiv a^2 \in \mathfrak{L}_\mu$. The second condition in (I) means that the "distance" from the function $p(t)$ to a^2 is at most $\Phi(a, \mu)$. The explanation of (II) is similar.

We first consider case (I). We may put

$$\Phi(a, \mu) = \inf \int_0^T [a^2 - p(t)]\, dt$$

over all $p(t) \in \Gamma^{(\mu)}$ such that $p(t) \leqslant a^2$. Indeed, let $\Phi(a, \mu)$ be defined in this way; let $p(t)$ satisfy (I) but $p(t) \notin \mathfrak{L}_\mu$. Set

$$p_\lambda(t) = (1-\lambda)\, p(t) + \lambda a^2 \qquad (0 \leqslant \lambda \leqslant 1).$$

Clearly,

$$\max p_\lambda(t) = a^2, \quad \int_0^T [-p_\lambda(t) + a^2]\, dt = (1-\lambda) \int_0^T [-p(t) + a^2]\, dt.$$

As λ varies from 0 to 1, $p_\lambda(t)$ varies from $p(t)$ to a^2. Since $p(t)\notin\mathfrak{L}_\mu$ and $a^2\in\mathfrak{L}_\mu$, there is some λ' for which $p_{\lambda'}(t)$ lies on the boundary of $\mathfrak{L}_\mu$, i. e., $p_{\lambda'}(t)\in\Gamma^{(\mu)}$. Clearly, $\lambda'\neq 0, 1$, and by the definition of $\Phi(a,\mu)$, *

$$\int_0^T [a^2-p_{\lambda'}(t)]\,dt \geqslant \Phi(a,\mu), \text{ i. e., } \int_0^T [a^2-p(t)]\,dt > \Phi(a,\mu),$$

a contradiction. Thus $p(t)\in\mathfrak{L}_\mu$.

Define an integer n by

$$\left.\begin{array}{l} n=0 \quad \text{if} \quad a^2\leqslant 0; \\ \dfrac{n\pi}{T}<a<\dfrac{(n+1)\pi}{T} \quad \text{if} \quad a^2>0 \quad (a>0). \end{array}\right\} \tag{4.72}$$

By the comparison theorem of §3.2, the conditions $p(t)\underset{\neq}{\leqslant} a^2$ and $p(t)\in\Gamma^{(\mu)}$ imply that $p(t)\in\Gamma_0^{(\mu)}\cup\Gamma_1^{(\mu)}\cup\ldots\cup\Gamma_n^{(\mu)}$. Denote

$$\Phi_k(a,\mu)=\inf\int_0^T [a^2-p(t)]\,dt \qquad (p(t)\in\Gamma_k^{(\mu)},\ p(t)\leqslant a^2).$$

Then

$$\Phi(a,\mu)=\min\Phi_k(a,\mu) \qquad (k=0, 1, \ldots, n). \tag{4.73}$$

To determine $\Phi_k(a,\mu)$, we proceed as follows. First let $n\geqslant 1$, so that $a>\pi/T$. Let $p(t)\in\Gamma_k^{(\mu)}$ and let $y(t)$ be the corresponding solution.

The solution $y(t)$ satisfies the following conditions:

a) y and $\dot y$ are absolutely continuous; $\ddot y/y$ is summable;

b) $y(t_0)=y(t_1)=\ldots=y(t_k)=0$, $t_0<t_1<\ldots<t_k=t_0+T$; for all other t, $y(t)\neq 0$;

c) $\dot y(t_0+T)=(-1)^k e^{\mu T}\dot y(t_0)$, $\dot y(t_i)\neq 0$ $\qquad (i=0, 1, \ldots, k)$;

d$_1$) $a^2+\dfrac{\ddot y}{y}\geqslant 0$ almost everywhere.

Conversely, any function $y(t)$ satisfying these four conditions defines a function $p(t)=-\ddot y/y\in\Gamma_k^{(\mu)}$ such that $p(t)\leqslant a^2$. Therefore

$$\Phi_k(a,\mu)=\inf\int_0^T\left(a^2+\frac{\ddot y}{y}\right)dt$$

over all y satisfying conditions (a), (b), (c), (d$_1$). Denote $\tau_i=(t_i, t_{i+1})$, $q_i=(-1)^i\dot y(t_i)$. By the comparison theorem (§3.2), $\tau_i>\pi/a$. It follows from the solution of the variational problem (I) in §4.3 that

$$\int_{t_0}^{t_0+T}\left(a^2+\frac{\ddot y}{y}\right)dt=\sum_{i=0}^{k-1}\int_{\tau_i}\left(a^2+\frac{\ddot y}{y}\right)dt\geqslant$$

$$\geqslant\sum_{i=0}^{k-1} a\left[a\tau_i-\pi+\frac{\left(\ln\dfrac{q_{i+1}}{q_i}\right)^2}{a\tau_i-\pi}\right]=a\left[aT-k\pi+\sum_{i=0}^{k-1}\frac{\left(\ln\dfrac{q_{i+1}}{q_i}\right)^2}{a\tau_i-\pi}\right].$$

* If $\lambda=1$, then $a^2\in\Gamma^{(\mu)}$, so that $a^2=-\mu^2$, $\Phi=0$. This case is excluded, for then the criterion becomes trivial.

Using the inequality*

$$\sum_i \frac{s_i^2}{t_i} \geqslant \frac{\left(\sum_i s_i\right)^2}{\sum_i t_i} \qquad (t_i > 0),$$

which is an equality for $s_i = \text{const} \cdot t_i$, we see that

$$\int_{t_0}^{t_0+T} \left(a^2 + \frac{\ddot{y}}{y}\right) dt \geqslant a\left(aT - k\pi + \frac{\mu^2 T^2}{aT - k\pi}\right).$$

By the variational problem (I) of §4.3, this must be an equality for some function. Therefore

$$\Phi_k(a, \mu) = a\left(aT - k\pi + \frac{\mu^2 T^2}{aT - k\pi}\right). \qquad (4.74)$$

Here $a^2 > \pi^2/T^2$; $k = 1, 2, \ldots$ It is easy to see, however, that (4.74) remains valid for $a^2 \leqslant \pi^2/T^2$, $k = 0$, as well. Indeed, in that case it follows from the condition $p(t) \in \Gamma^{(\mu)}$ that $p(t) \in \Gamma_0^{(\mu)} \subset \mathscr{H}_0$.

Let

$$y = e^{\mu t} u(t)$$

be the corresponding solution, $u(t+T) = u(t) \neq 0$. Integrating by parts, we have

$$\int_0^T [a^2 - p(t)]\, dt = \int_0^T \left(a^2 + \frac{\ddot{y}}{y}\right) dt = \int_0^T \left(a^2 + \frac{\mu^2 u + 2\mu \dot{u} + \ddot{u}}{u}\right) dt =$$

$$= (a^2 + \mu^2) T + \int_0^T \frac{\dot{u}^2}{u^2}\, dt \geqslant (a^2 + \mu^2) T.$$

Equality will hold for $u(t) = \text{const}$. Thus (4.74) is also valid for $a^2 \leqslant \pi^2/T^2$, $k = 0$. $\Phi(a, \mu)$ is found from (4.73):

$$\Phi(a, \mu) = \min_{k=0,1,\ldots,n} a\left(aT - k\pi + \frac{\mu^2 T^2}{aT - k\pi}\right).$$

Clearly,

$$\text{if } a^2 \leqslant \frac{\pi^2}{T^2} \quad n = k = 0, \quad \Phi(a, \mu) = \Phi_0(a, \mu) = (a^2 + \mu^2) T. \qquad (4.75)$$

Let $a^2 > \pi^2/T^2$. The function

$$f(u) = u + \frac{c^2}{u}$$

* Proof: $\left(\sum_i s_i\right)^2 = \left(\sum_i \sqrt{t_i} \frac{s_i}{\sqrt{t_i}}\right)^2 \leqslant \sum t_i \cdot \sum \frac{s_i^2}{t_i}.$

assumes its minimum value at $u=c$, $f(c)=2c$. In our case, $u=aT-k\pi$, $c=\mu T$. Therefore,

$$\Phi(a, \mu) \geqslant 2a\mu T.$$

Define an integer k_0 by

$$aT-k_0\pi \leqslant \mu T < aT-(k_0-1)\pi. \tag{4.76}$$

Elementary arguments now show that

$$\left.\begin{array}{ll} \text{if } n<k_0 & \Phi(a, \mu)=\Phi_n(a, \mu), \\ \text{if } n \geqslant k_0 & \Phi(a, \mu)=\min(\Phi_{k_0}, \Phi_{k_0-1}). \end{array}\right\} \tag{4.77}$$

We have thus proved

Theorem I. Let $\mathfrak{P}(\sigma, a^2)$ *be the class of T-periodic functions* $p(t)$, *Lebesgue-integrable on* $[0, T]$, *such that*

$$p_{\text{av}}=\frac{1}{T}\int_0^T p(t)\,dt=\sigma, \quad p(t) \leqslant a^2 \quad (a^2 \gtreqless 0).$$

Define an integer n *by (4.72) and the function* $\Phi(a, \mu)$ *by either (4.74), (4.73) or (4.74) — (4.77).*

Then a sufficient condition for the real parts of the characteristic exponents of any Hill's equation (3.1) with $p(t) \in \mathfrak{P}(\sigma, a^2)$ *to be bounded above by a prescribed number* μ *such that* $\mu^2+a^2>0$ *is that*

$$T(a^2-\sigma) \leqslant \Phi(a, \mu). \tag{4.77a}$$

Additional reasoning, using the structure of $\mathfrak{L}^3$ and the fact that $\Phi(a,\mu)$ is not merely a lower bound but an infimum, will show that (4.77a) is not only sufficient but also necessary.

This theorem is useful in estimating the characteristic exponents for equation (3.1). Indeed, the condition of the theorem means that

$$\mu^2T^2 \geqslant (aT-k\pi)\left(k\pi-\frac{1}{a}\int_0^T p(t)\,dt\right), \; k=0, 1, 2, \ldots, n. \tag{4.78}$$

By the definition of n, these inequalities will surely hold for any μ if

$$\int_0^T p(t)\,dt > an\pi. \tag{4.79}$$

Thus, condition (4.79) implies that the characteristic exponents of equation (3.1) are pure imaginary. Otherwise, if

$$\int_0^T p(t)\,dt \leqslant an\pi, \tag{4.80}$$

we can define a number

$$\tilde{\mu} = \frac{1}{T} \max \sqrt{(aT - k\pi)\left(k\pi - \frac{1}{a}\int_0^T p(t)\,dt\right)}, \tag{4.81}$$

where the maximum is evaluated over all $0 \leqslant k \leqslant n$ for which both factors in the radicand are positive.

Inequalities (4.78) will hold if $\mu \geqslant \tilde{\mu}$. We may thus rephrase Theorem I as follows.

Theoreom Ia. If $a^2 > 0$ and inequality (4.79) is satisfied, the characteristic exponents of equation (3.1) are pure imaginary.

Otherwise, i.e., if (4.80) is satisfied, the characteristic exponents of equation (3.1) are either pure imaginary or real and bounded above by $\tilde{\mu}$, where $\tilde{\mu}$ is defined by (4.81), so that $y(t) = O(e^{\tilde{\mu}t})$.

The first part of Theorem Ia is also a corollary of Test 5 of §4.4 (in this case, $c = a$). It follows easily from (4.81) that

$$\tilde{\mu} \leqslant \frac{1}{2aT}\int_0^T [a^2 - p(t)]\,dt.$$

Thus, if $a^2 > 0$ we have a rather coarse but very simple estimate* for the solutions of equation (3.1):

$$y(t) = O\left[\exp \frac{t}{2aT}\int_0^T [a^2 - p(t)]\,dt\right]. \tag{4.82}$$

If $a^2 \leqslant 0$, it follows from (4.81) or directly from Theorem I that

$$\tilde{\mu} = \sqrt{\frac{1}{T}\int_0^T [-p(t)]\,dt}$$

and so

$$y(t) = O\left\{\exp\left[t\sqrt{\frac{1}{T}\int_0^T [-p(t)]\,dt}\right]\right\}. \tag{4.83}$$

The same estimate is valid if $a^2 \leqslant \pi^2/T^2$, with the reservation that if the expression in the radicand becomes negative then $y(t) = O(1)$ (pure imaginary characteristic exponents).

As mentioned at the beginning of this subsection, these estimates, in conjunction with Theorem Ia, enable us to estimate the characteristic exponents of equation (4.68).

We shall now discuss the stability of the trivial solution of equation (4.68). If $\mu_0 < 0$ (see (4.69)), the trivial solution is unstable, since otherwise, as follows from (4.69) and (4.71), all solutions $y(t)$ of equation (3.1) would tend to 0 as $t \to +\infty$, and this is impossible.

If $\mu_0 = 0$ the trivial solution of equation (3.1) is stable if and only if the same holds for equation (4.68), and any stability test for the former is automatically a stability test for the latter.

* This estimate may also be established by a more direct method (see § II.4.3).

If $\mu_0 > 0$, we can deduce the following result from Theorem I.

Test I for equation (4.68). Define $p(t)$ and μ_0 for equation (4.68) by the formulas (4.70) and (4.71); assume that $\mu_0 > 0$. If the condition

$$T(a^2-\sigma) \leqslant \Phi(a, \mu)$$

of Theorem I holds for $p(t)$ and $\mu = \mu_0$, then the trivial solution of equation (4.68) is stable.

We now consider case (II). As in case (I), one shows that

$$\Psi(b, \mu) = \inf \int_0^T [p(t)-b^2]\, dt$$

for all $p(t) \in \Gamma^{(\mu)}$ such that $\min p(t) = b^2$. Indeed, let $\Psi(b, \mu)$ be defined in this way, and let $p(t)$ be a function satisfying (II) such that $p(t) \notin \mathfrak{L}_\mu$. Let

$$p_\lambda(t) = (1-\lambda)\, p(t) + \lambda b^2 \qquad (0 \leqslant \lambda \leqslant 1).$$

It is clear that

$$\min p_\lambda(t) = b^2.$$

As λ varies from 0 to 1, the function $p_\lambda(t)$ varies from $p(t)$ to b^2. Since $p_\lambda(t) \notin \mathfrak{L}_\mu$ and $b^2 \in \mathfrak{L}_\mu$, it follows that for some λ $(0 < \lambda < 1)$* the function $p_\lambda(t)$ is on the boundary. For this function

$$\int_0^T [p_\lambda(t)-b^2]\, dt = (1-\lambda) \int_0^T [p(t)-b^2]\, dt < \Psi(b, \mu),$$

contradicting the choice of $\Psi(b, \mu)$. Thus $p(t) \in \mathfrak{L}_\mu$. As before, define an integer n by

$$n = 0 \quad \text{if} \quad b^2 \leqslant 0, \quad \frac{n\pi}{T} \leqslant b < \frac{(n+1)\pi}{T} \quad \text{if} \quad b^2 > 0 \qquad (b > 0). \qquad (4.84)$$

Let $p(t) \in \Gamma^{(\mu)}$. We claim that if $p(t) \not\equiv b^2$, then

$$p(t) \in \Gamma^{(\mu)}_{n+1} \cup \Gamma^{(\mu)}_{n+2} \cup \ldots$$

This is obvious if $b^2 > 0$, since by the comparison theorem of §2.2 the solutions of equation (3.1) will have larger zeros on $[t_0, t_0+T)$ than the solutions of equation (3.1) with $p(t) \equiv b^2$, so that n will be larger. To show that the same holds for $b^2 \leqslant 0$, we prove a lemma which may itself be used to estimate characteristic exponents.

Lemma. Given equation (3.1), with $p(t) \in \mathcal{H}_0$ ($p(t) \not\equiv \text{const}$), let μ be the corresponding characteristic exponent. We may assume without loss of generality that $\mu > 0$. If $\min p(t) = -c^2 \leqslant 0$, then $\mu < c$.

* The case $\lambda = 1$ is excluded, for if $\lambda = 1$ we have $b^2 = -\mu^2$, $\Phi = 0$, and the test is trivial.

Proof. Since $p(t) \in \mathcal{H}_0$, there is a solution $y = e^{\mu t} u(t)$, where $u(t+T) = u(t) > 0$ and

$$p(t) = -\frac{\ddot{y}}{y} = -\frac{\mu^2 u + 2\mu\dot{u} + \ddot{u}}{u} \underset{\not\equiv}{\geqslant} -c^2, \quad c^2 - \mu^2 - 2\mu\frac{\dot{u}}{u} - \frac{\ddot{u}}{u} \underset{\not\equiv}{\geqslant} 0.$$

Integrating this inequality from 0 to T, we find

$$(c^2 - \mu^2) T - \int_0^T \frac{\ddot{u}}{u} dt > 0.$$

Since

$$\int_0^T \frac{\ddot{u}}{u} dt = \int_0^T \frac{\dot{u}^2}{u^2} dt > 0 \qquad (u \not\equiv \text{const}),$$

it follows that $(c^2 - \mu^2) T > 0$, i. e., $\mu < c$, Q. E. D.

It follows from the lemma and from the first condition in (I) (p. 741) that if $b^2 \leqslant 0$ then $p(t) \notin \Gamma_0^{(\mu)}$, i. e.,

$$p(t) \in \Gamma_1^{(\mu)} \cup \Gamma_2^{(\mu)} \cup \ldots$$

Thus the above assertion is also true when $b^2 \leqslant 0$. Now set

$$\Psi_k(b, \mu) = \inf \int_0^T [p(t) - b^2]\, dt \tag{4.85}$$

where the infimum extends over all $p(t) \in \Gamma_k^{(\mu)}$ such that $\min p(t) = b^2$. Then

$$\Psi(b, \mu) = \min_{k \geqslant n+1} \Psi_k(b, \mu). \tag{4.86}$$

Let $p(t) \in \Gamma_k^{(\mu)}$. The corresponding equation (3.1) has a solution

$$y = e^{\mu t} u(t),$$

where $u(t+T) = (-1)^k u(t)$. Let $y(t_0) = y(t_1) = \ldots = y(t_k) = 0$ $(t_k = t_0 + T)$. It is easy to see that then

$$\dot{y}(t_0 + T) = (-1)^k e^{\mu T} \dot{y}(t_0).$$

We see that, as before,

$$\Psi_k(b, \mu) = \inf \int_0^T \left(-\frac{\ddot{y}}{y} - b^2\right) dt$$

where the infimum is taken over all y satisfying conditions (a), (b), (c) and

(d$_2$) $$-\frac{\ddot{y}}{y} - b^2 \geqslant 0.$$

Denote

$$\tau_i = (t_{i-1}, t_i), \quad q_i = (-1)^i \dot{y}(t_i).$$

By the comparison theorem /87/, if $b^2 > 0$ we have $\tau_i < \pi/b$; thus $b^2 < \pi^2/\tau_i^2$ (it is possible that $b^2 \leqslant 0$). It follows from the solution of the variational problem (II) in §4.3 that

$$\int_{t_0}^{t_0+T} \left(-\frac{\ddot{y}}{y} - b^2\right) dt = \sum_i \int_{\tau_i} \left(-\frac{\ddot{y}}{y} - b^2\right) dt > F\left(\tau_i, \frac{q_i}{q_{i-1}}\right),$$

where

$$F(\tau, q) = \frac{b(q + q^{-1} + 2\cos b\tau)}{\sin b\tau}.$$

Therefore

$$\Psi_k(b, \mu) = \min \sum_{i=1}^{k} F\left(\tau_i, \frac{q_i}{q_{i-1}}\right)$$

over all $\tau_i > 0$ such that $\tau_1 + \ldots + \tau_k = T$, and all

$$q_i \qquad (i = 1, 2, \ldots, k-1; \; q_0 = 1, \; q_k = e^{\mu T}).$$

Now denote

$$\frac{q_i}{q_{i-1}} = e^{\mu_i T}, \quad F_1(\tau_i, \mu_i) = F(\tau_i, e^{\mu_i T}) = \frac{2b(\operatorname{ch} \mu_i T + \cos b\tau_i)}{\sin b\tau_i}.$$

Then we infer from (4.85) that

$$\Psi_k(b, \mu) = \min \sum_{i=1}^{k} F_1(\tau_i, \mu_i)$$

over all τ_i, μ_i satisfying the conditions $\tau_i > 0$, $b^2 < \pi^2/\tau_i^2$,

$$\tau_1 + \ldots + \tau_k = T, \quad \mu_1 + \ldots + \mu_k = \mu.$$

The set of values $(\tau_1, \ldots, \tau_k; \mu_1, \ldots, \mu_k)$ figuring in the problem is an open $(2k-2)$-dimensional domain whose boundary is the set of all points $(\tau_1, \ldots, \tau_k; \mu_1, \ldots, \mu_k)$ such that either $\tau_i = 0$, or $\tau_i = \pi/b$ (if $b > 0$) or $\mu_i = \pm\infty$. For these values we have $F_1 = \infty$. Thus the infimum is achieved at some interior point of the domain. At this point we have Lagrange's equations, which are for this case

$$\frac{1 + \cos b\tau_i \operatorname{ch} \mu_i T}{\sin^2 b\tau_i} = \lambda', \quad b\frac{\operatorname{sh} \mu_i T}{\sin b\tau_i} = \lambda'', \tag{4.87}$$

where λ' and λ'' are real constants (b is real or pure imaginary). Eliminating μ_i from these equations and setting

$$x_i = \frac{1}{b}\sin b\tau_i,$$

we obtain

$$x_i^2\left[(-b^2+\lambda''^2+\lambda' b^2)-x_i^2(\lambda''^2 b^2+\lambda'^2 b^4)\right]=0.$$

Since $x_i \neq 0$ and for real b we have $0<b\tau_i<\pi$, it follows from the above equation that

$$x_i = \frac{1}{b}\sin b\tau_i$$

is uniquely determined. If $b^2 \leqslant 0$, we deduce that $\tau_1 = \ldots = \tau_k$. If $b^2>0$, it follows that τ_i may take on at most two values:

$$\tau_i = \tilde{\tau} \quad \text{or} \quad \tau_i = \frac{\pi}{b}-\tilde{\tau}.$$

In either case, it follows from the second equation of (4.87) that $\mu_1 = \ldots = \mu_k$. Then the first equation of (4.82) tells us that $\cos b\tau_i$ is uniquely determined. It follows that $\tau_1 = \ldots = \tau_k$ when $b^2>0$ as well. Therefore, $\mu_i = \mu/k$, $\tau_i = T/k$, and

$$\Psi_k(b,\ \mu) = kF_1\left(\frac{T}{k},\ \frac{\mu}{k}\right) = \frac{2bk\left(\operatorname{ch}\frac{\mu T}{k}+\cos\frac{bT}{k}\right)}{\sin\frac{bT}{k}},$$

so that by (4.81)

$$\Psi(b,\ \mu) = \min_{k \geqslant n+1} \frac{2bk\left(\operatorname{ch}\frac{\mu T}{k}+\cos\frac{bT}{k}\right)}{\sin\frac{bT}{k}}. \tag{4.88}$$

This minimum is clearly attained at some value of k, since $\Psi_k(b,\ \mu) \sim \frac{4k^2}{T}$ as $k \to \infty$; in other words, from some k onward, $\Psi_k(b,\ \mu)$ increases with increasing k. We have thus proved

Theorem II. Let $\Omega(\sigma,\ b^2)$ *be the class of T-periodic functions* $p(t)$, *Lebesgue-integrable on* $[0,\ T]$, *for which*

$$p_{\text{av}} = \frac{1}{T}\int_0^T p(t)\,dt = \sigma, \quad p(t) \geqslant b^2 \quad (b^2 \gtreqless 0).$$

Define an integer n *and function* $\Psi(b,\ \mu)$ *by (4.84), (4.88).*

Then a sufficient condition for the characteristic exponents of any equation (4.68) with $p(t) \in \Omega(\sigma,\ b^2)$ *to be either pure imaginary or real and less than a prescribed number* $\mu \geqslant 0$ *such that* $b^2+\mu^2>0$ *is that*

$$T(\sigma-b^2) \leqslant \Psi(b,\ \mu).$$

Using the structure of the space $\mathfrak{L}^3$ and the fact that $\Psi(b, \mu)$ is the infimum, one can show that the inequality $T(\sigma - b^2) \leqslant \Psi(b, \mu)$ is not only sufficient but also necessary for the conclusion to hold.

Since the infimum in (4.85) is not achieved (this follows from the solution of the variational problem (II) in §4.3), the assumptions of the theorem guarantee that $p(t) \in \mathfrak{L}_\mu - \Gamma^{(\mu)}$, i. e., that $y = O(e^{(\mu-\varepsilon)t})$ for some $\varepsilon > 0$.

As before, it is readily seen that the theorem may be used to estimate characteristic exponents. Indeed, the condition of the theorem means that

$$\operatorname{ch}\frac{\mu T}{k} \geqslant \frac{\sin\frac{bT}{k}}{2bk}\int_0^T [p(t) - b^2]\,dt - \cos\frac{bT}{k} \tag{4.89}$$

for $k = n+1,\ n+2,\ \ldots$ These inequalities will surely hold for any μ if

$$\frac{1}{2bk}\sin\frac{bT}{k}\int_0^T [p(t) - b^2]\,dt - \cos\frac{bT}{k} \leqslant 1.$$

It is easy to see that, by the definition of n, these inequalities are equivalent to

$$\int_0^T [p(t) - b^2]\,dt \leqslant 2b(n+1)\operatorname{ctg}\frac{bT}{2(n+1)}. \tag{4.90}$$

We may thus take this last inequality as a stability test. It is easy to see that this test also follows from Test 4 of §4.4. If the reverse inequality

$$\int_0^T [p(t) - b^2]\,dt > 2b(n+1)\operatorname{ctg}\frac{bT}{2(n+1)} \tag{4.91}$$

is valid, we can define a real number $\tilde{\mu}$ by

$$\tilde{\mu} = \max\frac{k}{T}\operatorname{Arch}\left\{\frac{1}{2bk}\sin\frac{bT}{k}\int_0^T [p(t) - b^2]\,dt - \cos\frac{bT}{k}\right\}, \tag{4.92}$$

where the maximum extends over all $k \geqslant n+1$ such that the expression in square brackets in (4.92) is greater than unity. Inequalities (4.89) will be satisfied if $\mu \geqslant \tilde{\mu}$. Theorem II may now be rephrased as follows.

Theorem IIa. If inequality (4.90) is satisfied, then all characteristic exponents of equation (3.1) are pure imaginary. If the reverse inequality (4.91) is satisfied, the characteristic exponents are either pure imaginary or real and less than $\tilde{\mu}$, where $\tilde{\mu}$ is less than (4.92), so that

$$y(t) = O(e^{(\tilde{\mu}-\varepsilon)t})$$

for some $\varepsilon > 0$.

Theorem II implies

Test 2 for equation (4.68). Define $p(t)$ and μ_0 for equation (4.68) by (4.70) and (4.71), assume that $\mu_0 > 0$. If the assumptions of Theorem II are satisfied for $p(t)$ and $\mu = \mu_0$, then the trivial solution of equation (4.68) is stable and the corresponding solutions satisfy the estimate

$$x(t) = O(e^{-\varepsilon t})$$

for some $\varepsilon > 0$.

We pay specific attention to the case $b = 0$. In this case Theorem I (or Ia) implies a complement to Lyapunov's test $p(t) \geqslant 0,\ T\int_0^T p(t)dt \leqslant 4$:

Assume that in equation (3.1) $p(t) \geqslant 0$ and

$$J = \frac{T}{4}\int_0^T p(t)\,dt > 1.$$

Let

$$\tilde{\mu} = \frac{2}{T}\max\left[k \operatorname{Arch}\frac{\sqrt{J}}{k}\right]$$

where the maximum extends over all integers k such that $1 \leqslant k < \sqrt{J}$. Then $y(t) = O(e^{(\tilde{\mu}-\varepsilon)t})$ for some $\varepsilon > 0$.

It follows from the proof that these estimates (Theorems Ia and IIa) cannot be improved without additional assumptions.

§5. SOME OTHER STABILITY CONDITIONS FOR HILL'S EQUATION

5.1. Stability tests for Hill's equation based on the distance between zeros of solutions

The method of proving stability tests described in this subsection is based on Theorem I of §3.2. One consequence of this theorem is that $p(t) \in \mathfrak{G}_0$ if and only if the following two conditions hold: 1) all solutions of Hill's equation are oscillating, 2) the distance between consecutive zeros of any solution is greater than $T + \varepsilon$, where $\varepsilon = \varepsilon[p(t)] > 0$. Thus, our problem reduces to finding lower bounds for the distance between zeros of solutions. This method has been used by many authors since Zhukovskii first applied it /147/. The most complete results to date have been achieved by Komlenko /159a/. We present the most important ones.

As before, $[q(t)]_+$ will denote the positive part of a function $q(t)$:

$$[q(t)]_+ = \frac{1}{2}[|q(t)| + q(t)].$$

Set

$$C(t, s)=\begin{cases}\frac{1}{\sqrt{\mu}}\sin[\sqrt{\mu}(t-s)] & \text{if } \mu>0,\\ t-s & \text{if } \mu=0,\\ \frac{1}{\sqrt{|\mu|}}\operatorname{sh}[\sqrt{|\mu|}(t-s)] & \text{if } \mu<0.\end{cases} \tag{5.1}$$

This is the Cauchy function of the equation

$$\ddot{y}+\mu y=f(t). \tag{5.2}$$

In other words, the solution of equation (5.2) with initial conditions

$$y(t_0)=0, \quad \dot{y}(t_0)=0$$

is

$$y(t)=\int_{t_0}^{t} C(t, s) f(s)\, ds. \tag{5.3}$$

We shall make considerable use of the following lemma, concerning the existence of a positive solution of equation (3.1) on a prescribed interval (t_0, t_1).

Lemma (Komlenko /159a/). Suppose that for some $\mu\leqslant\pi^2(t_1-t_0)^{-2}$

$$\int_{t_0}^{t_1} C(t_1, s) C(s, t_0)[p(s)-\mu]_+ ds < C(t_1, t_0). \tag{5.4}$$

Then there exists a solution $y(t)$ *of equation (3.1) such that*

$$y(t)>0 \quad \text{for} \quad t_0\leqslant t\leqslant t_1, \tag{5.5}$$

and any solution $y(t)$ *of equation (3.1) has at most one zero in the closed interval* $[t_0, t_1]$.

Proof. We consider the auxiliary equations

$$\ddot{y}(t)+\mu y(t)+[p(t)-\mu]_+ y(t)=0. \tag{5.6}$$

Let $y_0(t)$ be a solution satisfying the initial conditions $y_0(t_0)=0$, $\dot{y}_0(t_0)=1$. Since $C(t, s)$ is the Cauchy function of equation (5.2), $y_0(t)$ satisfies the integral equation

$$y(t)=C(t, t_0)-\int_{t_0}^{t} C(t, s)[p(s)-\mu]_+ y(s)\, ds. \tag{5.7}$$

To this integral equation we apply Theorem I of §2.2 (p. 668); in the notation of equation (2.24),

$$\varkappa(t)=C(t, t_0), \quad K(t, s, \xi)=-C(t, s)[p(s)-\mu]_+\xi.$$

If $\mu \leqslant \pi^2 (t_1 - t_0)^{-2}$ $(t_1 \geqslant t \geqslant s \geqslant t_0)$, it follows from (5.1) that $C(t, s) \geqslant 0$. Therefore $\partial K/\partial \xi \leqslant 0$ and we may set $P^{\text{in}} \equiv 0$, $Q^{\text{in}} \equiv 0$, $P^{\text{d}} = Q^{\text{d}} \equiv K$ in (2.26), (2.27). Set

$$\xi_0''(t) = C(t, t_0).$$

The function $\xi_0'(t)$ will be defined below. Noting that $C(t, t_0) > 0$ for $t_0 < t \leqslant t_1$, we deduce via elementary arguments that for any fixed $s \in [t_0, t_1]$ the function $F(t) = C(t, s) C(s, t_0) C(t, t_0)^{-1}$ is monotone increasing in t. Consequently, inequality (5.4) implies

$$\int_{t_0}^{t} \frac{C(t, s)\, C(s, t_0)}{C(t, t_0)} [p(s) - \mu]_+ \, ds < 1 \qquad (t_0 < t \leqslant t_1).$$

Using this inequality, we see that the function

$$\xi_0'(t) = C(t, t_0) - \int_{t_0}^{t} C(t, s)\, C(s, a)\, [p(s) - \mu]_+ \, ds \tag{5.8}$$

is positive on $(t_0, t_1]$ and $\xi'(t_0) = 0$.

We now show that inequalities (2.28) of Theorem I, §2.2, are satisfied. Indeed, substituting the functions chosen above for $\varkappa$, P^{in}, P^{d}, Q^{in}, Q^{d}, we have

$$\xi_0''(t) - \varkappa(t) - \int_{t_0}^{t} Q^{\text{d}}(t, s, \xi_0'(s))\, ds - \int_{t_0}^{t} Q^{\text{in}}(t, s, \xi_0''(s))\, ds =$$
$$= \int_{t_0}^{t} C(t, s)\, [p(s) - \mu]_+ \xi_0'(s)\, ds \geqslant 0,$$

since $C(t, s) \geqslant 0$, $\xi'(s) \geqslant 0$. Similarly,

$$\xi_0'(t) - \varkappa(t) - \int_{0}^{t} P^{\text{in}}(t, s, \xi_0'(s))\, ds - \int_{0}^{t} P^{\text{d}}(t, s, \xi_0''(s))\, ds =$$
$$= \xi_0'(t) - C(t, t_0) + \int_{0}^{t} C(t, s)\, [p(s) - \mu]_+ C(s, t_0)\, ds = 0$$

by (5.8). Thus all the assumptions of Theorem I in §2.2 are valid. Hence, by (2.30) $(k = 0)$,

$$\xi_0'(t) \leqslant y(t) \leqslant \xi_0''(t) = C(t, t_0) \quad \text{for} \quad t_0 \leqslant t \leqslant t_1.$$

Since $\xi_0'(t) > 0$ for $t_0 < t \leqslant t_1$, it follows by the continuity of the solution of equation (5.6) with respect to initial conditions that there exists a solution $y_1(t)$ of equation (5.6) such that $y_1(t) > 0$ for $t_0 \leqslant t \leqslant t_1$.

Let us compare equations (5.6) and (3.1) by means of Sturm theory /87/. The coefficients of these equations satisfy the inequality

$$\mu + [p(t) - \mu]_+ \geqslant p(t) \quad \text{for} \quad t_0 \leqslant t \leqslant t_1.$$

By Sturm's theorem, the solution $y(t)$ of equation (3.1) with initial conditions $y(t_0)=y_1(t_0)$, $\dot{y}(t_0)=\dot{y}_1(t_0)$ has the property $y(t)>y_1(t)$ for $t_0\leqslant t\leqslant t_1$, so that (5.5) holds.

Between any two zeros of one solution of equation (3.1) there is always a zero of any other solution. It therefore follows from (5.5) that no solution $y(t)$ of equation (3.1) has more than one zero on $[t_0, t_1]$. This completes the proof of the lemma.

We now return to Hill's equation (3.1).

Test 8 *(Komlenko /159a/). Assume that in (3.1)* $p(t)\not\equiv 0$, $p(t)\not\equiv \pi^2/T^2$ *and*

$$\int_0^T p(t)\,dt\geqslant 0. \tag{5.8a}$$

If there exists $\mu<\pi^2/T^2$ *such that for any* $a\in[0, T]$

$$\int_a^{a+T} C(a+T, t)C(t, a)[p(t)-\mu]_+\,dt<C(a+T, a), \tag{5.9}$$

where $C(t, s)$ *is of the form (5.1), then all solutions of the equation are bounded and* $p(t)\in\mathfrak{S}_0$.

Proof. Condition (5.8a) guarantees that the equation is oscillating (p. 694).; It follows from (5.9) by continuity that a similar inequality holds with $a+T$ replaced on both sides of the inequality by $a+T+\varepsilon$, where $\varepsilon>0$ is sufficiently small. By the lemma, any solution $y(t)$ of equation (3.1) has at most one zero on the interval $[a, a+T+\varepsilon]$. Let $t_1>t_0$ be two consecutive zeros of $y(t)$. Setting $a=t_0$, we see that $t_1-t_0>T+\varepsilon$. Therefore, as shown previously, $p(t)\in\mathfrak{S}_0$, Q. E. D.

This proof makes use of Theorem 1 of §3.2. However, the fact that all the solutions are bounded may be deduced by quite elementary means, using only the Floquet-Lyapunov theorem. To this end, we show that if $|A|\geqslant 1$ and the equation is oscillating, then there exists a solution $y(t)$ of equation (3.1) which has zeros $t_0\leqslant t_1$ (i. e., $y(t_0)=y(t_1)=0$) such that $t_1-t_0\leqslant T$. If $\mathbf{x}(t)$ is a solution-vector of equation (3.2), then

$$\mathbf{x}(t)=\mathbf{F}(t)\,e^{t\mathbf{K}}\mathbf{x}(0),$$

where $\mathbf{F}(t+T)=\pm\mathbf{F}(t)$ and $\mathbf{K}$ are real matrices. Let $|A|\geqslant 1$, where A is the Lyapunov characteristic constant, so that $\mathbf{K}$ has real eigenvalues $\pm\lambda$ (which are equal if $|A|=1$). Here $\lambda\geqslant 0$. Let $\mathbf{x}_0$ be a real eigenvector, $\mathbf{K}\mathbf{x}_0=\lambda\mathbf{x}_0$, and let $\mathbf{x}(0)=\mathbf{x}_0$. Then

$$\mathbf{x}(t)=e^{\lambda t}\mathbf{F}(t)\,\mathbf{x}(0).$$

The second component of this vector is a solution $y(t)$ of equation (3.1). Consequently, if $|A|\geqslant 1$ then equation (3.1) has a solution

$$y(t)=e^{\lambda t}\varphi(t),$$

where $\lambda \geqslant 0$ and $\varphi(t+T) = \pm\varphi(t)$ is a real function. If all solutions are oscillating, then $y(t_0) = 0$ for some t_0. It is evident from the form of $y(t)$ that $y(t_0+T) = 0$, and this proves our assertion.

Since the equation in our case is oscillating (by (5.8a)) and $t_1 - t_0 > T + \varepsilon$ for any solution $y(t)$, it follows that $|A| < 1$, as claimed.

We now show that Test 8 implies Test 4 of §4.4 for $n = 0$.

Using (5.1), one readily shows that the left-hand side of inequality (5.9) is bounded above by

$$\Phi(\mu) \int_0^T [p(t) - \mu]_+ \, dt,$$

where

$$\Phi(\mu) = \begin{cases} \sin \dfrac{T\sqrt{\mu}}{2} & \text{if } \mu > 0, \\ \dfrac{1}{4} T^2 & \text{if } \mu = 0, \\ \operatorname{sh} \dfrac{T\sqrt{|\mu|}}{2} & \text{if } \mu < 0. \end{cases}$$

We thus deduce from Test 8:

Corollary. Let

$$p(t) \not\equiv 0, \quad p(t) \not\equiv \frac{\pi^2}{T^2}; \quad \int_0^T p(t)\, dt \geqslant 0$$

in (3.1), and suppose that for some $\mu < \pi^2/T^2$

$$\int_0^T [p(t) - \mu]_+ \, dt \leqslant \begin{cases} 2\sqrt{\mu} \operatorname{ctg} \dfrac{T\sqrt{\mu}}{2} & \text{if } \mu > 0, \\ \dfrac{4}{T} & \text{if } \mu = 0, \\ 2\sqrt{|\mu|} \operatorname{cth} \dfrac{T\sqrt{|\mu|}}{2} & \text{if } \mu < 0. \end{cases}$$

Then $p(t) \in \mathfrak{S}_0$.

This is precisely Test 4 for $n = 0$.

Using more sophisticated arguments, one can also prove Test 4 for arbitrary n via estimation of the zeros of solutions (see /159a/).

5.2. Survey of other stability tests for Hill's equation

In this subsection we describe (without proofs) a few typical results, generally established by variational methods; for details the reader is referred to the original papers.

The Swedish mathematician Borg /234a/ established several tests based on the solution of variational problems in spaces $\mathfrak{L} = \{p(t)\}$ of the coefficients of Hill's equation, with norms other than (4.1).

Borg considers, apart from the case prescribed in §4.5, two other cases, allowing $p(t)$ to be in $L_2(0, T)$ or $C[0, T]$. The norms are defined in the usual way: in the first case,

$$\|p\| = \left(\frac{1}{T}\int_0^T |p(t)|^2\,dt\right)^{1/2}$$

and in the second ($p(t)$ is a continuous function)

$$\|p\| = \max_{0 \leqslant t \leqslant T} |p(t)|.$$

Write the coefficient $p(t)$ as

$$p(t) = \alpha + \beta\psi(t),$$

where $\|\psi(t)\| = 1$. As in §4.5, one defines the concept of a stability point (α, β) in each case. Borg determines the sets of stability points explicitly in /234a/. In $C[0, T]$, if $T = \pi$ these domains are bounded by the curves

$$\sqrt{\alpha+\beta}\,\mathrm{tg}\,\frac{\pi\sqrt{\alpha+\beta}}{4(n+1)} = \sqrt{\alpha-\beta}\,\mathrm{ctg}\,\frac{\pi\sqrt{\alpha-\beta}}{4(n+1)} \quad (n = 0, 1, 2, \ldots)$$

through the points $\alpha = n^2$, $\beta = 0$, and contain the intervals $n^2 < \alpha < (n+1)^2$, $\beta = 0$ on the α-axis.

For the space $L_2(0, T)$ the boundaries are expressed in terms of elliptic integrals of the first and second kinds. Borg's test for the domain $\mathcal{O}_0$ is

$$p(t) \geqslant 0, \quad \int_0^T p(t)^2\,dt < \frac{63.03}{T^3}.$$

As in §4.1, the first inequality may be replaced by

$$p(t) \not\equiv 0, \quad \int_0^T p(t)\,dt \geqslant 0$$

or in general, any other condition for oscillation.

These results of Borg are based on his solutions to some special variational problems.

Neigauz and Lidskii /188/ established the following stability test by a variational method:

Let

$$b^2 \leqslant p(t) \leqslant a^2, \quad \frac{n\pi}{T} \leqslant b \leqslant \frac{(n+1)\pi}{T} \qquad (n = 0, 1, 2, \ldots).$$

If

$$\int_0^T [p(t) - b^2]\,dt < 2(n+1)(a^2 - b^2)\,\nu_0,$$

where ν_0 *is the smallest positive root of the equation*

$$b \operatorname{ctg}\left[\frac{bT}{2(n+1)} - b\nu\right] = a \operatorname{tg} a\nu,$$

then the trivial solution of equation (3.1) is stable.

This test is a particular case of the tests in §4.6 but is sometimes more convenient to use. It clearly covers all functions $p(t)$ in $\mathfrak{S}_n$.

Krein /166d/ develops methods yielding stability tests for the trivial solution of Hill's equation, based on estimates for the eigenvalues of periodic and semiperiodic boundary-value problems. Considering the boundary-value problem

$$\ddot{y} + \lambda p(t)\, y = 0, \quad y(0) = y(T) = 0,$$

Krein deduced the following test by estimating the first eigenvalue:

Denote

$$p^{+}(t) = \frac{1}{2}\left[p(t) + |p(t)|\right], \quad p_{\text{av}} = \frac{1}{T}\int_0^T p(t)\, dt.$$

If $p_{\text{av}} \geqslant 0$, $p(t) \leqslant H$ *and*

$$\varkappa\left(\frac{1}{H} p_{\text{av}}^{+}\right) > \frac{1}{4H}\left[T p_{\text{av}}^{+}\right]^2,$$

where $\varkappa(\tau)$ *is the smallest positive root of the transcendental equation*

$$\sqrt{\varkappa}\, \operatorname{tg} \sqrt{\varkappa} = \frac{\tau}{1-\tau} \qquad (0 \leqslant \tau \leqslant 1),$$

then the trivial solution of equation (3.1) is stable and $p(t) \in \mathfrak{S}_0$, *if* $p(t) \not\equiv 0$.

This test may also be derived from the results of §4.6, but is often more convenient to use.

The following table of values of $\varkappa(\tau)$ is due to Litovchin and Shvartsman /219/.

τ	$\varkappa$	τ	$\varkappa$	τ	$\varkappa$	τ	$\varkappa$	τ	$\varkappa$	τ	$\varkappa$	τ	$\varkappa$
0	0	0.16	0.179	0.32	0.405	0.48	0.697	0.62	1.029	0.76	1.458	0.90	2.001
0.02	0.020	0.18	0.204	0.34	0.438	0.50	0.740	0.64	1.083	0.78	1.528	0.92	2.090
0.04	0.041	0.20	0.230	0.36	0.472	0.52	0.784	0.66	1.140	0.80	1.600	0.94	2.181
0.06	0.062	0.22	0.257	0.38	0.506	0.54	0.830	0.68	1.199	0.82	1.674	0.96	2.274
0.08	0.084	0.24	0.285	0.40	0.542	0.56	0.878	0.70	1.260	0.84	1.751	0.98	2.370
0.10	0.107	0.26	0.314	0.42	0.579	0.58	0.927	0.72	1.323	0.86	1.832	1.00	2.468
0.12	0.130	0.28	0.343	0.44	0.617	0.60	0.977	0.74	1.388	0.88	1.915		
0.14	0.154	0.30	0.374	0.46	0.656								

For sufficiently small τ, the series expansion of $\varkappa(\tau)$ is

$$\varkappa=\tau+\frac{2}{3}\tau^2+\frac{19}{45}\tau^3+\frac{236}{945}\tau^4+\dots$$

The function $\varkappa(\tau)$ satisfies the inequalities

$$\frac{\tau}{1-\frac{\pi^2-4}{\pi^2}\tau}\leqslant\varkappa(\tau)\leqslant\frac{\tau}{1-\frac{2}{3}\tau}.$$

Using these inequalities, Shvartsman /219/ proved the following simple test:

If

$$p(t)\leqslant H,\quad 0\leqslant T\int_0^T p(t)\,dt\leqslant 4+\frac{16(\pi^2-4)}{\pi^2 HT^2},$$

then the trivial solution of equation (3.1) is stable (and $p(t)\in\mathfrak{S}_0$ if $p(t)\not\equiv 0$).

Lyapunov's classical paper /180b/ contains many useful transformations of Hill's equation, one of which we demonstrate. At that time, stability tests were available only for $p(t)\geqslant 0$, and so Lyapunov needed these transformations for the treatment of Hill's equation with coefficient of variable sign. (If $p(t)\leqslant 0$, then clearly $p(t)\in\mathcal{H}_0$; if $p(t)\geqslant 0$, Lyapunov was able to apply known stability tests.) Lyapunov's transformation applied to Hill's equation produces another instance of Hill's equation; it is defined by

$$y(t)=w(t)\,y_1(\tau),\quad \tau=\int_0^t\frac{dt_1}{w(t_1)^2},$$

where $w(t)$ is a positive T-periodic function with continuous second derivative (or, in a more general version, absolutely continuous first derivative).

Applying this transformation to equation (3.1), we again get Hill's equation

$$\frac{d^2y_1}{d\tau^2}+p_1(\tau)y_1=0,$$

where

$$p_1(\tau)=w^3\left[\ddot{w}+pw\right]_{t=t(\tau)}.$$

$p_1(\tau)$ is a periodic function of τ with period

$$T_1=\int_0^T\frac{dt}{w(t)^2}.$$

The characteristic constant A is not affected by the transformation. Since $w>0$, the solutions $y(t)$ and $y_1(\tau)$ have the same number of zeros over one period, so that $p(t)$ and $p_1(\tau)$ belong to stability (or instability) domains with the same subscript.

In subsections 15 to 24 of /180b/, Lyapunov takes the function $w(t)$ as

$$w = \exp\left[-\int_0^t v(t_1)\,dt_1\right],$$

where v is some T-periodic function with continuous derivative $\dot{v}$ and vanishing mean value, $v_{av}=0$. With this choice one has

$$p_1(\tau) = (p - \dot{v} + v^2)\exp\left(-4\int_0^t v\,dt\right).$$

Lyapunov shows that if $p_{av} \geqslant 0$ and we set

$$v = \int_0^t (p - p_{av})\,dt,$$

then

$$p_1(t) = (p_{av} + v^2)\exp\left(-4\int_0^t v\,dt\right).$$

An example will illustrate the advantage of using this transformation. If $p_{av} + v^2 \leqslant 0$, then $p_1(\tau) \leqslant 0$, and so $p_1(\tau) \in \mathcal{H}_0$. Thus also $p(t) \in \mathcal{H}_0$. Following /180b/, we obtain the following test for instability of Hill's equation:

$$p_{av} + \max_{0 \leqslant t \leqslant T}\left[\int_0^t (p - p_{av})\,dt\right]^2 \leqslant 0.$$

Another result proved in /180b/ is as follows. Set

$$M = \frac{1}{T}\max_{0 \leqslant t \leqslant T}\left|\int_0^T [p(t_1 + t) - p_{av}]\,t_1\,dt_1\right|.$$

a) *If* $p_{av} < 0$, $M^2 + p_{av} \leqslant 0$, *then the trivial solution of Hill's equation (3.1) is unstable.*

b) *If* $p_{av} \geqslant 0$, $MT \leqslant \zeta$, *where* ζ *is a positive root of the equation*

$$\left(\frac{T}{M}p_{av} + \zeta\right)(e^{\zeta} - 1)^2 = \pi^2\zeta,$$

then the trivial solution of Hill's equation (3.1) is stable (and $p(t) \in \mathfrak{S}_0$ *if* $p(t) \not\equiv 0$*).*

In particular, if the function $p(t)$ in Hill's equation has the property

$$p_{av} = 0 \quad \text{and} \quad \left|\int_0^T p(t_1 + t)\,t_1\,dt_1\right| \leqslant \ln(1 + \pi),$$

then the trivial solution of equation (3.1) is stable (and $p(t) \in \mathfrak{S}_0$ if $p(t) \not\equiv 0$).

§6. OSCILLATION THEOREM

6.1. Geometric interpretation of the oscillation theorem

Consider a canonical equation with Hamiltonian dependent on a parameter λ:

$$\tilde{\mathbf{J}}\frac{d\mathbf{x}}{dt}=\mathbf{H}_\lambda(t)\,\mathbf{x}. \tag{6.1}$$

As always in this book,

$$\tilde{\mathbf{J}}=\begin{Vmatrix}0 & 1\\ -1 & 0\end{Vmatrix},\quad \mathbf{H}_\lambda(t)=\begin{Vmatrix}\alpha(t,\lambda) & \beta(t,\lambda)\\ \beta(t,\lambda) & \gamma(t,\lambda)\end{Vmatrix}, \tag{6.2}$$

where $\alpha(t,\lambda)$, $\beta(t,\lambda)$, $\gamma(t,\lambda)$ are real T-periodic functions of t. We may assume without loss of generality that λ varies in the interval $-\infty<\lambda<+\infty$. We shall assume that $\mathbf{H}_\lambda(t)$ is continuous in λ in the metric of $\mathfrak{L}^3$ (see §1.6), and wherever necessary we shall assume that the derivative $\partial\mathbf{H}_\lambda/\partial\lambda$ exists and $\partial\mathbf{H}_\lambda/\partial\lambda\in\mathfrak{L}^3$.

The substitution specified at the beginning of §3 reduces Hill's equations

$$\ddot{y}+\lambda p(t)\,y=0, \tag{6.3}$$

$$\ddot{y}+[\lambda p(t)+q(t)]\,y=0 \tag{6.4}$$

to the form of equation (6.1). Lyapunov, Haupt, Hamel and other authors (see /180d, e, 255, 256, 114a, 156/), studying equations (6.3), (6.4) and using analytical methods, proved various theorems of the same type, which we shall refer to henceforth as oscillation theorems. The model $\mathfrak{R}^3$ constructed in §1 for the Hamiltonian space $\mathfrak{L}^3=\{\mathbf{H}(t)\}$, together with estimates for the rotation function as derived in §2, provide a highly transparent geometric interpretation for these theorems. This geometric approach, besides lending the theorems considerable visual clarity and elegance, enables one to augment their conclusions in various respects and to extend them to the general equation (6.1), in which the Hamiltonian may involve the parameter λ nonlinearly.

We shall say that *the oscillation theorem holds for equation* (6.1) if there exists a countable sequence

$$\ldots<\lambda_{-2}^{-}\leqslant\lambda_{-2}^{+}<\lambda_{-1}^{-}\leqslant\lambda_{-1}^{+}<\lambda_{0}^{-}\leqslant\lambda_{0}^{+}<\lambda_{1}^{-}\leqslant\lambda_{1}^{+}<\lambda_{2}^{-}\leqslant\lambda_{2}^{+}<\ldots, \tag{6.5}$$

infinite in both directions, which satisfies the following three conditions:

(I) The λ's with even subscripts are the eigenvalues of the boundary-value problem $\mathbf{x}(T)=\mathbf{x}(0)$; those with odd subscripts are the eigenvalues of the boundary-value problem $\mathbf{x}(T)=-\mathbf{x}(0)$. *

* This means that there is a solution $\mathbf{x}(t)\not\equiv\mathbf{0}$ of equation (6.1) such that $\mathbf{x}(T)=\mathbf{x}(0)$ or $\mathbf{x}(T)=-\mathbf{x}(0)$.

(II) If $\lambda_n^- \neq \lambda_n^+$, then at $\lambda = \lambda_n^+$ $(\lambda = \lambda_n^-)$ there is one periodic (antiperiodic) solution of period T. If $\lambda_n^- = \lambda_n^+$, all solutions are periodic (even n) or antiperiodic (odd n) with period T. The rotation in time T of any solution-vectors corresponding to an eigenvalue with subscript n is $n\pi$.

(III) The intervals $\lambda_n^+ < \lambda < \lambda_{n+1}^-$ are stability intervals (for these λ's all solutions are bounded on $(-\infty, +\infty)$), the intervals $\lambda_n^- < \lambda < \lambda_n^+$ are instability intervals (for $\lambda_n^- < \lambda < \lambda_n^+$ there are two linearly independent solutions $\mathbf{x}_1(t) = \mathbf{f}_1(t) \exp \varkappa t$, $\mathbf{x}_2(t) = \mathbf{f}_2(t) \exp(-\varkappa t)$, where $\varkappa > 0$ and $\mathbf{f}_1(t)$, $\mathbf{f}_2(t)$ are real vector-functions, $\mathbf{f}_j(t+T) = (-1)^n \mathbf{f}_j(t)$, $j = 1, 2$).

If the sequence (6.5) extends only to $+\infty$, i.e., there are only finitely many terms with negative subscripts, we shall say that *the oscillation theorem holds for* $\lambda \to +\infty$. If the sequence extends only to $-\infty$, we shall say that *the oscillation theorem holds for* $\lambda \to -\infty$.

Consider our model $\mathfrak{R}^3$ of the function space $\mathfrak{L}^3$ (§1.6). Recalling the properties of the Hamiltonians in the sets $\mathcal{O}_n$, $\mathcal{H}_n$, $\Pi_n^{\bullet +}$, $\Pi_n^{\bullet -}$, $\Pi_n^{\bullet\bullet}$, we at once arrive at the following geometric interpretation of conditions (I) — (III):

(A) As λ varies from $-\infty$ to $+\infty$, the point representing $\mathbf{H}_\lambda(t)$ in $\mathfrak{R}^3$ (see Figure 56 and Theorem II, §1.9) crosses $\mathfrak{R}^3$ from $-\infty$ to $+\infty$, crossing each of the surfaces $\tilde{\Pi}_n = \tilde{\Pi}_n^{\bullet +} \cup \tilde{\Pi}_n^{\bullet\bullet} \cup \tilde{\Pi}_n^{\bullet}$ from left to right.

(B) $\lambda = \lambda_n^+$ if and only if $\mathbf{H}_\lambda(t) \in \Pi_n^{\bullet +}$; $\lambda = \lambda_n^-$ if and only if $\mathbf{H}_\lambda(t) \in \Pi_n^{\bullet -}$; if $\lambda_n^+ = \lambda_n^-$, then $\mathbf{H}_\lambda(t) \in \Pi_n^{\bullet\bullet}$.

(C) If $\lambda_n^+ < \lambda < \lambda_{n+1}^-$, the Hamiltonian $\mathbf{H}_\lambda(t)$ belongs to the n-th stability domain $\mathcal{O}_n$; if $\lambda_n^- < \lambda < \lambda_n^+$ it belongs to the n-th instability domain $\mathcal{H}_n$.

Phrased in more rigorous terms, the first part of property (A) means that for arbitrarily large $n_0 > 0$ there exist λ', λ'' such that $\mathbf{H}_{\lambda'}(t) \in \mathcal{O}_{-n_0}$ and $\mathbf{H}_{\lambda''}(t) \in \mathcal{O}_{n_0}$, and $\lambda' \to -\infty$, $\lambda'' \to +\infty$ as $n_0 \to +\infty$. The rigorous meaning of the words "crossing each of the surfaces $\tilde{\Pi}_n$ from left to right" in condition (A) is interpreted in the natural way, as done in §4.1 for the phrase "to the left (right) of the surface $\tilde{\Pi}_n$."

It follows from the results of §1 that conditions (I), (II), (III) (in conjunction) are equivalent to (A), (B), (C).

As in §1, let $\psi_\lambda(\chi)$ denote the rotation function of equation (6.1), i.e., the rotation in the time from $t = 0$ to $t = T$ of a solution-vector $\mathbf{x}_\lambda(t)$ for which $\arg \mathbf{x}_\lambda(0) = \chi$ $(0 \leqslant \chi < \pi)$. We assume that $\mathbf{x}_\lambda(0)$ is independent of λ.

It follows from Theorem II of §1.9 (see also Figure 61) that conditions (A), (B), (C) (and hence also (I), (II), (III)) will hold if:

(I_ψ) $\psi_\lambda(\chi) \to +\infty$ as $\lambda \to +\infty$ for at least one χ.
(II_ψ) $\psi_\lambda(\chi) \to -\infty$ as $\lambda \to -\infty$ for at least one χ.

(III_ψ)) $\psi_\lambda(\chi)$ is a monotonically increasing function of λ for every χ $(0 \leqslant \chi < \pi)$.

Since $|\psi_\lambda(\chi_1) - \psi_\lambda(\chi_2)| < \pi$, it follows that if condition (I_ψ) holds then $\psi_\lambda(\chi) \to +\infty$ for any χ $(0 \leqslant \chi < \pi)$. A similar assertion is true with regard to property (II_ψ).

Another consequence of Theorem II in §1.9 is that if conditions (I_ψ) and (III_ψ) are valid, the oscillation theorem is valid for $\lambda \to +\infty$; if (II_ψ) and (III_ψ) are satisfied, the oscillation theorem is valid for $\lambda \to -\infty$.

If only condition (III_ψ) holds, the oscillation theorem may fail for both $\lambda \to +\infty$ and $\lambda \to -\infty$ (an example will be given below). By the previous discussion, there nevertheless exists a set (infinite, finite or empty) of numbers $\lambda_n^{\pm}$ satisfying (6.5), (I), (II), (III). If the set is empty, the entire λ-axis constitutes a single stability or instability interval.

We shall need the following device below. Apply to equation (6.1) the substitution

$$\mathbf{x}_1 = \mathbf{F}_\lambda(t)\,\mathbf{x},$$

where $\mathbf{F}_\lambda(t+T) = \pm\mathbf{F}_\lambda(t)$, $\det \mathbf{F}_\lambda(t) = 1$, and $\mathbf{F}_\lambda(t)$ is an absolutely continuous matrix-function, continuous in λ in the metric of $\mathfrak{L}^3$. Suppose that the result is a canonical equation

$$\mathbf{J}\frac{d\mathbf{x}_1}{dt} = \mathbf{H}_\lambda^{(1)}(t)\,\mathbf{x}_1.$$

If this equation satisfies one of properties (I_ψ), (II_ψ), the appropriate property will also hold for the original equation. If the new equation satisfies the oscillation theorem (whether for $\lambda \to +\infty$, for $\lambda \to -\infty$, or for both), the same is true of the original equation.

Indeed, let $\mathbf{F}_\lambda(t) \in \Omega_m$ (see §1.7). It is clear that m is the same for all λ (it is defined by $\varphi_x = m\pi$, where φ_x is the rotation in time T of a vector $\mathbf{F}_\lambda(t)\mathbf{c}$; m is independent of λ, since φ_x is a continuous function of λ). Suppose that for some λ the Hamiltonian $\mathbf{H}_\lambda(t)$ belongs to one of the sets $\mathcal{H}_n$, $\mathfrak{G}_n$, Π_n^{*+}, Π_n^{*-}, Π_n^{**}. By Theorem II of §1.10, the Hamiltonian $\mathbf{H}_\lambda^{(1)}(t)$ belongs to the analogous set with subscript $n+m$. This implies our assertion.

6.2. Formulation of oscillation theorems in different cases

The results of §2 immediately suggest the following

Theorem I. Let $h_{\min}(t,\lambda) \leqslant h_{\max}(t,\lambda)$ *be the eigenvalues of the Hamiltonian* $\mathbf{H}_\lambda(t)$ *or, in general, the roots of an equation*

$$\det[\mathbf{H}_\lambda(t) - \mathbf{H}_0] = 0,$$

where $\mathbf{H}_0$ *is some real matrix such that* $\mathbf{H}_0 = \mathbf{H}_0^* > 0$, $\det \mathbf{H}_0 = 1$. *Let*

$$\lim_{\lambda\to\infty}\int_0^T h_{\min}(t,\lambda)\,dt = +\infty, \tag{6.6}$$

$$\lim_{\lambda\to-\infty}\int_0^T h_{\max}(t,\lambda)\,dt = -\infty. \tag{6.7}$$

In addition, suppose that $\partial\mathbf{H}_\lambda(t)/\partial\lambda$ *exists almost everywhere on* $[0, T]$ *and*

$$\frac{\partial\mathbf{H}_\lambda(t)}{\partial\lambda} > 0 \qquad (-\infty < \lambda < +\infty) \tag{6.8}$$

almost everywhere on $[0, T]$.

Then the oscillation theorem is valid. If only (6.6) and (6.8) hold, the oscillation theorem is valid for $\lambda\to+\infty$; *if only (6.7) and (6.8) hold, the oscillation theorem is valid for* $\lambda\to-\infty$.

Proof. It follows from formula (2.66) (see §2.4) that

$$\frac{\partial\psi_\lambda(\chi)}{\partial\lambda}=\frac{1}{(\mathbf{x}_\lambda(T),\ \mathbf{x}_\lambda(T))}\int_0^T\left(\frac{\partial\mathbf{H}_\lambda(t)}{\partial\lambda}\mathbf{x}_\lambda(t),\ \mathbf{x}_\lambda(t)\right)dt. \tag{6.9}$$

Thus (6.8) implies condition (III_ψ).

Let $\mathbf{H}_0=\mathbf{I}_2$. Then by (2.5)

$$\int_0^T h_{\min}(t,\ \lambda)\,dt\leqslant\psi_\lambda(\chi)\leqslant\int_0^T h_{\max}(t,\ \lambda)\,dt.$$

Thus (6.6) implies (I_ψ), while (6.7) implies (II_ψ). Thus the theorem is proved for the case $\mathbf{H}_0=\mathbf{I}_2$.

Let $\mathbf{H}_0\neq\mathbf{I}_2$. The substitution $\mathbf{x}_1=\mathbf{H}_0^{1/2}\mathbf{x}$ gives the equation

$$\tilde{\mathbf{J}}\frac{d\mathbf{x}_1}{dt}=\mathbf{H}'_\lambda(t)\,\mathbf{x}_1, \tag{6.9a}$$

where

$$\mathbf{H}'_\lambda(t)=\tilde{\mathbf{J}}^{-1}\mathbf{H}_0^{1/2}\tilde{\mathbf{J}}\mathbf{H}'_\lambda(t)\,\mathbf{H}_0^{-1/2}.$$

For any $r\times r$ matrix $\mathbf{X}$ with determinant unity (hence also for $\mathbf{X}=\mathbf{H}_0^{1/2}$), we have $\tilde{\mathbf{J}}^{-1}\mathbf{X}\tilde{\mathbf{J}}=\mathbf{X}^{-1}$. Therefore

$$\mathbf{H}'_\lambda(t)=\mathbf{H}_0^{-1/2}\mathbf{H}_\lambda(t)\,\mathbf{H}_0^{-1/2},$$

so that $h_{\min}(t),h_{\max}(t)$ are the eigenvalues of $\mathbf{H}'_\lambda(t)$. Let $\psi'_\lambda(\chi)$ be the rotation function of equation (6.9a). By the previous case, conditions (6.6) and (6.7) imply conditions (I_ψ), (II_ψ), respectively, for the function $\psi'_\lambda(\chi)$. By Lemma II of §1.9, $|\psi'_\lambda(\chi)-\psi_\lambda(\chi)|\leqslant\text{const}$. Thus conditions (6.6), (6.7) imply (I_ψ), (II_ψ), respectively, for $\psi_\lambda(\chi)$, and the proof is complete.

Instead of (I_ψ), (II_ψ), consider the following conditions:

(I_G) For any sufficiently large number Λ_0 and integer $n>0$, there exist a number $\Lambda_n\geqslant\Lambda_0$ and a Hamiltonian $\mathbf{G}_n(t)\in\Pi_n^{**}$ such that $\mathbf{H}_{\Lambda_n}(t)\geqslant\mathbf{G}_n(t)$ $(0\leqslant t\leqslant T)$.

(II_G) For any sufficiently large number Λ_0 and integer $n>0$, there exist a number $\Lambda_n\leqslant-\Lambda_0$ and a Hamiltonian $G_{-n}(t)\in\Pi_{-n}^{**}$ such that $\mathbf{H}_{\Lambda_n}(t)\leqslant\mathbf{G}_{-n}(t)$ $(0\leqslant t\leqslant T)$.

The Hamiltonians $\mathbf{G}_n(t)\in\Pi_n^{**}$ were characterized in Theorem I of §1.10. If $\mathbf{H}_{\Lambda_n}(t)\geqslant\mathbf{G}_n(t)$ ($\mathbf{H}_{\Lambda_n}(t)\leqslant\mathbf{G}_n(t)$), it follows from Theorem I of §2.4 and Theorem II of §1.9 that the point in $\mathfrak{R}^3$ representing $\mathbf{H}_{\Lambda_n}(t)$ is to the "right" of the surface $\tilde{\Pi}_n$ (to the "left" of the surface $\tilde{\Pi}_{-n}$). Therefore, if the oscillation theorem holds for $\lambda\to+\infty$ ($\lambda\to-\infty$), condition (I_G) (condition (II_G)) is satisfied. If condition (III_ψ) holds, then conditions (I_G) and (I_ψ) ((II_G) and (II_ψ)) are equivalent. We have thus proved

Theorem II. *Condition* (I_G) *is necessary and, in conjunction with (6.8), also sufficient, for the oscillation theorem to hold for* $\lambda \to +\infty$. *Condition* (I_G) *is necessary and, in conjunction with (6.8), also sufficient, for the oscillation theorem to hold for* $\lambda \to -\infty$.

We now assume a weaker condition than (6.8):

$$\frac{\partial \mathbf{H}_\lambda(t)}{\partial t} \geqslant 0 \qquad (-\infty < \lambda < +\infty,\ 0 \leqslant t \leqslant T). \tag{6.10}$$

It follows from (6.9) that then

$$\frac{\partial \psi_\lambda(\chi)}{\partial \lambda} \geqslant 0. \tag{6.11}$$

Let us see what additional assumptions are required for condition (III_ψ) to hold. Suppose that (III_ψ) is false, so that

$$\frac{\partial \psi_\lambda(\chi)}{\partial \lambda} = 0$$

for some χ and all λ in some interval $\lambda' < \lambda < \lambda''$. Then, almost everywhere on $[0, T]$,

$$\left(\frac{\partial \mathbf{H}_\lambda(t)}{\partial \lambda} \mathbf{x}_\lambda(t),\ \mathbf{x}_\lambda(t)\right) = 0$$

and by (6.10),

$$\frac{\partial \mathbf{H}_\lambda(t)}{\partial \lambda} \mathbf{x}_\lambda(t) = 0.$$

Differentiating (6.1) with respect to λ and recalling that $\mathbf{x}_\lambda(0)$ does not depend on λ, we see via the last condition that

$$\frac{\partial \mathbf{x}_\lambda(t)}{\partial \lambda} \equiv 0 \qquad (0 \leqslant t \leqslant T, \quad \lambda' < \lambda < \lambda''). \tag{6.12}$$

Thus, if condition (III_ψ) is not satisfied, *there exist a vector* $\mathbf{x}_0$ ($\arg \mathbf{x}_0 = \chi$) *independent of* λ, *and an interval* $\lambda' < \lambda < \lambda''$ *such that equation (6.1) has a solution* $\mathbf{x} = \mathbf{x}_\lambda(t) \neq 0$ ($\mathbf{x}(0) = \mathbf{x}_0$), *independent of* λ *(i.e., satisfying (6.12)), for which*

$$\frac{\partial \mathbf{H}_\lambda(t)}{\partial \lambda} \mathbf{x}_\lambda(t) = 0 \qquad (0 \leqslant t \leqslant T, \quad \lambda' < \lambda < \lambda''). \tag{6.13}$$

If the italicized conditions are satisfied, we shall say that the singular case holds. We have thus shown that in the nonsingular case condition (III_ψ) must hold. Hence:

Theorem III. *Theorems I and II remain valid with condition (6.8) replaced by (6.10), provided that the singular case does not hold.*

If (6.10) holds and for every λ condition (6.8) is valid on a set of positive measure in $[0, T]$ (generally depending on λ), the singular case will not hold. Indeed, it then follows from (6.9) that for every λ and χ

$$\frac{\partial \psi_\lambda(\chi)}{\partial \lambda} > 0,$$

and this is impossible in the singular case (see (6.9) and (6.13)).

Even if the above conditions hold not for all λ but for a few isolated values, $\psi_\lambda(\chi)$ is a monotonically increasing function and the singular case does not hold.

If only condition (6.10) is satisfied, all we can say is that (6.11) is true; conditions (I_ψ), (II_ψ) need not hold. In this case, the point of $\mathfrak{R}^3$ representing $\mathbf{H}_\lambda(t)$ may be on a surface $\tilde{\Pi}_n$ in certain intervals $\lambda' < \lambda < \lambda''$. Now suppose that in addition $\mathbf{H}_\lambda(t)$ is an analytic function of λ. The values of λ for which $\mathbf{H}_\lambda \in \Pi_n$ are the roots of the equation

$$\det[\mathbf{X}(T, \lambda) - (-1)^n \mathbf{I}_2] = 0. \tag{6.14}$$

The function on the left is analytic in λ and therefore the roots cannot fill out whole intervals, except in the case when the left-hand side of (6.14) vanishes identically.

We have thus proved

Theorem IV. Let $\mathbf{H}_\lambda(t)$ be an analytic function of λ and assume that condition (6.10) holds. Then either $\mathbf{H}_\lambda(t) \in \Pi_n$, or $\mathbf{H}_\lambda(t) \in \mathfrak{S}_n$, or $\mathbf{H}_\lambda(t) \in \mathcal{H}_n$ for some n and all λ, or there exists a (finite or infinite) set of values $\lambda_n^\pm$ such that conditions (I), (II), (III) are satisfied.

Remark. Under the assumptions of Theorem IV, if condition (I_ψ) or (II_ψ) is satisfied, then the oscillation theorem is valid for $\lambda \to +\infty$ or $\lambda \to -\infty$, respectively.

Indeed, if, say, condition (I_ψ) holds, the point representing $\mathbf{H}_\lambda(t)$ in $\mathfrak{R}^3$ will cross all the surfaces $\tilde{\Pi}_n$ as $\lambda \to +\infty$, for all sufficiently large n. By Theorem IV, therefore, the only possibility is that there exists an infinite set of numbers $\lambda_n^\pm$ such that $\lambda_n^\pm \to +\infty$ as $n \to \infty$.

Consider the linear case,

$$\mathbf{H}_\lambda(t) = \lambda \mathbf{H}_1(t) + \mathbf{H}_2(t), \tag{6.15}$$

where

$$\mathbf{H}_1(t) \geqslant 0 \qquad (0 \leqslant t \leqslant T). \tag{6.16}$$

The conclusion of Theorem IV is clearly valid in this case.

Suppose that condition (6.16) holds and

$$\mathbf{H}_1(t) > 0 \tag{6.17}$$

on a set of positive measure in $[0, T]$. Let $h(t)$ be the minimum eigenvalue of $\mathbf{H}_1(t)$ and $h_{min}(t, \lambda) \leqslant h_{max}(t, \lambda)$ the eigenvalues of $\mathbf{H}_\lambda(t)$. Then by (6.15)

$$h_{\min}(t, \lambda) = \lambda\left[h(t) + O\left(\frac{1}{\lambda}\right)\right] \quad \text{if} \quad \lambda > 0,$$
$$h_{\max}(t, \lambda) = -\lambda\left[h(t) + O\left(\frac{1}{\lambda}\right)\right] \quad \text{if} \quad \lambda < 0.$$

It follows from (6.16), (6.17) that

$$\int_0^T h(t)\,dt > 0.$$

Thus conditions (6.6), (6.7) are satisfied. By Theorem I:

Theorem V. In the linear case (6.15), if inequality (6.16) is satisfied and inequality (6.17) holds on a set of positive measure, then the oscillation theorem is valid.

We claim that the condition (6.17) in this theorem cannot be replaced by

$$\int_0^T \mathbf{H}_1(t)\,dt > 0. \tag{6.18}$$

Let $\mathbf{H}_2(t) \equiv 0$ and let $\mathbf{H}(t)$ be a piecewise-constant matrix such that $\mathbf{H}(t) = \mathbf{H}_i$ in the interval (t_{i-1}, t_i), where

$$0 = t_0 < t_1 < \ldots < t_{n-1} < t_n = T.$$

In addition, assume that $\det \mathbf{H}_i = 0$. The motion on the plane $\{\mathbf{x}\}$ for $t_{i-1} < t < t_i$ takes place along straight lines. Thus, the motion in the whole interval $0 \leqslant t \leqslant T$ describes a polygonal line with n sides. The rotation over each side is at most π, so that the whole rotation is $|\varphi_x| < n\pi$. The same holds for $\mathbf{H}_\lambda(t) = \lambda\mathbf{H}(t)$. Thus, the oscillation theorem cannot hold for a piecewise-constant matrix $\mathbf{H}_\lambda(t) = \lambda\mathbf{H}(t)$ such that $\det \mathbf{H}_\lambda(t) \equiv 0$. If all the matrices $\mathbf{H}_i$ are in $\tilde{\Pi}_+$ (in the notation of §1.4), then

$$\mathbf{H}_i \geqslant 0, \quad \int_0^T \mathbf{H}(t)\,dt = \sum_i (t_i - t_{i-1})\,\mathbf{H}_i > 0.$$

The second inequality is obvious from geometric considerations; if $\mathbf{H}_i \in \tilde{\Pi}_+$, then

$$\sum_i (t_i - t_{i-1})\,\mathbf{H}_i \in \tilde{\mathcal{O}}_+.$$

Thus, if $\mathbf{H}_\lambda(t) = \lambda\mathbf{H}(t)$, where $\mathbf{H}(t)$ is a piecewise-constant matrix satisfying inequalities (6.16) and (6.18), then the oscillation theorem is not valid (there can be at most finitely many numbers $\lambda_n^\pm$).

We now consider equations (6.3) and (6.4).

Equation (6.4) may be reduced to an equation of type (6.1) (see the beginning of §3 and formulas (3.1), (3.2)) in which

$$\mathbf{x} = \begin{pmatrix} \dot{y} \\ y \end{pmatrix}, \quad \mathbf{H}_\lambda(t) = \begin{Vmatrix} 1 & 0 \\ 0 & \lambda p(t) + q(t) \end{Vmatrix}. \tag{6.19}$$

Theorem VI. Let $p(t) \underset{\not\equiv}{\geqslant} 0$ *in equation (6.4). Then the oscillation theorem holds for* $\lambda \to \infty$.

Proof. For $\lambda > 0$, we set

$$\mathbf{x}_1 = \begin{pmatrix} \frac{\dot{y}}{\sqrt{\lambda}} \\ y \end{pmatrix}$$

and so reduce equation (6.4) to a canonical equation with Hamiltonian

$$\mathbf{H}_\lambda^{(1)} = \begin{Vmatrix} \sqrt{\lambda} & 0 \\ 0 & \sqrt{\lambda}\, p(t) + \frac{q(t)}{\sqrt{\lambda}} \end{Vmatrix}.$$

It is clear that condition (6.6) holds for this Hamiltonian. Thus the corresponding rotation function $\psi_\lambda^{(1)}(\chi)$ satisfies (I_ψ). Since $\mathbf{x}_1(t) = \mathbf{L}_\lambda \mathbf{x}(t)$, where $\mathbf{L}_\lambda$ is a nonsingular linear transformation and $\mathbf{x}(t)$ is the vector defined in (6.19), it follows from Lemma I of §1.9 that $|\psi_\lambda^{(1)}(\chi) - \psi_\lambda(\chi)| < \pi$. Consequently, condition (I_ψ) also holds for equation (6.1) with the Hamiltonian $\mathbf{H}_\lambda$ of (6.19). This Hamiltonian clearly satisfies (6.10). The singular case cannot hold, for in that case

$$y(t) \not\equiv 0, \quad p(t)\, y(t) \equiv 0, \quad \ddot{y} + q(t)\, y = 0. \qquad (6.20)$$

Since any solution of the last equation may vanish at most finitely many times, it follows from the second relation of (6.20) that $p(t) \equiv 0$ almost everywhere. This contradiction shows that the singular case is impossible. Consequently, condition (III_ψ) holds. Thus equation (6.1) with Hamiltonian (6.19) satisfies the oscillation theorem for $\lambda \to +\infty$, as claimed.

Theorem VI implies certain propositions proved by other means in § VII. 1.6.

Consider equation (6.3) with a function $p(t)$ of a variable sign. We may assume without loss of generality that

$$p_{\mathrm{av}} = \frac{1}{T} \int_0^T p(t)\, dt \geqslant 0.$$

Theorem VII. Suppose that the interval $[0, T]$ *contains a subinterval* Δ' *on which* $p(t) \geqslant \varepsilon > 0$ *and a subinterval* Δ'' *on which* $p(t) \leqslant -\varepsilon < 0$. *Then equation (6.3) satisfies the oscillation theorem (for both* $\lambda \to +\infty$ *and* $\lambda \to -\infty$*), except that condition (II) must be modified as follows:*

(II′) *Let* $p_{\mathrm{av}} > 0$. *If* $\lambda_n^- \neq \lambda_n^+$, *then at* $\lambda = \lambda_n^+$ *and* $\lambda = \lambda_n^-$ *there is exactly one periodic solution of period* T. *If* $\lambda_n^- = \lambda_n^+$ *all the solutions are periodic (if* n *is even) or antiperiodic (if* n *is odd). We then have* $\lambda_0^+ = 0$, $\lambda_0^- < 0$. *The rotation of any solution-vector* $\mathbf{x} = \begin{pmatrix} \dot{y} \\ y \end{pmatrix}$ *corresponding to eigenvalues* $\lambda_{\pm n}^+$, $\lambda_{\pm n}^-$ $(n \geqslant 0)$ *in time* T *is* $n\pi$.

Let $p_{\mathrm{av}} = 0$. *Then* $\lambda_0^- = \lambda_0^+ = 0$ *and these conclusions are valid for all eigenvalues* λ_n^+, λ_n^-, $n \neq 0$.

Proof. Since $p(t) \geqslant \varepsilon > 0$ on Δ', it follows from the comparison theorem that any solution $y(t)$ of equation (6.3) has arbitrarily many zeros on Δ' for sufficiently large λ.

As shown in §3, as t varies from $t=t'$ to t'', where $t' < t''$ are two consecutive zeros of $y(t)$, then the vector $\mathbf{x}(t)$ with components $\dot{y}(t)$, $y(t)$ turns through the angle π in the positive sense. This vector cannot turn through an angle greater than π in the negative sense (see (3.1)). Consequently, condition (I_ψ) holds for equation (6.1) with Hamiltonian (6.19) (for $q\equiv 0$).

Since $p(t) \leqslant -\varepsilon < 0$ on Δ'', analogous arguments show that for equation (6.1) with Hamiltonian (6.19) we have $\psi_\lambda(\chi) \to +\infty$ as $\lambda \to -\infty$.

Set

$$S(t) = \int_0^t [p(t) - p_{\mathrm{av}}]\,dt \qquad (S(t+T) \equiv S(t)),$$

$$z = -S(t)\,y - \frac{1}{\lambda}\dot{y}, \quad \mathbf{x}_2 = \begin{pmatrix} y \\ z \end{pmatrix}.$$

The vectors $\mathbf{x}$ and $\mathbf{x}_2$ satisfy the relation

$$\mathbf{x}_2 = \mathbf{P}_\lambda(t)\,\mathbf{x}, \text{ where } \mathbf{P}_\lambda(t) = \begin{Vmatrix} 0 & 1 \\ -\lambda^{-1} & -S(t) \end{Vmatrix}.$$

Since $\det \mathbf{P}_\lambda(t)$ is independent of t, we have a canonical equation for $\mathbf{x}_2$ (see § 1.10):

$$\tilde{\mathbf{J}}\frac{d\mathbf{x}_2}{dt} = \mathbf{H}_\lambda^{(2)}(t)\,\mathbf{x}_2. \tag{6.21}$$

It is readily seen that

$$\mathbf{H}_\lambda^{(2)}(t) = \lambda \begin{Vmatrix} S^2 & S \\ S & 1 \end{Vmatrix} + \begin{Vmatrix} p_{\mathrm{av}} & 0 \\ 0 & 0 \end{Vmatrix}. \tag{6.22}$$

We now show that equation (6.21) satisfies conditions (I_ψ), (II_ψ).

If $\mathbf{X}_\lambda^{(2)}(t)$ and $\mathbf{X}_\lambda(t)$ are the matrizants of equations (6.21) and (6.1) (with Hamiltonian (6.19) for $q=0$), respectively, then

$$\mathbf{X}_\lambda^{(2)}(t) = \mathbf{P}_\lambda(t)\,\mathbf{X}_\lambda(t)\,\mathbf{P}_\lambda(0)^{-1}.$$

Let $\mathbf{c}$ be a constant vector $(\arg \mathbf{c} = \chi)$; set

$$\mathbf{x}_2 = \mathbf{X}_\lambda^{(2)}(t)\,\mathbf{c}, \quad \mathbf{x} = \mathbf{X}_\lambda(t)\,\mathbf{c},$$

$$\mathbf{u} = \mathbf{P}_\lambda(0)\,\mathbf{X}_\lambda(t)\,\mathbf{c}_\lambda, \quad \mathbf{c}_\lambda = \mathbf{P}_\lambda(0)^{-1}\mathbf{c}.$$

The rotation functions of equations (6.21) and (6.1) are, respectively, *

$$\psi_\lambda^{(2)}(\chi) = \varphi_{\mathbf{x}_2}, \quad \psi_\lambda(\chi) = \varphi_{\mathbf{x}}.$$

We have

$$\mathbf{x}_2 = \mathbf{F}_\lambda(t)\,\mathbf{u}, \text{ where } \mathbf{F}_\lambda(t) = \mathbf{P}_\lambda(t)\,\mathbf{P}_\lambda(0)^{-1}.$$

* As before, $\varphi_{\mathbf{z}}$ denotes the rotation of a vector $\mathbf{z} = \mathbf{z}(t)$ in the time from $t=0$ to $t=T$.

Joining the function $S(t)$ continuously to the identically zero function, we thereby join $\mathbf{F}_\lambda(t)$ to $\mathbf{I}_2$ without changing the endpoints: $\mathbf{F}_\lambda(T)=\mathbf{F}_\lambda(0)=\mathbf{I}_2$. Consequently,

$$\varphi_{x_2}=\varphi_u.$$

Set $\mathbf{v}=\mathbf{X}_\lambda(t)\mathbf{c}_\lambda$, $\vartheta_\mathbf{x}(t)=\operatorname{Arg}\mathbf{x}(t)$, $\vartheta_\mathbf{v}(t)=\operatorname{Arg}\mathbf{v}(t)$, where the arguments are defined by continuity, and

$$|\vartheta_\mathbf{x}(0)-\vartheta_\mathbf{v}(0)|<2\pi.$$

Then

$$|\vartheta_\mathbf{x}(t)-\vartheta_\mathbf{v}(t)|<2\pi,$$

since otherwise we would have

$$\vartheta_\mathbf{x}(t)-\vartheta_\mathbf{v}(t)=\pm 2\pi$$

for some $t\,(0<t\leqslant T)$. This would imply that $\mathbf{x}$ and $\mathbf{v}$ are linearly dependent at time t, and since they are solutions of the same equation it would follow that $\vartheta_\mathbf{x}(t)-\vartheta_\mathbf{v}(t)\equiv 2\pi$, which is impossible. Thus the rotation functions $\varphi_\mathbf{x}=\vartheta_\mathbf{x}(T)-\vartheta_\mathbf{x}(0)$ and $\varphi_\mathbf{v}=\vartheta_\mathbf{v}(T)-\vartheta_\mathbf{v}(0)$ satisfy the inequality $|\varphi_\mathbf{x}-\varphi_\mathbf{v}|<4\pi$.

Let $\lambda>0$. Since $\det \mathbf{P}_\lambda(0)>0$, it follows from Lemma I, §1.9, that

$$|\varphi_\mathbf{v}-\varphi_u|<\pi.$$

Consequently, $|\varphi_\mathbf{x}-\varphi_u|<5\pi$, so that

$$|\psi_\lambda^{(2)}(\chi)-\psi_\lambda(\chi)|<5\pi.$$

Thus condition (I_ψ) holds for (6.21).

Let $\lambda<0$. Denote

$$\mathbf{C}=\begin{Vmatrix}1 & 0\\ 0 & -1\end{Vmatrix},\quad \mathbf{w}=\mathbf{C}^{-1}\mathbf{P}_\lambda(t)\,\mathbf{v}.$$

Since $\det\,[\mathbf{C}^{-1}\mathbf{P}_\lambda(t)]>0$, it follows from Lemma I, §1.9, that

$$|\varphi_\mathbf{v}-\varphi_w|<\pi.$$

Consequently, $|\varphi_\mathbf{x}-\varphi_w|<5\pi$. Since, as shown above, $\varphi_\mathbf{x}\to+\infty$ as $\lambda\to-\infty$, it follows that also $\varphi_w\to+\infty$ as $\lambda\to-\infty$. We have $\mathbf{x}_2=\mathbf{C}\mathbf{w}$. Since $\mathbf{C}$ defines the operator of reflection in the horizontal axis, it follows that

$$\varphi_{x_2}\to-\infty \text{ as } \lambda\to-\infty.$$

Thus condition (II_ψ) holds for equation (6.21).

By (6.22),

$$\frac{\partial \mathbf{H}_\lambda^{(2)}}{\partial \lambda} \geqslant 0.$$

By the remark following Theorem IV, the oscillation theorem is valid for (6.21).

Again, let $\lambda > 0$, so that $\det \mathbf{P}_\lambda(t) > 0$. Since the solutions of equations (6.1) (with Hamiltonian (6.19), $q=0$) and (6.21) are related by $\mathbf{x}_2 = \mathbf{P}_\lambda(t)\,\mathbf{x}$, it follows by the remark at the end of §6.1 that equation (6.1) also satisfies the oscillation theorem for $\lambda \to \infty$.

The case $\lambda < 0$ is reduced to the case $\lambda > 0$ by substituting $-p(t)$ for $p(t)$ in (6.3). (At no point in the above arguments did we assume that $p_{av} \geqslant 0$.) Thus, as $\lambda < 0$ is decreased and $\lambda \to -\infty$, the Hamiltonian $\mathbf{H}_\lambda(t)$ (which is simply (6.19) with $q \equiv 0$) crosses $\mathfrak{L}^3$ (see Figure 68) "to the right," crossing each of the surfaces Π_n "from left to right." It follows directly from equation (6.3) that $\mathbf{H}_\lambda(t) \in \Pi_0^{\bullet +}$ at $\lambda = 0$.

Thus, as λ varies from $\lambda = -\infty$ to $\lambda = \infty$, the Hamiltonian $\mathbf{H}_\lambda(t)$ crosses $\mathfrak{L}^3$ (Figure 68) from "the point at infinity on the right" to the set $\Pi_0^{\bullet +}$ $-\infty < \lambda \leqslant 0$ and then $(0 \leqslant \lambda < \infty)$ back from $\Pi_0^{\bullet +}$ to "the point at infinity on the right." (One should, of course, understand this statement as referring to the point representing the Hamiltonian $\mathbf{H}_\lambda(t)$ in $\mathfrak{R}^3$, the phrase "to the right" being given its usual rigorous meaning.)

The behavior of the curve $\mathbf{H}_\lambda(t)$ in the neighborhood of the surface $\Pi_0^{\bullet +}$ may be described as follows. If $\lambda \neq 0$ the Hamiltonians $\mathbf{H}_\lambda(t)$ and $\mathbf{H}_\lambda^{(2)}(t)$ belong to the same one of the sets Π, $\mathcal{G}$ or $\mathcal{H}$. Let $p_{av} = 0$. It follows from (6.22) that $\mathbf{H}_\lambda^{(2)}(t) \geqslant 0$ for small $\lambda \geqslant 0$ and $\mathbf{H}_0^{(2)}(t) \in \Pi_0^{\bullet\bullet}$. By the comparison theorem of §2.4, as the Hamiltonian increases, so does the rotation function. This property, in conjunction with Theorem II of §1.9, implies that $\mathbf{H}_\lambda^{(2)}(t) \in \Pi_0^{\bullet +} \cup \mathcal{G}_0$ for small $\lambda > 0$. Since $\mathbf{H}_\lambda(t)$ is analytic in λ, it may belong to Π only for isolated λ values. Therefore $\mathbf{H}_\lambda^{(2)}(t) \in \mathcal{G}_0$ for small $\lambda > 0$. Consequently, $\mathbf{H}_\lambda(t) \in \mathcal{G}_0$ for small $\lambda > 0$. In similar fashion one shows, using the condition $p_{av} = 0$, that $\mathbf{H}_\lambda^{(2)}(t) \in \mathcal{G}_{-1}$ for small $\lambda < 0$. Consequently, $\mathbf{H}_\lambda(t) \in \mathcal{G}$ for small $\lambda < 0$, and since $\mathbf{H}_0(t) \in \Pi_0^{\bullet +}$ we see that $\mathbf{H}_\lambda(t) \in \mathcal{G}_0$ for small $\lambda < 0$. Thus, if $p_{av} = 0$ and λ is varied in the neighborhood of $\lambda = 0$, the point $\mathbf{H}_\lambda(t) \in \mathfrak{L}^3$ first $(\lambda > 0)$ moves from $\mathcal{G}_0$ to the surface $\Pi_0^{\bullet +}$, "grazing" the latter at $\lambda = 0$, and then $(\lambda > 0)$ returning to the domain $\mathcal{G}_0$. We have thus shown that the oscillation theorem is valid here, with the modification of condition (II) as in Theorem VII.

Now let $p_{av} > 0$. Using the table at the end of §1.10 (p. 658), we deduce from (6.22) that $\mathbf{H}_0^{(2)}(t) \in \Pi_0^{\bullet +}$. It follows from (6.22) that $\mathbf{H}_\lambda^{(2)}(t) \geqslant \mathbf{H}_0^{(2)}(t)$ for $\lambda > 0$. Reasoning as before, and using Theorem II of §1.9 and the comparison theorem (§2.4), we see that $\mathbf{H}_\lambda^{(2)}(t) \in \mathcal{H}_0$ for small $\lambda < 0$ and $\mathbf{H}_\lambda^{(2)}(t) \in \mathcal{G}_0$ for small $\lambda > 0$. Since $\mathbf{H}_\lambda(t) \in \Pi_0^{\bullet +}$ for $\lambda = 0$, this implies that $\mathbf{H}_\lambda(t) \in \mathcal{G}_0$ for small $\lambda > 0$ and $\mathbf{H}_\lambda(t) \in \mathcal{H}_0$ for small $\lambda < 0$. Thus, if $p_{av} > 0$ the Hamiltonian $\mathbf{H}_\lambda(t)$ is in $\mathcal{G}_0$ for small $\lambda > 0$, crosses the surface $\Pi_0^{\bullet +}$ at $\lambda = 0$, enters the domain $\mathcal{H}_0$ for small $\lambda < 0$, then crosses $\Pi_0^{\bullet +}$ again for the value of $\lambda = \lambda_0^- < 0$ at which $\mathbf{H}_\lambda^{(2)}(t) \in \Pi_0^{\bullet -}$, and re-enters $\mathcal{G}_0$ for small $\lambda - \lambda_0^- < 0$ (see Figure 68). As λ decreases further, the trajectory $\mathbf{H}_\lambda(t)$ goes off to the "right". In other words, the oscillation theorem is true with the desired modification of (II), and the proof is complete.

Remark. It is evident from the proof that in the cases $p_{av}=0$ and $p_{av}>0$ the curves $A(\lambda)$ (see § VII. 1) have the form shown in Figure 39. (The indices of the stability and instability domains in that figure correspond to the domains containing $\mathbf{H}_\lambda^{(2)}(t)$ but not to those containing $\mathbf{H}_\lambda(t)$.) Theorem VII also provides the proofs for certain propositions stated without proof in §§1.4, 1.5 of Chapter VII.

Reasoning along similar lines (the arguments are a little more cumbersome), one can show that the oscillation theorem is valid ($\lambda\to+\infty$ and $\lambda\to-\infty$) for equation (6.4) with $p(t)$ of variable sign, provided that $q(t)\leqslant \mathrm{const}$ and either $p_{av}\neq 0$, or $p_{av}=0$, $q_{av}\leqslant 0$.

In this case equation (6.21) is obtained by the substitution

$$z=-S_\lambda(t)\,y-\frac{1}{\lambda}\dot{y},\quad \mathbf{x}_2=\begin{pmatrix} y\\ z\end{pmatrix},$$

where

$$S_\lambda(t)=\int_0^t\left\{p(t)-p_{av}+\frac{1}{\lambda}[q(t)-q_{av}]\right\}dt.$$

If $q_{av}\leqslant 0$, we again obtain $\partial\mathbf{H}_\lambda^{(2)}/\partial\lambda\leqslant 0$. If $p_{av}\neq 0$, the substitution $\lambda\to\lambda+\lambda_0$ (for a suitable constant λ_0) brings us back to the case $q_{av}\leqslant 0$.

One corollary of Theorems VI and VII is that under the above assumptions the real spectrum defined by equation (6.3) or (6.4) and the condition

$$\int_0^\infty |y|^2\,dt<\infty$$

or, in general,

$$\int_0^\infty |y|^\beta\,dt<\infty$$

for any $\beta>0$, is the union of an infinite set of intervals $\lambda_n^-<\lambda<\lambda_n^+$ which extends to $+\infty$ (under the assumptions of Theorem VI) or to infinity in both directions (under the assumptions of Theorem VII). The real spectrum defined by equation (6.3) or (6.4) and the condition

$$|y(t)|<\mathrm{const}\ \text{ for }\ -\infty<t<+\infty,$$

is the union of the intervals $\lambda_n^+<\lambda<\lambda_{n+1}^-$.

6.3. Asymptotic formulas for boundaries of stability zones

We shall establish asymptotic formulas for the eigenvalues $\lambda_n^\pm$ of the vector equation

$$\mathbf{J}\frac{d\mathbf{x}}{dt}=(\lambda\mathbf{H}_1+\mathbf{H}_2)\,\mathbf{x}, \tag{6.23}$$

where

$$\mathbf{H}_1=\begin{Vmatrix}\alpha_1(t) & \beta_1(t)\\ \beta_1(t) & \gamma_1(t)\end{Vmatrix}>0,\quad \mathbf{H}_2=\begin{Vmatrix}\alpha_2(t) & \beta_2(t)\\ \beta_2(t) & \gamma_2(t)\end{Vmatrix}\qquad (0\leqslant t\leqslant T),$$

on the assumption that $\mathbf{H}_1(t)$, $\mathbf{H}_2(t)$ are absolutely continuous functions.

We shall use an asymptotic method due to Rapoport /77, 198a, b/, omitting some readily verified intermediate steps.

Denote

$$\mathbf{A}_1=\tilde{\mathbf{J}}^{-1}\mathbf{H}_1,\quad \mathbf{A}_2=\tilde{\mathbf{J}}^{-1}\mathbf{H}_2.$$

We know that $\mathbf{H}_1>\mathbf{0}$ if and only if

$$\alpha_1>0,\quad \gamma_1>0,\quad \alpha_1\gamma_1-\beta_1^2>0\qquad (0\leqslant t\leqslant T). \tag{6.24}$$

Using a substitution of the form

$$\mathbf{x}=\left[\mathbf{R}_0(t)+\frac{1}{\lambda}\mathbf{R}_1(t)\right]\mathbf{y} \tag{6.25}$$

where $\mathbf{R}_0(t)$, $\mathbf{R}_1(t)$ are suitably chosen matrices, we shall try to reduce equation (6.23) to the form of

$$\frac{d\mathbf{y}}{dt}=\left(\mathbf{B}+\frac{1}{\lambda}\mathbf{C}\right)\mathbf{y}, \tag{6.26}$$

where $\mathbf{B}=\mathbf{B}(t,\lambda)$ is a diagonal matrix and $\mathbf{C}=\mathbf{C}(t,\lambda)=O(1)$ as $\lambda\to\infty$. In fact, it is necessary and sufficient that the matrices $\mathbf{R}_0(t)$ and $\mathbf{R}_1(t)$ satisfy the equation

$$\frac{d}{dt}\left(\mathbf{R}_0+\frac{1}{\lambda}\mathbf{R}_1\right)=(\lambda\mathbf{A}_1+\mathbf{A}_2)\left(\mathbf{R}_0+\frac{1}{\lambda}\mathbf{R}_1\right)-\left(\mathbf{R}_0+\frac{1}{\lambda}\mathbf{R}_1\right)\left(\mathbf{B}+\frac{1}{\lambda}\mathbf{C}\right). \tag{6.27}$$

Setting $\mathbf{B}=\lambda\mathbf{B}_1(t)+\mathbf{B}_2(t)$, where $\mathbf{B}_1$ and $\mathbf{B}_2$ are diagonal matrices, we see that equation (6.27) is equivalent to three equations:

$$\mathbf{A}_1\mathbf{R}_0-\mathbf{R}_0\mathbf{B}_1=0, \tag{6.28}$$

$$\mathbf{A}_1\mathbf{R}_1-\mathbf{R}_1\mathbf{B}_1=\dot{\mathbf{R}}_0-\mathbf{A}_2\mathbf{R}_0+\mathbf{R}_0\mathbf{B}_2, \tag{6.29}$$

$$\left(\mathbf{R}_0+\frac{1}{\lambda}\mathbf{R}_1\right)\mathbf{C}=\mathbf{A}_2\mathbf{R}_1-\mathbf{R}_1\mathbf{B}_2-\dot{\mathbf{R}}_1. \tag{6.30}$$

The eigenvalues of the matrix $\mathbf{A}_1$ are $\pm ib$, where

$$b=\sqrt{\alpha_1\gamma_1-\beta_1^2}\,. \tag{6.31}$$

By assumption, these numbers are distinct, so that there exist a nonsingular matrix $\mathbf{R}_0$ and a diagonal matrix $\mathbf{B}_1$ satisfying equation (6.28):

$$\mathbf{R}_0=\begin{Vmatrix}\beta_1-ib & \gamma_1\\ -\alpha_1 & -\beta_1+ib\end{Vmatrix},\quad \mathbf{B}_1=\begin{Vmatrix}ib & 0\\ 0 & -ib\end{Vmatrix}. \tag{6.32}$$

To see whether equation (6.29) is solvable, we substitute

$$\mathbf{R}_1 = \mathbf{R}_0\mathbf{S}. \tag{6.33}$$

This gives an equation

$$\mathbf{B}_1\mathbf{S} - \mathbf{S}\mathbf{B}_1 = \mathbf{R}_2 - \mathbf{B}_2, \tag{6.34}$$

where

$$\mathbf{R}_2 = \mathbf{R}_0^{-1}\dot{\mathbf{R}}_0 - \mathbf{R}_0^{-1}\mathbf{A}_2\mathbf{R}_0. \tag{6.35}$$

From equation (6.34) we easily determine $\mathbf{S}$ and the diagonal matrix $\mathbf{B}_2$. It is easy to see that equation (6.34) is solvable only if $\mathbf{B}_2$ is equal to the diagonal part of $\mathbf{R}_2$. In that case the off-diagonal elements of $\mathbf{S}$ are determined from equation (6.34); the diagonal elements are arbitrary. Thus equation (6.34) and, with it, equation (6.29), are solvable. For sufficiently small λ the inverse $\left(\mathbf{R}_0 + \frac{1}{\lambda}\mathbf{R}_1\right)^{-1}$ exists; we may thus find $\mathbf{C}$ from equation (6.30) and $\mathbf{C} = O(1)$ as $\lambda \to \infty$.

Denote

$$\Delta = \det \mathbf{R}_0 = 2b(b + i\beta_1). \tag{6.36}$$

Since $\mathrm{Tr}\left[\mathbf{R}_0^{-1}\dot{\mathbf{R}}_0\right] = \dot{\Delta}/\Delta$, we have

$$\mathrm{Tr}\left[\mathbf{R}_0^{-1}\dot{\mathbf{R}}_0 - \frac{\dot{\Delta}}{2\Delta}\mathbf{I}_2\right] = 0,$$

i. e.,

$$\mathbf{R}_0^{-1}\dot{\mathbf{R}}_0 = \begin{Vmatrix} \varphi & * \\ * & -\varphi \end{Vmatrix} + \frac{\dot{\Delta}}{2\Delta}\mathbf{I}_2. \tag{6.37}$$

Here and below, asterisks will stand for elements which need not be specified. It is readily checked that

$$\varphi = \frac{1}{4}\left(\frac{\dot{\alpha}_1}{\alpha_1} - \frac{\dot{\gamma}_1}{\gamma_1}\right) - i\varphi_1, \quad \varphi_1 = \frac{\beta_1}{4b}\left(\frac{\dot{\alpha}_1}{\alpha_1} - \frac{\dot{\gamma}_1}{\gamma_1}\right). \tag{6.38}$$

Since $\mathrm{Tr}\left[\mathbf{R}_0^{-1}\mathbf{A}_2\mathbf{R}_0\right] = \mathrm{Tr}\,\mathbf{A}_2 = 0$, it follows that

$$\mathbf{R}_0^{-1}\mathbf{A}_2\mathbf{R}_0 = \begin{Vmatrix} \psi & * \\ * & -\psi \end{Vmatrix}, \tag{6.39}$$

and it is easily verified that

$$\psi = i\psi_1, \text{ where } \psi_1 = \frac{1}{2b}(\alpha_1\gamma_2 - 2\beta_1\beta_2 + \alpha_2\gamma_1). \tag{6.40}$$

Consequently,

$$\mathbf{B}_2 = \begin{Vmatrix} \varphi-\psi & 0 \\ 0 & \psi-\varphi \end{Vmatrix} + \frac{\dot{\Delta}}{2\Delta}\mathbf{I}_2. \tag{6.41}$$

The substitution

$$\mathbf{y} = e^{\int_0^t \mathbf{B}\,dt}\,\mathbf{z} \tag{6.42}$$

reduces equation (6.26) to the form

$$\frac{d\mathbf{z}}{dt} = \frac{1}{\lambda}\mathbf{D}(t, \lambda)\,\mathbf{z}, \quad \mathbf{D}(t, \lambda) = e^{-\int_0^t \mathbf{B}\,dt}\,\mathbf{C}(t, \lambda)\,e^{\int_0^t \mathbf{B}\,dt}. \tag{6.43}$$

It follows from the formulas (6.32) for $\mathbf{B}_1 = \mathbf{B}_1(t)$ and from (6.41) that

$$e^{\pm\int_0^t \mathbf{B}\,dt} = O(1)$$

as $\lambda \to \infty$, i. e., $\mathbf{D}(t, \lambda) = O(1)$ as $\lambda \to \infty$. Hence the matrizant $\mathbf{Z}(t, \lambda)$ $(\mathbf{Z}(0, \lambda) = \mathbf{I}_2)$ of equation (6.43) may be expressed as

$$\mathbf{Z}(t, \lambda) = \mathbf{I}_2 + \frac{1}{\lambda}\mathbf{Z}_1(t, \lambda) + \frac{1}{\lambda^2}\mathbf{Z}_2(t, \lambda) + \ldots, \tag{6.44}$$

where the coefficients $\mathbf{Z}_l(t, \lambda)$ are determined successively from

$$\mathbf{Z}_l(t, \lambda) = \int_0^t \mathbf{D}(t_1, \lambda)\,\mathbf{Z}_{l-1}(t_1, \lambda)\,dt_1 \qquad (\mathbf{Z}_0 = \mathbf{I}_2;\ l = 1, 2, \ldots). \tag{6.45}$$

By (6.30),

$$\mathbf{R}_0\left(\mathbf{I}_2 + \frac{1}{\lambda}\mathbf{R}_0^{-1}\mathbf{R}_1\right)\mathbf{C} = \mathbf{A}_2\mathbf{R}_1 - \mathbf{R}_1\mathbf{B}_2 - \dot{\mathbf{R}}_1,$$

$$\mathbf{C} = \left(\mathbf{I}_2 + \frac{1}{\lambda}\mathbf{R}_0^{-1}\mathbf{R}_1\right)^{-1}(\mathbf{R}_0^{-1}\mathbf{A}_2\mathbf{R}_1 - \mathbf{R}_0^{-1}\mathbf{R}_1\mathbf{B}_2 - \mathbf{R}_0^{-1}\dot{\mathbf{R}}_1) = \mathbf{C}_0 + \frac{1}{\lambda}\mathbf{C}_1 + \frac{1}{\lambda^2}\mathbf{C}_2 + \ldots,$$

where

$$\mathbf{C}_0 = \mathbf{R}_0^{-1}\mathbf{A}_2\mathbf{R}_1 - \mathbf{R}_0^{-1}\mathbf{R}_1\mathbf{B}_2 - \mathbf{R}_0^{-1}\dot{\mathbf{R}}_1.$$

Clearly, $\mathbf{Z}_l(t, \lambda) = O(1)$ as $\lambda \to \infty$ and the series (6.44) is uniformly convergent for all t, λ in the domain

$$0 \leqslant t \leqslant T < \infty, \quad \lambda_0 < \lambda < \infty. \tag{6.46}$$

It follows from (6.25), (6.42) that, if λ is sufficiently large, the matrizant $\mathbf{X}(t, \lambda)$ $(\mathbf{X}(0, \lambda) = \mathbf{I}_2)$ of equation (6.23) may be expressed as

$$\mathbf{X}(t, \lambda) = \left[\mathbf{R}_0(t) + \frac{1}{\lambda}\mathbf{R}_1(t)\right] e^{\int_0^t \mathbf{B}\,dt}\,\mathbf{Z}(t, \lambda)\left[\mathbf{R}_0(0) + \frac{1}{\lambda}\mathbf{R}_1(0)\right]^{-1}. \tag{6.47}$$

By (6.32), $\mathbf{R}_0(t+T)\equiv\mathbf{R}_0(t)$. We may assume that the diagonal elements of the matrix $\mathbf{S}=\mathbf{S}(t)$ are zero. Then $\mathbf{S}(t+T)\equiv\mathbf{S}(t)$, $\mathbf{R}_1(t+T)\equiv\mathbf{R}(t)$. On the boundary of the stability domains, the roots of the characteristic equation

$$\det[\mathbf{X}(T,\lambda)-\rho\mathbf{I}_2]=0$$

are $\rho=\pm1$. Thus the equation for the eigenvalues is

$$\det\left[\mathbf{I}_2+\frac{1}{\lambda}\mathbf{Z}_1(T,\lambda)+\frac{1}{\lambda^2}\mathbf{Z}_2(T,\lambda)+\ldots\pm e^{-\int_0^T \mathbf{B}\,dt}\right]=0. \qquad (6.48)$$

Using the fact that $\Delta(t+T)\equiv\Delta(t)$, $\alpha_1(t+T)=\alpha_1(t)$, $\gamma_1(t+T)=\gamma_1(t)$, we then obtain

$$\int_0^T \mathbf{B}\,dt=\begin{Vmatrix} i\int_0^T(\lambda b+\varphi_1+\psi)\,dt & 0 \\ 0 & -i\int_0^T(\lambda b+\varphi_1+\psi_1)\,dt \end{Vmatrix}.$$

The first approximation to the equation

$$\det\left[\mathbf{I}_2\pm e^{-\int_0^T \mathbf{B}\,dt}\right]=0$$

gives

$$\lambda=\lambda_n^0=\frac{n\pi+b_2}{b_1},\quad b_1=\int_0^T b\,dt,\quad b_2=\int_0^T(\varphi_1+\psi_1)\,dt. \qquad (6.49)$$

It is easy to see that equation (6.48) has a solution

$$\lambda=\lambda_n^0+\frac{k}{n}+O\left(\frac{1}{n^2}\right). \qquad (6.50)$$

Substituting (6.50) into (6.48), we see that k satisfies the equation

$$\det\left[\mathbf{Z}_1(T,\lambda_n^0)-\frac{k\pi}{b_1}\int_0^T \mathbf{B}_1\,dt\right]=0, \qquad (6.51)$$

where

$$\left.\begin{aligned} \mathbf{Z}_1(t,\lambda)&=\int_0^T \mathbf{D}_0(t_1,\lambda)\,dt_1,\\ \mathbf{D}_0(t,\lambda)&=e^{-\int_0^t \mathbf{B}\,dt}\,\mathbf{C}_0 e^{\int_0^t \mathbf{B}\,dt}. \end{aligned}\right\} \qquad (6.52)$$

Let k^-, k^+ denote the roots of the quadratic equation (6.51).

Inserting this into (6.50), we obtain asymptotic formulas for $\lambda_n^{\pm}$:

$$\lambda_n^{\pm} = \frac{n\pi}{b_1} + \frac{b_2}{b_1} + \frac{k^{\pm}}{n} + O\left(\frac{1}{n^2}\right), \tag{6.53}$$

where b_1, b_2 are defined by

$$b_1 = \int_0^T \sqrt{\alpha_1\gamma_1 - \beta_1^2}\,dt,$$

$$b_2 = \int_0^T \left[\frac{\beta_1}{4b_1}\left(\frac{\dot{\alpha}_1}{\alpha_1} - \frac{\dot{\gamma}_1}{\gamma_1}\right) + \frac{1}{2\sqrt{\alpha_1\gamma_1 - \beta_1^2}}(\alpha_1\gamma_2 - 2\beta_1\beta_2 + \alpha_2\gamma_1)\right] dt.$$

If $\mathbf{H}_2 = 0$, it can be shown, under quite broad assumptions (the matrix $\mathbf{H}_1(t)$ is only L-integrable on $[0, T]$), that although the above method is no longer applicable,

$$\lim \frac{\lambda_n^{\pm}}{n} = \frac{\pi}{b_1}$$

(see Krein /166d/).

6.4. Boundary-value problems for canonical equations whose coefficients are nonlinear in λ

Let us consider a canonical equation (6.1), defined for $0 \leqslant t \leqslant T$, whose Hamiltonian $\mathbf{H}_\lambda(t)$ depends on a real parameter λ. We recall that Hill's equation (6.3), (6.4) may be brought to the form of (6.1), with the vector $\mathbf{x}$ and Hamiltonian $\mathbf{H}_\lambda(t)$ defined as in (6.19).

Consider the boundary-value problem consisting of equation (6.1) and the boundary conditions

$$\mathbf{x}(T) = \mathbf{C}\mathbf{x}(0) \qquad (\det \mathbf{C} = 1) \tag{6.54}$$

(where $\mathbf{C}$ is a real 2×2 matrix) or the conditions

$$\mathbf{x}(T) = \rho\mathbf{x}(0) \qquad (|\rho| = 1,\ \rho \neq \pm 1), \tag{6.55}$$

or the Sturm boundary conditions

$$(\mathbf{x}(0),\ \mathbf{u}) = 0, \quad (\mathbf{x}(T),\ \mathbf{v}) = 0, \tag{6.56}$$

where $\mathbf{u}$, $\mathbf{v}$ are real vectors.

There is an essential difference between these boundary-value problems and the more familiar selfadjoint boundary-value problems: the coefficients of equation (6.1) are not linear in the parameter λ, while in the case of Hill's equation (6.3) or (6.4) the function $p(t)$ need not be of fixed sign.

We define the spectrum of problem (6.1), (6.54) (problem (6.1), (6.55) or problem (6.1), (6.56)) to be the set of all real λ for which there exists a nontrivial solution.

The eigenvalues $\lambda_{2n}^{\pm}(\lambda_{2n+1}^{\pm})$ (6.5) in the oscillation theorem (see §6.1, (I)) are the eigenvalues of the problems (6.1), (6.54) for $\mathbf{C}=\mathbf{I}_2$ $(\mathbf{C}=-\mathbf{I}_2)$.

We shall show that combined use of the geometric constructions of §1 and of the results of §2 yields analogs of the oscillation theorems for the boundary conditions (6.54)—(6.56). This geometric approach to boundary-value problems is also effective in far more general situations.

We first consider condition (6.54) together with a canonical equation

$$\tilde{\mathbf{J}}\frac{d\mathbf{x}}{dt}=\mathbf{H}(t)\,\mathbf{x}, \qquad (6.57)$$

where $\mathbf{H}(t)\in\mathfrak{L}^3$ is independent of λ. Let $\mathbf{X}(t)$ be the matrizant of equation (6.57). Problem (6.54), (6.57) has a nontrivial solution if and only if

$$\det[\mathbf{X}(T)-\mathbf{C}]=0. \qquad (6.58)$$

As usual, we wish to characterize the set of Hamiltonians in $\mathfrak{L}^3$ for which (6.58) holds; more precisely, to characterize the set of representative points on the model $\mathfrak{R}^3$ of $\mathfrak{L}^3$. First assume that $\mathbf{C}=\mathbf{I}$. Recall that the set of matrices $\mathfrak{G}$ in $\mathbf{X}$ such that

$$\tilde{\mathbf{J}}\dot{\mathbf{X}}=\mathbf{H}(t)\,\mathbf{X}$$

is the image of the cone $\tilde{\Pi}^+\subset\tilde{\mathfrak{R}}_0^+$; thus this set is the deformed cone

$$\tilde{\Pi}^+=\tilde{\Pi}_+^+\cup\tilde{\Pi}_-^+\cup\{\mathbf{I}\}\subset\mathfrak{G}$$

(this cone is hatched, together with $\tilde{\Pi}^-$, in Figure 52, p. 628). Thus the corresponding curves $\mathbf{X}(t)$ $(0\leqslant t\leqslant T)$ have initial point at $\mathbf{I}\in\mathfrak{G}$ and endpoint on the cone $\tilde{\Pi}^+\subset\mathfrak{G}$. Going over to the sets $\tilde{\mathfrak{R}}_0^+$, $\tilde{\mathfrak{R}}_0^-$, then to $\tilde{\mathfrak{R}}_{2n}$, $\tilde{\mathfrak{R}}_{2n-1}$, and finally to $\mathfrak{R}^3$ (see Figures 53, 54, 55, 56), we see that the corresponding curves begin at the point $\tilde{\Pi}_0^{\bullet\bullet}$ and end on the surfaces $\tilde{\Pi}_{2n}$. Thus the set of points in $\mathfrak{R}^3$ representing Hamiltonians which satisfy (6.58) is

$$\bigcup_{n=-\infty}^{+\infty}\tilde{\Pi}_{2n}, \qquad (6.59)$$

and the set of Hamiltonians itself is

$$\bigcup_{n=-\infty}^{+\infty}\Pi_{2n}.$$

This is the set of "eigen-Hamiltonians" of the boundary-value problem (6.57), (6.54) for $\mathbf{C}=\mathbf{I}$. It is the union of countably many closed connected subsets Π_{2n}.

Similarly, one shows that the set of "eigen-Hamiltonians" of the boundary-value problem (6.57), (6.54) for $\mathbf{C}=-\mathbf{I}$ is

$$\bigcup_{n=-\infty}^{+\infty} \Pi_{2n+1}.$$

This is the union of countably many closed connected subsets Π_{2n+1}.

Let $\mathbf{C} \neq \mathbf{I}$, $\mathbf{C} \neq -\mathbf{I}$. The transformation

$$\mathbf{X}' = \mathbf{C}^{-1}\mathbf{X},$$

is obviously a homeomorphism in $\mathfrak{G}$. The set of matrices $\mathbf{X}$ such that

$$\det [\mathbf{X} - \mathbf{C}] = 0 \tag{6.60}$$

is mapped onto the set of matrices $\mathbf{X}'$ such that

$$\det [\mathbf{X}' - \mathbf{I}] = 0,$$

which is represented in Figure 52 by the hatched deformed cone $\tilde{\Pi}^+$ in $\mathfrak{G} = \{\mathbf{X}'\}$ with center at $\mathbf{X}' = \mathbf{I}$. The point $\mathbf{X}' = \mathbf{I}$ corresponds to $\mathbf{X} = \mathbf{C}$. Thus, (6.60) is a "deformed cone" of the same type as $\tilde{\Pi}^+$, but its center is at $\mathbf{C}$. Applying the mapping from $\mathfrak{G}$ to the universal covering space $\mathfrak{R}^3$, we see that the corresponding set $\tilde{\Pi}(\mathbf{C})$ has the same form as the set (6.59) — a countable union of closed deformed cones $\tilde{\Pi}_{2n}(\mathbf{C})$. It is the image of (6.59) under a rigid displacement which takes the center $\tilde{\Pi}_0^{**}$ of $\tilde{\Pi}_0$ onto the point of $\tilde{\mathfrak{R}}_0^+$ (or of $\tilde{\mathfrak{R}}_0^-$) which is the image of $\mathbf{C}$ under the mapping demonstrated in Figure 52.

The boundary-value problem (6.1), (6.54) is solvable for all points λ, and only for these points, such that the point in $\mathfrak{R}^3$ representing $\mathbf{H}_\lambda(t)$ lies on one of the surfaces $\tilde{\Pi}_{2n}(\mathbf{C})$.

Hence the following proposition:

If conditions (I_ψ), (II_ψ) *of* §6.1 *are satisfied and* $\mathbf{H}_\lambda(t)$ *is analytic in* λ, *then the spectrum of problem (6.1), (6.54) is a countable set of eigenvalues* λ_n $(n = 0, \pm 1, \pm 2, \ldots)$ *such that*

$$\lambda_n \to +\infty, \quad \lambda_{-n} \to -\infty \quad \text{as} \quad n \to \infty. \tag{6.61}$$

Effective conditions guaranteeing the validity of conditions (I_ψ), (II_ψ) were established in §6.2, among them conditions (6.6) and (6.7).

Under the same assumptions, the spectrum of problem (6.1), (6.56) is also a set of eigenvalues λ_n satisfying (6.61). Indeed, let $\mathbf{x}(\lambda, t)$ be a solution of equation (6.1) satisfying the initial condition $\mathbf{x}(\lambda, 0) = \mathbf{x}_0$, where $\mathbf{x}_0$ is a constant vector such that $(\mathbf{x}_0, \mathbf{u}) = 0$.

If conditions (I_ψ) and (II_ψ) are satisfied, then

$$\operatorname{Arg} \mathbf{x}(\lambda, T) \to +\infty \quad \text{as} \quad \lambda \to +\infty,$$
$$\operatorname{Arg} \mathbf{x}(\lambda, T) \to -\infty \quad \text{as} \quad \lambda \to -\infty,$$

where $\operatorname{Arg} \mathbf{x}(\lambda, t)$ is a continuous branch of the argument of the vector $\mathbf{x}(\lambda, t)$.

The eigenvalues λ_n are the numbers for which this vector lies on the straight line $(\mathbf{x}, \mathbf{v}) = 0$. Consequently, there must be infinitely many, and there are values λ which are not eigenvalues. The eigenvalues are the roots of the equation $(\mathbf{x}(\lambda, t), \mathbf{v})=0$, whose left-hand side is analytic in λ (since by assumption $\mathbf{H}_\lambda(t)$ is an analytic function of λ). Consequently, the spectrum is a countable set of eigenvalues λ_n satisfying (6.61), as claimed.

Now consider the boundary conditions (6.55). Reasoning as before, we deduce that the corresponding set in $\mathfrak{G}$ is the set of matrices $\mathbf{X}$ with an eigenvalue ρ. This set separates into two surfaces, one in $\tilde{\mathfrak{G}}^+$ (for which ρ is an eigenvalue of the first kind) and the other in $\tilde{\mathfrak{G}}^-$ (ρ an eigenvalue of the second kind). The pre-images of these surfaces in $\tilde{\mathfrak{R}}^3$ (see §§1.4, 1.5) are the sheets of the hyperboloid $x^2+y^2-z^2=-\varphi^2$, where $\rho=e^{i\varphi T}$. The mapping onto $\tilde{\mathfrak{R}}_0^+$, $\tilde{\mathfrak{R}}_0^-$ (Figure 54) deforms these surfaces, producing surfaces lying in $\tilde{\mathfrak{G}}_{0-}^-$ or $\tilde{\mathfrak{G}}_{0+}^-$ in $\tilde{\mathfrak{R}}_0^-$, and in $\tilde{\mathfrak{G}}_{0-}^+$ or $\tilde{\mathfrak{G}}_{0+}^+$ in $\tilde{\mathfrak{R}}_0^+$, homeomorphic to the surfaces $\tilde{\Gamma}_{0+}^-\subset\tilde{\mathfrak{R}}_0^-$ and $\tilde{\Gamma}_{0+}^+\subset\tilde{\mathfrak{R}}_0^+$. Thus, any curve in $\tilde{\mathfrak{R}}_0^-(\tilde{\mathfrak{R}}_0^+)$ whose initial point is in $\tilde{\Gamma}_{0-}^-(\tilde{\Gamma}_{0-}^+)$ and endpoint in $\tilde{\Gamma}_{0+}^-(\tilde{\Gamma}_{0+}^+)$ must cut this surface. The corresponding set in $\tilde{\mathfrak{R}}^3$ is a countable union of surfaces of this type, each of which is in the set $\tilde{\mathfrak{G}}_n$. Consequently, the set of "eigen-Hamiltonians" for problem (6.57), (6.55) is the union of countably many closed "surfaces, " one surface in each domain $\mathfrak{G}_n$ each homeomorphic to a linear space. These surfaces are shown in Figure 56 by thin lines.

Now consider the boundary-value problem (6.1), (6.55), assuming again for simplicity's sake that $\mathbf{H}_\lambda(t)$ is analytic in λ. Suppose that conditions (I_ψ), (II_ψ) are satisfied. As λ varies from $\lambda=-\infty$ to $\lambda=+\infty$, the point representing $\mathbf{H}_\lambda(t)$ in $\mathfrak{R}^3$ crosses $\mathfrak{R}^3$ "from left to right" (in the usual sense). Thus the point representing $\mathbf{H}_\lambda(t)$ "crosses" the above "eigen-surfaces" infinitely many times. There are therefore infinitely many eigenvalues, and there are values $\mathfrak{R}^3$ which are not eigenvalues. Since $\mathbf{H}_\lambda(t)$ is analytic in λ, the eigenvalues can constitute at most a countable set. Thus, if conditions (I_ψ), (II_ψ) are satisfied, the spectrum of problem (6.1), (6.55) is a countable set of eigenvalues satisfying (6.5). We summarize the result as follows.

Theorem I. Let $\mathbf{H}_\lambda(t)$ be an analytic function of λ and suppose that conditions (I_ψ), (II_ψ) of Theorem I in § 6.1 are satisfied. Then the spectrum of each of problems (6.1), (6.5), or (6.1), (6.55), or (6.1), (6.56) is a countable set of eigenvalues λ_n $(n=0, \pm 1, \pm 2, \ldots)$ satisfying (6.61).

Now suppose that condition (III_ψ) is also satisfied. Then as λ increases the multiplier of the first kind will move along the unit circle in the positive sense (see § III. 1.7). Hence it follows that the eigenvalues λ_n of problem (6.1), (6.56) may be indexed in such a way that $\lambda_{2n}^+<\lambda_n<\lambda_{2n+1}^-$ (if $\operatorname{Im}\rho>0$) or $\lambda_{2n-1}^+<\lambda_n<\lambda_{2n}^-$ (if $\operatorname{Im}\rho<0$), where λ_n^+, λ_n^- are the eigenvalues (6.5) in the oscillation theorem.

The conclusions of Theorem I are no longer valid if we replace one of the boundary conditions (6.54), (6.55), (6.56) by a nonselfadjoint condition such as

$$\mathbf{x}(T)=\varkappa\mathbf{x}(0)$$

with real $\varkappa\neq 1$. Indeed, it follows from the results of §1 that the corresponding surfaces in $\mathfrak{R}^3$ (deformed one-sheeted hyperboloids in $\mathfrak{R}_{2n}$ if $\varkappa>0$ and in $\mathfrak{R}_{2n+1}$ if $\varkappa<0$) have the property that the point of $\mathfrak{R}^3$ representing $\mathbf{H}_\lambda(t)$ either crosses none of these surfaces as λ goes from $\lambda=-\infty$ to $\lambda=+\infty$ or crosses only finitely many of them.

We shall show now that problem (6.1), (6.54) may be reduced to a similar problem with $\mathbf{C}=\pm\mathbf{I}_2$. It was shown in §1.4 that if $\operatorname{Tr}\mathbf{C}\geqslant 0$ there exists a real logarithm $\ln\mathbf{C}$, and $\operatorname{Tr}\ln\mathbf{C}=0$. Set

$$\mathbf{K}=\frac{1}{T}\ln\mathbf{C}, \quad \text{if } \operatorname{Tr}\mathbf{C}\geqslant 0,$$

$$\mathbf{K}=\frac{1}{T}\ln(-\mathbf{C}), \quad \text{if } \operatorname{Tr}\mathbf{C}<0,$$

and substitute in (6.1)

$$\mathbf{x}=e^{t\mathbf{K}}\mathbf{x}_1.$$

This brings the boundary-value problem (6.1), (6.54) to the form

$$\tilde{\mathbf{J}}\frac{d\mathbf{x}_1}{dt}=\mathbf{G}_\lambda(t)\,\mathbf{x}_1, \tag{6.62}$$

$$\mathbf{x}(T)=\pm\mathbf{x}(0) \tag{6.63}$$

(plus if $\operatorname{Tr}\mathbf{C}\geqslant 0$, minus if $\operatorname{Tr}\mathbf{C}<0$). Since $\operatorname{Tr}\mathbf{K}=0$, it follows that

$$\mathbf{K}=\tilde{\mathbf{J}}^{-1}\mathbf{L}, \quad \text{where} \quad \mathbf{L}^*=\mathbf{L}.$$

Hence we easily deduce that $\mathbf{G}_\lambda(t)$ may be written

$$\mathbf{G}_\lambda(t)=-\mathbf{L}+(e^{t\mathbf{K}})^*\,\mathbf{H}_\lambda(t)\,e^{t\mathbf{K}}.$$

Therefore $\mathbf{G}_\lambda(t)^*=\mathbf{G}_\lambda(t)$, so that (6.62) is a canonical equation. It follows from the condition $\partial\mathbf{H}_\lambda/\partial\lambda\geqslant\mathbf{0}$ that $\partial\mathbf{G}_\lambda/\partial\lambda\geqslant 0$, and conversely. The boundary-value problem (6.62), (6.63) was studied in §6.1. It is readily checked that the rotations of the vectors $\mathbf{x}$ and $\mathbf{x}_1$ satisfy the relation

$$\frac{\partial\varphi_x}{\partial\lambda}=\frac{|\mathbf{x}_1(T,\lambda)|^2}{|\mathbf{x}(T,\lambda)|^2}\frac{\partial\varphi_{x_1}}{\partial\lambda}. \tag{6.64}$$

Using the above substitution and this last formula, one readily proves analogs of Theorems II – VII (§6.2) for boundary conditions (6.54).

For example, Theorem IV of §6.1 and Theorem II of §1.10 imply

Theorem II. Let $\mathbf{H}_\lambda(t)$ be an analytic function of λ (for any real λ) and $\partial H_\lambda(t)/\partial\lambda\geqslant\mathbf{0}$. Then the spectrum of the problem

$$\tilde{\mathbf{J}}\frac{d\mathbf{x}}{dt}=\mathbf{H}_\lambda(t)\,\mathbf{x}, \quad \mathbf{x}(T)=\pm\mathbf{C}\mathbf{x}(0) \qquad (\text{where } \operatorname{Tr}\mathbf{C}\geqslant 0) \tag{6.65}$$

is either empty, or the entire real axis $-\infty<\lambda<+\infty$, or a finite or infinite set of values (6.5). In that case:

1. *Eigenvalues λ with even subscripts are the eigenvalues of the problem $\mathbf{x}(T)=\mathbf{C}\mathbf{x}(0)$, those with odd subscripts the eigenvalues of the problem $\mathbf{x}(T)=-\mathbf{C}\mathbf{x}(0)$.*
2. *If $\lambda_n^+\neq\lambda_n^-$, $\mathbf{x}_n(t)$ is a corresponding eigenvector and $\mathbf{y}_n(t)=\exp(t\mathbf{K})\,\mathbf{x}_n(t)$, where $\mathbf{K}=T^{-1}\ln\mathbf{C}$, then the rotation of the vector $\mathbf{y}_n(t)$ from time $t=0$ to*

$t=T$ is $n\pi$. For all other solutions $\mathbf{x}(t)$, the rotation of the vector $\mathbf{y}(t)=\exp(t\mathbf{K})\mathbf{x}(t)$ is less than $n\pi$ if $\lambda=\lambda_n^-$ and greater than $n\pi$ if $\lambda=\lambda_n^+$.

3. *If $\lambda_n^-=\lambda_n^+$, then for all solutions $\mathbf{x}(t)$ the rotation of the vector $\mathbf{y}(t)=\exp(t\mathbf{K})\mathbf{x}(t)$ is $n\pi$. (In this case there are two linearly independent eigenvectors.)*

4. *If $\lambda_n^+<\lambda<\lambda_n^-$, the roots of the equation*

$$\det[\mathbf{X}(T,\ \lambda)-\rho\mathbf{C}]=0$$

lie on the unit circle $(\rho_{1,2}=\exp(\pm i\varphi))$, but if $\lambda_n^-<\lambda<\lambda_n^+$ they all lie on the real axis $(\rho_{1,2}=r^{\pm 1})$. Here φ, r are real, and $\mathbf{X}(t,\lambda)$ is the matrizant of equation (6.1).

The above method may also be utilized to establish various results concerning the relative positions of the eigenvalues $\lambda_n^{\pm}=\lambda_n^{\pm}(\mathbf{C})$ of boundary-value problems (6.65) with different matrices $\mathbf{C}$ (see /166d/).

APPENDIX 1

In this appendix we prove Theorems I, II (§ VIII. 4. 3) concerning the solution of the special variational problems formulated in the statements of these theorems. The proof uses no material not discussed in the text. Appendix 2 will present a simpler proof, due to Kazarinov, of these theorems, based on the Pontryagin maximum principle.

We first prove two auxiliary propositions.

Lemma I. *Let $y_1, \dot{y}_1, y_2, \dot{y}_2$ be absolutely continuous functions, $y_1(t_0) = y_2(t_0) \geqslant 0$, $\dot{y}_1(t_0) = \dot{y}_2(t_0)$, on an interval Δ containing the point t_0; suppose that*

$$\frac{\ddot{y}_1(t)}{y_1(t)} > \frac{\ddot{y}_2(t)}{y_2(t)} \text{ almost everywhere on } \Delta,\ y_1(t) > 0,\ y_2(t) > 0,$$

and if $y_1(t_0) = y_2(t_0) = 0$, then $\lim y_1(t)/y_2(t) = 1$ as $t \to t_0$. Then for all $t \in \Delta$, $t \neq t_0$,

$$y_1(t) > y_2(t).$$

Proof. Let $t \in \Delta$, $t > t_0$ (the proof for $t < t_0$ is similar). Set $\varphi(t) = \dot{y}_1 y_2 - \dot{y}_2 y_1$. Then

$$\dot{\varphi}(t) = y_1 y_2 \left(\frac{\ddot{y}_1}{y_1} - \frac{\ddot{y}_2}{y_2} \right) > 0, \quad \varphi(t_0) = 0.$$

Consequently,

$$\varphi(t) > 0, \quad \frac{\dot{y}_1}{y_1} - \frac{\dot{y}_2}{y_2} = \frac{\varphi}{y_1 y_2}.$$

Let $t_0 < t_1 < t$. Integrating the last inequality, we obtain

$$\frac{y_2(t_1)\, y_1(t)}{y_1(t_1)\, y_2(t)} = \exp \int_{t_1}^{t} \frac{\varphi}{y_1 y_2}\, dt > 1.$$

(This device is necessary in case $y_1(t_0) = y_2(t_0) = 0$, when we cannot simply set $t_1 = t_0$.) Letting $t_1 \to t_0$, we obtain $y_1(t) > y_2(t)$ for $t \in \Delta$, proving the lemma.

Lemma I will be used frequently in the sequel without special mention.

Definition 1. Functions $y_1(t)$ and $y_2(t)$ such that

$$y_1(t_0) = y_2(t_0), \quad \dot{y}_1(t_0) = \dot{y}_2(t_0)$$

are said to be joined smoothly* or simply joined at the point $t = t_0$.

* [Russian: "glued" or "pasted" together — Trans.]

We shall constantly be concerned with solutions of equations of the form

$$\ddot{y}+gy=0.$$

For brevity, we introduce the following

Definition 2. Any solution of the equation $\ddot{y}+gy=0$ will be called a *g*-curve and denoted by $y=f_g(t)$.

Lemma II. Let $y(t)>0$ be a function joined together from three g-curves:

$$y(t)=\begin{cases} f_{k_1}(t) & \text{if} \quad t_1\leqslant t\leqslant \xi, \\ f_g(t) & \text{if} \quad \xi\leqslant t\leqslant \eta, \\ f_{k_2}(t) & \text{if} \quad \eta\leqslant t\leqslant t_2 \end{cases} \tag{1}$$

(Figure 76). For any g_1 sufficiently close to g, we can construct a new joined function

$$y_1(t)=\begin{cases} f_{k_1}(t) & \text{if} \quad t_1\leqslant t\leqslant \xi_1, \\ f_{g_1}(t) & \text{if} \quad \xi_1\leqslant t\leqslant \eta_1, \\ f_{k_2}(t) & \text{if} \quad \eta_1\leqslant t\leqslant t_2. \end{cases}$$

If $g_1>g$, then $y_1(t)\geqslant y(t)$; if $g_1<g$, then $y_1(t)\leqslant y(t)$; the numbers ξ_1 and η_1 depend continuously on g_1 and $\xi_1\rightarrow\xi$, $\eta_1\rightarrow\eta$ as $g_1\rightarrow g$. The following three types of contact are possible:

external contact ($k_1>g$, $k_2>g$) — Figure 76a,

internal contact ($k_1<g$, $k_2<g$) — Figure 76b,

mixed contact ($k_1>g$, $k_2<g$ or $k_1<g$, $k_2>g$) — Figure 76c and d.

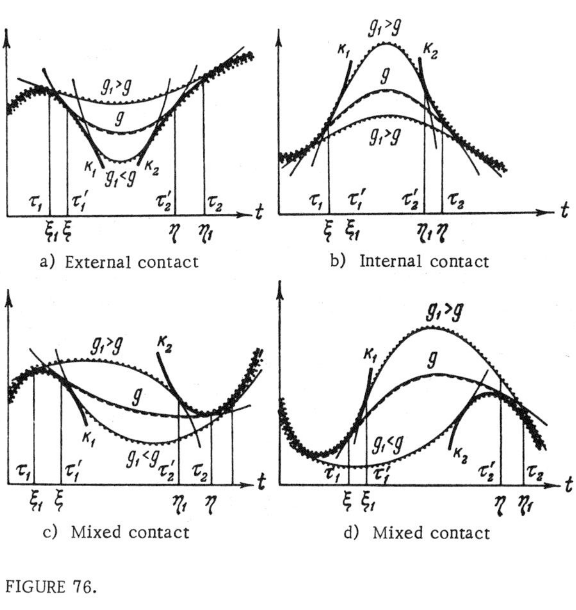

FIGURE 76.

Proof. The positions of the g-, k_1- and k_2-curves as shown in Figure 76a—d follow from Lemma I. For example, consider the case of internal contact, $g_1 > g$. The proof is the same for all other cases. Construct a curve through $t=\xi$ such that

$$y=\tilde{f}_{g_1}(t) > 0, \quad \tilde{f}_{g_1}(\xi)=f_g(\xi), \quad \dot{\tilde{f}}_{g_1}(\xi)=\dot{f}_g(\xi). \qquad (2)$$

By Lemma I, $\tilde{f}_{g_1}(t) < f_g(t)$, as long as $\tilde{f}_{g_1} > 0$. Take an arbitrary point $t_0 > 0$ such that $\tilde{f}_{g_1}(t-t_0) < f_g(t)$ for all t for which $\tilde{f}_{g_1}(t-t_0) \geqslant 0$. A fortiori,

$$\tilde{f}_{g_1}(t-t_0) < f_{k_1}(t) \text{ and } \tilde{f}_{g_1}(t-t_0) < f_{k_2}(t).$$

Consider the function $y=\lambda\tilde{f}_{g_1}(t-t_0)$. As λ increases from $\lambda=1$, this curve touches the curve $y=f_{k_1}(t)$ or the curve $y=f_{k_2}(t)$. If t_0 is sufficiently small, the first curve touched is $y=f_{k_1}(t)$. Let $\mathfrak{E}$ denote the set of all such points t_0. $\mathfrak{E}$ is clearly bounded. Let $t^*=\sup\mathfrak{E}$. As λ increases from $\lambda=1$, the curve $y=\lambda\tilde{f}_{g_1}(t-t^*)$ will touch both curves $y=f_{k_1}(t)$ and $y=f_{k_2}(t)$ at the same time at $\lambda=\lambda^*$, say. Denote

$$f_{g_1}(t)=\lambda^*\tilde{f}_{g_1}(t-t^*).$$

Let ξ_1 and η_1 be the values of t at which the curves touch. In the case of internal contact, we have $\xi_1 > \xi$ and $\eta_1 < \eta$. Indeed, otherwise it would follow that at the intersection points $t=t_1^*$ and $t=t_2^* > t_1^*$ of the curves $y=f_g(t)$ and $y=f_{g_1}(t)$

$$f_g(t_1^*)=f_{g_1}(t_1^*), \quad \dot{f}_{g_1}(t_1^*) < \dot{f}_g(t_1^*), \quad f_g(t_2^*)=f_{g_1}(t_2^*), \quad \dot{f}_{g_1}(t_2^*)=\dot{f}_g(t_2^*).$$

Now set $\varphi_g=\dot{f}_g f_{g_1}-\dot{f}_{g_1} f_g$; then

$$\varphi_g(t_1^*) > 0, \quad \varphi_g(t_2^*) < 0, \quad \dot{\varphi}_g=f_g f_{g_1}(g_1-g) > 0.$$

This contradiction shows that $\xi_1 > \xi$, $\eta_1 < \eta$. Therefore

$$y_1(t) > y(t).$$

Obvious arguments show that $\xi_1 \to \xi$, $\eta_1 \to \eta$ as $g_1 \to g$ and that ξ_1, η_1 depend continuously on g_1. This proves Lemma II.

Throughout this appendix we shall use the following notation. In the notation of Lemma II, let $h(t)=y_1(t)-y(t)$. We let (τ_1, τ_2) denote the interval on which $h(t)\neq 0$ and (τ_1', τ_2') the interval on which $h(t)=f_{g_1}(t)-f_g(t)$ $(\tau_1 < \tau_1' < \tau_2' < \tau_2)$. For external contact we have $\tau_1=\xi_1$, $\tau_1'=\xi$, $\tau_2'=\eta$, $\tau_2=\eta_1$, for internal contact $\tau_1=\xi$, $\tau_1'=\xi_1$, $\tau_2'=\eta_1$, $\tau_2=\eta$, and so on.

Remark to Lemma II. The function $|h(t)|/y(t)$ is increasing on (τ_1, τ_1') and decreasing on (τ_2', τ_2). On (τ_1', τ_2') we have

$$\frac{d}{dt}\left\{[f_g(t)]^2 \frac{d}{dt}\left[\frac{|h|}{f_g(t)}\right]\right\} \leqslant 0.$$

Proof. To fix ideas, let $g_1 > g$ and $h(t) \geqslant 0$. It follows from the proof of Lemma I that on (τ_1, τ_1')

$$\frac{h(t)}{y(t)} = \frac{y_1(t)}{y(t)} - 1 = \exp \int_{\tau_1}^{t} \frac{\varphi}{y_1 y_2} \, dt,$$

where $\varphi = \dot{y}_1 y - \dot{y} y_1$ for $t > \tau_1$. Thus $h(t)/y(t)$ is increasing on (τ_1, τ_1'). The proof that $h(t)/y(t)$ is decreasing on (τ_2', τ_2) is similar. On (τ_1', τ_2'),

$$\frac{h(t)}{y(t)} = \frac{f_{g_1}}{f_g} - 1,$$

$$\left(\frac{h}{f_g}\right)' = \frac{\dot{f}_{g_1} f_g - f_{g_1} \dot{f}_g}{f_g^2}, \quad \frac{d}{dt}\left[f_g^2 \frac{d}{dt}\left(\frac{h}{f_g}\right)\right] = -f_g f_{g_1}(g_1 - g) \leqslant 0.$$

This proves our assertion.

Let $y(t)$ be a function joined together from finitely many g-curves, which satisfies conditions (a'') of § VIII. 4.3. Suppose that we vary $y(t)$, using Lemma II to replace $f_g(t)$ on one of the intervals by, say, $f_{g_1}(t)$, at the same time slightly changing the interval. Let $y_1(t)$ be the resulting function, which is also assumed to satisfy conditions (a''). Denote $\Delta J = J(y_1) - J(y)$. We have

$$\Delta J = \int_0^{\tau} \left(\frac{\ddot{y}_1}{y_1} - \frac{\ddot{y}}{y}\right) dt = \int_{\tau_1}^{\tau_2} \frac{\ddot{h} y_1 - h \ddot{y}_1}{y y_1} \, dt = \int_{\tau_1}^{\tau_2} \frac{\ddot{h}}{y} \, dt - \int_{\tau_1}^{\tau_2} \frac{\ddot{y}_1}{y y_1} h \, dt,$$

$$\int_{\tau_1}^{\tau_2} \frac{\ddot{h}}{y} \, dt = \left[\frac{\dot{h}}{y} - \left(\frac{1}{y}\right)^{\cdot} h\right]_{\tau_1}^{\tau_2} + \int_{\tau_1}^{\tau_2} h \frac{d^2}{dt^2}\left(\frac{1}{y}\right) dt.$$

Since $h(\tau_1) = h(\tau_2) = \dot{h}(\tau_1) = \dot{h}(\tau_2) = 0$, it follows that for $\tau_1 \neq 0$, $\tau_2 \neq \tau$,

$$\int_{\tau_1}^{\tau_2} \frac{\ddot{h}}{y} \, dt = \int_{\tau_1}^{\tau_2} h \frac{d^2}{dt^2}\left(\frac{1}{y}\right) dt,$$

$$\Delta J = \int_{\tau_1}^{\tau_2} \left[\frac{d^2}{dt^2}\left(\frac{1}{y}\right) - \frac{\ddot{y} + \ddot{h}}{y(y+h)}\right] h \, dt. \tag{3}$$

If $\tau_1 = 0$ or $\tau_2 = \tau$ we have $y = 0$, and the above reasoning is invalid as it stands. Suppose, say, that $\tau_1 = 0$. We must prove that

$$\lim_{\varepsilon \to 0} \left[\frac{\dot{h}}{y} - h \frac{d}{dt}\left(\frac{1}{y}\right)\right]_{t=\varepsilon} = 0.$$

Since

$$\frac{\dot{h}}{y} - h \frac{d}{dt}\left(\frac{1}{y}\right) = \frac{y \dot{y}_1 - \dot{y} y_1}{y y_1} \frac{y_1}{y} \quad \text{and} \quad \lim_{\varepsilon \to 0} \frac{y_1(\varepsilon)}{y(\varepsilon)} = \frac{\dot{y}_1(0)}{\dot{y}(0)} = 1,$$

it will suffice to show that

$$\lim_{\varepsilon \to 0} \frac{y\dot{y}_1 - \dot{y}y_1}{yy_1} = 0.$$

We have

$$|y(\varepsilon)\dot{y}_1(\varepsilon) - \dot{y}(\varepsilon)y_1(\varepsilon)| = \left|\int_0^\varepsilon (y\ddot{y}_1 - \ddot{y}y_1)\,dt\right| =$$
$$= \left|\int_0^\varepsilon yy_1\left(\frac{\ddot{y}_1}{y_1} - \frac{\ddot{y}}{y}\right)dt\right| \leqslant \int_0^\varepsilon yy_1\left(\left|\frac{\ddot{y}_1}{y_1}\right| + \left|\frac{\ddot{y}}{y}\right|\right)dt \leqslant y(\varepsilon)y_1(\varepsilon)\left(\left|\frac{\ddot{y}_1}{y_1}\right| + \left|\frac{\ddot{y}}{y}\right|\right)\varepsilon,$$

since $\ddot{y}_1/y_1$ and $\ddot{y}/y$ are constants for sufficiently small ε, * and $y(t)$, $y_1(t)$ are increasing functions for $0 \leqslant t \leqslant \varepsilon$. Thus,

$$\lim_{\varepsilon \to 0} \frac{y\dot{y}_1 - \dot{y}y_1}{yy_1} = 0,$$

and so (3) is true in any case.

Let $y(t)$ and $y_1(t)$ be defined on the interval (τ_1, τ_2) by (1) and (2), respectively. We claim that the difference between ΔJ and the quantity

$$\delta_1 J = \int_{\tau_1}^{\tau_2} \left[\frac{d^2}{dt^2}\left(\frac{1}{f_g}\right) - \frac{\ddot{f}_g}{f_g^2}\right] h\,dt \tag{4}$$

is infinitesimal relative to $\delta_1 J$ if $\delta_1 J \neq 0$. Recalling that

$$f_g = Ae^{\sqrt{-g}t} + Be^{-\sqrt{-g}t}, \tag{5}$$

we bring (4) (the details are omitted) to the form

$$\delta_1 J = G\int_{\tau_1}^{\tau_2} \frac{h}{f_g^3}\,dt, \tag{6}$$

where

$$G = 8gAB. \tag{7}$$

If $g > 0$, the numbers A and B are complex conjugates. If $g < 0$, A and B are real. Since A and B depend on g, G need not vanish when $g = 0$. If $g > 0, G = 8g|A|^2 > 0$, and so by continuity $G \geqslant 0$ for $g = 0$. If $g < 0$, G may have any sign. If $G = 0$ the function $y = f_g(t)$ satisfies Euler's equation

$$\frac{d}{dt}\left(\frac{\dot{y}}{y}\right) = 0.$$

* $\ddot{y}_1/y_1$ and $\ddot{y}/y$ are step functions.

Let $G \neq 0$, and choose $\varepsilon > 0$ arbitrarily small. We claim that if g_1 is sufficiently close to g then

$$|\Delta J - \delta_1 J| < \varepsilon |\delta_1 J|. \tag{8}$$

We have

$$|\Delta J - \delta_1 J| \leqslant \left| \int_{\tau_1}^{\tau_1'} \right| + \left| \int_{\tau_1'}^{\tau_2'} \right| + \left| \int_{\tau_2'}^{\tau_2} \right|. \tag{9}$$

The integrand in this expression is

$$\left\{ \left[\frac{d^2}{dt^2}\left(\frac{1}{y}\right) - \frac{\ddot{y}+\ddot{h}}{y(y+h)} - \frac{d^2}{dt^2}\left(\frac{1}{f_g}\right) + \frac{\ddot{f}_g}{f_g^2} \right] y \right\} \frac{h}{y}. \tag{10}$$

Over the interval (τ_1', τ_2') we have $y = f_g$, $y + h = f_{g_1}$, and the expression in curly brackets is equal to

$$\left[\frac{g_1 f_{g_1}}{f_g f_{g_1}} - \frac{g f_g}{f_g^2} \right] f_g = g_1 - g.$$

Therefore,

$$\left| \int_{\tau_1'}^{\tau_2'} \right| = |g_1 - g| \int_{\tau_1'}^{\tau_2'} \left| \frac{h}{f_g} \right| dt.$$

Since

$$\int_{\tau_1'}^{\tau_2'} \left| \frac{h}{f_g} \right| dt \leqslant \max f_g^2 \cdot \int_{\tau_1'}^{\tau_2'} \left| \frac{h}{f_g^3} \right| dt < \frac{\max f_g^2}{|G|} |\delta_1 J|,$$

it follows that if g_1 is sufficiently close to g,

$$\left| \int_{\tau_1'}^{\tau_2'} \right| < \frac{\varepsilon}{2} |\delta_1 J|. \tag{11}$$

Now consider the intervals (τ_1, τ_1') and (τ_2', τ_2). The expression in curly brackets in (10) is bounded. This is obvious if $\tau_1 \neq 0$, $\tau_2 \neq \tau$. But if $\tau_1 = 0$ or $\tau_2 = \tau$ we have $y = f_g$, $y + h = f_{k_1}$ or f_{k_2} on the appropriate interval. In that case the expression in question becomes

$$\frac{\ddot{f}_g}{f_g} - \frac{\ddot{f}_{k_{1,2}}}{f_{k_{1,2}}} = -g + k_{1,2}.$$

Thus,

$$\left|\int_{\tau_1}^{\tau_1'}\right|+\left|\int_{\tau_2'}^{\tau_2}\right|\leqslant G\left(\int_{\tau_1}^{\tau_1'}\frac{|h|}{y}\,dt+\int_{\tau_2'}^{\tau_2}\frac{|h|}{y}\,dt\right)\ (\mathbf{C}=\text{const}>0).$$

By the remark to Lemma II,

$$\left|\int_{\tau_1}^{\tau_1'}\right|+\left|\int_{\tau_2'}^{\tau_2}\right|\leqslant \mathbf{C}\left[\left|\frac{h}{y}\right|_{\tau_1'}(\tau_1'-\tau_1)+\left|\frac{h}{y}\right|_{\tau_2'}(\tau_2-\tau_2')\right].$$

Denote

$$\sigma=\max\left[(\tau_1'-\tau_1),\ (\tau_2-\tau_2')\right].$$

Recall (see Lemma II) that $\sigma\to 0$ as $g_1\to g$. Thus

$$\left|\int_{\tau_1}^{\tau_1'}\right|+\left|\int_{\tau_2'}^{\tau_2}\right|\leqslant \mathbf{C}\sigma\left(\left|\frac{h}{y}\right|_{\tau_1'}+\left|\frac{h}{y}\right|_{\tau_2'}\right). \tag{12}$$

In order to estimate this expression, we consider the integral

$$\int_{\tau_1'}^{\tau_2'}\frac{|h|}{f_g^3}\,dt.$$

We transform to a new variable:

$$\frac{dt}{f_g^2}=ds,\quad s=\int_{\tau_1'}^{t}\frac{dt}{f_g^2}.$$

This gives the integrand

$$h_1(s)=\frac{|h(t)|}{f_g(t)}.$$

The function $h_1(s)$ is convex on the interval $(s_1,\ s_2)$ corresponding to $(\tau_1',\ \tau_2')$. Indeed,

$$\frac{d^2h_1(s)}{ds^2}=\frac{d}{dt}\left[\frac{dh_1}{dt}\frac{dt}{ds}\right]\frac{dt}{ds}=\frac{d}{dt}\left[f_g^2\frac{d}{dt}\left(\frac{|h|}{f_g}\right)\right]f_g^2\leqslant 0$$

by the remark to Lemma II. Therefore,

$$\int_{\tau_1'}^{\tau_2'}\frac{|h|}{f_g^3}\,dt=\int_{s_1}^{s_2}h_1(s)\,ds\geqslant\frac{1}{2}\left[h_1(s_1)+h_1(s_2)\right](s_2-s_1)=\frac{1}{2}\left(\left|\frac{h}{f_g}\right|_{\tau_1'}+\left|\frac{h}{f_g}\right|_{\tau_2'}\right)\int_{\tau_1'}^{\tau_2'}\frac{dt}{f_g^2}.$$

Since $y(\tau_1')=f_g(\tau_1')$, $y(\tau_2')=f_g(\tau_2')$, it follows that

$$\left|\frac{h}{y}\right|_{\tau_1'}+\left|\frac{h}{y}\right|_{\tau_2'}\leqslant 2\left(\int_{\tau_1'}^{\tau_2'}\frac{dt}{f_g^2}\right)^{-1}\int_{\tau_1'}^{\tau_2'}\frac{|h|}{f_g^3}\,dt.$$

Combining this estimate with (12) and (6), we see that

$$\left|\int_{\tau_1}^{\tau_1'}\right|+\left|\int_{\tau_1'}^{\tau_2}\right|<\frac{\varepsilon}{2}|\delta_1 J|.$$

Together with (11), this gives us the required estimate (8). Note that if $\tau_1=0$ or $\tau_2=\tau$, then

$$\int_{\tau_1'}^{\tau_2'}\frac{dt}{f_g^2}\to\infty \quad \text{as } g_1\to g,$$

but this causes no trouble.

We have proved, in particular, that if $G\neq 0$ the sign of ΔJ is the same as that of $\delta_1 J$ if g_1 is sufficiently close to g. Formula (6) and Lemma II imply

Lemma III. If $G\neq 0$ and the variation is subjected to the conditions stipulated above, then the sign of ΔJ is the same as that of G if $g'>g$ $(h>0)$ and opposite to that of G if $g'<g$ $(h<0)$.

We can now proceed directly to the proof of Theorems I and II of § VIII. 4. 3, attempting the determination of $\inf J(y)$ and $\sup J(y)$. We first assume that a^2 and b^2 are finite.

I) Determination of $\inf J(y)$. Denote $p(t)=-\ddot{y}/y$. First let $p(t)$ be a step function. Thus, there is a partition of the interval $(0,\tau)$ into $n\geqslant 3$ intervals

$$0=t_0\leqslant t_1\leqslant t_2\leqslant\ldots\leqslant t_{n-1}\leqslant t_n=\tau,$$

such that $p(t)=p_i=\text{const}$ on (t_i, t_{i+1}). Let $y(t)$ be a solution of the equation

$$\ddot{y}+p(t)y=0$$

with initial conditions $y(0)=0$, $\dot{y}(0)=1$ (y and $\dot{y}$ are continuous), satisfying conditions (a''), (b''), (c'') of § VIII. 4. 3.

The conditions $y(\tau)=0$, $\dot{y}(\tau)=-q$ are two equations for the numbers p_i, t_i. Condition (c'') implies that $a^2\geqslant p_i\geqslant -b^2$, the condition $y(t)>0$ $(0<t<\tau)$ is equivalent to $y(t)\geqslant 0$ $(0<t<\tau)$* and is an inequality imposing restrictions on t_i, p_i. The set of all numbers $t_1, t_2, \ldots, t_{n-1}$; $p_1, \ldots, p_n$, satisfying

* If $y(t)\geqslant 0$ for $0<t<\tau$, then $y(t)>0$ for $0<t<\tau$, since otherwise $y(t)=0$ at some point would imply $\dot{y}(t)=0$ at that point, so that $y(t)\equiv 0$, which is impossible.

all these conditions is closed and bounded, * the function

$$J(y) = -\sum_{i=0}^{n-1} p_{i+1}(t_{i+1}-t_i)$$

is continuous on this set (n is fixed), and so it attains a minimum there. Let $y=y_0(t)$ be the corresponding function; $y_0(t)$ is joined together from p_i-curves. If the numbers corresponding to $y_0(t)$ are t_i, p_i, then $\Delta J \geqslant 0$ for all numbers t_i', p_i' in the above set sufficiently close to t_i, p_i.

Consider an inner interval (t_i, t_{i+1}). Let $p_i = g$. By Lemma II, if we vary only the numbers g, t_i, t_{i+1}, then for any g' sufficiently close to g there are numbers t_i', t_{i+1}' such that the new function $y=y_1(t)$ will satisfy conditions (a''), (b''), (c'') and $\ddot{y}_1/y_1$ will be a step function with n distinct values.

Suppose that $g>0$ and $g>-b^2$. Then $G=g|A|^2>0$. Taking $g'<g$ sufficiently close to g, we can maintain the conditions

$$a^2 \geqslant p_i \geqslant -b^2,$$

and it follows from Lemma II that $h<0$. Then by (6) (or Lemma III), $\Delta J<0$. Thus it cannot be true that $g>0$. Hence, if b is pure imaginary $(-b^2>0)$ then $p(t)=-b^2$ on all inner intervals.

Now let $-b^2 \leqslant 0$ and $0 \geqslant g > -b^2$. If $AB \neq 0$, we may always ensure by increasing or decreasing g that $\Delta J<0$. Thus, if $0>g>-b^2$ then $AB=0$.

Let $g=-b^2<0$. In this case we can only increase g, and then $h>0$. If $AB>0$ it again follows that $\Delta J<0$, and so $AB=0$.

Thus, on the inner intervals either $p_i=-b^2$ and $AB<0$ (the function $y=y_0(t)$ is a boundary curve on these intervals) or $p_i \leqslant 0$ and $AB=0$ ($y=y_0(t)$ satisfies Euler's equation on these intervals).

We now show that both kinds of intervals cannot subsist together. Suppose the contrary. Then at the "join" of two intervals of different kinds we must have

$$Ae^{b\xi}+Be^{-b\xi}=Ce^{D\xi}, \quad Ae^{b\xi}-Be^{-b\xi}=\frac{1}{b}CDe^{D\xi}.$$

Squaring both equalities and subtracting, we obtain

$$4AB = C^2e^{2D\xi}\left(1-\frac{D^2}{b^2}\right). \tag{13}$$

Since $y=Ce^{Dt}$ must be an admissible function, we have $D^2 \leqslant b^2$, so that $AB \geqslant 0$ — contradiction. Thus, either $y=y_0(t)$ satisfies Euler's equation ($y_0=Ce^{Dt}$) on all inner intervals, or $p_i=-b^2<0$ and $AB<0$ on all the inner intervals. In other words, there is no point in speaking of different inner intervals — we have just three: two outer and one inner.

* The set of admissible numbers $t_1, \ldots, t_{n-1}$; $p_1, \ldots, p_n$ is the subset of the polyhedron $0 \leqslant t_1 \leqslant \ldots \leqslant t_{n-1} \leqslant \tau$, $a^2 \geqslant p_i \geqslant -b^2$ defined by the conditions $y(\tau)=0$, $\dot{y}(t)=-q$, $y(t) \geqslant 0$. Since the left-hand sides of these conditions are continuous functions of $t_1, \ldots, t_{n-1}$; $p_1, \ldots, p_n$, the set is closed (it is clearly bounded).

Consider the outer intervals. We claim that here $p_i = a^2$. Consider the interval at $t=0$. Suppose that there

$$p_i = g < a^2, \quad y = \frac{1}{\sqrt{g}} \sin \sqrt{g}\, t$$

(Figure 77). This is a special case of mixed contact ($\xi = 0$). If $g > 0$ then $G > 0$, $h < 0$, $\Delta J < 0$. Thus the case $g > 0$ is impossible. Let $g = -k^2 \leqslant 0$ and

$$y = A_1 e^{kt} + B_1 e^{-kt}.$$

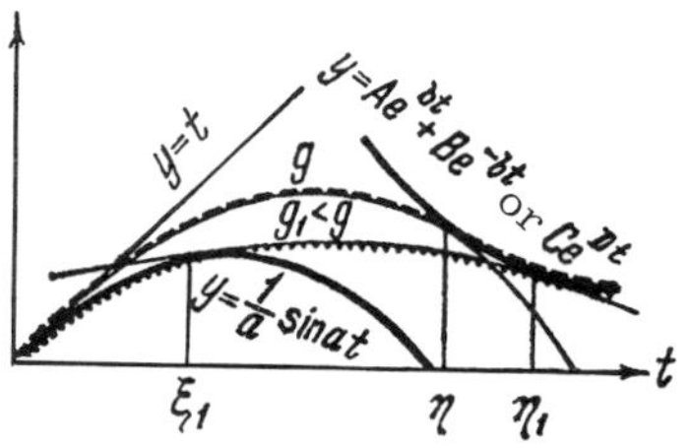

FIGURE 77.

We have

$$A_1 e^{k\eta} + B_1 e^{-k\eta} = Ae^{b\eta} + Be^{-b\eta}, \quad \frac{k}{b}(A_1 e^{k\eta} + B_1 e^{-k\eta}) = Ae^{b\eta} - Be^{-b\eta}$$

(the case that $y_0 = Ce^{Dt}$ on the inner interval is obtained formally by setting $A = C$, $B = 0$, $b = D$). Squaring both equalities and subtracting, we get

$$4A_1B_1 + \left(1 - \frac{k^2}{b^2}\right)(A_1 e^{k\eta} - B_1 e^{-k\eta})^2 = 4AB.$$

If $k < b$,

$$A_1B_1 < AB < 0, \quad G = -8k^2 A_1 B_1 > 0, \quad h < 0, \quad \Delta J < 0,$$

and so this case is impossible. If $k = b$ (and then $A_1B_1 = AB$) the curves

$$y = \frac{1}{\sqrt{g}} \sin \sqrt{g}\, t \text{ and } y = Ae^{bt} + Be^{-bt} \quad (\text{or } y = Ce^{Dt})$$

coincide. This case may be treated as that in which an outer interval degenerates to a point.

We have thus shown that $y = y_0(t)$ is joined together from a^2-curves on the outer intervals and a $(-b^2)$-curve or a curve $y = Ce^{Dt}$ on the inner interval. The latter alternative may hold only if $b^2 \geqslant 0$. We shall now see how to distinguish these cases. Construct two sine curves (a^2-curves) on the outer intervals. It will be shown below that these two curves may always be joined smoothly by a curve $y=Ce^{Dt}$. We claim that if the latter is admissible (i. e., $D^2\leqslant b^2$), then the resulting function is $y_0(t)$. Indeed, suppose that on the inner interval

$$y_0=Ae^{bt}+Be^{-bt}\equiv f_{-b^2}(t)$$
$$(AB<0)$$

(Figure 78). As in Lemma II, we "transfer" it to the point $t=\xi$. This gives

$$y=\lambda f_{-b^2}(t-t')=(\lambda Ae^{-bt'})e^{-bt}+(\lambda Be^{bt'})e^{-bt}.$$

This curve is joined smoothly to $y=Ce^{Dt}$ at the point $t=\xi$. By (13), we obtain

$$(\lambda Ae^{-bt'})(\lambda Be^{bt'})=\lambda^2 AB=C^2e^{2D\xi}\left(1-\frac{D^2}{b^2}\right)\geqslant 0$$

contradicting the condition $AB<0$.

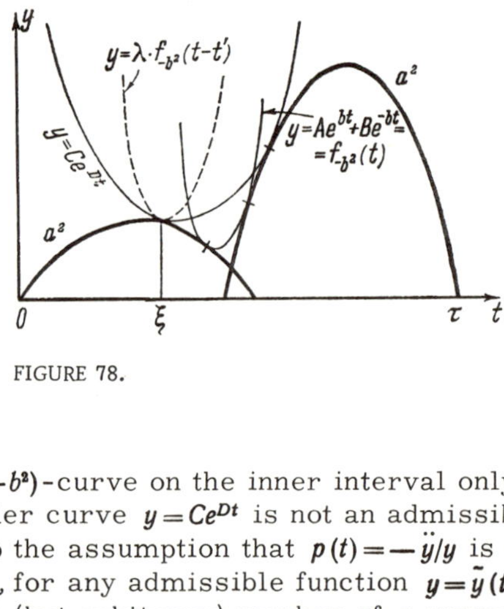

FIGURE 78.

Thus $y_0(t)$ is a $(-b^2)$-curve on the inner interval only if the function $y_0(t)$ consisting of an Euler curve $y=Ce^{Dt}$ is not an admissible function.

We can now drop the assumption that $p(t)=-\ddot{y}/y$ is a step function. We have shown that, for any admissible function $y=\tilde{y}(t)\not\equiv y_0(t)$ joined together from a finite (but arbitrary) number of g-curves, $J(y)>J(y_0)$. Now any summable function $p(t)$ may be approximated arbitrarily well in the sense of the norm

$$\|p\|=\int_0^\tau |p|\,dt$$

by a step function $\tilde{p}$, and the functional $J(y)=-\int_0^\tau p\,dt$ depends continuously on p. Therefore, if there is a function y such that $J(y)<J(y_0)$, then there is

also a step function $\tilde{p}=-\ddot{\tilde{y}}/\tilde{y}$ such that $J(\tilde{y})<J(y_0)$, which is impossible. Thus $J(y)\geqslant J(y_0)$ for any admissible function, and we have proved Theorem I of § VIII. 4. 3.

The proof is considerably simpler if we are required to find the infimum over the functions for which

$$-\frac{\ddot{y}}{y}\geqslant 0.$$

For such functions ($b^2\leqslant 0$) it is always the second case that holds.

II) Determination of $\sup J(y)$. * The reasoning is analogous. Let $y_0(t)$ be a function for which the supremum in question is attained in the class of functions $y(t)$ such that $p(t)=-\ddot{y}/y$ is a step function with $n\geqslant 3$ "steps."

We first examine the case $b^2\leqslant 0$. The same method as before shows that $y_0(t)$ has the property $p_i=a^2$ on all inner intervals. Indeed, if $0<p_i<a^2$ on some inner interval, we could increase p_i so that, by Lemma II, $h(t)\geqslant 0$, and by Lemma III $\Delta J>0$, which is impossible. (In this case $G=8p_i|A|^2$.) The same is true if $p_i=0$, $G>0$. If $p_i=0$, $G=0$, the function $y_0=\text{const}$ is a solution of Euler's equation. Since on the inner interval

$$\int_{\tau_1}^{\tau_2}\frac{\ddot{y}}{y}\,dt=\int_{\tau_1}^{\tau_2}\frac{\dot{y}^2}{y^2}\,dt,$$

it follows that for any variation,

$$\Delta J(y_0)=J(y_0+h)-J(y_0)=J(y_0+h)>0.$$

Thus $p_i=a^2$ on all inner intervals, and we can merge them, considering only three intervals: one inner and two outer.

We claim that on the outer intervals $p_i=-b^2$. Suppose that $p_i>-b^2$ on the right-hand outer interval, say. Construct a $(-b^2)$-curve at the point $t=\tau$, $y=y_0$ such that $y(\tau)=0$, $\dot{y}(\tau)=-q$ (Figure 79); the p_i-curve will join the a^2- and $(-b^2)$-curves at the points $t=\xi, t=\eta=\tau$. This is a particular case of mixed contact. By Lemma II, there exists a p_i'-curve, $p_i'>p_i$, joining the a^2- and $(-b^2)$-curves at points $t=\xi_1<\xi$ and $t=\eta_1<\eta$. We then have $h(t)>0$, $G=8p_i|A|^2$, and by Lemma III, $\Delta J>0$, contradicting the assumption $J(y_0)=\max$ and showing that $p_i=-b^2$ on the outer intervals.

The case $b^2>0$ is much more difficult. Here it may happen that $p_i<0$ on an inner interval. If at the same time $AB<0$, then $G=8p_i AB>0$, and by increasing p_i we again deduce from Lemma III that $\Delta J>0$. Thus the case $p_i<0$, $AB<0$ is impossible. If $p_i<0$, $AB=0$, then $y_0(t)$ satisfies Euler's equation on the relevant interval. Then $\Delta J>0$ for any variations. Indeed, in our case,

$$J=\int_{\tau_1}^{\tau_2}F(y,\dot{y})\,dt=\int_{\tau_1}^{\tau_2}\frac{\dot{y}^2}{y^2}\,dt.$$

* Recall that $J(y)=\int_0^{\tau}\frac{\ddot{y}}{y}\,dt$. As before, the function $y(t)$ satisfies conditions (a"), (b"), (c") of § VIII.4.3.

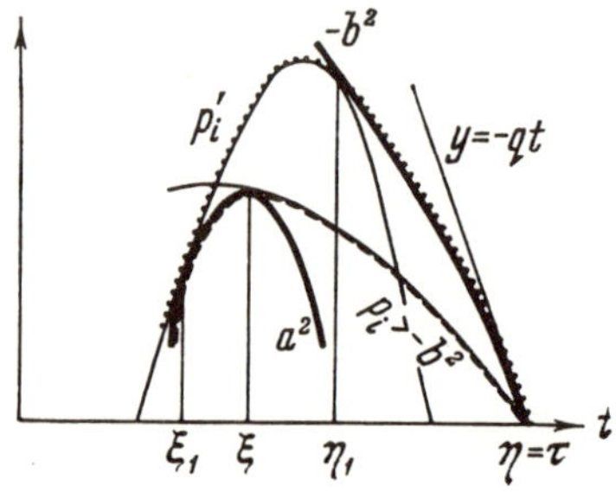

FIGURE 79.

The extremal $y_0 = C_0 e^{D_0 t}$ may obviously be included in the field of extremals $y = Ce^{D_0 t}$. Legendre's condition

$$\frac{d^2}{dt^2} F_{\dot{y}\dot{y}} = \frac{2}{y^2} > 0$$

is satisfied, and so there is a strong minimum on the extremal. The same conclusion may be drawn by consideration of the second variation. Carrying out the usual transformations, we see that

$$\delta^2 J = \frac{2}{c^2} \int_{\tau_1}^{\tau_2} \left[\frac{d}{dt} (e^{-D_t} h)\right]^2 dt > 0$$

for any variation $h(t)$ such that $h(\tau_i) = \dot{h}(\tau_i) = 0$. Thus the case $p_i < 0$, $AB = 0$ is also impossible. Now let $p_i < 0$, $AB > 0$. Then $G = 8p_i AB < 0$. If $p_i > -b^2$, we ensure by decreasing p_i that $h(t) < 0$ and by Lemma III, $\Delta J > 0$. But if $p_i = -b^2$, we cannot decrease p_i, and so $\Delta J < 0$ for any admissible variations.

We have thus shown that on the inner intervals either $p_i = a^2$, or $p_i = -b^2$ and $AB > 0$. On the outer intervals $p_i = -b^2$, as before. Any such function yields a relative maximum of our functional. To pick out the function that gives an absolute maximum, we must compare the various relative maxima. It can be shown by rather tedious computations that the absolute maximum is achieved for a function with one inner interval, as asserted in Theorem II. Since the only case used in § VIII. 4 with $b^2 > 0$ assumes that $a^2 = \infty$, we give the details for this situation only.

At this stage, we have a function $y_0(t)$ such that: on the outer intervals, $\ddot{y}_0/y_0 = b^2$; on the inner intervals, $\ddot{y}_0/y_0 = b^2$ and $AB > 0$, or else $\ddot{y}_0/y_0 = -a^2$ (Figure 80).

Let $\varphi_1, \varphi_2, \ldots, \varphi_k$ be the $(-b^2)$-curves entering into the curve $y = y_0(t)$.

Leaving the curves $y = \varphi_i(t)$ unchanged, we now join them by new a^2-curves, with larger constants a^2, retaining the notation $y_0(t)$ for the resulting function. An increase in a^2 implies an increase in $J(y_0)$ (since $G = 8a^2 |A|^2 > 0$, $h > 0$, $\Delta J > 0$). If we let $a^2 \to \infty$, the function $y_0(t)$ will tend to the function $y = \varphi(t)$ joined together (but not smoothly!) from the curves $\varphi_1, \varphi_2, \ldots, \varphi_k$.

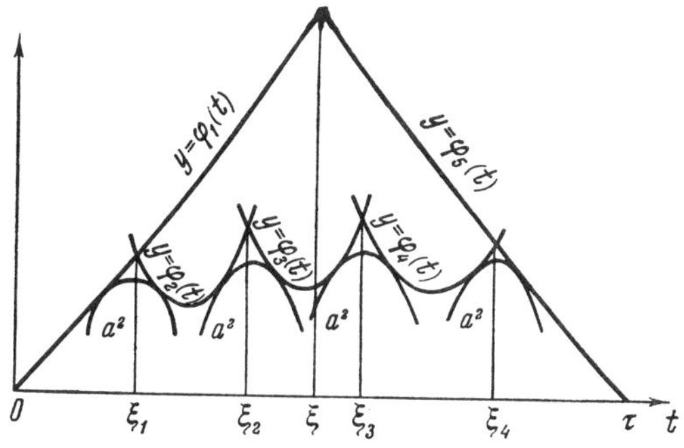

FIGURE 80.

Let $\xi_1, \xi_2, \ldots, \xi_{k-1}$ denote the points at which these curves intersect (Figure 80). The function $\varphi(t)$ is not admissible, as $\varphi(t)$ is discontinuous at the points $t=\xi_i$. Let us study the behavior of $J(y_0)$ as $a^2\to\infty$. Let

$$\sigma_i=(\xi_i^-,\ \xi_i^+)$$

denote the interval containing ξ_i on which $\ddot{y}_0/y_0=-a^2$, and $\sigma=(0,\tau)$. We have

$$\int_0^\tau \frac{\ddot{y}_0}{y_0}dt=\int_{\sigma-\Sigma\sigma_i}+\int_{\Sigma\sigma_i};\quad \int_{\sigma-\Sigma\sigma_i}=b^2\left[\tau-\sum_i(\xi_i^+-\xi_i^-)\right],$$

$$\int_{\sigma_i}\frac{\ddot{y}_0}{y_0}\,dt=\frac{1}{y_0(\xi_i')}\int_{\sigma_i}\ddot{y}_0\,dt=\frac{\dot{y}_0(\xi_i^+)-\dot{y}_0(\xi_i^-)}{y_0(\xi_i')},$$

where $\xi_i^-\leqslant\xi_i'\leqslant\xi_i^+$. As $a^2\to\infty$ we have $\xi_i^\pm\to\xi_i$ and

$$\frac{\dot{y}_0(\xi_i^+)-\dot{y}_0(\xi_i^-)}{y_0(\xi_i')}\to\frac{\dot{\varphi}_{i+1}(\xi_i)}{\varphi_{i+1}(\xi_i)}-\frac{\dot{\varphi}_i(\xi_i)}{\varphi_i(\xi_i)}.$$

Thus, as $a^2\to\infty$

$$\int_0^\tau\frac{\ddot{y}_0}{y_0}\,dt\to b^2\tau+S,$$

where

$$S=[\Phi_2(\xi_1)-\Phi_1(\xi_1)]+[\Phi_3(\xi_2)-\Phi_2(\xi_2)]+\ldots+[\Phi_k(\xi_{k-1})-\Phi_{k-1}(\xi_{k-1})],$$

$$\Phi_i(t)=\frac{\dot{\varphi}_i(t)}{\varphi_i(t)}.$$

Let ξ denote the point of intersection of the two outer curves

$$y=\varphi_1(t)\text{ and }y=\varphi_k(t).$$

We claim that

$$S \leqslant \Phi_k(\xi) - \Phi_1(\xi). \tag{14}$$

This is equivalent to

$$[\Phi_1(\xi_1) - \Phi_1(\xi)] + [\Phi_2(\xi_2) - \Phi_2(\xi_1)] + \cdots + \\ + [\Phi_{k-1}(\xi_{k-1}) - \Phi_{k-1}(\xi_{k-2})] + [\Phi_k(\xi) - \Phi_k(\xi_{k-1})] \geqslant 0.$$

It will suffice to prove that all the bracketed terms are nonnegative; in other words, we must show that $\Phi_i(t)$ is an increasing function on all the inner intervals $(i = 2, \ldots, k-1)$, but a decreasing function on the outer intervals $(i = 1, k)$. Recalling that

$$\varphi_i(t) = Ae^{bt} + Be^{-bt},$$

we see that

$$\dot{\Phi}_i(t) = \frac{4b^2 AB}{(Ae^{bt} + Be^{-bt})^2}.$$

Since $AB > 0$ on the inner intervals, it follows that $\Phi_i(t)$ is indeed an increasing function there. On the left outer interval we have $\varphi_1(0) = 0$, so that $A = -B$ and $AB < 0$. On the right outer interval, $\varphi_k(\tau) = 0$, so that

$$Ae^{b\tau} = -Be^{-b\tau} \text{ and } AB < 0.$$

Thus $\Phi_i(t)$ is a decreasing function on the outer intervals.

We have proved thus that as $a^2 \to \infty$ the integral

$$\int_0^\tau \frac{\ddot{y}_0}{y_0}\, dt$$

tends monotonically (increasing) to a number bounded above by

$$b_2\tau + \frac{\dot{\varphi}_k(\xi)}{\varphi_k(\xi)} - \frac{\dot{\varphi}_1(\xi)}{\varphi_1(\xi)}. \tag{15}$$

If there is only one inner interval, the limit of

$$\int_0^\tau \frac{\ddot{y}_0}{y_0}\, dt$$

is precisely (15).

Note that (15) is independent of n, the number of sine curves joined together to form the original function $y_0(t)$ (φ_1 and φ_{k+1} are the left- and rightmost sine curves in $y = y_0(t)$). Consequently, the quantity (15) is

$$\sup \int_0^\tau \frac{\ddot{\tilde{y}}}{\tilde{y}}\, dt$$

over all functions $\tilde{y}$ satisfying conditions (a''), (b''), (c'') of § VIII. 4.3 for $a^2 = \infty$ such that $\ddot{\tilde{y}}/\tilde{y}$ is a step function with an arbitrary but finite number of steps. It is clear that (15) is then also

$$\sup \int_0^\tau \frac{\ddot{y}}{y}\, dt$$

over all admissible functions, i. e., those satisfying conditions (a''), (b''), (c''), with $a^2 = \infty$ in (c''). * It is clear that the supremum is not achieved. We denote

$$y_1(t) = \varphi_1(t), \quad y_2 = \varphi_k(t),$$

and then

$$y_1(t) = \frac{1}{b}\operatorname{sh} bt, \quad y_2(t) = \frac{q}{b}\operatorname{sh} b(t-\tau). \tag{16}$$

We can now write (15) as

$$b^2\tau + \frac{\dot{y}_2(\xi)}{y_2(\xi)} - \frac{\dot{y}_1(\xi)}{y_1(\xi)}. \tag{17}$$

If $b^2 \leqslant 0$, the limiting behavior as $a^2 \to \infty$ is treated in exactly the same way. In this case it is natural to write the function (16) as

$$y_1(t) = \frac{1}{b_1}\sin b_1 t, \quad y_2(t) = \frac{q}{b_1}\sin b_1(\tau - t), \tag{18}$$

where $b_1^2 = -b^2 \geqslant 0$. As before, the number (17) is

$$\sup \int_0^\tau \frac{\ddot{y}}{y}\, dt$$

over the appropriate class of functions. This completes the proof of Theorem II of § VIII. 4. 3.

* Cf. the analogous argument at the end of the proof of Theorem I of § VIII.4.3.

APPENDIX 2

Here we present Kazarinov's proof of Theorems I and II of § VIII. 4.3, based on the Pontryagin maximum principle. * This proof is simpler than that given in Appendix 1, but makes additional demands on the reader, who should be familiar with Chapters 1 and 2 of the book /76c/ and some elementary material from functional analysis.

2.1. Statement of the problem. Existence of optimizing functions

The notation used below differs somewhat from that of § VIII. 4. We therefore reformulate the problem.

Consider equation (VIII. 3. 1),

$$\ddot{y}+p(t)\,y=0, \tag{1}$$

where $p(t)\in L(0,\ \pi)$.

Let $a,\ b,\ q$ be real numbers such that $q>0,\ \ a^2>1>b$. Throughout the sequel we shall assume without special mention that these inequalities hold.

Let $\mathcal{P}^{(a,\ b,\ q)}$ denote the class of functions $p(t)$ satisfying the following conditions:

(A) Equation (1) has a solution $y(t)$ such that

$$y(0)=y(\pi)=0,\quad \dot{y}(0)=1,\quad \dot{y}(\pi)=-q$$

and $y(t)>0$ for $0<t<\pi$.

(B) Almost everywhere on $[0,\ \pi]$,

$$b\leqslant p(t)\leqslant a^2. \tag{2}$$

We define a functional

$$J(p)=\int_0^\pi p(t)\,dt \tag{3}$$

and consider the following problems:

Problem I. To find $\inf J(p)$ *over all functions* $p\in\mathcal{P}^{(a,\ b,\ q)}$.

Problem II. To find $\sup J(p)$ *over all functions* $p\in\mathcal{P}^{(a,\ b,\ q)}$.

* Kazarinov, Yu. F. On an extremum problem. — Vestnik Leningrad. Univ. 4, No. 19 (1971). (Russian)

Conditions (a''), (b''), (c'') of § VIII. 4.3 are equivalent to conditions (A), (B).

The solution of Problems I, II is provided by Theorems I, II of § VIII.4.3, which we now proceed to prove. The solution will make use of the Pontryagin maximum principle, and we shall therefore call the function $p(t)$ in (1) a control.

If the class $\mathcal{P}^{(a,b,q)}$ is not empty, standard arguments show that it contains controls for which the infimum and supremum of $J(p)$ are achieved. We shall prove this, say for sup $J(p)$.

Let $p_k \in \mathcal{P}^{(a,b,q)}$, $J(p_k) \to \sup J(p)$ as $k \to \infty$. Since a bounded set of controls is weakly compact in the space $L_2[0, \pi]$, we may assume that the sequence p_k converges weakly to some function $p^* \in L_2[0, \pi]$. This means that $\int_0^\pi p_k f\,dt \to \int_0^\pi p^* f\,dt$ as $k \to \infty$, for any function $f \in L_2[0, \pi]$. It follows from (2) that $b \leqslant p^*(t) \leqslant a^2$ almost everywhere. Indeed, suppose that $p^*(t) > a^2$ on a set E of positive measure. Integrating the inequality $p_k(t) \leqslant a^2$ and letting $k \to \infty$ (which is legitimate since the sequence is weakly convergent), we obtain

$$\int_E p^*\,dt \leqslant a^2 \int_E dt.$$

This contradicts the assumption that $p^*(t) > a^2$ on E. Similarly one shows that $p^*(t) \geqslant b$ almost everywhere. Let $y^*(t)$ be a solution of equation (1) for $p = p^*$, $y^*(0) = 0$, $\dot{y}^*(0) = 1$. By the theorem of § II. 1.7,

$$y_k(t) \to y^*(t), \quad \dot{y}_k(t) \to \dot{y}^*(t) \quad \text{as} \quad k \to \infty, \tag{4}$$

uniformly on $[0, \pi]$. Setting $t = \pi$ in (4), we obtain $y^*(\pi) = 0$, $\dot{y}^*(\pi) = -q$.

Moreover, it follows from (4) that $y^*(t) \geqslant 0$ on $[0, \pi]$ and (1) is true almost everywhere on $[0, \pi]$. For any solution $y(t) \not\equiv 0$ of equation (1) and any $0 < t_0 < \pi$, it cannot be true that both $y(t_0) = 0$, $\dot{y}(t_0) = 0$. Therefore $y^*(t) > 0$ for $0 < t < \pi$. We have thus shown that $y^*(t)$ satisfies conditions (A), so that $p^*(t) \in \mathcal{P}^{(a,b,q)}$.

Since the above sequence is weakly convergent, it follows that $J(p_k) \to J(p^*)$, so that

$$J(p^*) = \sup_{\mathcal{P}^{(a,b,q)}} J(p),$$

as claimed.

2.2. Necessary conditions for optimality

We rewrite equation (1) as a system

$$\dot{y}_1 = y_2, \quad \dot{y}_2 = -p(t)\,y_1, \tag{5}$$

where $y = y_1$, $\dot{y} = y_2$. Replace (A) by the equivalent conditions

$$y_1(0) = y_1(\pi) = 0, \quad -qy_2(0) + y_2(\pi) = 0. \tag{6}$$

(By "equivalent" here we mean that system (5) with boundary conditions (6), on the one hand, and equation (1) with conditions (A), on the other, define the same set of controls $p(t)$.)

The problems will be solved with the help of the maximum principle /76c/. Following the general rule /76c/, we define the Hamiltonian*

$$H(p,\ y_1,\ y_2,\ \psi_1,\ \psi_2)=\psi_0 p+\psi_1 y_2-\psi_2 p y_1, \tag{7}$$

where $\psi_0=\text{const}$, and $\psi_1,\ \psi_2$ are a solution of the system $\dot{\psi}_1=-\frac{\partial H}{\partial y_1},\ \dot{\psi}_2=-\frac{\partial H}{\partial y_2}$, or

$$\dot{\psi}_1=p\psi_2,\quad \dot{\psi}_2=-\psi_1. \tag{8}$$

Setting $\psi=\psi_2$, we deduce from (8) that $\psi_1=-\dot{\psi}$ and

$$\ddot{\psi}+p(t)\psi=0. \tag{9}$$

The transversality conditions corresponding to the boundary conditions (6) are

$$\dot{\psi}(0)+q\dot{\psi}(\pi)=0. \tag{10}$$

Theorems 3 and 6 of /76c/ imply the following proposition. A necessary condition for a control $p(t)$ to be optimal is that there exist a nonzero function $\psi(t)$, satisfying (9) and (10), such that the Hamiltonian (7) is constant on the control:

$$\max_{b\leqslant u\leqslant a^2} H(u,\ y_1,\ y_2,\ \psi_1,\ \psi_2)=H[p(t),\ y_1(t),\ y_2(t),\ \psi_1(t),\ \psi_2(t)], \tag{11}$$

and moreover $\psi_0\leqslant 0$ for Problem I, $\psi_0\geqslant 0$ for Problem II.

Condition (11) becomes

$$\max_{b\leqslant u\leqslant a^2} u[\psi_0-\psi(t)y(t)]=p(t)[\psi_0-\psi(t)y(t)]. \tag{12}$$

Let $\xi(t)=(y(t),\ \psi(t))$ denote a trajectory in the $(y,\ \psi)$-plane, where $y(t)$ and $\psi(t)$ satisfy equations (1) and (9) and also satisfy the conditions of the maximum principle as formulated above. We shall call such trajectories ξ-*extremals*. The following properties of ξ-extremals will be needed:

$$\psi\dot{y}-\dot{\psi}y=R, \tag{13}$$

$$p(\psi_0-\psi y)-\dot{\psi}\dot{y}=S, \tag{14}$$

where R and S are constants for each ξ-extremal.

Condition (13) follows from the fact that $\psi(t)$ and $p(t)$ are solutions of equation (1). Condition (14) means that the Hamiltonian is constant on an optimal control. We observe moreover that for such ξ-extremals the function $y(t)$ satisfies conditions (A).

* [Not to be confused with the "Hamiltonians" of previous chapters! — Trans.]

Denote

$$z^{(1)}=\psi y, \quad z^{(2)}=\dot{z}^{(1)}=\dot{\psi} y+\psi\dot{y}. \tag{15}$$

Then $\dot{z}^{(2)}=-2pz^{(1)}+2\dot{\psi}\dot{y}$. Using property (14), we get the system

$$\dot{z}^{(1)}=z^{(2)}, \quad \dot{z}^{(2)}=-4pz^{(1)}+2(p\psi_0-S). \tag{16}$$

In the new notation, condition (12) becomes

$$\max_{b\leqslant u\leqslant a^2} u\,[\psi_0-z^{(1)}(t)]=p(t)\,[\psi_0-z^{(1)}(t)].$$

Hence it follows that

$$\left.\begin{aligned} p(t)&=a^2, &&\text{if} \quad z^{(1)}(t)\leqslant\psi_0,\\ p(t)&=b, &&\text{if} \quad z^{(1)}(t)>\psi_0. \end{aligned}\right\} \tag{17}$$

In addition, since $p(t)\in\mathcal{P}^{(a,b,q)}$, it follows that $b\leqslant p(t)\leqslant a^2$ if $z^{(1)}(t)=\psi_0$.

A trajectory $z(t)=(z^{(1)}(t), z^{(2)}(t))$ on the $(z^{(1)}, z^{(2)})$-plane, defined for $0\leqslant t\leqslant\pi$ and satisfying (16) and (17), will be called a *z-extremal*. In the sequel we shall consider only z-extremals defined in terms of some ξ-extremal as in (15).

It is easy to see that for a constant control $p=\tilde{p}\neq 0$ system (16) has the integral

$$\tilde{p}(z^{(2)})^2+4(\tilde{p}z^{(1)}-\tilde{p}\psi_0+S)^2=\text{const}. \tag{18}$$

If $\tilde{p}=0$, the integral of system (16) is

$$(z^{(2)})^2+4Sz^{(1)}=\text{const}. \tag{19}$$

We shall show that the z-extremals have the following properties:

(I) Any trajectory $z(t)$ is closed, since

$$z^{(1)}(0)=z^{(1)}(\pi)=0, \quad z^{(2)}(0)=z^{(2)}(\pi)=R.$$

(II) $z^{(1)}(t)$ cannot vanish identically on any interval $\Delta\subset[0,\pi]$ of positive length.

(III) If $\psi_0\neq 0$ and at some time the z-extremal passes through the point $(\psi_0, 0)$, then $\psi_0^{-1}S\geqslant 0$.

(IV) If $\psi_0=0$, $S=0$, then $z^{(1)}(t)$ and $z^{(2)}(t)$ cannot vanish at the same time $(0\leqslant t\leqslant\pi)$.

Property (I) follows from (15), (13) and condition (A). Property (II) is obvious, as $y(t)\not\equiv 0$, $\psi(t)\not\equiv 0$ are solutions of equation (1). Property (IV) follows directly from (16) and property (II). To prove (III), we deduce from (14), (15) that if

$$z^{(1)}=\psi y=\psi_0\neq 0, \quad z^{(2)}=\psi\dot{y}+\dot{\psi}y=0,$$

then $\dfrac{S}{\psi_0}=-\dfrac{\dot{\psi}\dot{y}}{\psi_0}=\left(\dfrac{\dot{y}}{y}\right)^2\geqslant 0$, as required.

Lemma. (a) *Let* $\Delta \subset [0, \pi]$ *be a "stationary" interval of positive length, i.e., an interval on which* $z^{(1)}(t) \equiv \text{const}$. *Then as* t *varies from* $t=0$ *to* $t=\pi$ *the point* $z(t)$ *describes an ellipse defined by (18) with* $\tilde{p}=a^2$, $\psi_0>0$, *starting from the point* $z(0)=(0, R)$ *and returning to it at* $t=\pi$, *with a "delay" in the stationary interval (i.e., for* $t\in\Delta$*) at the point* $(\psi_0, 0)$ *(see Figure 81).*

(b) *Suppose that* $z(t)$ *has no stationary interval in* $[0, \pi]$. *Then either* $z^{(1)}(t) \leqslant \psi_0$ *and* $p(t) \equiv a^2$ *(then the z-extremal is the ellipse (18) with* $\tilde{p}=a^2$, $\psi_0 \geqslant 0$*) or* $z^{(1)}(t) \geqslant \psi_0 \geqslant 0$ *and* $p(t) \equiv b > 0$ *(then the z-extremal is the ellipse (18) with* $p \equiv b$*), or the z-extremal intersects the straight line* $z^{(1)}=\psi_0$ *at* $z^{(2)} \neq 0$, *with no delay there. In the last case, the z-extremal is the union of the ellipse (18) with* $\tilde{p}=a^2$ *in the domain* $z^{(1)}<\psi_0$ *and the curve (18) or (19) for* $\tilde{p}=b$ *in the domain* $z^{(1)}>\psi_0$ *(see Figure 82).*

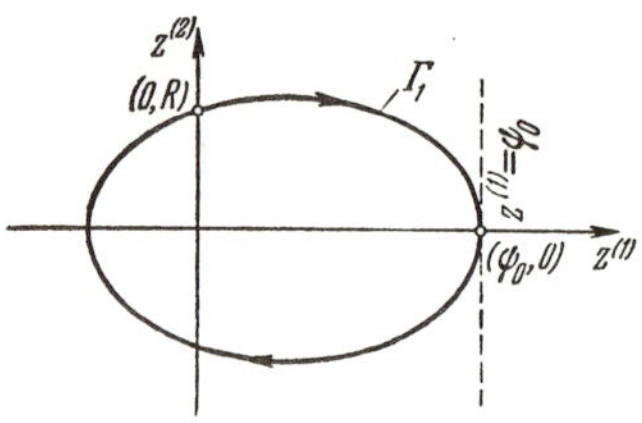

FIGURE 81.

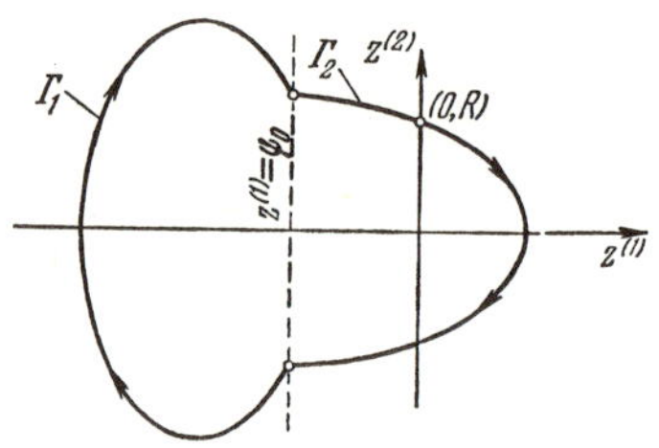

FIGURE 82.

Proof. There may be intervals on the t-axis, $t\in[0,\pi]$, on which $z^{(1)}(t)<\psi_0$ and intervals on which $z^{(1)}(t)>\psi_0$. Because of (17), in intervals of the first type $z(t)$ describes an ellipse Γ_1 defined by (18), $\tilde{p}=a^2$; in intervals of the second type, it moves along a second-order curve (18) or (19), $\tilde{p}=b$. By property (I), the curve $z(t)$ $(0\leqslant t\leqslant\pi)$ begins and ends at the point $z_0=(0, R)$. Any curve (18) or (19) intersects the straight line $z^{(1)}=\psi_0$ in at most two points, symmetric about the $z^{(1)}$-axis. Thus the trajectory Γ of the point $z(t)$ is made up of two second-order curves Γ_1 and Γ_2, one of which may degenerate to a point. The moving point $z(t)$ may be "delayed" at the "joins" of these curves.

Since $z^{(1)}(t)=\psi(t)y(t)$, the functions $\psi(t)\not\equiv 0$, $y(t)\not\equiv 0$ are solutions of equation (1) and $y(t)>0$ for $0<t<\pi$, it follows that $\psi(t)$ and, therefore, also $z^{(1)}(t)$ have only one zero on the interval $(0, \pi)$. Moreover, $z^{(2)}=\dot{z}^{(1)}$, so that $\dot{z}^{(1)}>0$ for $z^{(2)}>0$ and $\dot{z}^{(1)}<0$ for $z^{(2)}<0$. Thus, as t varies from $t=0$ to $t=\pi$ the point $z(t)$ describes a closed curve Γ in one direction, completing one revolution. Note that Γ is symmetric about the $z^{(1)}$-axis, since the curves (18) and (19) have this property.

The set of all t such that $z^{(1)}(t)=\psi_0$ may be made up of finitely many points and closed intervals, since its complement is the union of finitely many open intervals. Since $z^{(2)}=\dot{z}^{(1)}$, it follows that on the stationary intervals $z^{(2)}(t)\equiv 0$. Consequently, on these same intervals we have

$z(t) \equiv (\psi_0, 0)$. It is logically possible that there may be several stationary intervals (the situation would be as shown in Figure 83). We shall show, however, that this can never occur.

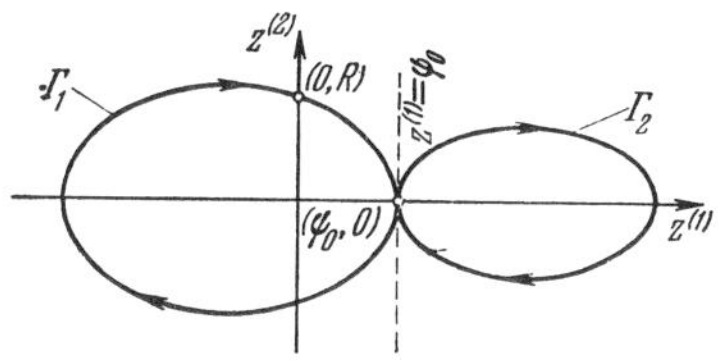

FIGURE 83.

It follows from (16) that on any stationary interval

$$p(t)\,\psi_0 + S \equiv 0.$$

We claim that $\psi_0 \neq 0$. Assuming the contrary, we see that $S = 0$ and $z^{(1)}(t) = z^{(2)}(t) \equiv 0$ on the stationary interval. This contradicts property (IV) of a z-extremal.

Since $\psi_0 \neq 0$, it follows that on the stationary interval

$$p(t) = -\frac{S}{\psi_0} \leqslant 0. \tag{20}$$

(We have used property (III).) Since $p(t) \geqslant b$, we have $b \leqslant 0$. Hence, for all $t \in [0, \pi]$,

$$z^{(1)}(t) \leqslant \psi_0. \tag{21}$$

Indeed, otherwise it follows from (17) and the inequality $b \leqslant 0$ that the curve Γ_2 is a parabola or a hyperbola tangent to the straight line $z^{(1)} = \psi_0$ at the point $(\psi_0, 0)$, but this curve cannot be closed. Consequently, the z-extremal cannot lie on the domain $z^{(1)} > \psi_0$, and this proves (21).

Since $z^{(1)}(0) = 0$, it follows from (21) that $\max z^{(1)}(t) \geqslant 0$, $\psi_0 \geqslant 0$, and since $\psi_0 \neq 0$ this gives $\psi_0 > 0$. This proves (a).

Suppose now that there are no stationary intervals on $[0, \pi]$. There are three possibilities: (i) the z-extremal lies entirely in the half-plane $z^{(1)} \leqslant \psi_0$, and then (by (17)) $p(t) \equiv a^2$; (ii) $z(t)$ is in the half-plane $z^{(1)} \geqslant \psi_0$ for all t, and then $p(t) \equiv b$ (and since the z-extremal is closed this is possible only if $b > 0$); or (iii) there are finitely many consecutive intervals in $[0, \pi]$ on which, alternately, $p(t) = a^2, p(t) = b$. We claim that in the latter case the z-extremal intersects the straight line $z^{(1)} = \psi_0$ at some $z^{(2)} \neq 0$. In other words, we wish to reject the case like that shown in Figure 83, except that now the moving point $z(t)$ "skips" the point $(\psi_0, 0)$ without a stationary interval. Thus the only possible case will be that shown in Figure 82, as asserted in part (b) of the lemma.

Suppose the contrary: there exists a time τ such that $z^{(1)}(t) < \psi_0$ on some interval (t_1, τ) and $z^{(1)}(t) > \psi_0$ on some interval (τ, t_2). By (17), when $t > \tau$ the z-extremal describes some closed second-order curve Γ_2 (i. e., ellipse), and $p(t) = b$ throughout this period. Hence $b > 0$, for if $b \leqslant 0$ the curve Γ_2 is not closed, contrary to (I). In view of the directions of motion implied by (16) (see Figure 83),

$$\left.\begin{aligned} \dot{z}^{(2)}(\tau-0) \leqslant 0, \quad \dot{z}^{(2)}(\tau+0) \geqslant 0, \\ z^{(1)}(\tau) = \psi_0, \quad z^{(2)}(\tau) = 0. \end{aligned}\right\} \tag{22}$$

By (16) and (22),

$$\begin{aligned} \dot{z}^{(2)}(\tau-0) &= -2(\psi_0 a^2 + S) \leqslant 0, \\ \dot{z}^{(2)}(\tau+0) &= -2(\psi_0 b + S) \geqslant 0. \end{aligned}$$

Consequently, there exists $\tilde{p}$ $(b \leqslant \tilde{p} \leqslant a^2)$ such that

$$\psi_0 \tilde{p} + S = 0.$$

If $\psi_0 = 0$, then

$$S = 0, \quad z^{(1)}(\tau) = 0, \quad z^{(2)}(\tau) = 0,$$

and this is impossible by property (IV) of the z-extremal. Therefore,

$$\psi_0 \neq 0 \text{ and } \tilde{p} = -\psi_0^{-1} S \leqslant 0$$

(by III)). This contradicts the inequalities $0 < b \leqslant \tilde{p}$, showing that the case illustrated in Figure 83 (no stationary interval) is impossible. This completes the proof.

2.3. Solution of Problem II

For Problem II, we have $\psi_0 \geqslant 0$ in the maximum condition (12). By the lemma (§2.2) we have either (a) or (b).

Assume first that case (a) is realized. Then $\psi_0 > 0$ and the z-extremal has a stationary interval in $[0, \pi]$.

During the time $[0, \pi]$, the point $z(t)$ describes the entire ellipse (18) (see Figure 81), so that $[0, \pi]$ may be divided into three subintervals Δ_1, Δ_2, Δ_3. On Δ_1 the moving point $z(t)$ leaves the point $(0, R)$ and moves along the ellipse Γ_1 (i.e., along the ellipse (18) for $\tilde{p} = a^2$) until the coordinate $z^{(1)}$ reaches ψ_0. The (open) interval Δ_2 is stationary: $z(t) \equiv (\psi_0, 0)$. On this interval (by part (a) of the lemma) we have $p(t) \equiv -\omega^2 \leqslant 0$. (We have denoted the quantity S/ψ_0, which is nonnegative by part (a) of the lemma, by ω^2.) On Δ_3 the point $z(t)$ returns to its original position along Γ_1. Thus:

$$\left.\begin{aligned} p(t) = a^2 \quad & \text{for } t \in \Delta_1 \text{ and } t \in \Delta_3, \\ p(t) \equiv -\frac{S}{\psi_0} = -\omega^2 \quad & \text{for } t \in \Delta_2. \end{aligned}\right\} \tag{23}$$

Let τ' and τ'' be the initial point and endpoint of the stationary interval Δ_2, $\delta = \tau'' - \tau'$. Since the time required to describe the whole ellipse (18) is $\frac{\pi}{a}$, it follows that

$$\tau' + (\pi - \tau'') = \frac{\pi}{a},$$

so that

$$\delta = \tau'' - \tau' = \pi - \frac{\pi}{a}. \tag{24}$$

Let $y(t)$ be the solution of equation (1) satisfying conditions (A), corresponding to the control $p(t)$. We have

$$y(t) = \begin{cases} \frac{\dot{y}(0)}{a} \sin at & (t \in \Delta_1), \\ c_1 e^{\omega t} + c_2 e^{-\omega t} & (t \in \Delta_2), \\ -\frac{\dot{y}(\pi)}{a} \sin a(\pi - t) & (t \in \Delta_3). \end{cases} \tag{25}$$

It remains to determine τ', τ'' and $\omega > 0$ such that the "joining" conditions hold for $y(t)$, i. e., $y(t)$ and $\dot{y}(t)$ are continuous at τ' and τ''. Elementary calculations give

$$q = -\frac{\dot{y}(\pi)}{\dot{y}(0)} = e^{\pm \omega S}. \tag{26}$$

We must take the plus sign in (26) if $q > 1$, the minus sign if $q < 1$. By (24),

$$\omega = \frac{|\ln q|}{\pi \left(1 - \frac{1}{a}\right)}. \tag{27}$$

Elementary arguments, using the joining conditions, show that on the stationary interval

$$y(t) = ce^{\pm \omega t} \quad \text{for} \quad t \in \Delta_2. \tag{28}$$

We may assume without loss of generality that the function $y(t)$ is normalized so that

$$\dot{y}(0) = 1.$$

We then obtain from (25), (28):

$$y(t) = \begin{cases} \frac{1}{a} \sin at & \text{if} \quad t \in [0, \tau'], \\ ce^{\pm \omega t} & \text{if} \quad t \in [\tau', \tau''], \\ \frac{q}{a} \sin a(\pi - t) & \text{if} \quad t \in [\tau'', \pi]. \end{cases} \tag{29}$$

Recall that ω is defined by (27), and c, τ', τ'' are determined by the joining conditions in conjunction with (24).

We have shown that when the z-extremal defined by an optimal control has a stationary interval the solution $y(t)$ has the form (29). Note that if

$$p = p(t)$$

is the control corresponding to the function (29), then

$$J(p) = a^2\tau' - \omega^2(\tau'' - \tau') + a^2(\pi - \tau''),$$

so that, by (24),

$$J(p) = a\pi - \pi\omega^2\left(1 - \frac{1}{a}\right). \tag{30}$$

We claim that if $b < -\omega^2$, i. e., the control $p(t)$ corresponding to (29) is admissible, then this is indeed an optimal control.

Suppose that the optimal control p' has no stationary intervals. By the lemma, we must have case (b): $p' = a^2$ on some set Δ' and $p' = b$ on the complement $[0, \pi] - \Delta'$, so that

$$J(p') = \int_{\Delta'} a^2\,dt + \int_{\Delta''} b\,dt.$$

Since the time required to describe the ellipse (18)($\breve{p} = a^2$) is $\frac{\pi}{a}$, we have

$$\int_{\Delta} a^2\,dt < a^2\frac{\pi}{a}.$$

Since

$$\Delta' \cup \Delta'' = [0, \pi] \text{ and } b \leqslant -\omega^2,$$

it follows that

$$\int_{\Delta''} b\,dt \leqslant \int_{\Delta''} (-\omega^2)\,dt < -\omega^2\left[\pi - \frac{\pi}{a}\right].$$

Therefore,

$$J(p') < a^2\frac{\pi}{a} - a^2\left[\pi - \frac{\pi}{a}\right] = J(p).$$

Consequently, p' cannot be optimal. Thus, if $b < -\omega^2$ the optimal control has stationary intervals: we have case (a), and the solution has the form (29). Thus, if $b \leqslant -\omega^2$ and the control $p(t)$ corresponding to (29) is admissible, then it is also optimal. In this case the z-extremal of the optimal control has a stationary interval.

Suppose now that $b>-\omega^2$, so that the control in question is not admissible. Then, as shown previously, the z-extremal of the optimal control has no stationary interval. By the lemma, we must have case (b) and the situation is that shown in Figure 82. This means that the solution defined by the optimal control has the form

$$y(t)=\begin{cases}\frac{1}{a}\sin at & \text{if} \quad t\in[0,\tau'],\\ c_1\frac{\sin\sqrt{b}\,t}{\sqrt{b}}+c_2\cos\sqrt{b}\,t & \text{if} \quad t\in[\tau',\tau''],\\ \frac{1}{a}\sin a(\pi-t) & \text{if} \quad t\in[\tau'',\pi].\end{cases} \tag{31}$$

Here b need not be positive. (If $b=0$, write t in place of $\sin\sqrt{b}\,t/\sqrt{b}$.) The constants c_1, c_2, τ', τ'' in (31) are determined by the joining conditions.

We summarize the result as follows:

Theorem I. Suppose that the class $\mathcal{P}^{(a,b,q)}$ *is not empty. If* $b\leqslant-\omega^2$, *i.e., the function* $p(t)$ *corresponding to the solution (29) is in* $\mathcal{P}^{(a,b,q)}$, *then* $\sup J(p)$ *over* $\mathcal{P}^{(a,b,q)}$ *is achieved for the function*

$$p(t)=-\frac{y''}{y},$$

where $y=y(t)$ *has the form (29) with* ω *as defined by (27). If* $b>\omega^2$, *so that* $p(t)$ *is not admissible, then* $\sup J(p)$ *over* $\mathcal{P}^{(a,b,q)}$ *is achieved for the function*

$$p(t)=-\frac{y''}{y},$$

where y *has the form (31). In either case, the constants* c, τ', τ''; c_1, c_2, τ', τ'' *are determined by joining conditions — continuity of* $y(t)$ *and* $\dot{y}(t)$ *at the points* τ' *and* τ''. *The class* $\mathcal{P}^{(a,b,q)}$ *is nonempty if and only if the joining equations have real solutions, and in either case we must have* $0\leqslant\tau'\leqslant\tau''<\pi$.

It is easy to see that this theorem is equivalent to Theorem I of § VIII.4.3.

We have assumed hitherto that a and b are finite. Now consider the case $b=-\infty$, so that $\mathcal{P}^{(a,-\infty,q)}$ is the class of functions

$$p(t)\in L(0,\pi),$$

satisfying conditions (A) and the condition

$$p(t)\leqslant a^2, \tag{32}$$

almost everywhere on $[0,\pi]$. The following theorem is readily deduced from Theorem I:

Theorem II. $\sup J(p)$ *over the class* $\mathcal{P}^{(a,-\infty,q)}$ *is achieved for*

$$p(t)=-\frac{y''}{y},$$

where $y=y(t)$ *has the form of (29), with* ω *defined by (27) and the numbers* c, τ', τ'' *determined by the joining conditions (continuity of* $y(t)$ *and* $\dot{y}(t)$ *at* τ' *and* τ''; *the joining equations are always solvable). We then have*

$$\sup_{\mathcal{P}^{(a,-\infty,q)}} J(p)=\pi a-\frac{(\ln q)^2}{\pi\left(1-\frac{1}{a}\right)}.$$

2.4. Solution of Problem I

For Problem I, we have $\psi_0 \leqslant 0$ in the maximum condition (12). It follows from the lemma of §2.2 that case (b) holds.

If $\psi_0 < 0$, the point $z_0 = (0, R)$, corresponding to zeros of the function $y(t)$, lies in the half-plane $z^{(1)} > \psi_0$. Thus, if $\psi_0 < 0$ the interval $(0, \pi)$ contains exactly two switching points of the control $p(t)$, at times τ' and τ'' $(\tau' < \tau'')$, with $p(t) = a^2$ on (τ', τ'') and $p(t) = b$ elsewhere.

If $\psi_0 = 0$, then either $\tau' = 0$ or $\tau'' = \pi$. We have thus proved

Theorem III. Suppose that the class $\mathfrak{P}^{(a, b, q)}$ *is not empty. Then* $\inf J(p)$ *over* $\mathfrak{P}^{(a, b, q)}$ *is achieved for the function*

$$p(t) = -\frac{y''}{y},$$

where

$$y(t) = \begin{cases} \frac{1}{\sqrt{b}} \sin \sqrt{b}\, t & \text{if } 0 \leqslant t \leqslant \tau', \\ c_1 \sin at + c_2 \cos at & \text{if } \tau' < t \leqslant \tau'', \\ \frac{q}{\sqrt{b}} \sin \sqrt{b} (\pi - t) & \text{if } \tau'' < t \leqslant \pi. \end{cases} \tag{33}$$

The numbers c_1, c_2, τ', τ'' *are determined by the joining conditions (continuity of* $y(t)$ *and* $\dot{y}(t)$ *at* τ' *and* τ''*). The class* $\mathfrak{P}^{(a, b, q)}$ *is nonempty if and only if the joining equations have real solutions* c_1, c_2, τ', τ'' *such that* $0 \leqslant \tau' \leqslant \tau'' \leqslant \pi$.

It is easy to see that this theorem is equivalent to Theorem II of § VIII. 4.3.

As before, the case $b \leqslant 0$ is not excluded in Theorem III; the expression $\sin \sqrt{b}\, t / \sqrt{b}$ should then be understood in the natural way.

We consider the class $a^2 = \infty$, i. e., the class $\mathfrak{P}^{(\infty, b, q)}$ of all functions satisfying conditions (A) and the condition

$$p(t) \geqslant b. \tag{34}$$

Letting $a \to \infty$ and using the special form of the function (33), we obtain the following solution of Problem I for the case $a = \infty$:

Theorem IV. $\inf J(p)$ *over* $\mathfrak{P}^{(\infty, b, q)}$ *is not achieved. An optimizing sequence is provided by the sequence of functions (33) with*

$$a = a_n \to \infty, \quad \tau' = \tau'_n, \quad \tau'' = \tau''_n, \quad c_1 = c_n^{(1)}, \quad c_2 = c_n^{(2)},$$

where the numbers $c_n^{(1)}, c_n^{(2)}, \tau'_n, \tau''_n$ *are determined by joining conditions. The joining equations have solutions for sufficiently large* a_n*. As* $n \to \infty$ *the numbers* τ'_n, τ''_n *tend to a common limit* τ^* *which is the unique root in* $(0, \pi)$ *of the equation*

$$\frac{1}{\sqrt{b}} \sin \sqrt{b}\, \tau = \frac{q}{\sqrt{b}} \sin \sqrt{b} (\pi - \tau).$$

The actual value of $\inf J(p)$ *is*

$$\inf_{\mathfrak{P}^{(\infty, b, q)}} J(p) = b\pi + \sqrt{b} [\operatorname{ctg} \sqrt{b} (\pi - \tau^*) + \operatorname{ctg} \sqrt{b}\, \tau^*].$$

BRIEF BIBLIOGRAPHICAL NOTES

Chapter I

§ 1. There is a wealth of good textbooks of linear algebra. We recommend in particular Gantmakher /21/, Gel'fand /23/, Mal'tsev /63/, Smirnov /85a/ (Volume III, Part I), Shilov /103a,b/ and Chapter I of Bulgakov /15/.

§2.1, §§ 2 and 3. The theory of power series in one or several matrix variables was developed by Lappo-Danilevskii /58a, b/. The fundamental idea of his work is the use of the theory of functions of matrices as a tool for studying systems of linear differential equations with analytic coefficients.

§2.3. The general formulas for the Lagrange-Sylvester interpolation polynomials may be found in Gantmakher /21/ (Chapter V, §§2, 3) and Bulgakov /15/ (§1.29).

§2. Some material on functions of matrices is included in Bulgakov /15/ (Chapter I, B), Gantmakher /21/ (Chapter V), Mal'tsev /63/ (Chapter IV, §4).

§3. The exposition follows the papers /224p, s/.

Chapter II

§1. The reader may find additional material in Stepanov /87/, Petrovskii /74/, Pontryagin /76b/, Coddington and Levinson /49/, Lefschetz /59/. A survey of the literature and a voluminous bibliography may be found in Cesari /100/.

§2. The main results are due to Floquet /106/ and Lyapunov (Chapter III of his doctoral dissertation /60/). The theory of reducible systems is due to Erugin /40a/.

§2.8. A survey of the use of Lyapunov's second method in stability analysis of the trivial solution of systems of linear differential equations with periodic coefficients is given in §3 of Rubanovskii /82/.

§§3.1 — 3.5 and 3.7. These are some of the results in Chapter III of Lyapunov's dissertation /60/ and Poincaré's celebrated memoir /111a/.

§3.9. Theorem II (in another version) was first published by Cesari /237/. Related results were obtained by Cesari and Hale /238/, Hale /252a/ and Gambill /249a/. The proofs used in these papers use a special variant of the method of successive approximations, based on Poincaré's method for eliminating secular terms in successive approximations. The exposition of §3.9 follows our paper /2241/.

§§ 4.1 — 4.3. The results are taken from /224e/.

§4.4. A more detailed account is given in /224v/.

§4.5. This subsection presents results due to G.G. Durnov, extending the theorems of /224v/ to a vector equation (4.64).

Chapter III

§1. The elements of the theory of Hamiltonian equations are due to M.G. Krein /166f/. The account given here, however, is more elementary than Krein's.

§1.3. What we call Krein's theorem is Theorem 1.2 of /166f/; the proof given here is different.

§1.4. A short proof of this fundamental theorem was published by Krein /166a/, and a more detailed version in /166f/. A detailed proof may also be found in Gel'fand and Lidskii /132/.

§§1.6 and 1.7. The reader may find similar formulas in /166f/ (§ 4) and /132/ (§ 2).

§3. The authors' original proof of the Krein-Gel'fand-Lidskii theorem was different. Our proof follows /224m/.

§3.2. The converse of Krein's theorem (§1.4) was proved by Gel'fand and Lidskii /132/. It was thus shown that Krein's sufficient condition /166a,f/ for strong stability of canonical (Hamiltonian) systems is also necessary. Our proof departs from /132/, following /224n/ (§1). Another proof of the sufficient and necessary conditions for strong stability (in a somewhat different version) was obtained by Moser /278a/. Certain physicists were also led to this theorem for fourth-order Hamiltonian systems, in connection with the theory of strong-focusing synchrotons (see /168/, p.281). Certain points in the main result of Diliberto /245a/ also have a bearing on the theorem. It was Krein /166f/, Gel'fand and Lidskii /132/ who first reduced the investigation of canonical systems to study of the real symplectic group, thus leading to many theoretically important conclusions.

The stability of solutions of Hamiltonian equations with unbounded periodic operator coefficients has been studied by Derguzov /141a,b,c,d,e/, Fomin /212a/ and /143/.

Generalizations to the case of differential equations in Banach spaces were given in the monograph of Daletskii and Krein /32/.

The Derguzov-Fomin theory referred to above, which is analogous to the finite-dimensional theory, refers to a class of partial differential equations arising in many different fields of physics and engineering. The center of gravity of this trend has now shifted toward the study of symmetric and unitary operators in spaces with indefinite metrics. A major role is now being played here by such facts as the difference between the spectra of similar operators under unbounded similarity transformations. Under various assumptions, the authors divide the spectrum into spectra of the first and second kinds and prove the Krein-Gel'fand-Lidskii necessary and sufficient conditions for strong stability.

§4. §§4.1 — 4.3 present results of Krein and Lyubarskii /167b/, with some complements and methodological changes.

§§ 4.4 — 4.6 give the results of /224p/. Similar material may be found in Breus /126a/.

§§ 5.1 — 5.4. The reader may find a more detailed account in /224m,u/. For different definitions of arguments on the symplectic group and their equivalence, see /224u/ (§§ 1 and 2).

§ 5.3. The proof of the Gel'fand-Lidskii theorem is one of the main results in /132/. The existence of stability (instability) domains in the space of Hamiltonians was first established for second-order equations in /224a/, where some of their properties were indicated. For the case of complex coefficients, see also /224t/. Lidskii and Frolov /178a/ consider a real selfadjoint system of $2k$ differential equations with periodic coefficients, namely $\mathbf{Q}(t)\dot{\mathbf{y}}-\left[\mathbf{S}(t)-\frac{1}{2}\dot{\mathbf{Q}}(t)\right]\mathbf{y}=\mathbf{0}$, where $\mathbf{Q}(t)$ is nonsingular and skew-symmetric, $\mathbf{S}(t)$ symmetric. If $\mathbf{Q}(t)$ is constant this is a canonical system. It is proved that in contrast to the case of canonical (and Hamiltonian) systems the set of all strongly stable systems is the union of a finite number (2^{k+1}) of open simply-connected components. Many results similar to these were proved independently for Hamiltonian systems by Coppel and Howe /241/. Lidskii and Frolov /178c/ consider complex selfadjoint systems $\mathbf{Q}(t)\dot{\mathbf{y}}-\left[\mathbf{S}(t)-\frac{1}{2}\dot{\mathbf{Q}}(t)\right]\mathbf{y}=\mathbf{0}$ and prove that the set of strongly stable systems of this type is the union of C_n^q connected components, where q is the number of eigenvalues of $\mathbf{Q}(t)$ in the upper half-plane and n the number of equations.

§5.4. Concerning different arguments on the symplectic group and the relations between them, see /224u/.

§5.5. This subsection gives a brief account of the main ideas of /224m/.

§6. The first theorems on "directional convexity" of stability domains were proved for second-order equations in /224c/. The exposition follows /224x/ for the most part.

§6.3. A proof that the stability domains are directionally convex was given in the important special case that $\mathbf{H}_1(t)$ and $\mathbf{H}_2(t)$ are scalar matrices by Neigauz and Lidskii /188/. Under the same assumption, these authors establish a stability test for systems of Hill equations similar to Zhukovskii's test, and establish other "best possible" stability tests.

§6.6. Another version of the necessary and sufficient condition for vector equations (6.78), similar to Theorem I, was cited without proof in Krein /166e/ (p.660).

§6.8. The theorem (without effective estimates for the number ε_0, and also subject to some additional though insignificant restrictions) was proved by Krein (see /166f/, Theorems 2.2 and 6.4).

§ 7. Some other conditions for stability and instability of systems (7.13) and equations (7.22) may be found in the papers /86, 204g,h, 224k,x/, Valeev /129a,b,c,d/, Gokhberg and Krein /137/, Karaseva /154/, Krein /166f/, Neigauz and Lidskii /188/, Rappoport /198b/, Hale /252a/, Massa /271/, Meadows /276/, Sandberg /286a,b/ and in the books Gokhberg and Krein /30/, Erugin /40b,c/, Hale /98/, Cesari /100/, Arscott /105/, Roseau /112/.

Further results relating to the infinite-dimensional case may be found in Daletskii and Krein /32/.

Chapter IV

The exposition of §§1, 2 is based on /213/. Concerning matrix perturbation theory, see also Vishik and Lyusternik /130/.

§§ 2 – 5. Various topics connected with the definition and computation of characteristic exponents of systems of linear differential equations are considered in the books Erugin /40b, c/, Malkin /62b/, Cesari /100/, and the papers Artem'ev /120b/, Shtokalo /221/, Shimanov /220a, b, c/, Valeev /129a, c, d/, Hukuhara /261/, Hale /252f, h, j/, Gambill /239b, c/, Rozenvasser /199a, b/. A survey of methods for calculating characteristic exponents may be found in §6 of Rubanovskii /82/. V.I. Ponomarenko and I.N. Markina took part in the calculations for the example of §2.5.

§4.5. The method based on computation of the exponent matrix was developed by Shtokalo /104, 221/, Erugin /40b, c/, Hukuhara /261/ and others.

§4.4. The propositions about the radius of convergence of the series (4.28) were proved in /224s/.

§§ 4.5 and 4.6. Formulas (4.50), (4.51) are presented in /224p/, in which it is shown that for canonical systems this method may be used "without leaving the realm of canonical systems," i.e., in such a way that the matrix $\mathbf{K}(\varepsilon)$ is Hamiltonian. In conjunction with the theorem of §3.1, this enables one to establish (generally via finitely many operations) whether the trivial solution of the original system is stable or unstable. (Note that the original canonical system is clearly critical in Lyapunov's sense.)

§§5.2 – 5.4. There is a tremendous literature on Hill's equation (see /86/, /100/). A bibliography on Hill's and Mathieu's equations may be found in McLachlan /61/.

Concerning various practical methods for determining the characteristic exponents for systems with a small parameter, see Lion /269/, Stevens /294/, Golokvoschyus /133/. Valeev /129/ suggests methods for determining the solutions and characteristic exponents, based on active use of Laplace transforms. The importance of these methods is that they are applicable for a broad range of equations, including equations with retarded arguments and equations with almost-periodic coefficients.

§6. The method of small parameters is widely used in the theory of nonlinear oscillations. Originating in celestial mechanics, the method was first formulated by Poincaré /111b/. Its further development is mainly linked with the Soviet school of research. Of the fundamental work we mention Papaleksi and Mandel'shtam /64/, Andronov, Vitt and Khaikin /3/, Krylov, Bogolyubov and Mitropol'skii /10, 11, 12, 55, 66a, b, c/, Bulgakov /15/, Malkin /62a, b/. The reader may find many other results in the recently published book Hale /98/.

§§6.4 and 6.5. Another, similar approach to the stability of periodic solutions of quasilinear nonautonomous systems with several degrees of freedom may be found in Volk /131/, Shimanov /220a, b, c/, Blekhman /123a, b Kushul' /172a, b/, Proskuryakov /195a, b, c, d/ and Plotnikova /194a, b/.

Chapter V

§1. For a broader definition of critical frequency, see /224n/. The present definition agrees with that of /224n/ for a one-parameter class of perturbed systems (1.2).

A survey of the results of Malkin, one of the founders of the mathematical theory of parametric resonance, may be found in Rubanovskii /82/ (§7).

§1.1. Formula (1.7) for the critical frequencies is due to Krein /166f/ (Theorem 10.1).

§1.2. The concept of the "kind" of a critical frequency was introduced in §3 of the survey article /168/.

§§1.4 and 1.5. A series of related results was first established, under certain special assumptions, by Malkin /62b/ (§§ 5 – 8). Formulas (1.29) were first proved in /224o/.

§1.11. Formulas (1.49) were derived, under certain additional assumptions, by Malkin /62b/ (Chapter IV, §8).

§§3.1 – 3.4. The results are taken from /224z/.

§§ 3.6 – 3.8. The formulas for the second approximation were first derived by Pittel' /191b/. The complement to these formulas is due to Chugaev /217/. The example in §3.8 is due to Pittel'.

§§4.1 and 4.2. See /204r/, /225/.

§4.3. The fact that a stable system may become unstable owing to friction was apparently first noticed by Schmidt and Weidenhammer /288a/, and then confirmed rigorously in Valeev /129i/ and /213/. The example at the end of §4.3 is due to Fomin.

The theory of parameteric resonance of elastic systems with infinitely many degrees of freedom (with distributed parameters) is presented in Fomin /212b, c, d, e/ and Derguzov /141/.

Chapter VI

§1. Contains the material of /193b/.

§2.1. Follows the account in Yuzefovich /222/.

§§ 2.2 and 2.3. A more detailed account may be found in the papers /204m, n, o, 205/ and in §§ 3 and 4 of the book /86d/.

§2.4. The L-instability domain was determined for other parameters by Zatepyakin /149b/.

§3. Here we consider only one of the many problems. Of the monographs on the dynamic stability of elastic systems, we mention Bolotin /13a, b/, Vol'mir /19/, Gol'denblat /28a, b/.

§4. For more details, see /204r, t/.

§§ 1 — 4. Concerning parametric resonance in torsional vibrations of crankshafts, see the paper by Zatepyakin /149a/. Using these methods, Pittel' /191a/ has constructed the dynamic instability domains for flexural vibrations of the blades of a helicopter tail rotor.

§§ 5 and 6. For additional material, see Koroza /160a, b, c, d/, and also /161, 162a, b, c, d, e/.

Chapter VII

Introduction. More about Lyapunov's method for estimating the characteristic constant may be found in the surveys Smirnov /85b/, /86a/ and Rubanovskii /82/ (§5).

§§1.1 — 1.4. For the most part, the account follows Krein /166g/. There are a number of methodological differences; in particular, the proof of property (III) of the function $A(\lambda)$ is different. We might mention that the proof for Hill's equation of property (III), which is basic for the entire theory of stability zones, had long caused considerable difficulties. It was first established by Krein, using highly sophisticated techniques of functional analysis. An elementary proof was first given in /166g/.

§2. These are results of /224f/. The case of an odd function $p(t)$ in Hill's equation was first considered by Lyapunov in /180d, g/, and an important additional contribution was made by Rubanovskii /201a/. Further development of Lyapunov's method may be followed in the papers /204b, c, d, e, f, g, h, j, k, l/.

§3.7. The results are taken from /204a — h/.

For the extension of Lyapunov's method to higher-order systems, see /224k/, /204g, h/, to one higher-order equation — see Los' /179a, b, c, d, e/ and Zadiraka and Los' /148a, b/. Other results for systems of two equations may be found in /204b, c, d, e, f, h, j, k, l/.

Chapter VIII

The exposition of §§ 1 — 4 in this chapter is based on /224a, b, c, d, e, g/.

§2. The first test was established by Zhukovskii /147/. The first analogs of Lyapunov's test for stability domains other than $\mathcal{G}_0$ are due to Gusarova /139a, b, c/, but the estimates in her tests proved to be inferior.

§4. The first important results similar to those presented here are due to Borg /234a/. Borg's theorem is presented in §4.5 in a sharper version (see Kozinets /158/). Many of the tests of §4, and also many other analogous tests, were established differently by Komlenko /159/.

§4.3. Similar variational problems are considered in Gernet /24/. There, however, constraints are imposed only on the first derivative, in contradistinction to our condition (c).

§5. A series of stability tests for Hill's equation may be found in the survey article /86a/. Stability tests involving $\max p(t)$ and $\min p(t)$ were first obtained by Krein /166d/. (The proofs were based on estimates for the eigenvalues of periodic and antiperiodic boundary-value problems.) Other results may be found in /166e/. Related results were proved by Karaseva /154a, b, c/. Also worthy of note are the original results of Levin /174a, b, c/. In conclusion, we mention that some sufficient conditions for stability of the trivial solution of second-order linear equations with bounded (but not necessarily periodic) coefficients may be found in Zhukovskii's survey article /41/.

§6. The account is based on the article /224/ [sic]. The first results about stability zones for Hill's equation are apparently due to Lyapunov; they were published without proofs in /180d, e/. There are no proofs of Lyapunov's assertions concerning functions $p(t)$ of variable sign in his papers, and a considerable period elapsed until Krein published the first complete proofs in /166a, f, g/. Other earlier work concerning the "oscillation theorem" for Hill's equation may be found in Hamel /225/, Haupt /256/, Kramers /264/, Schwerdtfeger /291/.

Appendices

Appendix 1 is a detailed account of results reported briefly in /224a/.

Appendix 2 is borrowed from Yu. F. Kazarinov's paper: On an extremum problem. — Vestnik Leningrad. Gos. Univ., No. 19 (1971), 27 – 34.

BIBLIOGRAPHY

Textbooks, Monographs, Surveys

1. Ince, E.L.
   Ordinary Differential Equations. 4th edition. — New York, Dover, 1953.
2. Andronov, A.A.
   Collected Works. — Moscow, USSR Acad. Sci., 1956. (Russian)
3. Andronov, A.A., A.A. Vitt, and S.E. Khaikin.
   Theory of Oscillations. 2nd edition. — Moscow, Fizmatgiz, 1959. (Russian)
4. Arnol'd, V.L.
   Lectures on Classical Mechanics. — Moscow State Univ., 1968. (Russian)
5. Artem'ev, N.A.
   Elements of the Qualitative Theory of Ordinary Differential Equations. Part 1. — Leningrad State Univ., 1941. (Russian)
6. Barbashin, E.A.
   a) Introduction to the Theory of Stability. — Moscow, "Nauka," 1964. (Russian)
   b) Lyapunov Functions. — Moscow, "Nauka," 1971. (Russian)
7. Beilin, E.A. and G.Yu. Dzhanelidze.
   Survey of literature on dynamic stability of elastic systems. — Prikl. Mat. Mekh. **16**, No.5 (1952), 635–648. (Russian)
8. Bellman, R.
   a) Stability Theory of Differential Equations. — New York, Dover, 1953.
   b) Introduction to Matrix Analysis. — New York, McGraw-Hill, 1960.
9. Belyaev, N.M.
   Stability of prismatic bars under the effect of variable longitudinal forces. — In: Inzhenernye sooruzheniya i stroitel'naya mekhanika. Moscow, "Put'," 1924. (Russian)
10. Bogolyubov, N.N.
    Selected Works in Three Volumes, **1**. — Kiev, "Naukova Dumka," 1969. (Russian)
11. Bogolyubov, N.N. and Yu.A. Mitropol'skii.
    Asymptotic Methods in the Theory of Nonlinear Oscillations. — Moscow, Fizmatgiz, 1958. (Russian)
12. Bogolyubov, N.N., Yu.A. Mitropol'skii, and A.M. Samoilenko.
    The Method of Accelerated Convergence in Nonlinear Mechanics. — Kiev, "Naukova Dumka," 1969. (Russian)
13. Bolotin, V.V.
    a) Dynamic Stability of Elastic Systems. — Moscow, Gostekhizdat, 1956. (Russian)
    b) Nonconservative Problems in the Theory of Elastic Stability. — Moscow, Fizmatgiz, 1961. (Russian)
14. Bondarenko, G.V.
    Hill's Equation and its Application to Vibrational Problems of Engineering. — Moscow, USSR Acad. Sci., 1936. (Russian)
15. Bulgakov, B.V.
    Oscillations. — Moscow, Gostekhizdat, 1954. (Russian)
16. Bylov, B.F., R.E. Vinograd, D.M. Grobman, and V.V. Nemytskii.
    The Theory of Lyapunov Exponents and its Application to Stability Problems. — Moscow, "Nauka," 1966. (Russian)

17. Vainberg, M.M. and V.A. Trenogin.
Theory of Branching of Solutions of Nonlinear Equations. — Moscow, "Nauka," 1969. (Russian)
18. Vlasov, V.Z.
Thin-Walled Elastic Bars, 2nd edition. — Moscow, Fizmatgiz, 1959. (Russian)
19. Vol'mir, A.S.
Stability of Elastic Systems. — Moscow, Fizmatgiz, 1963. (Russian)
20. Galerkin, B.G.
Bars and plates. — Vestnik Inzh. Tekh. 19 (1915). (Russian)
21. Gantmakher, F.R.
Theory of Matrices. — Moscow, Gostekhizdat, 1953. (Russian)
22. Gantmakher, F.R. and M.G. Krein.
Oscillation Matrices and Small Oscillations of Mechanical Systems, 2nd edition. — Moscow, Gostekhizdat, 1950. (Russian)
23. Gel'fand, I.M.
Lectures on Linear Algebra. — Moscow, Gostekhizdat, 1948. (Russian)
24. Gernet, N.N.
On the Fundamental Elementary Problem of the Calculus of Variations. — St. Petersburg, 1913. (Russian)
25. Glazman, I.M. and Yu.I. Lyubich.
Finite-Dimensional Linear Analysis. — Moscow, "Nauka," 1969. (Russian)
26. Goloskokov, E.G. and A.P. Filippov.
Nonstationary Oscillations of Mechanical Systems. — Kiev, "Naukova Dumka," 1966. (Russian)
27. Golubev, V.V.
Lectures on the Analytical Theory of Differential Equations, 2nd edition. — Moscow, Gostekhizdat, 1950. (Russian)
28. Gol'denblat, I.I.
a) Current Problems of Oscillations and Stability of Engineering Structures. — Moscow, Stroiizdat, 1947. (Russian)
b) Dynamic Stability of Structures. — Moscow, Stroiizdat, 1948. (Russian)
29. Goncharov, V.L.
Theory of Interpolation and Approximation of Functions, 2nd edition. — Moscow, Gostekhizdat, 1954. (Russian)
30. Gokhberg, I.Ts. and M.G. Krein.
The Theory of Volterra Operators in Hilbert Space and its Applications. — Moscow, "Nauka," 1967. (Russian)
31. Goursat, E.
A Course in Mathematical Analysis, Vol. II. — New York, Dover, 1959.
32. Daletskii, Yu.L. and M.G. Krein.
Stability of Solutions of Differential Equations in Banach Space. — Moscow, "Nauka," 1970. (Russian)
33. Dunford, N. and J.T. Schwartz.
Linear Operators, Vol. 1. General Theory. — New York, Wiley (Interscience), 1958.
34. Demidovich, B.P.
Lectures on the Mathematical Theory of Stability. — Moscow, "Nauka," 1967. (Russian)
35. Jones, H.
Theory of Brillouin Zones and Electronic States in Crystals. — New York, Wiley (Interscience), 1960.
36. Dimentberg, F.M., K.T. Shatalov, and A.A. Gusarov.
Oscillations of Machinery. — Moscow, "Mashinostroenie," 1964. (Russian)
37. Dmitriev, F.D.
Collapse of Engineering Structures. — Moscow, Gosstroiizdat, 1953. (Russian)
38. Dolginov, A.I.
Resonance in Electrical Circuits and Systems. — Moscow, Gosenergoizdat, 1957. (Russian)
39. Duboshin, G.N.
a) Elements of the Theory of Stability of Motion. — Moscow State Univ., 1952. (Russian)
b) Celestial Mechanics. Fundamental Problems and Methods. — Moscow, Fizmatgiz. 1962. (Russian)
c) Celestial Mechanics. Analytical and Qualitative Methods. — Moscow, "Nauka," 1964. (Russian)

40. Erugin, N.P.
   a) Reducible Systems. — Trudy Mat. Inst. Steklov 12 (1946). (Russian)
   b) The Lappo-Danilevskii Method in the Theory of Linear Differential Equations. — Leningrad State Univ., 1956. (Russian)
   c) Linear Systems of Ordinary Differential Equations with Periodic and Quasiperiodic Coefficients. — Minsk, BSSR Acad. Sci., 1963. (Russian)
   d) Lecture Notes for a General Course of Differential Equations. — Minsk, "Nauka i Tekhnika," 1970. (Russian)
41. Zhukovskii, V.I.
   Survey of literature on the stability of the trivial solution of a system of linear differential equations with variable coefficients. — Uchenye Zapiski Orekhovo-Zuevskogo Ped. Inst. 22, No.3(1964), 12—25. (Russian)
42. Siegel, C.L.
   Vorlesungen über Himmelsmechanik. — Berlin, Springer, 1956.
43. Zubov, V.I.
   a) Mathematical Methods for Investigation of Automatic Control Systems. — Moscow, Sudpromgiz, 1959. (Russian)
   b) Oscillations in Nonlinear and Control Systems. — Moscow, Sudpromgiz, 1962. (Russian)
44. Ishlinskii, A.Yu.
   Mechanics of Gyroscopic Systems. — Moscow, USSR Acad. Sci., 1963. (Russian)
45. Kamke, E.
   Differentialgleichungen; Lösungsmethoden und Lösungen, Part 2. Gewöhnliche Differentialgleichungen (3rd ed.). — New York, Chelsea, 1948.
46. Kantorovich, V.L. and G.P. Akilov.
   Functional Analysis in Normed Spaces. — Moscow, Fizmatgiz, 1959. (Russian)
47. Kantorovich, L.V. and V.I. Krylov.
   Approximate Methods of Higher Analysis, 5th edition. — Moscow, Fizmatgiz, 1962. (Russian)
48. Kauderer, H.
   Nichtlineare Mechanik. — Berlin, Springer, 1958.
49. Coddington, E.A. and N. Levinson.
   Theory of Ordinary Differential Equations. — New York, McGraw-Hill, 1955.
50. Kolmogorov, A.N. and S.V. Fomin.
   Elements of the Theory of Functions and Functional Analysis. — Moscow, "Nauka," 1968. (Russian)
51. Kononenko, V.O.
   Oscillating Systems with Narrow-Band Excitation. — Moscow, "Nauka," 1964. (Russian)
52. Krasnosel'skii, M.A.
   a) Positive Solutions of Operator Equations. — Moscow, Fizmatgiz, 1962. (Russian)
   b) The Translation Operator along Trajectories of Differential Equations. — Moscow, "Nauka," 1966. (Russian)
53. Krasnosel'skii, M.A., A.I. Perov, A.I. Povolotskii, and P.O. Zabreiko.
   Vector Fields on the Plane. — Moscow, Fizmatgiz, 1963. (Russian)
54. Krasovskii, N.N. Some Problems in the Theory of Stability of Motion. — Moscow, Fizmatgiz, 1959. (Russian)
55. Krylov, N.M. and N.N. Bogolyubov.
   Introduction to Nonlinear Mechanics. — Kiev, Ukr.SSR Acad. Sci., 1937. (Russian)
56. Kurosh, A.G.
   A Course of Higher Algebra. — Moscow, Gostekhizdat, 1946. (Russian)
57. Lavrent'ev, M.A. and L.A. Lyusternik.
   Elements of the Calculus of Variations, Vol.1, Part 2. — Moscow, ONTI, 1935. (Russian)
58. Lappo-Danilevskii, I.A.
   a) Theory of Functions of Matrices and Systems of Linear Differential Equations. — Moscow, ONTI, Gostekhizdat, 1934. (Russian)
   b) Application of Functions of Matrices to the Theory of Linear Systems of Ordinary Differential Equations. — Moscow, Gostekhizdat, 1957. (Russian)
59. Lefschetz, S.
   Differential Equations: Geometric Theory. — New York, Wiley (Interscience), 1963.

60. Lyapunov, A.M.
The General Problem of Stability of Motion. — Khar'kov, 1892; republished in: Collected Works, Vol.2, 7—263, 1956. (Russian)

61. McLachlan, N.W.
Theory and Application of Mathieu Functions. — New York, Dover, 1947.

62. Malkin, I.G.
a) The Methods of Lyapunov and Poincaré in the Theory of Nonlinear Oscillations. — Moscow, Gostekhizdat, 1949. (Russian)
b) Some Problems in the Theory of Nonlinear Oscillations. — Moscow, Gostekhizdat, 1956. (Russian)
c) Theory of Stability of Motion, 2nd edition. — Moscow, "Nauka," 1966. (Russian)

63. Mal'tsev, L.I. Elements of Linear Algebra, 2nd edition. — Moscow, Gostekhizdat, 1956. (Russian)

64. Mandel'shtam, L.I., N.D. Papaleksi, A.A. Andronov, A.A. Vitt, G.S. Gorelik, and S.E. Khaikin.
New Research in Nonlinear Oscillations. — Moscow, Gostekhizdat, 1936. (Russian)

65. Massera, J.L. and J.J. Schäffer. Linear Differential Equations and Function Spaces. — New York, Academic Press, 1966.

66. Mitropol'skii, Yu.A.
a) Problems of the Asymptotic Theory of Nonstationary Oscillations. — Moscow, "Nauka," 1964. (Russian)
b) Lectures on the Method of Averaging in Nonlinear Mechanics. — Kiev, "Naukova Dumka," 1966. (Russian)
c) The Method of Averaging in Nonlinear Mechanics. — Kiev, "Naukova Dumka," 1971. (Russian)

67. Mitropol'skii, Yu.A. and O.B. Lykova.
Lectures on the Method of Integral Manifolds. — Kiev, Ukr. SSR Acad. Sci., 1968. (Russian)

68. Mitropol'skii, Yu.A. and B.I. Moseenkov.
Lectures on the Application of Asymptotic Methods to the Solution of Partial Differential Equations. — Kiev, Inst. Math. Ukr. SSR Acad. Sci., 1968. (Russian)

69. Mikhailov, F.A.
Free oscillations of linear systems with variable parameters. — Trudy Moskov. Aviats. Inst., No.135, 1961. (Russian)

70. Moiseev, N.N.
Asymptotic Methods of Nonlinear Mechanics. — Moscow, "Nauka," 1969. (Russian)

71. Nemytskii, V.V.
Problems in the Qualitative Theory of Differential Equations. — Vestnik Moskov. Gos. Univ., Ser. Fiz. Mat. Estest. Nauk 8, No.5 (1952), 19—39. (Russian)

72. Nemytskii, V.V. and V.V. Stepanov.
Qualitative Theory of Differential Equations, 2nd edition. — Moscow, Gostekhizdat, 1949. (Russian)

73. Ostrowski, A.M.
Solution of Equations and Systems of Equations, 2nd edition. — New York, Academic Press, 1966.

74. Petrovskii, I.G.
Lectures on the Theory of Ordinary Differential Equations, 2nd edition. — Moscow, Gostekhizdat, 1947. (Russian)

75. Pliss, V.A.
Nonlocal Problems in the Theory of Oscillations. — Moscow "Nauka," 1964. (Russian)

76. Pontryagin, L.S.
a) Continuous Groups, 2nd edition. — Moscow, Gostekhizdat, 1954. (Russian)
b) Ordinary Differential Equations. — Moscow, Fizmatgiz, 1961. (Russian)
c) (with V.G. Boltyanskii, R.V. Gamkrelidze, and E.F. Mishchenko). Mathematical Theory of Optimal Processes. — Moscow, Fizmatgiz, 1961. (Russian)

77. Rapoport, I.M.
On Some Asymptotic Methods in the Theory of Differential Equations. — Kiev, Ukr. SSR Acad. Sci., 1954. (Russian)

78. Riesz, F. and B. Sz.-Nagy.
Functional Analysis. — New York, Ungar, 1955.

79. Rozenvasser, E.N.
Oscillations of Nonlinear Systems. Method of Integral Equations. — Moscow, "Nauka," 1969. (Russian)

80. [Same as /78/.]

81. Rocard, Y.
Dynamic Instability. — New York, Ungar, 1958.

82. Rubanovskii, V.N.
Stability of the trivial solution of systems of linear ordinary differential equations with periodic coefficients. — In: Itogi Nauki: General Mechanics, 1969. Moscow, All-Union Inst. Scient. Tech. Inform.(VINITI), 1971. (Russian)

83. Rumyantsev, V.V.
On the Stability of Steady Satellite Motion. — Mathematical Methods in Dynamics of Spacecraft. No. 4, Moscow, USSR Acad. Sci., Computation Center, 1967. (Russian)

84. Ryzhik, I.M. and I.S. Gradshtein.
Tables of Integrals, Sums, Series and Products, 3rd edition. — Moscow, Gostekhizdat, 1951. (Russian)

85. Smirnov, V.I.
a) A Course of Higher Mathematics, Vols. 1—5. — Moscow, Fizmatgiz, 1959. (Russian)
b) A survey of A.M. Lyapunov's scientific work. — Prikl. Mat. Mekh. **12**, No. 5 (1948), 479—560. (Russian)

86. Starzhinskii, V.M.
a) A survey of works on conditions of stability of the trivial solution of a system of linear differential equations with periodic coefficients. — Prikl. Mat. Mekh. **18**, No.4 (1954), 469—510. (Russian)
b) English translation of 186a/. — Am. Math. Soc., Translations, Ser.2, Vol.1 (1955), 189—237.
c) Chinese translation of /86a/. — Advancement in Math.**4**, No.3 (1958), 410—449.
d) Contributions to the Theory of Nonlinear Oscillations. — Moscow State Univ., Faculty of Mechanics and Mathematics. Part 1, 1970; Part 2, 1971; Part 3, 1972. (Russian)

87. Stepanov, V.V.
Course of Differential Equations, 6th edition. — Moscow, Gostekhizdat, 1953. (Russian)

88. Stoker, J.J.
Nonlinear Vibrations in Mechanical and Electrical Systems. — New York, Wiley (Interscience), 1950.

89. Strutt, M.J.O.
Lamé'sche-Mathieu'sche und verwandte Funktionen in Physik und Technik. — Berlin, Springer, 1932.

90. Timoshenko, S.P.
Stability of Elastic Systems. [Russian translation, 1955.]

91. Titchmarsh, E.C.
Theory of Functions, 2nd edition. — Oxford Univ. Press, 1939.

92. Whittaker, E.T. and G.N. Watson.
Course of Modern Analysis, 4th edition. — Cambridge Univ. Press, 1927.

93. Faure, R., A. Kaufmann, and M. Denis-Papin.
Mathématiques nouvelles, Vols. 1—2. — Paris, Dunod, 1964.

94. Friedrichs, K.O.
Perturbation of Spectra in Hilbert Space. — New York, Am. Math. Soc., 1967.

95. Fuks, B.A.
Theory of Analytic Functions of Several Complex Variables. — Moscow, Gostekhizdat, 1948. (Russian)

96. Hartman, P.
Ordinary Differential Equations. — New York, Wiley, 1964.

97. Hayashi, C.
Forced Oscillations in Non-linear Systems. — Osaka, Nippon Printing and Publishing Company, 1953.

98. Hale, J.K.
Oscillations in Nonlinear Systems. — New York, McGraw-Hill, 1963.

99. Chebotarev, N.G.
Theory of Algebraic Functions. — Moscow, Gostekhizdat, 1948. (Russian)

100. Cesari, L.
Asymptotic Behavior and Stability Problems in Ordinary Differential Equations. — Berlin, Springer, 1963.

101. Chelomei, V.N.
Dynamic Stability of Aircraft Structure Elements. — Moscow, "Aeroflot," 1939. (Russian)
102. Chetaev, N.G.
Stability of Motion, 2nd edition. — Moscow, Gostekhizdat, 1955. (Russian)
103. Shilov, G.E.
a) Introduction to the Theory of Linear Spaces, 2nd edition. — Moscow, Gostekhizdat, 1956. (Russian)
b) Mathematical Analysis; Special Course. — Moscow, Fizmatgiz, 1961. (Russian)
c) Mathematical Analysis; Finite-Dimensional Linear Spaces. — Moscow, "Nauka," 1969. (Russian)
104. Shtokalo, I.Z.
Linear Differential Equations with Variable Coefficients. — Kiev, Ukr. SSR Acad. Sci., 1960 (Russian)
105. Arscott, F.M.
Periodic Differential Equations. — Oxford, Pergamon Press, 1964.
106. Floquet, G.
Sur les équations différentielles linéaires à coefficients périodiques. — Ann. Ecole Norm., Ser.2, 12 (1883), 47–88.
107. Magnus, W. and S. Winkler.
Hill's Equation. — New York, Wiley (Interscience), 1966.
108. Mathieu, E.
Cours de physique mathématique, 1873.
109. Mettler, E.
a) Dynamic Buckling. Handbook of Engineering Mechanics. — New York, McGraw-Hill, 1962.
b) Stability and Vibration Problems of Mechanical Systems under Harmonic Excitation. Dynamic Stability of Structures. — Oxford, Pergamon Press, 1966.
110. Minorsky, N.
Introduction to Nonlinear Mechanics. — Ann. Arbor, Mich., Edwards, 1947.
111. Poincaré, H.
a) Sur le problème des trois corps et les équations de la dynamique. — Acta Math., 13 (1890), 5–270.
b) Les méthodes nouvelles de la mécanique céleste, Vols. I–III. Paris, 1892–1899.
112. Roseau, M.
Vibrations non linéaires et théorie de la stabilité. — Berlin, Springer, 1966.
113. Stodola, A.
Dampf- und Gasturbinen, 5th edition. — Berlin, 1922.

Articles

114. Adamov, N.V.
a) Sur quelques propriétés des intégrales d'une équation du second ordre à coefficients périodiques. — Compt. Rend. 197, No.22 (1933), 1280–1282.
b) Geometric meaning of Lyapunov's stability condition. — Dokl. Akad. Nauk SSSR **2**, Nos.5–6 (1935), 361–364. (Russian)
c) Some sufficient conditions for stability. — Dokl. Akad. Nauk SSSR **2**, No.7 (1935), 447–450. (Russian)
d) On oscillations of integrals of a second-order equation with periodic coefficients and some stability conditions. — Mat. Sb. **42**, No.6 (1935), 651–668. (Russian)
e) Some properties of transformations which leave the integral curve of a first-order equation invariant. — Dokl. Akad. Nauk SSSR **29**, Nos.8–9 (1940), 539–543. (Russian)
f) On some transformations which leave the integral curves of a first-order differential equation invariant. — Mat. Sb., Novaya Seriya **23**(65), No.2 (1948), 187–228.
115. Aizengendler, P.G.
Application of Newton's diagram to the problem of stability of periodic solutions. — Dokl. Akad. Nauk SSSR **179**, No.5 (1968), 1015–1018. (Russian)
116. Aizerman, M.A.
Sufficient conditions for stability of a class of dynamic systems with variable parameters. — Prikl. Mat. Mekh. **15**, No.3 (1951), 382–384. (Russian)
117. Aizerman, M.A. and F.R. Gantmakher.
a) Stability in the linear approximation of periodic solutions of a system of differential equations with discontinuous right-hand sides.— Dokl. Akad. Nauk SSSR **116**, No.4 (1957), 527–530. (Russian)

b) Stability in the linear approximation of a periodic solution of a system of differential equations with discontinuous right-hand sides. — Prikl. Mat. Mekh. **21**, No.5 (1957), 658–669. (Russian)

118. Akulenko, L.D.
Investigation of stability of resonant motions of some multifrequency systems. — Zh. Vychisl. Mat. i Mat. Fiz. 9, No. 1 (1969), 204–207. (Russian)

119. Arnol'd, V.I.
Small denominators and problems of stability of motion in classical and celestial mechanics. — Uspekhi Mat. Nauk 18, No.6 (1963), 91–192. (Russian)

120. Artem'ev, N.A.
a) Stabilité au sens de Liapounoff et nombre de solutions périodiques. — Compositio Math. **6** (1938), 78–92.
b) A method of determining characteristic exponents and its application to two problems of celestial mechanics. — Izv. Akad. Nauk SSSR Ser. Mat. 8, No.2 (1944), 61–100. (Russian)

121. Bainov, D. and G.V. Plotnikova.
On the stability of single-frequency periodic solutions of nonautonomous quasilinear systems with several degrees of freedom. — In: Nonlinear Vibration Problems. Państwowe Wydawnictwo Naukowe, pp.179–185. Warsaw, 1966.

122. Beyreuther, J.
Dynamic instability domains of almost-canonical systems. — Vestnik Moskov. Gos. Univ., ser.**1**, No.5 (1965), 80–84. (Russian)

123. Blekhman, I.I.
a) On the problem of stability of periodic solutions of quasilinear systems with many degrees of freedom. — Dokl. Akad. Nauk SSSR **104**, No.6 (1955), 809–812. (Russian)
b) On stability of periodic solutions of quasilinear autonomous systems with many degrees of freedom. — Dokl. Akad. Nauk SSSR **112**, No.2 (1957), 183–186. (Russian)

124. Blinov, I.N.
Analytic solutions of a linear system of differential equations with periodic coefficients depending on a parameter. — Differentsial'nye Uravneniya **1**, No.7 (1965), 880–885. (Russian)

125. Boyadzhiev, G. and B. Cheshankov.
Stability of periodic solutions of an autonomous system. — Godishnik Vissh. Tekhn. Uchebn. Zaved., Matem. **2**, No.2 (1965–1967), 113–126. (Bulgarian)

126. Breus, K.A.
a) On reducibility of a canonical system of differential equations with periodic coefficients. — Dokl. Akad. Nauk SSSR **123**, No.1 (1958), 21–24. (Russian)
b) On a class of linear differential equations with periodic coefficients. — Ukrain. Mat. Zh. **12**, No.4 (1960), 463–466. (Russian)
c) Canonical systems of differential equations with periodic coefficients. — Ukrain. Mat. Zh. **18**, No.1 (1966), 3–10. (Russian)

127. Bryuno, A.D.
a) On the formal stability of Hamiltonian systems. — Mat. Zametki **1**, No.3 (1967), 325–330. (Russian)
b) Instability in a Hamiltonian system and the distribution of asteroids. — Mat. Sb. **83**, No.2 (1970), 273–312. (Russian)
c) Analytical form of differential equations. — Trudy Moskov. Mat. Obshch. **25** (1971), 119–262; **26** (1972), 199–239. (Russian)

128. Burdina, V.I.
a) A criterion for boundedness of solutions of a system of second-order differential equations with periodic coefficients. — Dokl. Akad. Nauk SSSR 90, No.3 (1953), 329–332. (Russian)
b) Boundedness of solutions of a system of differential equations. — Dokl. Akad. Nauk SSSR **92**, No.4 (1953), 603–606. (Russian)

129. Valeev, K.G.
a) On the solution and characteristic exponents of solutions of some systems of linear differential equations with periodic coefficients. — Prikl. Mat. Mekh. **24**, No.4 (1960), 585–602. (Russian)
b) Hill's method in the theory of linear differential equations with periodic coefficients. — Prikl. Mat. Mekh. **24**, No.6 (1960), 979–987. (Russian)

c) On a method for solution of systems of linear differential equations with sinusoidal coefficients. — Izv. Vuzov, Radiofizika 3 (1960), 1113–1126. (Russian)

d) Hill's method in the theory of linear differential equations with periodic coefficients. Determination of characteristic exponents. — Prikl. Mat. Mekh. 25, No.2 (1961), 314–318. (Russian)

e) On the stability of solutions of a system of two linear differential equations of first order with periodic coefficients in the resonant case. — Prikl. Mat. Mekh. 25, No.4 (1961), 794–796. (Russian)

f) On the stability of solutions of second-order linear differential equations with sinusoidal coefficients. — Izv. Vuzov, Radiofizika 5, No. 4 (1962), 766–783. (Russian)

g) On convergence of series determining the boundaries of instability domains of solutions of a second-order linear differential equation with periodic coefficients. — Prikl. Mat. Mekh. 27, No.3 (1963), 565–572. (Russian)

h) On a system of linear differential equations with simple harmonic coefficients. — Izv. Akad. Nauk SSSR, Mekh. i Mashinostr., No.5 (1963), 203–205. (Russian)

i) On the danger of combination resonances. — Prikl. Mat. Mekh. 27, No.6 (1963), 1134–1142. (Russian)

j) Application of the Laplace transform to the solution and investigation of stability of linear delay-differential equations with variable coefficients. — Trudy Sem. Teor. Differentsial. Uravnenii s Otklon. Argumentom 1 (1967), 51–78. (Russian)

k) Investigation of oscillations in a quasilinear autonomous system in the resonant case. — Kiev, Prikl. Mekh. 5, No.4 (1969), 25–31. (Russian)

130. Vishik, M.I. and L.A. Lyusternik.

Solution of some perturbation problems in the case of matrices and selfadjoint and nonselfadjoint differential equations, I. — Uspekhi Mat. Nauk 15, No.3 (93) (1960), 3–80. (Russian)

131. Volk, I.M.

On the stability of periodic motions when the equations and their periodic solution are known only approximately. — Prikl. Mat. Mekh. 12, No.5 (1948), 647–650. (Russian)

132. Gel'fand, I.M. and V.B. Lidskii.

On the structure of stability domains of linear canonical systems of differential equations with periodic coefficients. — Uspekhi Mat. Nauk 10, No.1 (63) (1955), 3–40. (Russian)

133. Golokvoschyus, P.B.

a) Determination of the characteristic exponents of a system of two homogeneous linear differential equations with periodic coefficients containing a small parameter. — Dokl. Akad. Nauk BSSR 3 (1959), 361–367. (Russian)

b) Remarks on bounded and periodic solutions of a system of two linear differential equations with periodic coefficients integrable in closed form. — Izv. Vuzov., Matematika, No. 3 (1960), 113–117. (Russian)

c) Determination of characteristic exponents of a system of two differential equations with periodic coefficients which are analytic with respect to a small parameter. — Dokl. Akad. Nauk BSSR 4, No.6 (1960), 236–240. (Russian)

d) On bounded and periodic solutions of a certain system of two linear differential equations with periodic coefficients. — Uchen. Zap. Vil'n. Univ. Ser. Mat. Fiz. 33, No.9 (1960), 5–13. (Russian)

e) Sufficient conditions for the boundedness of all solutions of a system of two equations containing a small parameter. — Uchen. Zap. Vil'n. Univ. Ser. Mat. Fiz. 33, No.9 (1960), 21–35. (Russian)

f) Determination of the characteristic exponents of a system of two homogeneous linear differential equations. — Litovsk. Mat. Sb. 1, Nos.1–2 (1961), 59–74. (Russian)

g) Determination of the characteristic exponents of solutions of a class of systems of differential equations with periodic coefficients. — Litovsk. Mat. Sb. 3, No.1 (1963), 77–101. (Russian)

134. Gol'din, A.M.

On a certain criterion of Lyapunov. — Prikl. Mat. Mekh. 15, No. 3 (1951), 379–381. (Russian)

135. Gorbunov, A.D.

a) On conditions for asymptotic stability of the trivial solution of ordinary homogeneous linear differential equations. — Vestnik Moskov. Gos. Univ. 9 (1953), 49–55. (Russian)

b) Some problems in the qualitative theory of ordinary homogeneous linear differential equations with variable coefficients. — Uchen. Zap. Moskov. Gos. Univ., Matematika 7, No.165 (1954), 39–78. (Russian)

136. Gorelik, G.S.
Resonance phenomena in linear systems with periodically varying parameters. — Zh. Tekhn. Fiz. **4**, No.10 (1934), 1783—1817; **5**, No.2 (1935), 195—215. (Russian)

137. Gokhberg, I.Ts. and M.G. Krein.
On the effect of some transformations of the kernels of integral equations on the spectra of these equations. — Ukrain. Mat. Zh. **13**, No.3 (1961), 12—38. (Russian)

138. Gusarov, L.A.
a) On the boundedness of solutions of a second-order linear equation. — Dokl. Akad. Nauk SSSR **18**, No.2 (1949), 217—220. (Russian)
b) On the convergence of solutions of a linear second-order differential equation to zero. — Dokl. Akad. Nauk SSSR 21, No.1 (1950), 9—12. (Russian)

139. Gusarova, R.S.
a) On the boundedness of solutions of linear differential equations with periodic coefficients. — Prikl. Mat. Mekh. **13**, No.3 (1949), 211—246. (Russian)
b) Criteria for the boundedness of solutions of second-order linear equations. — Uspekhi Mat. Nauk **4**, No.3 (1949), 132—133. (Russian)
c) On the boundedness of solutions of a linear differential equation with periodic coefficients. — Prikl. Mat. Mekh. **14**, No.3 (1950), 313—314. (Russian)

140. Demidovich, B.P.
a) On stability in Lyapunov's sense of a linear system of ordinary differential equations. — Mat. Sb. 28 (70) (1951), 659—684. (Russian)
b) On some properties of the characteristic exponents of a system of ordinary linear differential equations with periodic coefficients. — Uchen. Zap. Moskov. Gos. Univ., Matematika **6**, No.163 (1952), 123—132. (Russian)

141. Derguzov, V.I.
a) On the stability of solutions of Hamiltonian equations in Holbert space with unbounded periodic operator coefficients. — Dokl. Akad. Nauk SSSR, Vol.152, No.6, 1963. (Russian)
b) On the stability of solutions of Hamiltonian equations with unbounded periodic operator coefficients.— Mat. Sb. **63 (105)**, No.4 (1964), 591—619. (Russian)
c) Sufficient conditions for the stability of Hamiltonian equations with unbounded periodic operator coefficients. — Mat. Sb. **64 (106)**, No.3 (1964), 419—435. (Russian)
d) Necessary conditions for the strong stability of Hamiltonian equations with unbounded periodic operator coefficients. — Vestnik Leningrad. Gos. Univ., No.19, Ser. Mat. Mekh. Astron., No.4 (1964), 18—30. (Russian)
e) Continuous dependence of the maximum exponent of exponential growth of solutions of a linear Hamiltonian equation with periodic operational coefficients. — In: Problemy matematicheskogo analiza, pp.120—134. Leningrad. Gos. Univ., 1966. (Russian)

142. Derguzov, V.I., V.N. Fomin, and V.A. Yakubovich.
Some mathematical methods for investigation of equations of dynamic stability of systems with many degrees of freedom. — In: Problemy ustoichivosti v stroitel'noi mekhanike, pp.378—389. Moscow, Izdatel'stvo Lit. po Stroitel'stvu, 1965. (Russian)

143. Derguzov, V.I. and V.A. Yakubovich.
Existence of solutions of linear Hamiltonian equations with unbounded operator coefficients. — Dokl. Akad. Nauk SSSR **151**, No.6 (1963), 1264—1267. (Russian)

144. El'shin, M.I.
On a method for computing the phase of a second-order linear equation. — Uchen. Zap. Moskov. Gos. Univ., Matematika, No.45 (1940), 97—107. (Russian)

145. Endovitskii, I.I.
Stability conditions for a linear differential equation with periodic coefficients. — Izv. Vuzov, Matematika, No.5 (1967), 28—33. (Russian)

146. Erugin, N.P.
a) On the stability of solutions of a system of homogeneous linear differential equations with periodic (and other) coefficients. — Prikl. Mat. Mekh. **23**, No.5 (1959), 818—825. (Russian)
b) Methods for investigation of the stability of solutions of linear systems of differential equations with periodic coefficients containing a small parameter. — Inzh.-Fiz. Zh. Akad. Nauk BSSR **3**, No.2 (1960), 115—127. (Russian)

c) Solution of existence problems for bounded solutions of a system of homogeneous linear differential equations with periodic coefficients on the basis of an integral substitution, Part I. — Izv. Akad. Nauk BSSR, Ser. Fiz.-Tekhn. Nauk, No.4 (1961), 24–29; Part 2, ibid. (1962), 5–12.

d) On periodic solutions of a homogeneous linear system of differential equations. — Dokl. Akad. Nauk BSSR **6**, No.7 (1962), 407–410. (Russian)

e) On periodic and bounded solutions of the equation $\ddot{x}+p(t)x=0$, $p(t+1)=p(t)$. — Izv. Akad. Nauk BSSR, Ser. Fiz.-Tekhn. Nauk, No.1 (1963), 5–13. (Russian)

f) On the radius of convergence of series representing a periodic solution of a linear system of differential equations as a function of the parameters. — Dokl. Akad. Nauk BSSR **7**, No.2 (1963), 73–75. (Russian)

147. Zhukovskii, N.E.

Finiteness conditions for integrals of the equation $d^2y/dx^2+py=0$. — Mat. Sb. **16**, No.3 (1892), 582–591. (Russian)

148. Zadiraka, K.V. and G.A. Los'.

a) On the problem of stability and instability of solutions of n-th order linear differential equations with periodic coefficients. — Ukrain. Mat. Zh. **18**, No.6 (1966), 117–122. (Russian)

b) New method of construction of the characteristic equation for second-order differential equations with periodic [coefficients]. — Proc. Fourth Conf. of Nonlinear Oscillations, Prague, pp.285–292, 1968.

c) Construction of solutions of second-order linear differential equations with periodic coefficients with the argument varying in an arbitrary interval. — Mat. Fiz. (Kiev), No.6 (1969), 83–93. (Russian)

149. Zatepyakin, M.M.

a) On the phenomenon of parametric resonance in torsional vibrations of crankshafts. — Ukrain. Mat. Zh. **14**, No.4 (1962), 403–407. (Russian)

b) On the stability of motion in a case of the three-body problem. — In: Priblizhennye metody reshenii differentsial'nykh uravnenii, pp.27–32. Kiev, Akad. Nauk Ukrain. SSR, 1963. (Russian)

c) Application of the method of a small parameter to calculation of the instability zones of a system of differential equations with periodic coefficients. — Proceedings, 6th Inter-Univ. Phys.-Math. Scient. Conf. (Far East), Vol.3, pp.87–92, 1967. (Russian)

150. Kalinin, S.V.

a) On the stability of periodic motions in the case of one root equal to zero. — Prikl. Mat. Mekh. **12**, No.5 (1948), 671–672. (Russian)

b) On the stability of periodic motions in the case of one root equal to zero. — Prikl. Mat. Mekh. **13**, No.3 (1949), 247–252. (Russian)

c) On the stability of periodic motions in the critical case of a characteristic equation with one pair of pure imaginary roots (simplified method). — Prikl. Mat. Mekh. **21**, No.1 (1957), 125–128. (Russian)

d) On the stability of periodic motions in the case of two zero roots. — Prikl. Mat. Mekh. **23**, No.5 (1959), 975–977. (Russian)

e) On the stability of periodic motions in some critical cases. — Izv. Akad. Nauk SSSR, Otdel. Tekhn. Nauk, Mekh. i Mashinostr., No.4 (1961), 79–82. (Russian)

f) On the stability of periodic motions in the case of several zero roots, pure imaginary roots, and roots with negative real parts. — Izv. Akad. Nauk SSSR, Otdel. Tekhn. Nauk, Mekh. i Mashinostr., No.6 (1961), 167–169. (Russian)

151. Kantorovich, L.V.

Functional analysis and applied mathematics. — Uspekhi Mat. Nauk **3**, No.6 (28) (1948), 89–185. (Russian)

152. Kamenkov, G.V.

a) On the problem of stability of motion in critical cases. — Prikl. Mat. Mekh. **29**, No.6 (1965), 1053–1069. (Russian)

b) On the stability of periodic motions. — Prikl. Mat. Mekh. **31**, No.1 (1967), 15–36. (Russian)

153. Kapralov, N.S.

On gyroscopic destabilization. — Uchen. Zap. Ryazan. Gos. Ped. Inst. **24** (1960), 33–35. (Russian)

154. Karaseva, T.M.
a) On a criterion for boundedness of solutions of Hill's differential equations. — Prikl. Mat. Mekh. **20**, No.4 (1956), 548–551. (Russian)
b) On a "sharp estimate" for the multipliers of second-order differential equations with periodic coefficients. — Dokl. Akad. Nauk SSSR **121**, No.1 (1958), 34–36. (Russian)
c) A boundedness criterion and a sharp estimate for the multipliers of solutions of Hill's equations. — Dokl. Akad. Nauk SSSR **127**, No.6 (1959), 1161–1163. (Russian)
d) On the rate of growth and the boundedness of solutions of second-order differential equations with periodic coefficients. — Izv. Akad. Nauk SSSR, Ser. Mat. **29**, No.1 (1965), 41–70. (Russian)

155. Kovalenko, K.R.
a) On the effect of the external and internal friction on the dynamic stability of bars. — Prikl. Mat. Mekh. **23**, No.2 (1959), 239–248. (Russian)
b) On a problem of G. Borg. — Ukrain. Mat. Zh. **13**, No.1 (1961), 95–100. (Russian)

156. Kovalenko, K.R. and M.G. Krein.
On some investigations of Lyapunov concerning differential equations with periodic coefficients. — Dokl. Akad. Nauk SSSR **75**, No.4 (1950), 495–498. (Russian)

157. Kovtun, I.I.
On the problem of stability of differential equations with periodic coefficients. — Trudy Sem. Differentsial. i Integral. Uravneniyam, Inst. Mat. Akad. Nauk Ukrain. SSR, No.1 (1969), 239–244. (Russian)

158. Kozinets, B.N.
On a stability test for Hill's equation. — Vestnik Leningrad. Gos. Univ., No.13, Ser. Mat. Mekh. Astron., No.3 (1964), 18–24. (Russian)

159. Komlenko, Yu.V.
a) On some criteria for nonoscillation and boundedness of solutions of linear differential equations. — Dokl. Akad. Nauk SSSR **164**, No.2 (1965), 270–272. (Russian)
b) On Chaplygin's method for the Cauchy problem. — Differentsial'nye Uravneniya **1**, No.7 (1965), 903–907. (Russian)
c) On some estimates for the interval of applicability of Chaplygin's theorem on differential inequalities and stability tests for differential equations with periodic coefficients. — Izv. Vuzov, Matematika, No.1 (50) (1966), 99–103. (Russian)
d) Conditions for solvability of some boundary-value problems for an ordinary linear second-order differential equation. — Dokl. Akad. Nauk SSSR **174**, No.5 (1967), 1018–1020. (Russian)

160. Koroza, V.I.
a) On a variational method for the boundary-value problem of wave propagation. — Dokl. Akad. Nauk SSSR **186**, No.4 (1969), 795–798. (Russian)
b) Stationary functionals for the problem of wave propagation in electromagnetic waveguides. — In: Uskoriteli, No.11. Atomizdat, 1969. (Russian)
c) The Ritz-Kantorovich method in the problem of wave propagation in irregular electromagnetic waveguides. — In: Uskoriteli, No.11. Atomizdat, 1969. (Russian)
d) Parametric resonance in propagation of electromagnetic waves in periodic moderators. — Radiotekhnika i Elektronika **15**, No.3 (1970), 450–455. (Russian)

161. Koroza, V.I., E.S. Masunov, and V.M. Starzhinskii.
Methods of the theory of oscillations in electrodynamics of waveguide systems. — Proceedings 5th Intern. Conf. on Nonlinear Oscillations, Vol.4, pp.258–264. Inst. Mat. Akad. Nauk Ukrain. SSR, Kiev, 1970. (Russian)

162. Koroza, V.I. and V.M. Starzhinskii.
a) Parameterresonanz beim Problem der Wellenausbreitung. — Wiss. Z. Techn. Hochsch. Karl-Marx-Stadt. **11**, No.3 (1969), 319–323.
b) Brillouin diagrams of one-dimensional periodic structures. — Vestnik Moskov. Gos. Univ., Ser.1, No.5 (1969), 110–116. (Russian)
c) On the theory of interaction of beams of charged particles with moderating systems. — Vestnik Moskov. Gos. Univ., Ser.1, No.6 (1970), 117–123. (Russian)
d) Some methods in the theory of oscillations in electrodynamics of wave systems. — Vestnik Moskov. Gos. Univ., Ser.1, No.1 (1971), 101–110. (Russian)

e) Parametric resonance in the problem of wave propagation in periodic structures. — Proceedings 5th Intern. Conf. on Nonlinear Oscillations, Vol.4, pp.248—257. Inst. Mat. Akad. Nauk Ukrain. SSR, Kiev, 1970. (Russian)

163. Kochin, N.E.
On torsional vibrations of crankshafts. — Prikl. Mat. Mekh., Staraya Seriya 2, No.1 (1934), 3—28; Collected Works, Vol.2, 507—535. Moscow, USSR Acad. Sci., 1949. (Russian)

164. Kotov, E.O.
On the analysis of linear systems with variable parameters, II. Case of rapid periodic variation of parameters. — Avtomat. i Telemekh. 28, No.7 (1967), 15—22. (Russian)

165. Krasnosel'skii, M.A.
On the stability of periodic solutions originating in equilibrium. — Dokl. Akad. Nauk SSSR 150 (1963), 463—466. (Russian)

166. Krein, M.G.
a) Generalization of some investigations of Lyapunov concerning linear differential equations with periodic coefficients. — Dokl. Akad. Nauk SSSR 73, No.3 (1950), 445—448. (Russian)
b) On the application of an algebraic proposition in the theory of monodromy matrices. — Uspekhi Mat. Nauk 6, No.1 (1951), 171—177. (Russian)
c) On the theory of entire matrix-functions of exponential type. — Ukrain. Mat. Zh. 3, No.2 (1951), 164—173. (Russian)
d) On some maximum and minimum problems for characteristic numbers and Lyapunov stability zones. — Prikl. Mat. Mekh. 15, No.3 (1951), 323—348. (Russian)
e) On criteria for stability and boundedness of solutions of periodic canonical systems. — Prikl. Mat. Mekh. 19, No.6 (1955), 641—680. (Russian)
f) Fundamental aspects of the theory of λ-zones of stability for a canonical system of linear differential equations with periodic coefficients. — In: Sb. pamyati A.A. Andronova, pp.413—498. Moscow, Akad. Nauk SSSR, 1955. (Russian)
g) On the characteristic function $A(\lambda)$ of a linear canonical system of second-order differential equations with periodic coefficients. — Prikl. Mat. Mekh. 21, No.3 (1957), pp.320—329. (Russian)
h) Introduction to the geometry of indefinite J-spaces and the theory of operators in these spaces. — In: Vtoraya letnyaya mat. shkola, pp.15—92. Kiev, "Naukova Dumka," 1965. (Russian)

167. Krein, M.G. and G.Ya. Lyubarskii.
a) On the theory of pass-bands of periodic waveguides. — Prikl. Mat. Mekh. 25, No.1 (1961), 24—37. (Russian)
b) On analytical properties of multipliers of periodic canonical differential systems of positive type. — Izv. Akad. Nauk SSSR, Ser. Mat. 26, No.4 (1962), 542—572. (Russian)

168. Krein, M.G. and V.A. Yakubovich.
Hamiltonian systems of linear differential equations with periodic coefficients. — Proceedings Intern. Symp. on Nonlinear Oscillations. Izv. Akad. Nauk Ukrain. SSR, 1 (1963), 277—305. (Russian)

169. Krylov, A.N.
On equilibrium modes of uprights under compression. — Izv. Akad. Nauk SSSR, OM i ON, No.7, 1931. (Russian)

170. Krylov, N.M. and N.N. Bogolyubov.
Investigation of resonance phenomena in transverse vibrations of bars under the effect of periodic normal forces applied to an end of the bar. — In: Issledovanie kolebanii konstruktsii, pp.25—42. Khar'kov-Kiev, Depart. Scient. and Techn. Information (ONTI), 1935. (Russian)

171. Kuptsov, N.P.
On a sufficient condition for stability of the trivial solution of a system of two linear differential equations. — Prikl. Mat. Mekh. 25, No.5 (1961), 791—793. (Russian)

172. Kushul', M.Ya.
a) On quasiharmonic systems close to systems with constant coefficients, the pure imaginary roots of whose fundamental equation have multiple elementary divisors. — Prikl. Mat. Mekh. 22, No.4 (1958), 519—533. (Russian)
b) Stability of periodic solutions of quasilinear elastic gyroscopic systems with distributed and lumped parameters. — Prikl. Mat. Mekh. 34, No.1 (1970), 115—126. (Russian)

173. Lavrent'ev, M. A. and A. Yu. Ishlinskii.
Dynamic modes of loss of stability in elastic systems. — Dokl. Akad. Nauk SSSR, Vol. 65, No. 6, 1949. (Russian)

174. Levin, A. Yu.
a) On the stability of solutions of second-order equations. — Dokl. Akad. Nauk SSSR **141**, No. 6(1961), 1298—1301. (Russian)
b) On a stability test. —Uspekhi Mat. Nauk. **17**, No. 3(105) (1962), 211—212. (Russian).
c) On the problem of the zeroth stability zone.— Dokl. Akad. Nauk SSSR **145**, No. 6(1962), 1221—1223. (Russian)

175. Levin, B. Ya.
An estimate for the growth of the integral of the Sturm-Liouville equation. — Trudy Odess'k. Derzhav. Univ., Matematika **2** (1938), 39—45. (Ukrainian)

176. Leonov, M. Ya.
a) On quasiharmonic oscillations.—Prikl. Mat. Mekh. **10**, Nos. 5—6(1946), 575—580. (Russian)
b) Some criteria for dynamic stability.—Prikl. Mat. Mekh. **12**, No. 6(1948), 737—748. (Russian)

177. Lidskii, V. B. and M. K. Neigauz.
On criteria for the stability of a system of differential equations.—Prikl. Mat. Mekh. **19**, No. 5(1955) 625—627. (Russian)

178. Lidskii, V.B. and P. A. Frolov.
a) On the topological structure of the stability domains of a selfadjoint system of differential equations with periodic coefficients. — Dokl. Akad. Nauk SSSR **161**, No. 4(1965), 764—766. (Russian)
b) Stability domains of complex selfadjoint systems of differential equations. — Dokl. Akad. Nauk SSSR **171**, No. 1(1966), 41—43. (Russian)
c) Structure of stability domains of a selfadjoint system of differential equations with periodic coefficients. — Mat. Sb. **71(113)**, No. 1(1966), 48—64. (Russian)

179. Los', G. A.
a) On determination of stability domains of a differential equation with periodic coefficients. — Ukrain. Mat. Zh. **18**, No. 4(1966), 110—116. (Russian)
b) A sufficient condition for stability of the trivial solution of a third-order differential equation with periodic coefficients. — Differentsial'nye Uravneniya **3**, No. 10(1967), 1707—1717. (Russian)
c) Theorems on the instability of solutions of a homogeneous linear differential equation of n-th order with periodic coefficients.— Differentsial'nye Uravneniya **4**, No. 4(1968), 625—630. (Russian)
d) Theorems on estimation of the invariant A_1 of the characteristic equation of a homogeneous linear differential equation of n-th order with ω-periodic coefficients.— Differentsial'nye Uravneniya **6**, No. 3(1970), 561—564. (Russian)
e) Solution of the problem of stability of a system of linear differential equations with π-periodic coefficients using the $\pi/4$-characteristic equation. — Differentsial'nye Uravneniya **6**, No. 3(1970), 565—568. (Russian)
f) The $\pi/4$-characteristic equation and stability of solutions of a system of homogeneous linear differential equations. — In: Proceedings 5th Intern. Conf. on Nonlinear Oscillations, Vol. 2, pp. 304—310. Kiev, 1970. (Russian)

180. Lyapunov, A. M.
a) On stability of motion in a special case of the three-body problem. — Soobshch. Khar'kovsk. Mat. Obshch., Vtoraya Seriya **2**, Nos. 1—2(1889), 1—94; Collected Works, Vol. 1, pp. 327—401. Moscow, 1954. (Russian)
b) On a problem concerning second-order linear differential equations with periodic coefficients. — Soobshch. Khar'kovsk. Mat. Obshch., Vtoraya Seriya **5**, Nos. 3 and 4, 5 and 6(1896—1897), 190—254; Collected Works, Vol. 2, pp. 332—386. Moscow, 1956. (Russian)
c) Sur une série relative à la théorie des équations différentielles linéaires à coefficients périodiques.— Compt. Rend. **123**, No. 26(1896), 1248—1252; Collected Works, Vol. 2, pp. 387—390. Moscow, 1956.
d) Sur une équation différentielle linéaire du second ordre. —Compt. Rend. **128**, No. 15(1899), 910—913; Collected Works, Vol. 2, pp. 401—403. Moscow, 1956.
e) Sur une équation transcendante et les équations différentielles linéaires du second ordre à coefficients périodiques. —Compt. Rend. **128**, No. 18(1899), 1085—1088; Collected Works, Vol. 2, pp. 404—406. Moscow, 1956.

f) Sur une série relative à la théorie d'une équation différentielle linéaire du second ordre. —Compt. Rend. **131**, No. 26(1900), 1185—1188; Collected Works, Vol. 2, pp. 407—408. Moscow, 1956.

g) Sur une série dans la théorie des équations différentielles linéaires du second ordre à coefficients périodiques. —Zap. Akad. Nauk Fiz. —Mat. Otdel., Vos'maya Seriya **13**, No. 2(1902), 1—70; Collected Works, Vol. 2, pp. 410—472. Moscow, 1956.

181. Makarov, S.M.

a) Investigation of the characteristic equation of a linear system of two first-order equations with periodic coefficients. —Prikl. Mat. Mekh. **15**, No. 3(1951), 373—378. (Russian)

b) Investigation of the characteristic equation of a linear system of two first order equations with periodic coefficients. —Trudy Kuibyshev. Aviats. Inst., No. 1(1952), 24—29. (Russian)

c) Some sufficient conditions for stability of motion for a linear system of two first-order equations.— Trudy Kuibyshev. Aviats. Inst., No.3(1957), 27—30. (Russian)

182. Malkin, I.G.

On the characteristic numbers of linear differential equations. —Prikl. Mat. Mekh. **16**, No. 1(1952), 3—14. (Russian)

183. Mandel'shtam, L.I. and N.D. Papaleksi.

On the parametric excitation of electrical oscillations. —Zh. Tekhn. Fiz., No. 3, 1934. (Russian). Also published in: Mandel'shtam, L.I. Complete Works, Vol. 2, Moscow, Akad. Nauk SSSR, 1947. (Russian).

184. Markhashov, L.M.

On the characteristic exponents of solutions of a second-order linear differential equation with periodic coefficients. —Prikl. Mat. Mekh. **23**, No. 6(1959), 1066—1073. (Russian)

185. Merman, G.A.

On the instability of a periodic solution of a canonical system with one degree of freedom in the case of principal resonance. —In: Problemy dvizhenii iskusstvennykh nebesnykh tel. Moscow, Akad. Nauk SSSR, 1963. (Russian)

186. Millionshchikov, V.M.

a) On the instability of characteristic exponents of statistically correct systems. —Mat. Zametki **2**, No. 3(1967), 315—318. (Russian)

b) Statistically correct systems. — Mat. Sb. **75(117)**, No.1(1968), 140—151. (Russian)

c) Metric theory of linear systems of differential equations. —Dokl. Akad. Nauk SSSR **179**, No. 1 (1968); 20—23. (Russian)

d) On the spectral theory of nonautonomous systems of differential equations. —Trudy Moskov. Mat. Obshch., **18**(1968), 146—186. (Russian)

187. Moiseenko, D.Ya.

On the stable boundedness of solutions of a second-order vector equation with periodic coefficients.— Ukrain. Mat. Zh. **21**, No. 1(1969), 117—120. (Russian)

188. Neigauz, M.G. and V.B. Lidskii.

On the boundedness of solutions of linear systems of differential equations with periodic coefficients. — Dokl. Akad. Nauk SSSR,77, No.1(1951), 25—28. (Russian)

189. Nugmanova, Sh.

On the stability of periodic motions. —Dokl. Akad. Nauk SSSR **42**, No. 5(1944), 206—208. (Russian)

190. Persidskii, K.P.

On an estimate for characteristic numbers. —Izv. Akad. Nauk Kazakh. SSR, No. 56, Ser. Mat. i Mekh., No. 2(1948). 36—45. (Russian)

191. Pittel', B.G.

a) On application of the method of a small parameter in a problem of dynamic stability. —In: Metody vychislenii, No. 1, pp. 66—75. Leningrad. Gos. Univ., 1963. (Russian)

b) An algorithm for construction of the boundaries of instability domains of linear Hamiltonian systems with periodic coefficients. —In: Metody vychislenii, No. 4, pp. 151—163. Leningrad. Gos. Univ., 1967. (Russian)

192. Pittel', B.G. and G.I. Yuzefovich.

Construction of dynamic instability domains of canonical systems with periodic coefficients. — Vestnik Leningrad, Gos. Univ., No. 1, Ser. Mat. Mekh. Astron., No. 1(1962), 89—101. (Russian)

193. Pittel', B.G. and V.A. Yakubovich.

a) Application of the theory of parametric resonance to explain the collapse of the Tacoma Narrows bridge. —Uspekhi Mat. Nauk **15**, No. 6(96) (1961), 183—184. (Russian)

b) Mathematical analysis of the stability of suspension bridges, illustrated by application to the Tacoma Narrows bridge.—Vestnik Leningrad. Gos. Univ., No. 1, Ser. Mat. Mekh. Astron., No. 1(1969), 80—91. (Russian)

194. Plotnikova, G.V.

a) On the stability of periodic solutions of nonautonomous quasilinear systems with one degree of freedom.—Prikl. Mat. Mekh. 27, No. 1(1963), 168—170. (Russian)

b) On the stability of periodic solutions of nonautonomous quasilinear systems with two degrees of freedom.—Prikl. Mat. Mekh. 29, No. 6(1965), 1084—1091. (Russian).

195. Proskuryakov, A.P.

a) Characteristic numbers of solutions of a second-order differential equation with periodic coefficients.—Prikl. Mat. Mekh. 10, Nos. 5—6(1946), 545—558. (Russian)

b) Stability of periodic solutions of quasilinear autonomous systems with one degree of freedom.—Prikl. Mat. Mekh. 27, No. 3(1963), 559—564. (Russian)

c) Stability of single-frequency periodic solutions of quasilinear autonomous systems with two degrees of freedom.—Prikl. Mat. Mekh. 29, No. 5(1965), 939—945. (Russian).

d) Stability of periodic solutions of quasilinear autonomous systems with several degrees of freedom.—Prikl. Mat. Mekh. 34, No. 1(1970), 105—114. (Russian)

196. Pyt'ev, Yu.P.

Stability domains of an equation with periodic coefficients.—Prikl. Mat. Mekh. 25, No. 2(1961), 294—302. (Russian)

197. Razumikhin, B.S.

Estimates of solutions of a system of differential equations of perturbed motion with variable coefficients.—Prikl. Mat. Mekh. 21, No. 1(1957), 119—120. (Russian)

198. Rapoport, I.M.

a) On linear differential equations with periodic coefficients.—Dokl. Akad. Nauk SSSR 76, No. 5(1951). 793—796. (Russian)

b) On the problem of stability of oscillations of a material system.—Dokl. Akad. Nauk SSSR 77, No. 1(1951), 25—28. (Russian)

199. Rozenvasser, E.N.

a) On forced oscillations and stability of quasiharmonic systems.—Prikl. Mat. Mekh. 25, No. 2(1961), 284—293. (Russian)

b) On a method for investigating the stability of a system of linear differential equations with periodic coefficients.—Dokl. Akad. Nauk SSSR 176, No. 4(1967), 783—786. (Russian)

200. Roitenberg, Ya.N.

On a method for constructing Lyapunov functions for linear systems with variable coefficients.—Prikl. Mat. Mekh. 22, No. 2(1958), 167—172. (Russian)

201. Rubanovskii, V.N.

a) On the problem of boundedness of solutions of Hill's equation.—Vestnik Moskov. Gos. Univ. Mat. Mekh., No. 1(1967), 99—102. (Russian)

b) On the problem of boundedness of solutions of periodic canonical systems of fourth order.—Vestnik Moskov. Gos. Univ. Mat. Mekh., No. 5(1968), 78—87. (Russian)

202. Sabirov, T.K.

a) On the problem of stability of small periodic solutions.—Dokl. Akad. Nauk SSSR 167, No. 4(1966), 755—757. (Russian)

b) On the stability of small periodic solutions in multidimensional degenerations.—Dokl. Akad. Nauk SSSR 181, No. 4(1968), 805—807. (Russian)

203. Sagitov, M.S.

On the stability of periodic motions in a certain critical case.—Izv. Akad. Nauk Uzbek. SSR, Ser. Tekhn. Nauk, No. 4(1966), 66—67. (Russian)

204. Starzhinskii, V.M.

a) Sufficient conditions for stability of a mechanical system with one degree of freedom.—Prikl. Mat. Mekh. 16, No. 3(1952), 369—374. (Russian)

b) On the stability of a mechanical system with one degree of freedom.—Prikl. Mat. Mekh. 17, No. 1(1953), 117—122. (Russian)

c) On the stability of the trivial solution of a second-order differential equation with periodic coefficients.—Inzh. Sb. 18(1954), 119—138. (Russian)

d) Remark on investigation of the stability of periodic motions.—Prikl. Mat. Mekh. 19, No. 1(1955), 119—120. (Russian)

e) On the stability of periodic motions in a certain special case. — Prikl. Mat. Mekh. 21, No. 5(1957), 720 — 722. (Russian)

f) On the problem of boundedness of solutions of a system of linear differential equations with periodic coefficients. — In: Proceedings 3rd All-Union Math. Congress, Vol. 4, pp. 37 — 39. 1959. (Russian)

g) On the stability of the trivial solution of linear systems with periodic coefficients. — Prikl. Mat. Mekh. 22, No. 5(1958), 646 — 656. (Russian)

h) On the stability of periodic motions, Part I. — Bul. Inst. Politehn. Iaşi 4(8), Nos. 3 — 4(1958), 19 — 68; Part V, 5(9), Nos. 1—2 (1959), 51—100. (Russian)

i) Torsional oscillations of crankshafts of looms. — Nauch. Dokl. Vyssh. Shkoly, Mashinostr. i Priborostr., No. 1(1959), 51 — 57. (Russian)

j) On Lyapunov's method for estimation of the characteristic constant. — Izv. Akad. Nauk SSSR, Otdel. Tekhn. Nauk, Mekh. i Mashinostr., No. 4 (1959), 46 — 55. (Russian)

k) On the stability of the trivial solution of two linear differential equations with periodic coefficients. — Prikl. Mat. Mekh. 24, No. 3(1960), 578 — 581. (Russian)

l) The stability of periodic motions in a special case. — Royal Aircraft Establishment, No. 8883(1960), 2 — 4.

m) Free, imperfectly elastic oscillating circuits. — Izv. Akad. Nauk SSSR, Otdel. Tekhn. Nauk, Mekh. i Mashinostr., No. 6(1961), 68 — 73. (Russian)

n) Free, perfectly elastic oscillating circuits. — Prikl. Mat. Mekh. 26, No. 1(1962), 172 — 181. (Russian)

o) Oscillating circuits. — Proceedings 1st Intern. Symp. on Nonlinear Oscillations, Vol.1, pp.446—455, Kiev, Akad. Nauk Ukrain. SSR, 1963. (Russian)

p) On the stability of periodic motions, Part I. — Amer. Math. Soc. Translations, Ser. 2, Vol. 33 (1963), 59 — 121; Part II, ibid., pp. 123 — 187.

q) On the stability of periodic regimes. — Nonlinear Vibration Problems 5(1964), pp. 360 — 369. 1964.

r) Parametric resonance in almost canonical systems. — Inzh. Zh. Mekh. Tverd. Tela 3, No. 3(1967), 174 — 180. (Russian)

s) On a variant of a method for determination of periodic solutions. — Inzh. Zh. Mekh. Tverd. Tela 4, No. 6(1968), 67 — 71. (Russian)

t) To the theory of parametric resonance. — Proc. Fourth Conf. of Nonlinear Oscillations, Prague, pp. 475 — 480. 1968.

u) Dynamic stability of thin-walled rods loaded with longitudinal periodic forces. — Proc. Fourth Conf. of Nonlinear Oscillations, Prague, pp. 467 — 474. 1968.

v) Anwendungen der Theorie linearer Differentialgleichungen mit periodischen Koeffizienten in der Mechanik. — Mitt. math. Gesellsch. DDR, No. 1 (1971).

w) Process of break-off of spring oscillations of a simple pendulum. — Mekh. Tverd. Tela, No.2(1971), 154 — 156. (Russian)

205. Starzhinskii, V.M. and J. Beyreuther.
Parameterresonanz in fastkanonischen Systemen. — Z. Angew. Math. Mech. 46, No.7(1966), 459 — 464. Translations.

206. Starzhinskii, V.M. and V.I. Koroza.
On the theory of periodic waveguides, Part I. — Bul. Inst. politehn. Iaşi, 15 (19), Nos. 3—4(1969); Part II 16(20), Nos. 3—4(1971), 31—37. (Russian)

207. Tonkov, E. L.
On a second-order periodic equation. — Dokl. Akad. Nauk SSSR 184, No. 2(1969), 296 — 299. (Russian)

208. Tonkov, E. L. and G. I. Yutkin.
Periodic solutions and stability of a linear differential equation with periodic coefficients. — Differentsial'nye Uravneniya 5, No. 11(1969), 1990 — 2001. (Russian)

209. Tribeleva, T. N.
On the stability of periodic motions. — Dokl. Akad. Nauk SSSR 133, No. 2(1960), 292 — 295. (Russian)

210. Uryupin, A. G.
Stability of flexural vibrations of coaxial shafts. — Izv. Akad. Nauk SSSR, Mekh. i Mashinostr., No. 1(1964), 180 — 182. (Russian).

211. Filippov, V.A.
On sufficient conditions of stability for linear systems. — Inzh. Zh., Mekh. Tverd. Tela 4, No. 4(1968), 50 — 55. (Russian)

212. Fomin, V.N.
a) On the stability of linear Hamiltonian equations with periodic coefficients in Hilbert space.— Vestnik Leningrad. Gos. Univ., No. 7, Ser. Mat. Mekh. Astron., No. 2(1964), 37—45. (Russian)
b) On the perturbation method in the theory of dynamic stability of systems with distributed parameters.— Dokl. Akad. Nauk SSSR **163**, No. 4(1965), 830—833. (Russian)
c) Parametric resonance of elastic systems with distributed parameters.— Dokl. Akad. Nauk SSSR **164**, No. 1(1965), 58—61. (Russian)
d) Parametric resonance of elastic systems with infinitely many degrees of freedom, Part I.— Vestnik Leningrad. Gos. Univ., No. 13, Ser. Mat. Mekh. Astron., No. 3(1965), 73—87; Part II, ibid., No. 13, Ser. Mat. Mekh. Astron., No. 4(1965), 74—86. (Russian)
e) Dynamic instability domains of parametrically excited systems with infinitely many degrees of freedom.— In: Problemy matematicheskogo analiza, pp. 135—165. Leningrad. Gos. Univ., 1966. (Russian)
f) Resonance of oscillations of linear systems under the effect of almost-periodic parametric excitation. — In: Problemy matematicheskogo analiza, pp. 28—79. Leningrad. Gos. Univ., 1969. (Russian)

213. Fomin, V.N. and V.A. Yakubovich.
Computation of characteristic exponents of linear systems with periodic coefficients.— In: Metody vychislenii, No. 3, pp. 76—104. Leningrad. Gos. Univ., 1966. (Russian)

214. Khazin, L.G.
Problems of the stability of Hamiltonian systems and resonances. — Inst. Prikl. Mat. Akad. Nauk SSSR, Preprint No. 20(1969), 23 pp.; Preprint No. 33(1969), 17 pp. (Russian)

215. Khapaev, M.M.
On the investigation of stability in the theory of nonlinear oscillations. — Mat. Zametki **3**, No. 3(1968), 307—318. (Russian)

216. Chetaev, N.G.
a) On a Cauchy problem. — Prikl. Mat. Mekh. **9**, No. 2(1945), 139—142. (Russian)
b) On the least characteristic number. — Prikl. Mat. Mekh. **9**, No. 3(1945), 193—196. (Russian)
c) On the sign of the least characteristic number. — Prikl. Mat. Mekh. **12**, No. 1(1948), 101—102. (Russian)
d) On the choice of parameters of a stable mechanical system. — Prikl. Mat. Mekh. **15**, No. 3(1951), 371—372. (Russian)
e) On some problems concerning the stability of nonsteady motions. — Prikl. Mat. Mekh. **24**, No. 1 (1960), 6—19. (Russian)
f) On the stability of structurally stable systems.— Prikl. Mat. Mekh. **24**, No. 1(1960), 20—22. (Russian)

217. Chugaev, V.V.
Stability of small oscillations of the Schuler vertical for an object moving along a terrestrial geodetic line. — Inzh. Zh. Mekh. Tverd. Tela **6**, No. 4(1970), 16—22. (Russian)

218. Sharshanov, A.A.
On the theory of linear differential equations with periodic coefficients. — In: Matematicheskaya fizika, pp. 172—193. Kiev, "Naukova Dumka," 1966. (Russian)

219. Shvartsman, A.P.
On the problem of boundedness of solutions of the differential equation $y''+p(x)y=0$. — Prikl. Mat. Mekh. **18**, No. 4(1954), 464—468. (Russian)

220. Shimanov, S.N.
a) On the theory of quasiharmonic oscillations. — Prikl. Mat. Mekh. **26**, No. 2(1952), 129—146. (Russian)
b) On the problem of determining the characteristic exponents of systems of linear differential equations with periodic coefficients. — Dokl. Akad. Nauk SSSR **109**, No. 6(1956), 1102—1105. (Russian)
c) On determination of the characteristic exponents of systems of linear differential equations with periodic coefficients. — Prikl. Mat. Mekh. **22**, No. 3 (1958), 382—385. (Russian)

221. Shtokalo, I.Z.
Criteria for stability and instability of solutions of linear differential equations with quasiperiodic coefficients. — Mat. Sb. **19**, No. 2(1946), 263—286. (Russian)

222. Yuzefovich, G. I.
On dynamic instability of circular saws. — Izv. Akad. Nauk SSSR, Otdel. Tekhn. Nauk, Mekh. i Mashinostr., No. 1(1961), 46—50. (Russian)

223. Yurovskii, A.V.
On some criteria for stability of integrals of systems of two linear differential equations with periodic coefficients. — Dokl. Akad. Nauk SSSR **62**, No. 5(1948), 595—598. (Russian)

224. Yakubovich, V.A.
a) On the boundedness of solutions of the equation: $y''+p(t)y=0,\quad p(t+\omega)=p(t)$. — Dokl. Akad. Nauk SSSR 74, No. 5(1950), 901—903. (Russian)
b) Criteria for stability of solutions of a system of two linear differential equations with periodic coefficients. — Uspekhi Mat. Nauk, No. 1(1951), 166—168. (Russian)
c) Criteria for stability for a system of two equations of canonical form with periodic coefficients. — Dokl. Akad. Nauk SSSR **78**, No. 2(1951), 221—224. (Russian)
d) Estimate of characteristic exponents and stability criteria for a second-order linear differential equation with periodic coefficients. — Dokl. Akad. Nauk SSSR **87**, No. 3(1952), 345—348. (Russian)
e) Estimates of the characteristic exponents of systems of linear differential equations with periodic coefficients. — Prikl. Mat. Mekh. **18**, No. 5(1954), 533—546. (Russian)
f) Lyapunov's method for determining the boundedness of solutions of equations $y''+p(t)y=0$, $p(t+\omega)=p(t)$, extended to the case of a function $p(t)$ of variable sign. — Prikl. Mat. Mekh. **18**, No. 6(1954), 705—718. (Russian)
g) Problems of the stability of solutions of a system of two linear differential equations of canonical form with periodic coefficients. — Mat. Sb. **37 (79)**, No. 1(1955), 21—68. (Russian)
h) On systems of differential equations of canonical form with periodic coefficients. — Dokl. Akad. Nauk SSSR **103**, No. 6(1955), 981—984. (Russian)
i) Structure of instability domains, stability and instability criteria for a system of differential equations of canonical form with periodic coefficients. — Uspekhi Mat. Nauk **6**, No. 4(1956), 191—192. (Russian)
j) On dependence of the eigenvalues of selfadjoint boundary-value problems for a system of two differential equations on the boundary conditions. — Vestnik Leningrad. Gos. Univ., No. 1, Ser. Mat. Mekh. Astron., No. 1(1957), 201—206. (Russian)
k) Extension of certain results of Lyapunov to linear canonical systems with periodic coefficients. — Prikl. Mat. Mekh. **21**, No. 4(1957), 491—502. (Russian)
l) Remarks on some works on systems of linear differential equations with periodic coefficients. — Prikl. Mat. Mekh. **21**, No. 5(1957), 707—713. (Russian)
m) Structure of the group of symplectic matrices and structure of the set of unstable canonical systems with periodic coefficients. — Mat. Sb. **44(86)**, No. 3(1958), 313—352. (Russian)
n) Critical frequencies of quasicanonical systems. — Vestnik Leningrad. Gos. Univ., No. 13, Ser. Mat. Mekh. Astron., No. 3 (1958), 35—63. (Russian)
o) On the dynamic stability of elastic systems. — Dokl. Akad. Nauk SSSR **121**, No. 4(1958), 602—605. (Russian)
p) The method of a small parameter for canonical systems with periodic coefficients. — Prikl. Mat. Mekh. **23**, No.1 (1959), 15—34. (Russian)
q) Oscillation properties of solutions of linear canonical systems of differential equations. — Dokl. Akad. Nauk SSSR **124**, No.3 (1959), 533—536. (Russian)
r) On some conditions of stability and instability for the system of differential equations with periodic coefficients. — In: Proc. Third All-Union Math.Congress, Vol.4, pp.41—42. 1959. (Russian)
s) On the radius of convergence of series in the method of a small parameter for linear differential equations with periodic coefficients. — Vestnik Leningrad. Gos. Univ., No.13, Ser. Mat. Mekh. i Astron., No.3 (1960), 81—89. (Russian)
t) Structure of the function space of complex canonical equations with periodic coefficients. — Dokl. Akad. Nauk SSSR **139**, No.1, pp.54—57. (Russian)
u) Arguments on the group of symplectic matrices. — Mat. Sb. **55 (97)**, No.3 (1961), 255—280. (Russian)
v) Conditions of unbounded stability for a certain second-order linear differential equation. — Vestnik Leningrad. Gos. Univ., No.19, Ser. Mat. Mekh. Astron., No.1 (1961), 83—92. (Russian)
w) Oscillation properties of solutions of canonical equations. — Mat. Sb. **56 (98)**, No.1 (1962), 3—42. (Russian)

x) On the properties of stability domains of linear Hamiltonian systems of differential equations with periodic coefficients. — Vestnik Leningrad. Gos. Univ., No.13, Ser. Mat. Mekh. Astron., No.3 (1962), 61–68. (Russian)

y) Upper and lower bounds for the solution of a homogeneous second-order differential equation. — In: Metody vychislenii, No.1, pp.30–44. Leningrad. Gos. Univ., 1963. (Russian)

z) Dynamic instability domains of Hamiltonian systems. — In: Metody vychislenii, No.3, pp.51–69. Leningrad. Gos. Univ., 1966. (Russian)

aa) Remark on the Floquet-Lyapunov theorem. — Vestnik Leningrad. Gos. Univ., No.1, Ser. Mat. Mekh. Astron., pp.88–92, 1970. (Russian)

225. Yakubovich, V.A. and V.M. Starzhinskii.
Parametric resonance of systems with several degrees of freedom. — In: Proc. Conf. on Mechanics, Kazan', pp.123–134, 1964. (Russian)

226. Abdel, K. and I.I. Rahmi.
Über den Resonanzfall bei gewöhnlichen linearen Differentialgleichungen N-ter Ordnung mit periodischen Koeffizienten. — Arch. Math. 3, No.1 (1967), 11–30.

227. Bailey, H.R. and L. Cesari.
Boundedness of solutions of linear differential systems with periodic coefficients. — Arch. Rational Mech. Anal. 1, No.3 (1958), 246–271.

228. Bellman, R.
On the stability of solutions of differential equations. — Rev. Ci. (Lima) **40**, Nos.5–42 (1958), 503–506.

229. Beyreuther, J.
a) Zur Darstellung von Instabilitätsbereichen bei einem mechanischen System mit mehreren Freiheitsgraden. — Z. Angew. Math. Mech. 45, No.6 (1965), 441–444.
b) Instabilitätsbereiche eines fastkanonischen Systems.— Math. Nachr. 32, Nos. 3–4 (1966), 131–137.

230. Beyreuther, J. and W.M. Starshinski.
Anwendung der Methode des kleinen Parameters in Matrizenschreibweise bei der Berechnung der charakteristischen Exponenten der Hillschen Gleichung. — Z. Angew. Math. Mech. 45, No.1 (1965), 57–59.

231. Bickart, T.A.
Periodically time-varying system: an instability criterion. — Proc. IEEE 55, No.11 (1967), 2057–2058.

232. Bóbr, St.
Eine Verallgemeinerung des v. Kochschen Satzes über die absolute Konvergenz der unendlichen Determinanten. — Math. Z. 10 (1921), 1–11.

233. Boman, J.
On the stability of differential equations with periodic coefficients. — Kungl. Tekn. Högsk. Handl. Stockholm, No.180 (1961), 1–21.

234. Borg, G.
a) Über die Stabilität gewisser Klassen von linearen Differentialgleichungen. — Ark. Mat. Astronom. Fys. 31A, No.1 (1944), 1–30.
b) On a Liapunoff criterion of stability. — Amer. J. Math. 71, No.1 (1949), 67–70.
c) Deux notes concernant la stabilité. — Publ. Sci. Tech. Ministère de l'Air (Paris), Actes du colloque intern. des vibrations non linéaires, No.281 (1953), 21–29.

235. Burton, T.A.
Linear differential equations with periodic coefficients. — Proc. Amer. Math. Soc. 17, No. 2 (1966), 327–329.

236. Caughey, T.K. and J.R. Dickerson.
Stability of linear dynamic systems with narrow-band parametric excitation. — Trans. ASME 34, No.3 (1967), 709–713.

237. Cesari, L.
Sulla stabilità delle soluzioni dei sistemi di equazioni differenziali lineari a coefficienti periodici. - Atti Accad. Italia, Mem. Cl. Sci. Fis. Mat. Natur. (6) 11 (1940), 633–695.

238. Cesari, L. and J.K. Hale.
Second-order linear differential systems with periodic *L*-Integrable coefficients. — Riv. Mat. Univ. Parma 5, Nos.1–3 (1954), 55–61.

239. Chizzetti, A.

a) Alcuni criteri sufficenti di stabilità per gli integrali di un'equazione differenziale lineare omogenea del 2 ordine. — Atti Accad. Naz. Lincei Rend. Cl. Sci. Fis. Mat. Natur. 33, No.5 (1962–63), 219–229.

b) Majoration formulae and sufficient criteria for the stability of the integrals of a linear homogeneous differential equation of n-th order. — Rend. Accad. Naz. Lincei, Ser. 8, No.7 (1963).

240. Christofer, P.A.T.

Stability of periodic solutions of a nonlinear differential equation arising in servomechanism theory. — College Aeronaut. Cranfield Rept., No.191 (1965), 1–17.

241. Coppel, W.A. and A. Howe.

a) On the stability of linear canonical systems with periodic coefficients. — J. Austral. Math. Soc. 5, No.2 (1965), 169–195.

b) Dynamical instability of linear canonical systems. — J. Austral. Math. Soc. 7, No.2 (1967), 247–251.

242. Counselman, C.C.

Asymptotic stability and generalized forces in periodic systems. — AIAA J. 6, No.7 (1968), 1383–1385.

243. Davison, E.J.

The stability of a second order linear periodic system. — Paper Amer. Soc. Mech. Engrs. (ASME), No. WA/Aut.-5 (1968), 1–4.

244. Dickerson, J.R.

Stability of linear systems with parametric excitation. — Trans. ASME E37, No.1 (1970), 228–230.

245. Diliberto, S.P.

a) On the stability of linear mechanical systems. — Proc. First Intern. Symp. on Nonlinear Oscillations, Vol.1, pp.189–203. Kiev, Akad. Nauk Ukrain. SSR, 1963.

b) A theorem of formal stability in the satellite orbit problem. — Astronom. J., Vol.71, No. 4 (1966).

246. Einaudi, R.

Sulle vibrazioni quassi-armoniche di un sistema dissipativo. — Atti Veneto 95, Part 2 (1966), 425–444.

247. Epstein, I.J.

On systems of linear differential equations with periodic coefficients: algebraic and topological aspects.—J. Differential Equations 4, No. 2 (1965), 206–221.

248. Fischer, U.

Stabilitätsbedingungen für die Lösung der Schwingungsgleichung mit stossartig veränderlicher Federzahl. — Wiss. Z. Techn. Hochsch. Otto von Guericke (Magdeburg) 11, No.1 (1967), 15–22.

249. Gambill, R.A.

a) Stability criteria for linear differential systems with periodic coefficients. — Riv. Mat. Univ. Parma 5, Nos.1–3 (1954), 169–181.

b) Criteria for parametric instability for linear differential systems with periodic coefficients. — Riv. Mat. Univ. Parma 6, Nos. 1–2 (1955), 37–43.

c) A fundamental system of real solutions for linear differential systems with periodic coefficients. — Riv. Mat. Univ. Parma 7, Nos. 3–5 (1956), 311–319.

250. Guggenheimer, H.

Hill's equations with coexisting periodic solutions. — J. Differential Equations 5, No.1 (1969), 159–166.

251. Haacke, W.

Über die Stabilität eines Systems von gewöhnlichen linearen Differentialgleichungen zweiter Ordnung mit konstanten Koeffizienten. — Math. Z. 56, No.1 (1952), 65–79.

252. Hale, J.K.

a) On boundedness of the solution of linear differential systems with periodic coefficients.—Riv. Mat. Univ. Parma 5, Nos. 1–3 (1954), 137–167.

b) On a class of linear differential equations with periodic coefficients. — Illinois J. Math. 1, No.1 (1958), 98–104.

c) Linear systems of first and second order differential equations with periodic coefficients. — Illinois J. Math. 2, No. 4A (1958), 586–592.

d) Periodic solutions of nonlinear systems of differential equations. — Riv. Mat. Univ. Parma 5, No.5 (1954), 281–311.

e) A short proof of a boundedness theorem for linear differential systems with periodic coefficients. — Arch. Rational Mech. Anal. 2, No.5 (1959), 429–434.

f) On the behavior of solutions of linear periodic differential systems near resonance points. — Ann. Math. Studies, No.45 (1960), 55–91.

g) On the stability of periodic solutions of weakly nonlinear periodic and autonomous differential systems. — Ann. Math. Studies, No.45 (1960), 91–115.

h) On the characteristic exponents of linear periodic differential systems. — Bol. Soc. Mat. Mexicana (1960), 58–66.

i) Integral manifolds of perturbed differential systems. — Ann. of Math. 73, No.3 (1961), 496–531.

j) On differential equations containing a small parameter. — In: Contributions to Differential Equations, Vol.1, No.2, pp.215–250. New York, Wiley, 1963.

253. Hale, J.K. and G. Seifert.
Bounded and almost periodic solutions of singularly perturbed equations. — J. Math. Anal. Appl. 3, No.1 (1961), 18–24.

254. Hale, J.K. and A.P. Stokes.

a) A discussion of differential equations on product spaces. — Bol. Soc. Mat. Mexicana (1960), 67–74.

b) Behavior of solutions near integral manifolds. — Ark. Mat. Mech. Anal. 6, No.2 (1960), 133–170.

c) Conditions for the stability of nonautonomous differential equations. — J. Math. Anal. Appl. 3, No.3 (1961), 496–531.

255. Hamel, G.
Über die lineare Differentialgleichung zweiter Ordnung mit periodischen Koeffizienten. — Math. Ann. 73, No.3 (1913), 371–412.

256. Haupt, O.
Über lineare homogene Differentialgleichungen zweiter Ordnung mit periodischen Koeffizienten. — Math. Ann. 79 (1919), 278–285.

257. Hill, G.W.
On the part of the motion of the lunar perigee which is a function of the mean motions of the sun and the moon. — Acta Math. 8 (1886), 1–36.

258. Hirai, K. and Y. Hamasaki.
The stability of a system with periodically varying parameters. — IEEE Trans. Circuit Theory 14, No.1 (1967), 56–60.

259. Hochstadt, H.

a) A class of stability criteria for Hill's equation. — Quart. Appl. Math. 20, No.1 (1962), 92–93.

b) A stability criterion for Hill's equation. — Proc. Amer. Math. Soc. 13, No.4 (1962), 601–603.

c) A stability estimate for differential equations with periodic coefficients. — Arch. Math. 15, Nos.4–5 (1964), 318–320.

260. Höhn, E.
Untersuchungen über die Stabilität von Hillschen Differentialgleichungen. — Z. Angew. Math. Mech. 45 (1965), 39–41.

261. Hukuhara, M.
Sur les équations différentielles linéaires à coefficients périodiques et contenant un paramètre. — J. Fac. Sci. Univ. Tokyo, Sect. 1, 7, Part 1, pp.69–85, 1954.

262. Infante, E.F.
Stability criteria for n-th order homogeneous differential equations. — In: Differential Equations and Dynamic Systems, pp.309–321. New York, Academic Press, 1967.

263. Janet, M.
Progrés recents sur la stabilité des mouvements périodiques. — In: Theory of Orbits in the Solar System and in Stellar Systems (International Astronomical Union, 25th Symposium), pp.1–33, New York, Academic Press, 1966.

264. Kramers, H.A.
Das Eigenwertproblem im eindimensionalen periodischen Kraftfeld. — Physica 2 (1935), 483–490.

265. Levi-Civita, T.

a) Sur les équations linéaires à coefficients périodiques et sur le moyen mouvement du noeud lunaire. — Ann. Sci. École Norm. Sup. 2-e Serie 28 (1911), 325–376.

b) Sopra alcuni criteri di instabilita. — Ann. Mat. Pura Appl., Ser. III 5 (1901), 221–307.

266. Levinson, N.
The stability of linear, real, periodic, self-adjoint systems of differential equations. — J. Math. Anal. Appl., No. 6 (1963), 473—482.
267. Levy, D.M. and J.B. Keller.
Instability intervals of Hill's equations. — Comm. Pure Appl. Math. 16, No.4 (1963), 469—476.
268. Lindh, K.G. and P.W. Likins.
Infinite determinant methods for stability analysis of periodic-coefficients differential equations. — AIAA J. 8, No.4 (1970), 680—686.
269. Lion, P.M.
Stability of linear periodic systems. — J. Franklin Inst. 281, No.1 (1966), 27—40.
270. Mangeron, D. and R. Chaléat.
Théorie générale de la synchronisation et applications, Note II. Conditions de stabilité dans la théorie générale de la synchronisation. — Bul. Inst. Politehn. Iaşi 11 (15), Nos.3—4 (1965), 301—306.
271. Massa, E.
Sulla instabilita delle vibrazioni ad un grado di liberta a parametri variabili periodicamente nel tempo e con resistenza viscosa. — Rend. Ist. Lomb. Sci. Lett., A. Sci. Mat. Fiz. Chim. e Geol. 98, No.1 (1964), 219—238.
272. Massaglia, F.
Sulle vibrazioni quasi-armoniche di un sistema dissipativo con elastita constante a tratti. — Atti Sem. Mat. Fis. Univ. Modena, Vol.7 (1953—1954), 167—181.
273. Massera, J.L. and J.J. Schäffer.
Linear differential equations and functional analysis, II. Equations with periodic coefficients.— Ann. of Math. 69, No. 1 (1959), 88—104.
274. Mathieu, E.
Mémoire sur le mouvement vibratoire d'une membrane de forme elliptique. — J. Math. 2—13 (1868), 137.
275. McGarvey, D.
Linear differential systems with periodic coefficients involving a large parameter. — J. Differential Equations 2, No.2 (1966), 115—142.
276. Meadows, H.E.
Solution of systems of linear ordinary differential equations with periodic coefficients. — Bell System Tech. J. 41, No.4 (1962), 1275—1294.
277. Mettler, E.
Schwingungs- und Stabilitätsprobleme bei mechanischen Systemen mit harmonischer Erregung.— Z. Angew. Math. Mech. 45, Nos. 7—8 (1965), 475—481.
278. Moser, J.
a) New aspects in the theory of stability of Hamiltonian systems. — Comm. Pure Appl. Math. 11, No.1 (1958), 81—114.
b) New results on the stability of periodic motions. — In: Proc. Intern. Congr. Math., August, 1962, pp.584—586. Uppsala, 1963.
279. Nishikawa, Y. and P.Y. Willems.
A method for stability investigation of a periodic dynamic system with many degrees of freedom. — J. Franklin Inst. 287, No.2 (1969), 143—157.
280. Opial, Z.
Sur un problème de stabilité. — Ann. Polon. Math. 5, No.2 (1958), 153—164.
281. Patry, I.
Über die linearen Differentialgleichungen mit sinusförmigen Koeffizienten. — Julis Verlag, Zurich, 1957, Prom., No.2618.
282. Putnam, C.R.
On the stability integrals of the Hill equation. — J. Soc. Indust. Appl. Math. 7, No.1 (1959), 101—106.
283. Richalet, J. and C. Lefèvre.
Systèmes lineáires d'ordre quelconque a coefficients périodiques. — Comptes Rendus AB 262, No.11 (1966), 662—665.
284. Roseau, M.
a) Sur la stabilité de la solution nulle de certains systèmes différentielles dans les cas critiques. — Comptes Rendus 257, No.22 (1963), 3273—3275.

b) Sur une méthode d'approximations succesives applicable à l'étude du problème de stabilité dans les cas critiques. — Comptes Rendus 257, No.23 (1963), 3549–3551.

c) Sur la stabilité des solutions périodiques de certains systèmes différentielles quasi linéaires. — Colloques Intern. du Centre National de la Recherche Scient., No.148 (1965), 145–166.

285. Salinas, B.R.
Sulla stabilita delle soluzioni per l'equazione differenziale del secondo ordine a coefficienti periodici. — Rend. Circ. Mat. Palermo 8, No.2 (1959), 206–224.

286. Sandberg, I.W.
a) On the stability of solutions of linear differential equations with periodic coefficients. — J. Soc. Indust. Appl. Math. 12, No.2 (1964), 487–496.
b) On the solutions of system of second-order differential equations with variable coefficients. — J. Soc. Indust. Appl. Math. A2, No.2 (1964), 192–198.

287. Schmidt, G.
a) On the stability of combination oscillation. — Proceedings of Vibration Problems 1, No.8 (1967), 34–45.
b) Instabilitätsbereiche bei rheolinearen Schwingungen. — Monatsb. Deutsch. Akad. Wiss. Berlin 9, Nos. 6–7 (1967), 405–411.
c) Zur Stabilität von Kombinationsschwingungen rotierender Wellen. — Monatsb. Deutsch. Akad. Wiss. Berlin 10, No.7 (1968), 475–477.
d) Resonanzlösungen nichtlinearer Schwingungsgleichungen und zur Stabilität der Resonanzlösungen nichtlinearer Schwingungsgleichungen. — Nova Acta Leopoldina 34, No.187 (1969), 3–68.

288. Schmidt, G. and F. Weidenhammer.
a) Instabilitäten gedämpfter rheolinearer Schwingungen. — Math. Nachr. 23, Nos. 4–5 (1961), 301–318.
b) Gedämpfte erzwungene Schwingungen in rheolinearen Systemen. — Math. Nachr. 27, Nos.3–4 (1964), 215–228.

289. Schumacher, H. and G. Schweitzer.
Zusammenhänge zwischen der Übertragungsfunktion und der Theorie von Floquet bei linearen Regelsystemen mit periodischen Parametern. — Z. Angew. Math. Mech. 48, No.8 (1968), 286–288.

290. Schweitzer, G.
a) Eine allgemeine Methode zur Untersuchung von Systemen mit periodisch sich ändernden Koeffizienten. — Arch. Electr. Übertrag. 19, No.9 (1965), 469–482.
b) Zur Stabilität eines parametererregten Schwingers. [No particulars given in Russian original.]

291. Schwerdtfeger, H.
The eigenvalue problem of Hill's equations. — J. Proc. Roy. Soc. New South Wales 71, No.4 (1945), 176–189.

292. Sibuya, Y.
Sur un système des équations différentielles ordinaires linéaires à coefficients périodiques et contenant des paramètres. — J. Fac. Sci. Univ. Tokyo Sect. 1, Vol.7 (1954), 229–241.

293. Sips, R.
Nouvelle méthode pour le calcul de l'exposant charactéristique de l'équation de Mathieu-Hill. — Bull. Acad. Roy. Belg. Cl. Sci. 51, No.2 (1965), 191–206.

294. Stevens, K.K.
On linear ordinary differential equations with periodic coefficients. — SIAM J. Appl. Math. 14, No.4 (1966), 782–795.

295. Strutt, M.J.O.
Der charakteristische Exponent der Hillschen Differentialgleichung. — Math. Ann., Vol.101 (1929), 559–569.

296. Thom, R.
Travaux de Moser sur la stabilité des mouvements périodiques. — Sémin. Bourbaki 16, No.1 (1963–1964), 264/01–264/13.

297. Van Vleck, F.S.
A note on the relation between periodic and orthogonal fundamental solutions of linear systems, I. — Amer. Math. Monthly 71, No.4 (1961), 406–408; II. — Ibid., No.7, pp.774–776.

298. Zlámal, M.
On a criterion for Lyapunov stability. — Czechoslovak. Math. J. 3(78); No.3 (1953), 257–264. (Russian)

*SUBJECT INDEX (Volumes 1 and 2)**

* Roman numerals I, II indicate Volume 1, 2, respectively.